INLAND FISHERIES MANAGEMENT
IN NORTH AMERICA
SECOND EDITION

INLAND FISHERIES MANAGEMENT IN NORTH AMERICA, 2ND EDITION

is a special project of and partly funded by the

Education Section and Fisheries Management Section,

American Fisheries Society

INLAND FISHERIES MANAGEMENT
IN NORTH AMERICA
SECOND EDITION

Edited by

Christopher C. Kohler

Fisheries Research Laboratory and Department of Zoology
Southern Illinois University at Carbondale

and

Wayne A. Hubert

U.S. Geological Survey
Wyoming Cooperative Fish and Wildlife Research Unit
University of Wyoming

American Fisheries Society
Bethesda, Maryland, USA
1999

Suggested Citation Formats

Entire book

Kohler, C. C., and W. A. Hubert, editors. 1999. Inland fisheries management in North
America, 2nd edition. American Fisheries Society, Bethesda, Maryland.

Chapter within the book

Rahel, F. J., R. T. Muth, and C. A. Carlson. 1999. Endangered species management. Pages
431–454 *in* C. C. Kohler and W. A. Hubert, editors. Inland fisheries management in
North America, 2nd edition. American Fisheries Society, Bethesda, Maryland.

Library of Congress Catalog Card Number 99-61579
ISBN 1-888569-13-1

Printed in the United States of America on acid-free paper.

American Fisheries Society
5410 Grosvenor Lane, Suite 110
Bethesda, Maryland 20814-2199
USA

CONTENTS

10 Stream Habitat Management
DONALD J. ORTH AND RAY J. WHITE

11 Lake and Reservoir Habitat Management
ROBERT C. SUMMERFELT

12 Maintenance of the Estuarine Environment
WILLIAM H. HERKE AND BARTON D. ROGERS

COMMUNITY MANIPULATIONS

13 Management of Introduced Fishes
HIRAM W. LI AND PETER B. MOYLE

14 Stocking for Sport Fisheries Enhancement
ROY C. HEIDINGER

15 Management of Undesirable Fish Species
RICHARD S. WYDOSKI AND ROBERT W. WILEY

16 Endangered Species Management
FRANK J. RAHEL, ROBERT T. MUTH, AND CLARENCE A. CARLSON

17 Managing Fisheries with Regulations
RICHARD L. NOBLE AND T. WAYNE JONES

COMMON MANAGEMENT PRACTICES

18 Coldwater Streams
J. S. GRIFFITH

19 Warmwater Streams
CHARLES F. RABENI AND ROBERT B. JACOBSON

CONTRIBUTORS

Frank J. Bulow (Chapter 21): Department of Biology, Wildlife, and Fisheries Science, Tennessee Tech University, Cookeville, Tennessee 38505, USA

Clarence A. Carlson (Chapter 16): Department of Fishery and Wildlife Biology, Colorado State University, Fort Collins, Colorado 80523, USA

Daniel J. Decker (Chapters 2 and 3): Associate Director for Research, Cornell University, 245 Roberts Hall, Ithaca, New York 14853-4203, USA

Stephen A. Flickinger (Chapter 21): Department of Fishery and Wildlife Biology, Colorado State University, Fort Collins, Colorado 80523, USA

J. S. Griffith (Chapter 18): Department of Biology, Idaho State University, Pocatello, Idaho 83209, USA

Daniel B. Hayes (Chapter 22): Department of Fish and Wildlife, Michigan State University, 13 Natural Resources Building, East Lansing, Michigan 48824, USA

Roy C. Heidinger (Chapter 14): Fisheries Research Laboratory, Southern Illinois University, Mail Code 6511, Carbondale, Illinois 62901-6511, USA

William H. Herke (Chapter 12): School of Forestry, Wildlife, and Fisheries, Louisiana State University, Baton Rouge, Louisiana 70803, USA

Wayne A. Hubert (Coeditor and Chapter 5): U.S. Geological Survey, Wyoming Cooperative Fish and Wildlife Research Unit, University of Wyoming, Laramie, Wyoming 82071-3166, USA

Daniel J. Isaak (Chapter 9): 954 McCue Street, #101, Laramie, Wyoming 82072, USA

Robert B. Jacobson (Chapter 19): U.S. Geological Survey, Columbia Environmental Research Center, 4200 New Haven Road, Columbia, Missouri 65201, USA

T. Wayne Jones (Chapter 17): Route 3, Box 321 E, Nashville, North Carolina 27856, USA

David J. Jude (Chapter 23): Center for Great Lakes and Aquatic Sciences, University of Michigan, 501 East University, Ann Arbor, Michigan 48109, USA

Christopher Kohler (Coeditor): Fisheries Research Laboratory and Department of Zoology, Southern Illinois University, Carbondale, Illinois 62901-6511, USA

Charles C. Krueger (Chapters 2 and 3): Cornell University, 206 D Fernow Hall, Ithaca, New York 14853, USA

Berton L. Lamb (Chapter 4): U.S. Geological Survey, Midcontinent Ecological Science Center, Biological Resources Division, Social, Economic, and Institutional Analysis Section, Fort Collins, Colorado 80525-3400, USA

Joseph Leach (Chapter 23): Lake Erie Fisheries Station, Rural Route 2, Wheatley, Ontario N0P 2P0, Canada

Hiram W. Li (Chapter 13): U.S. Geological Survey, Oregon Cooperative Fishery Research Unit, Oregon State University, 104 Nash Hall, Corvallis, Oregon 97331-3803, USA

Donna Lybecker (Chapter 4): U.S. Geological Survey, Midcontinent Ecological Science Center, Biological Resources Division, Social, Economic, and Institutional Analysis Section, Fort Collins, Colorado 80525-3400, USA

John R. Moring (Chapter 24): U.S. Geological Survey, Maine Cooperative Fish and Wildlife Research Unit, 5751 Murray Hall, Orono, Maine 04469-5751, USA

Peter B. Moyle (Chapter 13): Department of Wildlife, Fisheries and Conservation Biology, University of California, Davis, California 95616, USA

Robert T. Muth (Chapter 16): 2919 South Connor Street, Salt Lake City, Utah 84109, USA

John J. Ney (Chapter 7): Department of Fish and Wildlife, Virginia Polytechnic and State University, Mail Code 0321, Blacksburg, Virginia 24061, USA

Larry A. Nielsen (Chapter 1): School of Natural Resources, Pennsylvania State University, 113 Ferguson Building, State College, Pennsylvania 16802, USA

Richard L. Noble (Chapter 17): Fish and Wildlife Science, North Carolina State University, Campus Box 7646, Raleigh, North Carolina 27695, USA

Donald J. Orth (Chapter 10): Department of Fish and Wildlife, Virginia Polytechnic and State University, Mail Code 0321, Blacksburg, Virginia 24061, USA

Charles F. Rabeni (Chapter 19): U.S. Geological Survey, Missouri Cooperative Fish and Wildlife Research Unit, 302 Anheuser-Busch Natural Resources Building, Columbia, Missouri 65211-7240, USA

Frank J. Rahel (Chapter 16): Department of Zoology and Physiology, University of Wyoming, Box 3166, Laramie, Wyoming 82071, USA

Jerry L. Rasmussen (Chapter 20): U.S. Fish and Wildlife Service, 608 East Cherry Street, Columbia, Missouri 65201, USA

Barton D. Rogers (Chapter 12): School of Forestry, Wildlife, and Fisheries, Louisiana State University, Baton Rouge, Louisiana 70803, USA

Harold L. Schramm, Jr. (Chapter 5): U.S. Geological Survey, Mississippi Cooperative Fish and Wildlife Research Unit, Mississippi State University, Mail Stop 9691, Mississippi State, Mississippi 39762, USA

Robert J. Sheehan (Chapter 20): Fisheries Research Laboratory, Southern Illinois University, Carbondale, Illinois 62901, USA

Patricia A. Soranno (Chapter 22): Department of Fisheries and Wildlife, Michigan State University, 13 Natural Resources Building, East Lansing, Michigan 48824, USA

Robert C. Summerfelt (Chapter 11): Department of Animal Ecology, Iowa State University, Ames, Iowa 50011-3221, USA

William W. Taylor (Chapter 22): Department of Fisheries and Wildlife, Michigan State University, East Lansing, Michigan 48824, USA

Michael J. Van Den Avyle (Chapter 6): U.S. Geological Survey, Georgia Cooperative Fish and Wildlife Research Unit, University of Georgia, Athens, Georgia 30602, USA

A. Stephen Weithman (Chapter 8): Missouri Department of Conservation, Post Office Box 180, Jefferson City, Missouri 65101, USA

Thomas A. Wesche (Chapter 9): Department of Zoology and Physiology, University of Wyoming, Post Office Box 3166, Laramie, Wyoming 82071-3166, USA

Ray J. White (Chapter 10): 320 12th Avenue North, Edmunds, Washington 98020, USA

Robert W. Wiley (Chapter 15): Wyoming Game and Fish Department, 528 South Adams, Laramie, Wyoming 82070, USA

David W. Willis (Chapter 21): Department of Wildlife and Fisheries Science, South Dakota State University, Post Office Box 2140B, Brookings, South Dakota 57007, USA

Richard S. Wydoski (Chapter 15): U.S. Fish and Wildlife Service, Division of Federal Aid, Post Office Box 25486, Denver Federal Center, Denver, Colorado 80225, USA

LIST OF SPECIES

Fish Taxa

Alewife .... *Alosa pseudoharengus*
American eel .... *Anguilla rostrata*
American shad .... *Alosa sapidissima*
Apache trout .... *Oncorhynchus apache*
Arctic char .... *Salvelinus alpinus*
Arctic grayling .... *Thymallus arcticus*
Ash Meadows Armagosa pupfish .... *Cyprinodon nevadensis mionectes*
Ash Meadows killifish .... *Empetrichthys merriami*
Ash Meadows speckled dace .... *Rhinichthys osculus nevadensis*
Atlantic croaker .... *Micropogonias undulatus*
Atlantic salmon .... *Salmo salar*
Atlantic sturgeon .... *Acipenser oxyrhynchus*

Basses .... *Micropterus* spp.
Bighead carp .... *Hypophthalmichthys nobilis*
Bigmouth buffalo .... *Ictiobus cyprinellus*
Black basses .... *Micropterus* spp.
Black bullhead .... *Ameiurus melas*
Black carp .... *Mylopharyngodon piceus*
Black crappie .... *Pomoxis nigromaculatus*
Blackfin cisco .... *Coregonus nigripinnis*
Bloater .... *Coregonus hoyi*
Blue catfish .... *Ictalurus furcatus*
Blue pike .... *Stizostedion vitreum glaucum*
Blue sucker .... *Cycleptus elongatus*
Blue tilapia .... *Tilapia aurea*
Blueback herring .... *Alosa aestivalis*
Bluegill .... *Lepomis macrochirus*
Bonneville cutthroat trout .... *Oncorhynchus clarki utah*
Bonytail .... *Gila elegans*
Bowfin .... *Amia calva*
Brook stickleback .... *Culaea inconstans*
Brook trout .... *Salvelinus fontinalis*
Brown bullhead .... *Ameiurus nebulosus*

Brown trout .... *Salmo trutta*
Buffalos .... *Ictiobus* spp.
Bull trout .... *Salvelinus confluentus*
Bullhead catfishes .... *Ameiurus* spp.
Burbot .... *Lota lota*

California roach .... *Hesperoleucus symmetricus*
Carps .... Family Cyprinidae
Catfishes .... Family Ictaluridae
Chain pickerel .... *Esox niger*
Channel catfish .... *Ictalurus punctatus*
Chars .... *Salvelinus* spp.
Chinook salmon .... *Oncorhynchus tshawytscha*
Chum salmon .... *Oncorhynchus keta*
Cisco .... *Coregonus artedi*
Ciscoes .... *Coregonus* spp.
Coho salmon .... *Oncorhynchus kisutch*
Colorado pikeminnow .... *Ptychocheilus lucius*
Colorado River cutthroat trout .... *Oncorhynchus clarki pleuriticus*
Common carp .... *Cyprinus carpio*
Common shiner .... *Luxilus cornutus*
Crappies .... *Pomoxis* spp.
Creek chub .... *Semotilus atromaculatus*
Creek chubsucker .... *Erimyzon oblongus*
Cui-ui .... *Chasmistes cujus*
Cutthroat trout .... *Oncorhynchus clarki*

Daces .... Family Cyprinidae
Darters .... Family Percidae
Deepwater cisco .... *Coregonus johannae*
Deepwater sculpin .... *Myoxocephalus thompsoni*
Delta smelt .... *Hypomesus transpacificus*
Desert pupfish .... *Cyprinodon macularis*
Devils Hole pupfish .... *Cyprinodon diabolis*
Dolly Varden .... *Salvelinus malma*

Eastern mosquitofish .... *Gambusia holbrooki*
Emerald shiner .... *Notropis atherinoides*
Eulachon .... *Thaleichthys pacificus*

Fathead minnow .... *Pimephales promelas*
Flathead catfish .... *Pylodictis olivaris*
Florida largemouth bass .... *Micropterus salmoides floridanus*
Flounders .... Families Bothidae and Pleuronectidae
Freshwater drum .... *Aplodinotus grunniens*

Gars .... *Lepisosteus* spp.
Gila topminnow .... *Poeciliopsis occidentalis*
Gila trout .... *Oncorhynchus gilae*

Gizzard shad .... *Dorosoma cepedianum*
Golden shiner .... *Notemigonus crysoleucas*
Golden trout .... *Oncorhynchus aguabonita*
Goldeye .... *Hiodon alosoides*
Goldfish .... *Carassius auratus*
Grass carp .... *Ctenopharyngodon idella*
Graylings .... *Thymallus* spp.
Green sturgeon .... *Acipenser medirostris*
Green sunfish .... *Lepomis cyanellus*
Greenback cutthroat trout .... *Oncorhynchus clarki stomias*
Gulf menhaden .... *Brevoortia patronus*

Herrings .... Family Clupeidae
Humpback chub .... *Gila cypha*
Hybrid crappies .... *Pomoxis* spp.
Hybrid striped bass (white bass × striped bass) ...... *Morone chrysops × M. saxatilis*
Hybrid sunfishes .... *Lepomis* spp.

Inland silverside .... *Menidia beryllina*

Killifishes .... Family Cyprinodontidae
Kiyi .... *Coregonus kiyi*
Kokanee .... *Oncorhynchus nerka*

Lahontan cutthroat trout .... *Oncorhynchus clarki henshawi*
Lake herring .... *Coregonus artedi*
Lake sturgeon .... *Acipenser fulvescens*
Lake trout .... *Salvelinus namaycush*
Lake whitefish .... *Coregonus clupeaformis*
Lampreys .... Family Petromyzontidae
Largemouth bass .... *Micropterus salmoides*
Las Vegas dace .... *Rhinichthys deaconi*
Little Kern golden trout .... *Oncorhynchus aguabonita whitei*
Logperch .... *Percina caprodes*
Longear sunfish .... *Lepomis megalotis*
Longjaw cisco .... *Coregonus alpenae*
Longnose dace .... *Rhinichthys cataractae*
Longnose sucker .... *Catostomus catostomus*

Menhadens .... *Brevoortia* spp.
Minnows .... Family Cyprinidae
Mohave tui chub .... *Gila bicolor mohavensis*
Mooneye .... *Hiodon tergisus*
Mosquitofishes .... *Gambusia* spp.
Mottled sculpin .... *Cottus bairdi*
Mountain whitefish .... *Prosopium williamsoni*
Mozambique tilapia .... *Tilapia mossambica*
Mullets .... Family Mugilidae
Muskellunge .... *Esox masquinongy*

Nile perch .... *Lates niloticus*
Northern largemouth bass ....*Micropterus salmoides salmoides*
Northern pike .... *Esox lucius*
Northern pikeminnow .... *Ptychocheilus oregonensis*

Owens pupfish .... *Cyprinodon radiosus*
Owens sucker ....*Catostomus fumeiventris*
Owens tui chub .... *Gila bicolor snyderi*

Pacific salmon .... *Oncorhynchus* spp.
Paddlefish ....*Polyodon spathula*
Pahranagat spinedace .... *Lepidomeda altivelis*
Pahrump killifish ....*Empetrichthys latos latos*
Pahrump Ranch killifish .... *Empetrichthys latos pahrump*
Paiute cutthroat trout .... *Oncorhynchus clarki seleniris*
Pallid sturgeon .... *Scaphirhynchus albus*
Peacock bass .... *Cichla ocellaris*
Pecos gambusia .... *Gambusia nobilis*
Pecos pupfish .... *Cyprinodon pecosensis*
Pickerels .... *Esox* spp.
Pikeminnows .... *Ptychocheilus* spp.
Pink salmon .... *Oncorhynchus gorbuscha*
Prickly sculpin .... *Cottus asper*
Pumpkinseed ....*Lepomis gibbosus*
Pupfishes..*Cyprinodon* spp.

Quillback .... *Carpiodes cyprinus*

Rainbow smelt .... *Osmerus mordax*
Rainbow trout .... *Oncorhynchus mykiss*
Raycraft Ranch killifish .... *Empetrichthys latos concavus*
Razorback sucker .... *Xyrauchen texanus*
Red drum ....*Sciaenops ocellatus*
Red shiner...*Cyprinella lutrensis*
Redbreast sunfish .... *Lepomis auritus*
Redear sunfish .... *Lepomis microlophus*
Redside shiner ....*Richardsonius balteatus*
Rock bass...*Ambloplites rupestris*
Round goby .... *Neogobius melanostomus*
Ruffe .... *Gymnocephalus cernuus*

Sacramento blackfish .... *Orthodon microlepidotus*
Sacramento pikeminnow ....*Ptychocheilus grandis*
Sand seatrout .... *Cynoscion arenarius*
Sauger .... *Stizostedion canadense*
Saugeye (sauger × walleye hybrid) .... *Stizostedion canadense* × *S. vitreum*
Sculpins ....Family Cottidae
Sea lamprey ....*Petromyzon marinus*
Shads.. *Dorosoma* spp.

Shortnose cisco .... *Coregonus reighardi*
Shortnose sturgeon ....*Acipenser brevirostrum*
Shovelnose sturgeon ....*Scaphirhynchus platorynchus*
Silver carp.. *Hypophthalmichthys molitrix*
Slimy sculpin ....*Cottus cognatus*
Smallmouth bass....................................... *Micropterus dolomieu*
Smelts ....Family Osmeridae
Snail darter .... *Percina tanasi*
Snooks .... *Centropomus* spp.
Sockeye salmon .... *Oncorhynchus nerka*
Speckled dace .... *Rhinichthys osculus*
Spiny dogfish... *Squalus acanthias*
Splake (brook trout × lake trout hybrid) .... *Salvelinus fontinalis × S. namaycush*
Spoonhead sculpin...*Cottus ricei*
Spottail shiner .... *Notropis hudsonius*
Spotted bass .... *Micropterus punctulatus*
Spotted seatrout .... *Cynoscion nebulosus*
Steelhead .... *Oncorhynchus mykiss*
Striped bass..*Morone saxatilis*
Striped mullet ....*Mugil cephalus*
Sturgeons .... Family Acipenseridae
Suckers .... Family Catostomidae
Sunfishes.. *Lepomis* spp.
Surf smelt .... *Hypomesus pretiosus*
Swordtails .... *Xiphophorus* spp.

Threadfin shad ....*Dorosoma petenense*
Threespine stickleback .... *Gasterosteus aculeatus*
Tiger muskellunge (northern pike × muskellunge hybrid) .... *Esox lucius × E. masquinongy*
Tiger trout (brown trout × brook trout hybrid) .... *Salmo trutta × Salvelinus fontinalis*
Tilapias ....*Tilapia* spp.
Trout-perch ....*Percopsis omiscomaycus*
Trouts .... Family Salmonidae
Tubenose goby....................................... *Proterorhinus marmoratus*
Tui chub ....*Gila bicolor*

Utah chub .... *Gila atraria*

Walleye .... *Stizostedion vitreum*
Warmouth ....*Lepomis gulosus*
Western mosquitofish .... *Gambusia affinis*
Westslope cutthroat trout........................... *Oncorhynchus clarki lewisi*
White bass ....*Morone chrysops*
White catfish..*Ameiurus catus*
White crappie ....*Pomoxis annularis*
White perch ....*Morone americana*
White seabass .... *Atractoscion nobilis*
White sturgeon ....*Acipenser transmontanus*

White sucker .... *Catostomus commersoni*
Whitefishes .... Family Salmonidae
Winter flounder .... *Pleuronectes americanus*
Woundfin ....*Plagopterus argentissimus*

Yellow bullhead .... *Ameiurus natalis*
Yellow perch .... *Perca flavescens*

Nonfish Animal Taxa

Alligator .... *Alligator* sp.
Bald eagle .... *Haliaeetus leucocephalus*
Beaver .... *Castor canadensis*
Belted kingfisher .... *Ceryle alcyon*
Blue crab .... *Callinectes sapidus*
Blue crabs .... *Callinectes sapidus*
Brown shrimp .... *Panaeus aztectus*
Bullfrogs .... *Rana catesbeiana*

Common mergansers .... *Mergus merganser*
Cormorants .... Family Phalacrocoracidae
Crabs .... Suborder Brachyura

Double-crested cormorant .... *Phalacrocorax auritus*
Ducks .... Family Anatidae

European crayfish .... *Astacus* spp.

Great blue herons .... *Ardea herodias*
Grizzly bear .... *Ursus arctos*

Mink .... *Mustela vison*
Muskrats .... *Ondrata zibethica*
Mussels .... Family Unionidae

Opossum shrimp .... *Mysis relicta*
Otters .... *Lutra canadensis*
Oysters .... Family Ostreidae

Panaeid shrimps .... Family Panaeidae
Pelicans .... *Pelicanus* spp.

Quagga mussel .... *Dreissena bugensis*

Red swamp crawfish .... *Procambarus clarkii*
Rusty crayfish .... *Orconectes rusticus*

Sea lions .... Family Otariidae
Seals .... Families Phocidae, Otariidae
Signal crayfish .... *Pacifastacus leniusculus*

White river crawfish .... *Procambarus acutus*
White shrimp .... *Panaeus setiferus*

Zebra mussel .... *Dreissena polymorpha*

Plant Taxa

Alder .... *Alnus* spp.

Arrowhead .... *Vallisneria americana*

Bermuda grass .... *Cynodon dactylon*

Black spruce ....*Picea mariana*

Bluestems .... *Andropogon* spp.

Buffalo grass .... *Buchloe dactyloides*

Cottonwood .... *Populus* spp.

Douglas fir .... *Pseudotsuga menziesii*

Eurasian water milfoil .... *Myriophyllum spicatum*

Fir .... *Abies* spp.

Hydrilla .... *Hydrilla verticillata*

Juniper .... *Juniperus* spp.

Knotgrass .... *Polygonum aviculare*

Lodgepole pine ....*Pinus contorta*

Maiden cane ....*Panicum hemitomon*

Ponderosa pine ....*Pinus ponderosa*

Pondweed .... *Potamogeton* spp.

Purple loosestrife ....*Lythrum salicaria*

Reed canarygrass .... *Phalaris canariensis*

Smartweed .... *Polygonum* spp.

Spruce ....*Picea* spp.

Watercress .... Family Brassicaceae

Weeping lovegrass .... *Eragrostis curvula*

Western wheatgrass .... *Agropyron smithii*

White fir... *Abies concolor*

Widgeongrass .... *Ruppia maritima*

Wild millet..*Echinochloa* spp.

Willow .... *Salix* spp.

PREFACE

Inland Fisheries Management in North America was developed by fisheries scientists associated with the American Fisheries Society, particularly the Education and Management Sections, to describe the conceptual basis and current management practices being applied to manipulate freshwater and anadromous fisheries of North America. This revised edition expands and updates the text, and is devised for use in introductory university courses in fisheries management for juniors, seniors, or graduate students. It is assumed that students have a general background in ecology, limnology, ichthyology, and college-level mathematics.

The book is designed as an educational tool, not a reference work; however, each chapter is fully referenced to allow readers access to the specific references and to advanced, more detailed information. Accordingly, the book should be of use to the practicing fisheries manager. The book covers fishery assessments, habitat and community manipulations, and the common practices for managing stream, river, lake, and anadromous fisheries. Chapters on history; ecosystem management; management processing; communication with the public; introduced, undesirable, and endangered species; and the legal and regulatory framework provide the full context for modern fisheries management.

The text is divided into five sections. The first section (Introduction) provides a historical overview of inland fisheries management in North America followed by chapters detailing the decision processes, communications, legal considerations in fisheries management, and ecosystem management. The second section (Fishery Assessments) includes chapters that describe how biological statistics, including models of population dynamics, are used to assess the status of fisheries. A chapter on socioeconomic assessment is also included. In the third section (Habitat Manipulations), riverine, lacustrine, and estuarine environments are described along with specific measures that can be taken to maintain and improve the integrity of the habitats. Community Manipulations is the theme of the fourth section. Chapters cover the management of introduced species, use of stocking in fisheries management, controlling undesirable species, managing endangered species, and manipulating sport fisheries through the use of regulations. The fifth section (Common Management Practices) concludes the book with chapters on coldwater streams, warmwater streams, large rivers, farm ponds and small impoundments, natural lakes and large impoundments, the Great Lakes, and finally, anadromous stocks.

The editors are indebted to the following individuals for assistance in developing the text outline: David H. Bennett, Frank J. Bulow, Robert F. Carline, James S. Diana, Charles C. Krueger, Stephen P. Malvestuto, John R. Moring, Peter B. Moyle, Donald J.

Orth, Charles G. Scalet, Richard J. Strange, Michael J. Van Den Avyle, and Jimmy Winter. Although an attempt was made to develop a comprehensive textbook, the editors recognized that not every aspect of inland fisheries management can be covered within the confines of a single volume.

Each chapter was externally peer reviewed, and the criticisms were thoroughly considered by and constructive to the authors and editors. For the time and care they invested as peer reviewers, we thank: Kendall Adams, David D. Aday, Michael Allen, Stanley Anderson, Mark B. Bain, Charles R. Berry, John G. Baughman, Richard C. Bishop, Stephen B. Brandt, Walter R. Courtenay, Jr., Larry B. Crowder, Richard A. Cunjak, James T. Davis, William D. Davies, Joe G. Dillard, Harvey R. Doerkson, Lauren R. Donaldson, Don A. Duff, James R. Fazio, David L. Galat, Thomas Gengerke, Robert Gresswell, Bobby G. Grinstead, Vern Hacker, Arthur D. Hasler, Burchard Heede, Robert L. Hunt, Daniel Isaak, Robert M. Jenkins, Charles C. Krueger, Thomas J. Kwak, Boyd Kynard, Robert T. Lackey, Peter A. Larkin, Joseph G. Larscheid, William M. Lewis, Hiram Li, John J. Magnuson, Stephen P. Malvestuto, F. Joseph Margraf, Eugene Maughan, James K. Mayhew, Thomas S. McComish, Keith McLaughlin, L. Esteban Miranda, Brian R. Murphy, Fred P. Myer, Richard J. Neves, Anthony A. Nigro, Richard L. Noble, Gary Novinger, John Osborn, Vaughn L. Paragamian, William D. Pearson, R. Ben Peyton, Edwin P. Pister, James B. Reynolds, William E. Ricker, Thomas R. Russell, Charles G. Scalet, Dennis L. Scarnecchia, James C. Schmulbach, Harold Schramm, Jr., William L. Shelton, George R. Spangler, Jon G. Stanley, Lynne B. Starnes, Roy A. Stein, David K. Stevenson, Robert R. Stickney, Richard H. Stroud, David H. Wahl, David K. Whitehurst, James A. Williams, David W. Willis, Paul J. Wingate, and Alejandro Yáñez-Aranciba,

Several authors also wish to acknowledge the help they received as they created their respective chapters. Charles Krueger and Daniel Decker thank the following individuals for providing reviews of Chapters 2 and 3: G. A. Barnhart, T. L. Brown, A. Creamer, R. E. Eshenroder, D. McElveen, B. A. Knuth, L. A. Neilsen, P. Pister, C. P. Schneider, R. J. Wattendorf, and P. J. Wingate.

Berton Lamb and Donna Lybecker received helpful suggestions for Chapter 4 from E. R. Mancini, J. B. Gabbert, E. F. Tritschler, T. DeYoung, A. Locke, S. Mumme, C. W. B. Stubbs, N. P. Lovrich, and P. Eschmeyer.

Steve Weithman appreciates reviews for Chapter 8 by R. O. Anderson, E. K. Brown, N. Burkardt, N. Sexton, and D. J. Wittler.

Thomas Wesche appreciates the critical reviews of Chapter 9 by Q. Skinner and J. F. Orsborn.

Donald Orth and Ray White thank C. A. Dolloff, P. L. Angermeier, and W. M. Turner for reviews of Chapter 10; M. D. Lobb and J. Fanz for locating suitable photographs; and R. L. Hunt for his pioneering field evaluations on habitat development.

Robert Summerfelt is grateful to L. R. Mitzner and G. R. Ploskey for advice on Chapter 11.

William Herke and Barton Rogers thank the numerous respondents to their requests for information while preparing Chapter 12. An earlier draft was much improved as a result of critical reviews by H. Austin, D. Fruge, C. Levings, F. Perkins, R. Shaw, and T. Shirley. The same is true of reviews by students at Iowa State University, Louisiana State University, Montana State University, and the University of Massachusetts: M. Adkinson, C. Butson, Y. Lee, J. Mack, M. Mullins, S. Murphy, J. Serrano and four who wished to remain anonymous. Special thanks to A. Yáñez-Arancibia for providing information for Mexico, and appropriate references to the Spanish literature. The criticisms and suggestions of the above were most helpful, but not all were accepted in the final version; therefore, errors or omissions must remain the responsibility of the authors and editors.

Hiram Li and Peter Moyle thank K. Currens and D. Buchanan for review comments on Chapter 13.

Roy Heidinger appreciates the critical reviews of Chapter 14 by M. Marcinko and P. Shafland.

Robert Wiley and Richard Wydoski appreciate the helpful suggestions on Chapter 15 by J. W. Mullan and W. F. Sigler.

Frank Rahel, Clare Carlson, and Robert Muth extend gratitude to H. Rolston and R. J. Behnke for review comments as well as to O. Bray, J. Hamill, L. Lamb, and N. Kaufman for expert advice and unpublished material.

Richard Nobel and Wayne Jones thank R. O. Anderson, F. A. Harris, C. R. Inman, J. R. Jackson, and L. C. Redmond for providing review comments on Chapter 17.

Jack Griffith thanks D. W. Chapman, T. W. Hillman, and T. R. Angradi for reviewing Chapter 18.

Steve Flickinger and Frank Bulow thank the numerous respondents to their requests for information while preparing Chapter 21, the Alabama Division of Game and Fish for demonstrations used in some of the photographs, and T. C. Modde for supplying the basic drawing for the pond distribution map.

Dan Hayes, Bill Taylor, and Patricia A. Soranno thank S. Marod, J. Kocik, and R. Brown for their reviews of early drafts of Chapter 22. Assistance in literature searches by S. Marod are also gratefully acknowledged.

Chapter 23 by David Jude and Joseph Leach is contribution 600 from the Center for Great Lakes and Aquatic Sciences and No. 98-01 of the Ontario Ministry of Natural Resources. The authors also extend thanks to T. Wells, J. Christie, D. Ryder, and R. Eshenroder for review comments. The seemingly endless draft versions of the chapter were typed by B. McClellan. The authors are in her debt.

John Moring thanks the following individuals for providing review comments on Chapter 24: D. Buchanan, H. Lorz, N. McHugh, M. Stratton, D. Locke, S. Pierce, E. Baum, N. Dubé, A. Meister, W. Hauser, D. Jones, D. Sidelman, G. Thomas, D. Misitano, A. Knight, F. Ayer, R. Barnhart, J. Delabbio, B. Glebe, G. Allen, L. Flagg, T. Squiers, S. Ellis, and S. Wilson.

The editors gratefully acknowledge Teresa Cavitt and Nellore Collins for typing the entire volume (1st edition) on a single word processing program. This greatly facilitated the editorial process. We thank Allory Deiss, Thomas Lund, Carol Stevens, and Elizabeth Ono-Rahel with the graphics art shop at the University of Wyoming for the preparation of technical drawings.

We also appreciate the help provided by Beth Staehle and Amy Wassmann (1st edition) and Janet Harry and Eric Wurzbacher (2nd edition) of the Society's headquarters staff who contributed editorial improvements in the final manuscript and saw the book through production.

We acknowledge additional editorial assistance provided by Eva Silverfine to the 2nd edition. Her comments greatly improved many chapters.

Elizabeth Ono-Rahel designed the cover for the 2nd edition.

Funding for the book was provided by the Education Section and the Fisheries Management Section of the American Fisheries Society.

Any trade names and product vendors mentioned in this book are done so as a convenience to the readers. Specific citations do not imply endorsements by the American Fisheries Society, the authors and editors, or the employers of the book's contributors.

Christopher C. Kohler
Wayne A. Hubert

INTRODUCTION

Chapter 1

History of Inland Fisheries Management in North America

LARRY A. NIELSEN

1.1 INTRODUCTION

Fisheries management is a young profession. Its history—at least the part of greatest practical interest—can be recounted through the personal experience of older fisheries managers still working. After 44 years as a fisheries biologist, Woody Seaman described himself as "a 'starter' of new things. ...As you know, I was the first Chief of Fisheries, West Virginia Conservation Commission. Prior to this, it was only a hatchery program. I staffed up a group of fish biologists who tackled the fish management problems of the state" (Seaman 1988). Seaman was describing conditions in 1946, a time when many states and provinces were also just beginning their conservation programs and when many other starters like Woody Seaman were pushing the profession forward. These managers quickly developed most of the strategies and techniques used in fisheries management today. This rapid evolution of fisheries as a profession coincides with the accelerating development of the North American continent and with our increasing concern about the long-term consequences of human action.

The precedents of fisheries management, however, extend back to the origins of European settlement of North America. As is common with all subsets of human activity, the events and attitudes of society have affected fisheries and fisheries managers profoundly. The observation that fisheries management has a political and sociological context, as well as an ecological basis, is as accurate for earlier centuries as it is for today.

Understanding fisheries management, therefore, requires a familiarity with the ideas and events that have shaped the North American personality and landscape.

This chapter has two purposes. First, it connects fisheries management and societal development. It illustrates that fisheries are part of the larger society, validating the obvious fact that fisheries management is, and always will be, subject to the desires of the public and societal leadership. Second, the chapter introduces many of the concepts developed in detail in later chapters. It displays the rich variety of approaches and techniques that make up fisheries management.

The chapter in total provides a conceptual definition of fisheries management. A dictionary style definition might read, "the manipulation of aquatic organisms, aquatic environments, and their human users to produce sustained and ever increasing benefits

3

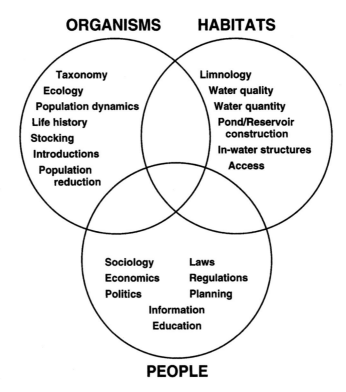

ORGANISMS   **HABITATS**

Taxonomy
Ecology
Population dynamics
Life history
Stocking
Introductions
Population
  reduction

Limnology
Water quality
Water quantity
Pond/Reservoir
  construction
In-water structures
Access

Sociology   Laws
Economics   Regulations
Politics   Planning
  Information
  Education

PEOPLE

Figure 1.1 Fisheries management depicted as three overlapping circles that represent the three interacting components of fisheries—aquatic organisms, aquatic habitats, and people.

for people." This conception is often illustrated as three overlapping circles that represent the three principal components of fisheries: organisms, habitats, and people (Figure 1.1). Each is important, each affects the other two, and each presents opportunities for enhancing the value of fisheries resources.

1.2   THE HISTORICAL BASIS FOR FISHERIES MANAGEMENT

Fisheries in North America are public resources. State, provincial, and federal governments hold the resources in trust for the general use of their citizens. Although this system is different from the private ownership of fisheries in much of Europe, public ownership in North America derives directly from early English practices.

Ownership of fisheries resources has changed through time to follow the pattern of government. When the feudal system of governance spread across Europe during the Middle Ages, feudal kings and land barons declared themselves owners of the land. They claimed ownership of wildlife as well, not necessarily because wildlife was valuable, but because they wished to keep weapons out of the hands of the people. Later, as the feudal system in England evolved to a national monarchy, royalty expanded their ownership to include all wildlife and fisheries, exercising their assumed right to assign both exclusive and nonexclusive franchises (permission to use) to private individuals (Bean 1977). This system also governed fisheries use and ownership in North America up to the time of the American Revolution.

After the United States became a sovereign nation, U.S. courts struck down the exclusive rights previously given by English royalty. They reasoned that under the English law principle, the king held fisheries as a public trust; moving such franchises to private, individual ownership violated that principle. After the revolution, all the rights and responsibilities of being the trustee of public fisheries were transferred to the new governments, specifically the state governments. Since that time, state governments in the United States have been the principal guardians of fisheries resources, vested with the major responsibilities for protecting and managing inland fisheries (Bean 1977).

In Canada, the responsibilities for managing inland fisheries were originally held by the provinces, operating as independent governments. Since the Confederation Act of 1867, however, by which the Canadian provinces agreed to join under a federal government, fisheries have been shared by provincial and federal governments. The provinces own most inland fisheries, holding them in trust and managing the distribution of fisheries benefits. The federal government holds the responsibility to protect inland fisheries. This dual authority over Canadian fisheries management, which has produced a complicated legal history, has evolved through time; most responsibility now lies with the provincial governments (Thompson 1974). Management of Mexican resources is conducted primarily by the federal government, which assumed control of wildlife resources in the 1910 revolution (Beltran 1972).

The question of how a public trustee should treat fisheries has also been answered by a tradition as old as government itself: fisheries are common property. Common properties are those resources owned by the entire populace, without restriction on who may use them and, at least in earlier times, on how they may be used. The principle of common property was established formally in 1608, when the Dutch statesman Hugo Grotius proffered the Doctrine of Freedom of the Seas. The open ocean, including its fishes, was declared the property of all people. This seemingly idealistic principle was actually just a statement of fact. In 1608, no one could control the oceans because defining boundaries and then protecting them was essentially impossible. Ownership was not necessary in any case because the wealth of the oceans was believed to be inexhaustible—the oceans held more fish than anyone could ever imagine needing or being able to catch (Nielsen 1976). The Doctrine of Freedom of the Seas has been modified throughout history but has remained the principle for the management of fishes, both marine and inland, as common property to this day.

The common property principle is a good one—under certain conditions. As long as the productivity of a fishery exceeds the demand for its products, common property provides an efficient and equitable allocation system. When demand gets too high, however, the productive capacity of the resource declines, and everyone suffers. For centuries, fisheries did supply all the food that people sought, and little or no attention was paid to managing the resource. Within the past century, however, the demand for fish has exceeded the available supply in place after place around the world.

Fisheries management was born of a need to balance the supply–demand equation. The modern history of fisheries is basically a chronicle of individual and governmental attempts to control the exploitation of common property fisheries. Over the past century, scientists and public officials have struggled to develop suitable objectives for managing fisheries—objectives that preserve the time-honored ideal of free access to fisheries and that preserve the productive capacity of fish populations. In pursuing these objectives, they have also developed the technical capacity to enhance fisheries productivity and to reduce the influence of other human activities on fisheries resources.

1.3   THE PRELUDE TO FISHERIES MANAGEMENT

To European colonists, North America truly was a new world. The continent was populated at such a low density and was so naturally productive that early colonists faced a land entirely beyond their experience. Resources were seemingly limitless. Of the Chesapeake Bay, Robert Beverly wrote in 1705:

> As for fish, both of fresh and saltwater, of shellfish, and others, no country can boast of more variety, greater plenty, or of better in their several kinds…In the spring of the year, herrings come up in such abundance into their brooks and fords to spawn that it is impossible to ride through without treading on them.…Thence it is that at this time of the year,…the rivers…stink of fish. (Quoted in Wharton 1957.)

It is little wonder that the principle of common property and an attendant lack of concern for conservation were the standards for conduct.

European settlers approached the new continent aggressively. The untamed landscape was viewed as an enemy that had to be subdued in order to provide a suitable human environment. Various explanations of this aggressiveness have been offered, including the Judeo–Christian ethic and the rise of democracy, but the prevailing forces must surely have been necessity and opportunity. The colonists were strictly utilitarian, and their early laws were developed to govern only the commercial aspects of natural resource use (Kawashima and Tone 1983).

Native Americans, in contrast, lived more harmoniously with nature. Natural resources, including fishes, were the base of their existence, and virtually all Native Americans lived largely by subsistence fishing, hunting, and gathering (Swanton 1946). Their mythology often ascribed human qualities to natural objects, thus compelling an ethical treatment of their environment. Where Native American populations were dense, as in coastal California, Native Americans restricted their fish harvests in ways that sustained high productivity for centuries (McEvoy 1986). In most areas, however, their population density was low; at the time of European colonization, fewer than 12 million humans occupied North America. The sparse population, coupled with the low-technology forms of agriculture and fish harvest Native Americans practiced, never made significant impacts on the continent's resources.

The aggressiveness of the colonists could not be sustained long without local problems. Local resource depletion and environmental problems, like those long experienced in Europe, developed rapidly in North America. Pastures were overgrazed, forests were overharvested, streams were overburdened with wastes, and fisheries were overexploited. Laws to correct and prevent local degradation became common in colonial governments, beginning with a 1652 Massachusetts law restricting fish catches (Stroud 1966). By the time of the American Revolution, hundreds of statutes restricted the times, places, and mechanisms for harvesting fishes, precursing the variety of similar laws in force today.

Figure 1.2 The westward colonization of North America used rivers as liquid highways. The Ohio River was perhaps the most important highway because its westward flow carried settlers and cargo across the rugged Appalachian Mountains. Drawing is courtesy of the U.S. Army Corps of Engineers, Huntington, West Virginia.

The answer to resource depletion and environmental degradation, however, was not yet to be found in good management. The United States' answer was immortalized in the words of Horace Greeley, "Go West, young man!" Westward expansion of the new nation was a dominant goal of the United States' leaders, most notably Thomas Jefferson. Jefferson envisioned a nation of small landowners—hardworking, stable, and dedicated to preserving the nation. Federal land policy, from the Louisiana Purchase in 1803 to the Homestead Act of 1862, reflected the Jeffersonian doctrine of expanding, subdividing, and settling the frontier (Cox 1985).

Water played a dominant role in western settlement as a liquid highway transporting people and goods across the eastern mountains. The Ohio River was the primary route of immigrants embarking at Pittsburgh for destinations in the Northwest Territory (Figure 1.2). River travel was difficult and dangerous, interrupted by portages around rapids, strandings in shallow water, and damage from floating debris. Waterway improvements, therefore, also became national priorities, and the United States' first federally supported public works were for improved navigation on the Ohio River (Smith 1971). By 1824, under the guise of improving national defense, the U.S. Army Corps of Engineers was actively and continuously modifying the Ohio and Mississippi rivers by digging canals, removing snags, and deepening the channels. Since then, water development projects have become a dominant feature in U.S. domestic policy and have modified the fisheries of virtually every major U.S. river system (see Chapter 20).

Fisheries in the nineteenth century were primarily subsistence and commercial. The subsistence use of fisheries is poorly recorded, but it certainly provided substantial food supplies for local communities. Commercial fisheries, which had been based on marine fishes in colonial times, began to grow in freshwaters. The Great Lakes commercial fisheries were well established by 1835, with expectations of much greater possible harvests as the century progressed (Whitaker 1892). The state of Iowa reported a commercial fish catch of more than 1.8 million kilograms in 1886, or close to 1.4 kg per person (Carlander 1954). The Mississippi River was well used for commercial and subsistence fishing, although not everyone agreed that the river offered either palatable food or enjoyable sport (see Box 1.1).

Giving the western lands to eastern migrants also meant taking it away from Native Americans. During the first half of the nineteenth century, scores of treaties between Native American nations and the U.S. government traded millions of hectares desired by settlers for more distant lands currently not desired. As the frontier pushed westward, so did the treaty lands, removing Native Americans progressively farther from their homes and disrupting their ties with the land and its resources. The continuing efforts of the U.S. government to accommodate western settlement progressively stressed relocation, separation, and assimilation of Native Americans, policies that did not improve Native American lifestyles or protect their rights. During this time, however, a body of Native American law developed and a series of treaty-based promises accumulated. Modern interpretation of these laws and promises shows that in many cases Native Americans retain ownership of the natural resources, including fishes, on their aboriginal and treaty lands (Busiahn 1985). Recognition of those rights is reshaping the governance and use of many fisheries today and will continue to do so for the foreseeable future (see Chapter 4).

1.4   THE BIRTH OF FISHERIES MANAGEMENT

The completion of the U.S. transcontinental railway in 1869 signified both practical and symbolic ends to the frontier ethic. The continent had been tamed by the railroad, which allowed fast and convenient transportation across the formerly treacherous landscape. The question was no longer if the land would yield to human development, but how that development would proceed. The last half of the nineteenth century represents the period when North Americans of European descent began to exert their influence broadly on natural resources, including fisheries, and when they developed a unique New World character tied closely to the character of their environment.

1.4.1 The Cultural Basis for Natural Resource Management

The wilderness of the New World had a deep effect on its settlers. While carving homesteads out of the wilderness, the pioneers also developed a kinship with the expansiveness and wildness of the land. After 1850, that kinship flowered into a North American personality, reflected in many aspects of human endeavor. The culture emphasized the natural environment, represented in philosophy by the transcendentalism of Ralph Waldo Emerson; in art by the Hudson River school of landscape artists, most notably Thomas Cole; and in literature by romanticists such as William Cullen Bryant (Figure 1.3). The worldwide recognition of this cultural growth legitimized the place of nature in the arts, and it welded an appreciation for the natural environment to the growing surge of land development.

BOX 1.1   LIFE ON THE MISSISSIPPI

The Mississippi River was a dominant force in the lives of riverside communities. Harriet Bell Carlander (1954) chronicled the history of the river's fisheries, quoting examples of the wonder, inspiration, and indignation of river watchers.

Thomas Jefferson was an early champion of the Mississippi, as he recorded in his *Notes on Virginia*:

> The Mississippi will be one of the principal channels of future commerce for the country westward of the Allegheny....This river yields turtle of a peculiar kind, perch, trout, gar, pike, mullets, herrings, carp, spatula fish of fifty pound weight, catfish of one hundred pounds weight, buffalo fish and sturgeon.

Estwick Evans swam in the Mississippi in 1818 and noted the sediment in the water with mixed emotion:

> It is, however, not very unpalatable and is, I think, not unwholesome. The fish in the river are numerous and large; but they are too fat to be delicate.

Mark Twain quotes an English sea captain who visited the Mississippi in 1837 and was less ambivalent:

> It contains the coarsest and most uneatable of fish such as catfish and such genus....There are no pleasing associations connected with the great common sewer of the western America which pours out its mud into the Mexican Gulf, polluting the clear blue sea for many miles beyond its mouth.

Charles Lanman, a New Englander who wrote about the river in 1856, also was dubious about catfishes:

> ...This fish is distinguished for its many deformities and is a great favorite with all persons who have a fancy for muddy water. In the Mississippi they are frequently taken weighing upwards of one hundred pounds...but it has always seemed to us that it requires a very powerful stomach to eat a piece from one of the mammoths of the western waters.

The editor of the *Muscatine Journal* assessed fishing in 1869:

> Fishing parties are fashionable now-a-days. For our part we should prefer lighter employment, such as sawing wood or carrying a hod to the seventh story of a building.

The truth about the Mississippi, however, is probably best recorded by Mel Ellis, writing in the *Milwaukee Journal* in 1949:

> If you haven't fished Ol' Man Mississip, forget about any preconceived notions you may have as far as rivers are concerned. Because Ol' Man River isn't really a river at all. In fact, he's a hundred rivers and a thousand lakes and more sloughs than you could explore in a lifetime. He is creeks, bayous, ditches, puddles, and thousands and thousands of impenetrable lotus beds that break big yellow flowers out above green pads.

This linkage between the human and the natural landscapes was first tied to resource management through the writings of George Perkins Marsh. Marsh's 1849 book, *Man and Nature,* showed how human activity affected the physical environment and how both could be improved by applying scientific and aesthetic principles. He denounced the destructive effects of human exploitation and anticipated the development of the natural sciences of biology, geology, and chemistry (Nash 1987). His work aimed landscape design in a new naturalistic direction, one that would be implemented in full form through the genius of Frederick Law Olmsted, the architect of New York's Central Park and of the first U.S. national parks.

While Marsh was changing landscape architecture, a transformation also occurred in natural science. Until the mid-1800s, natural science had been an avocation. Amateur scientists collected bizarre and grotesque animals or spent their lives classifying animals, plants, and geological specimens. The application of physical and biological sciences to human endeavors, however, was like the touch of flame to oil. The pace and magnitude of human affairs exploded. As depicted in the United States' 1876 Centennial Exposition in Philadelphia, the industrial revolution unfolded in an array of machinery and engines. Natural science radiated into useful specialties. The application of zoology to agriculture was mandated in the United States by the creation of land grant colleges in 1863 (these same universities house many of the U.S. fisheries programs today). The science of ecology began to take form at about the same time; the encyclopedic observations of Canadian and U.S. naturalists provided the basis for its theoretical development (Fry and Legendre 1966). Charles Darwin's *Origin of Species*, published in 1859, was the precursor to ecology, a term formally defined by Ernst Haeckel in 1866 (Egerton 1976).

Fisheries and other natural resources were also subject to the modernization brought by the industrial revolution. Modifications of the landscape were more massive and ubiquitous, multiplying and merging the local environmental impacts of previous generations. Power dams were built on many rivers; by 1849 on the Connecticut River, these dams had blocked the spawning migrations of Atlantic salmon and American shad. The combination of mill dams, deforestation, and pollution had virtually exterminated Atlantic salmon from the St. Lawrence system by the mid-1800s. Throughout the continent, the use of waterways for power, transportation, mining, and waste disposal and the use of watersheds for rapid exploitation of timber and minerals and for crops were destroying the capacity of aquatic environments to sustain fish populations.

1.4.2 The First Steps for Fisheries Conservation

Fisheries exploitation had always been a local industry, but the industrial revolution of the late 1800s allowed rapid expansion of exploitation for an ever growing market. The demand for fishes increased as the human population grew and as spreading railroad lines made possible rapid transportation of fishes to distant cities. Fish processors developed better canning and refrigeration techniques, which led to massive exploitation of concentrated fish stocks. On the Sacramento River in California, for example, the chinook salmon fishery expanded from 2,500 cases of canned fish in 1874 to 200,000 cases in 1882, caught by 1,500 boats and pro-

Figure 1.3 The harmony between the North American personality and its wild natural environment is the subject of this 1849 painting by Asher Durand. Entitled *Kindred Spirits*, it depicts Thomas Cole, the foremost natural landscape artist of the time, and William Cullen Bryant, a naturalistic poet and journalist, in a setting like those that inspired them. Photograph is courtesy of the New York Public Library.

cessed in 21 canneries. Fishers responded by improving their own technology. Steam replaced sail as the power for fishing vessels in the 1880s, permitting larger vessels and more reliable fishing schedules. The effectiveness of fishing equipment advanced regularly—for example, the otter trawl and the deep trap net were developed in the late 1860s.

Concern for the decline of fisheries was building throughout this period. In addition to a general society-wide interest in natural resources, fishers, scientists, and government officials began to doubt that fisheries—in either marine or freshwaters—were inexhaustible. The Sacramento River salmon, for example, responded to heavy exploitation by producing regularly declining spawning runs; after 1882, the fish harvests declined as well. The United States institutionalized its concern in 1871 by creating the U.S. Fish Commission for the express purpose of investigating the decline in commercial fisheries (Allard 1978). Similar actions were underway around the world, with national investigations in Russia, Germany, and England during this time (Nielsen 1976). Provinces and states began establishing their own fisheries agencies, beginning with Quebec in 1857, New York in 1868, and California in 1870. In 1870, a small group of fish culturists formed the American Fish Culturists' Association (now the American Fisheries Society) to promote fish culture as a cure for the widespread destruction of fisheries (Thompson 1970).

Although the modern techniques of fisheries management were used in rudimentary form during the late 1800s, most people believed that the future belonged to fish culture. In its second year of operation, the U.S. Fish Commission was given the added task of raising fishes and distributing them throughout the United States for the promotion of commercial fisheries. Spencer Baird, the first U.S. Fish Commissioner, hired Livingston Stone in 1872 to carry American shad to the West Coast and, once there, to establish salmon culture stations for Pacific salmon eggs to be transported back East.

These early fish culture operations were enormously successful, and they spawned an era of unrestrained enthusiasm for raising and stocking fishes throughout the continent (Regier and Applegate 1972). Striped bass were shipped from New Jersey to California in 1879 and developed into a commercial fishery in California by 1889 (Craig 1930). Rainbow trout were first distributed into the eastern United States in 1880, and by 1896 many states east of the Rocky Mountains boasted self-sustaining populations of rainbow trout (Wood 1953). European introductions were equally popular; common carp was imported from Germany in 1877 and brown trout in 1883. Missouri built its first hatchery in 1881 for the production and distribution of common carp (Callison 1981). The belief that fish culture could sustain commercial fisheries was the established position of the U.S. government throughout this period. A 1900 law, for example, required fishers in Alaska to establish sockeye salmon hatcheries on every river they fished (Roppel 1982). The technical improvements of the time also enhanced fish stocking activities. Long-distance transportation of fishes was accomplished in specialized railway cars, first used in 1873 and finally retired in 1947 (Leonard, no date). The initial transfers used milk cans to carry eggs, but soon the cars were outfitted with specialized fish-holding cans (Fernow pails) and later with tanks carrying ice and aeration devices (Figure 1.4).

Fish rescue operations provided another major source of fishes for stocking and redistribution, especially in the midwestern United States. Annual flooding of the Mississippi River and its major tributaries stranded multitudes of fishes as the waters receded. State officials reasoned that the death of these fishes reduced fish abundance in the river. In 1876, therefore, Iowa began an annual program of rescuing fish, a program that continued until 1930 (Carlander 1954). Most of the fishes were returned to the

Figure 1.4 Railroad cars, like this one, were specially adapted to transport fish and fish eggs across the continent. Long-distance transport by rail continued well into the 1940s. Photograph is courtesy of the U.S. Fish and Wildlife Service.

main channel of the river, but game fishes were kept for distribution into other waters throughout the region. The work was difficult—weeks of seining in shallow, silt-bottomed ponds in hot weather. The death of one young worker after several years of fish rescue work prompted an infamous epitaph on his Iowa gravestone (Figure 1.5).

Fishing regulations proliferated during the latter nineteenth century, generally in response to the declining fisheries and the desire to protect stocked fishes. Most laws regulated either the seasons or methods of commercial fishing. Based on the implicit belief that spawners were needed to assure future yields, closed seasons were implemented to protect spawning fishes. Regulation of fish-catching methods, however, was usually politically motivated, designed to restrict the effectiveness of some fishers while enhancing the effectiveness of others (Nielsen 1976). Riddled with loopholes and contradictions, such regulations were usually ineffective for fish conservation. Early commercial fishing laws also provided no means for enforcement, other than the efforts of ordinary police. The few regulations that did exist for sportfishing were even less likely to be enforced. Among the first law enforcement officers specifically hired for natural resource work were those in California, authorized in 1876, and in New York, authorized in 1883.

At the end of the nineteenth century, fisheries were experiencing the worst conditions in history. Fish populations had been badly depleted throughout the developed world. Environmental degradation tied to industrial frenzy, and overfishing tied to insatiable demand, had affected freshwater and anadromous fisheries to the point that some commercially exploited fishes were extinct and many were economically useless. But a new attitude was pervading the continent. The respect for nature that was intrinsic in the North American heritage was being directed toward fisheries. Unlimited exploitation had been benign on the frontier, but people recognized that the closure of the frontier necessitated a conservative attitude. Attempts to control exploitation were largely political and insufficient, but the scientific basis for more rational management was beginning to develop. The new century would harbor a new science and new approaches to the use and conservation of natural resources.

Figure 1.5 The dreadful truth about fisheries work is shown on this gravestone standing in Manchester, Iowa. The young man who died was not actually a hatchery worker but a member of one of the fish rescue crews that seined fishes along the Mississippi River.

1.5 THE SCIENTIFIC MANAGEMENT OF FISHERIES

Most historians mark the beginning of the North American conservation movement with the Governors' Conference of 1908. The U.S. president, Theodore Roosevelt, invited governors of all states and territories, federal legislators, the justices of the Supreme Court, cabinet officers, and selected authorities to the White House to discuss the conservation of natural resources. The conference serves well as a milestone marking the transition of the conservation movement from a collection of separate ideas and actions into a national priority.

Natural resource conservation received this attention partly because of the acknowledged need for better management, but mostly because it was swept along in a new national outlook. The twentieth century began with an obsession for efficiency in the conduct of human affairs—the progressive movement. Born of the waste and corruption of the concluding century, the progressive movement presumed that direct manipulation of human affairs, often aided by the best scientific expertise, could produce a better society—even a perfect one. The philosophy was the force behind the work of Gifford Pinchot, Chief of the U.S. Forest Service under President Roosevelt. Known in conservation as the founder of modern forestry, Pinchot was a zealous proponent of efficiency in all aspects of resource use and societal development—so zealous that his forceful rhetoric has been called "the gospel of efficiency." He popularized the notion that natural resources should be used for the long term, conserving their capacity to produce human value indefinitely. The other part of that philosophy, however, held that fully using the annual production of renewable resources was an obligation in the service of the present human population, as he observed for coal:

> The first principle of conservation is development, the use of the natural resources now existing on this continent for the benefit of the people who live here now. There may be just as much waste in neglecting the development and use of certain natural resources as there is in their destruction. We have a limited supply of coal, and only a limited supply. Whether it is to last for a hundred or a hundred and fifty or a thousand years, the coal is limited in amount, unless through geological changes which we shall not live to see, there will never be any more of it than there is now. But coal is in a sense the vital essence of our civilization. If it can be preserved, if the life of the mines can be extended, if by preventing waste there can be more coal left in this country after we of this generation have made every needed use of this source of power, then we shall have deserved well of our descendants. (Pinchot 1910)

Fish populations under the progressive philosophy served the primary purpose of providing food and the secondary purpose of providing economic value, much like agricultural crops (Bower 1910). For example, at Roosevelt's 1908 Governors' Conference, only two speakers mentioned fisheries, both extolling the virtues of commercial fisheries as thriving industries that provided a wholesome food supply (Smith 1971). The idea that natural resources were crops to be planted, managed, and harvested would later evolve into the founding principle of wildlife management (Leopold 1933) and would dominate the thinking of fisheries scientists for the first half of the twentieth century.

The efficient use of fish populations became known as maximum sustainable yield, or MSY. During the early twentieth century, the MSY concept was developed independently several times. E. S. Russell, a British scientist studying marine fisheries, presented it most effectively in his classic book, *The Overfishing Problem* (Russell 1942). He related the present abundance of a fish population to additions via growth, recruitment, and immigration and to losses via natural mortality, fishing mortality, and emigration. Russell's simple mathematics showed that the greatest long-term yield of fish was achieved by allowing small fish to grow before harvesting them. The "vital statistics" of a fish population—growth, recruitment, natural mortality, fishing mortality, immigration, and emigration rates—became the standard descriptors of fish population dynamics and remain so to this day (see Chapter 6).

Logical arguments like Russell's and political philosophies like Pinchot's were powerful, but they also needed a strong scientific basis. The science of ecology answered that need. Ecology, which began in earnest in the late 1800s, developed rapidly during the 1920s and 1930s (McIntosh 1976). Much of the early work in ecology centered on aquatic environments, relating the lives of organisms to the physical and chemical characteristics of the habitat. Significant in that early work was Stephen A.

Forbes, Director of the Illinois State Laboratory of National History in the 1880s. His classic paper, "The Lake as a Microcosm," explains the attraction of aquatic ecosystems and the need for comprehensive information:

> It forms a little world within itself—a microcosm within which all the elemental forces are at work and the play of life goes in full, but on so small a scale as to bring it easily within the mental grasp. Nowhere can one see more clearly illustrated what may be called the *sensibility* of such an organic complex, expressed by the fact that whatever affects any species belonging to it, must have its influence of some sort upon the whole assemblage. He will thus be made to see the impossibility of studying completely any form out of relation to the other forms; the necessity for taking a comprehensive survey of the whole as a condition to a satisfactory understanding of any part. If one wishes to become acquainted with the black bass, for example, he will learn but little if he limits himself to that species. He must evidently study also the species upon which it depends for its existence, and the various conditions upon which *these* depend. He must likewise study the species with which it comes in competition, and the entire system of conditions affecting their prosperity; and by the time he has studied all these sufficiently he will find that he has run through the whole complicated mechanism of the aquatic life of the locality, both animal and vegetable, of which his species forms but a single element. (Forbes 1925)

1.5.1 The Scientific Basis for Fisheries Management

Ecology and fisheries science grew in mutualistic fashion, with ecology supplying hypotheses and principles and fisheries science supplying natural laboratories for testing them. Fisheries science grew exponentially in this exciting environment, with early scientists concentrating on gathering information in three areas critical to fisheries management.

First, scientists needed to describe and survey the fishes and invertebrates in important waters before the management of fisheries could even be considered. Building on classical studies by naturalists such as Louis Agassiz, many prominent ichthyologists, including Tarleton Bean, David Starr Jordan, Barton Evermann, and J. R. Dymond, conducted faunal surveys for state, provincial, and federal agencies in the early 1900s. Conservation agencies and universities carefully documented those surveys in elaborate reports complete with taxonomic keys and full-color drawings. The first volume of the *Roosevelt Wild Life Annals*, for example, published by the New York State College of Forestry, contains a 300-page report of the fishes and fishery of Oneida Lake (Figure 1.6; Adams and Hankinson 1928). This style of reporting was largely replaced by the next generation of zoologists, who prepared standardized taxonomic references for the continent or for large ecological regions. Works on the fishes of the Great Lakes by Carl Hubbs and Karl Lagler (1941) and on the freshwater invertebrates of North America by Ronald Pennak (1957), for example, became major references for biological surveyors.

The second type of information desired by fisheries scientists related to the physicochemical characteristics of important waters. Work in this area had developed a strong theoretical base through the work of European scientists like Francois Forel, who named the new field limnology in 1869 (Egerton 1976). Based on the ecological link between organisms and their environments and on the concern for the devastating pollution of the early 1900s, physical and chemical limnology became important components of

Figure 1.6 Agency reports were the first textbooks on fisheries science, ecology, and management. Drawings like this, from an intensive survey of Oneida Lake, New York, provided scientifically accurate records of fish species.

fisheries management (Osburn 1933). Realizing the need to survey the conditions of lakes and streams, researchers blanketed the continent with meters, water bottles, and nets. The most famous field-workers were Edward Birge and Chancey Juday, limnologists for the University of Wisconsin who surveyed waters throughout North America in the early 1900s (Beckel 1987). Their efforts were matched in Canada by Donald Rawson, who used his extensive field experience to create indices predicting fish production based on lake characteristics (Northcote and Larkin 1966). Large-scale physicochemical surveys are seldom conducted by fisheries managers today and are now handled by specialists in water management. Nevertheless, the idea remains firmly planted in fisheries management that water quality determines the type and productivity of a fishery.

The third set of information desired by early fisheries scientists concerned the life history and ecology of fishes. Scientists recognized that management of fish populations required knowledge of the critical elements in a fish's life and of the factors that altered the vital statistics of the population. Therefore, management agencies began to sponsor directly research on fish ecology (Larkin 1979). In 1908, the Canadian government opened the Pacific Biological Station at Nanaimo, British Columbia, to advance both marine and freshwater research. Similar laboratories were founded throughout Canada and the United States, providing an institutional home for fisheries research. As knowledge grew, the study of changes in fish abundance based on reproduction, growth, and death rates (both natural and from fishing) became known as population dynamics. Understanding of fisheries ecology and population dynamics grew in this environment so massively that it defies a brief synopsis. The list of outstanding scientists working in these areas is a virtual who's who of fisheries (Box 1.2).

This growing wealth of facts and theories regularly confirmed the experiences of fishers and government officials that habitat destruction and unlimited fishing could have monumental impacts on fish populations. The corroboration of practical knowl-

BOX 1.2   A FISHERIES HALL OF FAME

Fisheries science developed so rapidly during the mid-1900s that a complete history would
fill many volumes. A partial story, however, may be told in the accomplishments of those
scientists honored to receive the American Fisheries Society's Award of Excellence. (Bio-
graphical information was compiled by Yanin Walker of the American Fisheries Society.)

1969	William E. Ricker	Population dynamics theory and quantitative fisheries computation methods
1970	Stanislas F. Snieszko	Fish diseases, especially bacteriological, and fish health management
1971	F. E. J. Fry	Dynamics of exploited fish populations and environmental effects on fishes
1972	Ralph O. Hile	Population dynamics of Great Lakes fishes and fisheries statistics
1973	Carl L. Hubbs	Taxonomy, distribution, ecology, life history, and evolution of fishes
1974	Clarence M. Tarzwell	Pollution biology, water quality requirements of fishes, and toxicty testing
1975	Robert R. Miller	Taxonomy and evolution of freshwater fishes and protection of endangered fishes
1976	Alfred W. H. Needler	Marine mollusk culture and management of marine fisheries
1977	Arthur D. Hasler	Fish ecology, fish behavior, and environmental effects on fishes
1978	Peter Doudoroff	Toxicity testing of fishes and fish physiology and energetics
1979	Kenneth D. Carlander	Life history of freshwater fishes and fish population management
1980	Reeve M. Bailey	Taxonomy, distribution, ecology, and nomenclature of freshwater fishes
1981	Gunnar Svardson	Genetics of northern European freshwater fishes, especially the whitefish group
1982	Douglas G. Chapman	Population dynamics of fishes and marine mammals
1983	Peter A. Larkin	Population dynamics of Pacific salmon and fish ecology
1984	James L. McHugh	Dynamics of exploited fish populations and management of North Atlantic fishes
1985	Lauren R. Donaldson	Culture of Pacific salmon, especially nutrition and genetics
1986	William B. Scott	Taxonomy, distribution, and ecology of freshwater and marine fishes
1987	David Cushing	Dynamics of exploited marine fishes and productivity of the sea
1988	Clark Hubbs	Fish ecology in relation to environmental conditions and endangered fish management
1989	John Cairns, Jr.	Environmental toxicology and biological monitoring
1990	John A. Gulland	Populations dynamics of marine fishes
1991	Kenneth Wolf	Fish virology, cell culture, and fish parasitology
1992	Henry A. Regier	Impacts of use and development on fishery resources and rehabilitation of damaged ecosystems
1993	Raymond Beverton	Fish population dynamics
1994	Shelby D. Gerking	Fish ecology and fish production dynamics
1995	William M. Lewis	Warmwater aquaculture
1996	Thomas Northcote	Fish ecology, especially salmonids
1997	William C. Leggett	Ecology of lake fishes
1998	Carl Bond	General ichthyology and fish biology

edge by the new sciences of ecology and limnology thus provoked the flowering of fisheries management to a form much as we see it today. Most of the techniques and approaches used today in fisheries management were developed during the first half of the twentieth century.

1.5.2 The Management of Fish Populations

The exploration of the principles of population dynamics of marine fishes by Russell (1942) and others was readily transferred to inland fisheries. Population dynamics became a fundamental concern to scientists and managers, and developments occurred rapidly throughout North America. In the Pacific Northwest, concern for Pacific salmon fisheries fueled the pioneering work of William Ricker and Earle Forester at Canada's Nanaimo laboratory. They demonstrated the individuality of salmon stocks in different rivers and the relationships between the abundance of spawning fish and the abundance of the subsequent generation (Ricker's stock–recruitment relationship; Ricker 1958). Ricker also invented and published computational techniques for measuring the vital statistics of fish populations; Chapter 6 in this book relies directly on Ricker's 6 decades of population dynamics research.

In the southeastern United States, the field of population dynamics was developing from an empirical approach. In the 1930s Homer Swingle, an extension entomologist at Auburn University in Alabama, began work on the possibility of raising fish effectively in farm ponds (Figure 1.7). His work expanded quickly into a continuing series of pond experiments relating the species, sizes, and densities of fishes to the sustained quality of fishing. Swingle called his concept of sustainable quality fishing the "balance" of the fish population, monitored by a series of ratios comparing the relative abundance of predators and prey (Swingle 1950). Swingle's ideas have guided two generations of fisheries managers and, like Ricker's computational techniques, form the basis for the practical fisheries statistics described in Chapters 7 and 21.

The mideastern and midwestern areas of the continent have contributed many management innovations. Because of the abundance of natural waters and the high density of human populations, fisheries have always been important recreational and commercial resources in these regions. The dominant fishes of these waters, however, often have not been the species desired by anglers. Managers in these areas, therefore, pioneered techniques for the removal or control of undesired fishes. Commercial netting to reduce the abundance of unwanted fishes was used commonly in many states. In Minnesota, large-scale netting began in the 1920s and continued for several decades (Johnson 1948). Commercial fishing is still used successfully today to control undesirable fishes in several situations, but it has never produced the large increases in desired fishes that were expected. Fisheries managers, therefore, turned increasingly to chemical fish poisons. First used in 1913, chemical poisons became common tools after World War II as a larger variety and supply of chemicals became available (Cummings 1975). The use of fish poisons is now greatly restricted, but a few poisons still are important parts of specific management plans (see Chapter 15).

Fish stocking changed in important ways in the mid-1900s. Stocking remained a favorite tool for fisheries managers, but the promiscuous introductions of the late 1800s were supplanted by a more scientific, or at least conscientious, approach. Increasing knowledge of fish ecology and an increasing body of stocking experience allowed

Figure 1.7 Homer Swingle pioneered the empirical approach to fisheries management. Working first with a few ponds and later with hundreds, Swingle and his colleagues at Auburn University discovered combinations of species, sizes, and densities of fishes that provided quality pond fishing.

fisheries managers to choose stocking locations where growth and survival were probable (Wood 1953). This knowledge was first employed systematically in 1927 by George Embody, a New York fisheries scientist, who published tables relating trout stocking densities to stream characteristics. Stocking eggs and larval fishes was largely abandoned in the 1920s; instead, fishes were raised in hatcheries for extended periods and were stocked when larger and more likely to survive. This idea was rapidly expanded to include stocking fishes at sizes desired by anglers, creating the put-and-take fisheries so common today (Chapter 14).

1.5.3 The Management of Habitats

Since earliest times, fisheries managers have modified aquatic habitats. Among the earliest laws governing fisheries in Europe and the United States were those that restricted dams. Dams that blocked entire rivers in order to catch fish, called fixed engines in England, were first prohibited in the Magna Carta (Nielsen 1976). Interest in habitat improvements in North America followed closely on the investigation of life history requirements of fishes and the development of the ecological concept of carrying capacity. Once these habitat-related limits on the abundance or size of fishes were outlined, managers began developing ways to remove them.

Initial attempts to increase carrying capacity focused on the living space available for fishes. In 1938 Carl Hubbs and Ralph Eschmeyer published their handbook, *The Improvement of Lakes for Fishing*, the first comprehensive description of lake habitat

Figure 1.8 Reservoir construction has provided millions of hectares of fishing water in North America. Most construction occurred between 1930 and 1960, when agencies like the Tennessee Valley Authority built hundreds of major impoundments. Photography is courtesy of the Tennessee Valley Authority.

management (Hubbs and Eschmeyer 1938). They emphasized the addition of artificial structures, especially brush piles, as a way to increase the standing crop of fishes. Managers had long been using analogous measures in coldwater streams, placing various low dams and shoreline obstructions where they would diversify the depth and velocity of water (Cooper 1970). Although the effectiveness of such structures has been evaluated only rarely, the manipulation of habitat diversity by adding structures is enormously popular, and, therefore, continuously employed (see Chapters 10 and 11).

Where natural lakes were scarce, the preferred method for expanding fisheries habitat was to build artificial lakes. Farm ponds and other small impoundments have always graced the human landscape, but the environmental problems of the early twentieth century greatly stimulated their construction. When the extended droughts of the 1920s settled in on the Great Plains, farming practices had already produced a land stripped of its natural protection from erosion and with little water available for irrigation. In 1934, the U.S. Soil Erosion Service (now part of the Natural Resource Conservation Service) was created to rectify the causes of the dust bowl. It developed a cost-sharing program for building farm ponds. The purpose was to raise the water table, but the ponds also provided fisheries habitat. As an incentive to build ponds, the U.S. Fish and Wildlife Service supplied farmers with free fishes for stocking. Thus began nationwide programs of farm pond construction and fish stocking that have produced over 5 million small fishing ponds in the United States. Agency-sponsored fish stocking is greatly reduced today, but small impoundments are still primary opportunities for fishing (see Chapter 21).

Reservoir construction had similar beginnings. Fewer than 100 reservoirs (larger than 200 ha) existed in North America in 1900. Under the impetus of burgeoning human development, the U.S. government accelerated the construction of large reservoirs in the 1930s (Figure 1.8). The development of the Tennessee River valley, for example, began in 1933 as part of a federal program to revitalize the region's economic and social well-being through natural resource conservation. By 1941, the Tennessee Valley Authority had built seven major impoundments, providing navigation, flood control, hydroelectric energy—and fishing (TVA 1983). More than 1,300 large reservoirs had been built in the United States by 1970, mostly in the southeast and upper midwest (Jenkins 1970). Because the most desirable sites for reservoir construction have been used, the rate of construction has slowed markedly, and few new reservoirs are being built today. Although fishing was generally a secondary objective for most reservoir construction, the nearly 4 million hectares in major impoundments are intensively managed for fisheries today (Chapter 22).

1.5.4 The Management of Fisheries Users

Early in the twentieth century, the importance of recreational fisheries began to grow. This expansion caused demands for regulations on the competing harvests of commercial fishers. Exhortations for governmental restrictions on environmental degradation and on overfishing became common after World War I. In response to massive depletion of inland fish stocks for food during the war, the U.S. government passed the Black Bass Act of 1926. This act, which prohibited interstate movement of bass taken in violation of state laws, effectively eliminated most large commercial markets for freshwater predatory fishes in the United States. Since that time, commercial fishing in freshwaters has been regularly reduced and is now a minor part of most inland fisheries management programs.

Regulation of recreational fishing, in contrast, has fluctuated widely. The earlier regulatory emphasis on closed seasons and areas was supplemented with new techniques based on studies of fish population dynamics. Once managers began to understand the value of higher growth rates and lower mortality rates, they tried to improve those rates through regulations. Highly restrictive fishing regulations, including minimum size limits, fishing equipment restrictions, and daily catch (creel) limits, were implemented broadly and uniformly by state agencies (Redmond 1986). In the early 1940s, however, the regulatory pendulum swung in the other direction, because several new studies showed the inappropriateness of restrictive regulations in many waters. State agencies responded by liberalizing regulations, again in a largely uniform manner. The pendulum has reversed direction again, and stricter fishing regulations have returned. Because regulations represent one of the few sure methods of decisively manipulating fisheries, they are popular among managers (see Chapter 17).

The decades from 1900 to 1950 might well be called the golden age of fisheries management. The continent had discovered the value of recreational fisheries and had backed the research and management to improve them. Scientific information grew rapidly and, when combined with the experience of previous years, provided fertile ground for the proliferation of management techniques. Maximum sustainable yield was firmly established as the goal for fisheries management, whether for commercial or recreational fisheries. Managers stocked and poisoned fish, they built and modified water bodies, and they regulated fish harvest with the single aim of providing the greatest sustained quantity of fish. Fisheries management now had a well-developed kit of tools, and the significant changes after 1950 would come in its purposes rather than its techniques.

1.6   MODERN FISHERIES MANAGEMENT

The modern era of fisheries management began after the end of World War II. State and provincial agencies grew rapidly by enlarging management, fish culture, and law enforcement staffs. Spurred on by the trend for higher education and by the increasing use of fisheries resources for recreation, governments invested generously in education, research, and management. Along with this investment came an outpouring of data that has allowed fisheries management to develop as the multifaceted profession it is today.

Perhaps the greatest stimulus to inland fisheries management in the United States was the passage in 1950 of the Federal Aid in Sport Fish Restoration Act, popularly known as the Dingell–Johnson Act. Patterned after the earlier Pittman–Robertson Act for wildlife, the Dingell–Johnson Act created a 10% excise tax on specified fishing equipment. The tax was dispersed to state fisheries agencies to support the creation and improvement of recreational fisheries. From its inception through 1985, the Dingell–Johnson program had dispersed over US$480 million for fisheries development and research. In 1985, the Dingell–Johnson program was expanded significantly, increasing the range of items taxed, adding marine fuel taxes to the program, and authorizing development of marine, as well as freshwater, recreational fisheries. This important legislation, known as the Wallop–Breaux Act, has more than doubled the annual funds available, with $332 million dispersed in 1992 alone.

The initial decades of the twentieth century had been dominated by the concept of fisheries as crops, with the single objective of achieving the highest physical yield, or MSY. Since 1950, focus on MSY has been challenged repeatedly, and possible objectives for management have been continually expanded. This expansion has not been a sequential process, with each new idea following neatly behind the previous. For most modern viewpoints, the antecedents extend well back into the last century. Their rise to widespread acceptance, however, has followed a somewhat identifiable path.

The initial challenge to MSY was the idea that producing physical yields was really secondary to the more universal objective of producing economic value. Biologists had learned that physical yield could be maximized (theoretically) by regulating the total fish catch via a quota or similar means. Economists now reasoned that if the effort used to catch those fish was so great that the cost equaled the revenue, no economic value (profit) would exist. This idea was expressed eloquently by Michael Graham, a leading British fisheries scientist, in his classic essay *The Fish Gate*: "Fisheries that are unlimited become unprofitable" (Graham 1943). The result of such economic inefficiency was a different type of common property dilemma: although the natural resource was preserved, the dominant human benefit—economic gain—would be lost if fishing effort was unrestricted.

In the 1950s, fisheries economists began to point out that MSY should be replaced by the concept of maximizing profit, alias maximizing net economic revenue (MNER). Management for MNER required a different approach. Scientists such as Milner Schaefer, who analyzed the Pacific tuna fisheries, and Anthony Scott, working in British Columbia, stated that not only must the fish harvest be regulated but also the effort used to acquire that harvest must be limited. The fundamental economic premise was that the fish harvest must be allocated to particular fishing units—individual fishers, corporations, or nations—in order to keep the fishing effort low. The concept, which is usually called limited entry, grants some persons the right to fish while excluding everyone else. This economic principle is not universally accepted, but it is being increasingly used for important commercial fisheries and is the basis for the 200-mi limit that most coastal nations now claim over their marine fisheries.

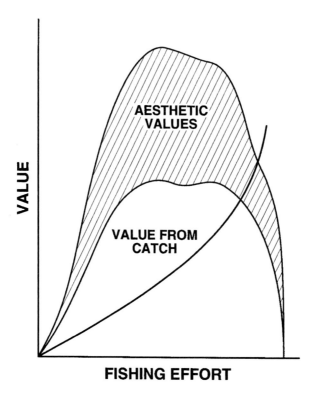

Figure 1.9   The addition of nonbiological concerns to the objectives of fisheries management is shown in this figure from James McFadden's keynote address to the American Fisheries Society in 1968.

While economic concerns were challenging MSY in commercial fisheries, MSY was also being challenged as a sufficient objective for recreational fisheries. The quality of recreational fishing had been measured traditionally as the number and size of harvested fishes. Nevertheless, anglers had always admitted that other parts of the fishing experience were at least as important as the catch itself. People value companionship and pleasant surroundings in their fishing experiences, and their preference for catch might vary from one large fish to many small fish. The idea that such sociological concerns should be part of fisheries management gained popularity in the 1960s as public opinion became more important in directing government decisions. The addition of aesthetic values to the relationship between fishing effort and fish harvest was formalized by James McFadden (1969) in a modern classic of fisheries literature (Figure 1.9). Since then, the development of socioeconomic principles for fisheries management has been a high priority for management agencies (see Chapter 8).

The third major addition to the objectives of fisheries management that challenged MSY was the result of continuing advances in ecological science. Because fisheries are components of the productivity of aquatic ecosystems, ecological research has continuously improved the theoretical foundation of fisheries management. In the 1970s, for example, ecologists supplied the notion that the management of single fish species must be replaced by multiple-species management. The yield of predatory fish (e.g., largemouth bass) depends on the condition of the food base; when the food base is also

---

BOX 1.3   R.I.P. FOR MSY

Maximum sustainable yield (MSY) served the fisheries profession well for many decades. In the 1970s, however, the concept of optimum sustainable yield supplanted MSY as the dominant fisheries objective. In the keynote address at the 1976 American Fisheries Society meeting, Peter Larkin (1977) laid MSY to a gentle rest:

Whatever lies ahead in the development of new concepts for harvesting the resources of the world's fresh waters and oceans, it is certain that the concept of maximum sustained yield will alone not be sufficient. The concept has served an important service. It arrived just in time to curb many fisheries problems. To appreciate what MSY has done, we need only ask what the world's fisheries would have looked like today if the concept had not been developed and advocated with such fervor. The fish, I'm sure, would shudder to think of it. Like the hero of a western movie, MSY rode in off the range, caught the villains at their work, and established order of a sort. But it's now time for MSY to ride off into the sunset. The world today is too complex for the rough justice of a guy on a horse with a six-shooter. We urgently need the same kind of morality, but we also need much more sophistication.

Accordingly, I tender the following epitaph:

MSY
1930s–1970s
Here lies the concept, MSY.
It advocated yields too high,
And didn't spell out how to slice the pie.
We bury it with the best of wishes,
Especially on behalf of fishes.
We don't know yet what will take its place,
But hope it's as good for the human race.

R.I.P.

---

exploited (e.g., bluegill), the pair of species must be managed together, not separately. The idea that fisheries must be thought of as communities, or at least as interacting groups of populations, has become firmly rooted through a series of symposia focusing on multispecies fisheries (e.g., Clepper 1979). Since the days of Forbes, ecology has been and will continue to be the basic science that contributes most to the understanding of fisheries management.

The accretion of additional concerns—economic, sociological, and ecological—into the management of fisheries forced MSY off its throne in the 1970s (see Box 1.3). It has been replaced by a new guiding principle: management for optimum sustainable yield, or OSY. Optimum sustainable yield was formalized in a 1975 symposium (Roedel 1975) that assessed management from a variety of viewpoints. The basic tenets of OSY are that the appropriate goal for fisheries management includes a broad range of considerations (not just maximizing physical yield) and that a unique management goal exists for each fishery. Optimum sustainable yield thus greatly complicates fisheries management. Defining the OSY for a fishery is much more difficult than defining the MSY because fishery-specific information is needed about biological, ecological, economic, and sociological aspects of fishery use. Optimum sustainable yield, however, is also much more realistic in that it recognizes the diversity of aquatic ecosystems and the diversity of human needs in relation to them.

The job of the fisheries manager has evolved rapidly. What the manager does today has been conditioned as much by the events of the past 2 decades as by the accumulated history of fisheries. The remainder of this chapter describes some of the substantive areas of fisheries work today, specifically as they relate to later chapters in the book.

1.6.1 The Management of Habitats

The growth of the environmental movement in the 1960s changed the world of the fisheries manager. United States law had required fisheries managers to assess the impacts of federal development projects since the 1930s (the Fish and Wildlife Coordination Act), but recent decades have greatly expanded and defined that general charge (see Chapter 4). Of greatest significance were the U.S. National Environmental Policy Act of 1969, which created the process of developing environmental impact statements, and the 1970 amendments to the Canadian Fisheries Act, which broadly prohibits harm to aquatic habitats (Pearse et al. 1985). Fisheries managers are now required to review thousands of government-sponsored developments each year, and many states have passed legislation requiring impact analysis for state-regulated projects.

The need to predict the relationship between habitat modification and the effects on fisheries has given rise to entirely new specialties concerned with habitat analysis (see Chapter 10). Habitat analysis has been concentrated in the western half of the continent, where water supplies are low and where fisheries compete directly with consumptive uses of water. The most significant efforts have helped define the water flow conditions needed to sustain fish life (usually called instream flow). Scientists have created methods for expressing the relationships between instream flow and habitat conditions and have determined the specific habitat conditions needed by dozens of fish species.

Environmental laws have also allowed fisheries managers to address the ecological principle that land uses throughout a watershed affect the well-being of fish (see Chapter 9). Fisheries managers have advanced well past the point of merely predicting changes. They now prescribe remedies for the anticipated problems in the design, construction, and operation of development projects (Swanson 1979). Mitigation, as this process is called, allows the manager to participate in the planning process for land and water projects. The manager can then help developers make good choices—eliminating avoidable damages by redesign or improved operation and gaining suitable restitution for the damages that are unavoidable.

The legal framework for maintaining fisheries, along with their increasing economic and social value, has also allowed for the small- and large-scale rehabilitation of habitats. Stream reconstruction in the West and wetland reconstruction in the East are widely practiced by public agencies and private companies. Earlier techniques of stream improvement have been augmented so that now entirely new stream channels can be constructed, kilometers of stream bottoms can be removed, washed, and replaced, and existing dams and other structures can be removed (see Chapter 10). Although lake habitat is not as easily remodeled as stream habitat, the methods available far exceed those of previous decades. New reservoirs are constructed with outlets designed to enhance fisheries, and a seemingly endless supply of devices is available for improving water quality and fisheries habitat (see Chapter 11).

1.6.2 The Management of Aquatic Organisms

As ecological knowledge has increased, the breadth of interest in management of fishes and other aquatic organisms has similarly increased. The concentration on a limited number of predatory fishes has been replaced by the recognition that the entire aquatic community is valuable and needs management attention.

Endangered and rare species are now managed primarily by fisheries and wildlife professionals. Although the addition of this responsibility to their more traditional roles has been resisted by some managers, fisheries agencies are the natural homes for endangered species concerns. Fisheries managers have the skills and experience to manipulate ecosystems to enhance endangered species, just as they manipulate other fisheries organisms. Today the management of endangered species is recapitulating the entire course of fisheries management, rapidly moving through the stages of life history descriptions, distribution studies, and habitat assessments to the stage of prescribing how to remove species from the danger of extinction (see Chapter 16).

Management interest has also broadened to incorporate the nonharvested species of a typical aquatic environment. Attention is now being directed toward smaller fishes that provide the food for predators and toward competing fishes that reduce the reproduction, growth, and survival of desired predators (see Chapters 12 and 13). Concern has also developed for other species that may not enter in any way into a fishery. Conservation biology, as this interest is called, concentrates on preserving and enhancing the diversity of all species. Although initially stronger in Europe than in North America (Maitland 1974), the conservation of aquatic biodiversity is now a standard part of agency management, from the U.S. Forest Service to state fish and game departments.

Aquaculture remains one of the most valuable management techniques in fisheries today (Stroud 1986)—as well as one of the most controversial. It still serves traditional purposes of stocking fish for angling, but it also serves faunal restoration efforts and endangered species management. As the use of natural waters for commercial fishing has declined, aquaculture has developed rapidly as the source for food fish. At the same time, serious questions have arisen about the wisdom of relying on fish culture as a long-term solution to fisheries problems. Scientists and managers, therefore, will be forced to confront many significant issues related to aquaculture, including genetic engineering, the effects of hatchery-reared fish on wild stocks, and the use of public waters for commercial fish farming.

1.6.3 The Management of Fisheries Users

The human component in fisheries management, so long de-emphasized, is now a major element in all management activities. Optimum sustainable yield demands that the needs and desires of fisheries users be discovered and incorporated into management plans. Consequently, creel surveys and attitude assessments are now continuing activities of state and federal agencies. Methods for assessing user demands and for comparing the relative value of various fisheries are evolving rapidly in the face of both professional and legal demands for information (see Chapter 7).

Now that fisheries agencies know more about the demand for fisheries, they have changed their notions about how to serve society. Fisheries programs now typically include urban fisheries and fisheries accessible to physically impaired anglers. Aquatic education, which opens fishing to people without a family tradition of angling, is increasingly important in fisheries agencies. A more representative balance also is being achieved in the allocation of fishes between commercial and recreational fisheries. Comanagement of fisheries, by which traditional regulatory agencies relinquish some authority to local governments, Native American groups, and conservation organizations, is beginning and seems highly likely to increase in the future. The utility of socioeconomic information to guide management will continue to expand as techniques improve and as the success of incorporating such information is established.

As fisheries management has become more complex—more quantitative, more political, more scientifically sophisticated, and more diverse—the need for effective planning of fisheries programs has become apparent. Beginning in the 1970s, the concept of comprehensive planning has become increasingly popular as a way to anticipate the future. An increasing number of state, provincial, and federal agencies are functioning under strategic plans (which describe what could be done) and operational plans (which describe how to do it). The emphasis is now placed on setting specific objectives for fisheries management and monitoring the achievement of those objectives (see Chapter 2).

1.7   CONCLUSION

And just what is this profession we call fisheries management? Viewpoints about management are as diverse as the collection of human activities that fall under the rubric of what we call fisheries. As the profession has evolved and radiated in recent decades, the precepts that fisheries management was applied biology, applied ecology, or even applied economics have all proven too restrictive. A dictionary definition is not really important; a principle for guiding the management of fisheries, however, is necessary. Most fisheries professionals would agree that the principle objective of fisheries management is to provide people with a sustained, high, and ever increasing benefit from their use of living aquatic resources. In the pursuit of that principle, fisheries managers manipulate all aspects of the natural and human ecosystem. Where people and water meet, fisheries exist; where people and water could meet, potential fisheries exist; and wherever fisheries, real or potential, exist, fisheries management can make them better.

1.8   REFERENCES

Adams, C. C., and T. L. Hankinson. 1928. The ecology and economics of Oneida Lake fish. Roosevelt Wild Life Annals 1(34):235–548, Roosevelt Wild Life Forest Experiment Station, Syracuse, New York.
Allard, D. C., Jr. 1978. Spencer Fullerton Baird and the U.S. Fish Commission. Arno Press, New York.
Bean, M. J. 1977. The evolution of national wildlife law. U.S. Council on Environmental Quality, Washington, D.C.

Beckel, A. L. 1987. Breaking new waters. A century of limnology at the University of Wisconsin. Transactions of the Wisconsin Academy of Sciences, Arts and Letters, Special Issue, Madison.

Beltran, E. 1972. Programs for renewable natural resources in Mexico. Transactions of the North American Wildlife and Natural Resources Conference 37:4–18.

Bower, S. 1910. Fishery conservation. Transactions of the American Fisheries Society 40:95–100.

Busiahn, T. R. 1985. An introduction to native peoples' fisheries issues in North America. Fisheries 9(5):8–11.

Callison, C. 1981. Men and wildlife in Missouri. Missouri Department of Conservation, Jefferson City.

Carlander, H. B. 1954. History of fish and fishing in the upper Mississippi River. Upper Mississippi River Conservation Committee, Rock Island, Illinois.

Clepper, H., editor. 1979. Predator–prey systems in fisheries management. Sport Fishing Institute, Washington, D.C.

Cooper, E. L. 1970. Management of trout streams. Pages 153–162 in N. G. Benson, editor. A century of fisheries in North America. American Fisheries Society, Special Publication 7, Bethesda, Maryland.

Cox, T. R. 1985. Americans and their forests. Romanticism, progress, and science in the late Nineteenth Century. Journal of Forest History 29:156–168.

Craig, J. A. 1930. An analysis of the catch statistics of the striped bass (Roccus lineatus) fishery of California. California Department of Fish and Game, Sacramento.

Cummings, K. B. 1975. History of fish toxicants in the United States. Pages 5–21 in P. H. Eschmeyer, editor. Rehabilitation of fish populations with toxicants: a symposium. American Fisheries Society, North Central Division, Special Publication 4, Bethesda, Maryland.

Egerton, F. N. 1976. Ecological studies and observations before 1900. Pages 311–351 in B. J. Taylor and T. J. White, editors. Issues and ideas in America. University of Oklahoma Press, Stillwater.

Forbes, S. A. 1925. The lake as a microcosm. Illinois Natural History Survey Bulletin 15:527–550.

Fry, F. E. J., and V. Legendre. 1966. Ontario and Quebec. Pages 487–519 in D. G. Frey, editor. Limnology in North America. University of Wisconsin Press, Madison.

Graham, M. 1943. The fish gate. Faber and Faber, London.

Hubbs, C. L., and R. W. Eschmeyer. 1938. The improvement of lakes for fishing. Michigan Department of Conservation, Institute for Fisheries Research, Bulletin 2, Lansing.

Hubbs, C. L., and K. F. Lagler. 1941. Guide to the fishes of the Great Lakes and tributary waters. Cranbrook Institute of Science, Bloomfield Hills, Michigan.

Jenkins, R. M. 1970. Reservoir fish management. Pages 173–182 in N. G. Benson, editor. A century of fisheries in North America. American Fisheries Society, Special Publication 7, Bethesda, Maryland.

Johnson, R. E. 1948. Maintenance of natural population balance. Proceedings, Convention of the International Association of Game, Fish and Conservation Commissioners 38:35–42.

Kawashima, Y., and R. Tone. 1983. Environmental policy in early America: a survey of colonial statutes. Journal of Forest History 27:168–179.

Larkin, P. A. 1977. An epitaph for the concept of maximum sustained yield. Transactions of the American Fisheries Society 106:1–11.

Larkin, P. A. 1979. Maybe you can't get there from here: foreshortened history of research in relation to management of Pacific salmon. Journal of the Fisheries Research Board of Canada 38:98–106.

Leonard, J. R. No date. The fish car era. U.S. Fish, and Wildlife Service, Washington, D.C.

Leopold, A. 1933. Game management. Scribner, New York.

Maitland, P. S. 1974. The conservation of freshwater fishes in the British Isles. Biological Conservation 7(1):7–14.

McEvoy, A. F. 1986. The fisherman's problem. Ecology and law in California fisheries 1850–1980. Cambridge University Press, Cambridge, UK.

McFadden, J. G. 1969. Trends in freshwater sport fisheries of North America. Transactions of the American Fisheries Society 98:136–150.

McIntosh, R. P. 1976. Ecology since 1900. Pages 353–372 in B. J. Taylor and T. J. White, editors. Issues and ideas in America. University of Oklahoma Press, Stillwater.

Nash, R. 1987. Aldo Leopold's intellectual heritage. Pages 63–90 in J. B. Callicott, editor. Companion to A Sand County Almanac. University of Wisconsin Press, Madison.

Nielsen, L. A. 1976. The evaluation of fisheries management philosophy. U.S. National Marine Fisheries Service Marine Fisheries Review 38(12):15–23.

Northcote, T. G., and P. A. Larkin. 1966. Western Canada. Pages 451–486 *in* D. G. Frey, editor. Limnology in North America. University of Wisconsin Press, Madison.

Osburn, R. C. 1933. Some important principles of fish conservation. Transactions of the American Fisheries Society 63:91–97.

Pearse, P. H., F. Bertrand, and J. W. MacLaren. 1985. Currents of change. Environment Canada, Inquiry on Federal Water Policy, Final Report, Ottawa.

Pinchot, G. 1910. The fight for conservation. Doubleday, New York.

Redmond, L. C. 1986. The history and development of warmwater fish harvest regulations. Pages 186–195 *in* G. E. Hall and M. J. Van Den Avyle, editors. Reservoir fisheries management: strategies for the 80's. American Fisheries Society, Southern Division, Reservoir Committee, Bethesda, Maryland.

Regier, H. A., and V. C. Applegate. 1972. Historical review of the management approach to exploitation and introduction in SCOL lakes. Journal of the Fisheries Research Board of Canada 29:683–692.

Ricker, W. E. 1958. Handbook of computations for biological statistics of fish populations. Fisheries Research Board of Canada Bulletin 119.

Roedel, P. M., editor. 1975. Optimum sustainable yield as a concept in fisheries management. American Fisheries Society, Special Publication 9, Bethesda, Maryland.

Roppel, P. 1982. Alaska's salmon hatcheries, 1891–1959. Alaska Historical Commission, Studies in History 20, Juneau.

Russell, E. S. 1942. The overfishing problem. Cambridge University Press, Cambridge, UK.

Seaman, E. A. 1988. Silent letters released. Seven decades of one life. Vantage Press, New York.

Smith, F. E. 1971. Land and water, 1492–1900. Chelsea House Publications, New York.

Stroud, R. H. 1966. Lakes, streams, and other inland waters. Pages 57–73 *in* H. E. Clepper, editor. Origins of American conservation. Ronald Press, New York.

Stroud, R. H., editor. 1986. Fish culture in fisheries management. American Fisheries Society, Fish Culture Section and Fisheries Management Section, Bethesda, Maryland.

Swanson, G. A., technical coordinator. 1979. The mitigation symposium: national workshop on mitigating losses of fish and wildlife habitats. U.S. Forest Service General Technical Report RM-65.

Swanton, J. R. 1946. The Indians of the southeastern United States. Smithsonian Institution, Bureau of American Ethnology, Bulletin 137, Washington, D.C.

Swingle, H. S. 1950. Relationships and dynamics of balanced and unbalanced fish populations. Alabama Agricultural Experiment Station, Auburn University, Bulletin 274.

TVA (Tennessee Valley Authority). 1983. The first fifty years: changed land, changed lives. TVA, Knoxville.

Thompson, P. C. 1974. Institutional constraints in fisheries management. Journal of the Fisheries Research Board of Canada 31:1965–1981.

Thompson, P. E. 1970. The first fifty years–the exciting ones. Pages 1–12 *in* N. G. Benson, editor. A century of fisheries in North America. American Fisheries Society, Special Publication 7, Bethesda, Maryland.

Wharton, J. 1957. The bounty of the Chesapeake: fishing in colonial Virginia. University Press of Virginia, Charlottesville.

Whitaker, H. 1892. Early history of the fisheries of the Great Lakes. Transactions of the American Fisheries Society 21:163–179.

Wood, E. M. 1953. A century of American fish culture, 1853–1953. Progressive Fish-Culturist 15:147–162.

Chapter 2

The Process of Fisheries Management

CHARLES C. KRUEGER AND DANIEL J. DECKER

2.1 INTRODUCTION

Over the twentieth century, fisheries management in North America has evolved institutionally, socially, and conceptually. Modern management incorporates not only scientific understanding about fishes and their habitats but also considers economics, aesthetics, user attitudes and desires, and the interests of environmentalists and the general public. Considerable efforts are being made by management agencies to integrate these concerns into a comprehensive approach to management. Here, fisheries management is defined as

> the use of all types of information (ecological, economic, political, and sociocultural) in decision making that results in actions (e.g., regulations) to achieve goals established for fish resources.

Central to the management process is the evaluation of actions taken to achieve goals. Evaluation allows learning from past management and results in the development of an information base for future decision making.

Effective management requires information about fishes and their habitats as well as about society which uses fishes or otherwise benefits from fisheries management (e.g., businesses). Unfortunately, natural resource agencies do not have the comprehensive information needed to make perfect decisions. Yet every day decisions must be made by agencies based on whatever information is available. Scientific research plays a critical role in providing sound information for decision making. The fisheries management process itself can also increase the information available by encouraging careful observation of the effects of today's decisions so that new information can be acquired and used in future decisions. Learning as we manage is the key to success!

Who manages commercial, recreational, and tribal fisheries and fish populations? Who makes decisions and implements actions? Many government agencies have authority to implement specific actions to provide for the wise use and conservation of our aquatic resources. As a general rule, in most areas of North America either the state or provincial governments have direct authority over fisheries resources. Some situations exist, however, where authority is shared or vested in federal or tribal governments (see Chapter 4).

The institutional complexity surrounding fisheries management increases when natural resources are shared along state or provincial boundaries or between countries. For example, state, provincial, tribal, and federal agencies each have authority over specific aspects of Great Lakes fisheries management (see Chapter 23). To facilitate cooperation among these agencies, in 1955 the Great Lakes Fishery Commission was created by a treaty between the federal governments of Canada and the United States. The complications caused by multiple agencies sharing responsibility for management increase when agencies have different missions or purposes. Coordination of each agency's actions is essential if effective management is to occur. All involved parties must effectively communicate and cooperate with each other (see Chapter 3).

Fisheries management is not only the concern of government agencies; it allows for, and often requires, involvement by members of the public. Historically, management agencies focused mostly on the desires of recreational or commercial interests who sometimes formally organized into groups or associations to communicate their concerns more effectively to fisheries agencies. Since the 1970s, more specialized groups (e.g., bass anglers and trout anglers) have emerged, dramatically increasing the number and sophistication of public interest groups that interact with agencies. These groups often represent interests distinct from each other, interests that may or may not overlap. In addition, groups representing environmental concerns have formed with goals much different from those of the traditional fishing public. For example, nonconsumptive users of lakes and streams who are not anglers often have environmental values (e.g., preservationist) that differ from those of sport and commercial users. Also, in some locations in North America, aboriginal peoples have special rights as users, and are often outside the jurisdiction of state or provincial management agencies. Native American tribes may have comanagement responsibilities (e.g., McDaniels et al. 1994; Pinkerton 1994). Thus, resource agencies have the complicated and challenging task of managing fisheries for an array of users who represent different sets of values.

Who are the beneficiaries of fisheries management that should be considered in decision making? Recently, the concept of beneficiaries has broadened considerably beyond the traditional clients (e.g., anglers) to include all those who have a stake in fisheries management decisions. The term stakeholder has emerged to represent any citizen potentially affected by or having a vested interest in an issue, program, action, or decision leading to an action (Decker et al. 1996). The stakeholder approach to management recognizes a larger set of beneficiaries of management (including the fish populations themselves and future human generations) than did traditional concepts of constituencies, clients, or customers (McEwan 1997).

What are the products of fisheries management? What are the goals that management attempts to achieve? Is it an angler's catch of largemouth bass from a lake in Mexico? Is it the successful restoration of Atlantic salmon to streams in the Maritime Provinces? Is it the sustained commercial harvest of channel catfish from the Mississippi River? Successful fisheries management can achieve all of these outcomes. These events may be thought of as the products of a management agency and should reflect achievement of goals. Products of management may be not only the catching of fish but, more fundamentally, simply providing the opportunity to go fishing. People enjoy knowing such opportunities exist close to where they live, even if they can participate only rarely. In the case of restoring fish populations such as Atlantic salmon, many people can appreciate when a native species has regained a secure, stable status in a watershed, even if they never see or fish for the species themselves. Goals and products of management are closely related concepts that are derived from the interests of the public. Goals are the focus of fisheries management activities.

Meeting today's challenges of fisheries management depends on the willingness of managers to adopt a comprehensive concept of fisheries management. The concept used must be general enough to be applicable to management of an urban fishery in New York City or a wilderness alpine lake in British Columbia. The success of a fisheries professional depends on a thorough understanding of the management process and on the ability to communicate effectively with other fisheries professionals and public sectors (see Chapter 3; Pringle 1985).

In this chapter, the concept of fisheries management is presented as a cyclic, self-correcting, goal-oriented process. The chapter is organized to describe the environment within which management must function, the team of people who conduct management, and the elements of the management process. The idealized process presented here does not necessarily portray fisheries management as it has occurred in the past; rather, it provides a conceptual approach to fisheries management for the future. The importance of a goal-oriented approach to management combined with a feedback or evaluation step has been recently emphasized for commercial fisheries (Stephenson and Lane 1995), for restoration management (Hobbs and Norton 1996), and in fish and wildlife texts (Scalet et al. 1996).

Herein we describe the entire fisheries management process. Many elements of the process are similar to past descriptions of natural resource planning (e.g., Anderson and Hurley 1980; Crowe 1983) and business management (e.g., Odiorne 1979) and follow closely an earlier general description of natural resource management (Krueger et al. 1986). However, the concepts described here include not only the steps for the development of a strategic plan for management of a fishery but also the actual implementation of such a plan.

2.2 MANAGEMENT ENVIRONMENT: THE FORCES THAT SHAPE MANAGEMENT

The combination of ecological, political, economic, and sociocultural factors that influence the management process can be thought of as the special environment set within space and time wherein fisheries management is conducted (Figure 2.1). Understanding each component of the management environment is essential for effective fisheries management. For example, the ecological characteristics of this environment control potential fish production and yield, such as the number of walleye that can be caught from a lake on a sustained basis. The sociocultural aspects will limit the types of actions that a manager can take (Ross 1997). For example, it may be a socially unacceptable action for state government to condemn private property for the improvement of stream watersheds. The components and dimensions of the management environment are described below.

The ecological component of the management environment includes the ecosystem within which fishes and other species of interest complete their life cycles. This component includes aquatic and terrestrial, biotic and abiotic environmental factors, such as forage availability, abundance of predators, and quality and quantity of habitats required for different life stages. Manipulation of these environmental factors has been a traditional focus of fisheries management. However, the ecological component also includes genetic factors, such as traits that affect migration, habitat preferences,

Management Environment

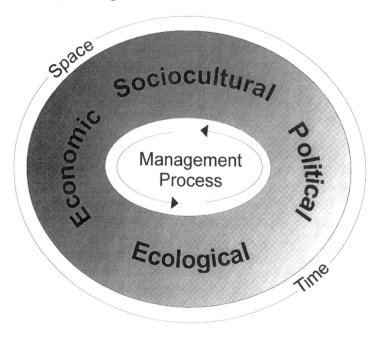

Figure 2.1 Representation of the management environment. The ecological, economic, political, and sociocultural components categorize the forces that shape management in space and time.

and survival. Increasing knowledge about genetics has focused attention on the genetic aspects of fisheries management and has allowed agencies to pursue exciting new ways to manage fisheries resources more effectively (e.g., Ryman and Utter 1987; Utter 1991). Both environmental and genetic factors control the annual production of fishes and other aquatic organisms within bodies of water. Managers who understand the potential for fish production within a system can realistically assess fishing opportunities and identify ways to sustain or protect fisheries resources. The ecological component establishes the bounds within which the other components can be considered.

The economic component of the management environment includes the marketplace and nonmarket forces that influence the monetary valuation of fisheries resources. The economic component has long been an important consideration in making decisions about commercial fisheries (e.g., Christy and Scott 1965; Clark 1976; Anderson 1986), and since the 1970s has gained recognition as an important consideration for sport fisheries management (e.g., Talhelm 1987; Chapter 8). This component also includes the economics of fisheries agencies, such as license fee structures (e.g., nonresident angler fees and trout stamps) and annual budgets available to conduct management activities. As ecological factors limit the productivity of aquatic ecosystems, agency budgets limit the magnitude and type of management actions that may be considered for implementation. Before making a decision, a manager should consider the entire economic component to reflect on the costs and benefits to society of different fisheries policies.

The political component of the management environment includes laws and official policies of government, as well as the personal values of government employees who interpret and enforce such directives. Laws and policies of government are often well-defined by constitutions, statutes, administrative codes, and published agency management plans; they establish in general terms the mission of an agency (see Chapter 4). In contrast, the political climate of government is constantly changing as employees assume new responsibilities, new employees are hired, and new politicians are elected. Both aspects, defined laws and the shifting political climate, must be considered if current management practices are to change. For example, before a new program can be proposed, managers must first determine whether their agency has legal jurisdiction to implement the program and whether the political climate in their agency or government is "right" for the new proposal. The political component includes the branches of government that surround a fisheries agency (executive, legislative, and judicial); these branches each can exercise an important decision-making function for the agency. If fishery-related conflicts occur that are not resolved, then a binding legal solution to the conflict may develop through the legislative or judicial branches.

The sociocultural component includes traditions, values, norms, religions, and philosophies of the society within which management is conducted. This component provides the primary motivation for fisheries management to occur. Fisheries management is conducted because the end products of the process are believed to have value to society. Study of this component will help managers understand, and possibly influence, fishery-related values held by the public. These sociocultural values influence the choice of management goals and actions (Brown and Manfredo 1987), especially in favor of high-value species such as walleye or lake trout or against undesirable species (see Chapter 15). In some instances, the fisheries profession works to change or influence society's values about fisheries resources. For example, the American Fisheries Society published a book about common carp to enhance the general public's awareness of the potential recreational value of this species (Cooper 1987). As fisheries managers, our interpretation of ecological, economic, and political components is strongly influenced by our personal identification with particular values and philosophies within the sociocultural component (Nielsen 1985). Our values influence what we do and how we choose to do it (Barber and Taylor 1990). Understanding sociocultural values will help managers understand the needs and motivations of different public sectors and help managers to communicate effectively with the public.

Separation of the management environment into ecological, economic, political, and sociocultural components provides a useful categorization of the forces that shape fisheries management. Clearly the components of the management environment described here are not independent of each other. For example, the economic and political components are derivatives of the sociocultural component but are important enough to warrant their separate consideration. Managers faced with a fishery decision should first consider how each component of the management environment may affect the problem and its potential solutions rather than plunging headlong into a favorite mode of thinking, such as pursuing an ecological solution. On the other hand, factors other than ecology can dangerously drive decision making to exclude ecological considerations to the detriment of the resource. For example, decision making for fisheries management of chinook salmon stocks in the Pacific Northwest has been influenced more strongly by sociocultural and political factors than by ecological considerations, despite the presence of adequate biological data and analytical capabilities (Fraidenburg and Lincoln 1985; NRC 1996).

Key dimensions of the management environment are space and time. Space delimits the geographical area of focus for management, such as a particular lake, river, watershed, state, or province (e.g., Roos 1991; Boon et al. 1992). The habitats used during the various stages of a fish's life history help define the spatial area of concern for management. For example, Atlantic salmon management must consider the watersheds and associated streams that support juvenile rearing and adult spawning as well as the ocean migratory routes and feeding grounds in the North Atlantic. The political component will define the legal jurisdictional area that a management agency can affect (see Chapter 4). Time dictates the urgency of decision making for management: when actions should be implemented and evaluated and when success in management can be expected. The life history of fish species also strongly affects the temporal dimensions of management. For example, the long life of species such as lake sturgeon or lake trout necessitates that the achievement of restoration goals not be expected for several decades or more. This ecological reality provides special challenges to management (Krueger et al. 1995). A sense of immediate urgency is often placed on managers by the sociocultural or political components because of the high value given certain species or the fisheries they support or because there is risk of extinction and listing under the U.S. Endangered Species Act (see Chapter 16). Because of the time dimension, the option of not making a decision also constitutes a distinct type of decision that has associated risks and benefits that must be considered.

2.3   MANAGEMENT TEAM APPROACH

Each component of the management environment should be considered during agency decision making. Fisheries managers are expected to anticipate the economic and cultural effects of proposed actions, for example, new regulations, as well as the biological effects of these regulations on fish populations. It is difficult for an individual fisheries manager to be educated sufficiently in all disciplines to make all of these predictions. Rather, a team approach should be used to provide input to decision making (Harville 1985; Coughlan and Armour 1992). Teams most often will function in an advisory capacity with final decision-making authority vested at some level in the hierarchy of an agency (e.g., the regional manager, bureau chief, or the director of the agency). If a team is delegated the authority for decision making, then members of the team must be willing to accept responsibility for their decisions and be held accountable for those decisions.

The team should include individuals having general fisheries science backgrounds and those having a special focus in one or more of the core areas of ecology, economics, government, or social science (Figure 2.2). Fisheries managers should have a sufficiently broad educational background to communicate effectively with all team members. The team should also include representatives from all agencies that share jurisdiction for a fishery and possibly from other nonfisheries, environmental, or land management agencies (e.g., in the United States, the Forest Service, Bureau of Land Management, and Environmental Protection Agency). Leaders from key public sectors often constitute important, additional members of a management team. Public partici-

P.O. BOX 818
WILSON, NY 14172

**Toward Total Economic Valuation of
Great Lakes Fishery Resources**

RICHARD C. BISHOP

*Department of Agricultural Economics, University of Wisconsin
Madison, Wisconsin 53706, USA*

KEVIN J. BOYLE

*Department of Agricultural and Resource Economics, University of Maine
Orono, Maine 04469, USA*

MICHAEL P. WELSH

*Department of Agricultural Economics, Oklahoma State University
Stillwater, Oklahoma 74078, USA*

Figure 2.2 The fisheries management team includes members representing the different components of the management environment.

pation can produce helpful insights on potential economic and cultural effects that may result from new management programs. Representation on the team from the disciplines of ecology, economics, government, and social sciences, from agencies that have jurisdiction, and from the public will help ensure proper guidance of the fisheries management process. The actual composition of the team will likely vary depending on which step of the management process is being conducted. In the following section (2.4), suggestions for the composition of the team are made for each step of the management process.

Members of management teams, who represent a wide variety of backgrounds, will likely approach the challenges of management quite differently. This diversity is the strength behind the team approach to fisheries management but can also be challenging to orchestrate effectively (Miller 1984; Chapter 5). Team members must be able to understand a diversity of viewpoints if the strengths of a team are to be channeled into innovative management strategies.

The team approach to fisheries management is valid at all spatial scales of management, from local areas within a state or province to large international fisheries. For example, a local manager may be solely responsible for a set of lakes and streams in the region. Managers in this case must recognize that the concerns of disciplines outside their own training are relevant and must be considered in decision making. Such managers should seek advice from individuals who have needed expertise. Such expertise may be found within agencies and universities and may be located by referral through professional organizations such as the American Fisheries Society.

2.4 THE FISHERIES MANAGEMENT PROCESS

2.4.1 Process Summary

The fisheries management process has five steps: (1) choice of goals, (2) selection of objectives, (3) identification of problems, (4) choice and implementation of actions, and (5) evaluation of actions (Figure 2.3). Decision making occurs within each of these steps, from choice of goals for a program to deciding upon the best actions to implement. The steps of the process require the use of an information base that contains knowledge about each component of the management environment. Managers use and contribute to the information base as each step is executed.

Goals provide long-term statements about what fisheries programs are to achieve. Objectives specify measurable, expected outcomes that indicate achievement or progress toward attainment of goals and state when the outcomes are to be achieved. Problem identification determines what factors impede achievement of goals and objectives. Actions are the activities chosen and implemented to overcome the problems. Evaluation determines whether the actions implemented helped to solve the problems and achieve goals and objectives. Revision of management programs occurs based on an assessment of the information gained in the evaluation step. The term management program refers to the application of the management process to achieve goals related to a fishery (e.g., the management program to rehabilitate lake trout populations in Lake Ontario). After completion of the evaluation step, management then returns to the next most appropriate step in the process—management redefines goals, chooses new objectives, identifies new problems, or implements alternative actions. Below, we provide an actual example of this process, followed by descriptions of the conceptual details of each step.

2.4.2 Example of the Management Process

Lake trout was originally one of the most important species native to the Great Lakes. From the 1930s, lake trout populations declined sharply due to predation by the exotic sea lamprey, overfishing, and habitat degradation (see Chapter 23). In Lake Ontario, lake trout were presumed to be extinct around 1960. In 1983, a plan for lake trout rehabilitation in Lake Ontario was established between fisheries agencies in the United States and Canada (Schneider et al. 1983). The plan documented the past and future direction of lake trout management. Implementation of the plan and the program that resulted illustrates the fisheries management process as described above (Figure 2.3). All block quotations are from Schneider et al. (1983:11). The goal of the lake trout management program in Lake Ontario is

> To rehabilitate the lake trout population of Lake Ontario such that the adult spawning stock(s) encompasses several year classes, sustains itself at a relatively stable level by natural reproduction, and produces a usable annual surplus.

This goal clearly and concisely states that the purpose of this program is to restore naturally reproducing lake trout populations to Lake Ontario in sufficient numbers and quality to permit a fishery.

An interim objective was then formulated to measure achievement of the rehabilitation goal stated above.

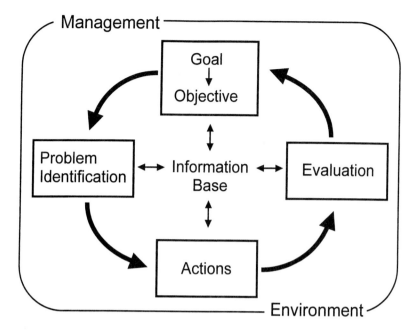

Figure 2.3 The steps of the fisheries management process as they cycle around the information base and are set within the management environment.

> By the year 2000, demonstrate that rehabilitation is feasible by developing a Lake Ontario lake trout stock consisting of 0.5 to 1.0 million adult fish with adult females that average 7.5 years of age and produce 100,000 yearlings annually.

This objective specifies measurable expected outcomes that can later be evaluated and also defines a specific date by which this objective is to be achieved. The development of this objective required scientific knowledge from the information base about the productive capacity of Lake Ontario and an understanding of lake trout population dynamics. This objective was intended as a checkpoint on the way to goal achievement but does not define full population rehabilitation. An ultimate objective was also established to define and quantify final achievement of the goal.

> To develop a lake trout population in Lake Ontario of 0.5 to 1.0 million adults that produce 2 to 3 million yearlings annually and provide 450,000 kg of usable surplus.

This objective is unusual because a date is not specified for its achievement. The agencies involved decided that a realistic date could not be specified until achievement of the first objective. The time required to achieve the first objective would provide an important piece of information for setting a completion date for the second objective.

Several problems were identified that would prevent achievement of these objectives, for example:

- sea lamprey predation on lake trout;
- overharvest of stocked lake trout;
- environmental degradation of Lake Ontario; and
- lack of optimal strains and procedures for stocking lake trout.

Management actions were then developed and implemented to solve these problems. Strategies were identified (e.g., to reduce lake trout mortality), and specific actions were chosen to accomplish the strategy.

1. Sea lamprey control. Since 1971, agents of the Great Lakes Fishery Commission, the Department of Fisheries and Oceans (Canada) and the U.S. Fish and Wildlife Service have maintained an active sea lamprey control program primarily through the application of selective lampricides into Lake Ontario tributary streams. (Strategy: increase lake trout population by reduction of mortality.)

2. Sport fisheries regulation. In 1988, the state of New York instituted more restrictive angler regulations (slot length limit) that reduced by 50% the sportfishing mortality of lake trout that had reached spawning size. (Strategy: increase lake trout population by reduction of mortality.)

3. Water quality legislation. Agencies in both countries continue to support legislation to reduce the eutrophication and pollution of Lake Ontario. (Strategy: protect and restore lake trout habitat.)

4. Lake trout stocking. Approximately 1.0 million lake trout of several genetically different strains were marked and stocked into Lake Ontario annually. The marking program has permitted determination of the strains that survive best and identification of improved stocking procedures (Elrod et al. 1995). (Strategy: increase lake trout population by means of improved stocking methods.)

Evaluation of the program has been ongoing and includes field studies and annual assessment meetings by the fisheries agencies of the United States and Canada. For example, the natural production of yearlings and the size and age of the adult spawning stock is evaluated annually by trawling and gill netting conducted by the Biological Resources Division of the U.S. Geological Survey, New York State Department of Environmental Conservation, and Ontario Ministry of Natural Resources. Assessment is carried out by comparing the results of field studies to the program's objectives during the annual meetings of the Lake Ontario Committee organized by the Great Lakes Fishery Commission. This committee includes representatives of the management agencies that are directly involved in fisheries management and who established the goal for lake trout management. The management program is annually revised and fine-tuned based on discussions during the meetings. Major program reviews took place in 1990 and 1997, and in 1997, a new objective was developed.

> By the year 2017, sustain the density of wild lake trout in U.S. waters at a catch of 26 age-2 fish in bottom trawling conducted during July, and in Canadian waters increase the abundance of wild age-2 lake trout above their 1997 levels.

This new objective was developed because in 1994 the first substantial natural reproduction (a real sign of success!) was detected by standardized assessment trawling in Lake Ontario.

2.4.3 Goals: Why We Manage Fisheries

The first step in the development of a management program is to determine the goals that are to be achieved (Figure 2.3). Goals are long-term, broad statements of intent that define the purpose of management but not how the goals will be achieved

(Barber and Taylor 1990). Goals are statements that explain why agencies manage a particular fishery resource. Consequently, goals set the frame of reference in which management actions are chosen and subsequently implemented.

Goals that are specified within agencies' programs will be hierarchically organized. The highest level may be determined legislatively and are often termed mission or vision statements for an agency. For example, Wisconsin statutes establish that conservation law and the Department of Natural Resources were instituted "to provide an adequate and flexible system for the protection, development and use of forests, fish and game, lakes, streams, plant life, flowers and other outdoor resources" (Wisconsin State Statutes 23.09). Following this statement is an administrative code that specifies "the goal of fish management is to provide opportunities for the optimum use and enjoyment of Wisconsin's aquatic resources, both sport and commercial" (Wisconsin Administrative Code of Natural Resources 1.01(2)). Although useful in establishment of the authority for an agency, such legislative mandates are usually so general as to help only to bound goals defined lower in the hierarchy. Nested within these broad, general statements are goals that direct specific management programs. The goal statement described earlier for lake trout in Lake Ontario is an example of a lower-level goal that gives real direction to a fisheries management program.

Determination of goals is usually a difficult, complex process that must be done carefully because of the long-term effects that goals have on the future direction of management programs. Both fisheries professionals and representatives from the public should be involved in discussions leading to the choice of goals. Senior-level agency administrators responsible for policy must always be included in the process of determining goals (Crowe 1983). Their involvement will help them to understand a program so that they can later provide the essential internal agency support needed for program implementation. Inclusion of regional or district managers in the process will help ensure that goals will be realistic with respect to the resource and the public. In some cases, fisheries scientists from other governmental agencies or universities may be helpful in providing specialized information or viewpoints.

The public should always be involved in the goal determination process (Anderson and Hurley 1980). If programs are to succeed, they must reflect the values and interests of society (Yarbrough 1987). Public participation ensures that management will reflect public interests. Participation also provides the opportunity for agencies to communicate new ideas and information being used to help guide future program direction. Fisheries programs that include goals derived with public input are more likely to be successful because the public will better understand the program, will have helped shape its direction, and, therefore, will support the management efforts of the agency. Management programs that result from public participation are politically defensible, especially when the goal determination process has been documented (e.g., meeting minutes). Forums for this type of public involvement include task forces, steering committees, advisory councils, review committees, surveys, and public hearings (Anderson and Hurley 1980).

A major challenge of public participation is effective communication between the fisheries agency and public groups (see Chapter 3). Various media, communication, and extension specialists can be of substantial assistance in bridging the gap between fisheries professionals and the public. Discussions about goals are often difficult to initiate with either agency personnel or the public because there is a natural tendency to discuss management actions (e.g., stock more fish) rather than to ensure consensus about what management is to achieve (Giles 1981). To counteract this tendency, the management process should be reviewed at the start of discussions to help participants understand the significance of the definition of goals.

Administrators, regional biologists, resource scientists, and public groups will provide important linkages to the information base (Figure 2.3; see Section 2.4.8). Ecological information provided by scientific studies will help define which goals are biologically feasible. Economic information will help predict the effects of achieving different goals on businesses, jobs, markets, and the standard of living. Political information will identify which goals are legally appropriate for the agency to pursue. Sociocultural information can identify which groups of people will be affected by programs designed to achieve different goals or how different groups may react to a set of goals.

Simple procedures do not exist to help determine goals that are optimally compatible with ecological realities and the sociocultural values of different public groups. Philosophers and social scientists need to provide methods to define, interpret, and understand the values of various groups that make up the public sector. These methods would help managers identify where values among groups overlap or are in conflict. In addition, a process is needed to identify values that are of importance to society in general. The identification of these values would provide a general definition of the "public interest," which fisheries agencies are often required by legislation to serve. This type of analysis and identification could then be combined with decision theory, such as is used in businesses, to help choose goals for fisheries management (Stephenson and Lane 1995; Malvestuto and Hudgins 1996). The process of goal determination should be a top priority for study by philosophers and scientists interested in fisheries management (see Scarnecchia 1988).

Realistically, management of fisheries resources cannot wait until a decision theory that includes cultural values is developed and tested. Management goals must be chosen. Basic questions such as, where are we? and, where do we want to go? will be helpful in choosing goals (Crowe 1983; see Box 2.1). Recognition of unique opportunities to develop, enhance, or protect fisheries should occur at this step. The conservation-minded manager often must judge the degree to which biological problems can be safely balanced against pressing economic and social desires (Olver et al. 1995). Though the goal determination process may be imperfectly defined, fisheries managers must consider the long-term well-being of fisheries resources and the use of those resources by future generations. These considerations are not always fully represented in discussions about management goals. Occasionally, the well-being of fisheries resources is assured, yet conflicts over resource allocation occur among user groups. In this case, the manager may assume a more neutral role as a facilitator to resolve conflicts and to allocate the resource fairly. Managers, however, should not simply be mediators among special interest groups without regard for fishery resources.

2.4.4 Objectives: Management Targets with Deadlines

The second step in the management process is to choose objectives that will define progress toward achievement of goals (Figure 2.3). Objectives have two important characteristics: they are measurable and they have a specified time period in which to

---

Box 2.1   Goals for Urban and Wilderness Fisheries— An Exercise

Nine major lakes occur within the city limits of Minneapolis and St. Paul, Minne-sota. Most of their shorelines are part of an extensive city park system that provides easy access. Outboard motors are prohibited on some lakes. The nine city lakes are fertile and productive and have comparatively hard water for this midcontinental region. The fish communities include northern pike, muskellunge, walleye, largemouth bass, bluegill, pumpkinseed, common shiner, and fathead minnow. In addition to angling, the lakes and their shorelines are used for picnics, jogging, bird watching, sailing, cross-country ski-ing, and ice skating. The larger metropolitan area of seven counties has about two million residents and more than 100 lakes other than those in the cities.

Develop a goal or a set of goals around which an urban fishery management pro-gram could develop for the nine lakes. What types of anglers would you identify as key users? Would your goals address the interests of any special angler groups? What prod-ucts of the fishery will the anglers enjoy as the result of management? Are there unique opportunities for management in this urban situation? What members of the public might have problems with your goals?

Now develop fisheries management goals for Utah's High Uinta Wilderness Area. This region has over 250 limited-access, infertile, alpine lakes that provide trout and Arctic grayling fishing. The region is surrounded by an area that has a low-density human population. Salt Lake City is a one to two hour drive away from the area and has a metro-politan population of over 600,000. A review of Kennedy and Brown (1976; their paper used the earlier designation of Uinta Primitive Area) will be useful to you as you consider and develop possible goal statements.

How do your goals for the Uinta area contrast with those developed for the urban Minnesota fishery? Would you use a different approach for gaining public input about goals for these two fisheries?

---

be accomplished. Objectives are the criteria by which agencies can determine their progress toward goal achievement over time (Barber and Taylor 1990; Box 1996). Objectives provide a measurable definition of successful management. They should be specific, attainable, and realistic.

Choice of the measures used to define objectives requires several consider-ations. First, measures must relate to the management goal and thus be related to the anticipated products of management (see Box 2.2). Second, they must be real-istic, or attainable. Similar to the goal-setting process, an assessment of ecologi-cal, economic, political, and sociocultural information is required to ensure that the objectives are reasonable relative to the management environment. Third, the feasibility, both logistically and economically, of conducting studies to measure the progress toward achievement of the objectives must be ensured. Such measure-ment is a requirement for the fifth step in the management process—evaluation (Figure 2.3). Money from an agency's budget will have to be available for these studies. Fourth, chosen variables must be measurable with sufficient precision to permit statistical comparison before and after management actions are taken. Avail-able technology and budget will determine the degree of precision attainable by evaluation studies.

Management objectives related to a commercial fishery may be defined more rigorously than those related to a recreational fishery or ecosystem objective. For commercial fisheries concepts have been developed to define and estimate sustainable yield (see Chapter 6) and economic yield. These concepts can help establish goals and measurable objectives for commercial fisheries management (e.g., Parsons 1993; King 1995). Measurable objectives that gauge the satisfaction of recreational anglers are more difficult to define and evaluate (Larkin 1980; Holland and Ditton 1992). Objectives could include measures of resource use such as the proportion of anglers who catch their legal limit. Alternatively, recreational satisfaction could be measured directly through sociological surveys of the users. Objectives that quantify goals related to population reestablishment or endangered species management will most likely be measures of population characteristics, such as illustrated above for lake trout in Lake Ontario.

Objectives should be defined with the help of several types of fisheries professionals from several disciplines. Applied ecologists, sociologists, economists, statisticians, planners, and budget analysts either from within or external to an agency may be helpful. This team must also include field managers who are responsible for implementation of the management program. Typically, senior-level administrators need not be involved as intimately as they were in the goal determination process (Crowe 1983). Direct public input or information from sociological surveys can help guide managers in their selection of parameters that best define recreational fisheries objectives.

Box 2.2   Measures that Could Be Used for Objectives

- Catch per unit effort

- Number of fish caught

- Number of trophy fish caught

- Number of angler hours

- Number of angler trips

- Proportion of anglers catching limits

- Measures of variation related to species, size, or number caught per trip

- Measures of angler satisfaction

- Value of landed commercial catch (maximum economic yield)

- Kilograms of fish per year (maximum sustained yield)

- Spawning stock abundance (species restoration management)

- Year class recruitment index (species restoration management)

- Number of eggs deposited (species restoration management)

2.4.5 Problem Identification: What Can Prevent Success

The next step in fisheries management is identification of problems that could prevent achievement of objectives and goals. If the management team has answered, where do we want to go? then they should ask, what will prevent us from getting there? For example, sea lamprey predation was considered to be an important problem preventing achievement of the lake trout restoration goal in Lake Ontario (Figure 2.4). One approach to this step is to seek answers to questions about the potential problems that could prevent attainment of goals, for example: Is stream habitat quality reducing trout abundance? Is prey abundance sufficient to support an increase in predators? Do markets exist to accommodate expansion of commercial harvest? Do nonnative species affect the viability of native species populations? Do anglers know about the fishing opportunities available? Based on answers to questions such as these, a group of problems can be identified that will be the focus for the next step—management actions (Figure 2.3).

Management not guided by goals or objectives usually will focus on immediate problems and thus will start with problem identification instead of goal selection. Management agencies often feel tremendous pressure to respond immediately with actions demanded by the public or politicians to solve specific, short-term problems. If issue-oriented management is followed, then problems are identified and actions are taken but without goals, objectives, or evaluation in mind. Issue-oriented management can result in management actions that are unfocused and often contradictory in purpose. Nevertheless, issues or problems will be raised not only by the public and politicians but also by personnel within fisheries agencies. How should the manager respond? When discussions begin with problems, the manager should redirect attention back to the goals and objectives of programs and thus focus the discussion on how the specific issues prevent achievement of goals and objectives.

Problem identification in goal-oriented management should be conducted by a team that represents disciplines related to the measures specified in the objectives. This group must also include regional or local field managers responsible for program implementation. Resource scientists external to the agency can often provide helpful expertise not represented by agency personnel. The team assembled for problem identification within the lake trout program for Lake Ontario included state, provincial, and federal agency personnel and scientists from universities specializing in hatchery production, sea lamprey control, population dynamics, and genetics. Involvement of specialists in identifying problems ensures that a broad information base is used to identify and analyze the problems. Public involvement may also occur at this step, especially when problems related to user conflicts need to be identified. The discussions that occur within the group should be conducted in an open, creative atmosphere where all ideas are considered.

In preparation for the next step, problems identified must be examined to determine which have the greatest effect on the objectives and offer the most potential for resolution. Limited agency budgets will usually prevent all problems from being addressed initially. A final subset of problems needs to be established; however, all the problems identified should be documented because the agency may need to reconsider them later.

Figure 2.4 Lake trout mortality caused by sea lampreys is considered an important problem preventing reestablishment of lake trout populations in Lake Ontario.

2.4.6 Actions: Solutions to Problems

The fourth step in the management process is to identify, choose, and implement the strategies and actions required to solve problems (Figure 2.3). Actions are the specific tactics (sometimes called techniques) used by an agency to solve problems that prevent achievement of goals and objectives. If we have answered the question, where do we want to be? (goals and objectives) and have identified, what will prevent us from getting there? then this step answers the question, how will we get there? (Crowe 1983). Sea lamprey control, stocking, and angler regulation are examples of the actions chosen and implemented to begin achievement of the lake trout restoration goal in Lake Ontario (Figure 2.5). Actions are unfortunately what many public groups and some fisheries professionals define as management. Instead, management includes the entire goal-oriented process set within the management environment (Figures 2.1 and 2.3). Discussions about proposed actions should always be linked back to goals, objectives, and the problems intended to be solved (Cowx 1994; AFS 1995).

Strategies define the purpose and state the desired outcomes of actions. Often, specific problems are part of a larger, more general problem. In this case, strategies can be developed that address the general problem and join together many specific problems. It is critical to first determine strategies before implementing actions. For example, the reduction of lake trout mortality is a strategy that links and solves the problems created by sea lamprey predation and overharvest by anglers. Most importantly, the identification of strategies often results in the identification of additional, less obvious solutions or actions. Too often actions are chosen that simply address the symptoms of problems and not their causes (Meffe 1992). For example, a manager might respond to an overharvest problem by simply choosing the obvious action of reducing angler creel limits. However, if the strategy chosen is to reduce all forms of mortality, then a much broader set of actions will be considered, such as size limits,

Figure 2.5 Stocking of yearling lake trout and regulation of angler harvest were two important actions chosen and implemented as a part of lake trout management in Lake Ontario.

seasons, fish refuges, reduction of predation and other natural causes of mortality, and the establishment of a voluntary catch-and-release program. Once strategies are established, actions should be chosen and implemented.

Actions may address any component of the management environment—ecological, economic, political, or sociocultural. Ecological actions include what is often thought of as the traditional domain of fisheries management (Sigler and Sigler 1990). Traditional actions by fisheries agencies include regulation of resource use (e.g., season or creel limits; Chapter 17), population manipulation (e.g., stocking or chemical control of sea lamprey; Chapters 14 and 15), and habitat manipulation (e.g., stream improvement; Chapter 10). Management actions, however, include a much broader array than these examples. Economic solutions for a commercial fishery could include the establishment of monthly catch quotas to prevent temporary market gluts that drive product prices down. Political solutions could include an increase in an agency budget to purchase and protect critical spawning habitats or the passage of new legislation to give authority to an agency to manage fisheries through limited entry. Sociocultural actions may be related to the education of resource users and include programs about outdoor ethics or how to harvest underutilized resources (e.g., how to catch and cook common carp). Sociocultural actions could also focus on encouraging landowners to provide anglers access to their lands or encouraging them to adopt wise land use practices to reduce soil erosion and stream siltation in watersheds.

Choice of management actions can follow a qualitative approach based on experience, a quantitative analytical process, or a combination both. Often the local fisheries manager will have an intuitive sense based on past experience of how the resource and its users will respond to a particular action. Having public groups suggest actions that would be acceptable to them can effectively augment this experience. In other cases, sufficient information will be available to construct a simulation model of the management environment to assist in the decision-making process (see Walters 1986). Such models will have as input variables specific management actions and have as output indicators the measures specified in the objectives (e.g., catch per unit effort). The model is used to predict output indicators for several different combinations of actions in order to identify the best set of actions by which to achieve objectives and goals. Regardless of the process used, usually all actions cannot be implemented because of monetary and personnel constraints. Actions should be based on the greatest cost effectiveness and the greatest potential for solving problems.

The team of individuals involved in selecting management actions will be similar to the team that identified problems in the previous management step. It is important to include local managers responsible for program implementation and who have experience with the specific resource and its users. Others who know the agency's conservation laws and represent law enforcement will also be helpful when actions related to regulation of users are to be considered. If simulation models will be used, then people with resource modeling expertise must be included. Public involvement often can be helpful at this stage. In some situations, several actions may be equally effective in addressing a particular strategy. Public input can help managers choose actions most socially acceptable. For example, after several meetings with the public, the New York State Department of Environmental Conservation chose size limit and season regulations over more restrictive creel limits as the regulations most acceptable to the public for the reduction of angler-induced lake trout mortality in Lake Ontario.

The team involved in choosing strategies and actions may conclude that the problems related to the achievement of goals are insurmountable. If this results, the management process must then return to redefine goals and objectives. Alternatively, the group may decide that the problems and associated strategies cannot be clearly identified without more information, and they may recommend as the first management action that studies be conducted for further problem definition.

After management actions are chosen, they must be implemented. For example, fish must be propagated and stocked. New laws that govern anglers must be passed. Extension publications must be written and distributed. The implementation of actions requires the allocation of money and personnel from agency budgets. Control of implementation is often termed operational or tactical planning (Crowe 1983). Each year a tactical plan for Lake Ontario is developed among the state, provincial, and federal agencies that determines, for example, which tributary streams will be chemically treated to control sea lampreys, how many lake trout will be produced by hatcheries, and where these fish will be stocked. After money has been allocated, the actions can be implemented, and the components of the management environment evaluated for their response.

2.4.7 Evaluation: Information for Program Improvement

The fifth step in the management process is evaluation of the management environment's response to the actions implemented and an assessment of that response in terms of the entire management process (Figure 2.3). This step answers the question, did we achieve our goals and objectives? (Crowe 1983). This step has four parts: measurement of the effects of actions, comparison of measurements with the objectives, assessment of the comparison, and revision, continuation, or termination of particular management programs.

First, the effects of our actions must be measured. The focus of such measurement should be the outcomes stated in the objectives. For example, evaluation studies are annually conducted to estimate how many yearling lake trout are produced naturally in Lake Ontario (Figure 2.6). Choice of this parameter comes directly from the program's

Figure 2.6 The RV *Seth Green* operated by the New York State Department of Environmental Conservation is used to measure the parameters stated in the management objectives for lake trout in Lake Ontario.

objectives. Careful statistical design of sampling studies prior to the initiation of management actions is essential to detect differences over time. Obtaining necessary measurements is the focus of many of the field studies conducted by state, provincial, and federal fisheries agencies and often occur simultaneously with implementation of actions rather than just on the date specified in the objectives. These measurements provide continuous feedback information to guide management.

Second, the measurements must be compared with those stated in the objectives. For instance, the question must be answered as to whether 26 age-2 wild lake trout were caught last July by trawling in Lake Ontario (see the new objective for 2017 stated in section 2.4.2).

Third, an assessment of the management program must follow. Decisions must be made as to what to do next. If objectives were successfully met, the agency must decide whether to maintain, expand, reduce, or terminate the current management program. Failure to achieve objectives, however, often stimulates in-depth analysis that is informative and helps guide future management. We often learn more from our mistakes than from our successes. For example, if few or no yearling lake trout are captured in Lake Ontario, it must be determined why. Is the new objective unrealistic? Were spawning substrates unsuitable? Were adults unable to locate spawning reefs? Was stocking and sea lamprey control ineffective? Assessment of the lake trout program occurs at an annual meeting of all the agencies (Figure 2.7).

Fourth, revision of the management program is implemented based on the decisions made in the assessment step above. The management process now cycles to the next most appropriate step. Goals may need to be revised. Objectives may need to be redefined with different time frames. New problems may have arisen. New strategies may need to be developed, and different actions may need to be taken.

The evaluation step allows learning to occur and is crucial to the management process. This step provides the information required for program redirection. Evaluation is the step that gives management the capability to correct earlier errors and to adapt to changes in the management environment. Evaluation permits analysis of the past so that management can improve in the future (Stephenson and Lane 1995). To choose to conduct the evaluation step means that limited resources within agency budgets will need to be directed to evaluation, thus reducing the amount available for management actions. Nevertheless, the long-term benefits of evaluation will more than counterbalance budget and personnel expenditures and the temporary delay in implementation of other management actions.

2.4.8 Information Base: Resources for Effective Decision Making

The information base assists decision making and is used during each step of the fisheries management process (Figure 2.3). The evaluation step causes the information base to grow. Also, research conducted by scientists provides an important contribution to the information base. The information base includes not only the published results of scientific research but also unpublished data and the experiences of fisheries managers and the public. Ecological, economic, political, and sociocultural knowledge is included in the information base. For example, the information base could include published accounts in both professional journals and local newspapers about the economic benefits brought about by lake trout rehabilitation, as well as the anecdotal observations by a hatchery manager that lake trout yearlings reared at high densities in raceways appear in poor condition. Agencies, universities, private consultants, and industries can contribute to the information

Figure 2.7   Assessment of annual evaluations occurs at the Lake Ontario Committee meetings attended by New York, Ontario, and the two federal fisheries agencies. The meetings are organized by the Great Lakes Fishery Commission. (Pictured here, Carlos Fetterolf, Jr., former Executive Secretary, Great Lakes Fishery Commission.)

base available for fisheries management. Access to the information base is often facilitated when the management teams are made up of individuals with a variety of perspectives—scientists, managers, and members of the public. The teams can then access a broad information base and provide recommendations concerning the appropriateness of goals and objectives, identification of problems, actions to consider, and the evaluation of a management program. Impartial development of these recommendations is critical so that informed choices can be made among management options. A management team's unrestricted access to information and freedom to develop recommendations must be vigorously defended and be unconstrained by political and sociocultural control (Hutchings et al. 1997).

2.5   SOUND RATIONALE FOR PROGRAMS: KEY TO SURVIVAL

Management programs developed through the process described above should have a positive effect on fisheries resources. Programs, however, may be canceled prematurely by legislative actions on agency budgets or by a lack of public support. Therefore, defensibility is an important characteristic for program survival. Use of a goal-oriented approach to management will help ensure that sound rationales exist for fisheries management programs. Fisheries managers and the management team will then be able to explain the rationale behind past and planned agency decisions and subsequent actions.

Defensibility is especially enhanced when public involvement has occurred throughout the development of the management program. In this chapter, we have identified some key steps in the management process at which direct public involvement will assist and improve fisheries management. In addition, agency personnel need to communicate with (including listening to) the public about the current status of management programs. An informed public is a most effective ally when a fisheries management agency needs support to maintain its programs.

2.6   CONFLICT RESOLUTION

2.6.1 Resolving Conflicts: Part of the Fisheries Manager's Job

Don't be surprised but be ready—conflicts will inevitably occur in a fisheries management career! Understanding the nature of conflicts will help a fisheries manager anticipate and respond to them appropriately (Carpenter and Kennedy 1988; Gale 1992). All change, including new directions for management programs, will produce conflict among resource users, between an agency and its stakeholders, and sometimes within an agency. Conflicts can occur between anglers and other aquatic resources users, such as canoeists, municipal water supply authorities, and industry. Tensions can also exist within fisheries, such as between commercial and sport fishing or between those who fly-fish and those who use bait. Stakeholders in a fishery may demand an agency to stock nonnative fishes which may be in conflict with long-term agency goals to restore native species. Unfortunately, conflicts among fisheries professionals also occur and can be between employee and employer, between central versus regional offices, among state agencies, among university researchers, among states, and between countries. A good manager must learn to manage to prevent conflicts and, when conflicts exist, to negotiate these conflicts to a positive end. Conflict prevention by using sound communication with stakeholders (Chapter 3) and by making sure to involve stakeholders appropriately in programs is a key approach. Not all conflicts, however, can be prevented. The fisheries manager must be able to resolve conflicts efficiently when they occur. A short excellent text on this subject that every manager should read is *Getting to Yes—Negotiating Agreement without Giving In* (Fisher and Ury 1991). Most of the material below is drawn from this reference.

2.6.2 Good Negotiation Procedures: Three Essential Characteristics

Three criteria should be used to judge negotiation procedures used to resolve fisheries management conflicts. First, the procedure should produce wise agreements. Resolution of a conflict should produce an agreement that best meets the interests of all parties and should be durable. Second, negotiation procedures should exhibit a level of efficiency—they should not require an extraordinary amount of time to reach a wise agreement. Third, the negotiation procedures used to resolve conflicts should improve, or at least not damage, relationships among parties. Usually some link exists among the parties that will probably cause them to continue to interact in the future. Damaged relationships today may affect future negotiation outcomes tomorrow. The negotiation procedure used should meet these three criteria.

2.6.3 Positional Bargaining: Poor Decisions, Lengthy Negotiations, and Damaged Relationships

One generally bad approach to conflict resolution is called positional bargaining. With this approach, each side states a position, argues for it, and then makes concessions to reach a compromise. Because concessions will be made, the best strategy to take is to state initially an extreme position so that after concessions the agreement will be satisfactory. Most often the end result is a situation where one side views themselves as winning and the other side as losing. Unfortunately, many agreements reached by this process results in both sides losing. This approach is common when buying used cars—"This car is pretty bad, but I will give you $1000 for it." Positional bargaining often characterizes the outset of conflicts between user groups ("Double the stocking!") and a fisheries management agency. How an agency responds to the initial position by the public will establish whether positional bargaining is used to resolve the conflict or whether an alternative approach is used.

What is wrong with positional bargaining? Central to this approach are the stated positions between the parties. Nowhere in this process is there an identification of the underlying interests or real concerns of the parties. Instead, a regular sequence of attack and defense of the stated positions characterizes this approach. Because little or no attention is given to meeting the interests of the other side, positional bargaining encourages a selfish attitude toward proposed solutions. Negotiators often develop a personal identification with a stated position, and thus an attack on the position by the other side is interpreted as an attack on the person who stated the position. A strategy for success is to take an extreme position and then drag out the negotiations and wear out the opposition to gain substantial concessions. Thus, success with this approach often requires dishonesty or deceit about what you want (take an extreme position as opposed to an honest one), is time inefficient (drag negotiations out), and endangers your future relationship with the other side (bend others to your rigid will). Thus, positional bargaining will not produce wise and durable agreements that meet the interests of both parties, it is time inefficient, and it hurts relationships. Because the criteria stated above for judging negotiation procedures are violated, fisheries managers must choose a different approach.

2.6.4 Principled Negotiation: A Focus on Interests, Not Positions

What is an alternative, better approach to conflict resolution?—principled negotiation (Fisher and Ury 1991). Principled negotiation focuses on the interests of all parties (as opposed to positions) and chooses the best solution to meet those interests among many potential options. The approach has four basic elements:

- people—separate the people from the problem;
- interests—focus on interests, not positions;
- options—generate a variety of possibilities before deciding what to do; and
- criteria—insist that solutions be based on objective criteria.

These elements are incorporated throughout a cyclic process of analysis, planning, and discussion.

First, separate the people from the problem. Conflicts bring out strong emotions in people that get in the way of wise agreements or solutions. The emotions of all sides must be understood and respected and should guide communication. Understand all sides' perception of the conflict, their personal feelings, fears, hopes, and dreams. Their perceptions, even if ill founded, are their terms of reality and must be addressed in the negotiations. Remember, understanding their point of view is not the same as agreeing with it. Negotiations create anxiety or fear because of the perception of a win–lose outcome—parties to the negotiations fear being the losers. Fear breeds resentment and then anger, and anger causes fear…and on it goes! Try to create the perception of a win–win outcome. Last, learn to communicate based on an understanding of all sides (see Chapter 3). Rehearse what will be said and listen to it through the participants' ears. The goal is to speak to be understood. Say the same concept several ways (repetition) and concisely (succinctly). If possible, try to build a relationship with all parties before the negotiations begin. Success with the "people element" will have occurred if the parties negotiate as though the problem is the adversary rather than each other.

Second, focus on interests, not positions. The needs, desires, concerns, and fears of all sides must be identified. A key concept here is to explore the reasons behind any stated position—this is the interest. For example, an agency states they want to reduce the lake trout bag limit from three fish to one fish in Lake Ontario. Anglers counter that they will not allow the agency to reduce the bag limit. In this case, the agency's interest may be that it needs to reduce lake trout mortality so fish can reach maturity and reproduce. The interest of the private anglers may be that they don't want their fishing success reduced. The charter fishing industry's interest may be that it's worried about its business. Mutual or shared interests must then be identified as well as those that are opposed or mutually exclusive. Considerable rapport can often be gained among the parties when a clear statement of each of their interests is made. When negotiations are tense, some tensions will be eased if each party knows that its concerns are understood. Knowing the other parties' interests is a key piece of information required to develop options that can meet some of the interests of all sides.

Third, invent options for mutual gain. The ideal is to generate options that address the interests of more than one party in the negotiations. Options should be created by brainstorming or inventing as many options as possible without attempting to judge the merits of each. The purpose is to increase and stimulate the development of creative options and not to look for a single answer. Once a list is generated, then each option should be evaluated (see below) for its potential to produce mutual gain among the parties.

Fourth, insist on objective criteria. Almost always, some interests will conflict. Choices ultimately will have to be made that favor one interest over another. Criteria, standards, or procedures must be determined by which to make these decisions. If possible, gain agreement among the parties at the start of a negotiation on the standards to be used. Types of criteria could be past precedence (past settlements of fisheries issues), external, independent scientific judgement (e.g., for issues over commercial quotas), professional standards (e.g., government salary ratings), minimization of costs, moral standards (e.g., fairness), equal treatment (a 50 to 50 split), and reciprocity. One well-known approach used by parents to allocate the last piece of cake between two children is "one cuts and the other chooses." This approach could be used to solve a conflict between commercial and sport fisheries over zone regulations. If zones between the two activities must be chosen, then one group could set the boundaries for the zones and let the other group choose an equal number of sport and commercial zones.

These four elements must be carefully woven into the negotiation process of analysis, planning, and discussion. Analysis is knowing or defining the current status of the conflict, gathering all possible information about all sides, knowing your interests and those of the other parties, and knowing the time constraints that may be driving the need for resolution. Planning is next and requires choosing tactics to use to solve problems, setting your own priorities, identifying mutual interests shared among parties, and generating options for solutions. Analysis and planning are done before meeting with the other parties. Discussion is the actual communication or negotiation with the other parties. Sometimes having a professional facilitator who is neutral can be helpful for this discussion phase. A facilitator familiar with group decision-making techniques (e.g., Coughlan and Armour 1992) can help guide a group efficiently toward resolution of a conflict. Usually several meetings will have to take place before agreement will occur. Thus, after each discussion step, analysis and planning should occur before the next meeting.

Conflicts occur because fishes and the fisheries they support are important to many people in different ways. Attempting to resolve conflicts through positional bargaining can damage relationships and often results in less than satisfying solutions. Principled negotiation procedures have been developed to produce wise, durable agreements in a time-efficient manner and to provide an opportunity to build relationships. Because conflicts are a part of the life of a fisheries manager, conflicts should be accepted as a characteristic of the profession and approached with confidence using the negotiation techniques presented here. Conflicts often provide exceptional opportunities to promote better understanding among groups and to educate others concerning the benefits of wise stewardship of natural resources.

2.7   ADAPTIVE MANAGEMENT: LEARNING FROM PAST ACTIONS

The management process described above is based on the fundamental premise that uncertainty exists for every decision because of insufficient information. Science may help reduce uncertainty in the future; however, it will not result in risk-free management with perfectly predictable results (Ludwig et al. 1993). Assessment of the probabilities that different outcomes will occur from a management action will help the management team understand the level of uncertainty associated with any decision option (Francis and Shotton 1997). Uncertainty of outcomes arises from a lack of knowledge about the ecological and social processes that affect fisheries as well as from stochastic events. Evaluation as a step in the process was emphasized because it permits learning from the responses within the management environment.

The characteristics of the management process described in this chapter are similar to the concept of adaptive management presented by Walters (1986). Walters and others view management actions as primary tools for whole-system study of the management environment. Adaptive management is in contrast to the more typical incremental changes that occur via the actions implemented by many agencies (Lindblom 1959). The adaptive management concept promotes aggressive experimentation with management actions coupled with careful evaluation. The primary purpose of evaluation is to understand, or "adaptively learn," about the full range of responses in the management environment that are possible as various actions are implemented. The predictability of these responses becomes a central aid to decision making in management. With this approach and the knowledge that the management process is inherently uncertain, unexpected outcomes of management can be viewed as opportunities to learn rather than failures to predict (Lee 1993).

Application of adaptive management will rapidly provide the information needed for better assessment of risk in management options. This is especially true when compared with the traditional incremental approach used by agencies or the reductionist approach of some scientists. Nevertheless, aggressive experimentation with management actions incurs the real risk of negatively affecting fisheries resources (at least temporarily), which could upset users and result in a political liability for a program. To avoid such risk, an agency should seek public approval before using an adaptive management approach.

2.8   ECOSYSTEM MANAGEMENT: A PARADIGM SHIFT

Natural resources management, including for fisheries, is shifting from commodity production of a single resource to management of whole systems for a variety of purposes (Vogt et al. 1997). In many cases, goals have shifted from maximum sustained yield to sustaining ecosystem diversity and function. Fundamentally, this approach has developed from the recognition that a single species cannot be managed effectively without understanding its interconnectedness with other species and ecosystem processes. Thus, single species have become less of a focus and ecosystems themselves are emerging as management units.

The interconnectedness of species within ecosystem processes may be illustrated by Pacific salmon. Salmon management in Canada and the United States has historically focused on the notion that the abundance of a spawning stock will establish the abundance of the next year-class of salmon (stock–recruitment relationship; Chapter 6). Thus, fisheries have been regulated to allow sufficient numbers (termed escapement) of adult fish to move into spawning streams to ensure adequate spawning and sustained yields. However, the role that salmon and their carcasses (Pacific salmon die after spawning) play in the upstream transport of marine nutrients (nitrogen and phosphorous) has recently become better understood (e.g., Larkin and Slaney 1997). These nutrients affect primary production in streams, which, in turn, affects secondary production (such as by invertebrates), which determines the availability of food for young salmon and other species. Without adequate escapement of adults, spawning streams exhibit nutrient loss and oligotrophication over time, which seriously affects a stream's ability to support young salmon. Salmon escapement can no longer be seen simply as an issue of sufficient egg deposition but instead must also be viewed in terms of the effects of salmon carcasses on nutrient cycling and stream ecosystems. Evidence of the shift from single-species to an ecosystem focus can be seen in the emphasis of watershed approaches to management (Naiman 1992; Chapter 9) and research focused on land–water linkages (e.g., France 1997). Ecosystem management is described further in Chapter 5.

2.9   CONCLUSION

Fisheries management was defined at the start of this chapter as the use of all types of information (ecological, economic, political, and sociocultural) in decision making that results in actions (e.g., regulations) to achieve goals established for fish resources. Learning as we manage fisheries through evaluation of programs has been stressed throughout this chapter. Adoption of a goal-focused, cyclic process for fisheries management will prevent haphazard construction of programs. Budgets are always limited, yet as human populations increase, fisheries resources demand more management attention than in the past. Society simply cannot afford the luxury or potential waste that can be caused by the unfocused evolution of fisheries management programs. To stay within the process, the fisheries manager must always evaluate each decision in terms of a program's goals and objectives and the strategies that actions are to address.

The conduct of management demands day-to-day decision making. The fisheries manager must recognize that past decisions will often affect the types of choices or opportunities that will be available in the future. Each decision must be made with an eye to its potential effects on future decisions. For example, the decision to introduce a species may be irrevocable. The new species may establish naturalized populations or a new group of anglers may develop who specialize in catching a species that must be continually stocked.

The personal wisdom and skill required for effective decision making in fisheries management is not something gained simply by completion of a course or reading a textbook. The aspiring fisheries professional must recognize that management effectiveness requires practical experience in an agency and a lifelong commitment to studying the forces that affect management. The professional must also work hard to develop skills to communicate effectively with fellow professionals and the public. Citations within this chapter have been given to assist the reader in the continued study of the processes of management.

Fisheries management offers an exciting profession to those who enjoy direct involvement with management of a portion of the Earth's natural resources. Decisions made today by a manager may ensure conservation and wise use of fishes for the future. This aspect of direct involvement in management decisions helps distinguish the fisheries manager in an agency from the fisheries scientist in a university. The scientist's role is primarily to discover, interpret, and provide information. Managers within agencies are responsible for decision making that actually affects fisheries resources.

The foci of the management process are fishes and people. Fishes, and the aquatic communities of which they are a part, provide delightful and intriguing subjects for study and management. Equally fascinating is understanding the economic, political, and sociocultural forces caused by the many interactions between fishes and people. These interactions inextricably weave fishes and people together within the process of fisheries management. Thus, the best fisheries managers are those who enjoy, understand, and work effectively with both fishes and people.

2.10 REFERENCES

AFS (American Fisheries Society). 1995. Special fishing regulations for managing freshwater sport fisheries: an American Fisheries Society position statement. Fisheries 20(12):32–33.

Anderson, K. H., and F. B. Hurley, Jr. 1980. Wildlife program planning. Pages 455–471 in S. D. Schemnitz, editor. Wildlife management techniques manual, 4th edition. The Wildlife Society, Washington, D.C.

Anderson, L. G. 1986. The economics of fisheries management. The Johns Hopkins University Press, Baltimore, Maryland.

Barber, W. E., and J. N. Taylor. 1990. The importance of goals, objectives, and values in the fisheries management process and organization: a review. North American Journal of Fisheries Management 10:365–373.

Boon, P. J., P. Calow, and G. E. Petts, editors. 1992. River conservation and management. Wiley, New York.

Box, J. 1996. Setting objectives and defining outputs for ecological restoration and habitat creation. Restoration Ecology 4:427–432.

Brown, P. J., and M. J. Manfredo. 1987. Social values defined. Pages 12–23 in D. J. Decker and G. R. Goff, editors. Valuing wildlife—economic and social perspectives. Westview Press, Boulder, Colorado.

Carpenter, S. L., and W. J. D. Kennedy. 1988. Managing public disputes. Jossey-Bass Publishers, San Francisco.

Christy, F. T., Jr., and A. Scott. 1965. The common wealth in ocean fisheries. The Johns Hopkins University Press, Baltimore, Maryland.

Clark, C. W. 1976. Mathematical bioeconomics: the optimal management of renewable resources. Wiley, New York.

Cooper, E. L., editor. 1987. Carp in North America. American Fisheries Society, Bethesda, Maryland.

Coughlan, B. A., and C. L. Armour. 1992. Group decision-making techniques for natural resource management applications. U.S. Fish and Wildlife Service, Resource Publication 185, Washington, D.C.

Cowx, I. G. 1994. Stocking strategies. Fisheries Management and Ecology 1:15–30.

Crowe, D. M. 1983. Comprehensive planning for wildlife resources. Wyoming Game and Fish Department, Cheyenne.

Decker, D. J., C. C. Krueger, R. A. Baer, Jr., B. A. Knuth, and M. E. Richmond. 1996. From clients to stakeholders: a philosophical shift for fish and wildlife management. Human Dimensions of Wildlife 1:70–82.

Elrod, J. H., and six coauthors. 1995. Lake trout rehabilitation in Lake Ontario. Journal of Great Lakes
    Research 21(Supplement 1):83–107.
Fisher, R., and W. Ury. 1991. Getting to yes -negotiating agreement without giving in, 2nd edition.
    Penguin Books, New York.
Fraidenburg, M. E., and R. H. Lincoln. 1985. Wild chinook salmon management: an international con-
    servation challenge. North American Journal of Fisheries Management 5:311–329.
France, R. 1997. Land-water linkages: influences of riparian deforestation on lake thermocline depth
    and possible consequences for cold stenotherms. Canadian Journal of Fisheries and Aquatic Sci-
    ences 54:1299–1305.
Francis, R. I. C. C., and R. Shotton. 1997. "Risk" in fisheries management: a review. Canadian Journal
    of Fisheries and Aquatic Sciences 54:1699–1715.
Gale, R. P. 1992. Is there a fisheries management revolution in your future? Fisheries 17(5):14–19.
Giles, R. H. 1981. Assessing landowner objectives for wildlife. Pages 112–129 in R. T. Dumke, G. V.
    Burger, and J. R. March, editors. Wildlife management on private lands. The Wildlife Society,
    Wisconsin Chapter, Madison.
Harville, J. P. 1985. Expanding horizons for fishery management. Fisheries 10(5):14–20.
Hobbs, R. J., and D. A. Norton. 1996. Towards a conceptual framework for restoration ecology. Resto-
    ration Ecology 4:93–110.
Holland, S. M., and R. B. Ditton. 1992. Fishing trip satisfaction: a typology of anglers. North American
    Journal of Fisheries Management 12:28–33.
Hutchings, J. A., C. Walters, and R. L. Haedrich. 1997. Is scientific inquiry incompatible with govern-
    ment information control? Canadian Journal of Fisheries and Aquatic Sciences 54:1198–1210.
Kennedy, J. J., and P. J. Brown. 1976. Attitudes and behavior of fishermen in Utah's Uinta Primitive
    Area. Fisheries 1(6):15–17,30–31.
King, M. 1995. Fisheries biology, assessment and management. Fishing News Books, London.
Krueger, C. C., D. J. Decker, and T. A. Gavin. 1986. A concept of natural resources management: an
    application to unicorns. Transactions of the Northeast Section of the Wildlife Society 43:50–56.
Krueger, C. C., M. L. Jones, and W. W. Taylor. 1995. Restoration of lake trout in the Great Lakes:
    challenges and strategies for future management. Journal of Great Lakes Research 21(Supplement
    1):547–558.
Larkin, G. A., and P. A. Slaney 1997. Implications of trends in marine-derived nutrient influx to south
    coastal British Columbia salmonid production. Fisheries 22(11):16–24.
Larkin, P. A. 1980. Objectives of management. Pages 245–262 in R. T. Lackey and L. A. Nielsen, edi-
    tors. Fisheries management. Blackwell Scientific Publications, Cambridge, Massachusetts.
Lee, K. N. 1993. Compass and gyroscope–integrating science and politics for the environment. Island
    Press, Washington, D.C.
Lindblom, C. E. 1959. The science of "muddling through". Public Administration Review 19:79–88.
Ludwig, D., R. Hilborn, and C. Walters. 1993. Uncertainty, resource exploitation, and conservation:
    lessons from history. Science 260:17,36.
Malvestuto, S. P., and M. D. Hudgins. 1996. Optimum yield for recreational fisheries management.
    Fisheries 21(6):6–17.
McDaniels, T. L., M. Healey, and R. K. Paisley. 1994. Cooperative fisheries management involving First
    Nations in British Columbia: an adaptive approach to strategy design. Canadian Journal of Fisher-
    ies and Aquatic Sciences 51:2115–2125.
McEwan, D. 1997. Customers, constituents, and the public trust. Fisheries 22(12):4.
Meffe, G. K. 1992. Techno-arrogance and halfway technologies: salmon hatcheries on the Pacific coast
    of North America. Conservation Biology 6:350–354.
Miller, A. 1984. Professional collaboration in environmental management: the effectiveness of expert
    groups. Journal of Environmental Management 16:365–388.
Naiman, R. J., editor. 1992. Watershed management –balancing sustainability and environmental change.
    Springer-Verlag, New York.
Nielsen, L. A. 1985. Philosophies for managing competitive fishing. Fisheries 10(3):5–7.
NRC (National Research Council, Committee on Protection and Management of Pacific Northwest
    Anadromous Salmonids). 1996. Upstream–salmon and society in the Pacific Northwest. National
    Academy Press, Washington, D.C.

Odiorne, G. S. 1979. MBO II a system of managerial leadership for the 80s. Fearon Pitman Publishers, Inc., Belmont, California.

Olver, C. H., B. J. Shuter, and C. K. Minns. 1995. Toward a definition of conservation principles for fisheries management. Canadian Journal of Fisheries and Aquatic Sciences 52:1584–1594.

Parsons, L. S. 1993. Management of marine fisheries in Canada. Canadian Bulletin of Fisheries and Aquatic Sciences 225.

Pinkerton, E. W. 1994. Local fisheries co-management: a review of international experiences and their implications for salmon management in British Columbia. Canadian Journal of Fisheries and Aquatic Sciences 51:2363–2378.

Pringle, J. D. 1985. The human factor in fishery resource management. Canadian Journal of Fisheries and Aquatic Sciences 42:389–392.

Roos, J. F. 1991. Restoring the Fraser River salmon. The Pacific Salmon Commission, Vancouver.

Ross, M. R. 1997. Fisheries conservation and management. Prentice-Hall, Englewood Cliffs, New Jersey.

Ryman, N., and F. Utter, editors. 1987. Population genetics and fishery management. University of Washington Press, Seattle.

Scalet, C. G., L. D. Flake, and D. W. Willis. 1996. Introduction to wildlife and fisheries –an integrated approach. Freeman, New York.

Scarnecchia, D. L. 1988. Salmon management and the search for values. Canadian Journal of Fisheries and Aquatic Sciences 45:2042–2050.

Schneider, C. P., D. P. Kolenosky, and D. B. Goldthwaite. 1983. A joint plan for the rehabilitation of lake trout in Lake Ontario. Lake Ontario Committee, Great Lakes Fishery Commission, Ann Arbor, Michigan.

Sigler, W. F., and J. W. Sigler. 1990. Recreational fisheries–management, theory, and application. University of Nevada Press, Las Vegas.

Stephenson, R. L., and D. E. Lane. 1995. Fisheries management science: a plea for conceptual change. Canadian Journal of Fisheries and Aquatic Sciences 52:2051–2056.

Talhelm, D. R., editor. 1987. Social assessment of fisheries resources-proceedings. Transactions of the American Fisheries Society 116:289–540.

Utter, F. M. 1991. Biochemical genetics and fishery management: an historical perspective. Journal of Fish Biology 39(Supplement A):1–20.

Vogt, K. A., and 10 coauthors. 1997. Ecosystems–balancing science with management. Springer-Verlag, New York.

Walters, C. 1986. Adaptive management of renewable resources. Macmillan Publishing Company, New York.

Yarbrough, C. J. 1987. Using political theory in fishery management. Transactions of the American Fisheries Society 116:532–536.

Chapter 3

Communication for Effective Fisheries Management

DANIEL J. DECKER AND CHARLES C. KRUEGER

3.1 INTRODUCTION

Communication is a process that each of us engages in daily—at school, on the job, and at home. We are all highly experienced in the process of communication, either as the originators of messages or recipients of information that bombards us from every direction. Think about it for a second. Every person is involved in hundreds to thousands of communication events daily. Our routine exposure to communications includes face-to-face conversations, telephone calls, letters, meetings, television commercials and newscasts, radio programs, newspapers and magazines, posters, billboards, "junk" mail, graffiti, and even course lectures and textbooks like this one, all of which are intended to communicate certain messages to us in the hopes that we will buy, believe, or behave in a certain way. Some communications are very effective, some are not effective at all. Ever wonder why? Communication professionals have amassed a wealth of information on the communication process that can be useful to fisheries managers. Effective communication among staff of a natural resources agency, and between staff and the public, is essential to the fisheries management process presented in Chapter 2.

In this chapter we will use a term borrowed from program evaluation specialists to refer to all people interested in fisheries management—stakeholders. As the name implies, a stakeholder is someone who has a stake in an issue or program that addresses a problem, interest, or concern of that person. We feel this term is more appropriate as a general descriptor than others commonly used to describe people interested in fisheries management—terms such as constituent, which really means a supporter, although not all those interested in management are supporters; client, which really means someone who receives a service, usually for a fee, although not all those interested in management pay for it; user, which means someone who actually uses the resource, like an angler, although not all those interested in management personally use the resource; and public, which from a public relations standpoint is accurate but lacks intuitive appeal as a descriptor. Thus, we have settled on stakeholder as the best general term for people with an interest in fisheries management.

Stakeholders in fisheries management include a wide variety of people, such as direct users, people who benefit from others' use (e.g., business operators who sell supplies and services to anglers), people who incur costs because of fisheries resources (e.g., riparian landowners who suffer trespass problems from anglers), staff of related natural resource agencies, people who might not even realize initially that they have a stake in the resource (e.g., local government officials who will eventually have to deal with providing expanded services in an area with a developing recreational fishery, such as occurred in conjunction with the Great Lakes salmonid fishery), and even people who are opposed to sport fishing. These and other stakeholders, as individuals and groups, may all be involved in any single fisheries management issue. Effective communication with them must occur to develop and implement fisheries management programs. This represents a real challenge for the fisheries manager but can contribute to an interesting and exciting career.

The primary purpose of this chapter is to increase your awareness of the communication process and to help you understand its role in fisheries management. Communication is viewed broadly here. The discussion of the management process (Chapter 2) should have started you thinking about the role of communication in fisheries management. Aspiring fisheries managers need to become aware of the direct relationship between effective communication and effective fisheries management early in their academic training, while they still have an opportunity to enroll in courses that can help develop and hone communication skills. These skills may become the most important fisheries management techniques used by a professional manager. Some reasons why this is the case will be discussed. Minimally, aspiring fisheries professionals should recognize that management is as much working with people as with fish. Dealing effectively with people is achieved by the skillful use of communication.

3.2   FISHERIES MANAGEMENT IN TRANSITION

Fisheries management is in a period of transition that began in the 1970s (Harville 1985). Two major changes associated with the transition have heightened the importance of communication in fisheries management. First, the broad goals of fisheries management have moved toward optimum sustainable yield (OSY). Based on OSY philosophy, some optimal set of stakeholders' preferences and other economic and sociocultural benefits are to be achieved (Roedel 1975; see Chapter 1). Such benefits are defined by people in terms of their desires, values, preferences, and satisfactions. For fisheries managers to set goals to produce economic and sociocultural benefits, they must engage in communication that will give them a clear understanding of the kinds of benefits desired by different user groups and other stakeholders in the fishery. Managers also need to be able to communicate to stakeholders their plans to provide the benefits desired. As a result, the changing management philosophy has redefined and expanded the roles and responsibilities, and therefore the academic training and expertise required, of contemporary fisheries managers.

The second major change associated with the transition in fisheries management is greater public involvement in the management decision-making process. The management process presented in Chapter 2 reflects this philosophy. The U.S. Fishery Conservation and Management Act of 1976 mandated inclusion of active public participation throughout the planning and decision-making process for marine fisheries management. Although this act is outside the context of inland fisheries management, the notion of including various groups of interested stakeholders in decision making has spilled over into the philosophy and conduct of fisheries management in general. Fisheries managers cannot operate without public involvement in policy and management decision making. As Harville (1985:15) remarked, "While enlightened fish and wildlife agencies long have recognized the need to involve their user-publics in review and dissemination of agency policy,...our constituency now demands active public participation in the formulation of that policy." This participation requires effective, two-way communication with stakeholders, particularly resource users. Public involvement in decision making is discussed further in section 3.4.

In addition to changes in fisheries management, the characteristics of the stakeholders have changed, further complicating communication. They are more diverse in their interests and more sophisticated in their ecological understanding and political activities. Today's stakeholders have greater expectations for clear and complete communication, expectations that are certain to challenge a manager's abilities.

3.3 FUNDAMENTALS OF COMMUNICATION

Every natural resource professional, whether employed by a university, consulting firm, foundation, or governmental agency should understand the fundamentals of communication. Understanding communication is a *requirement* for managers. Texts on this subject have been written especially for the natural resources professional. For example, the book *Public Relations and Communications for Natural Resource Managers* (Fazio and Gilbert 1986) is full of practical information on communication theory and practice and should be in every manager's personal library. If managers develop a working understanding of this subject, they will be able to facilitate the kinds of communication needed to apply the concept of fisheries management described in Chapter 2. This section introduces fisheries managers to the general process of communication and addresses the ways that managers can use communication in fisheries management.

3.3.1 Communication: The Process of Interaction

The term communication is used in a variety of ways and with different meanings. A definition that reflects the notion of communication used in this chapter is the interchange of information (e.g., data, beliefs, or insights) by written and spoken words, visual illustrations, and actions for the purpose of developing mutual understanding between the sender and receiver about a situation, concept, or event. A principal element of this definition is the concept of interchange among people. There should be two-way communication between a fisheries management agency and its stakeholders. Listening to stakeholders' opinions and observations relative to a management issue is as important as telling them the agency's viewpoints. Building effective feedback, or listening mechanisms, is essential to the two-way communication needed for meaningful public involvement in decision making. A simple model of the communication process has five basic components: communicator, message, channel, audience, and feedback (Figure 3.1). To simplify the discussion below, fisheries managers can be thought of as the communicators and stakeholders as the audience; however, in two-way communication these roles often switch (response arrow in Figure 3.1).

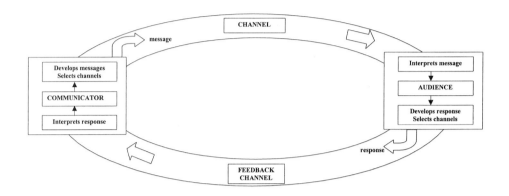

Figure 3.1 Model representing the five basic components of the process of communication.

The communicator (i.e., source of a message) needs to be perceived as being technically sound and credible if the message is to be accepted and acted upon by the audience (Figure 3.1). Communication is enhanced by a climate of belief on the part of the audience. This climate is created by the credible past performance of the communicator. Trustworthiness and honesty are particularly important components of credibility. Credible past performance reflects an earnest desire to serve the audience and helps the audience to have confidence in the communicator (Cutlip et al. 1985). When a fisheries manager or agency does not hold the confidence of stakeholders, the desired communication will not take place.

Managers as communicators should know their audience—its preferences, beliefs, attitudes, communication habits, and other traits pertinent for effective interchange. Good communicators have skills in selecting and structuring messages, selecting and using channels (the means by which messages are delivered to the intended audience), and soliciting and interpreting feedback. Fisheries managers, not the public, are primarily responsible for effective communication.

The message is the information that the manager wants the audience to understand, usually to accept, and often to act upon (Figure 3.1). Consequently messages should be clearly interpreted and easily understood by the recipients and perceived as being useful to them. Two aspects of messages are important (Cutlip et al. 1985).

1. Content. The message must have meaning for the recipients and must be relevant; people tend to listen to messages that offer them something tangible and useful. Usually the content of a message largely determines the audience (i.e., those who will listen).

2. Clarity. The message must be put in simple terms. Words must mean the same thing to the audience as they do to the communicator. Complex fishery issues must be simplified and clarified.

Possibly the most difficult task for a communicator is that of developing meaningful ways to express messages. This task is challenging because to do it well requires that the manager transform fisheries management ideas into words and illustrations that can be correctly interpreted by the public. Managers have to recognize that the text and illustrations they use to convey their message may not be interpreted in the manner they intended. Fisheries managers cannot simply create messages that are clear to them and be confident that they have executed their communication responsibilities adequately. Instead, they must carefully and skillfully develop ways to express their messages so that their audiences will understand.

Basically, a "good" message should be (1) matched to the mental, social, economic, and physical capabilities of the audience; (2) significant—economically, socially, or aesthetically—to the needs, interests, and values of the audience; (3) concise, that is, containing no irrelevant material; (4) timely, especially when seasonal factors are important and issues are current; (5) balanced, that is, containing factual material covering both sides of an issue; (6) applicable so that the audience can use the recommendations presented; and (7) manageable so that the message can be handled skillfully within the constraints imposed by the time and resources available for communication.

The channel is the way in which a message reaches the intended audience (Figure 3.1). Channels of communication can be thought of as bridges over which the message travels, connecting the communicator and the audience. Some examples of channels commonly used by fisheries managers are meetings (and other face-to-face situations), letters, newsletters, newspapers, magazines, radio announcements and broadcasts, and reports. Established channels of communication should be used as much as possible, particularly channels that the audience uses frequently and trusts. Although new channels may be difficult to create, at times fisheries managers may find it necessary to do so because existing channels to which they have access are inadequate or inappropriate.

Surveys can be used to identify the channels used by different groups. For example, a fisheries manager may want to reach charter boat captains with a message. The manager may discover through a survey of readers that the agency magazine it has used in the past reaches only 5% of the captains. As a result, a special newsletter may have to be developed for direct mailing just to this group to be sure that messages reach them in an effective and timely fashion.

Fisheries managers should use multiple channels to reach an audience rather than rely entirely on a single channel for most communication purposes. Use of channels in "parallel" rather than in "series" is advisable. In other words, use several channels that reach the audience directly instead of one channel to reach another channel that then reaches the audience. For example, to get a message to charter boat captains, don't rely on a general news release to be picked up and interpreted accurately by a wire service and then, in turn, to be discovered (perhaps) and edited further by an editor of a charter boat captains' association newsletter. Each link in the series has potential for problems. Instead, send out the general news release but prepare another release for a targeted mailing. Better yet, hand deliver the latter release to the newsletter editor. The disadvantage of the first approach is that information "gatekeepers" usually exist within channels, and they can alter the message in such a sequence. Using several channels has multiple advantages, such as the reinforcement of ideas brought about by receiving a message more than once, the added credibility associated with receiving a message from more than one channel, and the increased likelihood of reaching more people.

Overall the selection of channels should be approached with attention to the following considerations: specific goal of the message, content of the message, characteristics of the audience, channels available that will reach the audience, personnel available with the right skills needed to use a particular channel, channels that can be combined and used simultaneously, cost of channel use relative to its effectiveness, and time constraints related to the urgency of message.

Most of the important considerations relative to the audience have been presented in the discussion of the other components of the communication process (Figure 3.1). Given that a message reaches the intended audience (and they read or listen to it), possibly the most important consideration is the ability of the audience to interpret the message as the communicator had anticipated. If the message was developed with this essential purpose in mind, then communication probably will be effective and the audience will gain a new or better understanding of the situation through correct interpretation. However, understanding a message does not mean the audience will agree with it.

Feedback requires either being receptive to or, preferably, seeking the audience's reactions and responses to the message (Figure 3.1). Obtaining accurate and comprehensive feedback is important in fisheries management for two reasons. First, the manager needs confirmation that the message was received. Second, reactions to the message need to be assessed so decisions can be made about the need for additional communication efforts. Feedback can be thought of as either intentional or unintentional. Intentional feedback is the type that stakeholders engage in to make certain the agency knows their interpretation of the content of the message they received and their opinions about that message. Unintentional feedback is information the manager receives about stakeholders' reactions through observations, informants, and other means. This kind of feedback may be revealing and useful for assessing the need for additional communication efforts.

Fisheries managers should not leave the feedback element to chance. Obtaining feedback is simply too important. Rather, they should develop mechanisms to solicit feedback from the audience. This approach has two distinct advantages. First, managers are more likely to obtain accurate, timely, and representative information if the feedback mechanisms are developed carefully (e.g., a survey of the angling public). Second, seeking input has an important public relations value. By doing so, managers are telling their stakeholders that they are not merely being "talked at" but that their opinions are valued enough to be sought. Mechanisms that are used to encourage feedback may be as simple as the address of an agency contact person on a brochure or as formal as surveys and public hearings. This topic is discussed further in section 3.4.2.

Contemporary fisheries managers need to have a basic understanding of the communication process and a willingness to develop and fine-tune their communication skills. The ability to speak well in public or to write interesting media releases must be combined with a good understanding of communication as a process to apply these speaking and writing skills effectively. When contemplating communication, managers should keep the entire process in mind to improve the probability of engaging in an effective interchange.

3.3.2 External Communication: Requirement for Good Public Relations

Experience has shown that good public relations is a prerequisite to success in managing fisheries resources. A definition of public relations applicable to fisheries managers is the planned effort to influence public opinion through good character and responsible performance coupled with mutually satisfactory two-way communication (Fazio and Gilbert 1986). This definition of public relations may give a slightly narrower view of the communication function than that proposed for managers in this chapter. The goal of communication in the context of the fisheries management process is more than the development of a favorable public opinion climate for the agency; it is also the development of informed groups who can contribute to decision making through public involvement processes. Developing and maintaining good public relations, though, is a responsibility of all agency personnel involved in fisheries management. Fazio and Gilbert (1986) discussed in detail seven principles of public relations, the essence of which are outlined in Box 3.1.

External public relations must be carefully planned and executed to be effective. External communication can be multipurpose: (1) to listen to stakeholders' desires for a fishery's resources, (2) to inform them of the status of the management environment (Chapter 2) and what future options appear available, (3) to assess their preferences, (4) to gain their support for proposed actions, and (5) to evaluate their reactions to management actions. Consequently, ways to accomplish effective communication vary with the purpose of the communication and the particular characteristics of the stakeholders who are the audience. Methods also depend on the particular situation. For example, if a message needs to reach a group of anglers in a short time, the channel of choice may be first-class mail, telephone, teleconference, or a special meeting, depending on the geographic distribution of the target audience and the amount of funds available to contact its members. On the other hand, if there is no urgency about the message, some combination of a news release, article in an agency magazine, or an agenda item for a regularly scheduled meeting may suffice.

Box 3.1   Principles of Public Relations

Principle 1   Every agency action makes an impression on its stakeholders.

Principle 2   Good public relations is a prerequisite to success in agency programs.

Principle 3   The public is actually many different groups of people.

Principle 4   Truth and honesty are essential to credible public relations.

Principle 5   Proactive communication is more effective than reactive communication.

Principle 6   Communication is the key to good public relations.

Principle 7   Planning comprehensive communication strategy is essential.

A number of barriers to communication may exist in any management situation, but four in particular have needlessly persisted in fisheries management and should be avoided. These barriers are assumptions that some fisheries managers have accepted and have used as reasons for not working more diligently on communication. They are presented here so you can recognize and avoid them.

1.   Assumption: Anglers have little concern for either the resource base or the future of the fishery.

> Response: Most fisheries managers who work to communicate effectively with anglers and other users discover that anglers care about the resource base and, given an informed choice, will choose the side of conservation (Pringle 1985; Walters 1986).

2.   Assumption: The public is unable or unwilling to understand the ecological concepts underlying fisheries management deliberations.

> Response: This assumption reflects an attitude of professional elitism that can seriously impede sound resource management. Although many people may not be able to understand readily some aspects of fisheries management, the opinion leaders of stakeholder groups are typically both intelligent and highly motivated to learn. If just these people understand the basis for management and lend their support, others who may not fully understand the alternative often will trust the leaders and also be supportive of management. Thus, fisheries managers should communicate the ecological bases for their management deliberations.

3.   Assumption: The public will oppose almost any fisheries management program the agency suggests.

> Response: This assumption emanates from an us-versus-them attitude that sees management as a contest of wills between the agency and the public rather than a team approach to solve problems important to the agency and the public (Fisher and Ury 1981; Chapter 2). Managers who understand that stakeholders care about fisheries resources will find many concerns they share in common.

4.   Assumption: The public has nothing to offer the management process; fisheries managers are professionally trained people who know best how to manage the resources.

> Response: The public is made up of many individuals and groups of stakeholders having diverse and extensive experience. Many people are astute observers who often have important insights and may be able to teach managers valuable lessons. To gain these helpful insights, managers must have a receptive attitude about open interchange with their stakeholders.

Fisheries managers need to remember that contact with the public is not equivalent to effective communication. In other words, simply reaching stakeholders with a message does not mean that they heard or saw it, interpreted it correctly, or put it in the context of management as managers had intended. Managers should not be satisfied that a piece of information simply gets into the hands of the people who they want to have it. Rather, managers should want people to be able to interpret, understand, and use that information to participate meaningfully and responsibly in the fisheries management process. In essence, successful managers view themselves as educators of their stakeholders.

3.3.3 Internal Communication: Interchange of Information within Agencies

Professionals inside one's own agency are sometimes disregarded, particularly by inexperienced managers, when communication outside the agency is planned. Inevitably, the importance of good internal relations and attendant communication become apparent. In some instances, internal communication is more important and challenging than external communication.

A principle that managers soon learn is that good internal relations among professionals is a prerequisite for good external public relations. Many examples of resource management difficulties and failures can be traced to poor communication within agencies that led to external public relations problems. The individual fisheries manager must be mindful not to work independently within an agency. Support for a fisheries management program is essential from administrators, supervisors, other managers, and technicians. Program support by other agency personnel results from understanding the elements and rationale of current management.

Poor internal communication can have disastrous results for a program. The effort devoted to developing the support of a skeptical outside interest group can be undone by someone else in the agency making a disparaging remark about the program at the wrong time. The reaction of the outside skeptic is to ask if agency staff cannot even agree on the merits of this program, why should we support it? In a brief moment the public relations work of weeks or possibly months may be negated. There is no guarantee that managers can avoid this scenario by engaging in purposeful internal communication; however, they can significantly reduce the probability of such happening if they plan and execute communication to establish broad agency support for proposals.

Managers should be careful not to overlook any group of people potentially important to a program when considering internal communication (e.g., conservation law enforcement personnel). In Chapter 2, important agency personnel were identified for each step in the management process. Several general categories of agency personnel exist, including administrators, law enforcement officers, researchers, office staff, seasonal staff, licensing agents, commissioners, and advisory boards (Fazio and Gilbert 1986). Field staff represent one category of individuals with whom it is absolutely essential to develop continuous communication. Many fisheries management agencies have a decentralized organizational structure, and most managers and technicians are located in regional offices. What commonly develops is tension between field staff and central office staff. This counterproductive atmosphere occurs largely because of poor communication. Regardless of the cause of this situation, if a program proposal is to move ahead, be accepted, and have a reasonable chance for success, the manager has to pay careful and continuous attention to internal communication. To ignore this reality of management is to court failure.

Internal communication can be pursued in several ways. Seeking input from all relevant groups within an agency before establishing policies can contribute substantially to the development of internal support. Proposals must be explained clearly and reactions to them need to be addressed. Some suggestions will be incorporated and some will not, but even people whose ideas are not fully accepted will feel better if they know their suggestions have been considered honestly and rejected for understandable reasons. The process of explaining why an idea was not adopted will help the person(s) who offered the idea learn even more about the plan. If handled skillfully, these learning episodes can help establish strong internal support for a management program.

A special type of communication is that which occurs among agencies. Among-agency communication is essential if effective fisheries management is to occur where jurisdiction is shared by two or more agencies. Communication among agencies often verges on diplomacy. Special kinds of sensitivities must be considered when representatives of agencies that share jurisdictions interact with one another. Especially important is developing a thorough understanding of each agency's jurisdiction over fisheries resources. Agencies often jealously guard their authority over resources. Communication strategies must be designed to facilitate the cooperation needed. Remember that communication with other agencies is actually communication with individuals. Be sensitive to them as people and fellow professionals. Correcting a communication blunder is much more difficult and time consuming than doing the planning necessary ahead of time to avoid problems.

3.3.4 Adoption of New Ideas and Practices

The following discussion of innovation–adoption and adoption–diffusion is a brief introduction to a body of thinking (Rogers and Shoemaker 1971; Rogers 1983) that can serve the manager in planning comprehensive communication efforts in support of fisheries management. Although the concepts are presented in terms of communica-

Figure 3.2   Innovation–adoption in the angler's world.

Box 3.2   Adoption of Downrigger Trolling by Anglers

The innovation–adoption process can be illustrated by examining the use of new technology by anglers. Let us go through the process as it has occurred probably thousands of times since the early 1970s as the Great Lakes salmonid fishery has developed. An angler with no previous experience in fishing for trout and salmon in a large body of water hears from other anglers at a tackle shop or reads in a fishing magazine about the tremendous catches of salmon and trout being made using moderate-size boats (5–7 m long) equipped with fish-finders and cannon-ball downriggers (awareness stage). The stories and pictures are alluring, but uncertainty about necessary skills and equipment slows enthusiasm for the potential of this new activity. This person begins buying more fishing magazines and going to exhibitions and boat shows to learn more about boats, motors, downriggers, and sonar (interest stage). After accumulating additional knowledge about this type of fishing and talking to people who have tried it, the angler soon forms some opinions about its suitability for his or her own enjoyment. It is likely that the angler will try to find someone with whom to go fishing a few times or will perhaps go out on a charter boat to gain firsthand experience (evaluation stage). If this experience–evaluation period is positive and reinforces the notion that this type of fishing experience is right (including being affordable), the angler might next invest in the minimum equipment required and use this limited-investment outfit for a season or two (trial stage). If the angler is successful and enjoys this new kind of fishing, and anticipates that its rewards will continue, he or she may invest more into the existing outfit or upgrade to an even more elaborate outfit that suits his or her needs as they have developed (adoption stage).

---

tion with external sectors, they apply also to internal agency communication. Both theories were developed in the context of voluntary behavior, so their applicability to some areas of management, such as acceptance of new regulations, may not be valid.

3.3.4.1 Innovation–Adoption: Idea Acceptance by Individuals

Innovation-adoption theory proposes that new ideas or practices (i.e., innovations) are seldom adopted as the result of a single decision by an individual. Rather, a person usually goes through a process that can be thought of as a series of stages, each stage having a characteristic set of information-seeking behaviors and decisions and each leading to a progressively advanced stage in the process. The final stage is to adopt the idea or reject it as being inappropriate for adoption at the time. Thus, the process is a series of stages by which a person (1) becomes aware of the idea, (2) develops an interest in it, (3) evaluates it for personal application, (4) tries it on a limited basis, and (5) finally adopts the idea or, alternatively, rejects the idea as being inappropriate (Figure 3.2; Box 3.2).

At each stage in the process, information is important for an individual's continued progress to the next stages. Research has shown that people rely on different kinds and sources of information at different stages in the process. For example, mass media are of greatest importance in the awareness and interest stages, opinions of friends and neighbors are most influential in the evaluation and trial stages, and personal experience gained during the trial stage has the greatest bear-

---

Box 3.3   Adoption–Diffusion of Sonar by Anglers

The adoption–diffusion process operates constantly among boat anglers. The 1980s and 1990s have witnessed a proliferation of technological advancements in equipment to aid anglers in their pursuit of salmonids, largemouth bass, walleye, and many other fishes. With the development of each new product, such as sonar or fish-finders (a relatively new product in the 1960s), the innovators are the first to purchase the gadget, regardless of cost. The early adopters tend to wait a while and see what kinds of refinements are made to the equipment. They are assessing the applicability and advantages of this new development compared with the old way of fishing. Given a positive assessment, the early adopters will obtain the new fish-finder and will readily communicate its successful use to members of their peer group. The early majority will eventually be convinced to make similar purchases to increase their catch rate and to be in step with the new wave of technologically advanced anglers. The late majority will follow suit and typically will have access to a greater variety of models and price ranges as the market grows and more manufacturers are competing for a share of the market. Some anglers will be very late in adopting the now not-so-new technology or may never adopt it (nonadopters). Their reasons for not adopting may be as pragmatic as the inability to afford the technology or as value oriented as wanting to continue a traditional approach to fishing (e.g., sonar provides an unfair advantage; anglers should spend time "learning" a lake without this equipment).

---

ing on the ultimate decision to adopt or reject. Knowledge of the particular information needs and preferred sources of information will help a manager in determining message content and in selecting communication channels as stakeholders go through each stage of the adoption process.

3.3.4.2 Adoption–Diffusion: Idea Acceptance by Society

The adoption–diffusion theory is an explanation of how an idea gains acceptance throughout a social system (Box 3.3). This theory can help fisheries managers understand how proposals for a new regulation, such as the establishment of a slot limit for lake trout in Lake Ontario, gains acceptance among anglers. The adoption–diffusion theory also helps managers have realistic expectations for the rate and extent of acceptance of a new idea. Basically, the theory explains that a new idea is adopted, or accepted, at differential rates by people in a social system. Five categories of people will be discussed: innovators, early adopters, early majority, late majority, and nonadopters (Figure 3.3; Table 3.1). There are, of course, exceptions to categorizations, so it is important not to make assumptions when interacting with the individual.

The first people to adopt a new practice are the experimenters or, in terms used by sociologists, the innovators. These people tend to be more highly educated, more affluent, and more willing to take risks. Usually they are not the local opinion leaders in a social system, but the opinion leaders keep an eye on the innovators to identify "good" ideas. Innovators tend to be information seekers; they are willing and able to access primary sources of information, such as agency personnel and university researchers.

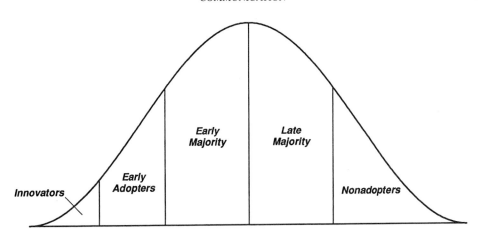

Figure 3.3 Adoption–diffusion: a relative frequency of the different types of people regarding their tendency to adopt an innovation or idea.

The early adopters are very influential at the local level, operating within a variety of social groupings, such as fishing associations. These people do not necessarily hold elected offices in organizations. Nevertheless, they are looked to and typically sought out for their opinions regarding new ideas. Like the innovators, early adopters are well educated and informed. They constantly seek information from a variety of sources, so they usually are the willing recipients of any communication offered about management.

The early majority is made up of many informal leaders. They tend to be older and more cautious than are early adopters. These people are considered wiser and more conservative and are similar in many respects to the majority of people in the social system. They have some influence when it comes to the adoption of ideas by the majority. Members of the early majority may hold offices in state and local groups, such as angling clubs and associations of charter boat operators. Thus, these people must be kept informed of any developments relative to fisheries management. Opinion leaders of the early majority are the people who are sought by individuals during the evaluation stage of the innovation-adoption process. Opinion leaders of both the early adopters and early majority represent those people most likely to be directly involved in the process of fisheries management.

The late majority are the people who take a wait-and-see posture before making a commitment to adopting a new idea. They rely on the early majority for much of their information and for their formation of opinions about an idea. These people tend to be the followers, the rank-and-file members of angler organizations or the people who never affiliate with any organization. For sake of efficiency, a minimum amount of time should be spent targeting communication with the late majority because the pay-off for the effort will be low. Energy is better spent on the early adopters and early majority who will be the key influencers of the late majority anyway. This is not to suggest that the late majority should be ignored, but the amount of attention they receive should be tempered by the knowledge of how they react to and use information emanating directly from an agency.

Table 3.1 Characteristics of different categories of adopters. Table adapted from Fazio and Gilbert (1986).

Adopter category	Salient personality characteristics	Profile	Communications behavior	Social relationships
Innovators (first 2.5% to try out idea)	Venturesome; willing to accept risks	Youngest age; highest socio-economic status; largest and most specialized fishing operations	Closest contact with scientific information sources; most interaction with other innovators; greatest use of impersonal sources	Some opinion leadership; very cosmopolitan
Early adopters (next 13.5%)	Respectable; regarded by many others as the social role model	High socioeconomic status; large, specialized fishing operations	Greatest contact with local agents of change (innovators)	Greatest opinion leadership in most social systems; very local
Early majority (next 34%)	Deliberate; willing to consider innovations after peers have adopted them	Above-average socio-economic status; average-size fishing operations	Considerable contact with agents of change (innovators) and early adopters	Some opinion leadership
Late majority (next 35%)	Skeptical; overwhelming pressure from peers needed before adoption occurs	Below-average socio-economic status; small fishing operation with little specialization	Contact with peers who are mainly early or late majority adopters; less use of mass media	Little opinion leadership
Nonadopters (last 15%)	Traditional; oriented to the past	Oldest; lowest socio-economic status; smallest fishing operation with little specialization	Contact with neighbors, friends, and relatives with similar values	Very little opinion leadership; isolated

The nonadopters are people who either do not adopt or are exceptionally slow to adopt new ideas. These people depend almost entirely on peers for their information. From a strictly pragmatic standpoint, time is not well spent trying to reach the nonadopters. From an ethical standpoint, managers have a responsibility to offer them access to information about fisheries resources and their management. Again, knowledge of this category of people suggests that the best way to reach them, for all parties involved, is through their peers in the other categories; thus, this provides another reason to do an effective job in educating the latter groups.

3.4 PUBLIC INVOLVEMENT IN DECISION MAKING

The trend toward greater public participation in management decision making brings the importance of effective communication front and center. Public involvement generally produces better decisions and increased public acceptance of those decisions but seldom makes the process any easier (Peyton and Talhelm 1984). Public involvement is not an easy out for managers; it will not make a decision for the manager (Heberlein 1976). Nevertheless, communication between stakeholders and man-

Box 3.4   Public Involvement and Communication

Seven basic elements of a comprehensive public involvement program have been identified (Peyton and Talhelm 1984):

1. a staff that understands the dynamics of public behaviors, attitude formation, and group decision making;

2. a staff that is aware of existing public preferences, values, and systems of belief pertaining to fisheries resources.

3. a continuous monitoring system to anticipate and detect changes in public behaviors, needs, preferences, values, and perspectives;

4. an ongoing information system to develop and maintain adequate public awareness;

5. a structured public involvement system to provide opportunity for appropriate public participation in management;

6. a coordinated, credible public image; and

7. a network of stakeholders who are capable of maximizing the agency's attempts to facilitate public involvement in fisheries management.

agers will ensure that the manager understands societal goals for the fishery resource and interprets those goals correctly as specific objectives and actions are developed. Similarly, communication between managers and stakeholders is essential to improve the latter group's understanding and support of how managers are translating the goals into objectives and objectives into actions.

Managers will find that the communication required for public involvement can be difficult. However, without effective communication fisheries management can become a frustrating, unrewarding, and often unproductive undertaking. Poor communication about management programs can result in confrontations between stakeholders and agency representatives during meetings and in the pursuit of political solutions by the public. Ultimately, uninformed or misinformed stakeholders may circumvent the management process (from the fisheries manager's perspective) through nonparticipation, litigation, and various levels of noncompliance with regulations required to meet program goals. Recently, the use of ballot initiatives to set policy and decide management actions has increased, circumventing traditional decision-making processes. The probability and severity of these unfortunate outcomes can be greatly diminished through carefully planned and executed communication efforts designed as public involvement mechanisms (Box 3.4).

3.4.1 Four Functions of Public Involvement

Identifying the purpose or purposes of public involvement is important to the development of effective communication mechanisms. Four functions of public involvement may be described as follows (Heberlein 1976):

- Informational—disseminating information to and obtaining it from stake-holders;
- Interactive—working together on an issue or problem with information going back and forth between managers and stakeholders freely and rapidly;
- Assurance—making sure that a group of stakeholders knows that its views have been considered in the program-planning process; and
- Legalistic—involving stakeholders to satisfy legal requirements or social norms and to allow input from any sources that may have been inadvertently overlooked.

The goal of the legalistic function of public involvement is to demonstrate to others (i.e., besides the key stakeholders) that there have been open mechanisms for public involvement. This type of public involvement should be well documented.

3.4.2 Integration of Public Involvement into Management

Public involvement efforts will vary depending on an agency's purpose and type of management program. We identify three categories of management; each requires a different approach to public involvement (from Peyton and Talhelm 1984).

1. Mandated management. Some management programs are mandated by legislation, administrative codes, commission rulings, or court orders. In this case, the political or legal component of the management environment is dominant in decision making. When these mandates are clear and not a matter of interpretation, the public education system of the agency must simply communicate these mandates to the public. When options are provided or the mandates must be interpreted, involvement of the affected stakeholders is necessary. The agency should also be monitoring changes in public perceptions, needs, and preferences that may lead to changes in or creation of mandates.

2. Ecological management. Many fisheries management decisions are constrained by ecological or biological considerations. Sometimes, the survival of the fishery resource requires decisions based on biological aspects. Little flexibility for alternative approaches may be available to the manager. The ecological component of the management environment is the dominant force in decision making. In this case, stakeholders must be educated to understand the ecological constraints imposed on management. Care must be taken not to treat an issue as this category of management when, in fact, it belongs to the category listed below. Managers must be alert not to attempt to defend decisions by overstating or misrepresenting the biological aspects of an issue. To do so could quickly lead to diminished credibility and trust among important stakeholders.

3. Issue resolution management. Most decisions facing a fisheries management agency involve issue resolution management. In this category, programs are not clearly or precisely mandated, nor are they ecologically constrained to the point where no flexibility or consideration of options are possible. The sociological and economic components of the management environment strongly influence decision making. Issue resolution management situations are confounded by conflicting pressures from stakeholders with disparate views about the best management objectives and

Box 3.5   Possible Outcomes of Public Involvement

Peyton and Talhelm (1984) report that public involvement can be used to

1. educate the public in the processes of effective public participation and about fisheries resources and management;

2. change public behaviors and attitudes;

3. supplement staff resources and provide additional expertise or volunteer efforts;

4. prevent anticipated conflict by involving expected opponents in the decision-making process;

5. use existing public opinion leaders and their influence to get broader public acceptance of decisions;

6. legitimize agency decisions among stakeholders;

7. present an image of citizen participation without compromising agency autonomy (tokenism);

8. resolve conflict (e.g., in matters of resource allocation); and

9. achieve better representation of various stakeholder and resource interests in decision making.

Peyton and Talhelm noted that although some of the outcomes listed above may seem to represent more noble purposes (e.g., 9, achieving better representation) than others (e.g., 7, tokenism), an ethical case could be developed for the judicial balance of all of the above in a public involvement effort related to a fisheries management program.

approaches. Managers typically have to deal with consideration of different priorities among stakeholders, inadequate data, and insufficient time lines for decisions and action. Carefully designed public involvement can help resolve conflicts among user groups and between user groups and the agency.

It is often difficult for the fisheries manager to determine the most effective means to obtain information about public needs and meet requirements for public involvement. Many possible methods exist, but the decision about which approach(es) to employ is seldom straightforward. Each method can be rated in terms of its effectiveness in serving the four functions of public involvement (Fazio and Gilbert 1986; Table 3.2). In every case, the conscientious manager should determine why public involvement is needed and what result is desired from public involvement (Heberlein 1976; see Box 3.5).

3.4.3 Benefits from Public Involvement

What can public involvement bring to the decision-making process to benefit management? Involvement of stakeholders in management is intended to lead to better decisions and decisions that stakeholders support. Specifically, public involvement can result in the following benefits (from Fazio and Gilbert 1986).

Table 3.2 Effectiveness of various methods in fostering public involvement and in serving the four functions of public involvement. Table adapted from Fazio and Gilbert (1986).

| Form of public involvement | Function of public involvement | | | | | |
| | Informational | | Interactive | Assurance | Legalistic | Overall effectiveness |
	To give	To receive				
Open public meetings	Good	Poor	Poor	Fair	Yes	Poor
Workshops (small)	Excellent	Excellent	Excellent	Excellent	Yes	Potentially good
Presentations to groups	Good	Fair	Fair	Fair	Yes	No clear assurance
Ad hoc committees	Good	Good	Excellent	Excellent	Yes	Potentially good
Advisory groups	Good	Good	Excellent	Excellent	Yes	Potentially good
Key contacts	Excellent	Excellent	Excellent	Excellent	No	No clear assurance
Analysis of incoming mail	Poor	Good	Poor	Poor	Yes	Poor
Direct mail from agency to public	Excellent	Poor	Fair	Good	No	Potentially good
Questionnaires and surveys	Poor	Excellent	Poor	Fair	Yes	Potentially excellent
Behavioral observation	Poor	Excellent	Poor	Poor	No	Potentially excellent
Reports from key staff	Poor	Good	Poor	Poor	No	No clear assurance
News releases and mass media	Good	Poor	Poor	Poor	Yes	Potentially fair to good
Analysis of mass media	Poor	Fair	Poor	Poor	Yes	Potentially fair to good
Day-to-day public contacts	Good	Good	Excellent	Fair	No	Poor
Nominal group process	Poor	Excellent	Poor	Excellent	Yes	Potentially good
Delphi technique	Fair	Excellent	Fair	Excellent	Yes	Potentially excellent

1. New factual data may be brought to light about the resources involved.

2. Preferences, conflicts, and unknown social complexities will be made known.

3. People will accord greater respect and confidence to an agency that openly invites their participation in management.

4. Involvement reduces the likelihood that management decisions will be reversed in court, by new legislation, or by superiors.

The ultimate result desired from public involvement is the development of the best possible fisheries programs to serve the public interest and ensure the long-term well-being of fisheries resources.

Figure 3.4 Examples of survey forms used to gain information from anglers–a critical step in fisheries management.

3.5 STUDIES OF STAKEHOLDERS: A SPECIAL TYPE OF COMMUNICATION

Identifying values, attitudes, satisfactions, and preferences of stakeholders often requires systematic studies of stakeholder groups. Comprehensive studies that use in-depth interviewing and observation techniques in combination with attitude and opinion surveys can effectively yield valuable information about stakeholders. Studies of stakeholders are important to the communication process in two ways. First, such studies will systematically obtain information for an agency from a targeted group of stakeholders, such as steelhead anglers who use drift boats in Oregon or largemouth bass anglers in Florida. Improved knowledge of the varying tastes, preferences, expectations, beliefs, economic characteristics, and lifestyles of stakeholders helps managers understand them. The effective communicator, whether an educator, diplomat, or fisheries manager, learns to tailor messages and to select methods to deliver information based on key characteristics of the intended audience. Second, the process of conducting a study communicates to those contacted, and others who learn of the effort, that fisheries managers are interested in their opinions, observations, and preferences. This concept is an important message to send to stakeholders, and it can become even more powerful if the people in the study observe that the information they provided was used as input to decision making (Figure 3.4).

3.6 CONCLUSION

Managers can learn much about communication that will facilitate the process of fisheries management. Managers need to understand (1) the elements of the communication process; (2) the process individuals go through in adopting or accepting new ideas and the kinds of communications that people use in each stage of the process; (3) the way new ideas gain acceptance throughout a social system and the influence that other people have in forming opinions about and accepting new ideas; and (4) the communication challenges associated with public involvement processes.

Fisheries managers should keep in mind certain basic premises presented in this chapter.

1. Contact does not mean communication.

2. Communication is a two-way interchange leading to mutual understanding.

3. The public comprises many individuals and groups of people having a stake in fisheries management. Communication must occur with professionals inside agencies and externally with the public. Understanding each group's characteristics is important for effective communication to occur.

4. Public relations is important for support of agency programs, and effective communication is the vehicle of public relations. Credible performance in fisheries management is the foundation of good public relations and necessary for effective two-way communication with public groups.

5. Public involvement is usually a requisite of contemporary fisheries management and requires effective use of communication skills to realize its full benefit.

Managers should expect that regardless of how hard they try to avoid conflicts with other people over management, conflicts will occur. Knowing this beforehand will allow fisheries managers to deal rationally with conflict situations and manage them to positive resolutions. Courses and textbooks are available that describe techniques to help managers resolve conflicts (e.g., Fisher and Ury 1981; Chapter 2).

Finally, communication with all individuals and groups external and internal to agencies is a primary responsibility of the fisheries manager. Ability to communicate effectively will be directly related to the effort put into learning more about the concepts discussed in this chapter and cumulative experience in applying them. Skill in communication is a requirement for a professional to be effective in the fisheries management process. Clearly, fishes are not the only focus of the management process. A fisheries manager must communicate effectively with many types of people.

3.7 REFERENCES

Cutlip, S. M., A. H. Center, and G. M. Broom. 1985. Effective public relations. Prentice-Hall, Englewood Cliffs, New Jersey.
Fazio, J. R., and D. L. Gilbert. 1986. Public relations and communications for natural resource managers, 2nd edition. Kendall/Hunt, Dubuque, Iowa.
Fisher, R., and W. Ury. 1981. Getting to yes - negotiating agreement without giving in. Penguin Books, New York.

Harville, J. P. 1985. Expanding horizons for fishery management. Fisheries 10(5):14–20.

Heberlein, T. A. 1976. Principles of public involvement. University of Wisconsin, Cooperative Extension Programs, Rural and Community Development, Madison.

Peyton, A. B., and D. L. Talhelm. 1984. Expanding public involvement programs in the Michigan Department of Natural Resources: a recommendation of the Great Lakes Fisheries Advisory Committee. Michigan Department of Natural Resources, Fisheries Division, Lansing.

Pringle, J. D. 1985. The human factor in fishery resource management. Canadian Journal of Fisheries and Aquatic Sciences 42:389–392.

Roedel, P. M., editor. 1975. Optimum sustainable yield as a concept in fisheries management. American Fisheries Society, Special Publication 9, Bethesda, Maryland.

Rogers, E. M. 1983. Diffusion of innovations, 3rd edition. The Free Press, New York.

Rogers, E. M., and F. F. Shoemaker. 1971. Communication of innovations: a cross-cultural approach. The Free Press, New York.

Walters, C. 1986. Adaptive management of renewable resources. Macmillan, New York.

Chapter 4

Legal Considerations in Inland Fisheries Management

BERTON LEE LAMB AND DONNA LYBECKER

4.1 INTRODUCTION

Sharon Douglas is a fictitious midlevel manager with an agency of the United States. In this capacity, she supervises fisheries and wildlife biologists who are routinely and frequently called upon to consult with other federal and state agencies, as well as with private citizens, on the economic development of water resources. These consultations are often neither easy nor pleasant. The biologists often must tell citizens, agency representatives, and decision makers that their favored projects are unwise as advocated because the projects are harmful to aquatic life. Sometimes the projects must be modified or even abandoned to protect society's aquatic resources. Douglas has found that her staff biologists are frequently not well prepared to participate in this arduous task, which requires not only scientific knowledge but also a healthy measure of diplomacy.

Douglas has also limitations in her subordinates that reflect a misunderstanding of how law works, especially in regulatory processes. Although she is a fictitious supervisor, the problem she faces is real. Often managers in fish and wildlife agencies must build an understanding of the law and how it works without formal legal training. That is not to say that law is so complex that it is incomprehensible to those without a law degree. Rather, the law is more alive than most professionals recognize, and an understanding of the law grows with experience and observation.

4.1.1 A Brief History or North American Governments and Their Laws

Understanding the law begins with an appreciation of government. When first introduced to the natural sciences, people are often attracted by the rigor of analysis and comprehensive perspective provided by the "laws of science." Later they realize that even these scientific laws are never absolute. Experience teaches the scientist to have a healthy skepticism and to be careful about accepting any claim. The same can be said for understanding government, politics, and law.

It is important to recognize that each nation has its own particular culture that broadly determines people's attitudes and values toward government, political processes, and law; this is called political culture. Consider political culture as the environment in which political institutions live and evolve. In North America, nations divide their political institutions into three parts: executive, legislative, and judicial. Beyond this similarity, there are vast differences among North American nations in terms of political history, beliefs about appropriate political behavior, and the nature of lawmaking. For example, in the United States a person convicted of a crime may appeal that decision to the highest court in the land. That court's decision sets a precedent for future cases. Lower courts must follow the precedent. This is not the system in Mexico, where each case is decided solely on the basis of the written statute as passed by the legislature. Each nation has its own view of law depending on its political culture, and that culture is based in large measure on the history of the nation.

Mexico, Canada, and the United States were all colonies, and thus all have similarities to one another. Canada and the United States were both colonized by Great Britain and thus inherited the British common law legal system. The Canadian province of Quebec and the state of Louisiana also use the French civil law legal system. In addition, both Canada and the United States have strong democratic traditions. But these two nations also differ in ways that profoundly affect how natural resource management is conducted. Not the least of these differences is found in the executive branch. In the United States, there is an executive branch that consists of the president, the Executive Office of the President (including the Council on Environmental Quality), the president's cabinet, and the executive departments. For inland fisheries management, the important departments are the Department of the Interior, Department of Commerce, Environmental Protection Agency, Department of Defense, and Department of State.

In Canada there is no separate executive branch. The Canadian government is a parliamentary system in which executive functions are an extension of the legislature. Formally, the executive power is vested in the British monarch; however, the executive power through which Canada is governed is exercised almost entirely by or on the advice of the Canadian prime minister and his or her cabinet (Ward 1987). In this respect, the executive functions are an extension of the legislature; in day-to-day practice, however, the prime minister performs most functions apart from the legislature. Of course there are bureaucracies, which are led by ministers who are also elected members of the legislature. The federal ministries that affect inland fisheries management are Environment Canada and the Department of Fisheries and Oceans (DFO).

As in the United States, Canada and Mexico have federal systems that provide for responsibilities in fisheries management to be shared between the central and regional or local governments. In Mexico and the United States, these responsibilities are shared with their states. The United States suffered a major civil war to determine the role of the states versus the federal government, and Mexico's system was shaped, in part, by its revolution of 1917. Canada has not experienced the trauma of civil war and is still evolving a system in which lawmaking as well as enforcement is shared between the federal government and the provinces. Canada's constitution is embodied in a series of documents beginning with the British North America Act (1867 [U.K.] R.S.C. 1970, Appendix II, No. 5), with which the British implemented a plan for confederation that had been worked out in a series of conferences (Hogg 1985). The process of developing a constitution was completed in 1982 when the British North America Act was adopted with minor adjustments along with the Charter of Rights and Freedoms (Constitution Act 1982). The federal system is continuing to develop in all three countries.

Among democracies, the functions of the executive branch are probably nowhere more fundamentally different than among the nations of North America. Within Mexico, the decision-making structure is extremely hierarchical. The president exercises much influence and is expected to exercise authority over the entire system (Camp 1993). Additionally, Mexico has had a strong one-party government in which the Institutional Revolutionary Party (PRI) has enjoyed virtually complete control. Although the PRI does not dominate policy making in the legislature or executive branches, it does play a very visible role (Camp 1993). From its inception in 1929, the PRI did not lose a national election until 1997, when it lost control of the lower house of Congress (the Chamber of Deputies). This, in addition to opposition parties' successes in recent state and local elections including Mexico City's mayoralty, shows the increasing degree of democracy found in the Mexican system. Nonetheless, although Mexico is technically and increasingly democratic, political power is still strongly centralized.

The United States exhibits a rather unique political system with its two-party competition and its checks-and-balances relationship among the three branches of government. The president is very powerful, but that power is limited by the other branches. Canada's parliamentary system is radically different in terms of the chief executive: the prime minister is an elected member of the parliament chosen by the majority party to organize a government. Thus, the executive and legislative duties are formally interconnected, giving the prime minister substantial control of the legislature. The three patterns exhibited in North America are essentially repeated in the organization of each country's state and provincial executives.

In North America, the legislative branch is made up of the most important lawmakers. Although the legislature is not the only source of law, a statute created by a legislature is one of the most important types of law. In Mexico, the majority party formulates policy, and, because it controls the legislature, it ensures the passage of statues reflecting party preferences. Most Canadian governments obtain majorities from the electoral success of one party. That party can control legislation. When strong third parties arise, some governments are sustained by coalitions among members of several parties (Ward 1987). In coalition governments, bargaining among parties is required to maintain a majority. In the United States, the limited cohesiveness of the majority party means it is frequently unable to guarantee legislative success, and statutes are passed through a process of bargaining among members of the legislature.

In Mexico, courts—those denizens of the judicial branch—do not interpret the law; they literally apply statutes created by the legislature. Yet, with the 1994 Judicial Reform, judges have been given more independence and power to protect the constitution. In other words, there is more room for interpretation, although Mexico is still very restricted to statutes created by the legislature (Comment 1995). Comparing the judicial branches of North American countries illustrates one of the common misunderstandings of law in Canada and the United States—the importance of the heritage imparted by English common law. In those nations, the law grows and develops with each decision a court makes, as well as with legislative changes. Law in Canada and the Unites States has three sources: legislative statutes, court opinions, and common practice (i.e., common law).

The Canadian province of Quebec and the state of Louisiana use civil law, based on the French Napoleonic Code. The right of Quebec citizens to maintain their legal system was laid down in the Quebec Act of 1774 and confirmed in the British North American Act of 1867 (Ward 1987). It is important to note that Canadian federal jurisdiction also covers topics explicitly laid out in the civil law.

4.1.2 The Role of Law in Conflicts

Many beginning professionals in fish and wildlife biology see the law as a set of static principles held in place for all time, unless superseded by new laws. This view ignores the dynamic nature of law in resolving conflicts. In fact, it is in this conflict resolution process that most resource managers uncomfortably find themselves (see Chapters 2 and 3).

Learning that there is little court interpretation of statutes in Mexico might lead to the conclusion that enforcement of laws is simple. Nothing could be further from the truth. Law is applied according to political culture. In Mexico, that culture requires bargaining and compromise among the many agencies and organizations designated by statutes to enforce the laws. The political cultures of the United States and Canada also put a premium on compromise. The three countries perform this bargaining differently, but all use negotiation to protect the natural environment. Negotiation occurs at all levels.

It is in the context of negotiation that law serves to resolve conflicts. Law is a guide to negotiation and promotes resolution of conflict. Courts provide a formal remedy when bargaining fails. Where the law says little about how society should operate, conflict may be resolved by resorting to force. Happily, violence is rarely a part of fisheries management.

4.1.3 The Role of Law in Fisheries Management

There are many ways in which law affects fisheries management. Four of the most important avenues are treated here: (1) prescribing rules to resolve conflicts, (2) balancing the powers of government branches, (3) defining the powers of the central government, and (4) describing the boundary between legal and political issues.

The first role of law is in prescribing rules for resolving conflicts. If Sharon Douglas' employees have a poor understanding of law, they probably believe they will end up in court whenever they are involved in a dispute. Whether Ms. Douglas is supervising an office in Mexico, Canada, or the United States, this assumption is probably wrong. Her employees *will* participate in negotiations and enforcement actions outside the courtroom. Law does more than just guide conflict into the judicial system; it also guides the behavior of agencies, sets their missions, constrains their actions, and defines their powers.

The second role concerns balancing the power of legislative and executive branches. In the United States, the balance between the legislative and executive branches is a kind of dynamic equilibrium that oscillates over time. Which of the two branches decides the priorities of fisheries protection? Which plans? Which budgets? Which branch sets the standards for successful protection? All these questions must be answered by the executive branch, legislative branch, or both.

The third role of law in fisheries management is defining the powers of the central government as opposed to those of the states and provinces. The first level of law in this regard is the constitution. Each nation has a constitution defining basic government powers. The constitution clarifies such matters as who has the power to regulate trade between the states or provinces. Who owns the land, water, or air—and who owns the fish—may be set forth in the constitution, statutes, or court cases. All of these issues are subject to litigation. For example, how far can a Canadian province go in regulating the harvest of inland fish populations? Is such regulation under the control of federal, provincial, or Native jurisdiction?

The fourth role of law is deciding what is a legal question and what is a political one. In democracies, almost every issue is open for debate. However, some issues are regarded as beyond partisan politics. In the United States, the individual states are understood to own the fish and wildlife populations within their boundaries; this is no longer a subject for serious partisan debate. What is debated in the political arena is the degree of control that comes with ownership. For example, treaties and federal statutes can limit state prerogatives, but the extent of those limits is often a political question. It was once thought that the states owned the water, but this has become a political question where the federal government controls the protection of habitat for endangered species, manages the water on national forests and Native American reservations, and oversees interstate transfers.

4.1.3.1 Who Owns Fishes Under the Law?

Most fisheries managers are trained to think in terms of rules and regulations governing the taking of fishes, for example, size limits, closed seasons, closed areas, and permissible equipment (see Chapter 17). In these cases, the agencies are enforcers of clear rules. The result of this perspective is that legal problems may be seen as cut and dried. But regulations must have some basis in ownership or formal authority.

Certainly, fishes are owned by some jurisdiction, such as a nation, state or province, or Native American government. In that ownership capacity, there is a trust responsibility to protect the fishery for the people (see Chapter 1). In the United States, that responsibility can result in curtailing water development, limiting commercial harvests of fishes, or regulating land use to preserve a fishery. Ownership also means that the government can regulate the taking of sport fishes and enforce civil or criminal penalties for illegal takings. The police power of states and provinces means that they can abridge the rights of private property owners to protect a fishery resource. In Mexico and Canada, the federal government owns the inland fishery; in the United States, it is under the jurisdiction of the individual states.

4.2 ALLOCATION OF WATER TO FISHERIES PURPOSES

The regulation of fishing pressure and the enhancement of fisheries through hatchery propagation are not enough to provide the public with an adequate inland fishery resource. Especially in the face of extensive land and water development, what is needed is the protection of habitat. Suitable habitat is a necessity for maintaining fish populations, and although fishing pressure can eliminate a population at a particular site, a lack of adequate habitat can eliminate an entire fishery.

Protection of habitat enters the realm of environmental values. Water development is the major impediment to maintenance of fish habitat. Water development (often equated to economic development) may take the form of diverting water for agriculture, hydropower production, municipal water supply, or navigation. Generally, water development cannot proceed without water rights. Because the states and provinces own the water, they may regulate its use; hence, the right to use water is granted by the states and provinces. However, in recent years the federal governments of Canada and the United States have exerted increasing authority over these water allocations. This development has raised the specter of conflict over which jurisdiction has the power to regulate the use of water. Controlling the use of water means controlling the quality of fish habitat. In both the United States and Canada, two basic legal doctrines control the allocation of water: the riparian and appropriation doctrines. We also review doctrines and policies that affect water allocation.

4.2.1 Riparian Doctrine

Where rainfall is plentiful it is common to find a water rights doctrine reflecting that abundance. The idea of the doctrine is simple: persons owning land that abuts a surface body of water have the right to use that water; persons whose lands are not contiguous to a body of water have no such right. Typically, these nonriparian landowners must rely on groundwater. The simplicity of the riparian doctrine belies the confusion that arises when water becomes scarce or many users compete for a limited supply.

When water is scarce, the riparian doctrine applies one of two cardinal principles. The first of these principles is known as the natural flow rule, by which a landowner is entitled to use the water in a stream or lake so long as the water body remains substantially undiminished in either quantity or quality. This rule seems to maintain habitat because holders of water rights cannot diminish the flow of a stream or level of a lake. However, because the doctrine is laxly enforced, this protection is minimal. The natural flow rule developed in connection with domestic water uses and an industry powered by water wheels. Such uses either consumed little water or operated on a run-of-the-river basis (i.e., use of the water as it passes). When other uses began competing for water, the natural flow rule lost its appeal.

The other principle for dealing with scarcity in riparian law jurisdictions is the reasonable use rule. Under this principle, a riparian landowner may use water as long as that use is reasonable. This means that a reasonable use might completely dewater a stream or lake. What is a reasonable use of water? That question is answered in a relative fashion—reasonableness depends on other uses on the stream. If there is only one riparian owner on the stream, almost any use is reasonable. Between two competing users, a court must weigh each claim and determine which is relatively more reasonable. Should a third user arrive on the stream, the determination must be made again—also by a court weighing the relative merits of the three uses.

This system does not work well for the fisheries manager because the regulation of water rights requires special environmental, water regulation, or fisheries management legislation or a court finding that environmental protection is somehow required. Typically, there is so much water in riparian doctrine jurisdictions that fish habitat needs have not been an issue and few statutes exist. As a consequence, there is generally little legal leverage for the fisheries manager under the riparian doctrine.

Laws protecting fish habitat can modify the riparian doctrine. These laws have often stemmed from environmental protection statutes empowering agencies to promulgate regulations to protect stream environments. An example can be found in Iowa, where state agencies set streamflow levels below which water users may not divert water. When streamflows approach the protected low level, the state informs riparian owners to cease or reduce diversions so that the flow levels are maintained (see Trelease and Gould 1987).

4.2.2 Appropriation Doctrine

The maintenance of instream flows for fish habitat is also cumbersome under the appropriation doctrine, although it is more common. The appropriation doctrine developed as a way to allocate effectively a chronically scarce resource. The heart of the doctrine entails that the first person to put water to a "beneficial use" on a stream has the best right to use that water. The second person to apply water to a beneficial use has the second best water right, and so on: "first in time is first in right." The measure of a water right is not land ownership, as in the case of the riparian doctrine; rather, it is sustained beneficial use. Each jurisdiction has defined what constitutes beneficial use; included are diversion of water for agriculture, industry, municipal water supply, mining, and other developmental purposes. These water rights are private property that may be sold and moved. Historically, the concept of beneficial use excluded environmental protection. However, beginning in 1959 instream flows were recognized as beneficial uses of water under the appropriation doctrine (Lamb and Doerksen 1990).

Instream flows are accommodated under the appropriation doctrine in one of three ways: (1) appropriations of water are allowed for instream uses, (2) an amount of water is set aside that cannot be used for water rights, or (3) special requirements are placed on diversionary water rights.

First, it is possible for a province or state to declare instream flows for fish habitat to be a beneficial use and allow appropriation of water for that purpose. Such a program involves either a water management agency or a fish and wildlife agency appropriating water for use between two points in a stream, for a certain amount of flow, and for preferred species of fishes. This water right would be administered in order (i.e., time) of priority like any other right. An example of this system is found in Colorado, where the Water Resources Conservation Board, on the recommendation of the Division of Wildlife, can file for and hold an instream flow right. Because instream flow programs began in the late 1970s, these water rights are very junior in time. Old water rights are said to be senior, reflecting their more secure status; newer rights, or junior rights, are less frequently allowed to take water. However, under a provision of the appropriation doctrine a junior user is entitled to have the stream remain in the condition in which it was found. Therefore, a junior instream flow water right holder can prevent senior rights from being moved upstream.

Second, a way to protect aquatic habitat is through the reservation of water. This reservation is much like a water right. A model reservation system is found in Washington State, where the Department of Ecology sets aside a base flow of water from further appropriation. This agency set-aside is managed as a water right, but it can be revised by administrative action. A water right is held in perpetuity.

Third, some jurisdictions (such as California) allow the fish and wildlife agency to review water rights applications and to suggest adjustments. These conditions usually require that the water right holder not use water when flow levels fall below a specified volume (see Trelease and Gould 1987).

4.2.3 Mixed Doctrine Jurisdictions

Some jurisdictions, including California and Nebraska, use a mixture of the appropriation and riparian doctrines. One essential difference between the two doctrines is that under the riparian doctrine a water right cannot be lost by nonuse, whereas an appropriative right can be lost if not persistently applied to a beneficial use. This raises the specter of an instream flow appropriative right being injured, without recourse, by a long-dormant riparian right.

4.2.4 The Public Trust Doctrine

Under English common law as it has developed in the United States, governments hold fish and wildlife in trust for the people. The government may allow the use and taking of fish and wildlife so long as this management is mindful of the trust responsibility. The public trust doctrine is the "bedrock of modern wildlife regulation" (Veiluva 1981). It is under this doctrine that state legislatures have developed laws to control the use of the resources held in trust for the public.

Similarly, the government has a trust responsibility for other natural resources, including water. Acknowledgment of the public trust for water resources is growing slowly, but in some jurisdictions it has come to include the protection of fish and wildlife habitat, coastal access, and aesthetic characteristics. The test for the public trust doctrine is whether the government is cognizant of its trust responsibilities when making decisions about allocating resources to private uses. Failure to consider the public trust may result in a court reversing a natural resource management decision even years after the decision was made.

4.2.5 The Takings Doctrine

The Constitution of the United States (Amendment XIV, section 1 [1868]) provides that private property cannot be taken by the government without just compensation under due process of law. Thus, for example, when the government wishes to build a highway through a person's land, the government must pay the full market price for the land. This is called "a taking" of private property. However, when a government creates a statute or regulation to promote the public welfare, safety, or health, property rights may be infringed upon without constituting a taking.

For example, if a state were to pass legislation protecting wetlands from draining and filling to preserve the natural environment, landowners might not be able to sell their wetland properties to developers. This would mean a loss of value to the landowner, but it would not be a taking so long as the landowner still has some economic use of the land. This remaining economic use might be for activities as low in value as growing wild cranberries. Because legislation protecting beaches, wetlands, rivers, endangered species, and other environmental values has become commonplace, the law of taking is in a state of flux.

4.2.6 Federal Reserved Water Rights

Reserved water rights are an issue only in the United States. In 1908, the U.S. Supreme Court created a doctrine enabling the federal government to obtain water rights in appropriation doctrine states even though those water rights may have lain dormant for many years. The Supreme Court found that when the U.S. Congress reserved the land of an Indian reservation, it reserved with that land enough water to accomplish the purposes of the reservation. Succeeding cases have extended the reserved rights doctrine to include other federal enclaves, such as national parks, monuments, forests, wildlife refuges, recreation areas, and military reservations. The amount of water reserved is the minimum amount necessary to ensure that the purposes of the land reservation are not entirely defeated. In a national park, this may mean the entire natural flow of all rivers and streams. In national forests, it may mean the amount of water flow necessary to maintain stream channels. On wildlife refuges, it may mean a minimum flow or lake level sufficient to propagate fishes and waterfowl. In any event, the amount of water that is reserved depends entirely on the purposes of the reservation.

This doctrine is often controversial because it can cause long-time users of water to lose senior standing. If the date of the federal land reservation predates senior water rights, the reservation will become the senior user, adversely affecting the state-granted water rights. Such a situation causes a great deal of consternation among water management interests and has led to a long series of convoluted state and federal court cases. The final shape of this doctrine is yet to be determined.

4.2.7 Interstate Compacts

Under the federal system in the United States, each state administers water within its borders. However, rivers often cross state boundaries. When rivers flow between and among states, the administration of water resources becomes complicated by the plethora of jurisdictions controlling water use. When this happens, states may enter into interstate compacts, which are authorized under the U.S. Constitution (Article I, section 10, Clause 3). Each one operates as a treaty between the states and requires the consent of Congress. These compacts have been used to settle disputes over navigation, boundaries, fishing rights, consumptive use of water, and water quality (Doerksen and Wakefield 1975; Goldfarb 1988).

Not all conflicts between the states are resolved by compact. Other techniques include congressional apportionment and judicial apportionment. In disputes between the states, the U.S. Constitution (Article III, section 2, Clause 2) provides that the Supreme Court has original jurisdiction. The Supreme Court has also held that Congress can itself apportion water among the states contiguous to an interstate navigable stream (*Arizona v California*, 29 US 558 [1936]).

4.2.8 Canadian Federal Policy

Federal regulation affecting fish habitat has an effect on water allocation. In Canada, the DFO has determined that "fish habitats constitute healthy production systems for the nation's fisheries and, when the habitats are functioning well, Canada's fish stocks will continue to produce economic and social benefits throughout the country" (DFO

1986:12). Under the federal Fisheries Act of 1970 (R.S.C. 1985, c. F-14), habitat is defined as those parts of the environment on which fish depend either directly or indirectly for their life processes. The goal of the federal policy is an increase in the natural productive capacity of habitats for the nation's fisheries resources to benefit present and future generations of Canadians (DFO 1986). This goal leads to fish habitat conservation based on the guiding principle of "no net loss of the productive capacity of habitats." The policy also envisions rehabilitating productive habitat in selected areas and creating habitats where there is some social or economic benefit. In implementing the Fisheries Act, the DFO works closely with other agencies, including Environment Canada, the Department of Indian and Northern Affairs, and especially the provincial governments, to ensure compliance in matters related to habitat destruction, pollution control, habitat improvement, and fisheries research.

4.2.9 Mexican Federal Policy

The focus of the federal government in Mexico has been on economic development (Juergensmeyer and Blizzard 1973; Wright 1990; Simonian 1995). However, this has begun to change since the 1980s because of a growing awareness in Mexico of the importance of sound environmental policies in improving the long-term outlook for the national economy (Cabrera Acevedo 1978; Mumme et al. 1988). These policies affect water allocation and fish habitat. In 1988, the federal government enacted legislation greatly increasing the strength of the nation's environmental programs. This statute, the General Law of Ecological Equilibrium and Protection of the Environment (*Diario Oficial de la Federación*, 28 January 1988), which became effective on 1 March 1988, is administered by the Secretariat of Social Development. The ecological equilibrium statute provides greater authority to prevent and control environmental disruptions and allows state and municipal governments to share with the federal government in resource conservation. The law encourages joint approaches to solving environmental problems. Important provisions of the statute call for protecting natural areas and conserving plants and animals, as well as conducting conservation education in elementary schools. In May 1992, new legislation further reduced the role of the federal environmental authorities, giving local authorities greater discretion to interpret and apply Mexico's environmental regulations. The legislation also created the National Institute of Ecology, which is in charge of defining federal environmental policy, setting national ecological norms, and conducting research (Lewis 1992). Additionally, the 1994 passage of the North American Free Trade Agreement (NAFTA), has necessitated that Mexico increase its level of environmental concern.

4.2.10 United States Federal Policy

In the United States, as in Mexico and Canada, there are many federal statutes relating water allocation to management of inland fisheries resources. These include the Endangered Species Act, Anadromous Fish Conservation Act, Clean Water Act, National Environmental Policy Act, Federal Power Act, and the Fish and Wildlife Coordination Act. Among these, the Fish and Wildlife Coordination Act (FWCA) sets forth the national policy probably as well as any statute. This act requires that agencies undertaking water development activities—either building structures or issuing permits—consult with the U.S. Fish and Wildlife Service (FWS), National Marine Fisheries Service, and state fish and wildlife agencies. It does not require that the agency undertaking water development activities follow the recommendations of the wildlife agencies. Quite often, however, the views of the fish and wildlife agencies prevail in decision making, and projects are revised to take account of their concerns (Hamilton 1980; Clarke and McCool 1996).

4.3 FEDERAL REGULATION OF WATER DEVELOPMENT AND QUALITY

4.3.1 Canada and Mexico

Numerous Canadian federal statutes cover water pollution control. The most powerful of these is the Fisheries Act of 1970 (amended in 1985; Johnson and Beaulieu 1996). Section 33 of this act states that "no person shall deposit or permit the deposit of a deleterious substance of any type in water frequented by fish or in any place under any conditions where such deleterious substance or any other deleterious substance that results from the deposit of such deleterious substance may enter any such water." Judicial rulings have found that the stirring up of silt and sediment, which endangers fish eggs, is covered under this provision. The deleterious substance need not be discharged in harmful quantities to be prohibited; all that must be shown is that a substance is harmful to fishes in any concentration.

The Fisheries Act gives the Canadian federal government broad powers in controlling water pollution and other development activities in inland water bodies, although federal regulation must be tied to the harmful effects on fishes and must involve cooperation with the provinces. Approval of the DFO must be obtained for any project or undertaking that may possibly harm fish habitat. Proponents of a project submit a statement that includes a description of the proposed undertaking, any probable effects on fish habitat, and the plans for mitigation and compensation of harmful effects. The DFO can approve the project, impose conditions on approval, or reject the proposal. In practice, the negotiations can be long and arduous. The Fisheries Act is administered jointly by the DFO and Environment Canada. In addition, a federal environmental assessment and review process was established by the Environmental Assessment and Review Process Guidelines Order (SOR/84–467). Under this order the application for specific permits triggers an environmental assessment that must investigate all impacts within federal jurisdiction (Elder and Thompson 1992).

The Canada Water Act of 1970 (R.S.C. 1970, 1st Supp. c.5) authorizes the establishment of federal–provincial committees to consult on water resource issues, including designing specific projects for conservation, development, and use of any waters in which management is of significant national interest. The act also authorizes studies, quality tests, recommendation of water quality standards, and the design, construction, and operation of wastewater facilities by federal–provincial committees or designated

provincial agencies. The Canada Water Act sets forth a "polluter must pay" philosophy, emphasizing that the costs of pollution control should be considered part of production costs. It is one of several statutes governing the pollution of inland waters in
Canada (see Cram 1971).

The British North America Act of 1867 (see also Department of Justice, Canada,
A Consolidation of the Constitution Acts, 1867–1982 [1982]) grants the provinces
jurisdiction over local works and undertakings (section 92 [10]). These works include
the generation and distribution of energy. A provincial agency, such as Alberta's Department of the Environment, is responsible for approving hydropower development
plans. Federal approval is required under the Fisheries Act for any undertaking that
may have harmful effects on fishes, and under the Constitution Act for projects on
navigable waters (Hogg 1985).

Another important federal statute in Canada is the 1988 Canadian Environmental
Protection Act (amended in 1996 and 1998; R.S.C. 1998, c.16). This statute describes
the environment as an entity in itself, separate from humans and their dependence
upon the natural environment. Moreover, it states that the well-being of Canada is
dependent upon the protection of the environment (Morley 1995). The Canadian Environmental Protection Act is an example of consolidation of legislation. Many provinces have adopted consolidated general environmental codes that include environmental management provisions (Johnson and Beaulieu 1996).

The Mexican federal government has predominant jurisdiction over environmental issues. The Constitution of 1917 declares the ownership of all waters within the
nation's boundaries to be vested in the nation. Environmental laws are issued by the
federal government and administered by secretariás (secretariats) at the federal level.
These secretariats are equal to U.S. departments and Canadian ministries. Only minor
roles are delegated to the state or local governments in Mexico. In the administration
of all environmental legislation, enforcement has not been a priority (Mumme et al.
1988).

Environmental protection and pollution control in Mexico are addressed by the
1971 Federal Law for the Prevention and Control of Environmental Pollution (*Diario
Oficial de la Federación*, 23 March 1971), as amended in 1984 and 1987. In 1982, the
Federal Law for the Protection of the Environment (*Diario Oficial de la Federación*,
11 January 1982) was passed. These two laws were replaced, in large part, by the
massive 1988 General Law of Ecological Equilibrium and Protection of the Environment (Johnson and Beaulieu 1996). These laws established the framework for legislative control and regulation of the environment including pollution control, resource
management, and environmental education. The ecological equilibrium act is a detailed ecological plan that covers the rational use of water, the aquatic ecosystem, and
living resources (DuMars and Baltran del Rio 1988). In 1992, the ecological equilibrium act, in combination with the National Water Law (approved December 1992 and
amended in 1994), was used to set forth a legal framework to deal with water pollution
control in Mexico (Paule 1996). No authority for state or local governments was originally defined, although all levels of government, along with the Secretariats of Agriculture and Hydrologic Resources; Industry and Commerce; Fisheries, Oceans, and
Urban Development; and Ecology are involved. The National Program for Environ-

mental Protection (1990–1994) dictated basic environmental programs to be accomplished within Mexico during the 1990s. It provided goals, objectives, strategies, and actions to be taken for the execution of these programs by establishing ecological and environmental efficiency as a national priority (Lewis 1992).

The regulation of pollution is targeted toward urban areas and the larger lakes, principally Lake Chapala and Lake Pátzcauro (e.g., see Faudon et al. 1983). Federal involvement is usually in the form of investing in wastewater treatment facilities and encouraging industries to comply voluntarily with existing regulations (Friedrich Ebert Stiftung Foundation 1990).

Water allocation, and hence its development, is an important federal responsibility in Mexico. Several priority uses are defined in the ecological equilibrium act, including urban water supply, livestock watering, irrigation, electrical power generation, industrial uses, and aquaculture, as well as other uses. The National Water Plan was prepared in 1975 by the Secretariat of Agriculture and Hydrologic Resources to organize and encourage the development of policies for the social and economic development of Mexico's water resources. The goal of the plan was to develop water for economic purposes. The generation of electrical power is exclusively an activity of the federal government in Mexico, and any fisheries protection is worked out by the central government.

The central government of Mexico has been involved in large-scale water development projects. Beginning in 1947, river basin commissions under the control of the Secretariat of Agriculture and Hydrologic Resources were established in several regions to develop and administer large-scale irrigation projects. These commissions also became involved in flood control, hydropower development, comprehensive regional planning, and the coordination of the activities of other secretariats and agencies within their region (Barkin and King 1970).

4.3.2 The United States

The U.S. statute that seeks to integrate environmental protection policies, including fisheries management, is NEPA, passed in 1969. The statute has been effective in bringing federal agencies together to consider the environmental consequences of their actions by requiring that federal agencies prepare environmental impact statements for "major Federal actions significantly affecting the quality of the human environment." The courts have interpreted this to mean that federal agencies must give "at least as much automatic consideration to environmental factors" as they do to economic factors (*Zabel v Tabb*, 430 F2nd 199 [1970]). Furthermore, individuals can bring suits against agencies to force agencies to comply with NEPA (*Sierra Club v Morton*, 405 US 727 [1972]). The watchdog of the NEPA process is located within the Executive Office of the President, at the Council on Environmental Quality.

The FWCA was promulgated in response to the general deterioration of rivers caused by major urban development (Veiluva 1981). The idea behind the act is that forcing federal agencies to work together on construction projects will preserve fisheries. The FWCA specifies how the studies and recommendations of the FWS, National Marine Fisheries Service, and state fish and wildlife agencies will be reported to and considered by the U.S. Army Corps of Engineers (ACE), Bureau of Reclamation (as well as other agencies of the Department of the Interior), Department of Agriculture (especially the Natural Resource Conservation Service), and the Congress as they plan impoundment, diversion, dredging, and navigation projects. In short, the act requires that agencies with construction responsibilities consult with fish and wildlife agencies to determine how to compensate or mitigate for losses to fish and wildlife resources resulting from projects. The construction agencies are to give full consideration to the reports of these fish and wildlife agencies. Veiluva (1981) argued that FWCA is an ineffective mechanism for protecting fisheries because, at least into the late 1970s, federal agencies with construction responsibilities continued to issue permits and build projects over the objections of fish and wildlife agencies. More recently, the Department of the Interior, through the FWS, has developed joint working agreements with ACE and the Environmental Protection Agency to improve the consultation process.

Every country needs to build water supply projects for irrigation and municipal needs. A number of books have detailed how the United States government has subsidized these projects. Reisner's *Cadillac Desert* (1986), Hundley's *Water and the West* (1975), and Nadeau's *The Water Seekers* (1950) detail how the great cities of the American West—Denver, Los Angeles, San Francisco, Salt Lake City, and Phoenix—strove for and found water in pristine valleys, future national parks, wilderness areas, and productive farmlands and moved it hundreds of miles to slake the thirst of city dwellers. Also chronicled are stories of how federal agencies and Congress aided in the development of this water, not only for cities but also for farmers and ranchers. These books argue that subsidizing water development with federal tax dollars is no longer appropriate. Whatever the suitability of some water development, it is clear that the combined constituencies of city and farm are powerful influences on Congress when it considers water projects.

The FWCA affects federal hydropower licensing in the United States. The Federal Energy Regulatory Commission issues licenses to private parties who wish to develop hydroelectric power facilities. Under a number of authorities, including the Federal Power Act (16 U.S.C. 791a–828 [1982]), the commission may approve hydropower projects but must take into consideration the effects of these developments on fish and wildlife. The FWCA is a further reinforcement of the requirement to consult with fish and wildlife agencies before issuing a license to anyone wishing to generate electric power. The commission has routinely given weight to the opinions of the FWS, National Marine Fisheries Service, and state fish and wildlife agencies. This does not mean that the task is easy for fisheries managers faced with evaluating hydropower projects. In fact, the large number of hydropower license applications, dearth of information on fish species life requirements, and the federal incentive program encouraging hydropower projects have meant that fish and wildlife agencies are often overwhelmed.

The Clean Water Act has a long and fascinating history. It has evolved incrementally as the nation's understanding of pollution has developed. Together with other environmental statutes—the FWCA, NEPA, and aspects of the Federal Power Act—the Clean Water Act helps weave together a fabric of protections. The history of this extensive statute can be traced from the Water Quality Act of 1965 (33 U.S.C. 1151) to the present statute. The present law redefines pollution from what was considered conventional wisdom in 1965. In section 502, pollution is defined as the "man-made or man-induced alteration of the chemical, physical, biological, and radiological integrity of water." Goldfarb (1988) argues that what Congress meant by the "integrity of water" is its ecological stability. Before 1972, the public did not have a right to clean water but a right to water not polluted beyond an individual state's standards. Pollution was defined as excessive discharge (Goldfarb 1988). The goals of the act today include elimination of the discharge of pollutants into U.S. waters, protection of aquatic life, prohibition of the discharge of pollutants, and provision of financial assistance for the construction of wastewater treatment plants and research.

The Clean Water Act is an ambitious statute. Perhaps the most important and controversial aspect for fisheries management is section 404 because it affects state water allocation practices and influences a vast array of nonfederal water developments. This section requires developers to acquire a permit from ACE before discharging dredge or fill material into the "waters of the United States." Much broader than navigable waters, this description of the jurisdiction of ACE takes in almost all watercourses. Goldfarb (1988) points out that intrastate streams, freshwater wetlands, drainage ditches, mosquito canals, and intermittent streams are all covered by section 404. This has led to a pitched battle between environmentalists and developers over who really controls water allocation, the states or federal government. On the advice of fish and wildlife agencies, ACE has denied permits under this statute strictly on ecological grounds. Such permit denials may have the effect of superseding state-granted water rights.

Another similar statute is section 10 of the Rivers and Harbors Act of 1899 (33 C.F.R. 320 et seq.). This section prohibits the unauthorized construction in, or alteration of, any navigable water of the United States. Under the statute, ACE issues permits for "the construction of any structure in or over any navigable water…the excavation from or depositing of material…or the accomplishment of any other work affecting the course, location, condition, or capacity of (navigable) waters" (see Goldfarb 1988). Of course, section 10 is administered within the consultation requirements of the FWCA.

The FWS also operates under the Anadromous Fish Conservation Act and a host of other statutes, including the Federal Aid in Fish Restoration Act (16 U.S.C. 777–777k as amended). The Anadromous Fish Conservation Act provides that the FWS can plan and conduct research and develop programs affecting anadromous species whenever two or more states have a common interest in any river basin. Programs can be designed to conserve, enhance, or develop anadromous fisheries resources nationwide. The Federal Aid in Fish Restoration Act is often referred to as the Dingell–Johnson Act after its main sponsors; the later, amended act is known as the Wallop–Breaux Act (P.L. 98–369 [1984]) after its sponsors. The act is one of several providing federal aid to states for the management and restoration of fishes. Funds for this aid come from taxes on the sale of fishing equipment, boats, motors, and motor boat fuel.

Whenever a U.S. federal agency undertakes an action, such as building a water resource project or issuing a permit for construction, dredging and filling, or hydropower production, it must ask the FWS if any threatened or endangered species are present at the project site. This is a main provision of the Endangered Species Act. The act has four approaches to protecting endangered species: the listing of species, agency consultation, mandatory agency responsibility to conserve endangered species, and a ban on any "taking" of endangered species. The first step is to list endangered species, and no action can be taken under the act until this has been completed. Once a species is listed, it cannot be taken either through collection or harassment, and federal agencies are prohibited from actions adversely affecting the species. When the FWS makes a tentative determination that an endangered species is present at a project site, the agency undertaking an action must submit a biological assessment to the FWS. If the assessment identifies potential adverse effects, the FWS prepares a formal biological opinion setting forth whether or not the action is likely to jeopardize the species. The opinion may recommend no project or mitigation or conservation measures. If differences over a project's effects on the species cannot be resolved, a federal committee may grant an exception to the otherwise absolute ban against injuring endangered species.

The Endangered Species Act is a stringent law with controlling power over many projects. It is not administered as prohibitively as might be expected, however. Yaffee (1982) has demonstrated that throughout the history of the Endangered Species Act, the FWS has bargained rather than regulated the impacts on endangered species. Chapter 16 deals extensively with the implementation of this act.

4.4 ENVIRONMENTAL PROTECTION: STATE AND LOCAL REGULATION

4.4.1 State and Provincial Statutes

In a recent count, 30 states in the United States had "little NEPAs," or state environmental protection acts. These statutes are similar to the federal NEPA and are applicable to state and local projects. Typically, they require a state environmental impact statement. These little NEPAs are operative for state projects as long as the conditions placed on the projects by the state do not conflict with the paramount jurisdiction of the United States under a law passed by Congress. Generally speaking, when Congress has acted rightfully under the Constitution, the states may not contradict congressional intent. A good example would be a state's attempt to prevent construction of a hydropower project that has a permit from the Federal Energy Regulatory Commission. The state is likely to lose this battle because Congress has authorized the commission, under the commerce clause of the Constitution, to regulate hydropower (see Trelease and Gould 1987).

The authority of Canadian provinces to enact environmental protection legislation is based on the British North American Act of 1867. The policies therein were carried forward in the Canadian Constitution Act of 1982, giving the provinces jurisdiction over property and civil rights. Pollution control legislation exists in all the provinces in the form of a water resources, pollution control, or public health act, and provincial board administers these statutes. Under these statutes, activities that could cause pollution to waters within the province are regulated and licensed. The jurisdiction over and liability for activities that cause pollution in another province or in interprovincial waters have not been clearly determined (Morgan 1970; Hogg 1985).

Under a 1982 revision of the Federal Law for the Prevention and Control of Environmental Pollution, state and local governments in Mexico received the authority to enact environmental legislation. However, the actual power to take action has been slow due to the existing political culture and lack of funding (Simon 1997). In 1987, the federal government provided some modest funding for awareness campaigns and educational programs. Additionally, Mexico has received money from the World Bank to assist in decentralizing environmental legislation (Lewis 1992). The local environmental legislation that does exist tends to mirror the federal ecological equilibrium act of 1988. One example is the Ecological Law for the State of Chihuahua (*Folleto Anexo Periodico Oficial* 86, 26 October 1991), designed to preserve and restore ecological equilibrium and environmental protection (Lewis 1992). It provides for concurrent jurisdiction at the state and municipal levels. By 1996, 29 of the 31 Mexican states had their own environmental laws, which were equivalent to or more restrictive than the federal law (Paule 1996).

4.4.2 Health Codes

In the United States, individual states have enacted legislation for the protection of public health. These laws are enacted under state police powers and range from wide reaching to ineffective in their contribution to the protection of the natural environment. Control of pesticides and toxic substances has important ramifications for fisheries management. Just as in the United States, Canadian municipalities may construct and operate water and wastewater treatment facilities. Public health acts provide the provincial departments of health with the power to control unsanitary conditions as well as pesticides and toxic substances.

All the nations of North America have been painfully slow in recognizing the problem of toxic waste.

Mexico, too, has health codes. The Secretariat of Agriculture and Hydrologic Resources is responsible for the control of all wastewater discharges and the regulation of water quality for industrial use. The Secretariat of Health and Welfare has the authority to protect the public health under the Federal Law for the Protection of the Environment, which combines several older statutes. In 1990, according to Mexico's National Commission of Water, 30% of the population still lacked formal access to established potable water systems, and 51% had no sewer connection (Castro 1995).

4.4.3 State and Provincial Endangered Species Protection

In the United States, state fish and wildlife agencies provide information to the FWS, petition for changes in the status of endangered species, and help enforce the provisions of the federal Endangered Species Act. More than two-thirds of the states have some form of endangered species law or regulation. Most of these state statutes are copies of the federal law. Many states also have cooperative agreements with the FWS. Yaffee (1982) summarized the situation when he observed:

> The state agencies, whose historic constituency is hunters and fishermen, are by and large game animal agencies. Their professional traditions are those of conservation and management of wildlife to produce huntable surpluses. Most of their programs are financed principally by hunting and fishing license fees.

Because of the [Congress'] fear of their game animal bias, the states were not given a larger scale in the implementation of the [Endangered Species Act]. (Yaffee 1982)

Endangered species in Canada are protected by provincial legislation such as that of Ontario, which lists species of plants and animals that are considered endangered and makes it an offense to kill, injure, or destroy their habitats. Provincial fish and wildlife departments (Burton 1984) administer such legislation.

Mexico's General Law of Ecological Equilibrium and Protection of the Environment recognizes the need for protection of endangered species. Although this is a federal law, cooperation for its implementation is encouraged from all levels of government.

4.4.4 State and Provincial Fish and Wildlife Agencies

State fish and wildlife agencies in the United States enact their own fish and game harvest regulations, generate funds via licenses, designate special fishing areas, and enforce fish and game laws. In some states, game wardens are deputized law officers. States operate fish hatcheries and stocking programs, conduct fisheries research, and manage nongame populations. Some state agencies manage commercial fishing. All of this is done under the state's trust responsibility to protect the wildlife resources of the state, and state legislatures pass statutes to effect this purpose.

Less well understood is the role of state fish and wildlife departments in sharing information with and making recommendations to other state and federal programs. Programs with which state fish and wildlife agencies may interact range from the Federal Energy Regulatory Commission licensing program—for which the state agency must certify, pursuant to the Clean Water Act section 401, that any discharge will comply with state standards—to the state's instream flow program, for which the state agency must recommend streamflow levels (Lamb and Doerksen 1990).

In Canada, fishing regulations established by the provinces are enacted as federal regulations under the Fisheries Act. These regulations are administered by provincial fisheries agencies in Ontario, Manitoba, Saskatchewan, Alberta, Quebec, and British Columbia. The federal government directly manages the fisheries of Newfoundland, New Brunswick, Nova Scotia, Prince Edward Island, and the Northwest and Yukon territories.

In Mexico the Federal Fisheries Law of 1992 (*Diario Oficial de la Federación*, 25 June and 21 July 1992) establishes the national regulations for fishing. However, each of the states has a delegation to help with the administration of the law and the interpretation of the law for application in their state.

4.4.5 Land Use Planning, Special Districts, and Commissions

One of the most bewildering aspects of fisheries management is the plethora of state and local boards, districts, and commissions that have some sort of direct or indirect effect on fish habitat. These entities can range from the municipal planning board to water conservation districts and even to intergovernmental councils. There are many organizations that develop water or help local governments plan for economic and natural resource development. Examples in the United States are councils of government in which several cities and counties may be represented in joint planning efforts. Examples in Canada are the Canadian Council of Resource and Environmental Ministers, which was formed to enhance interprovince communications, as well as the Prairie Provinces Water Board, which helps allocate the water of interprovincial rivers. Many of these districts and commissions have public participation programs and planning and rule-making authority. In Mexico an example would be *Consejos de Cuenca*, or river basin management commissions, which provide for the inclusion of urban interests in basinwide management of water resources.

4.5 REGULATION OF THE HARVEST

Regulation is a tool of management (see Chapter 17). When to regulate, what to regulate, and how to regulate are all policy questions. In the United States, Canada, and Mexico, the power to regulate is derived from ownership of, trust responsibility for, or statutory authority to control the resource for the public. In the United States, this authority is referred to as police power, the power to control individuals for the good of society. It must be balanced against the rights of individuals—including property rights, religious freedom, and free speech. However, the power to regulate is not an end in itself. How the resource is regulated depends on management philosophy. The right to regulate may not change, but how states and provinces carry out their regulatory responsibilities depends on political culture and management decisions. Thus, regulation of fishing is guided by a management philosophy. That philosophy must be founded on the law and reflect good management practices in the face of many competing uses of the resource.

Traditionally, fish and wildlife agencies in the United States have concentrated on producing enough fish and game to provide adequate hunting and fishing experiences. The environmental movement has led to more emphasis on nongame species and management for ecological protection, but the primary emphasis of these agencies has remained service to a hunting and fishing constituency.

Competition occurs between environmental (conservation) and harvest uses, but there is international competition as well. Fisheries managers pay attention to international treaties as well as to the effect of high seas harvests of anadromous stocks by domestic and foreign commercial fishing enterprises. States or provinces do not regulate foreign fishing interests when fishing is outside a country's exclusive economic zone (188 nautical miles, or 348.2 km, beyond the territorial sea), but the effect of such fishing pressure must be counted in domestic regulations.

4.5.1 Regulation of Sport Fisheries

State and federal governments in the United States share power over fisheries resources. This sharing is an evolving, complicated concept arising out of conflict over states rights and federal powers. There is little doubt that the states do have the right to regulate the taking of wildlife within their borders (see Bean 1977). A classic case setting forth this right to regulate is *Geer v Connecticut* (161 US 519 [1896]), in which the U.S. Supreme Court found that the states have the right "to control and regulate the common property in game." Going further, the Supreme Court said "the wild game within a State belongs to the people in their collective sovereign capacity." However, in the same case the Supreme Court said that the states have this power "insofar as exercise may not be incompatible with, or restrained by, the rights conveyed to the Federal government by the Constitution."

Three rights conveyed to the federal government are important in limiting state power: the treaty power, commerce clause, and property clause. Briefly, the federal government has the exclusive right to conclude treaties with foreign powers, and treaties have the force of the Constitution; Congress can regulate interstate commerce, from which comes most of the nation's environmental legislation; and Congress has the right to manage federal property such as national forests, parks, and wildlife refuges. Finally, the federal government is also trustee for Native Americans. In this trust relation, it must protect the interests of various tribes. The states own the game within their boundaries, but the federal government can often tell the states how to behave.

A similar sort of shared power over fisheries resources exists in Canada. Besides the authority to prohibit the discharge of deleterious substances into waters frequented by fishes, the Fisheries Act grants the Minister of Fisheries and Oceans complete discretion to issue and determine conditions of fishing leases and licenses as well as authority to regulate a wide range of activities relating to fishing. In practice, the administration of fishing regulations is shared with the provinces (VanderZwaag 1983). Where anadromous fisheries are concerned, the federal agencies do indeed fully manage these fish stocks and administer the Fisheries Act. However, in all provinces, freshwater fisheries are managed by the provincial agencies, and those agencies actually administer the federal Fisheries Act. For example, Alberta has passed fisheries regulations pursuant to the federal Fisheries Act. The province administers the act subject to annual federal approval of regulations. Provincial administration of the Fisheries Act is accomplished through what might be best termed an administrative understanding between Alberta and the federal government.

Mexico's fishing regulations are more centralized. Although fisheries laws have existed since the 1920s, in 1972 the Federal Law on Fisheries Development (*Diario Oficial de la Federación*, March 1972) was enacted (Torres Garcia 1987). The 1992 Federal Fisheries Law is very similar to the 1939 Code of Fisheries, but it is more specific in dealing with interior waters (Ortiz Martinez and Guzman Arroyo 1995). The Secretariat of Fisheries was established in 1976 and elevated to cabinet status in 1982. This secretariat is responsible for the management and development of the nation's fisheries resources. The secretariat is responsible for overseeing the operation of the national fishing industry and has been primarily concerned with the development of Mexico's marine fishing industry. At the federal level, the secretariat consists of four divisions: fishery development, infrastructure, aquaculture, and the National Institute of Fish. Within this institute, 31 delegates represent the interests of the states. The secretariat also coordinates the activities of the National Fishery Collectives Commission, which is an advisory board consisting of representatives from other cabinet level departments and the fishing industry (Cicin-Sain et al. 1986).

4.5.2 Regulation of Commercial Fisheries

The government of Canada is also involved in the management of commercial fisheries. This responsibility too is shared informally with the provinces. Several acts affect federal financial involvement in the fishing industry. These acts, plus the DFO's complete discretion on fishing leases and licenses, form a strong regulatory base (VanderZwaag 1983).

The first significant legislation regulating fishing was enacted in 1857. This act dealt primarily with protecting the salmon fisheries of Canada's rivers and estuaries and developing the marine fishery resource. This legislation defined the types of gear and fishing practices allowed, established closed seasons and the licensing and leasing of fishing rights, and has provided the basis for the current policy for the regulation of fishing. The Minister of Fisheries and Oceans' licensing authority does not provide exclusive federal control of fishing because some provincial jurisdiction has been established by the courts under the constitutional role of the provinces to regulate property rights. As a result, the federal and provincial governments cooperate to regulate both commercial and sport fishing in inland waters. Provincial departments are responsible for developing regulations, although these must be enacted at the federal level (Scott and Neher 1981).

Mexico's history has been one of regulating and promoting the marine fishery. There are a number of important inland fisheries as well. An example of these is Lake Chapala near Guadalajara. As is the case elsewhere, the federal government regulates fishing cooperatives that harvest fishes from this lake.

In the United States, individual states regulate the commercial take. An example is the paddlefish fishery in Iowa, where the state sets the times and means of fishing, monitors the commercial harvest, and makes adjustments to ensure that the population remains viable. All the surrounding states also regulate the commercial take of paddlefish in the same way. This means that without close interstate cooperation—which is not consistent—maintaining a population by regulation of the timing and means of harvesting can be an uncertain process.

4.5.3 Regulation of Harvest by Native Americans

In the United States, Native American fishing rights are an area of great uncertainty. Some tribes have a right to virtually unregulated subsistence fishing, as in Alaska, and others have treaty rights. Though states can regulate some aspects of Native Americans' harvest, the federal government's trustee status makes it an important actor in this process.

The United States recognizes 507 Native American tribes and Alaska Native groups. As of 1993, these tribes and groups have a total population of 1.4 million. The federal government is the trustee for the lands and natural resources held by these tribes. The Bureau of Indian Affairs, in the U.S. Department of the Interior, is the government agency charged with representing Native American interests. In addition to this trust relationship with the federal government, tribes are, to a certain extent, sovereign nations that have binding treaties with the United States. Even though these nations are part of the states in which they are located, they are exempt via the treaties from many state laws, pay few state taxes, and may collect royalties for the extraction of natural resources.

In the area of fisheries management, states can and do regulate aspects of tribal fishing. Generally the tribes have treaties that protect some element of their sovereignty in regard to fishing. Tribes can control sport and commercial fishing on the reservations and under treaty may have relatively open access to fishing sites off the reservations. The limit that may be placed on the harvest by Native Americans is a matter of controversy. States have a responsibility to conserve viable fish populations, tribes have sovereign rights under treaties to take fishes, and the federal government has a trust responsibility to protect Native American natural resources. Any attempt to summarize the long line of court decisions that have tried to make sense out of this jumble would fail to be brief, but a fairly recent case is perhaps a good example of decisions about Native American fishing rights. In *Washington v Washington State Commercial Passenger Fishing Vessel Association* (99 Sup. Ct. 3055 [1979]), the Supreme Court seems to have upheld a kind of fair-share approach to Native American fishing rights off the reservation. The Supreme Court said that, at least for certain types of tribes with treaties, a 50-50 split of the harvestable fishes with nontribal fishers is a fair result. What is harvestable is in debate because some fish must be left in the stream for conservation and reproduction. Good introductions to the concepts regarding Native American fishing rights are provided by Landau (1980), Miklas and Shupe (1986), and Shupe (1986).

Native rights in Canadian law stem from aboriginal rights, the property rights of native peoples that are theirs by virtue of their occupation of certain lands from time immemorial. These rights were first recognized in the Royal Proclamation of 1763 ([U.K.], R.S.C. 1970, Appendix II, No. 1), when Britain forbade settlement on Native American lands and decreed that such lands be obtained only by purchase or cession. Treaties were also made in recognition of these rights, granting the right to unrestricted fishing and hunting at all times on all unoccupied government lands or any other lands to which native people may have a right of access. As a result, all native peoples can claim fishing rights, either under a treaty or through aboriginal rights. It is unclear to what degree aboriginal rights provide protection of fishing and hunting in the absence of a treaty. Native rights are under federal jurisdiction, and provincial laws can make only limited infringements on any rights granted under a treaty (Indian Act, R.S.C. 1960, c. 39, section 88). For example, certain sections of the Alberta Wildlife Act (R.S.C. 1984, c. W-9.1) specify that while native people may hunt year round for food, they cannot sell the meat or use automatic weapons.

At one time, the federal parliament could enact legislation limiting treaty rights. Such limitations of hunting rights were included in legislation and had been upheld by the courts (Cummings and Mickenberg 1972). However, as with other areas of law, this has changed in recent years as the duties of the federal government regarding Native Peoples have been defined. The Constitution Act, 1982 (section 35[1] 1982) provides protection of "existing aboriginal rights." The meaning of this and other provisions is still being worked out in the courts. A recent decision by the Supreme Court of Canada summarized the law as follows: "Federal power must be reconciled with

federal duty and the best way to achieve that reconciliation is to demand the justifica-
tion of any government regulation that infringes upon or denies aboriginal rights"
(*Ronald Edward Sparrow v Her Majesty the Queen, et al.* Supreme Court of Canada,
File No. 20311, 3 November 1990).

4.5.4 International Treaties

International treaties are negotiated by national governments subject to the provisions of their constitutions. In the United States, these provisions mean negotiation by
the president and subsequent ratification by a two-thirds vote of the U.S. Senate. There
are several treaties in force between the United States and Canada affecting fisheries;
many such treaties are also in force between the United States and Mexico (Rohn
1983; Paule 1996). Important treaties regulate the boundary waters between nations
and the harvest of marine fishes. Because treaties become the supreme laws of the
land—equivalent to the constitution—it is imperative that they are worked out carefully and that the states and provinces be cognizant of their provisions. To this end,
treaties frequently establish commissions that are responsible for approving and carrying out projects affecting treaty-protected resources. Such commissions include the
International Joint Commission covering boundary water issues for the United States
and Canada, the Great Lakes Fisheries Commission (Dworsky 1986; Sadler 1986),
and the Border Environmental Cooperation Commission (created shortly after passage
of NAFTA), covering boundary water issues between the United States and Mexico
(Johnson and Beaulieu 1996).

In addition, with the signing of NAFTA, the North American Agreement on Environmental Cooperation took effect. Among other measures, the agreement created the
North American Commission on Environmental Cooperation, which monitors the ecological effects of NAFTA in Canada, Mexico, and the United States (Johnson and
Beaulieu 1996).

4.6   CONCLUSION: THE BALANCING ACT

At the opening of this chapter, the problem faced by Sharon Douglas was presented. That fictitious problem might be summarized by this quote: "We were never
trained to work in an arena where everything is controversial and everything is political" (quoted by Booth 1988). The biologist who said this might well have added that
science is rarely the controlling factor in the decision-making process. Fisheries management is a bewildering mixture of the scientific, legal, and political arenas, where
the technically trained manager must integrate professional and pragmatic judgments
about policy outcomes.

4.6.1 The Essential Points of Law

There are six essential categories of law that affect the fisheries manager: constitutional law, treaties, native people's rights, harvest regulations, water allocation, and
property rights. Constitutional law means discerning those powers given to the federal

government and those reserved to the states or provinces; this is to say, who has jurisdiction? The answer to this question is evolving in every nation, but the trend seems to be toward central control.

Treaty making is a power of the federal government, and from this capacity has evolved a new field of the law affecting fisheries management. The combination of Native American water rights and fishing rights leads to complexity for any manager seeking to conserve and protect fish habitat and populations. The final answer on who controls this important resource in is not fully known and may not be known for some time. Treaties, of course, are also international. These formal international agreements are important considerations for the manager because their provisions may be paramount over other statutes.

The rights of Native Americans are a further intervening variable in fisheries management. These rights arise from treaties, aboriginal rights, and federal trusteeship. Native Americans have a right to fish for subsistence, religious and commercial purposes, and maintenance of a moderate living standard. These rights are to be balanced against state ownership responsibilities, including the duty to provide self-sustaining fish populations. These rights may be abridged by federal statute in both the United States and Canada—for example, in the case of an endangered species—but such abridgement is politically difficult to accomplish.

Fishing regulation is the responsibility of the states in the United States and of the federal governments in Mexico and Canada. In practice, fishing regulation is shared between and among jurisdictions in each country. Treaties, federal trust responsibilities, and the regulation of interstate commerce gives the federal government in the United States power to affect state regulation, whereas property rights give Canadian provinces power to affect federal regulation in that country.

Water allocation is important because it affects fish habitat. As one noted biologist has commented, "We've become so focused on preserving a species here and a species there that we are losing sight of the habitat" (quoted by Booth 1988). The water allocation systems of the North American continent historically have not considered habitat preservation as a beneficial use of water worthy of legal protection. This situation has been slowly changing, but taking an ecosystem view of water rights is not on the legal horizon. Most jurisdictions still give water in perpetuity to the first user who invests money in developing the resource. Environmental uses of the water are struggling to achieve a strong place in that system.

Property rights stand for a host of legal doctrines and policies that essentially tell landowners what they can or cannot do with their property. The states have a police power to regulate unsafe practices and may legislate land uses to protect the public. The federal power to regulate interstate commerce has extended to the control of pollution. The public trust doctrine means that states cannot dissolve their ownership of public land or natural resources without first considering their obligation to the public to protect those resources. All of these are limits on private property.

4.6.2 The Essential Points of Politics

It is often remarked that "politics is the art of the possible." Learning what can be done and how to accomplish goals is the key to the political art. Four areas of politics concern the fisheries manager.

First, electoral politics are of concern as they would be to any citizen. The process of selecting good leaders is important in a democracy. Second, legislative politics is sometimes important to fisheries management because the manager is called upon to report to the legislature, provide background for pending legislation, or provide expert advice on policy. It is important to understand these processes and to be attuned to the needs of the legislative branch to promote the democratic process. As public servants, managers have an obligation to be well informed and efficient in carrying out legislative programs.

Third, judicial politics is uniquely frustrating and complex. The court system provides a formal means of resolving disputes based on, in part, the rules of advocacy and fair play in debate. As such, there are many rules that seem foreign to the casual observer but that serve to assure each side protection until a judge or jury can make a decision. This is a form of politics that the fisheries manager will likely face in one of two capacities: expert witness or client. As an expert witness, the fisheries biologist is a servant of the court. The biologist will be called to testify based on special knowledge. Imparting that knowledge to the court means using expert scientific judgment to give the court enough information to decide. The adversary process can make this a formidable task. As a client, the biologist is the one who guides the preparation of the case. The client uses lawyers to make sense out of the trial process. It is the client whose interests are represented and who must instruct counsel on pertinent facts, agency policy, and procedures. It is this knowledge that enables lawyers to present an adequate case.

Fourth, implementation politics requires an additional skill. This political arena deals with the process for working out how statutes will be administered. The task of implementing statutes may sound straightforward, but it is not. Yaffee (1982) has demonstrated how even the most clear statute may require interagency bargaining at all phases of implementation. The fisheries manager must come to recognize these bargaining situations and plan for them. This means combining knowledge of both the law and past behavior of agencies to develop and carry through an effective bargaining position.

4.6.3 Expertise of Others

What should be evident from this discussion is that modern fisheries management is multidisciplinary. Certainly the manager—like the fictitious Sharon Douglas described at the beginning of this chapter—must have a strong knowledge base in a number of fields. But more than that, the manager must have a sense of how to seek out needed expertise to complete the administrative task.

It is imperative that the manager has an understanding of the relevant law and how it works. Knowledge of how law is made, interpreted, and enforced is crucial. When faced with a new legal problem, the manager must know where to turn to develop an understanding of the history of the statute or case, its precedents, and relevant procedures. Building this knowledge will lead the manager beyond the law itself into the fields of hydrology, meteorology, chemistry, physics, and other technical subjects. Because management deals with people and agencies, however, problems almost always involve other disciplines, including public administration and applied policy analysis. It is in these two areas that the manager comes to appreciate fully the legal considerations in fisheries management.

4.7 REFERENCES

Barkin, D., and T. King. 1970. Regional economic development: the river basin approach in Mexico. Cambridge University Press, London.

Bean, M. J. 1977. The evolution of national wildlife law. Praeger, New York.

Booth, W. 1988. Reintroducing a political animal. Science 241:156–159.

Burton, E. C. 1984. Enforcement of natural resources litigation. Carswell Company, Ottawa.

Cabrera Acevedo, L. 1978. Legal protection of the environment in Mexico. California Western International Law Journal 8:22–42.

Camp, R. A. 1993. Politics in Mexico. Oxford University Press, New York.

Castro, J. E. 1995. Decentralization and modernization in Mexico: the management of water services. Natural Resources Journal 35:461–487.

Cicin-Sain, B., M. K. Orbach, S. J. Sellers, and E. Mansanilla. 1986. Conflictual interdependence: U.S.–Mexican relations on fishery resources. Natural Resources Journal 26:769–792.

Clarke, J. N., and D. McCool. 1996. Staking out the terrain, 2nd edition. State University of New York Press, Albany.

Comment. 1995. Liberalismo contra democracia: recent judicial reform in Mexico. Harvard Law Review 108:1919–1936.

Cram, J. S. 1971. Water: Canadian needs and resources. Harvest House Ltd., Ottawa.

Cummings, P. A., and N. H. Mickenberg. 1972. Native rights in Canada. General Publishing, Toronto.

DFO (Department of Fisheries and Oceans). 1986. National policy on fisheries. Environment Canada, Ottawa.

Doerksen, H. R., and G. Wakefield. 1975. Columbia basin compact issues review. Pacific Northwest Regional Commission, Washington State Water Research Center, Pullman.

DuMars, C. T., and S. Baltran del Rio. 1988. A survey of the air and water quality laws of Mexico. Natural Resources Journal 28:787–813.

Dworsky, L. B. 1986. The Great Lakes: 1955–1985. Natural Resources Journal 26:291–336.

Elder, P. S., and D. Thompson. 1992. Rivers and Canadian environmental impact assessment. Rivers 3:45–52.

Faudon, E. E., E. F. Tritschler, and G. J. E. R. Michel. 1983. Lago de Chapala: investigacion actualizada 1983. Universidad de Guadalajara, Instituto de Geografia y Estadistica, Guadalajara, Mexico.

Friedrich Ebert Stiftung Foundation. 1990. Desarrollo y medio ambiente en Mexico, diagnostico, 1990. Fundacion Universo Vientiuno, Mexico DF.

Goldfarb, W. 1988. Water law, 2nd edition. Lewis Publishers, Chelsea, Michigan.

Hamilton, M. S. 1980. Summaries of selected federal statutes affecting environmental quality. Colorado State University, Cooperative Extension Service, Fort Collins.

Hogg, P. W. 1985. Constitutional law in Canada. The Carswell Company, Toronto.

Hundley, N. 1975. Water and the west: the Colorado River compact and the politics of water in the American west. The University of California Press, Berkeley.

Johnson, P. M., and A. Beaulieu. 1996. The environment and NAFTA: understanding and implementing the new continental law. Island Press, Washington, D.C.

Juergensmeyer, J., and E. Blizzard. 1973. Legal aspects of environmental control in Mexico: an analysis of Mexico's new environmental law. Pages 101–102 in A. Utton, editor. Pollution and international boundaries. University of New Mexico Press, Albuquerque.

Lamb, B. L., and H. R. Doerksen. 1990. Instream flow uses of water in the United States–water laws and methods for determining flow requirements. Pages 109–116 in J. E. Carr, E. B. Chase, R. W. Paulson, and D. W. Moody, editors. National water summary 1987. U.S. Geological Survey, Water Supply Paper 2350, Washington, D.C.

Landau, J. L. 1980. Empty victories: Indian treaty fishing rights in the Pacific Northwest. Environmental Law 10:413–456.

Lewis, T. N. 1992. Environmental law in Mexico. Denver Journal of International Law and Policy 21(1):159–184.

Miklas, C. L., and S. J. Shupe, editors. 1986. Indian water 1985. AILTP/American Indian Resources Institute, Oakland, California.

Morgan, F. 1970. Pollution-Canada's critical challenge. Ryerson Press and Maclean-Hunter Ltd., Toronto.

Morley, C. G. 1995. The legal framework for public participation in Canadian water management. Inland Waters Directorate, Ontario Region, Water Planning Management Branch, Burlington, Ontario.

Mumme, S. P., C. R. Bath, and V. J. Assetto. 1988. Political development and environmental policy in Mexico. Latin American Research Review 27:7–34.

Nadeau, R. A. 1950. The water seekers. Peregrine Smith, Santa Barbara, California.

Ortiz Martinez, J. M., and M. Guzman Arroyo. 1995. Legislation pesquera. Pages 231–248 *in* M. Guzman Arroyo, editor. La pesca en el lago chapala: hacia su ordenamiento y explotation racional. Universidad de Guadalahara and Comision Nacional del Agua, Guadalajara, Mexico.

Paule, A. 1996. Underground water: a fugitive at the border. Pace Environmental Law Review 13:1129–1171.

Reisner, M. 1986. Cadillac Desert: the American west and its disappearing water. Viking Penguin, New York.

Rohn, P. H. 1983. World treaty index. ABC-Clio Information Services, Santa Barbara, California.

Sadler, B. 1986. The management of Canada–U.S. boundary waters: retrospect and prospect. Natural Resources Journal 26:359–376.

Scott, A., and P. A. Neher. 1981. The public regulation of commercial fisheries in Canada. Minister of Supply and Services, Ottawa.

Shupe, S. J. 1986. Water in Indian country: from paper rights to a managed resource. University of Colorado Law Review 57:561–592.

Simon, J. 1997. Endangered Mexico: an environment on the edge. Sierra Club, San Francisco.

Simonian, L. 1995. Defending the land of the jaguar: a history of conservation in Mexico. University of Texas Press, Austin.

Torres Garcia, F. 1987. El marco legal de la pesca en Mexico. Pages 185–193 *in* Sintesis pesquera, 1982–1987. Secretaria Pesca, Mexico City.

Trelease, F., and G. A. Gould. 1987. Water law case book. West Publications, Minneapolis, Minnesota.

VanderZwaag, D. L. 1983. The fish feud. Lexington Books, Toronto.

Veiluva, M. 1981. The Fish and Wildlife Coordination Act in environmental litigation. Ecology Law Quarterly 9:489–517.

Ward, N. 1987. *Dawson vs the government of Canada*, 6th edition. University of Toronto Press, Toronto.

Wright, A. 1990. The death of Ramon Gonzalez: the modern agricultural dilemma. University of Texas Press, Austin.

Yaffee, S. L. 1982. Prohibitive policy: implementing the federal endangered species act. MIT Press, Cambridge, Massachusetts.

Chapter 5

Ecosystem Management

HAROLD L. SCHRAMM, JR. AND WAYNE A. HUBERT

5.1 INTRODUCTION

The term ecosystem management came into usage in the 1990s, but elements of ecosystem management originate from the 1930s and 1940s (Leopold 1949; Grumbine 1994). Ecosystem management encompasses concepts of biological diversity, ecological integrity, sustainability, watershed management, and landscape ecology; however, it is also a social issue and a philosophy for management. Lackey (1998) suggested that ecosystem management can be viewed as evolution or revolution: an evolution of the concepts of natural resource management or a radical revolution by a minority of people trying to increase support for preservationist values. Indeed, there are widely different interpretations of what ecosystem management is and what it may accomplish. Debate about ecosystem management will continue well into the twenty-first century because society is far from unified in establishing goals for natural resource management. We have taken the position that ecosystem management is evolutionary and is an incremental change in the way resource managers do business; however, the change is substantial. Ecosystem management will involve further evolution of philosophies and practices regarding the use and conservation of natural resources, and fisheries are part of the overall array.

What has fueled the evolution of ecosystem management? Three forces stand out: (1) the biodiversity crisis, (2) recognition of the interconnectedness of ecosystems in time and space, and (3) consideration of human values in the process of natural resource management. Biodiversity is the variety and variability among living organisms and the ecological complexes in which they occur (Grumbine 1992). Biodiversity includes not only different species, unique populations, and gene pools but also the structure and function of all the interdependent parts of ecosystems. In North America, hundreds of millions of dollars have been spent to save individual species, yet biodiversity continues to decline. There is increasing recognition that efforts to restore ecosystems which support imperiled species are necessary to preserve rare species and prevent endangerment of other species. The interconnectedness of ecosystems in time and space is a second stimulus for ecosystem management. Fisheries managers are well aware of the interconnectedness of ecosystems because activities on land ultimately affect the waters. Although recognized, this interconnectedness, or holistic landscape perspective, has been rarely addressed in fisheries management. Failure to address the larger ecosystem stems to a large extent from the limited authority of fisheries management agencies and the focus on single-species management driven by sport and commercial fishing interests. Consideration of human values is a third force driving ecosystem management. Human values are essential for setting policy, establishing laws, and making management decisions.

Half a century ago, Leopold (1949) stated that "a thing is right when it tends to preserve the integrity, stability, and beauty of the biotic community. It is wrong when it tends otherwise." This statement holds the philosophical basis for Leopold's ethic for managing the land. The land ethic requires a basic respect for all forms of life and concerted and serious efforts to minimize impacts of human actions on the other members of the biotic community (Beatley 1994). The obligations extend to individual organisms, species, and the larger ecosystems that sustain them. Over the last 50 years, this philosophy has grown and diverged in several ways. Nevertheless, people are recognizing the benefits, both tangible and intangible, that are tied to natural resources and the present and future benefits of stewardship of these resources.

5.2 DEFINITIONS OF ECOSYSTEM MANAGEMENT

Ecosystems can be viewed as having three primary attributes: (1) composition, (2) structure, and (3) function (Franklin 1981; Grumbine 1992, 1994, 1997). Ecosystem components (i.e., composition) are the habitat, which provides the physical and chemical conditions necessary for existence, and the inhabiting species in all their variety and richness. Structure refers to the patterns and organization of the life forms. Functions are the jobs the various organisms perform to maintain a system, such as the cycling of nutrients or the capture and flow of energy. Humans have intervened to alter the composition, structure, and function of ecosystems in an effort to attain values important to human survival and well-being. The actions of humans to manipulate ecosystems to achieve benefits is natural resource management.

All management is a process. The process of ecosystem management has a vision, or a goal, that includes long-term sustainability of ecosystems and human communities mediated by societal values. Ecosystem management is centered on the compatibility of long-term sustainability of ecosystems (and their attendant biological diversity) and human communities. Crucial in the ecosystem management process is recognition of human values in establishing management goals and objectives (see Chapter 2). Presenters at an ecosystem management symposium (1995 American Fisheries Society annual meeting) strongly supported the benefits of managing fisheries resources from an ecosystem perspective (Schramm and Hubert 1996). They suggested that ecosystem management includes both goals and processes. A strict definition of ecosystem management may not be necessary, but definitions can help us understand the components of ecosystem management and, thereby, aid in establishing goals and developing processes for achieving the goals. Several definitions have been proposed that encompass ecosystem management philosophy.

> [Ecosystem management is] a proactive approach to ensuring a sustainable economy and a sustainable environment. (Vice President Gore's National Performance Review; Interagency Ecosystem Management Task Force 1995:1)

[Ecosystem management is a management strategy to] preserve, restore, or where those are not possible, simulate, ecosystem integrity as defined by composition, structure, and function that also maintains the possibility of sustainable societies and economies. (Keystone Forum on Ecosystem Management; U.S. General Accounting Office 1994:38)

[Ecosystem management] integrates scientific knowledge of ecological relationships within a complex sociopolitical and values framework toward the general goal of protecting native ecosystem integrity over the long-term. (Grumbine 1994:31)

[Ecosystem management is] the application of ecological and social information, options, and constraints to achieve desired social benefits within a defined geographic area and over a specified period. (Lackey 1998:29)

Common to all of the definitions are the concepts ecosystem, sustainability, and societal[1] values. From this commonality of concepts, we can derive a goal or vision for ecosystem management:

To maintain ecosystems that will support diverse human societal benefits for present and future generations.

This goal is pragmatic. Human benefits, both tangible (e.g., fish, water, agricultural crops, and minerals) and intangible (e.g., aesthetics, the knowledge that species unseen or undescribed are being conserved), are included because there is no basis for management without providing benefits to humans (see Chapter 2).

Ecosystem integrity is a concept that is also included in the definitions of ecosystem management (U.S. General Accounting Office 1994; Grumbine 1994). Various definitions of ecological or biological integrity have been offered, two of which are "maintenance of the community structure and function characteristic of a particular locale or deemed satisfactory to society" (Cairns 1977:171) and "the capability of supporting and maintaining a balanced, integrated, adaptive community of organisms having a species composition and functional organization comparable to that of the natural habitat of the region" (Karr and Dudley 1981:55). Ecosystem integrity recognizes that ecosystems are adaptable, have a capacity to return to a functional state after perturbation, and have the ability to respond to stresses. Today, ecosystems do not exist without human influences (Regier 1993). Some human activities are compatible with ecosystem function and do not alter ecosystem integrity. Still other activities alter ecosystem structure or function, but the ecosystems are able to accommodate these stresses and return to a state similar to the unaltered state; thereby, ecosystem integrity is maintained. Of greatest concern are human activities that exceed the level of stress that can be accommodated by an ecosystem and lead to ecosystem disintegration. The concept of sustained use (e.g., sustainable societies or sustainable economies) can be compatible with ecosystem integrity. The adaptability of ecosystems allows for numerous uses by humans, as long as the uses do not exceed the ecosystem's capacity to maintain

---

[1]Throughout this chapter, society or social includes not only people, individually or in groups or cultures, but also economic and political concerns.

a fully integrated state (i.e., a state with complete structure and function). "Satisfactory to society" (see Cairn's definition above) is a matter of values and is subject to diverse influences. Society's values, and subsequently definitions of what is satisfactory to society, differ from nation to nation, among local, state and provincial jurisdictions, and over time.

5.3   IMPLEMENTATION OF ECOSYSTEM MANAGEMENT

5.3.1 A Collaborative Process

The fisheries management process described by Krueger and Decker (Chapter 2) is germane to ecosystem management. Ecosystem management mandates that the management process be used to the fullest extent and necessitates a team approach.

Because societal values are an important component of ecosystem management, consideration of human dimensions is essential. Processes for identifying societal values and incorporating them into natural resource management are beginning to emerge. For example, in *Reinventing Government*, Osborne and Gaebler (1992) acknowledged that cynicism about government runs deep, and they outlined several principles by which the public's interests can be inserted into the work of government; all of their principles apply to natural resources and fisheries management. They suggest that government should be catalytic and serve the function of guiding management rather than carrying out the management activities. They also suggested that the decision-making authority of government be returned to the community level, that government become mission driven instead of rule driven, and that the voice of the stakeholders must be incorporated into the governing process. Processes for ecosystem management are engendered in the principles presented by Osborne and Gaebler.

The spatial scale of ecosystem management requires the involvement of multiple agencies and jurisdictions, as well as private landowners and a wide array of stakeholders (O'Connell and Noss 1992; Montgomery et al. 1995). For example, many fisheries management issues are habitat problems (e.g., eutrophication and sedimentation), and the sources of the problems are land uses throughout a watershed. Fisheries management alternatives for streams can be identified to fix *symptoms* (e.g., stocking fish to supplement recruitment lost due to effects of sediment on spawning habitat or constructing instream habitat to enhance spawning sites temporarily), but long-term solutions require fixing the *sources* of the problems; that is, working on the watershed to reduce erosion and sediment delivery into streams. Watershed scale management requires a framework to structure watershed analysis and subsequent management activities (Montgomery et al. 1995). Such a framework involves consideration of landscape level ecological processes, history, current conditions, and definition of potential responses to management activities. Because watersheds can range from relatively small (<100 ha for a first-order stream) to extensive (41% of the land area of the continental United States for the Mississippi River basin), the magnitude may become overwhelming, thereby requiring stratification of the watershed into manageable units.

As fisheries managers look beyond the water, they are confronted with management issues that exceed their legislated authorities. Activities on watersheds are generally controlled by private landowners and regulated to some degree by municipal governments (e.g., zoning); state departments of agriculture, forestry, or transportation; or one or more federal government agencies. In the eastern two-thirds of the United States, most land is in private ownership. Landowners are independent, but they may be represented by organizations such as producer groups or a soil and water conservation district. Large watersheds commonly are under the jurisdiction of two or more states or provinces or even nations. Consequently, there is substantial jurisdictional fragmentation with many administrative players (Grumbine 1992) as well as numerous private interests to consider when managing at the watershed scale (O'Connell and Noss 1992). Some associations (e.g., commissions, councils, and organizations) of different interests or different governments have been formalized to facilitate the collaborative processes; comprehensive listings of these associations are available in the *Conservation Directory* published annually by the National Wildlife Federation.

From consideration of the social, spatial, and political dimensions of ecosystem level management, it should be apparent that collaborative processes are necessary to identify and solve issues. The benefits of collaborative processes are those of the management team approach described in Chapter 2. Effective participation in a collaborative process may pose a serious challenge for some fisheries scientists; they must realize that science can do the job it was designed to do only when we have resolved conflicts and can accept scientific data for what they are meant to be: tentative insight into universal relationships (Maser 1996).

5.3.2 Education and Communication

The ecosystems that result from a society intent on rapid and unplanned development for short-term profit differ from the ecosystems maintained by a society that understands the values of clean air, clean water, productive lands, and biological diversity. Although some societies or portions of societies are becoming more cognizant of the need to maintain ecosystem integrity, others are less concerned about the future. In North America, environmental consciousness appears to be growing (National Environmental Education and Training Foundation 1996), but it is questionable whether a majority of North Americans are sufficiently supportive of a healthy environment to maintain the integrity of our diverse ecosystems. Environmental education efforts are being incorporated into the programs of private, local, state, and federal entities responsible for natural resource management and are creating awareness for environmental issues. Although significant, these programs need to be broadened to reach the large majority of people.

Achieving increased environmental awareness and improving communication of resource management needs and strategies will be difficult. For fisheries managers, as for others engaged in resource management, this task is made more difficult because they must consider nontraditional stakeholders and a new market for the managers' services. Most fisheries management agencies have become adept at communicating with recreational anglers and commercial fishers. However, anglers are only 17% of the U.S. population (U.S. Fish and Wildlife Service 1997) and 21% of the Canadian population (Department of Fisheries and Oceans 1994); similar statistics are not available for Mexico. If the integrity of aquatic ecosystems depends on society's values, environmental awareness and public support are needed from a much larger proportion of the population. Therefore, education and communication efforts must not only increase, but the focus of these efforts and the message must change. For example, an acceptable message like "we need to do this to improve fishing for species A" might be replaced by "we need to do this to create a healthier ecosystem with good water quality that will support diverse and stable fish populations and produce better fishing for species A." Communication and education must be high priorities for management teams (see Chapter 3), which should include people skilled in communications. The business community fully understands the importance of marketing, and partnerships with them (e.g., sportfishing, tourism, and real estate industries) will be productive.

The need for communication within agencies, among agencies, and with the public is an integral component of ecosystem management (Grumbine 1997). Some fisheries management agencies have been daunted by, and possibly are resistant to, such change, but most are initiating programs to obtain public input in their decision-making processes (Wilde et al. 1996). Agencies are employing communications specialists and providing in-service training for their employees to communicate more effectively within agencies, among agencies, and with special interest groups and the general public. Employees trained in fisheries management or other ecologically based disciplines *and* in communications or education are becoming more prevalent in management agencies and are contributing to environmental education, effective internal and external communications, conflict resolution, and collaborative management. Fisheries managers will be expected to have skills far beyond fisheries science as ecosystem management principles and processes are adopted.

5.3.3 Opportunities

Many fisheries managers have been frustrated because the limited authority possessed by their employing agencies often prevents them from addressing the sources (rather than the symptoms) of the fisheries problems they face. Ecosystem management will require managers to work across institutional boundaries and to direct management to a broad geographic scale (Grumbine 1997). As a result, opportunities will exist to address and solve persistent fisheries and environmental problems that have been beyond the abilities of most managers constrained by politically based organizational and jurisdictional boundaries.

The habitats that fisheries managers work with today are often drastically different from the habitats present less than a century ago. Eutrophication of lakes and reservoirs is typically viewed with disdain by a large segment of the public. Although a natural process (Sawyer 1966), cultural influences have accelerated eutrophication. Eutrophication can be reversed and lakes returned to an improved state of ecological integrity. Projects to reduce nutrient inputs and restore lakes (supposedly to more "natural" conditions) are ongoing throughout the United States with the assistance of the U.S. Environmental Protection Agency's Clean Lakes Program. Similarly, cooperative efforts between the United States and Canada through the Great Lakes Water Quality Agreement have improved water quality in the Great Lakes. Lakes Okeechobee and Apopka in Florida are examples of ongoing eutrophication reduction efforts, and Lake Erie is a recent lake restoration success story (see Chapter 23). Although an expensive and slow process, restoration of eutrophic lakes has considerable public support because a large segment of the public values clean water.

Another fisheries management problem involving even larger spatial scales is the loss of habitat and fragmentation of large rivers as a result of construction of dams and reservoirs (see Chapter 20). From 1930 to 1970 dam building flourished as impoundments (reservoirs) were built for reliable sources of water for domestic, agricultural, and industrial use, electricity production (hydropower generation and cooling reservoirs), flood control, and commercial navigation. Reservoirs provided expanded recreational opportunities, much more than could be accommodated by the small streams before they were impounded. However, impoundments eliminated vast reaches of lotic habitats and reduced or extirpated fauna dependent on flowing-water habitats to complete their life cycle. Dams blocked migration and fragmented habitats, resulting in loss of fish stocks and genetic potential. Altered hydrographs affected spawning and recruitment of resident and migratory species. Seizing the opportunities to maximize the fisheries potential of reservoirs, fisheries managers relied heavily on stocking fishes adapted to the newly created lentic conditions, further stressing the native fauna. Restoration of streams and rivers affected by dams is a dilemma when managing for ecosystem integrity. On one side of the value equation are natural habitats and biodiversity lost 20–60 years ago (most of the people living in the Colorado, Columbia, or Tennessee river watersheds have never seen these rivers free flowing); on the other side are cheap electricity, inexpensive commerce, flood control, reliable water supplies, and bountiful recreational opportunities. Solving this dilemma will call for all the collaborative skills and ecological knowledge that fisheries scientists, natural resource management agencies, and stakeholders can muster. It will require ecosystem management.

5.3.4 Constraints

One constraint to ecosystem management is identifying the appropriate spatial scale for management. The spatial scope may be formidable: the ecosystem could be as large as the Great Lakes basin, Chesapeake Bay, or the North Pacific Ocean. Although management of large multijurisdictional resources faces many challenges, significant progress has been made in conserving or restoring fish stocks and the ecosystems that support them at large spatial scales (see Box 5.1).

The temporal scope of ecosystem management is also a problem because the success of management may not be realized for many years, maybe several generations. Society has become accustomed to relatively quick technological fixes. Many fisheries managers and their angling constituents are also accustomed to seeing relatively quick changes. For example, measurable changes in short-lived sport fish populations can be expected 2–4 years after implementation of an appropriately chosen harvest regulation. Ecosystem management generally will require even longer time frames and greater patience among all participants in the process.

Working in collaborative teams can be a long and tedious process. Effective participation in collaborative teams will require facilitation, negotiation, and conflict resolution skills currently possessed or accepted by few fisheries managers. Fisheries managers will be working with agencies, organizations, and stakeholders whose missions, perspectives, paradigms, and methods are foreign. Likely, effective partnerships will not just happen because different agencies or organizations are part of a management team; however, productive relationships among different concerns will evolve if mem-

Box 5.1   Chesapeake Bay Restoration *

Chesapeake Bay is the largest estuary in the United States. Its shoreline stretches 300 km through Maryland and Virginia, and the bay's 166,000 km^2 watershed is shared by six states and the District of Columbia. The Chesapeake Bay ecosystem provides habitat for more than 2,000 species of plants and animals.

With its abundant fish and shellfish resources, easy accessibility by water, and natural shipping ports, Chesapeake Bay has been a center of commerce and an area of high human growth since Captain John Smith arrived there in the seventeenth century. Although the consequences of intense human use and industrialization were recognized early, it wasn't until the 1970s that agencies such as the U.S. Environmental Protection Agency (EPA) began to focus research efforts on the bay as an ecosystem. By that time, conditions were at critical stages. Nearly all oyster reefs (which provided not only oysters but also important fish habitat and filtering of bay water) had disappeared due to intense oyster harvesting. Anoxic conditions in many parts of the estuary impeded aquatic life, and toxic chemicals were still being released into the bay's waters. Submerged aquatic vegetation was gone from most of the Bay, the striped bass population was declining to precariously low levels (see Chapter 24), and spawning runs of American shad had disappeared from major river systems.

In 1987, the District of Columbia, Maryland, Pennsylvania, Virginia, the EPA, and the Chesapeake Bay Commission (a tri-state legislative advisory commission) signed the Chesapeake Bay Agreement, initiating a coordinated restoration program for this invaluable ecosystem. The agreement specified control of point and nonpoint source pollution, management of human population growth in the watershed, restoration of living resources, creation of programs to foster increased citizens' enjoyment and responsibility for the bay, and provision for opportunities for citizen participation in decisions and programs affecting the bay.

Over the following 10 years, wastewater treatment plants installed advanced biological nutrient removal systems, phosphorous and nitrogen levels were reduced, best management practices were implemented on 648,000 ha of agricultural lands, urban runoff was curbed by innovative landscape and construction technologies (cooperatively developed by government agencies, citizen groups, and the private sector), industrial chemical releases declined 65%, submerged aquatic vegetation rebounded 70% from its 1984 low point, and more than 460 km of tributary streams were reopened to passage of anadromous fishes. Furthermore, many goals set for the year 2000 and beyond are on, or ahead of, schedule.

Because 90% of the Atlantic Coast striped bass historically were produced in Chesapeake Bay, restoration of Chesapeake Bay was necessary to provide essential spawning and nursery habitat. However, striped bass and many other fish stocks in Chesapeake Bay migrate to Atlantic coastal waters; hence, successful restoration and management of these species required collaboration with other states. In 1942, the 13 Atlantic states and the District of Columbia joined under an interstate compact (Atlantic States Marine Fisheries Commission, ASMFC) for the purposes of developing cooperative fisheries management and data-sharing programs. Before 1984, implementation of fisheries management plans developed through the ASMFC was voluntary and only marginally successful. In 1984, the Atlantic Striped Bass Conservation Act was signed into federal law and required states to implement and enforce the cooperative Atlantic Striped Bass Management Plan.

The combined effects of a cooperative Chesapeake Bay ecosystem restoration program and interstate and federal fisheries management have reversed the decline of Atlantic Coast striped bass. The striped bass population has increased almost 800% in the 15 years since 1982. Juvenile production has steadily increased, with reproduction indices in 1993 and 1996 higher than any since records began in 1954. The number of age-classes in the spawning stock, important for stable reproduction, have increased from 2 to 3 in the early 1980s to 6 in 1997. Likewise, other fisheries have begun to rebound. The Susquehanna River stock of American shad had declined to fewer than 6,000 fish. Through 15 years of intense management, habitat restoration, and fish passage development, the population increased to more than 200,000 fish in 1997.

Despite the biological, social, economic, and political complexity, the Chesapeake Bay states recognized that an ecosystem perspective was the only truly effective approach to restoring and managing such a large system. The enormous task of managing the bay and its resources from the ecosystem perspective involved multiple state and municipal governments along with the federal government and also required the support of and participation by the 15 million citizens of the bay region. Due to the commitment of each of these partners, sound planning, measurable goals, and dedicated financial resources, the state of the Chesapeake Bay has improved tremendously and is expected to continue to improve under such cooperative management.

* Written by Andrew J. Loftus, Chair of the Citizen's Advisory Committee to the Chesapeake Executive Council, 3116 Munz Drive, Suite A, Annapolis, Maryland 21403; and Marjorie S. Adkins, Director of the Chesapeake Regional Information Service, Alliance for the Chesapeake Bay, Post Office Box 1981, Richmond, Virginia 23218.

bers of a management team can discard destructive conflicts and focus on issues, goals, and objectives. Patience and confidence in achieving a better world (i.e., sustainable communities and ecosystems) are central to the process.

Policy must precede management. Government policy encouraging collaboration and cooperation within and among agencies and, most importantly, with the public will be necessary for the initiation of ecosystem management philosophies (Osborne and Gaebler 1992). Currently, collaborative processes are minimally applied through the National Environmental Policy Act, Endangered Species Act, Fishery Conservation and Management Act, and other legislation guiding state, provincial, and federal land management agencies. International commissions such as the Great Lakes Fishery Commission have been created to foster multinational collaboration (see Chapter 23).

Although diversity within management teams (see Chapter 2) can be a constraint, it provides opportunity. Diverse perspectives help ensure that appropriate goals and objectives are selected and that problems associated with the implementation of management actions are recognized in advance. The diversity of perspectives also provides a wider array of skills and techniques for addressing problems and implementing management plans.

Funding of ecosystem management is a significant issue for two reasons. First, ecosystem management requires substantially more funding than is currently appropriated for environmental conservation. By definition, ecosystem management is much broader than the more traditional, relatively simple fisheries management approach of maximizing fisheries output within the confines of a water body. Managing for ecological integrity, which for many ecosystems will involve ecosystem restoration, will be extremely expensive and require long-term funding commitments. Certainly, more staff will be needed by natural resource management agencies to carry out expanded responsibilities. These needs will include not only more biologists, researchers, and managers fulfilling rather traditional roles and responsibilities, but also administrative staff especially skilled to participate on management teams and personnel to fulfill new and important responsibilities (e.g., assessing social preferences and education).

At present, most funding for conservation and management of fish and wildlife resources in the United States comes from federal appropriations to land management agencies (the Bureau of Land Management, Bureau of Reclamation, National Park Service, Forest Service, and Fish and Wildlife Service) and the Department of Defense, fishing and hunting license revenues, and the Sport Fish and Wildlife Restoration Program. Conservation activities of federal, state, and provincial agencies are already constrained by insufficient funding. Funds for enhanced staffs will have to come from the public at large and will require substantial policy changes. In the United States, the Teaming with Wildlife initiative may be a way to obtain funding for nongame wildlife conservation. This national initiative, modeled after the Sport Fish and Wildlife Restoration Program, initially proposed to obtain funds from nonconsumptive wildlife recreationists and the industries that manufacture the goods they use. The congressional responses in 1998 and 1999 proposed instead that federal royalties from offshore oil and gas leases provide the funding rather than expanded user fees. The final outcome is still to be determined. Although a significant first step, a larger and broader funding base will be needed to accomplish ecosystem management. Missouri and Arkansas (Box 5.2) have established the necessary broad funding base and provide models for other states.

The second funding concern stems from traditional sources of funding and accountability. For at least the past 2 decades, many fisheries managers have struggled with integrating sport fish management with management for biodiversity and sensitive species. The issue has not been what should be done (i.e., management to ensure ecosystem integrity and conservation of natural resources for present and future generations) but accountability to the bill-paying stakeholders–anglers who buy licenses. For example, operational funds for most fisheries management agencies in the United States come largely or entirely from fishing license sales. Although many anglers are relatively supportive of environmental issues, they also expect a return on the investment they perceive they are making for recreational fishing. Given limited funding and staff, accountability to licensed anglers often means sport fishes take precedence over darters, pupfishes, mussels, and other nonsport species. Given traditional funding mechanisms, hard decisions will have to be made between maintaining or restoring native biodiversity and sport fish management. Implicit in ecosystem management philosophy is achieving recognition among diverse stakeholders (and, in some cases, resource managers) that biodiversity and sport fish management are compatible goals.

5.4   CONCLUSIONS

The inception of Optimum Sustained Yield (OSY) as the predominant fisheries management philosophy in the 1970s (see Chapter 1) marks movement by progressive fisheries managers to adopt an ecosystem management philosophy. Although OSY

Box 5.2   Funding Natural Resources that Benefit Everyone

The first of July in 1996 marked the dawn of a new era for the Arkansas Game and Fish Commission (AGFC) and the future of Arkansas's fish and wildlife resources. It was on this day that a one-eighth of a penny conservation sales tax went into effect, assuring the continuance and expansion of scientific fish and wildlife management practices in Arkansas. The statewide tax increased the AGFC's budget by almost 50% (from US$28 million per year to almost $45 million per year). These funds will be used exclusively to protect, enhance, and promote Arkansas's treasured natural resources.

Before passage of the Conservation Amendment, Arkansas voters were polled as to their opinion on AGFC's functions and programs and within what areas the public wanted to see the AGFC doing "more." Four topics stood above the rest: more enforcement, more public places to hunt and fish, more conservation efforts for endangered and nongame species, and more conservation education. As a result, the AGFC developed a plan for conservation based on these priorities and submitted it to the voters as a promise for the future.

Keeping with this promise, new tax monies were budgeted to hire additional wildlife officers and endangered and nongame species staff, to purchase land from willing sellers, to enhance AGFC-owned lakes and wildlife management areas, to develop new lakes and public fishing areas, to establish private land management programs, and to institute new and innovative conservation programs, such as "Hooked On Fishing—Not On Drugs." In one fell swoop, the new money enabled the AGFC to integrate funding for high-priority resource management, nongame and endangered species management, conservation education, public and private lands programs, and regulations enforcement based on a dynamic new source that is totally unattached to traditional funding sources.

The new money in effect enabled the AGFC to expand its constituent tent and include all stakeholders in comprehensive resource management. The inclusion of various stakeholders into the plan for conservation makes the management of biodiversity and sensitive species not only acceptable but understandably practical.

---

does not specify the need to manage at the ecosystem level, it clearly states the need to manage for diverse values: "a deliberate melding of biological, economic, social, and political values designed to produce the maximum benefit to society from a given stock of fish" (Roedel 1975:85). Although the OSY concept has been accepted by the fisheries management community, implementation has been slow, and many management activities are still directed at maximizing yields of single species (Malvestuto and Hudgins 1996). Most fisheries managers recognize the benefits of habitat management, and they recognize that good habitat benefits a broad diversity of fauna. Similarly, fisheries managers recognize that activities on the watershed affect life in the water. However, limited funding, constrained authority, and public (angler) pressure have resulted in a focus on sport fishes, sometimes with excessive management of single species.

Introduction of nonnative fishes to enhance sportfishing opportunities (see Chapter 13) is an example of attempting, and sometimes succeeding, in satisfying millions of anglers but failing to consider ecosystem management concepts. In the United States, the number of nonnative species stocked and managed for sportfishing ranges up to 30 among the states, and nonnative sport fishes account for at least 50% of the recreational fishery program in 18 states (Horak 1995). This situation is most prevalent in western states, some of which would have only three or four sport species without introductions. These introduced species have contributed to reductions or even extinctions of native fauna (Lassuy 1995). The dilemma currently faced by most, if not all, fisheries managers is managing for ecological integrity while maintaining sportfishing opportunities. On the social side of this issue is the challenge of increasing the public support for natural resource conservation without losing the support of traditional stakeholders in recreational, subsistence, and commercial fisheries.

As demonstrated by the restoration and management of Apache trout in Arizona, multiple agencies can work across jurisdictional boundaries to restore a declining species while building the support of anglers (Carmichael et al. 1995; Rinne and Janisch 1995). Affected by abusive grazing and logging on the watershed and competition, predation, and hybridization with nonnative salmonids, native Apache trout stocks declined and the species was listed as endangered under the Endangered Species Act (16 U.S.C.A. §§ 1531–1534). A recovery plan was developed and the species was downlisted to threatened status in 1975. Through cooperation among the Arizona Game and Fish Department, Trout Unlimited, Forest Service, Fish and Wildlife Service, and White Mountain Apache Tribe, instream and watershed habitat improvements have contributed to self-sustaining Apache trout populations, the abundance and the number of sites with Apache trout populations have increased, and the genetic integrity of Apache trout has been ensured by good hatchery practices and reduced (generally eliminated) stocking of nonnative salmonids into Apache trout habitats. A combination of stocking and restrictive harvest regulations have allowed the expansion of an Apache trout sport fishery while accomplishing the restoration of this threatened species. The enhanced sport fishery has generated the support of anglers for the restoration of Apache trout as well as conservation and restoration of other native Arizona fishes. The success of this ongoing restoration effort depends on holistic habitat management, long-term planning, substantial public involvement, and collaborative management; in other words, ecosystem management.

5.5 REFERENCES

Beatley, T. 1994. Ethical land use. Principles of policy and planning. Johns Hopkins University Press, Baltimore, Maryland.

Cairns, J. 1977. Quantification of biological integrity. Pages 171–187 in R. K. Ballentine and L. J. Guarraia, editors. The integrity of water. U.S. Environmental Protection Agency, Office of Water and Hazardous Materials. U.S. Government Printing Office, Washington, D.C.

Carmichael, G. J., J. N. Hanson, J. R. Novy, K. J. Meyer, and D. C. Morizot. 1995. Apache trout management: cultured fish, genetics, habitat improvements, and regulations. Pages 112–121 in Schramm and Piper (1995).

Department of Fisheries and Oceans. 1994. 1990 survey of recreational fishing in Canada. Economic and Commercial Analysis, Report 148, Economic and Policy Analysis Directorate, Ottawa.

Grumbine, R. E. 1992. Ghost bears. Exploring the biodiversity crisis. Island Press, Washington, D.C.

Grumbine, R. E. 1994. What is ecosystem management? Conservation Biology 8:27–38.

Grumbine, R. E. 1997. Reflections on "what is ecosystem management?" Conservation Biology 11:41–47.

Horak, D. 1995. Native and nonnative fish species used in state fisheries management programs in the United States. Pages 61–70 in Schramm and Piper (1995).

Interagency Ecosystem Management Task Force. 1995. The ecosystem approach: healthy ecosystems and sustainable economies. U.S. Government Printing Office, Washington, D.C.

Karr, J. R., and D. R. Dudley. 1981. Ecological perspective on water quality goals. Environmental Management 5:55–68.

Lackey, R. T. 1995. Ecosystem management: implications for fisheries management. Renewable Resources Journal 13(4):11–13.

Lackey, R. T. 1998. Seven pillars of ecosystem management. Landscape and Urban Planning 40:21–30.

Lassuy, D. R. 1995. Introduced species as a factor in extinction and endangerment of native fish species. Pages 391–396 in Schramm and Piper (1995).

Leopold, A. 1949. A Sand County almanac and sketches here and there. Oxford University Press, New York.

Malvestuto, S. P., and M. D. Hudgins. 1996. Optimum yield for recreational fisheries management. Fisheries 21(6):6–17.

Maser, C. 1996. Resolving environmental conflict. St. Lucie Press, Delray Beach, Florida.

Montgomery, D. R., G. E. Grant, and K. Sullivan. 1995. Watershed analysis as a framework for implementing ecosystem management. Water Resources Bulletin 31:369–386.

National Environmental Education, and Training Foundation. 1996. Report card: environmental attitudes and knowledge in America. Roper Starch, New York.

O'Connell, M. A., and R. F. Noss. 1992. Private land management for biodiversity conservation. Environmental Management 16:435–450.

Osborne, D., and T. Gaebler. 1992. Reinventing government. Plume, New York.

Regier, H. A. 1993. The notion of natural and cultural integrity. Pages 3–18 in S. Woodley, J. Kay, and G. Francis, editors. Ecological integrity and the management of ecosystems. St. Lucie Press, Delray Beach, Florida.

Rinne, J. N., and J. Janisch. 1995. Coldwater fish stocking and native fishes in Arizona: past, present, and future. Pages 397–406 in Schramm and Piper (1995).

Roedel, P. M. 1975. A summary and critique of the symposium on optimum sustainable yield. Pages 79–89 in P. M. Roedel, editor. Optimum sustainable yield as a concept in fisheries management. American Fisheries Society, Special Publication 9, Bethesda, Maryland.

Sawyer, C. N. 1966. Basic concepts of eutrophication. Journal of the Water Pollution Control Federation 38:737–744.

Schramm, H. L., Jr., and W. A. Hubert. 1996. Ecosystem management: implications for fisheries management. Fisheries 21(12):6–11.

Schramm, H. L. Jr., and R. G. Piper, editors. 1995. Uses and effects of cultured fishes in aquatic ecosystems. American Fisheries Society, Symposium 15, Bethesda, Maryland.

U.S. Fish and Wildlife Service. 1997. 1996 national survey of fishing, hunting, and wildlife-associated recreation. U.S. Government Printing Office, Washington, D.C.

U.S. General Accounting Office. 1994. Ecosystem management: additional actions needed to adequately test a promising approach. U.S. General Accounting Office, GAO/RECD-94–111, Washington, D.C.

Wilde, G. R., R. B. Ditton, S. R. Grimes, and R. K. Riechers. 1996. Status of human dimensions surveys sponsored by state and provincial fisheries management agencies in North America. Fisheries 13(11):12–17.

FISHERY ASSESSMENTS

Chapter 6

Dynamics of Exploited Fish Populations

MICHAEL J. VAN DEN AVYLE AND ROBERT S. HAYWARD

6.1 INTRODUCTION

Knowledge about the dynamics of fish populations is essential for developing management plans and evaluating management success. In the context of fisheries management, population dynamics includes estimation of changes in population numbers, composition, or biomass. Population dynamics can be subdivided into two areas having different applications in fisheries management. One, termed population assessment, focuses on estimating characteristics of fish populations and comparing those characteristics with other populations or with the same population over time. Assessments are important for describing fluctuations in abundance, measuring responses to exploitation or other perturbations, and defining fishery components that could be successfully managed. The second area of population dynamics is the use of mathematical models to predict future trends. Models have been used to forecast changes in fish population size and harvest by assuming certain levels of fishing effort. Estimates of parameters in the models are obtained from assessment studies. The most common application of forecasting techniques has been to predict yield of commercial fisheries, but several models have proven useful for predicting trends for sport fisheries. This chapter focuses on single-species population dynamics. Although fisheries management philosophy is moving toward multispecies approaches, analyses of single-species' dynamics are commonly used and serve as the basis for most multispecies approaches (see Hilborn and Walters 1992).

6.2 CHARACTERISTIC DYNAMICS OF INLAND FISH POPULATIONS

6.2.1 Population Biology and Regulation

Exploitation reduces the number of harvestable-size fish in a body of water, but its influences on the long-term stability of a population are rarely obvious. Models developed for predicting effects of harvest are based on the premise that a population, even when not fished, remains at or near some equilibrium level of abundance. Populations that fluctuate randomly or cyclicly above and below a long-term equilibrium level can also be regarded as stable.

Processes that regulate population size can be categorized as density dependent or density independent. Density-dependent factors, such as food availability, predation, cannibalism, diseases, parasites, and availability of spawning sites, vary with population size. These factors usually operate in a compensatory manner, so that extremes in population size are moderated by their action. For example, with increasing fish population density, food availability per fish declines, leading to slower growth and poorer condition of the surviving fish. These responses may, in turn, lead to increased vulnerability to predation or cause delayed sexual maturation, which would cause a decline in population size because of reduced rates of survival and reproduction. A low population density could lead to rapid growth and maturation, relatively high survival and reproduction, and increased population size.

Density-independent processes are not affected by population density. Factors such as water temperature, river flows, lake levels, drought, and other features of the environment may affect a population in ways that are not influenced by the number of fish present.

The relative importance of density-dependent and density-independent factors in regulating population size can vary among ecosystems and life stages of a species. In systems where the environment is relatively stable or undergoes recurring long-term cycles, density-dependent processes tend to produce an equilibrium level about which the population varies in response to density-independent environmental factors. Because oceans and large lakes provide relatively constant environments they often support fish populations that are regulated in this manner. An exception to this relative constancy is the well-known El Niño effect, which can substantially alter the thermal environment. Ecosystems with relatively unstable and unpredictable physical characteristics have fish populations that are regulated to a greater extent by density-independent factors. Examples of these systems are streams, rivers, and some reservoirs, which are subject to influence by storms, drawdowns, and temperature changes as well as other weather factors.

For many freshwater fishes, the period of reproduction, including spawning and egg incubation, occurs during a relatively short period in spring when weather conditions, water levels, turbidity, and other factors can be unstable. Hence, egg hatching success and the resultant number of young produced may be a function of density-independent factors. Survival of juveniles and adults is generally thought to be regulated primarily by density-dependent processes, although in many freshwater situations the most pronounced influence of increased population size is reduced growth. Subsequently, reduced growth can lead to increased mortality from predation because smaller fish are more vulnerable. Older fish are better capable of surviving or avoiding extremes of physical environmental factors than are eggs and larvae.

Most freshwater fish populations are characterized by considerable variation in the number of young produced annually. Such variation is most pronounced for species with brief spawning periods or those that spawn in variable, unpredictable environments. The extent to which variation in reproduction influences the adult population depends on the rate of survival of the group of fish spawned in a given year (termed a cohort or year-class) and the age structure of the population.

The overall population abundance of a long-lived species will show relatively less annual variation than will a short-lived species subjected to the same annual variation of reproduction.

A key requirement for management of a fishery is knowledge of the processes and factors that control survival of young fish to the age at which they are mature or reach a desired size. The relative abundance of a year-class (termed "year-class strength") at any early developmental stage may show no obvious relationship to either the abundance of the spawning population that produced it or the number of fish that eventually is added to the adult population; therefore, knowledge of the processes that regulate the dynamics of year-classes during their development is required to prescribe effective management programs.

The idea that year-class strength is established during some specific, relatively distinct phase of a species' life cycle has been applied to fish populations. The term critical period is defined as the time when natural regulatory factors determine the eventual abundance of a cohort. The concept that a critical period exists during the early life of fishes is consistent with the belief that natural processes of population regulation have their greatest influence on the youngest life stages.

The critical period has usually been postulated to occur during early larval development, at a time when the fish become reliant on exogenous food (Cushing and Harris 1973). Initially, larvae use energy contained in the yolk sac to develop functional mouth parts and become capable of swimming and foraging for food. At the point of transition to external food, larval energy reserves are low, and the fish are vulnerable to weather extremes, food shortages, and predation. Hence, biologists have often concluded that the number of fish surviving to juvenile and adult stages is functionally determined during the larval stage.

It is possible that a critical period may occur at later developmental stages. In temperate inland waters, seasonal changes in food availability can lead to a critical period during the first winter of life. This has been demonstrated for juvenile largemouth bass in reservoirs and ponds (Shelton et al. 1979; Miranda and Hubbard 1994a, 1994b; Ludsin and DeVries 1997). Young largemouth bass initially feed on zooplankton and other small aquatic invertebrates but switch to larger invertebrates and small fish as they grow. Individuals in a cohort that initially grow faster can realize an even greater growth and survival advantage because of their increased size and greater flexibility in prey use. During fall and winter, prey abundance can decline to an extent that food becomes limiting, especially to the smallest individuals that have the least flexibility in prey selection. Consequently, these individuals can have reduced condition and suffer greater losses to predators, disease, and other stresses. In these examples, population regulation is a function of density-dependent processes, but the intensity of their influence is expected to vary with severity of fall and winter weather conditions. In a study of smallmouth bass, Watt (1959) showed convincingly that age-0 fish that had not attained a critical size by October would not survive the winter.

Knowledge of the existence and timing of a critical period during a species' life cycle can provide guidance for making management decisions. Using largemouth bass as an example, one can see that in systems regulated like those described above, management efforts should not be directed at increasing the number of young during the fall because survival is regulated by food availability during winter. A more effective approach might be to use procedures that would enhance growth in summer and fall, thereby increasing overwinter survival. It is also clear from this example that estimates of cohort abundance should be made after, not before, the critical period if one intends to use the data to forecast year-class strength in a fishery.

6.2.2 Effects of Fishing on Population Dynamics

Unexploited stocks are typified by a high proportion of old fish, slow individual growth rates, and low rates of total annual mortality (Clady et al. 1975; Goedde and Coble 1981). The presence of old fish in poor body condition is often reflective of little or no exploitation. When unexploited populations are opened to fishing, length- and age-frequency distributions typically shift toward smaller and younger fish, mean age declines, and total mortality increases. For stocks that are naturally regulated by density-dependent processes, it is also expected that individual growth rates of surviving fish would increase after exploitation because of reduced intraspecific competition (Backiel and Le Cren 1967). At initial stages of exploitation, a population usually is relatively stable because the abundance of adult fish is not reduced to an extent that reproduction is affected. In fact, it is possible that the reduction in numbers of larger, older fish could lead to increased survival of young because of reduced cannibalism (Ricker 1954). Management objectives in these cases usually are directed at maximizing the recreational or economic benefit that can be obtained from each fish newly added to the population by reproduction. However, if harvest is further increased, the reproductive potential of the population may be reduced to an extent that the adult population declines substantially. At such times, management goals are adjusted to help assure adequate reproduction in the population.

Angler catch rates (number or weight of fish caught per unit of effort) often are high for newly exploited stocks but decline rapidly thereafter. Redmond (1974) estimated that during the first 3 d that five Missouri lakes (9–83 ha) were opened to angling, 39–66% of the largemouth bass populations was removed by anglers. Similarly, Goedde and Coble (1981) showed that 1 month's angling in a recently opened Wisconsin lake (5 ha) reduced the number of harvestable-size pumpkinseed, yellow perch, largemouth bass, and northern pike by 74, 86, 53, and 46%, respectively. High initial exploitation rates are partly a function of intense angling pressure, but unexploited stocks may contain a large proportion of naive fish that are highly vulnerable to exploitation.

The effects of exploitation on the abundance of mature fish in a population is determined by the extent to which rates of mortality and replacement by reproduction are altered. For a population that is not fished, all growth and reproduction are balanced by natural mortality (Ricker 1975), and the population size is expected to remain close to some equilibrium level. With the addition of harvest, mortality increases, and the number of mature fish is reduced. The long-term effects of harvest on the population are a function of the new rates of mortality, growth, and reproduction. An excessive rate of harvest may tip the balance, and fishing may steadily reduce a population to a level at which harvest is no longer economical or possible. More commonly, however, a new equilibrium population level is reached because the decreased abundance of mature fish (from harvest) allows the remaining fish to respond with (1) a greater rate of growth, (2) a reduced rate of natural mortality, or (3) greater rates of reproduction and survival of young (Ricker 1975). Fisheries managers use estimates of these rates and population size to determine appropriate levels of harvest for fish stocks.

Ricker (1954) developed generalized models of the relationships between abundance of adult fish (stock size) and the number of new fish surviving to reach an exploitable size or age (termed recruitment) for stocks that are regulated by density-dependent factors. Because populations vary in age structure, fecundity, and relative importance of density-dependent and density-independent mortality factors, there are many ways that populations can respond to a reduction in the number of adults. The possibilities range from direct proportionality between the number of adults and recruitment to total independence of these two measures. Stock–recruitment models have proven useful, primarily in marine fisheries, for predicting population responses to changes in exploitation and for estimating optimum levels of harvest. Application of stock–recruitment models to freshwater fisheries has been less common.

6.2.3 Quantification of Dynamics

A stock is the biological unit of interest in studies of fish population dynamics. Stocks are expected to respond differently to exploitation because of differences in growth or mortality rates. They can often be defined as geographically isolated, and biologists generally attempt to gather information for distinct stocks and manage each separately. The term stock is almost synonymous with biological usage of the term population, which is defined as a collection of interbreeding organisms having its own birth rate, death rate, sex composition, and age structure. The major distinction is that stock refers to the biological unit that is exploited; it may be a subset of a larger population or a collection of species that is exploited as a single unit.

Delineation of stocks and descriptions of their reproduction, behavior, and genetic characteristics are major areas of study in fisheries management; such studies are necessary to define the extent to which management actions may influence a particular fishery. Stocks of many inland fish species are easily defined spatially—populations that occur in isolated lakes or are geographically distant obviously represent different stocks. Marine fisheries managers often face a difficult task of stock identification because of species' life histories that include extensive migrations that lead to mixing of stocks. It is also difficult in some freshwater situations to know if a particular species within a body of water comprises one or more stocks. This problem is most common in large river systems, where the potential for evolution of distinct stocks is greater than it is in most lakes. Migratory behavior of adults may lead to aggregations of individuals from several stocks in one location. In this situation, management actions influencing the species' abundance and mortality at this location would affect several stocks simultaneously; conversely, management actions directed toward improving survival at one stock's spawning grounds might not have the expected influence on the overall abundance of adults because recruitment from other stocks is unaffected.

Traditionally, the most important biological statistics of fish populations have been population size, total mortality rates at successive ages, the fractions of total mortality attributable to natural mortality and fishing mortality, individual growth rates, recruitment rates, and the rate of surplus production (Ricker 1975). These parameters are needed to determine the greatest amount of biomass that can be harvested from a stock on a sustained basis.

For recreational fisheries typical of inland waters, management objectives usually are far more complicated than simply maximizing harvest, which suggests that managers need to collect information beyond the statistics listed above. Fisheries that illustrate this need emphasize catch-and-release, fishing-for-fun, or trophy angling, where success is measured in terms of recreational enjoyment rather than biomass harvested. Management objectives of inland fisheries frequently address the need for maintaining "balance" in systems regulated by density-dependent processes. Here, the size distribution of fish available to anglers can be more important than total harvest or yield. In such cases, aesthetic and economic values of a given fishery may not be related simply to stock biomass, meaning that other measurements will be needed to monitor management success. These measurements might include estimates of the number of trips or hours fished by anglers, economic benefits derived from a fishery, number of hours required to catch a trophy fish, number of fish caught and released, number of fishing licenses sold, and various indices of condition of the fish population itself. Methods for conducting angler surveys to determine fishing effort, catch rates, and harvest have been reported by Malvestuto (1996), and procedures for determining social and economic values are described in Smith (1983), Weithman (1986), and Chapter 8.

6.3 METHODS OF ESTIMATING POPULATION PARAMETERS

6.3.1 Estimation of Population Size

Estimates of population size often provide the information needed for making fisheries management decisions. Research or survey programs that track fluctuations in numbers of fish in a stock are used to identify influences of environmental factors and human exploitation and ultimately identify effective management strategies. As such, population monitoring activities often make up a significant proportion of a fisheries biologist's workload. This section introduces three commonly used methods of population estimation: counts on sample plots, mark and recapture, and decline in catch per unit effort. Otis et al. (1978), Seber (1982), White et al. (1982), Brownie et al. (1985) contain information on more advanced models.

6.3.1.1 Counts on Sample Plots

An estimate of population size can be obtained by determining the average density of animals per unit area in sample plots and multiplying this value by the total area covered by the population. Seber (1982) outlined three main steps in developing a sampling scheme of this type.

1. The size and shape of the sample area, or plot, should be determined. This choice will be a function of the behavior of the animals to be evaluated, physical features of the habitat, and practical constraints associated with the sampling gear. Plots can cover a standardized area and be shaped as squares, circles, or rectangles, termed quadrats, or plots could consist of nonoverlapping strips running through the population area, termed transects.

2. The number of plots to be sampled should be established in advance. Sampling of more than one plot is necessary to estimate sampling variance, and the desired level of precision of the population estimate can be used to determine the number of plots required.

3. The sample plots should be located randomly so that valid statistical estimates of sampling error can be calculated.

This method of estimation is used primarily when all members of the target population within each sample plot can be counted with reasonable certainty. For example, plots could be established by using nets to block off sections of a small stream, and fish could be counted following removal with toxicants or electrofishing. Another example is the use of a seine to block a standardized quadrat along the shoreline of a lake or reservoir, followed by application of toxicants, such as rotenone, to facilitate removal and counting of the fish.

Counts of fish made per unit of time or volume can also be used to estimate population size, provided that the steps outlined above are followed. For example, counts of larval fish or plankton samples can be expanded to estimate population size provided that samples are collected randomly and have a standard sample volume.

An estimate of the population size (N) in an area can be calculated from the individual plot counts as follows:

$$\hat{N} = \frac{A}{a}\bar{n},\tag{6.1}$$

where A is the size of the study area, a is the size of the plot (same unit of measure as A), and $\bar{n}$ is the average number of animals counted per sample plot. The variance, $V(\hat{N})$, of the population estimate is calculated as follows:

$$V(\hat{N}) = \frac{A^2}{a}\frac{V(n)}{s}\frac{(A - s \cdot a)}{A},\tag{6.2}$$

where $V(n)$ is $\sum_{i=1}^{s}(n_i - \bar{n})^2/(s-1)$; n_i is the number of animals counted in the ith plot; and s is the number of plots used (Cochran 1977).

An approximate 95% confidence interval for the true population size can be calculated as $\hat{N} \pm (t_{s-1,0.05})\left(\sqrt{V(\hat{N})}\right)$.

In designing a study, it is important to predetermine a desired level of precision to be achieved for estimates of important parameters. A convenient way of expressing the precision is to calculate a coefficient of variation, CV, which is defined as the square root of the variance of an estimate divided by the estimate itself. Thus, CV is a unitless measure of the relative amount of variation about an estimate. When using counts from sample plots to estimate population size, we define the coefficient of variation to be CV $= \sqrt{V(\hat{N})}/\hat{N}$. A coefficient of variation of 0.20 or less is usually judged to be adequate.

The number of plots sampled is a principal determinant of the precision of population estimates from simple random sampling designs. Prior to conducting fieldwork, a researcher should determine the number of plots that need to be sampled to achieve the target level of precision (Cochran 1977).

The above sampling procedures are termed simple random sampling when all potential sampling plots, transects, or intervals within their respective population areas or times have an equal chance of being included in the sample. Thus, every animal in the population has an equal chance of being included in the sample provided that the members of the population are randomly or uniformly distributed throughout the area. Fish populations, however, are rarely distributed randomly and more typically are aggregated in certain areas or times. In such cases, estimates from equation (6.1) are not biased but have poor precision because of extreme variability in counts among plots. If fish distribution patterns are known prior to conducting a population study, precision may be improved by subdividing the study area into zones, or strata, expected to have different fish densities and selecting sample plots at random within each stratum. This is termed stratified random sampling.

6.3.1.2 Mark and Recapture

The simplest mark–recapture technique of population estimation requires one sample period in which fish are collected, marked, and released and another period in which fish are collected and examined for marks. This method is the Petersen index (alternatively known as the Lincoln index), which is based on the assumption that the proportion of marked fish in the second sample estimates the proportion of marked fish in the total population. The estimator of population size is

$$\hat{N} = \frac{MC}{R}, \qquad (6.3)$$

where M is the number of fish initially marked and released, C is the number of fish collected and examined for marks in the second period, and R is the number of recaptures (i.e., previously marked fish) found in C. This estimate applies to the population present during the first sample period, not the recapture period.

The Petersen index can give biased estimates of population size when the number of fish sampled is low, but several modifications of equation (6.3) have been proposed to help correct this bias. Bailey's (1951) modification is

$$\hat{N} = \frac{M(C+1)}{(R+1)},$$

(6.4)

with variance

$$V(\hat{N}) = \frac{M^2(C+1)(C-R)}{(R+1)^2(R+2)}.$$

(6.5)

Bailey's modification is used in cases in which sampling during the recapture period is conducted with replacement, meaning that each fish is returned (re-placed) to the population after it is examined for marks and thus is eligible to be included in the sample again. Chapman (1951) recommended using

$$\hat{N} = \frac{(M+1)(C+1)}{(R+1)} - 1,$$

(6.6)

with variance

$$V(\hat{N}) = \frac{(M+1)(C+1)(M-R)(C-R)}{(R+1)^2(R+2)}.$$

(6.7)

This model is used when sampling during the recapture period is done without re-placement, as in cases in which anglers examine their catch for marks or when all fish collected in the recapture period are marked in a way different from the first mark and then released. Differences among population estimates obtained from equations (6.3), (6.4), and (6.6) would probably be of little significance in making fishery manage-ment decisions if the number of fish recaptured (R) exceeds 7.

Several important conditions or assumptions must be met to obtain valid esti-mates using the Petersen index or its modifications: (1) marked fish do not lose their marks prior to the recapture period; (2) marked fish are not overlooked in the recapture sample; (3) marked and unmarked fish are equally vulnerable to cap-ture in the recapture period; (4) marked and unmarked fish have equal mortality rates during the interval between the marking and recapture sample periods; (5) following release, marked animals become randomly mixed with the unmarked ones or recapture effort is distributed in proportion to the number of animals in different parts of the population area; and (6) there are no additions to the popula-tion during the study interval.

Assuring that these conditions are satisfied is one of the most difficult aspects of estimating population size with the Petersen method. Any factor causing underrepresentation of marked fish in the second sample will lead to overestimation of the population size. This could result from poor mark retention, failure to recognize all recaptures in the second sample, impaired survival of marked fish, and immigration of new (thus unmarked) animals before the recapture sampling. Conversely, any factor leading to overrepresentation of marked fish in the second sample, caused perhaps by increased susceptibility of marked animals to capture, will result in underestimation of the true population number.

Several methods may be used to establish confidence intervals for Petersen-type population estimates. These have been thoroughly developed for the most common sampling design—one in which sampling is done without replacement during the recapture period. For this design, the random variable is the ratio R/C, which estimates M/N for the population, and the distribution of R/C is hypergeometric. Unfortunately, neither tables nor explicit formulas are available for determining exact confidence intervals for the hypergeometric distribution. Consequently, various approximations based on the binomial, Poisson, or normal distributions have been used, depending on the magnitude of R/C and the values of M, C, and R for a particular study (Seber 1982).

Precision and accuracy of Petersen estimates are affected by the numbers of fish marked and subsequently checked for marks. Charts prepared by Robson and Regier (1964) can be used to determine values of M and C required to produce Petersen population estimates expected to differ from the true population number by no more than 50%, 25%, or 10% at the 95% level of confidence. They recommend the 50% level for preliminary surveys, 25% for management studies, and 10% for research evaluations. Use of the charts requires an initial guess of population size. A particular combination of M and C can be chosen as a function of the relative costs associated with marking fish and sampling for recaptures.

Mark–recapture methods of population estimation that use two or more sample periods for marking animals are termed multiple-census procedures. The simplest of these was originally described by Schnabel (1938). Fish are collected from the population, marked, and released for a series of samples; the numbers of recaptures and unmarked fish collected in each sample are recorded. Assumptions of the method are identical to those of the Petersen index except that no mortality is allowed during the study. Because of this requirement, the multiple census is most appropriate when sampling periods are closely spaced and restricted to a relatively short overall period so that the occurrence of mortality would not have a great influence on the validity of the population estimate. Table 6.1 illustrates typical data and computational procedures for this method.

The Schnabel population estimation formula is

$$\hat{N} = \frac{\sum_{t=1}^{n} C_t M_t}{\sum_{t=1}^{n} R_t}, \tag{6.8}$$

where the subscript t refers to the individual sample period and n is the number of periods.

Table 6.1 Data records and calculations for Schnabel (1938) multiple-census population estimate.

Sample period (t)	Number of fish captured			Total number of marked fish released prior to sample period (M)	C × M
	Marked (R)	Unmarked	Total (C)		
1	0	150	150	0	0
2	22	203	225	150	33,750
3	26	86	112	353	39,536
4	53	150	203	439	89,117
5	38	80	118	589	69,502
6	28	53	81	669	54,189
7	87	150	237	722	171,114
Total	254				457,208

For our example (Table 6.1),

$$\hat{N} = \frac{\sum_{t=1}^{7} C_t M_t}{\sum_{t=1}^{7} R_t} = \frac{457,208}{254} = 1,800.$$

Confidence limits are determined by first computing the variance of (1/N) because the inverse of N is more normally distributed than is N itself:

$$V(1/\hat{N}) = \frac{\sum_{t=1}^{n} R_t}{(\sum_{t=1}^{n} C_t M_t)^2}.$$ (6.9)

We next determine a 95% confidence interval for (1/N) as $(1/\hat{N}) \pm 1.96\sqrt{V(1/\hat{N})}$ and compute the inverses of the limits to find the confidence interval of N itself. For our example, $1/\hat{N} = 1/1,800 = 0.000556$, and

$$V(1/\hat{N}) = \frac{254}{(457,208)^2} = 1.215 \times 10^{-9}.$$

From these, the 95% confidence interval for (1/N) is 0.000488–0.000624, and by calculating the inverses of these limits, we obtain a confidence interval for N of 1,602–2,049.

For cases in which the total number of recaptures in the study is small, say less than 25, we do not expect (1/N) to be normally distributed. We then must calculate confidence limits by alternative methods. This can be done by using tables of the Poisson distribution (see Ricker 1975) to determine 95% limits for the total number of recaptures and then substituting these values into the denominator of equation (6.8) to determine limits for N.

6.3.1.3 Removal Methods

Population size can be estimated from data on fishing effort and catch rates. Several estimators have been developed, all of which are based on the theory that the number of fish caught per unit of effort will progressively decline as members of the population are removed. The most common methods assume that (1) all members of the target population are equally vulnerable to capture, (2) vulnerability to capture is constant over time, and (3) there are no additions to the population or losses other than those due to fishing during the study interval. Additionally, one must be able either to quantify fishing or sampling effort or to create a sampling situation in which equal effort is expended in consecutive sampling periods. Examples of ways that effort could be quantified include hours spent electrofishing, angler trips, vessel-days, seine hauls, or gill nets fished.

The Leslie and DeLury methods are used in cases in which sampling effort may vary among periods. These models, which are part of a general class of methods described by Schnute (1985), have been used mainly for large populations for which there is low probability that an individual fish will be caught during a single unit of effort. Typical applications have included commercial fisheries for which data on sampling effort and catch are obtained by monitoring the fishers, but Leslie models have recently been applied to reservoir stock assessments (Maceina et al. 1993, 1995).

The Leslie method of estimation assumes that the number of fish caught per unit effort during some time interval, t, is proportional to the number of fish present at the beginning of the interval:

$$\frac{C_t}{f_t} = qN_t, \tag{6.10}$$

where C_t is the catch during period t, f_t is the amount of fishing effort during period t, N_t is the number of fish present at beginning of period t, and q is the catchability coefficient.

Because the method assumes that the population is closed to additions or losses other than fishing, we can express N_t as a function of the original population size (N_0) minus the total number of fish caught and removed (K_t) prior to time t as

$$N_t = N_0 - K_t. \tag{6.11}$$

We can substitute this expression for N_t into equation (6.10) and obtain

$$\frac{C_t}{f_t} = qN_0 - qK_t, \tag{6.12}$$

which is a linear relationship of the form $Y = a + bX$, where $Y = C_t/f_t$ and $X = K_t$. A plot of catch per effort (C_t/f_t) versus cumulative catch (K_t) will approximate a straight line with slope (actually, the absolute value of the slope) equal to q and intercept of qN_0. We can use linear least-squares regression methods to estimate the slope and intercept, and then estimate the original population size as

$$\hat{N}_0 = \frac{\text{intercept}}{|\text{slope}|} = \frac{qN_0}{q}. \tag{6.13}$$

When the fraction (q) of a population that is taken by a given unit of fishing effort is small, say less than 0.02 (<2% of the population), DeLury's modification of the Leslie model is preferred. The DeLury method also is based on the premise that catch per effort is proportional to population size (equation (6.10) and assumes a closed population (other than losses due to removal), but uses a different expression of population decline:

$$N_t = N_0 e^{-qE_t}, \tag{6.14}$$

where E_t is the cumulative total effort expended prior to period t, and other variables are defined as before. This implies that the population declines in proportion to total effort, whereas the Leslie method assumes that the decline is a function of the total catch.

Substituting the expression of N_t from equation (6.14) into equation (6.10), we get $C_t / f_t = qN_0 e^{-qE_t}$. By taking the natural log of both sides, we obtain

$$\log_e (C_t / f_t) = \log_e (qN_0) - qE_t, \tag{6.15}$$

which is also of the form $Y = a + bX$, with $Y = \log_e (C/f)$ and $X = E_t$. So in this method, we can plot the natural log of catch per effort versus cumulative effort and again use linear least-squares regression to estimate the slope and intercept. The estimate of population size is

$$\hat{N}_0 = \frac{e^{\text{intercept}}}{|\text{slope}|}. \tag{6.16}$$

Confidence limits for population estimates obtained from the Leslie and Delury methods are calculated from intermediate statistics obtained when performing the least-squares regression, and they may be determined using Ricker (1975). Because regression techniques are used, these methods of estimation require a minimum of three sample periods. Precision can be improved by increasing the number of sample periods, but the influences of immigration or natural mortality on accuracy of the estimates could become significant if the duration of the study is extended.

Removal methods of population estimation are also used in situations in which the catchability of fish is high and equal effort is expended in each sample period. The most common example of this is sampling small streams, where sections are blocked off with nets, and fish are collected by making consecutive passes with electrofishing gear (e.g., Thompson and Rahel 1996). Each pass represents one sample period. Fish captured during each period can be released outside the sample reach or marked and then released back into the sampling area. Marking in this case can be used to remove fish from consideration in subsequent samples. A model for population estimation was described by Zippin (1956, 1958) as

$$\hat{N} = \frac{C}{1 - \hat{p}^s}, \tag{6.17}$$

where C is the total catch over all sample periods $\left(\sum_{t=1}^{s} C_t\right)$, s is the number of sample periods, and $\hat{p}$ is the probability that a fish escapes capture during a sample period (i.e., $\hat{p} = 1 - \hat{q}$, where $\hat{q}$ is the catchability coefficient as defined before). To calculate $\hat{N}$, we must first estimate $\hat{p}$ from experimental data using the equation

$$\frac{\hat{p}}{\hat{q}} - \frac{s\hat{p}^s}{(1 - \hat{p}^s)} = \frac{\sum_{t=1}^{s}(t-1)C_t}{C}. \tag{6.18}$$

Using an example from Seber (1982), we have the following catches made from three consecutive sampling efforts: $C_1 = 165$, $C_2 = 101$, and $C_3 = 54$. First, estimate $\hat{p}$ by iteratively solving equation (6.18):

$$\frac{\hat{p}}{\hat{q}} - \frac{3\hat{p}^3}{(1 - \hat{p}^3)} = \frac{(1-1)165 + (2-1)101 + (3-1)54}{320} = 0.65.$$

We then find $\hat{p} = 0.58$. Using equation (6.17), calculate $\hat{N} = 320/(1 - 0.58^3) = 400$. Charts originally published by Zippin (1956) and partially reproduced in Seber (1982) can be used to help solve equation (6.18).

A 95% confidence interval for N can be calculated as $\hat{N} \pm 1.96\sqrt{V(\hat{N})}$, where

$$V(\hat{N}) = \frac{\hat{N}(1 - \hat{p}^s)\hat{p}^s}{(1 - \hat{p}^s)^2 - (\hat{q}s)^2 \hat{p}^{s-1}}. \tag{6.19}$$

For our example, $V(\hat{N}) = 691.1$, and the confidence interval is 348–452.

One limitation of the Zippin method is that a large proportion of the population must be sampled to obtain reasonably accurate and precise estimates. This is especially restrictive as the population size decreases; thus, for small populations, it may be necessary to mark fish to simulate removal or to otherwise hold the collected fish for eventual release in order to avoid depleting the population. In cases in which catchability varies among individuals in the population, perhaps in relation to sex, age, or size (Thompson and Rahel 1996), it may be necessary to identify distinct segments of the population and estimate the size of each segment separately. White et al. (1982) described removal–depletion estimation methods useful in cases in which capture probability ($\hat{q}$) may change between sampling efforts.

6.3.2 Estimation of Mortality Rates

Sources of mortality in fish populations are usually placed into one of two categories: natural mortality, including losses to predation, diseases, and weather, or fishing mortality, which is mortality due to harvest (Table 6.2). The combined effect of natural and fishing mortalities is termed total mortality. For exploited stocks, we usually regard the life span of the fish as having a prerecruitment phase, when only natural mortality occurs, and a postrecruitment phase, when fishing and natural mortality occur. In this context, recruitment refers to the addition of fish to the exploited portion of the stock.

6.3.2.1 Estimation of Total Mortality

Methods used to estimate mortality rates vary in relation to assumptions that are made regarding temporal patterns of reproduction and survival rates. The simplest estimation procedures are based on three assumptions: (1) reproduction is constant from year to year; (2) survival is equal among all age-groups; and (3) survival is constant from year to year. Under these assumptions, the population will be in steady state; that is, the number of fish added equals the number dying annually, and the population's age composition would be stable. Hence, information on the population's age composition at any point in time would be representative of mortality rates over a much broader interval. A more realistic model for mortality of inland fish stocks would allow for annual variation in reproduction and survival rates. These variations would produce an unstable age structure in the population, meaning that samples would have to be collected annually for several years (at least) to estimate mortality rates.

For a population having a stable age distribution, the number of fish alive at any age could be described by the curve shown in Figure 6.1. The curve could also represent the decline in abundance of one year-class (or cohort) throughout its life, given the assumption of equal mortality among age-groups. The curve in Figure 6.1 is based on a mathematical model that assumes a constant proportion (Z) of the population (N) dies per unit of time (t):

$$\frac{dN}{dt} = -ZN. \tag{6.20}$$

Upon integration, we obtain an exponential equation for predicting the number of fish, among those currently present, that will still be present at some specified future time:

$$N_t = N_0 e^{-Zt}, \tag{6.21}$$

Table 6.2 Types of mortality rates and relationships among them.

Mortality rates	Symbol	Relationships
Instantaneous		
Total	Z	$A = 1 - e^{-Z}$
From fishing	F	$F = Z - M$
From natural causes	M	$M = Z - F$
Actual		
Total	A	$A = u + v$
From fishing (exploitation rate)[a]	u	$u = FA/Z$
From natural causes[a]	v	$v = MA/Z$

[a] Relationships shown are for cases in which the ratio of fishing mortality to natural mortality is similar throughout the year.

where N_t is the number alive at time t, N_0 is the number alive initially (at time t_0), Z is the force of total mortality (also known as the instantaneous total mortality rate), and t is the time elapsed since t_0.

Although any unit of time can be used with the above model, years are most commonly used so that the number of fish alive at any age t can be computed as a function of the number alive initially and the force of total mortality. If we let $t = 1$ year, the probability (S) that a fish survives the year can be expressed as

$$S = \frac{N_1}{N_0} = e^{-Z},$$
(6.22)

and the complement of survival, the annual mortality rate (A), is equal to $1 - S$ or $1 - e^{-Z}$.

In many applications, the exponential model is used to estimate the annual mortality rate of fish that are age 1 and older, and it sometimes is not applied until ages 2, 3, or older. Additionally, because many management efforts are focused on the postrecruitment phase of the species' life span, the exponential model has traditionally been used to describe mortality beginning with the youngest age-group that is exploited in the fishery. When the model includes the youngest exploited age-group, the assumption of constant reproduction is modified to become an assumption that the number of fish alive at the age first considered in the model is constant from year to year. If this age coincides with the age at recruitment, the assumption is one of constant recruitment.

Figure 6.1   Relationship between the number of fish alive and fish age for a population that has a stable age distribution and a constant proportion dying per unit of time.

Total mortality rates are often estimated from the age structure in samples taken from a population. If we assume that recruitment and survival rates are equal across years, the population age structure would be stable, and a random sample taken from the population at any time would always show the same age composition (allowing for random variation among samples). Thus, relationships among numbers of fish at specific ages in the sample can be used to estimate mortality rates.

In practical applications, it is first necessary to assure that the sample is representative of the entire population. Most sampling gears, including those used in commercial operations, are selective for certain sizes or ages of fish, and they would produce biased estimates. A typical problem is that young fish are underrepresented in the sample because either they are too small to be effectively sampled by the experimental gear or they occur in a different habitat than that sampled. In these cases, age-groups that are not fully vulnerable to the sampling gear are excluded from the mortality computations.

The simplest methods of estimation assume that mortality is equal among ages (in addition to the assumptions of constant recruitment and mortality among years). Consider the following hypothetical sample, for which we have data from a population in which the annual survival rate (S) is 0.60.

Age	1	2	3	4	5	6	7	8	9
Number sampled	100	180	140	84	50	30	18	11	6

It is obvious that age-1 fish are underrepresented in the sample, and closer inspection also shows that age-2 fish were inadequately sampled (i.e., 140/180 exceeds the actual survival rate). Thus, ages 3–9 would be used in further computations.

A more realistic situation is one in which we have the sample, but the survival rate is unknown (it is, of course, the object of the investigation). One way to evaluate gear selectivity in this case is to plot the number collected versus age. This plot should be similar to Figure 6.1, but in the example given above, we would find that the youngest age-groups fall below the line expected from the exponential model (Figure 6.2). An alternative method is to plot the sample results on semilog paper (Figure 6.3). This latter plot is known as a catch curve. Because we assume that recruitment is constant and that $N_t = N_0 e^{-Zt}$ (equation 6.21), a plot of the logarithms of the numbers sampled at each age (N_t) versus age (t) should be a descending straight line with a slope of $-Z$. In Figure 6.3 we see that the numbers collected for age 3 and older form a straight line whereas the points for ages 1 and 2 lie below an extension of this line, indicating they were not adequately sampled. The catch curve can be used to estimate mortality rates. By taking the natural log of both sides of equation (6.21), we obtain

$$\log_e(N_t) = \log_e(N_0) - Z(t), \tag{6.23}$$

which is of the form $Y = a + bX$, with $Y = \log_e(N_t)$ and $X = t$. Linear least-squares regression can then be used to estimate the annual instantaneous total mortality rate, Z. For example, assume we have collected the following data.

Age (t)	1	2	3	4	5	6
Number (N_t)	100	150	95	53	35	17

For ages 2–6, we regress $\log_e(N_t)$ versus t and obtain a slope of -0.54. Thus, $\hat{Z} = 0.54$, and $\hat{S} = 0.59$ (from $S = e^{-Z}$). By assuming constant recruitment and survival over time and equal survival among ages, we have determined that 59% of fish ages 2–6 survive annually (hence, total annual mortality is 41%).

The accuracy and precision of this estimate are affected by the number of age-groups included and the representativeness of the sample. Age-groups having fewer than five fish in the sample usually are excluded from the regression because of the extreme variation that could be introduced by the collection of, or failure to collect, a few individuals. This typically occurs for the oldest age-groups. In practice, the data set is truncated beginning with the youngest age-group having fewer than five fish in the sample. Because regression techniques are used, data for at least three age-groups that are fully vulnerable to the gear are required; precision will increase with the number of age-groups included. Confidence limits for Z are equal to the confidence limits for the slope of the regression; methods for computation are included in most statistical texts.

An alternative approach to constructing catch curves can be used when annual recruitment to a population is thought to vary substantially. This involves developing multiple catch curves, each based on a single cohort. Sometimes, despite varying annual recruitment, cohort mortality rates can be similar. The average of Z values derived from multiple catch curves should be more representative of the population but will require more years of data.

Other models that can be used to estimate survival in stable age populations are based on ratios among numbers of the various ages of fish collected. A method described by Robson and Chapman (1961) estimates survival as

$$\hat{S} = \frac{T}{n + T - 1},$$
(6.24)

where n is the total number of fish in the sample, beginning with the first age that is fully vulnerable to the sampling gear, and T is determined from the age distribution of the sample.

First, the sample data are coded so that the number collected in the first fully vulnerable age is labeled N_0, the number in the next oldest group is N_1, and so on, so that N_k is the number of fish in the oldest age-group in the sample. For our example, we have the following data.

Age	2	3	4	5	6
Coded age (x)	0	1	2	3	4
Number (N_x)	150	95	53	35	17

Mathematically,

$$T = \sum_{x=0}^{k} x(N_x) = 0(N_0) + 1(N_1) + 2(N_2) + 3(N_3) + 4(N_4),$$

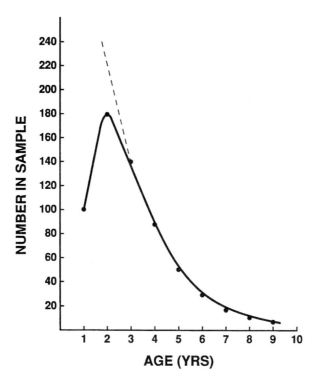

Figure 6.2   Plot of number of fish collected versus fish age. Figure illustrates underrepresentation of age-1 and age-2 fish relative to numbers expected from an exponential model (dashed line).

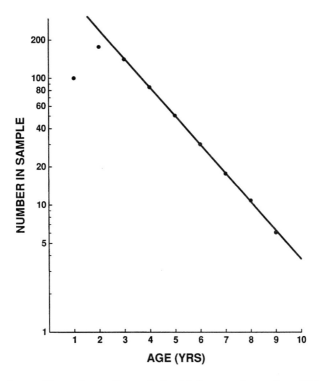

Figure 6.3   Catch curve illustrating the linear relationship between the number of fish collected and fish age when sampling results are plotted on a semilog scale. Ages 1 and 2 were underrepresented in the sample.

which, for the example, is

$$T = 0(150) + 1(95) + 2(53) + 3(35) + 4(17) = 374.$$

Hence, from equation (6.24), we have

$$\hat{S} = \frac{374}{350 + 374 - 1} = 0.52.$$

The variance of this estimator is

$$V(\hat{S}) = \hat{S}\left(\hat{S} - \frac{T-1}{n+T-2}\right), \tag{6.25}$$

which, for our example is

$$V(\hat{S}) = 0.52\left(0.52 - \frac{374-1}{350+374-2}\right) = 0.0018.$$

Methods for estimating mortality rates in populations that do not have stable age distributions can require considerable amounts of data and may be mathematically complicated.

6.3.2.2 Separation of Natural and Fishing Mortality

For exploited fish populations, it is often important to account for the influences of natural and fishing mortality separately. The total instantaneous mortality rate (Z) can be partitioned into instantaneous rates of fishing mortality (F) and natural mortality (M) from the relationship $Z = F + M$. Equation (6.21) can be modified as

$$N_t = N_0 e^{-(F+M)t} = N_0 e^{-Ft} e^{-Mt}. \tag{6.26}$$

Likewise, the actual total mortality rate (A) can be expressed as the sum of two components,

$$A = u + v, \tag{6.27}$$

where u is the expectation of death from fishing (also known as the rate of exploitation) and v is the expectation of death from natural causes. For cases in which fishing and natural mortality are similarly apportioned throughout the year, it is possible to equate

$$\frac{Z}{A} = \frac{F}{u} = \frac{M}{v}. \tag{6.28}$$

Upon rearrangement we find that

$$u = FA/Z, \quad \text{and} \tag{6.29}$$

$$v = MA/Z. \tag{6.30}$$

These relationships are useful if it can be reasonably assumed that the number of fish harvested in any given time period (perhaps a month) relative to the total number harvested annually is similar to the ratio of the number dying naturally in the same given period divided by the total number dying annually from natural causes.

Separation of total mortality into fishing and natural components is accomplished usually by first estimating total and fishing mortality and then estimating natural mortality as the difference. The exploitation rate, u, can be calculated by estimating the population size at the beginning of some time interval and then counting the number of fish harvested by fishers in the ensuing period. Provided that there is no immigration or other additions to the population, the number harvested divided by the initial population size is an estimate of u. If the number harvested is monitored for 1 year, the value of u estimates the annual exploitation rate. The precision and accuracy of the estimate are a function of methods used to estimate population size and number harvested.

Another way to estimate u is to release a known number of tagged fish at one point in time and then determine the proportion of tagged fish harvested in a year. Counts of number harvested are often obtained from anglers who report their catches of tagged fish; in many cases, monetary rewards are offered by management agencies to encourage complete reporting of harvested fish. The validity of tag reward procedures is based on assumptions that tagging does not affect the fish's vulnerability to angling or other sources of mortality, that the tagged fish are representative of the target population, that tags are not lost by the fish, and that all harvest of tagged fish is reported. Precision of the estimate is primarily a function of the number of fish tagged and released.

Another approach for estimating mortality components is based on the relationship $Z = M + F$. The instantaneous fishing mortality rate (F) can be related to the amount of fishing effort (f) as

$$F = qf, \tag{6.31}$$

where q is a catchability coefficient.

Because $Z = M + F$, we can substitute for F and obtain

$$Z = M + qf, \tag{6.32}$$

which is equivalent to a linear equation with slope q and intercept M. A hypothetical relationship between Z and f is illustrated in Figure 6.4. Thus, if we have annual estimates of Z and corresponding values of annual fishing effort, it is possible to estimate M using least-squares regression. The method requires several years of data (a minimum of three) and assumes that the catchability coefficient is constant over time and equal among age-groups. The time-specific annual estimates of Z could be obtained from age-specific catch data as described earlier in this section. The basis of this approach is that if natural mortality (M) is constant, any annual variation of total mortality (Z) will be due to variation in only fishing mortality (F), which, in turn, varies only in relation to effort (f).

6.3.2.3 Age-Specific Mortality

Virtual population analysis is a method commonly used to estimate age-specific fishing mortality rates (F_t), age-specific catchability (q_t), and annual recruitment for exploited populations. The material presented in this section is based on the more complete treatment of the topic by Hilborn and Walters (1992). Also known as cohort analysis, because parameters are estimated for individual cohorts, virtual population analysis has the advantage of being free of the sometimes unrealistic assumptions associated with previously described estimation procedures. Consider a population that is fished during only a brief period at the end of each year. For most of the year, only natural mortality would operate on the population. During the fishing period we could ignore natural mortality and consider only fishing mortality to be important, thereby separating the two types of mortality into distinct phases over time. If N_t is the number of fish alive in a cohort at the beginning of a year, and N_{t+1} is the number alive at the start of the next year (after the current year's fishing season), a model for cohort size during this interval is

$$N_t = N_{t+1} + C_t + D_t,\tag{6.33}$$

where C_t is the number of fishing deaths and D_t is the number of natural deaths that occurred between t and $t + 1$.

Natural deaths are considered proportional to the number in the cohort at the start of the 1-year period (N_t) such that

$$D_t = N_t v,\tag{6.34}$$

where v is the portion of N_t expected to die from natural causes before the start of the next fishing season. By rearranging equation (6.33) and substituting $N_t v$ for D_t, we obtain

$$N_t = \frac{N_{t+1} + C_t}{1 - v}.\tag{6.35}$$

Assumptions of virtual population analysis are that the expectation of death from natural causes (v) is known for all age-groups within cohorts and that an oldest living age-group (called the terminal age-group, for which $N_{t+1} = 0$) can be identified and is consistent over all cohorts.

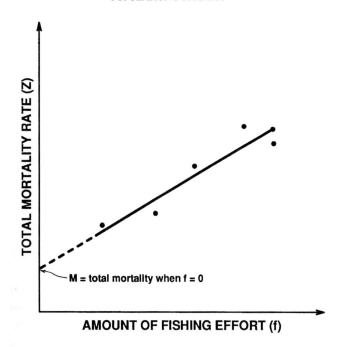

Figure 6.4 Illustration of a straight line (solid) fit to 5 years' estimates of fishing effort (*f*) and total instantaneous mortality (*Z*). The point at which an extension of this line (dashed) intercepts the vertical axis estimates the instantaneous natural mortality rate (*M*).

Assume that a population recruits to a fishery at age 2 and has a terminal age of 6 (i.e., 7-year-old fish are never caught). Next, consider the hypothetical catch-at-age data provided for a single cohort of that population (Table 6.3) and assume that *v* is 0.3 for all age-groups. Beginning with the terminal age-group and working backwards, numbers of fish in the cohort at the beginning of each year (N_t) can be successively back-calculated from equation (6.35): for age 6, N_t = (0 + 112)/0.7 = 160; for age 5, N_t = (160 + 2,245)/0.7 = 3,436; and so on. The cohort size each year just before the fishery begins is $N_t - D_t$, readily determined as N_t (1 − *v*). Exploitation rate (*u*) for each age-group in the cohort is calculated as $C_t /(N_t - D_t)$, and *F* for each age-group is calculated as $F_t = -\log_e(1 - u)$. If age-specific effective fishing effort (f_t) is known, then the catchability coefficient of each age-group (q_t) can be calculated as

$$q_t = \frac{F_t}{f_t}.$$ (6.36)

Values of q_t given in Table 6.3 assume 150 units of effective fishing effort for each age-group. Note that values of F_t and q_t could not be calculated for the terminal age-group. The number of fish from this cohort (spawned in 1992) that recruited to the fishery in 1994 was 22,903 according to this analysis. The information in Table 6.3 from catch-at-age data are typically developed for many cohorts in the population to gain insight into the dynamics of *F*, *q*, and annual recruitment.

Often, it is inappropriate to consider annual losses from a cohort due to fishing and natural mortality as separate in time. When both mortality sources are operating simultaneously, changes in number (N) over time (t) proceed according to the equation

$$N_{t+1} = N_t e^{-(F+M)},$$ (6.37)

which is identical to equation (6.26) except t is 1 year. Catch during any one year can be represented as

$$C_t = \frac{F}{F+M}(N_t - N_{t+1}).$$ (6.38)

To perform virtual population analysis in this setting, we would like to solve for N_t in terms of N_{t+1}, C_t, and M, as done previously in equation (6.35) when periods of natural and fishing mortality were separated in time. If we first solve for F we have

$$F = -\log_e \frac{N_t}{N_{t+1}},$$ (6.39)

and if we then substitute the right side of this equation for F in equation (6.38) we get,

$$C_t = \frac{M}{\log_e N_t - \log_e N_{t+1}}(N_t - N_{t+1}).$$ (6.40)

Although this is not exactly what we wanted, equation (6.40) can be used to determine values for N_t, when N_{t+1}, C_t, and M are known. The equation is solved iteratively by substituting values of N_{t+1}, C_t, and M until equality is obtained. Situations in which M operates solely during one part of the year and in combination with F during another part can be handled by appropriately combining the two approaches that have been described.

Table 6.3 Hypothetical catch-at-age data and calculations associated with virtual population analysis. It was assumed that the natural mortality rate (v) was 0.30. Values of the age-specific catchability coefficient, q_t, were calculated under the assumption that effective fishing effort, f, was 150 units for each age-group.

Year	Age (t)	Catch (C_t)	Cohort size at start of year (N_t)	Cohort size when fishing begins	Exploitation rate (u_t)	Instantaneous rate of fishing mortality (F_t)	Catchability coefficient (q_t)
1999	7	0	0	0			
1998	6	112	160	112	1.00		
1997	5	2,245	3,436	2,405	0.93	2.66	0.018
1996	4	4,545	11,401	7,981	0.57	0.84	0.006
1995	3	3,937	21,911	15,338	0.26	0.30	0.002
1994	2	992	32,719	22,903	0.04	0.04	<0.001

6.3.3 Growth Rates

Growth of fishes, as in many other poikilothermic animals, is indeterminate, meaning that individuals have no innate pattern of growth and can continue to increase in size throughout life. Because growth can be affected by food abundance, weather, competition, and many other factors, the measurement of growth rates is a common way fisheries biologists evaluate the effectiveness of management practices. Additionally, because the size of fish caught by recreational and commercial fishers greatly affects aesthetic and economic values of the catch, an understanding of growth dynamics in populations is important to predict fishery trends adequately. Mathematical models of fish growth are often incorporated into models of population dynamics to predict changes in stock biomass resulting from various harvest strategies.

Growth may be measured in terms of length (l) or weight (w) and is expressed in several ways:

1. absolute increase per unit time, that is, $l_2 - l_1$ or $w_2 - w_1$;
2. relative rate of increase per unit time, that is, $(l_2 - l_1)/l_1$ or $(w_2 - w_1)/w_1$; and
3. instantaneous rate of increase per unit time, that is,
   $\log_e l_2 - \log_e l_1$ or $\log_e w_2 - \log_e w_1$.

Estimates of these rates can be made from observations of length (or weight) of a cohort of fish at two or more points in time, from tagging studies, or from size-at-age data obtained from age and growth analyses.

These expressions assume linear increases of length or weight over time and should not be calculated for time periods where growth is typically nonlinear. Trends of length and weight of a cohort throughout life usually show an early period of rapid growth and a subsequent period of more gradual increase (Figure 6.5). Likewise, absolute, relative, and instantaneous growth rates computed from size at successive ages are initially low, increase to a maximum, and then decline with age.

6.3.3.1 Growth in Length

One useful model that mimics this pattern of declining growth rate with age was originally described by von Bertalanffy (1938). The von Bertalanffy model is based on the theory that the rate of change in length per unit of time (dl/dt) will get smaller and eventually become zero as a fish nears its maximum possible size (L_∞). Mathematically,

$$\frac{dl}{dt} = K(L_\infty - l_t), \tag{6.41}$$

where K is a growth parameter (not a rate in the sense defined earlier) and l_t is the length at time t. Equation (6.41) shows that the rate of increase in length is a constant proportion (K) of the difference between the maximum size and present length ($L_\infty - l_t$). Upon integration of equation (6.41), we obtain a predictive relationship:

$$l_t = L_\infty \left[1 - e^{-K(t-t_0)} \right],$$ (6.42)

where t is time (or age) in years and t_0 is the time at which l_t is 0.

A plot of this relationship shows a progressive increase in body size that asymptotically approaches L_∞ (Figure 6.6). Parameters of the model (L_∞, K, and t_0) are typically estimated from annual length-at-age data; procedures have been outlined by Gulland (1969) and Galluci and Quinn (1979).

A Walford plot provides a classical approach for obtaining values of L_∞ and K when fitting a von Bertalanffy growth model to a population's mean-length-at-age data. Across successive ages (usually excluding age 0), mean length at age $t + 1$ (vertical axis) is plotted against mean length at age t (horizontal axis). The plotted data should appear somewhat linear so that they can be well represented by a straight line by means of simple linear regression. The slope of the fitted line estimates e^{-K}, and

$$K = -\log_e(\text{slope}).$$ (6.43)

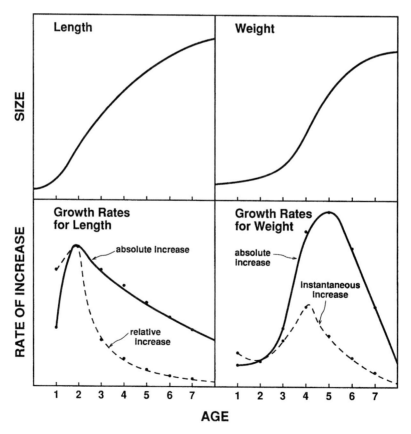

Figure 6.5   Typical trends of length, weight, and growth rates throughout the life of a fish.

The point of intersection between the fitted line and a 45° line (where l_t equals l_{t+1}) estimates L_∞, the theoretical upper length that fish in the population approach but never quite reach, even if they live many years. This upper length is determined from the fitted regression line as

$$L_\infty = \frac{\text{intercept}}{1 - \text{slope}}. \qquad (6.44)$$

Because the value of L_∞ is obtained using mean-length-at-age data, the estimate represents the length that the average fish in the population would reach if it lived to age infinity. Accordingly, one might occasionally find a fish longer than the estimated L_∞.

To determine t_0, a second regression analysis is done. Consider the linear form of the von Bertalanffy model, which is acquired by taking the natural logarithm of both sides of equation (6.42):

$$\log_e(L_\infty - l_t) = \log_e(L_\infty) + Kt_0 - Kt. \qquad (6.45)$$

Using the value of L_∞ determined from equation (6.44) and the mean-length-at-age data, values of t are plotted (on the horizontal axis) against corresponding values of $\log_e(L_\infty - l_t)$; a linear pattern with negative slope should be observed. The slope that results when linear regression analysis is run on these data provides another estimate of K (this value may differ slightly from the previous value of K estimated from equation (6.43). The intercept from the regression analysis will estimate $\log_e(L_\infty) + Kt_0$. Consequently, t_0 can be estimated as

$$t_0 = \frac{\text{intercept} - \log_e L_\infty}{K}. \qquad (6.46)$$

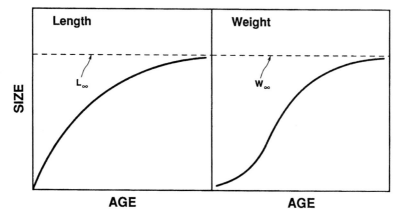

Figure 6.6   Growth trends predicted from von Bertalanffy models of length and weight.

As an example, consider the following mean-length-at-age data for a cohort of gizzard shad.

Age (years)	1	2	3	4	5	6	7	8	9
Mean length (mm)	133	176	202	240	249	256	268	271	276

A regression of l_{t+1} on l_t (Walford plot) gives a slope of 0.71 and an intercept of 83.2 mm. By applying equations (6.43) and (6.44), we find values of 0.34 for K and 286.8 mm for L_∞. Next, a regression of $\log_e (L_\infty - l_t)$ on t (using 286.8 mm for L_∞) gives a slope of -0.34 ($K = 0.34$) and an intercept of 5.36. Applying equation (6.46) leads to an estimate of -0.88 for t_0. Therefore, the fitted von Bertalanffy model is

$$l_t = 286.8\left[1 - e^{-0.34(t+0.88)}\right].$$

6.3.3.2 Growth in Weight

In many stock analyses, it is more desirable to model growth in weight because stock biomass is the product of population size times mean weight. The von Bertalanffy model can be converted to an expression for weight by first using the allometric relationship

$$W = aL^b, \tag{6.47}$$

where the parameters a and b describe the form of the weight–length relationship. Methods for estimating these parameters are outlined in Carlander (1969) and Ricker (1975). In many cases, b is near 3.0, and for computational ease a value of 3 for b is often used, so that only the parameter a is estimated. This produces a simple form of the von Bertalanffy weight model:

$$w_t = W_\infty\left[1 - e^{K(t-t_0)}\right]^3, \tag{6.48}$$

where W_∞ is aL_∞^3, and the other parameters are the same ones used in the length model.

A plot of w_t versus age shows an initial period of accelerating growth, an intermediate period when growth is approximately linear, and a final phase of decelerating growth as w_t approaches W_∞ (Figure 6.6). Ricker (1975) presents methods of predicting w_t when the value of b (equation 6.47) is not equal to 3.

6.3.3.3 The Gompertz Model

A second expression that can be used to describe fish growth in either length or weight over many years of life is the Gompertz model, which represents fish growth in weight quite well across all life stages, including age 0. The model for weight is

$$w_t = w_0 e^{G\left(1-e^{-gt}\right)}, \tag{6.49}$$

where w_t is weight at age t, w_0 is the weight of a fish at the beginning of the growth period, G is the instantaneous growth rate when t is 0 and w is w_0, and g is the instantaneous rate at which G decreases as t increases. Estimates of w_0, G, and g can be made using observations of fish weight for a series of age-groups (Ricklefs 1967; Hilborn and Walters 1992).

The Gompertz model has been used less frequently than the von Bertalanffy model. Ricker (1975) suggested that this is not due to its inferiority but because the von Bertalanffy model became more well-known after it was adopted for use in yield models. However, the Gompertz model is gaining popularity for describing growth in length of larval fishes (Michaletz 1997), which often slows as the juvenile stage is approached. When used to describe growth in length, only the upper portion of the S-shape Gompertz curve (beyond the inflection point) is used. Consequently, the model appears as

$$l_t = L_\infty e^{-e^{-(G-gt)}}, \tag{6.50}$$

where L_∞ is the theoretical asymptotic upper length (as in the von Bertalanffy model), and G and g are as previously described (but for growth in length).

To fit this model to length-at-age data, equation (6.50) can be linearized as

$$\log_e\left(\log_e\frac{L_\infty}{l_t}\right) = G - gt, \tag{6.51}$$

and a regression of $\log_e[\log_e(L_\infty/l_t)]$ versus age (t) is used to estimate G from the intercept and g from the slope. An iterative procedure can then be used whereby values of L_∞ are selected until the best linear fit to the data is acquired. A Gompertz model for weight can also be linearized so that linear regression can be applied.

6.4 PREDICTION OF FISHERY TRENDS

A major activity of fisheries managers has been to predict the effects of different amounts of fishing effort on the numbers and sizes of fish obtained on a continuing basis from a stock. Models are developed within constraints imposed by data availability, types of predictions desired, and mathematical and computational complexity. Models are tools that should be as simple as possible while providing appropriate types of predictions—emphasis should be placed on making decisions regarding feasible management practices rather than impossible ones. Formulation of any mathematical representation of population dynamics will require certain assumptions, and the validity of the assumptions will affect the accuracy and meaning of predictions from a model. Evidence of failure of assumptions does not necessarily mean that the model is useless; rather, the manager should recognize the limitations of the model and determine if such limitations will have a significant effect on decisions or recommendations that will be made (Gulland 1983; Johnson 1995). Models often can be used to explore the influences of different management options on a specific fishery even if the predicted values are known to be only approximately proportional to the actual values. Recent examples of the application of models for inland fisheries management include models of crappie populations in reservoirs (Colvin 1991), channel catfish in rivers (Gerhardt and Hubert 1991), and whitefish stocks in the Great Lakes (Walker et al. 1993) and models of the effects of harvest regulations on numerous species (Johnson et al. 1992; Luecke et al. 1994; Allen and Miranda 1995; Beamesderfer and North 1995).

Three general types of models have been used for fisheries predictions. Models of the first type, surplus production, consider trends in a population as a whole in relation to harvest. The influences of growth, mortality, and reproduction are combined into a model of overall population change (Shaefer 1968). The second type predicts the yield that would be obtained from a year-class of fish throughout its lifespan as a function of harvest practices. These are known as yield-per-recruit models because the predictions are usually expressed as the yield per fish that is newly recruited to the stock. Such models do not explicitly account for the effects of variable recruitment on yield. The most common of these models is known as the dynamic-pool yield model, which was first proposed by Beverton and Holt (1957). The third category of models treats each age-group of a population separately and sums yield predictions among ages and years to obtain an overall estimate of long-term yield. Such predictions are used as a reference, or index, in determining optimal management policies (Walters 1969). Age-structured models require estimates of age-specific rates of reproduction, growth, and mortality, meaning that data requirements are greater than those of surplus production or yield-per-recruit models, which rely primarily on data that are readily available from commercial fishery landings.

6.4.1 Surplus Production Models

Surplus production models can be used to predict yield for a fishery by using information on either stock abundance or fishing effort. Surplus production models do not explicitly consider the growth, reproduction, and mortality rates in a population; rather, they deal primarily with relationships among overall stock biomass, yield, and fishing effort. These models have been popular because their parameters can be estimated from commonly available statistics, such as annual records of commercial harvest and effort.

Biological assumptions of surplus production models are similar to those discussed for stock–recruitment relationships. Populations are assumed to be regulated by density-dependent processes that affect reproduction and control survival and growth early in life, prior to the size (or age) at which fish are harvestable. Survival after this time is assumed to be density independent and is not necessarily compensatory. An unexploited population would be expected to occur at some equilibrium level of abundance at which the number of fish added each year equals the number dying. If the stock is subjected to a fixed level of fishing mortality annually, the population is expected to reach a new equilibrium abundance at which the number of recruits produced annually exceeds the number of adults required to produce them. This excess of new recruits, termed surplus production, is available for exploitation. It is reasonable to expect that the greatest harvestable surplus will occur at some level of stock abundance that is less than the primitive equilibrium because at this equilibrium density-dependent regulation limits the number of recruits, and, by definition, there is no surplus production when recruitment exactly equals the stock size required for replacement.

Graham (1935) proposed a surplus production model based on the assumption that the annual change in biomass of a stock is proportional to the actual stock biomass and also to the difference between present stock size and the maximum biomass the habitat can support:

$$\frac{dB}{dt} = kB\frac{B_\infty - B}{B_\infty}, \tag{6.52}$$

where B is stock biomass, B_∞ is maximum biomass, k is the instantaneous rate of increase of stock biomass, and t is time in years (Ricker 1975).

This equation describes how the stock biomass would increase (dB/dt) following a reduction of biomass from B_∞ to B. If the reduction in biomass to B has been brought about by human harvest, and if the level of fishing mortality (or effort) operates in a manner that stock biomass remains at B from year to year, the stock is regarded as being at equilibrium, and the yield that could be obtained annually from the fishery is dB/dt. Hence, equation 6.52 can be rewritten as:

$$Y = kB\frac{B_\infty - B}{B}, \quad \text{or} \tag{6.53}$$

$$Y = kB - \frac{k}{B_\infty}(B^2), \tag{6.54}$$

where Y is the annual yield when the stock is at equilibrium biomass B.

The primary purpose of developing the model usually is to determine maximum sustainable yield, the level of fishing effort required to obtain maximum sustainable yield, and the stock size at which this would occur. Equation (6.54) is a parabolic relationship (Figure 6.7) with the greatest yield (Y_{max}) at a stock abundance (B) equal to $B_\infty/2$. By substituting B = $B_\infty/2$ into equation (6.54), the maximum sustainable yield is found to be

$$Y_{max} = \frac{kB_\infty}{4}, \tag{6.55}$$

showing that estimates of k and B are needed to determine the maximum sustainable yield. Data on equilibrium yield obtained at two or more different levels of fishing effort are used to estimate these parameters (Ricker 1975). Shaefer (1954) outlined methods for estimating the parameters for nonequilibrium situations.

6.4.2 Yield-per-Recruit Models

Dynamic-pool yield models are based on the premise that stock biomass varies with growth and mortality rates in the stock. Hence, yield is taken by humans from a changing, dynamic pool of available biomass that is increased by recruitment and diminished by mortality. Because biomass is the product of the average fish weight times the number of fish present, it is possible to predict yield (Y) as

$$Y = \int_{t_c}^{t_l} FN_t W_t dt. \qquad (6.56)$$

This indicates that total yield from a given cohort of fish in a stock is the integration of the force of fishing (F) times the number of fish alive at age t (N_t) and the average weight per fish at age t (W_t) for the period of life during which the cohort is exploited. This period begins at the age when fish are first exploited (t_c) and ends at the cohort's theoretical maximum age (t_l). If a stock is at a stable equilibrium, the total yield each year would equal the total harvest of a cohort during its life, so equation (6.56) also estimates the annual equilibrium yield in such cases (Shaefer 1968).

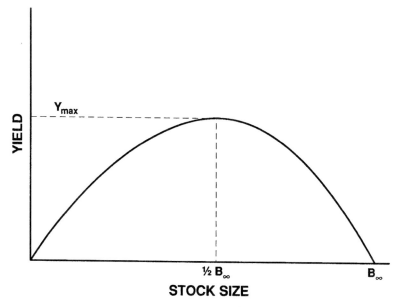

Figure 6.7 Relationship between yield and stock size for the surplus production model. The maximum sustainable yield (Y_{max}) theoretically occurs at a stock size equal to one-half of the maximum biomass (B) that the habitat can support.

Dynamic-pool models are useful for predicting yield where fishing effort and the minimum size (or age) of capture can be regulated. Mesh size restrictions in a commercial fishery or minimum length limits in a recreational fishery would affect the age at which a cohort is first exploited. Regulations that influence fishing effort, such as restrictions on numbers of licenses issued, or those that otherwise influence fishing mortality, such as closed seasons or refuge zones, would affect F. By simulating the effects of different levels of F and age at first harvest, one can evaluate the extent to which yield is influenced and determine levels that produce maximum yield.

Implementation of the model requires mathematical expressions for changes in numbers and average weight of a cohort during its life span. The formulation most commonly used follows that originated by Beverton and Holt (1957):

$$Y = FRe^{-M(t_c - t_r)} W_\infty \sum_{n=0}^{3} \left[\frac{U_n}{F + M + nk} e^{-nK(t_c - t_0)} \right], \tag{6.57}$$

t_r = the age, in years, when fish reach a size at which they could be harvested;

t_c = the age at which fish actually are first harvested ($t_c \geq t_r$);

t_0 = a constant introduced upon integration of equation (6.56), usually interpreted as the age at which fish length is 0;

F = annual instantaneous rate of fishing mortality, assumed equal for all fish older than age t_c;

M = annual instantaneous rate of natural mortality, assumed constant over time and equal for all fish older than age t_r;

R = recruitment, the number of fish alive in a cohort at age t_r;

W_∞, K = parameters of a von Bertalanffy growth model, assumed constant over time; and

U_n = a mathematical term used to simplify equation (6.57), where $U_0 = 1$, $U_1 = -3$, $U_2 = 3$, and $U_3 = -1$.

The model includes three distinct periods during a cohort's life span. The first period is the prerecruitment phase, when the fish are too young (small) to be harvested. This stage begins at age t_0 and ends at t_r, when the fish reach a size at which they could be harvested. The postrecruitment stage in the life cycle can include a preexploited phase as well as an exploited phase. The age at which the fish are first harvested (t_c) sometimes exceeds t_r because of economic or social factors. For example, fishing technology may be available to harvest small fish commercially, but dollar value per unit biomass may be low for these fish, and it thus would be more profitable to harvest fish at a larger size (older age). In a recreational fishery, t_r would correspond to the age at which fish reached a size that anglers would harvest in the absence of length limits, whereas t_c would correspond to the age at which fish reached a legally imposed size limit.

The rate of fishing mortality (F) and the age at first harvest (t_c) can be manipulated by fisheries managers, and the influences of these on yield can be predicted using equation (6.57). Estimates of the number of recruits entering the fishery are rare, and it has become customary to divide both sides of equation (6.57) by R so that the prediction of yield actually is yield per recruit rather than an estimate of total yield. Yield-per-recruit estimates (Y/R) can be used as an index of actual yield for stocks that have relatively stable (albeit unknown) levels of recruitment. Gulland (1983) suggested that yield-per-recruit models are best suited to providing guidance on how to manage specific cohorts in stocks that have variable, environmentally regulated recruitment levels. In these cases, data from some index of cohort abundance at the time of recruitment could be used to de-fine appropriate timing and levels of harvest.

In practice, fisheries managers are often more interested in how the yield-per-recruit predictions would vary across a range of management options, as op-posed to obtaining point estimates for specific combinations of parameters. In the dynamic-pool model, it is possible to have many combinations of F and t_c that would produce the same Y/R. After calculating yield for ranges of t_c and F in a given fishery, the results can be summarized by plotting isopleths that connect the same Y/R estimates (Figure 6.8). The plots then can be used to choose desirable levels of F and t_c for the fishery.

For example, suppose that the fishery represented in Figure 6.8 has been operating at point A, where F is 0.7, t_c is 1.5 years, and Y/R is 50. If our goal is to increase Y/R, it is clear that we could accomplish the most by altering t_c rather than increasing F. If F is maintained at 0.7, t_c should be increased to about 4.5 years to maximize yield. Thus, by delaying the age at first harvest by about 3 years, we would expect a three- to fourfold increase in Y/R.

If, however, the fishery was presently operating at point B on Figure 6.8, Y/R still equals 50, but F is 0.2 and t_c is 4. In order to increase Y/R from this point, the most effective approach will be to increase F, possibly by allowing or promoting increased effort or improved harvest techniques. An approximate fourfold increase in F would be required to approach the maximum yield.

Despite the inclusion of the adjective dynamic in the name dynamic-pool model, the suitability of using the yield predictions for making management deci-sions is dependent on several assumptions of stability or constancy of the stock's vital statistics. For example, the model assumes that natural mortality and growth rates remain constant over time and are independent of stock size. This may be a reasonable assumption for stable environments or for stocks where postrecruitment survival and growth are relatively constant from year to year. The model also assumes that the fishery has an equal influence on all ages of fish older than t_c and that rates of growth and natural mortality are not influenced by changes in the level of harvest.

6.4.3 Age-Structured Models

The basic structural unit of a fish population is the age-group. Each age-group has rates of growth, natural mortality, fishing mortality, and reproduction that can vary with time and differ from those of other age-groups. Age-structured models incorporate these differences in an attempt to provide realistic predictions of population behavior.

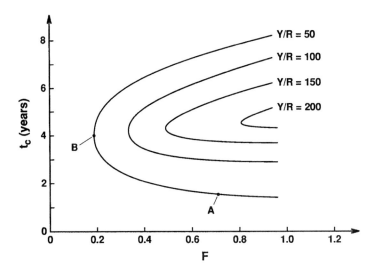

Figure 6.8 Summary of yield predictions obtained from the Beverton–Holt dynamic-pool yield model across a range of age at first capture (t_c) and fishing mortality (F). Lines are isopleths indicating combinations of t_c and F expected to produce the same yield per recruit (Y/R). Points A and B are discussed in text.

Population processes most frequently included in age-structured models are illustrated in Figure 6.9. This graph shows that the number of fish alive at any age is a function of the number alive in the previous year, less the numbers lost to fishing and natural causes. The number alive at age 0, however, is a function of the reproductive rates and age structure of the entire population; this feature is not included in the surplus production or yield-per-recruit models discussed earlier. Age-structured models of population number can be converted to biomass models by including a predictor of size at age, such as a von Bertalanffy or Gompertz growth model, or by using observed size-at-age data for a given population.

Walters (1969) formulated an age-structured model for fish population studies and illustrated its application for making management decisions in a brook trout fishery. He used an exponential model to predict the number (N) of brook trout in a year-class that would survive from year to year, given observed age-specific rates of instantaneous natural mortality (M) and hypothetical rates of instantaneous fishing mortality (F):

$$N_{i+1} = N_i e^{-(F_i + M_i)}. \tag{6.58}$$

The model was used to predict total population size and yield under conditions of either no fishing, a constant rate of fishing over time and among ages, or a periodic fishery.

Taylor (1981) claimed that fisheries models available prior to the 1980s were either too simplified or too complicated to be useful to inland fisheries managers, and he attempted to solve this problem by developing a computerized population simulator specifically aimed at inland fisheries. His model, termed a generalized inland fishery simulator, is fundamentally similar to the one proposed by Walters (1969), but it uses a different mathematical approach and can incorporate a number of population processes not considered by Walters. Examples of the application of Taylor's model have been reported by Taylor (1981) for rainbow trout, Zagar and Orth (1986) for largemouth bass, Johnson et al. (1992) for walleye, and Luecke et al. (1994) for lake trout.

Greater versatility for addressing management options of inland fisheries is achieved from age-structured models. However, a great deal of information is needed to estimate model parameters, and, in many cases, data are inadequate to develop the model for a specific fishery. According to Taylor (1981), the minimum data requirements are (1) the initial age structure or number of fish in each age-class, (2) the average length-at-age relationship for the population, (3) weight–length regression coefficients, and (4) natural and fishing mortality rates.

During the 1990s, there has been increased use of mathematical models that follow large numbers of individual animals simultaneously to describe population dynamics. Unlike the previously described population models, which implicitly assume that all individuals in the modeled population respond identically, individual-based models allow the study of individual variation as well as calculation of the average dynamics for a population (DeAngelis and Gross 1992; Chambers 1993). Actually, the development of individual-based models is an endpoint reached by the reduction of a population into its component parts. We may initially think that a population needs to be categorized according only to age-group but then find that the distinction of gender, size, or other subdivision is necessary to represent dynamics adequately (Crowder et al. 1992). Ultimately, this process leads to the modeling of each individual separately and using the results of numerous simulations to develop generalities about the subdivisions or the population as a whole. Some recent applications of individual-based models include evaluations of population dynamics of young-of-the-year striped bass (Rose and Cowan 1993), growth of young bluegills (Breck 1993), and contaminant accumulation in stocked lake trout (Madenjian and Carpenter 1993).

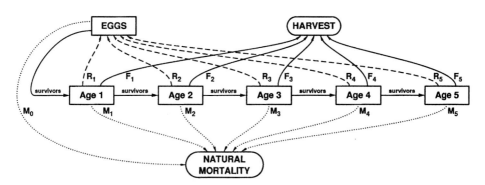

Figure 6.9 Representation of an age-structured population with maximum fish age of 5 years. Age-specific rates of reproduction (R_i), fishing mortality (F_i), and natural mortality (M_i) regulate survival from one age to the next.

6.5 CONCLUSION

Recommendations for optimal management of a fishery should be based on changes in population number, composition, and biomass that are expected to result from management efforts. Development of optimal management strategies requires assessment of fish population characteristics, including population size, mortality, and growth rates, as well as predictions of future trends by use of mathematical models. This chapter provides an introduction to basic methods for studying population dynamics; fisheries managers should consider the suitability of more advanced techniques in specific cases.

The development of models useful for inland fisheries management has been difficult because of complicated processes of fish population regulation and multiple management objectives. Models with broader applicability and greater predictive capability remain elusive targets for managing inland recreational fisheries on a sustained basis. Annual and within-year variation of rates of reproduction, growth, and survival often make it necessary to conduct population assessments over several years to obtain estimates needed for model development. In many cases, long-term studies are not feasible, and population statistics like those described in Chapter 7 are used in lieu of predictive models for making management decisions.

6.6 REFERENCES

Allen, M. S., and L. E. Miranda. 1995. An evaluation of the value of harvest restrictions in managing crappie fisheries. North American Journal of Fisheries Management 15:766–772.

Backiel, T., and E. D. Le Cren. 1967. Some density relationships for fish population parameters. Pages 261–293 in S. D. Gerking, editor. The biological basis of freshwater fish production. Blackwell Scientific Publications, Oxford, UK.

Bailey, N. J. J. 1951. On estimating the size of mobile populations from recapture data. Biometrika 38:293–306.

Beamesderfer, R. C. P., and J. A. North. 1995. Growth, natural mortality, and predicted response to fishing for largemouth bass and smallmouth bass populations in North America. North American Journal of Fisheries Management 15:688–704.

Beverton, R. J. H., and S. J. Holt. 1957. On the dynamics of exploited fish populations. Fishery Investigations, Series II, Marine Fisheries, Great Britain Ministry of Agriculture, Fisheries and Food 19.

Breck, J. E. 1993. Hurry up and wait: growth of young bluegills in ponds and in simulations with an individual-based model. Transactions of the American Fisheries Society 122:467–480.

Brownie, C., D. R. Anderson, K. P. Burnham, and D. S. Robson. 1985. Statistical inference from band recovery data–a handbook, 2nd edition. U.S. Fish and Wildlife Service Resource Publication 156.

Carlander, K. D. 1969. Handbook of freshwater fishery biology, volume 1. Iowa State University Press, Ames.

Chambers, R. C. 1993. Phenotypic variability in fish populations and its representation in individual-based models. Transactions of the American Fisheries Society 122:404–414.

Chapman, D. G. 1951. Some properties of the hypergeometric distribution with applications to zoological sample censuses. University of California Publications in Statistics 1:131–160.

Clady, M. D., D. E. Campbell, and G. P. Cooper. 1975. Effects of trophy angling on unexploited populations of smallmouth bass. Pages 425–129 in R. H. Stroud and H. Clepper, editors. Black bass biology and management. Sport Fishing Institute, Washington, D.C.

Cochran, W. G. 1977. Sampling techniques. Wiley, New York.

Colvin, M. A. 1991. Evaluation of minimum size limits and reduced daily limits on the crappie populations and fisheries in five large Missouri reservoirs. North American Journal of Fisheries Management 11:585–597.

Crowder, L. B., J. A. Rice, T. J. Miller, and E. A. Marshall. 1992. Empirical and theoretical approaches to size-based interactions and recruitment variability in fishes. Pages 237–255 in D. L. DeAngelis and L. J. Gross, editors. Individual-based models and approaches in ecology: populations, communities, and ecosystems. Routledge, Chapman, and Hall, New York.

Cushing, D. H., and J. G. Harris. 1973. Stock and recruitment and the problem of density-dependence. Rapports et Proces Verbaux des Reunions, Conseil Permanent International pour l'Exploration de la Mer 164:142–155.

DeAngelis, D. L., and L. J. Gross, editors. 1992. Individual-based models and approaches in ecology: populations, communities, and ecosystems. Routledge, Chapman, and Hall, New York.

Galluci, V. F., and T. J. Quinn. 1979. Reparameterizing, fitting, and testing a simple growth model. Transactions of the American Fisheries Society 108:14–25.

Gerhardt, D. R., and W. A. Hubert. 1991. Population dynamics of a lightly-exploited channel catfish stock in the Powder River system, Wyoming–Montana. North American Journal of Fisheries Management 11:200–205.

Goedde, L. E., and D. W. Coble. 1981. Effects of angling on a previously fished and an unfished warmwater fish community in two Wisconsin lakes. Transactions of the American Fisheries Society 110:594–603.

Graham, M. 1935. Modern theory of exploiting a fishery, and application to North Sea trawling. Journal du Conseil International pour l'Exploration de la Mer 10:264–274.

Gulland, J. A. 1969. Manual of methods for fish stock assessment, part 1. Fish population analysis. FAO (Food and Agriculture Organization of the United Nations) Manuals in Fisheries Science 4:1–154, Rome.

Gulland, J. A. 1983. Fish stock assessment. A manual of basic methods. Wiley, New York.

Hilborn, R., and C. J. Walters. 1992. Quantitative fisheries stock assessment: choice, dynamics, and uncertainty. Chapman and Hall, New York.

Johnson, B. M., and five coauthors. 1992. Forecasting effects of harvest regulations and stocking of walleyes on prey fish communities in Lake Mendota, Wisconsin. North American Journal of Fisheries Management 12:797–807.

Johnson, B. L. 1995. Applying computer simulation models as learning tools in fishery management. North American Journal of Fisheries Management 15:736–747.

Ludsin, S. A., and D. R. DeVries. 1997. First-year recruitment of largemouth bass: the interdependency of early life stages. Ecological Applications 9:1024–1038.

Luecke, C., T. C. Edwards, Jr., M. W. Wengert, Jr., S. Brayton, and R. Schneidervin. 1994. Simulated changes in lake trout yield, trophies, and forage consumption under various slot limits. North American Journal of Fisheries Management 14:14–21.

Maceina, M. J., S. J. Rider, and D. R. Lowery. 1993. Use of a catch-depletion method to estimate population density of age-0 largemouth bass in submersed vegetation. North American Journal of Fisheries Management 13:847–851.

Maceina, M. J., W. B. Wrenn, and D. R. Lowery. 1995. Estimating harvestable largemouth bass abundance in a reservoir with an electrofishing catch depletion technique. North American Journal of Fisheries Management 15:103–109.

Madenjian, C. P., and S. R. Carpenter. 1993. Simulation of the effects of time and size at stocking on PCB accumulation in lake trout. Transactions of the American Fisheries Society 122:492–499.

Malvestuto, S. P. 1996. Sampling the recreational fishery. Pages 591–623 in B. R. Murphy and D. W. Willis, editors. Fisheries techniques, 2nd edition. American Fisheries Society, Bethesda, Maryland.

Michaletz, P. H. 1997. Factors affecting abundance, growth, and survival of age-0 gizzard shad. Transactions of the American Fisheries Society 126:84–100.

Miranda, L. E., and W. D. Hubbard. 1994a. Length-dependent winter survival and lipid composition of age-0 largemouth bass in Bay Springs Reservoir, Mississippi. Transactions of the American Fisheries Society 123:80–87.

Miranda, L. E., and W. D. Hubbard. 1994b. Winter survival of age-0 largemouth bass relative to size, predators, and shelter. North American Journal of Fisheries Management 14:790–796.

Otis, D. L., K. P. Burnham, G. C. White, and D. R. Anderson. 1978. Statistical inference from capture data on closed animal populations. Wildlife Monographs 62:1–135.

Redmond, L. C. 1974. Prevention of overharvest of largemouth bass in Missouri impoundments. Pages 54–68 *in* J. L. Funk, editor. Symposium on overharvest and management of largemouth bass in small impoundments. American Fisheries Society, North Central Division, Special Publication 3, Bethesda, Maryland.

Ricker, W. E. 1954. Stock and recruitment. Journal of the Fisheries Research Board of Canada 11:559–623.

Ricker, W. E. 1975. Computation and interpretation of biological statistics of fish populations. Fisheries Research Board of Canada Bulletin 191.

Ricklefs, R. E. 1967. A graphical method of fitting equations to growth curves. Ecology 48:978–983.

Robson, D. S., and D. G. Chapman. 1961. Catch curves and mortality rates. Transactions of the American Fisheries Society 90:181–189.

Robson, D. S., and H. A. Regier. 1964. Sample size in Petersen mark-recapture experiments. Transactions of the American Fisheries Society 93:215–226.

Rose, K. A., and J. H. Cowan, Jr. 1993. Individual-based model of young-of-the-year striped bass population dynamics. I. Model description and baseline simulations. Transactions of the American Fisheries Society 122:415–438.

Schnabel, Z. E. 1938. The estimation of the total fish population of a lake. American Mathematical Monographs 45:348–368.

Schnute, J. 1985. A general theory for analysis of catch and effort data. Canadian Journal of Fisheries and Aquatic Sciences 42:414–429.

Seber, G. A. F. 1982. The estimation of animal abundance and related parameters, 2nd edition. Griffin, London.

Shaefer, M. B. 1954. Some aspects of the dynamics of populations important to the management of the commercial marine fisheries. Inter-American Tropical Tuna Commission Bulletin 1:27–56.

Shaefer, M. B. 1968. Methods of estimating effects of fishing on fish populations. Transactions of the American Fisheries Society 97:231–241.

Shelton, W. L., W. D. Davies, T. A. King, and T. J. Timmons. 1979. Variation in growth of the initial year class of largemouth bass in West Point Reservoir, Alabama–Georgia. Transactions of the American Fisheries Society 108:142–149.

Smith, C. L. 1983. Evaluating human factors. Pages 431–446 *in* L. A. Nielsen and D. L. Johnson, editors. Fisheries techniques. American Fisheries Society, Bethesda, Maryland.

Taylor, M. W. 1981. A generalized inland fisheries simulator for management biologists. North American Journal of Fisheries Management 1:60–72.

Thompson, P. D., and F. J. Rahel. 1996. Evaluation of depletion-removal electrofishing of brook trout in small Rocky Mountain streams. North American Journal of Fisheries Management 16:332–339.

von Bertalanffy, L. 1938. A quantitative theory of organic growth. Human Biology 10(2):181–213.

Walker, S. H., M. W. Prout, W. W. Taylor, and S. R. Winterstein. 1993. Population dynamics and management of lake whitefish stocks in Grand Traverse Bay, Lake Michigan. North American Journal of Fisheries Management 13:73–85.

Walters, C. J. 1969. A generalized computer simulation model for fish population studies. Transactions of the American Fisheries Society 98:505–512.

Watt, K. E. F. 1959. Studies on population productivity. II. Factors governing productivity in a population of smallmouth bass. Ecological Monographs 29:367–392.

Weithman, A. S. 1986. Measuring the value and benefits of reservoir fisheries programs. Pages 11–17 *in* G. E. Hall and M. J. Van Den Avyle, editors. Reservoir fisheries management: strategies for the 80's. American Fisheries Society, Southern Division, Reservoir Committee, Bethesda, Maryland.

White, G. C., D. R. Anderson, K. P. Burnham, and D. L. Otis. 1982. Capture-recapture and removal methods for sampling closed populations. Los Alamos National Laboratory, LA-8787-NERP, Los Alamos, New Mexico.

Zagar, A. J., and D. J. Orth. 1986. Evaluation of harvest regulations for largemouth bass populations in reservoirs: a computer simulation. Pages 218–226 *in* G. E. Hall and M. J. Van Den Avyle, editors. Reservoir fisheries management: strategies for the 80's. American Fisheries Society, Southern Division, Reservoir Committee, Bethesda, Maryland.

Zippin, C. 1956. An evaluation of the removal method of estimating animal populations. Biometrics 12:163–169.

Zippin, C. 1958. The removal method of population estimation. Journal of Wildlife Management 22:82–90.

Chapter 7

Practical Use of Biological Statistics

JOHN J. NEY

7.1 INTRODUCTION

Managers are often responsible for conducting routine surveys of fish communities. Fish are captured, counted, and measured for length and weight, and scales are removed for age analysis. The resulting data are conscientiously recorded, but too often they receive only superficial analysis: a small payoff for a large effort. If properly designed and consistently executed, surveys can yield information required for management of fish populations. Samples of fish can provide statistics concerning recruitment, growth, and survival that describe how the fish stock of interest is responding to physical, biological, and human factors that control its health and abundance.

Management decisions should always begin with a thorough assessment of available biological statistics. At the very least, this evaluation will identify aspects of population and community dynamics that may need more focused study. When biological statistics are sufficiently descriptive, they can be related to the socioeconomic concerns of the fishing clientele (Chapter 8) to determine options for enhancing angler satisfaction. Management decisions involving habitat modifications (Chapters 9–12), community manipulations (Chapters 13–16), or altered fishing regulations (Chapter 17) can be guided by their potential for success and the availability of resources.

7.2 ASSESSING INFORMATION NEEDS

The nature and amount of data required to evaluate a fishery periodically depend on the management intentions for that resource. The first step in designing a program (see Chapter 2) should be to formulate those intentions as a goal (e.g., to develop a trophy muskellunge fishery, to maintain balanced bluegill–largemouth bass populations, or to maximize sustainable crappie harvest). From this general goal, specific objectives with precise and measurable characteristics (e.g., within 3 years to grow muskellunge to 80 cm and to achieve 40% annual survival) are derived. The objectives, in turn, direct the search for biological data that accurately measure the desired characteristics.

The final step in identifying data needs is the choice of statistics to be used to estimate population or community parameters. In most instances, several alternative data collection methods and resultant statistics can provide a measure of each parameter. Selection of the best statistic is determined by a balance between the cost (effort) to obtain it and the amount of information that it provides.

The population statistics described in this chapter are simple to obtain, can be incorporated into monitoring studies, and provide information on the dynamics of target stocks. They are intended to provide explanations of how species are responding to their total environment in terms of the dynamic factors that dictate abundance—recruitment, growth, and survival (Chapter 6). Understanding why a population performs as it does often requires more in-depth study. Declining growth, for example, results from an inadequate amount of food available per consumer, which could be due either to too many consumers (including competing species) or poor food production. Sometimes, satisfactory explanations can be developed from the manager's knowledge of the system and the environmental influences acting upon it. Community level, rather than population level, statistics, particularly those describing predator–prey relations, also frequently provide explanations that guide management decisions.

7.3 BIOLOGICAL DATA

7.3.1 Sampling Design

Before implementing a sampling design, the manager should make a careful review of the recent literature to insure that the statistics to be gathered can be compared with existing information and that they will be as accurate as possible by current standards. Detailed discussions of sampling designs to monitor fish stocks, as well as techniques to evaluate angler harvest, are provided by Brown and Austen (1996) and Malvestuto (1996). The sampling design assures two attributes for these data—consistency and representativeness.

7.3.1.1 Consistency

Most uses of fisheries statistics are comparative. The health of a population or community may be evaluated by relating current statistics to those previously reported for fishes from the same system (temporal or trend comparisons). Statistics from several bodies of water may also be compared to assess relative performance (spatial comparisons). When adequate statistics have been obtained on the performance (abundance, recruitment, growth, or survival) of key species in a regional array of systems, standards may be developed to gauge how well fish are doing in a particular system. Valid comparison, whether over time or among different waters, requires that these data always be obtained in the same manner. It may be invalid, for example, to compare the relative abundance of crappies caught in gill nets and trap nets because crappies are much better at avoiding gill nets (Guy et al. 1996). Consistency in sampling design (timing, site, gear, and effort) does not ensure that the statistics being compared are true estimates of populations because the sampling designs can still be biased.

7.3.1.2 Representativeness

Bias occurs when the portion of the population sampled is not typical of the population as a whole. As a consequence, the statistics developed from sampling will not accurately estimate population parameters. The assumption that what can be collected is representative of the population is one of the most troubling assumptions to fisheries scientists. Several measures can be employed to evaluate the accuracy of biological samples and provide a degree of confidence in statistics (Chapter 6). However, very precise samples, for which the difference between one to the next is slight, might also not be accurate if an atypical segment of the population is consistently collected. If trap nets always catch a disproportionate share of young crappies, the age structures depicted by repeated trap-net collections could be very similar but still fail to reflect the real age distribution in the crappie population. A degree of precision (repeatability) is essential to have confidence in statistical estimates, but the only way to insure true representativeness, or accuracy, is to compare the sample statistic directly with the population parameter. Because all the individual fish that compose a population can rarely be examined (as in draining a pond), accuracy of sample statistics is usually evaluated indirectly by comparing estimates developed from two or more sampling methods. Precision can usually be improved by taking more or larger samples. Accuracy is dependent on sampling design to collect fish in proportion to their abundance (Brown and Austen 1996; Wilde and Fisher 1996).

7.3.2 Sources of Data

7.3.2.1 Assessment Survey Data

Fishes captured in the course of surveys provide simple, easily measured data, including length, weight, and appearance. Relationships between length and weight are indicative of the fish's growth and overall health or condition. Occurrence of sores, tumors, ectoparasites, and scars can also be noted as indicators of health. The frequency of evidence of disease in a sample can indicate the health of a population and has been used as a measure of water quality in U.S. streams (Leonard and Orth 1986). More refined methods of fish health assessment require autopsy of specimens and measurement of tissue, organ, and blood variables (Goede and Barton 1990).

Enumeration and measurement of the fishes captured (or a representative subset) can lead to several statistics that estimate population parameters. The species composition (by number and biomass) of the sample can be useful to evaluate predation, competition, and resource use within the fish community as well as water quality.

For a target species, the number of individuals captured with a given amount of fishing (or sampling) effort provides a measure of its absolute abundance or an estimate of its relative abundance for temporal or spatial comparisons. Relative abundance is best expressed as the number caught per unit of effort. Relative abundance of young fish is widely interpreted to assess recruitment.

Plotting the number of fish captured by length interval provides a length-frequency distribution. Where overlap of lengths between successive age-groups are minimal or can be discounted (as is particularly likely for younger fish), the length-frequency distribution will describe the age structure of the population. Under steady state conditions, the length-frequency distribution also describes the patterns of growth with age and can be used to estimate annual mortality rates by the catch curve method (Chapter 6).

When survey designs encompass several different habitats (e.g., depths or substrates), a record of the location of each catch can provide information on fish distribution by size, sex, and species. Analysis of spatial distribution data can help define habitat use.

7.3.2.2 Data from Anglers

The single most important measure of a fishery is the catch. Harvest (or yield) is that portion of the catch that is removed from the system. The number of anglers and the time they spend fishing vary greatly over time and among locations. Consequently, total fishing effort is routinely estimated and applied as the divisor to total catch to provide the mean catch-per-unit-effort statistic (e.g., kg/d or number/h) to describe average angler success for comparative purposes.

If a representative fraction of angler harvest is examined, population data can be obtained that are similar to data obtained by sampling procedures used by biologists. Fish length, weight, growth, size and age distributions, relative abundance, recruitment, mortality, and species composition can all be measured or estimated from hook-and-line samples. Fisheries managers increasingly recognize the recreational catch as an inexpensive and rapid alternative to obtaining samples. Although conventional creel surveys are often labor intensive, large catches can be quickly examined at fishing tournaments or by mail survey of tournament anglers. Dolman (1991) found good agreement in catch statistics (harvest rate and mean weight) for largemouth bass between postcard returns from tournament anglers and creel surveys for six Texas reservoirs. An alternative and complementary approach is to provide anglers with diaries in which to record details of their catch. The time required of fisheries managers varies among the different approaches. For example, in a study of largemouth bass in Lake Minnetonka, Minnesota, one person could examine 1 fish per hour during a creel survey, 5 per hour by electrofishing, 15 per hour through angler diaries, and 40 per hour at tournaments (Ebbers 1987).

However, there is a strong probability that fish taken by hook and line will not be representative of the population, and statistics derived from such samples will inaccurately estimate population parameters. Anglers may use techniques that are not only species selective but size selective as well. Comparison of biological and angler-generated sample statistics for largemouth bass indicate that bias will vary with the body of water and the type of angler. Although Ebbers (1987) reported strong similarity in length-frequency distributions, growth and mortality rates, and population estimates obtained by electroshocking and angler surveys, Gabelhouse and Willis (1986) found that tournament anglers caught a disproportionate number of intermediate-size large-mouth bass. In general, the angler catch should not be substituted for conventionally obtained samples to provide population statistics unless comparison to biological data from each sample shows no real differences.

7.4 STATISTICS FOR STOCK ASSESSMENT

Samples of fish can quickly be described by measuring lengths and weights of individual specimens. These basic data can be transformed into simple statistics to evaluate the two fundamental characteristics of fish stocks—well-being and abundance.

7.4.1 Weight–Length Relationships

The ratio of the weight of a fish to its length varies with the species, the fish's size, and the ecological conditions under which it feeds and expends energy to live. When species and size effects can be discounted, the ratio provides a measure of the fish's health or well-being. Comparison of weight–length ratios of fish in a body of water over time or between different bodies of water helps a manager assess how well fish are able to feed and grow. Weight–length ratios must be obtained from precise and accurate measurements. Under field conditions, accurate measurements of fish length are routine, but weight measurements will be inconsistent unless factors that can affect scale readings (e.g., wind, wave action, and fish movement) are controlled (Gutreuter and Krzoska 1994).

Because weight–length ratios are derived from individual specimens in a sample, the data show both averages and variance, essential properties for statistical tests comparing different groups. When populations are to be compared, weight–length ratios should always be taken in the same season—when tissue accumulation is neither extremely high nor low (e.g., avoid pre- and postspawning samples). The mid- to late growing season is preferred (Wege and Anderson 1978).

Weight–length ratios are usually expressed in whole numbers, called condition factors. Weight (W) of fish tends to increase as a cubic function of length (L), and the weight–length relationship can be expressed as a power curve:

$$W = aL^b,$$

where a and b are population-specific constants. For many populations, b will be close to 3. Condition (K) of an individual fish is sometimes expressed as

$$K = \frac{W \cdot X}{L^3},$$

where X is a scaling constant (100,000 for metric units and 10,000 for English units) to achieve integer status (see, for example, Anderson and Neumann 1996). In this form of expression, known as the Fulton type, the condition factor has achieved wide usage. Comparisons to other populations can be made from published values, such as those in the *Handbook of Freshwater Fishery Biology* (Carlander 1977). However, comparisons are not valid between species or even between length-groups within species because b is not truly 3, and fishes have different or changing body shapes.

In an effort to eliminate the size bias, Le Cren (1951) developed the relative condition factor (K_n) which uses the weight–length relation developed over all size-groups in a particular population:

$$K_n = \frac{W}{\hat{W}},$$

where W is weight and $\hat{W}$ is equal to aL^b. The relative condition factor expresses the deviation of an individual's weight from the average for fish of its length in that population. As such, its use is limited to within-population comparisons, as for seasonal effects or sexual differences in growth.

The concept of relative condition has been refined for interpopulation comparisons by replacing the population-specific weight-length relation with a standard for the species (Wege and Anderson 1978). The resulting condition factor is termed relative weight (W_r) and is determined as follows:

$$W_r = \frac{W}{W_s} \times 100,$$

where W_s is the standard weight for a specimen of the measured length. The standard weight-length relations are developed from available weight-length relations for the species (Wege and Anderson 1978; Murphy et al. 1990). The standard weight represents the 75th percentile weight at a given length for all populations surveyed. When W_r is 100 or greater, a specimen is considered to be in excellent condition. Relative weight may become the most commonly used index of condition because it enables direct comparison of different sizes and species of fishes and provides an instant benchmark for evaluating the well-being of a population without a literature search. Standard weight equations have been developed and published for at least 40 freshwater fish species (see Anderson and Neumann 1996).

Condition factors in their various forms describe the energy status of fish, which is itself dependent on food consumption and metabolic costs. Relative weight has been directly related to fat content in bluegill and fecundity in white crappie (Anderson and Neumann 1996). Within populations, condition factors can be used to monitor the influence of environmental change or human manipulations over time. For example, Colle and Shireman (1980) used condition to evaluate the response of centrarchids to hydrilla infestations in Florida lakes, and Bryan and Ney (1994) compared condition of tagged versus untagged brook trout. Analysis of condition factor dynamics can also identify life stages or seasons during which available food is inadequate. In interpopulation comparisons, poor condition factor may signal the need for more focused study on the amount of food available per consumer (a function of both food production and intensity of competition) or the environmental factors (such as cover and water quality) that affect its efficient use.

Feeding efficiency does not necessarily translate into somatic growth, and condition has a mixed record as a predictor of growth rate (Anderson and Neumann 1996). In a study of 10 Quebec lakes, Liao et al. (1995) found little evidence of a relationship between relative weight and growth for either pumpkinseed or golden shiner. Conversely, DiCenzo et al. (1995) reported that relative weight was positively related to growth rate in populations of Alabama spotted bass. Use of condition as a surrogate measure of growth rate in assessment surveys should be limited to situations in which a consistent positive relation has been demonstrated.

7.4.2 Abundance

Knowledge of the numerical abundance of a fish stock is a component of the information required for its management. When numbers are matched to weight-at-length data, total stock biomass (usually referred to as standing crop or standing stock) can be calculated. Methods to estimate absolute abundance (Chapter 6) frequently require more effort and expense than can be allocated in assessment surveys. Instead, the catch of fishes in a survey sample is related to the effort expended to collect that sample in a statistic known as relative abundance. Division of the catch (C, usually number but sometimes weight) by effort (f) yields catch per unit effort (C/f, or CPUE) and, in theory, removes the effect of variable effort in the measurement of abundance. Relative abundance is used to make temporal or spatial comparisons. If a high correlation can be demonstrated between relative and absolute abundance, CPUE may also be used to estimate actual stock size.

Sampling effort must be precisely measured and consistently expended for developing relative abundance statistics. A wide variety of effort units are in common use; most involve measures of the amount of gear deployed, duration of sampling, or area sampled, alone or in combination. Consistency of sampling effort is a requisite for valid comparisons; capture efficiency cannot be affected by how the sample was collected. Gear should be standardized in type, technical specifications (e.g., gill net mesh sizes or trawl throat dimensions), and manner of operation. Similar habitats should always be used as sampling locations. Sampling to compare relative abundances should also be conducted at the same time of year, but unusual weather or water conditions (e.g., storms or turbidity) likely to affect capture success should be avoided. Prolonged sampling effort can change fish behavior or otherwise cause the gear to become less effective. Catching efficiency can be altered by even slight changes in the sampling regime.

Application of the relative abundance statistic is predicated on the assumption that catch per unit effort is directly proportional to population size (N). For this to be true, catchability (q, the probability of catching an individual fish in one unit of effort) has to be constant (from Chapter 6, $C/f = qN$). In reality, catchability is never constant. Fish activity patterns, weather, and water quality change rapidly and influence capture success. Gear saturation (excess effort) also reduces catchability. These influences can usually be minimized by sampling under typical environmental conditions with a moderate amount of well-distributed effort. Sampling gear should be chosen neither to attract nor to repel the target species but rather to catch them by random encounter. In several commercial and sport fisheries, catchability has declined as abundance has increased (Bannerot and Austin 1983), causing catch per unit effort to stabilize. The opposite relationship appears to result from intense fishing pressure or from changes in angler behavior, but this increased catchability does not seem likely to occur in typical assessment surveys. The only means to validate the proportionality assumption is to compare relative and absolute abundance directly, and this is usually not possible. Sampling programs to obtain relative abundance statistics have been developed for most prominent freshwater species in different types of systems.

If a paired series of CPUE statistics and absolute abundance indices are highly correlated, a single measure of relative abundance can be used to predict actual population size. The technique is illustrated by Hall (1986), who compared shoreline electrofishing CPUE for largemouth bass with concurrent mark–recapture population estimates in 12 Ohio impoundments (Figure 7.1). The paired data were analyzed by linear regression with CPUE as the independent variable (X) and population density as the dependent variable (Y).

Relative abundance statistics obtained for a single stock over time can be used to assess whether stock size is changing. Analysis of temporal patterns of relative abundance is valuable both to identify trends and to evaluate phenomena (management actions, fishing pressure, or environmental alterations) that have the potential to cause change in population sizes. Although it is often possible to spot trends visually from graphs of CPUE versus time, short-term variability might mask trends. Linear regression, in this instance with CPUE as the dependent variable and time as the independent variable, should be applied to determine whether a trend actually exists (slope of the regression line is not zero), the direction and rate of change (the slope itself), and the strength (r^2) of the CPUE-versus-time relationship (Figure 7.2). Measures of the sus-

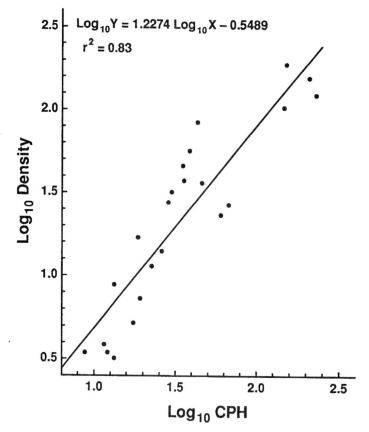

Figure 7.1 Estimated population density (mark-recapture estimates) as a function of relative abundance (electroshocking catch per hour, CPH) of largemouth bass in 12 Ohio impoundments. Both variables are transformed to $\log_{10}$ to increase linearity of the relationship. Figure is from Hall (1986).

pected agent of change in fish abundance (e.g., fishing pressure or nutrient concentration) may be available over the time period of assessment; these can then substitute for time as the independent variable in the regression.

Relative abundance of juvenile fish is frequently used to assess reproductive success and to project the future abundance of the adult stock. To be effective as a predictor of recruitment, relative abundance of juvenile fish must be positively related to their later abundance in the catchable-size stock. Larval and juvenile survival rates are highly variable within many fish populations because these life stages are particularly vulnerable to starvation and predation (Ploskey and Jenkins 1982). It is imperative that the validity of a recruitment index be established before its use. Comparison of the relative abundances of a cohort as juveniles and following recruitment by means of regression analysis can provide that assessment (Curtis et al. 1993).

Management agencies commonly employ standardized sampling schemes in assessment surveys (Nielsen and Johnson 1983). Relative abundances of species and life stages obtained by standard procedures can then be compared among similar systems to give an overview of density patterns, provide insight into influencing factors, develop management strategies, and adjust angler expectations. Although sampling effort is easily controlled, lakes and rivers are all unique and have inherent differences in the ease with which fishes may be taken from them by a particular method. For example, CPUE for gill-net samples from clear and turbid lakes will likely be affected by differential net avoidance, regardless of true population densities. Care must therefore be taken to limit intersystem comparisons to waters in which catchability should be similar.

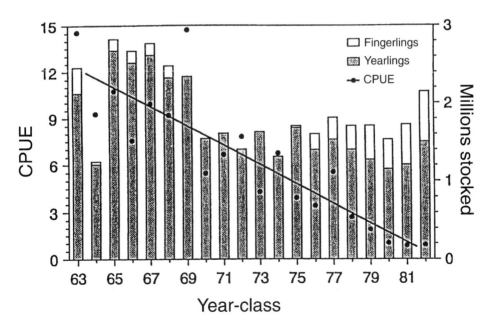

Figure 7.2   Survival of lake trout stocked in U.S. waters of Lake Superior, 1963–1982. Dots represent catch per unit effort (CPUE; number fish per 205-m gill net) for age-7 lake trout of each year-class. Solid line is CPUE regression. Figure is from Hansen et al. (1994).

7.4.3 Population Structure

When all ages or sizes of a species are taken in proportion to their true abundances, the sample is representative of the structure of the population. Analysis of the relative abundance of the various age-groups in the sample provides measures of the dynamic processes that govern population number and biomass. Size-group comparisons can be formulated in indices for rapid assessment of the harvest potential of the population. Coupling of size-structure indices (usually length based) with size-at-age statistics permits an understanding of how the size composition of the population is achieved.

The age-frequency distribution of a representative population sample describes the status of the successive cohorts. Changes in age-frequency distributions over time will identify instances of excessive mortality. Age-frequency distributions can also be used to describe the pattern of growth in length within populations. If growth conditions are assumed to be static, plotting means and standard deviations of lengths at successive ages will be sufficient; otherwise, particular cohorts must be tracked over their life spans by means of annual samples.

The age composition of a sample is best determined by aging individual fish from hard body parts that deposit annual rings (DeVries and Frie 1996). Aging fish from scales or other structures is a laborious process and often not possible where growth is year-round, such as in tropical waters. Consequently, length-frequency analysis, in which length modes (most frequent lengths) are assigned ages, has become a popular alternative to direct aging procedures.

Assignment of ages from length-frequency distributions is widely practiced but of limited utility. A length-frequency histogram is likely to show distinct modes for the youngest age-groups but less distinct or indistinct modes for older fish (see DeVries and Frie 1996:Figure 16.1). Short spawning periods and low variability in early growth rates cause these modes to be sharp and well separated. However, a multimodal length distribution can develop if spawning is prolonged or intermittent or if differential growth occurs (Shelton et al. 1979). The distance between true modes and the variability within each mode's associated distribution will determine the degree of resolution of individual modes in the overall length-frequency distribution. For normal distributions of length at age, only a single mode will be apparent for two adjacent age-groups if the difference between modes is less than twice the minimum standard deviation (Figure 7.3). Growth slows and becomes more variable as fish age, causing high overlap in distributions and smaller distances between modes. Both damping of modes and spurious modes confound the assignment of ages to length-frequency distributions. Resolution for early age-groups can be improved by proper choice of measurement interval for graphical display (Anderson and Neumann 1996), but assignment of ages by visual inspection will still likely be limited to the first few age-groups. Microcomputer programs that define overlapping distributions are available, but they require informed decisions on input parameters to minimize error and have not as yet received much use for inland, temperate-latitude fishes.

For comparative purposes, the length at age of only a single age-group may be adequate to characterize growth in the population. Kruse (1988) used total length at age 3 to assess growth rates of largemouth bass in Missouri impoundments and to clarify interpretation of the length-frequency distributions of these populations. It gen-

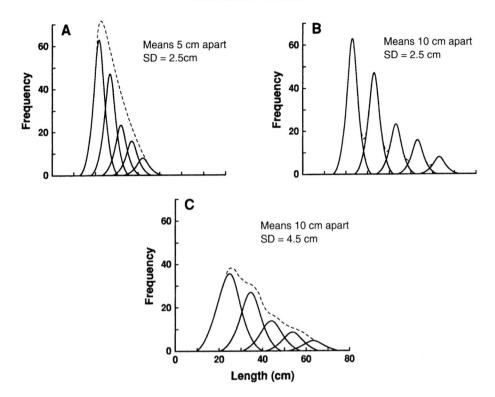

Figure 7.3 Effects of distribution about means on the distinctness of successive modes in length-frequency plots. Dashed curves are frequency plots for an infinite population. In (B), dashed curves follow solid curves very closely. Proportions of the population represented by each successive age-group are the same in all three panels. Figure is from Macdonald (1987).

erally will be best to choose an intermediate age to indicate both cumulated growth performance over several years and how rapidly fish are reaching harvestable size. The effort required to estimate mean total length at any single age should be a small fraction of that needed to determine length at each successive age in the population.

Length-frequency distributions are better suited to describe the status of a fishable population than its dynamics. The size distribution of fish in a representative sample can be quickly analyzed without reference to age composition to assess the balance of the population or community. As defined by Swingle (1950), a population is balanced when it can sustain a satisfactory (for management objectives) harvest of good-size fish in proportion to the productivity of the habitat. Balance therefore depends on the density of fish of various sizes in the population—both adequate numbers of catchable-size fish and sufficient numbers of smaller fish to provide replacement.

Indices of population balance derived from length-frequency distributions have been most widely used in small (<200 ha) midwestern impoundments with largemouth bass–bluegill fisheries. The basic index is proportional stock density (PSD) developed by Anderson (1976) as

$$PSD\ (\%) = \frac{number \geq quality\ size}{number \geq stock\ size} \times 100.$$

Minimum stock and quality lengths are defined as some length within 20–26% and 36–41%, respectively, of angling world record length (Anderson and Weithman 1978). These length categories correspond roughly to the minimum sizes at which anglers will first catch the species (stock length) and consider the specimen desirable (quality length). Gustafson (1988) has developed a confidence interval estimator for PSD, and Miranda (1993) has developed the methodology to determine minimum sample sizes to achieve particular levels of precision.

Two approaches can be used to determine the range of PSD values indicative of a population structure that meets management objectives for a balanced population of a particular species in a given system. For the empirical method, PSDs can be calculated for situations in which management objectives (e.g., harvest or predation intensity) are being met, and the range of these PSD values is then used as the standard for similar systems. Alternatively, a sample model can be employed to generate PSDs based on average or optimal annual growth and mortality rates and constant recruitment (Anderson and Weithman 1978). Desirable PSD values will vary with management objectives. For example, control of stunting in bluegill populations could require a much higher proportion of stock size largemouth bass than would be optimal to support a trophy largemouth bass fishery.

Generally, PSDs indicative of balance in a target species population are based on sustainable harvest of sizes preferred by anglers (Table 7.1). The relative stock density (RSD) index has been developed to assess better the size distribution within the quality length portion of the population. Although RSD can be the proportion of any size-group within the total sample of stock-size and larger fish, it is usually used to track subsets of the quality size group (RSD_{30}, for example, would refer to all fish over 30 cm). A standard categorization of quality size fish that is linked to the angling experience has been proposed (see Table 15.2 in Anderson and Neumann 1996). These additional length-groups are also percentages of world record lengths: preferred (minimum 45–55%), memorable (minimum 59–64%), and trophy (minimum 74–80%). Determination of RSDs of game fishes in these categories provides a readily understandable description of the fishing opportunity provided by the population. Precision can be increased by calculating specific RSDs as the proportion of total stock within each category length range rather than as the proportion of all fish longer than the minimum category length.

Table 7.1 Ranges of proportional stock density (PSD) values indicative of balance when the population supports a substantial fishery.

Species	PSD	Source
Largemouth bass	40–70	Anderson (1980)
Bluegill	20–40	Anderson (1980)
Yellow perch	30–50	Anderson and Weithman (1978)
Walleye	30–60	Anderson and Weithman (1978)
Northern pike	30–60	Anderson and Weithman (1978)
Muskellunge	30–60	Anderson and Weithman (1978)
Smallmouth bass	30–60	Anderson and Weithman (1978)

Although the PSD and RSD indices assess population structure in terms of angling potential, management actions should not be undertaken without additional information. Balance requires not only size but also number; density of the population should also always be estimated (most often this will be as CPUE). Management responses to adjust population size structure may also require knowledge of growth, mortality, and recruitment rates. A PSD value judged to be too high could be due to rapid growth, low mortality, or recruitment of an extremely strong year-class; the inverse is true for low PSD. The appropriate management response to rectify an extreme PSD might be quite different if the source of the perceived imbalance is growth rather than recruitment or survival. For example, imposition of a minimum length limit for white crappie was followed by an increase in PSD in two Texas reservoirs subject to overharvest of small, young fish; however, no change in PSD occurred in a third impoundment with a stunted white crappie population (Webb and Ott 1991).

Size selectivity in sampling can be a major impediment to obtaining an accurate length-frequency index, particularly in large systems with diverse habitats. Sampling bias associated with gear selectivity or differential seasonal distributions of fish can cause PSD values to vary severalfold. Comparison of the length distributions in samples collected with different gears or in successive samples can be used to identify and control bias (Carline et al. 1984).

A potentially greater obstacle to the general adoption of length-frequency indices in inland fisheries management is system instability. Most management alternatives to correct imbalance in fish populations assume that recruitment, growth, and mortality are density dependent. Although density dependence does appear to control these processes in small, central-latitude impoundments, abiotic factors may play a greater role in streams, in larger waters, or at higher latitudes. In particular, reproductive success and eventual recruitment may fluctuate widely from year to year in response to climatic factors. Extremes in year-class strength will be reflected in length-frequency indices. The PSDs of largemouth bass in an Ohio impoundment and of walleye in a natural Wisconsin lake varied threefold in one year, largely as a function of recruitment (Carline et al. 1984; Serns 1985). In these situations, the usefulness of structural indices is likely to be limited to tracking population changes beyond management control. Use of structural indices as an assessment tool for management has worked best in small waters with simple communities that are not overfished (Willis et al. 1993).

7.4.4 Angler Catch and Fishing Effort

The three fundamental descriptors of a fishery are the catch (C, and its subcomponent, harvest), fishing effort expended (f), and catch per unit of effort (C/f, or CPUE). Any two of these statistics can be used to calculate the third. All are typically estimated by the creel survey technique, which consists of interviews of anglers, inspection of the catch, and tabulation of hours spent fishing. For most bodies of water, the creel survey is a sample requiring statistical expansion over time and area to achieve total estimates of catch and effort in a particular period. Choice of sampling design and the method to contact anglers can strongly influence the resulting estimates; options are discussed by Malvestuto (1996).

Quantitative description of a catch usually requires interviewing anglers in addition to inspecting their fish to characterize the fraction discarded. Size and bag limits have long restricted angler harvest, and the rising popularity of catch-and-release fishing for many species is widening the discrepancy between what is caught and what is retained. However, the number and weight of fish harvested by fishing remains the most critical piece of information to the manager responsible for sustaining a balanced yield.

Fishing effort is best measured as the number of angler hours because trip length is highly variable. Total effort in time fished is frequently related to the size of the body of water (e.g., hours per surface hectare or stream kilometer) to describe the distribution of fishing pressure over time or among systems. Profiles of fishing pressure can prompt management actions to redistribute fishing effort to alleviate crowding, promote catch rates, and enhance angler satisfaction. In multispecies fisheries, effort should be partitioned by hours spent pursuing target species to avoid underestimating angler success.

The ratio of the catch of target species to the effort expended in their pursuit may be the single best indicator of fishing quality. However, fishing success as measured by CPUE should not be equated with angler satisfaction. The objectives for "going fishing" are varied, reflecting not only the angler's personal philosophy but also that individual's expectations for the total experience. Definition of angler satisfaction remains elusive, but a better approximation will be achieved by including social as well as fisheries metrics in the assessment (Chapter 8).

Catch per unit effort can be estimated from a representative sample of the angling population and used to derive total catch from total effort ($C = f[C/f]$) or total effort from total catch ($f = C/[C/f]$) if either of these statistics is not available. The CPUE statistic has also sometimes been used as an index of abundance for intensively fished species. Use of angler CPUE as an abundance estimator should be avoided because hook-and-line capture rate is influenced by many factors other than abundance, especially the interactive behavior of the fishers and their prey.

Creel surveys on large waters are laborious and expensive, but they are usually the only way to obtain accurate total catch and effort statistics. However, anglers and their catch can be monitored in a variety of less intensive ways to obtain useful statistics. Citation and tag return programs have both proved useful in this regard.

Rewarding successful anglers with certificates or other forms of recognition is an inexpensive way to monitor species catch on a state or regional basis. Citation programs usually recognize trophy fish; minimum species-specific weights are required for entry. Information gathered in this manner must be supplemented by biological and creel statistics to appraise the status of the whole population.

Tags recovered by anglers from previously marked fish (M = number of fish marked) are commonly used to estimate fishing mortality. When the number of tag returns (R) is totaled for the year following marking, the ratio R/M is a direct estimate of the annual exploitation rate (u) (i.e., the percent of the initial population that has been harvested). The exploitation rate can be used alone or in conjunction with the total mortality rate (A) to assess the effect of fishing on the population (Chapter 6).

The validity of mortality and abundance estimates based on angler tag returns depends on, among other things, the ability of all anglers to recognize tags and angler willingness to report tags, as well as the tag retention rate. The tag reporting rate can best be determined by comparing the number of tags observed in a creel survey with the number subsequently reported by interviewed anglers (Green et al. 1983). The percent of tags reported may double if rewards (e.g., cash or hats) are offered (Zale and Bain 1994). Double-tagging fish will generate an estimate of percentage tag loss (Muoneke 1992).

7.4.5 Fishery Productivity

The capacity of a body of water to support fish biomass (carrying capacity) and to provide a sustainable fish harvest is extremely useful information to the fisheries manager. Knowledge of system capacity allows the manager to adjust angler expectations and manipulate the harvest to make best use of the resource. However, direct determination of biomass carrying capacity and sustainable yield for individual waters generally requires detailed, long-term data.

Much research has been devoted to empirical prediction of fish standing stock and yield. A measure of standing stock or yield as the dependent variable is paired with an independent variable or variables without substantiation of a cause-and-effect relationship. Paired data are obtained for a set of waters where the data have been accurately measured and then analyzed by regression analysis (Ryder 1965).

The search for suitable predictors of fishery productivity has been long and intense because of the obvious benefits. Early efforts to predict fishery productivity in lakes focused on morphometric features such as surface area and mean depth. Predictive power was improved by using nutrient or other chemical (edaphic) variables as predictors (Hanson and Leggett 1982). The most well-known predictor of fishery productivity in natural lakes is the morphoedaphic index, which is the concentration of total dissolved solids divided by mean depth (Ryder et al. 1974). The morphoedaphic index remains among the best predictors, but it must be applied judiciously to lakes that meet particular climatic and chemical criteria. Biological variables such as phytoplankton or benthic invertebrate standing stock or chlorophyll-*a* activity have also demonstrated good predictive power for fishery productivity in lakes, but these variables may be difficult to measure accurately. Regression models that predict fishery productivity in North American streams are also numerous, but their usefulness to date has been limited by small sample size, narrow focus, complex (multivariate) structure and data requirements, and low predictive power.

The number of simple regression models that can predict fish standing stock or harvest potential in standing waters is substantial (Table 7.2) and can be expected to increase for both lakes and streams. Selection of a model to project fishery productivity in a particular system should be based on meeting three criteria: (1) feasibility of consistent and representative measurement of the predictor, (2) predictive power, and (3) similarity of the system to those in the regression data set.

Table 7.2   Single-variable predictors of fish standing stock or yield for North American waters that show good predictive power.

Predictor	Prediction	Waters tested	Source
Benthos standing stock (kg/ha)	Total yield (kg/ha/yr)	Northern natural lakes	Matuszek (1978)
Chlorophyll a (mg/m²)	Sport fish yield (kg/ha/yr)	Midwestern U.S. lakes and reservoirs	Jones and Hoyer (1982)
Morphoedaphic index	Total yield (kg/ha/yr); total standing stock (kg/ha)	Northern natural lakes; U.S. hydropower storage reservoirs	Ryder (1965); Jenkins (1982)
Surface area (km²)	Total yield (kg/yr)	Northern natural lakes	Youngs and Heimbuch (1982)
Total phosphorus (µ/L)	Total yield (kg/ha/yr); total standing stock (kg/ha)	Northern natural lakes Southern Appalachian reservoirs	Hanson and Leggett (1982); Yurk and Ney (1989)

7.5   STATISTICS FOR COMMUNITY ASSESSMENT

An aquatic community includes all plants and animals living in or closely associated with a body of water. In practice, community assessment refers to evaluation of any multispecies combination within the ecosystem. Community assessment statistics have been used for two general purposes by fisheries managers. Status of the relationship between prey (forage) and predatory fishes is described to determine community balance—the potential to provide a satisfactory angling harvest. Composition of the fish species assemblage is analyzed to assess the ability of the system to support a healthy aquatic biota. The statistics employed for each purpose are varied and still evolving. No single estimator of either prey–predator balance or water quality has been proven suitable for general application.

7.5.1 Prey–Predator Relations

The PSD index developed to assess balance within the population can be extended to visualize simultaneously the status of both the prey and predator components of the community. Prey PSD is plotted versus predator PSD in a tic-tac-toe grid (Figure 7.4). Parallel lines bound the PSD percentages indicative of balance (as determined by management objective; Gabelhouse 1984) for each trophic group. The central rectangle formed by the intersection of horizontal and vertical lines defines the desired state of mutual balance for prey and predators. The PSD grid offers a rapid visual status assessment. Originally developed to describe largemouth bass–bluegill relationships in ponds, it can be expanded to other predator–prey combinations (Guy and Willis 1991) or to encompass all predators and prey by weighing for the relative abundance of species in each category (Anderson and Weithman 1978). Drawbacks of the PSD graph are the same as for the single-species PSD index. Additional information on density, growth, recruitment, and mortality are required to explain the status graphically depicted; management actions should not be undertaken on the basis of graph results alone. Uncontrollable variability in population dynamics, particularly recruitment, will be reflected in extreme and potentially misleading annual changes in both prey and predator PSDs. Plots of PSD have been used primarily to track response to management actions in small, relatively simple ponds.

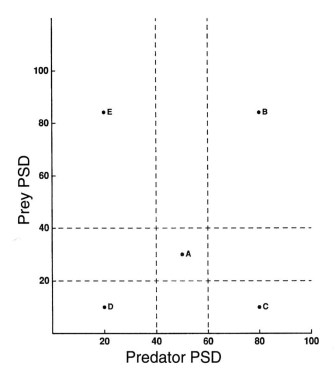

Figure 7.4 Tic-tac-toe grid comparing proportional stock densities (PSD) of predators and prey. Parallel lines bound the desired PSD ranges. Potential interpretations for the different combinations are A, mutual balance for satisfactory fishing; B, community comprises large, old specimens, indicative of an unfished population; C, stunted prey interfering with predator reproduction; D, overfishing of predators and stunting of prey; and E, high population of small predators excessively cropping young prey.

Because biomass is the product of numbers and weight, biomass ratios incorporate densities (absolute or relative abundance) of prey and predators directly. The ratios also serve as indicators of the adequacy of the food supply to meet predator demand (Ney 1990). Biomass ratios were first developed for small, southern U.S. ponds to indicate balance between largemouth bass and their forage (Swingle 1950). The F/C ratio is the total weight of forage species (usually predominately bluegill) divided by the total weight of predator species; a ratio of range 3 to 6 is desirable. The Y/C ratio is a refinement that recognizes size limitations on predation in which Y is the total weight of forage fish small enough to be eaten by the average-size adult predator and C is as before. Balance is indicated by Y/C ratios of 1 to 3. Use of both ratios has been largely limited to southern pond management (see Chapter 21). Desirable pond values do not apply in southern reservoirs, but appropriate values for these and other systems could conceivably be generated through empirical analysis of extensive abundance and harvest data sets.

Jenkins and Morais (1978) developed the available prey to predator ratio (AP:P) as an elaboration of Y/C to account for size limitations on prey consumption for predators of all lengths. The basis of the AP:P determination is a series of curvilinear equations that predict the maximum total length of a forage fish species that can be ingested by a largemouth bass of a given total length. Availability equations are derived from comparisons of mouth diameter of largemouth bass and maximum body depth of prey;

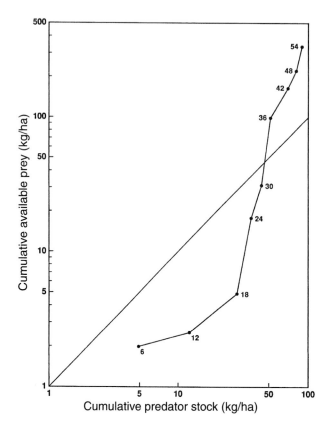

Figure 7.5   Logarithmic plot of prey standing stock morphologically available to the cumulative standing stock of predators at successive 6-cm total length intervals. Points below the 1:1 available prey to predator (AP:P) line indicate clear deficiencies in available prey for predators of these sizes. Figure modified from Jenkins (1979).

all predators are standardized in terms of largemouth bass length equivalents on the basis of their relative throat diameters. Using standing stock estimates and length distributions for all species, available prey biomass is plotted as a function of cumulative predator biomass for successively larger predators (Figure 7.5). Jenkins and Morais (1978) considered that an AP:P of 1:1 was indicative of prey sufficiency for southern reservoirs sampled in August. The AP:P has been used to identify periods and sizes at which predators encounter prey deficiencies in southern reservoirs (Timmons et al. 1980). The AP:P is most easily applied in these systems because standing stocks and size distributions of all species are routinely estimated by cove rotenone sampling. The AP:P approach can be adapted to consider behavioral and distributional effects in addition to size influences on prey availability (Ney 1990). The ingestibility limit equations can be used independent of abundance data to compare relative availability of different prey species or to suggest predator stocking strategies to maximize use of prey (Ney and Orth 1986). Alternatively, prey usage can be determined directly by examining predator diets. DiCenzo et al. (1996) computed an index of vulnerability (defined as the percent of a prey population subject to predation) of gizzard shad to largemouth bass based on the observation that gizzard shad larger than 203 mm in total length were rarely eaten.

It is now recognized that top carnivores in lakes can influence the abundance and composition of lower trophic levels (Carpenter et al. 1985). By consuming planktivorous forage fishes, piscivores relax the predation pressure on zooplankton. The abundance of large zooplankters (e.g., *Daphnia* spp.), which are positively selected by forage fishes, should therefore be directly related to the intensity of predation by piscivores. Mills et al. (1987) confirmed that mean zooplankton size increases with the ratio of predator to prey fishes in New York lakes and advocated zooplankton size as an index of predator–prey balance. They suggested monitoring zooplankton by vertical-haul sampling in spring and again in summer, the latter sample providing a gauge of the abundance of age-0 fishes. Annual changes in mean zooplankton size might be used to assess quickly the success of game and forage fish introductions or the impact of major perturbations, such as winterkill.

7.5.2 Water Quality Indicators

Characterization of the fish assemblage to indicate the well-being of an aquatic ecosystem is attractive because fish represent aquatic life to the general public. Fish, as the dominant vertebrates in most systems, also integrate the effects of multiple influences on the water resource. Most attempts to use fishes as ecosystem indicators have been directed at flowing waters, where disturbances are potentially more severe and identifiable than in lakes.

The single measurement most commonly used to describe a taxonomic group (such as fishes) within a community is diversity. Although ecologists disagree on the exact definition of diversity, the consensus is that it is a function of both the number of species present (richness) and the equitability of the distribution of individuals within these species (evenness). Diversity indices that simultaneously consider both richness and evenness in mathematical formulations are included in Table 7.3. High values of diversity indices have generally been interpreted to indicate relatively unspoiled systems, with diversity inverse to the degree of degradation. However, it is now recognized that diversity can actually increase with disturbance of the system, and that diversity is not inevitably linked with system stability (Angermeier and Karr 1994).

Similarity indices (Table 7.3) compare diversity between two areas. Although their use in aquatic ecosystems has been limited, they have particular potential for contrasting sites that differ only in exposure to degradation; for example, upstream and downstream locations with a point-source discharge between them. Of course, the assumption of uniformity of sites is difficult to verify because a stream is a physical and biological continuum, progressively changing from its headwaters to its mouth. The merits of the many diversity and similarity indices are critically reviewed by Washington (1984).

The indicator organism approach to assessing ecosystem degradation is now widely applied in a formal manner, although subjective appraisal based on the presence or absence of tolerant (e.g., common carp) and intolerant (e.g., rainbow trout) fish species have been made for decades. Karr (1981) proposed the index of biotic integrity (IBI) to assess stream degradation from measurable attributes of the fish assemblage that can be easily derived from a representative sample. As applied to midwestern streams in agricultural areas, the IBI consists of 12 attributes in three categories: species composition, trophic composition, and fish health and abundance (Table 7.4). Spe-

Table 7.3 Diversity indices frequently used to characterize fish communities and similarity indices that have potential for intersite comparisons.

Name of index	Formulation [a]	Source
Diversity indices		
Margalef's diversity index (*D*)	$D = \dfrac{S - 1}{\log_e N}$	Margalef 1958
Shannon–Weaver diversity index (H')	$H' = \displaystyle\sum_{i=1}^{S} \dfrac{n_i}{N} \log_e \dfrac{n_i}{N}$	Shannon and Weaver 1949
Similarity indices		
Sorenson's similarity index (SI)	$SI = \dfrac{2C}{A + B}$	Sorenson 1948
Percentage similarity (PS)	$PS = 100 - 0.5 \displaystyle\sum_{i=1}^{S} \left| a_i - b_i \right|$	Whittaker 1952

[a] Symbols are number of species in sample A (*A*); number of species in sample B (*B*); number of species common to both samples (*C*); percentage of sample A represented by the species (*a*); percentage of sample B represented by the species (*b*); number of individuals in the sample (*N*); number of species in the sample (*S*); and number of individuals of species *i* in the sample (n_i).

cies composition attributes focus on overall richness and richness within major taxonomic groups as well as the occurrences of notably tolerant and intolerant species. Food habits of the fish assemblage as categorized by tropic composition are products of the diversity and productivity of the lower trophic levels in the community. Fish abundance and fish health reflect system productivity and habitat stability. A fish sample is assigned one, three, or five points for each attribute by comparison to expectations for a pristine stream of similar size in the same region. Total scores assign stream health to one of six classes ranging from excellent (pristine) to extremely degraded (no fish). Both IBI metrics and scoring criteria are frequently modified to indicate environmental quality accurately in different regions and waterbodies (Fausch et al. 1990). Still most often used to evaluate streams, IBIs are now being developed for standing waters. The IBI for the littoral zone of the Great Lakes includes metrics based on biomass proportions as well as numerical percentages (Minns et al. 1994). The reservoir fish assemblage index is used by the Tennessee Valley Authority to make annual environmental quality assessments of its main-stem reservoirs (Hickman and McDonough 1996).

Development of an effective IBI first requires knowledge of the structure and function of regional fish communities and of species' tolerances. Ideally, nondegraded systems will be available for selection of appropriate metrics. This was not possible in the generation of the reservoir fish assemblage index because reservoirs by their nature are artificial systems. Once metrics and scoring criteria are developed, the IBI must be field tested for validation.

Table 7.4 Fish community metrics and scoring criteria proposed for the index of biotic integrity. Table taken from Leonard and Orth (1986).

Attribute category and metric	Scoring criteria		
	5 (best)	3 (fair)	1 (worst)
Species composition			
Total number of fish species			
Species richness and composition of darter species			
Species richness and composition of centrarchid species		Varies with stream size and region	
Species richness and composition of sucker species			
Presence of intolerant species			
Proportion of individuals that are green sunfish	<5%	5–20%	>20%
Trophic composition			
Proportion of individuals that are omnivores	<20%	20–45%	>45%
Proportion of individuals that are insectivorous cyprinids	>45%	20–45%	<20%
Proportion of individuals that are top carnivores	>5%	1–5%	<1%
Fish abundance and health			
Number of individuals in sample		Varies with stream size	
Proportion of individuals that are hybrids	0	0–1%	>1%
Proportion of fish with disease or anomalies	0	0–1%	>1%

It is also critical that sampling effort is adequate to achieve accuracy in scoring the metrics. Species composition metrics can be particularly demanding if the fish assemblage includes species with limited distribution or low abundance. Angermeier and Smogor (1995) estimated that stream electrofishing surveys may require longitudinal sampling of the equivalent of 105 stream widths to capture 95% of the species present. Comparative use of the IBI should be limited to systems that are physically similar. For example, headwater streams may have lower species richness than like-size tributaries to larger rivers (Osborne et al. 1992). Despite these cautions, the simplicity and practical relevance of the IBI have made it a popular tool for assessing the health of aquatic ecosystems.

7.6 CONCLUSION

The population and community statistics described in this chapter are not inclusive, and many approaches to estimating key parameters are still evolving. It is critical, therefore, that the manager give careful consideration to the choice of statistics in the design of assessment and creel surveys. Rarely will a single biological statistic provide adequate information for decision making. Rather, the sampling program should be designed to provide an array of statistics that will describe the response of the population and community to the multiple factors that influence its performance. Indices of growth, abundance, recruitment, survival, harvest, and population and community structure may all be needed to develop an understanding of that response.

Choice of parameters to be estimated should be dictated by the potential uses of the information to be gained. Selection of the appropriate statistic to estimate each parameter should then be based on the criteria of (1) effort required, (2) consistency with which the statistic can be measured and thus confidently compared, and (3) the probable representativeness of the sample and consequent accuracy of the estimator. Evaluation of how well statistics meet these criteria will require knowledge of their performance for the target organisms in similar systems. Although biological descriptors of fisheries resources are essential to effective management, fisheries statistics are not so routine or universal that a standard recipe formula can be expected to provide them.

7.7 REFERENCES

Anderson, R. O. 1976. Management of small warmwater impoundments. Fisheries 1(6):5–7, 26–28.

Anderson, R. O. 1980. Proportional stock density (*PSD*) and relative weight (*Wr*): interpretive indices for fish populations and communities. Pages 27–33 *in* S. Gloss and B. Shupp, editors. Practical fisheries management: more with less in the 1980s. Proceedings of the American Fisheries Society, New York Chapter, Ithaca.

Anderson, R. O., and A. S. Weithman. 1978. The concept of balance for coolwater fish populations. Pages 371–381 *in* R. L. Kendall, editor. Selected coolwater fishes of North America. American Fisheries Society, Special Publication 11, Bethesda, Maryland.

Anderson, R. O., and R. M. Neumann. 1996. Length, weight, and associated structural indices. Pages 447–482 *in* Murphy and Willis (1996).

Angermeier, P. L., and J. R. Karr. 1994. Biological integrity versus biological diversity as policy directives. BioScience 44:690–697.

Angermeier, P. L., and R. A. Smogor. 1995. Estimating number of species and relative abundance in stream-fish communities: effects of sampling effort and discontinuous spatial distributions. Canadian Journal of Fisheries and Aquatic Sciences 52:936–949.

Bannerot, S. P., and C. B. Austin. 1983. Using frequency distribution of catch per unit effort to measure fish stock abundance. Transactions of the American Fisheries Society 112:608–617.

Brown, M. L., and D. J. Austen. 1996. Data management and statistical techniques. Pages 17–62 *in* Murphy and Willis (1996).

Bryan, R. D., and J. J. Ney. 1994. Visible implant tag retention by and effects on condition of a stream population of brook trout. North American Journal of Fisheries Management 14:216–219.

Carlander, K. D. 1977. Handbook of freshwater fishery biology, volume 2. Iowa State University Press, Ames.

Carline, R. F., B. L. Johnson, and T. J. Hall. 1984. Estimation and interpretation of proportional stock density for fish populations in Ohio impoundments. North American Journal of Fisheries Management 4:139–154.

Carpenter, S. R., J. F. Kitchell, and J. R. Hodgson. 1985. Cascading trophic interaction and lake productivity. Bioscience 35:635–639.

Colle, D. E., and J. V. Shireman. 1980. Coefficients of condition for largemouth bass, bluegill, and redear sunfish in hydrilla-infested lakes. Transactions of the American Fisheries Society 109:521–531.

Curtis, G. L., C. R. Bronte, and J. H. Selgeby. 1993. Forecasting contribution of lake whitefish year classes to a Lake Superior commercial fishery from estimates of yearling abundance. North American Journal of Fisheries Management 13:349–352.

DeVries, D. R., and R. V. Frie. 1996. Determination of age and growth. Pages 483–507 *in* Murphy and Willis (1996).

DiCenzo, V. J., M. J. Maceina, and W. C. Reeves. 1995. Factors related to growth and condition of the Alabama subspecies of spotted bass in reservoirs. North American Journal of Fisheries Management 15:794–798.

DiCenzo, V. J., M. J. Maceina, and M. R. Stimpert. 1996. Relations between reservoir trophic state and gizzard shad population characteristics in Alabama reservoirs. North American Journal of Fisheries Management 16:888–895.

Dolman, W. B. 1991. Comparison of bass club tournament reports and creel survey data from Texas reservoirs. North American Journal of Fisheries Management 11:185–189.

Ebbers, M. A. 1987. Vital statistics of a largemouth bass population in Minnesota from electrofishing and angler-supplied data. North American Journal of Fisheries Management 7:252–259.

Fausch, K. D., J. Lyons, J. R. Karr, and P. L. Angermeier. 1990. Fish communities as indicators of environmental degradation. Pages 123–144 in S. M. Adams, editor. Biological indicators of stress in fish. American Fisheries Society, Symposium 8, Bethesda, Maryland.

Gabelhouse, D. W., Jr. 1984. A length-categorization system to assess fish stocks. North American Journal of Fisheries Management 4:273–283.

Gabelhouse, D. W., Jr., and D. W. Willis. 1986. Biases and utility of angler catch data for assessing size structure and density of largemouth bass. North American Journal of Fisheries Management 6:481–489.

Goede, R. W., and B. A. Barton. 1990. Organismic indices and an autopsy-based assessment as indicators of health and condition of fish. Pages 93–108 in S. M. Adams, editor. Biological indicators of stress in fish. American Fisheries Society, Symposium 8, Bethesda, Maryland.

Green, A. W., G. C. Matlock, and J. E. Weaver. 1983. A method for directly estimating the tag-reporting rate of anglers. Transactions of the American Fisheries Society 112:412–415.

Gustafson, K. A. 1988. Approximating confidence intervals for indices of fish population size structure. North American Journal of Fisheries Management 8:139–141.

Gutreuter, S., and D. J. Krzoska. 1994. Quantifying precision of in situ length and weight measurements of fish. North American Journal of Fisheries Management 14:318–322.

Guy, C. S., and D. W. Willis. 1991. Evaluation of largemouth bass–yellow perch communities in small South Dakota impoundments. North American Journal of Fisheries Management 11:43–49.

Guy, C. S., D. W. Willis, and R. D. Schultz. 1996. Comparison of catch per unit effort and size structure of white crappie collected with trap nets and gill nets. North American Journal of Fisheries Management 16:947–951.

Hall, T. J. 1986. Electrofishing catch per hour as an indicator of largemouth bass density in Ohio impoundments. North American Journal of Fisheries Management 6:397–400.

Hansen, M. J., and five coauthors. 1994. Declining survival of lake trout during 1963–1986 in U.S. waters of Lake Superior. North American Journal of Fisheries Management 14:395–402.

Hanson, J. M., and W. C. Leggett. 1982. Empirical prediction of fish biomass and yield. Canadian Journal of Fisheries and Aquatic Sciences 39:257–263.

Hickman, G. D., and T. A. McDonough. 1996. Assessing the reservoir fish assemblage index: a potential measure of reservoir quality. Pages 85–97 in L. E. Miranda and D. R. DeVries, editors. Multidimensional approaches to reservoir fisheries management. American Fisheries Society, Symposium 16, Bethesda, Maryland.

Jenkins, R. M. 1979. Predator–prey relations in reservoirs. Pages 123–134 in H. Clepper, editor. Predator–prey systems in fisheries management. Sport Fishing Institute, Washington, D.C.

Jenkins, R. M. 1982. The morphoedaphic index, and reservoir fish production. Transactions of the American Fisheries Society 111:133–140.

Jenkins, R. M., and D. I. Morais. 1978. Prey–predator relations in the predator-stocking-evaluation reservoirs. Proceedings of the Annual Conference Southeastern Association of Fish and Wildlife Agencies 30 (1976):141–157.

Jones, J. R., and M. V. Hoyer. 1982. Sportfish harvest predicted by summer chlorophyll-a concentrations in midwestern lakes and reservoirs. Transactions of the American Fisheries Society 111:176–179.

Karr, J. R. 1981. Assessment of biotic integrity using fish communities. Fisheries 6(6):21–27.

Kruse, J. S. 1988. Guidelines for assessing largemouth bass fisheries in large impoundments. Missouri Department of Conservation, Final Report, Jefferson City.

Le Cren, E. D. 1951. The length-weight relationship and seasonal cycles in gonad weight and condition in the perch Perca fluviatilis. Journal of Animal Ecology 20:201–219.

Leonard, P. M., and D. J. Orth. 1986. Application and testing of an index of biotic integrity in small, coolwater streams. Transactions of the American Fisheries Society 115:401–414.

Liao, H., C. L. Pierce, D. H. Wahl, J. B. Rasmussen, and W. C. Leggett. 1995. Relative weight (Wr) as a field assessment tool: relationships with growth, prey biomass, and environmental conditions. Transactions of the American Fisheries Society 124:397–400.

Macdonald, P. D. M. 1987. Analysis of length-frequency distributions. Pages 371–384 in R. C. Summerfelt and G. R. Hall, editors. Age and growth of fish. Iowa State University Press, Ames.

Malvestuto, S. P. 1996. Sampling the recreational creel. Pages 591–623 in Murphy and Willis (1996).

Margalef, R. 1958. Information theory in ecology. General Systems Bulletin 3:36–71, University of Louisville, Systems Science Institute, Louisville, Kentucky.

Matuszek, J. M. 1978. Empirical predictions of fish yields of large North American lakes. Transactions of the American Fisheries Society 107:385–394.

Mills, E. L., D. M. Green, and A. Schiavone. 1987. Use of zooplankton size to assess the community structure of fish populations in freshwater lakes. North American Journal of Fisheries Management 7:369–378.

Minns, C. K., V. W. Cairns, R. G. Randall, and J. E. Moore. 1994. An index of biotic integrity (IBI) for fish assemblages in the littoral zone of Great Lakes areas of concern. Canadian Journal of Fisheries and Aquatic Sciences 51:1804–1822.

Miranda, L. E. 1993. Sample sizes for estimating and comparing proportion-based indices. North American Journal of Fisheries Management 13:383–386.

Muoneke, M. I. 1992. Loss of Floy anchor tags from white bass. North American Journal of Fisheries Management 12:819–824.

Murphy, B. R., M. L. Brown, and T. A. Springer. 1990. Evaluation of the relative weight (W_r) index, with application to walleye. North American Journal of Fisheries Management 10:85–97.

Murphy, B. R., and D. W. Willis, editors. 1996. Fisheries techniques, 2nd edition. American Fisheries Society, Bethesda, Maryland.

Ney, J. J. 1990. Tropic economics in fisheries: assessment of demand/supply relationships between predators and prey. Reviews in Aquatic Sciences 2:55–81.

Ney, J. J., and D. J. Orth. 1986. Coping with future shock: matching predator stocking programs to prey abundance. Pages 81–92 in R. H. Stroud, editor. Fish culture in fisheries management. American Fisheries Society, Fish Culture Section and Fisheries Management Section, Bethesda, Maryland.

Nielsen, L. A., and D. L. Johnson, editors. 1983. Fisheries techniques. American Fisheries Society, Bethesda, Maryland.

Osborne, L. L., and six coauthors. 1992. Influences of stream location in a drainage network on the index of biotic integrity. Transactions of the American Fisheries Society 21:635–643.

Ploskey, G. R., and R. M. Jenkins. 1982. Biomass model of reservoir fish and fish-food interactions with implications for management. North American Journal of Fisheries Management 2:105–121.

Ryder, R. A. 1965. A method for estimating the potential fish production of north-temperate lakes. Transactions of the American Fisheries Society 94:214-218.

Ryder, R. A., S. R. Kerr, K. H. Loftus, and H. A. Regier. 1974. The morphoedaphic index, a fish yield estimator—review and evaluation. Journal of the Fisheries Research Board of Canada 31:663-688.

Serns, S. L. 1985. Proportional stock density index—is it a useful tool for assessing fish populations in northern latitudes? Wisconsin Department of Natural Resources Research Report 132, Madison.

Shannon, C. E., and W. Weaver. 1949. The mathematical theory of communication. University of Illinois Press, Urbana.

Shelton, W. L., W. D. Davies, T. A. King, and T. J. Timmons. 1979. Variation in the growth of the initial year class of largemouth bass in West Point Reservoir, Alabama and Georgia. Transactions of the American Fisheries Society 108:142–149.

Sorenson, T. 1948. A method establishing groups of equal amplitude in plant society based on similarity of species content. Koneglige Danske Videnskabernes Selskab 5:1–34.

Swingle, H. S. 1950. Relationships and dynamics of balanced and unbalanced fish populations. Agricultural Experiment Station, Bulletin 274, Auburn University, Alabama.

Timmons, T. J., W. L. Shelton, and W. D. Davies. 1980. Differential growth of largemouth bass in West Point Reservoir, Alabama-Georgia. Transactions of the American Fisheries Society 109:176–186.

Washington, H. G. 1984. Diversity, biotic and similarity indices. A review with special relevance to aquatic ecosystems. Water Research 18:653–694.

Webb, M. A., and R. A. Ott, Jr. 1991. Effects of length and bag limits on population structure and harvest of white crappies in three Texas reservoirs. North American Journal of Fisheries Management 11:614–622.

Wege, G. J., and R. O. Anderson. 1978. Relative weight (Wr): a new index of condition for largemouth bass. Pages 79–81 in G. D. Novinger and J. G. Dillard, editors. New approaches to the management of small impoundments. American Fisheries Society, North Central Division, Special Publication 5, Bethesda, Maryland.

Whittaker, R. H. 1952. A study of summer foliage insect communities in the Great Smokey Mountains. Ecological Monographs 22:1–47.

Wilde, G. R., and W. L. Fisher. 1996. Reservoir fisheries sampling and experimental design. Pages 397–409 *in* L. L. Miranda and D. R. DeVries, editors. Multidimensional approaches to reservoir fisheries management. American Fisheries Society, Symposium 16, Bethesda, Maryland.

Willis, D. W., B. R. Murphy, and C. S. Guy. 1993. Stock density indices: development, use and limitations. Reviews in Fisheries Science 1:203–222.

Youngs, W. D., and D. G. Heimbuch. 1982. Another consideration of the morphoedaphic index. Transactions of the American Fisheries Society 111:151–153.

Yurk, J. J., and J. J. Ney. 1989. Analysis of phosphorus-fishery productivity relationships in southern Appalachian reservoirs: can lakes be too clean for fish? Lake and Reservoir Management 5:83–90.

Zale, A. V., and M. B. Bain. 1994. Estimating tag-reporting rates with postcards as tag surrogates. North American Journal of Fisheries Management 14:208–211.

Chapter 8

Socioeconomic Benefits of Fisheries

A. STEPHEN WEITHMAN

8.1 INTRODUCTION

Careful consideration of the full range of fisheries benefits began when the fisheries profession evolved from a guiding philosophy of maximum sustained yield (MSY) to a philosophy of optimum sustained yield (OSY; see Chapter 1). For MSY, the term yield can be defined as weight of fish harvested. For OSY, however, yield refers to all socioeconomic, as well as biological, benefits associated with fisheries, including the joy of the sport. Malvestuto and Hudgins (1996) discuss optimum yield and four associated management components, biological, sociocultural, economic, and human health, to cover considerations associated with fisheries management. This chapter identifies socioeconomic benefits of fisheries, outlines methods for measuring benefits, and discusses management implications of socioeconomic data.

Modern fisheries management requires a mastery of more than the technical problems associated with fishes and water. Habitat and fish populations have been studied extensively, but today this traditional approach is too limited. Because fisheries are managed for people, an effort must also be made to understand angler and nonangler attitudes, preferences, characteristics, and needs. If fisheries managers accept this challenge of incorporating social sciences into the evaluation of recreational and commercial fisheries benefits, we will improve our ability to justify important management programs.

8.2 IDENTIFICATION OF FISHERIES BENEFITS

Fisheries provide a myriad of benefits to society. Through the use of an excellent historical perspective and classification system, Steinhoff et al. (1987) presented the benefits of wildlife; similar benefits are applicable to fisheries. This section identifies groups of people that receive benefits from fisheries and develops a logical framework for categorizing the benefits these groups receive.

First, consider the angler's perspective. When anglers plan a fishing trip, they often envision the location, company of friends or family, fishing gear, and fish they will catch. Fishing is just one type of recreation that individuals can choose. Reasons for fishing are personal; however, angler expectations are related to realizing certain tangible or intangible benefits in addition to harvesting fish (Figure 8.1). Individual

Figure 8.1   The benefits of fishing include more than just the catch. (Photograph provided courtesy of Missouri Department of Conservation.)

anglers repeat their fishing trips if their expectations have been reasonably fulfilled. Thus, the experience has value. Most people, with the exception of tournament anglers and people who depend on anglers for sales revenue, do not think of recreational fishing as a business, nor are they concerned about what happens to the money spent on recreational fishing.

The perspective of charter operators or fishing guides is different. Recreational fishing is their business (Figure 8.2). Benefits to those who provide direct services include employment, income, and the satisfaction of helping anglers realize their expectations by showing them where and how to catch fish. The people who provide these services are a part of the community and the local economy. They continue to provide their services as long as they (1) make money, (2) are successful at guiding anglers to fish, and (3) enjoy what they are doing. Lichtkoppler (1997) provides a characterization of Ohio's Lake Erie charter fishing industry from 1985 to 1994.

Some businesses, including marinas, boat and bait dealerships, and fishing tackle wholesalers and retailers, depend heavily upon fishing-related expenditures. Benefits to those who provide goods and indirect services include employment, income, and

Figure 8.2   To the charter boat operator, fishing is a way of life. (Photograph provided courtesy of Missouri Department of Conservation.)

the satisfaction of providing needed goods and services to anglers. The people who provide these goods and services are also a part of the community and local and regional economies. They continue to provide goods and services as long as doing so is profitable and there is a demand. Other businesses, such as restaurants, service stations, and motels, that derive a portion of their income directly or indirectly from anglers could also be included in this group.

Commercial fishers have yet another perspective. For them, fishing is a means of support and a way of life. Their benefits include employment, income, and the satisfaction of providing a source of nutritious food. As with the charter operators and other local businesses, commercial fishers are a part of the community and local economy, and they continue to fish as long as it can be done profitably and they enjoy what they are doing.

Native American fishers must also be considered. Their fishing rights have been legally protected by treaties based on cultural requirements and precedent and permit use of the resources as subsistence fisheries and as a source of income. Benefits to Native American fishers include a way of life, a tradition that is maintained, and a source of food and cultural pride. They continue to fish as long as their rights are upheld and the rewards are worth the effort.

Fisheries benefits depend upon the perspective and circumstances of the participant, given that the recreational, commercial, and Native American fishers have access to a place to fish. The quality of fishing, which is based on the species, size, number,

Table 8.1 Socioeconomic benefits of fisheries.

Values	Impacts
Social benefits	
Cultural	Quality of life
Societal	Social well-being
Psychological	
Physiological	
Economic benefits	
User	Direct
Consumptive	Indirect
Nonconsumptive	Induced
Indirect	
Nonuser	
Option	
Existence	

and diversity of fishes caught by fishers with a distinct set of attitudes about those fish, dictates the magnitude of benefits available for anglers and businesses. Fisheries management is the process of allocating fisheries resources between current and future generations of sport, commercial, and Native American fishers (Lichatowich 1992).

All fisheries-related benefits can be considered social or economic, but an important distinction must be made between two types of benefits: values and impacts (Table 8.1). Socioeconomic values represent the importance that people place on the resource (Rockland 1985)—that is, both the satisfaction that people derive from a resource (intangible) and the worth they place on it (tangible). Values describe what people receive related to their expenses—for recreational anglers, the satisfaction of the trip (which can be converted to dollars as described in section 8.2.1); for commercial fishers, profit. As an example, values include the opportunity to fish. For recreational anglers, tangible values (the worth of a trip) could include money spent for bait, tackle, and other equipment, services, gasoline, lodging, and meals. For commercial fishers, money is invested in equipment and wages. Socioeconomic impacts, in contrast to values, represent the social and economic effects that are generated by the use of the resource (Rockland 1985). In other words, impacts represent the effects on the community and local, regional, and national economies in terms of jobs, income, and tax receipts.

8.2.1 Socioeconomic Values

Socioeconomic values can be subdivided into social values that are held and economic values that are assigned. Held values are usually associated more with ideas, behaviors, outcomes, and experiences, whereas assigned values are usually associated more with goods, services, and opportunities.

Brown and Manfredo (1987) identified four categories of social values that are important considerations for fisheries managers—cultural, societal, psychological, and physiological. Cultural and societal values are more generic, pertaining to nations and communities, whereas psychological and physiological values relate to individual anglers. Cultural values represent a collective feeling toward fishes and wildlife. An example of a modern cultural value is the view that fish can provide recreation through catch-and-release fishing in addition to serving as a source of food. Today, subcultural values have evolved around the use of specific methods for harvesting particular species, such as snagging for paddlefish, fly-fishing for rainbow trout, or bait-casting for largemouth bass. Societal values are based on relationships among people as part of a family or community. The togetherness of a family fishing trip would be representative of societal values. Psychological values are those that relate to the satisfaction, motives, or attitudes associated with the use or knowledge of the existence of a fishery. An example would be an angler's perception of the quality of fishing. Physiological values relate to improvements in human health (better conditioning and reduction of stress) related to the sport of fishing.

Two types of economic values associated with fisheries have been identified—user and nonuser (Bishop 1987; Rockland 1985). User values result from direct or indirect participation. Nonuser values are potential (option) or intrinsic (existence) values of the resource.

User values can be further subdivided into consumptive, nonconsumptive, and indirect use values (Bishop 1987; Table 8.1). An example of consumptive use is harvest (removal) of fish by an angler. The fact that people go fishing despite the time and money required indicates there are values associated with participation. If expenses exceed perceived values, people might change location, species, or method or could stop fishing altogether and substitute other activities. Willingness to pay, which includes actual expenditures and excess value (benefits that exceed monetary cost) to users, is an appropriate measure of the economic value of a recreational fishery. An equivalent measure for commercial fisheries would be individual, transferable quotas, which grant individuals the salable rights to harvest specified amounts of fish (Anderson 1991).

Nonconsumptive uses of fisheries resources include sight-seeing, fish watching, snorkeling, scuba diving, visiting aquaria, studying nature, and photographing fishes and their habitat. These unlicensed uses do not deplete fisheries resources but are important. Just as with consumptive use, money is spent to participate, and values are realized or participation in the activities ceases. Measurement of values is a little more difficult because many nonconsumptive fisheries uses occur in conjunction with recreational activities such as camping, swimming, or boating.

Indirect users do not actually come into contact with the resource or habitat. Indirect uses are limited to such things as reading about fishes, seeing pictures of fishes, and watching television programs about fishes. Values are derived from these activities because people still invest time and money in them.

Nonuser values can be subdivided into option and existence values. An option value means that individuals are leaving the door open to future participation although they currently choose not to participate. A type of option value, the quasi-option value, allows postponement of the decision to participate until more information is gathered. Existence values occur when people value a fishery despite the fact that they do not currently use the resource or ever plan to in the future. Bequest value is an existence value in which a person wants a particular resource to exist for future generations. Loomis and White (1996) discuss option, existence, and bequest values in addition to use values for rare and endangered fishes. Altruistic value is an existence value for which the motivation is to allow for the survival of a species.

8.2.2 Socioeconomic Impacts

Socioeconomic impacts are the effects of money spent by recreationists (primarily by anglers but also by nonconsumptive and indirect users) and commercial fishers on local, regional, and national economies. The initial or obvious effects are economic, but social changes are also likely to result (Table 8.1).

The total economic impact within a defined region exceeds actual expenditures by recreationists because some of the money spent in a local economy continues to circulate in that area. A portion of money leaves the area to pay for goods, services, and taxes; money that remains is used to pay wages or represents profit. Wages and profits are then spent by residents in the local economy, resulting in another round of economic impact. Money spent by recreationists is said to have a direct economic impact. Indirect economic impacts result from businesses that procure materials locally to produce goods and provide services to meet the needs of recreationists. Money spent by people who earn wages or profits in the region is said to have an induced economic impact (Table 8.1).

An example illustrating the different economic impacts follows. Assume anglers enter a bait-and-tackle shop and buy fishing rods and reels, lures, and night crawlers. The direct economic impact occurs when anglers exchange money for merchandise. An indirect economic impact is the purchase of materials by businesses that manufacture the fishing rods and lures. Wages paid by those businesses and the wholesaler who provides night crawlers are also indirect economic impacts. All products and services have an indirect economic impact as long as they are produced and provided within the region of interest. If the reel is imported from another region, however, the money spent to bring it to the region is lost to the regional economy and no longer represents a local indirect economic impact. Induced economic impacts occur when clerks in the bait-and-tackle shop spend their wages locally. A similar set of direct, indirect, and induced economic impacts can be envisioned for a commercial fishing operation.

Economic impacts can be quantified by determining the number of jobs created, wages paid, sales, or profits. These figures depend upon a definition of the region of interest, which could range from an individual community to a county, state, multistate, or national area. Most economic impacts can be converted to dollars for comparison (measurement of the impacts will be discussed in section 8.3.2).

Social impacts are more elusive and are not as readily quantifiable. They relate to quality of life and social well-being. Quality of life and social well-being often have been used interchangeably in the literature. Measures of well-being track changes in people's quality of life. Social impacts are reflected by changes in social relationships within the community and the social organization of a fishery's production, distribution, and marketing systems (Vanderpool 1987). Social impacts due to changes in fisheries management are most likely to result from changes in business conditions or employment. Improvements or changes in fisheries can cause jobs to be created, transferred, or lost. Changes in employment and business can result in population shifts and the need for more or fewer housing and public services.

Creation of jobs, especially in environmentally compatible industries, is definitely a benefit to people and their communities. Employment provides people with a sense of worth and identity. Adequate employment strengthens community cohesion and allows for development of social institutions such as schools and churches. Loss of jobs has the exact opposite social effect. Changes in types of employment can be either positive or negative. For instance, if satisfying, career-oriented positions are replaced with minimum-wage jobs, the effect on individuals and the community will be negative. Besides career considerations, income is another factor related to employment. More income generally creates greater social well-being.

8.3   MEASUREMENT OF FISHERIES BENEFITS

In this section two general categories of benefit assessment techniques—nonmonetary and monetary—are discussed, and a variety of approaches for each technique are reviewed (Table 8.2). Total value assessment, a third category, is a synthesis of nonmonetary and monetary evaluations. The focus of this section is on assessment techniques for fisheries that have broad acceptance by sociologists and economists as well as biologists. Some methods target a specific set of benefits, whereas other methods provide a comprehensive assessment. Methods discussed hold the greatest promise for measuring socioeconomic benefits associated with recreational and commercial fisheries.

Almost every method of assessing fisheries benefits requires survey data from people who fish or who in some other way might be affected by a change in a fishery (Table 8.2). The three most common approaches for collecting information are on-site,

Table 8.2   Methods of assessing fisheries benefits.

Benefit measurement	Approach
Nonmonetary	
Social well-being measurement	Angler survey on changes in a fishery
Psychophysical measurement	Angler survey on aesthetic appeal
Multi-attribute choice approach	Angler survey on a variety of fishery characteristics
Attitude measurement	Angler survey on factors that affect the quality of fishing
Social impact assessment	Projection of changes that will likely result from a new policy or program
Monetary	
Economic impact assessment	Angler expenditures used as input for regional economic models
Economic value assessment	Angler expenditures used as input for travel cost or contingent valuation
Total economic valuation	Angler expenditures plus nonuser values used as input for contingent valuation
Combination	
Total value assessment	Comprehensive evaluation that combines nonmonetary and monetary evaluations to determine social and economic values

mail, and telephone surveys. Each method has advantages and disadvantages that must be considered for individual projects. Thorough reviews are available to help decide how to collect survey data (Malvestuto 1983).

8.3.1 Nonmonetary Methods for Assessing Social Benefits

Nonmonetary assessments are conducted because of the need for information about social and environmental effects related to commercial and recreational fisheries resulting from improvements due to effective management or declines due to habitat degradation. Use of social sciences and the advent of nonmonetary methods in solving fisheries problems began in earnest in the 1970s after economic valuations gained acceptance. Gregory (1987) provides additional details and references on social benefit measurement.

The social well-being measurement approach assesses the effects of a fishery project or policy on the well-being of individuals to determine its overall effects on society. Theoretically, an index of well-being should reflect individuals' observations on changes in their quality of life and hence changes in the community and in social relationships. A typical social assessment would describe current conditions and identify groups of people who would be affected by a change in the fishery. Interviews can be as basic as asking people how they feel about a possible or actual change in a fishery and how seriously it would affect them. An example would be assessing acceptance of a new regulation by recreational fishers; the outcome would depend on their preferences for keeping fish to eat compared with catch-and-release fishing. Other indicators of well-being within the community include employment of women, crime, and divorce rates (Gregory 1987).

Psychophysical measures determine the sensitivity of people to physical features of a fishery environment. Sites can be ranked based on aesthetic appeal, or a single site can be judged after modification to the environment. Rating scales are used to assess personal preferences of some sample population for the site modification that has taken place. This approach has been applied in studies on the importance of landscape features to study participants and for assessing the scenic beauty of forest environments. This technique could be used to evaluate the sensory impression of the fishery setting to anglers (Figure 8.3).

The multi-attribute choice approach analyzes preferences to aid in making decisions on multifaceted fishery problems. The procedure involves selecting relevant attributes or characteristics of a fishery, specifying how the attributes will be measured, identifying all possible outcomes, weighting the attributes based on relative importance, and determining the effect a management policy will have on each attribute so a single, overall score can be computed.

Walker et al. (1983) used the multi-attribute approach to analyze trade-offs with respect to allocation and production of wild and hatchery coho salmon in Oregon. As a result of the analysis, they selected the most effective management policy of the 12 proposed, determined that harvest rate was the most important decision variable, and found that the number of smolts released should be increased only up to the point that smolt survival would be affected by additional releases.

Figure 8.3 Point sources of pollution are not only deleterious to the integrity of the offended aquatic ecosystem but severely detract from the aesthetic appeal of the site and thus diminish the angling experience. (Photograph provided courtesy of Missouri Department of Conservation.)

Attitude measurements focus on the response of anglers to any of a number of factors that can influence the quality of a fishing trip. Attitude measures address indirect behavioral associations with environmental conditions, such as the setting or quality of the experience (Gregory 1987). Numerical scales list conditions used to rate angler response.

Many researchers have recently focused on angler attitudes as being important to recreational fisheries management. For example, an investigation of Missouri angler attitudes revealed the importance of the opportunity to be outdoors, the setting, companionship, and obviously the catching of fish (Weithman and Anderson 1978b). Another example of attitude measurement was the development of indices by Weithman and Anderson (1978a) and Weithman and Katti (1979) to measure the quality of fishing. Weithman and Anderson (1978a) developed equations to rate fishing quality based on angler attitudes about the importance of species, size, number, and diversity of fishes caught. The accuracy and precision of these indices were tested by Weithman and Katti (1979) by calculating index values and comparing the results to actual angler preferences between two fish species or groups of fishes. After angler preferences were determined, higher index values for a fishing trip indicated greater enjoyment for anglers. Further work in this area has been directed at asking anglers to rate fishing quality on a 10-point scale for a particular day compared with their normal success. Mean values of this numerical rating also appear to be quite sensitive in predicting the level of enjoyment an angler experiences based on the species, size, and number of fish they catch.

Social impact assessments determine economic and social costs and benefits associated with development of fisheries policies and programs. This type of assessment is more comprehensive than the other nonmonetary techniques. The emphasis is on understanding and projecting likely changes in social relationships, social structures and institutions, and normative systems and world views (Vanderpool 1987). A social impact assessment should occur before management changes in a fishery and should include an analysis of historical, cultural, economic, ecological, and demographic dimensions. This approach is especially applicable to evaluation of commercial fisheries, although not exclusively so, compared with the other nonmonetary measures. Social impact assessments in fisheries have been extremely limited, but based on applications in other natural resource areas they hold great potential.

8.3.2 Monetary Methods for Assessing Economic Benefits

Monetary methods have been used for a number of years to evaluate the economic impacts and values of recreational and commercial fisheries. This section discusses accepted valuation techniques and their use in conducting three types of economic benefit assessment: (1) economic impact, (2) economic value, and (3) total economic valuation (Table 8.2). Several researchers have studied valuation techniques in recent years and have presented specific information that would allow the reader to develop similar models and analyses for assessing economic benefits (Dwyer et al. 1977; Rockland 1986; Talhelm and Libby 1987). In general, impacts are best determined by using economic models to track angler expenditures through the economy, and values are best estimated by using travel cost or contingent valuation models (section 8.3.2.2) to determine to what extent values derived by anglers exceed their expenses.

8.3.2.1 Economic Impact Assessment

Economic impact assessments are conducted to determine the effect of fishing-related expenditures on the local, regional, or national economy. Such an assessment is used to quantify socioeconomic impacts identified in section 8.2.2, including direct, indirect, and induced impacts. An economic impact assessment is the process of summing all three types of impacts with respect to the effects of recreationists' and commercial fishers' expenditures on income, employment, and taxes in the region of interest.

Rockland (1986) discussed four models that can be used in the analysis of the economic impacts of recreational fisheries: (1) economic base, (2) econometric, (3) input–output, and (4) modified input–output. Given relevant employment data and angler expenditures, it is possible to estimate employment, earnings, business output, and the resulting socioeconomic impacts. Rockland discussed the strengths, weaknesses, and recommendations for application of the four models.

Data on gross expenditures by anglers in the study area are needed as an input for an economic impact assessment. Multipliers (for output, employment, or income) can be applied to expenditures to determine the full impact on the economy if, and only if, they are properly developed and specified (Hushak 1987). These multipliers express the total amount of income generated in an economy by recreationists spending money by measuring the circulation of these expenditures through individual components of the economy. Multipliers reflect general relationships. Labor-intensive industries, such as amusement and recreation, have high multiplier values. Much of the revenue from these industries is converted to employee wages—money that remains, in part, in the local economy to be spent again. Industries that market products, such as grocery stores, have low multiplier values. Much of their revenue pays the cost of goods supplied by wholesalers outside the community and is lost to the local economy.

Economic impact assessments have been conducted for a number of recreational activities. Two fisheries management examples are a determination of the value of a salmonid fishery in New York (Brown 1976) and a rainbow trout fishery in Missouri (Weithman and Haas 1982). Economic impact assessments have also been conducted for commercial fisheries, such as a salmonid fishery (Huppert and Fight 1991), and with respect to natural resource damage assessments, such as the *Exxon Valdez* oil spill (Hanemann and Strand 1993; Cohen 1995).

8.3.2.2 Economic Value Assessment

Economic value assessments determine the value derived by participants from recreational fishing. This approach is used to quantify economic values (identified in section 8.2.1) for consumptive users of the resource. Although many valuation techniques have been proposed, two in particular deserve further discussion because of their applicability to fisheries evaluations—travel cost and contingent valuation (Dwyer et al. 1977). Both methods have been designed to estimate consumer surplus, which is the value that anglers receive in excess of their expenditures.

The travel cost and contingent valuation methods allow a more complete and accurate analysis of value provided by fisheries. They take into account shifts in demand for recreation due to changes in the quantity and quality of recreation provided. This type of evaluation is the best measure of the value of management programs. In addition, an economic evaluation allows for comparisons with other fisheries or other activities that compete with fisheries for limited resources.

Both methods are based on the relationship between angler preference for, and costs associated with, particular fisheries. The travel cost method depends on observed days of fishing and actual angler expenses; contingent valuation is based on an estimate of days fished and angler response to hypothetical questions about the value of the fishery. Two important assumptions are involved in applying these methods. The travel cost method depends on the assumption that anglers would respond the same to a fee for fishing as they would to an increase in their travel expenses. The basic assumption for the contingent valuation method is that an angler's response to a hypothetical question about the value of a fishery would actually fit that angler's behavior in a real market situation. Literature is available that outlines the methods of applying the travel cost (Hansen 1986) and contingent valuation (Hoehn 1987) techniques.

Both methods have been used repeatedly to evaluate benefits from outdoor recreation. A travel cost model was used by Burt and Brewer (1974) to predict recreational benefits at a proposed reservoir in Missouri; Knetsch et al. (1976) used a travel cost model to estimate benefits for a number of reservoirs in California. Adamowicz et al. (1989) used data from recreational hunters in Alberta and employed a travel cost model to estimate consumer surplus. Morey et al. (1993) compared a variety of travel cost models, including partial demand, single-site demand, nested-logit, and three-level nested-logit, to measure consumer surplus of Atlantic salmon anglers on the Penobscot River. A nested-logit model implies that the vector of random components associated with a trip, such as site choice or number of fishing trips in a season, is drawn from a nonnormal, extreme value distribution. The contingent valuation method was used by Oster (1977) to determine willingness to pay to clean up the Merrimack River; by Walsh et al. (1978) to determine recreational benefits from high-mountain reservoirs in Colorado; by Connelly and Brown (1991) to determine the net economic value of New York's freshwater recreational fisheries; and by Teasley et al. (1994) to estimate revenue capture potential associated with public area recreation. Loomis (1989) examined the reliability of the contingent valuation method by comparing responses from the general population and responses from people who visited specific recreation sites. Gan and Luzar (1993) successfully used conjoint analysis, a modification of the contingent valuation method, to determine willingness to pay for waterfowl hunting in Louisiana.

Use of the travel cost and contingent valuation methods should be emphasized in future valuation of recreational fisheries. Studies already completed provide good examples for future applications, but there is room for improvement. Some areas currently under investigation include incorporating travel time and grouping participants in the travel cost model, refining questioning techniques for contingent valuation methods, and determining the importance of site quality and substitute sites and choosing an appropriate functional form to estimate recreation demand for both types of models.

Incorporation of travel time into travel cost models is an important consideration because time is considered an expense of the trip (Wilman 1980). However, because travel time is usually correlated with distance traveled or total expense, inclusion in the model has been difficult. Travel time is taken into account most often by assuming a trade-off between money and time based on some function of the average wage rate (Cesario 1976). McConnell and Strand (1981) developed a method to set the cost of travel time by use of an independent estimate of travel time and round-trip distance to the site. Two problems are apparent: (1) travel time might not be a liability in every instance because some people derive benefits during the trip by sight-seeing, and (2) if trips for which the site of interest is not the only destination are included in the valuation, site benefits will be overestimated. Few attempts have been made to separate benefits provided by primary and secondary destinations (Dwyer et al. 1977). Mendelsohn et al. (1992) successfully valued trips to Bryce Canyon National Park by identifying unique multiple-destination trips as additional sites.

The travel cost model uses data on distance traveled from an angler's residence to the site under evaluation to produce a demand curve. Grouping anglers into categories for consideration and comparison based upon point of origin has become an art. Binkley and Hanemann (1976) recommended categorizing people according to areas within concentric rings surrounding the site being evaluated. Others, however, have separated participants by county of origin or geographic areas with relatively homogeneous populations to make their evaluation. Another approach has been to consider individual responses and ignore aggregation over a particular area (Brown and Nawas 1973). Count data models such as the Poisson and negative binomial have become popular for travel cost analysis (Hellerstein 1991). Haab and McConnell (1996) recently devel-

Figure 8.4   For many, the setting for fishing is as important or is more important than the expectation to catch fish.

oped a count data model for consumer demand that deals with nonparticipants in the recreation of interest who live in close proximity to the resource. How anglers are grouped is a matter of choice for individual researchers, but each type of grouping can provide valid estimates. Sample selection bias can present serious methodological problems, but methods for correction have been offered by Bockstael et al. (1990).

Site quality, including factors such as congestion, aesthetics, and fishing quality, can be an important variable for consideration in predicting visitation by anglers in both types of models (Figure 8.4). Much research has been conducted on the effects of congestion as they relate to the management of natural resources. Overcrowding, while not a problem at most fisheries, can affect the quality of the experience at sites near some metropolitan areas and on opening day for certain species. Cesario (1969) approached the question of site quality by observing actual behavior of recreationists. Smith and Desvousges (1985) examined the effects of water quality as a site attribute that could influence recreational demand. Fishing quality might be the most important site quality variable, yet little consideration has been given to this topic. Stevens (1966) included catch rates as an indicator of fishing quality. The overall quality-of-fishing index proposed by Weithman and Katti (1979) could be used as one indicator of site quality. Englin and Mendelsohn (1991) used travel cost analysis to determine preference estimates of the recreational value of old-growth forest, clear-cuts, and other types of forest management. Further investigation of site variables is needed.

Visitation to one reservoir or stream is affected by the availability of other nearby fishing sites. Likewise, another important factor that can affect visitation is previous experience at any given recreation site (Adamowicz 1994). Substitute sites can play a major role in the development of realistic models to predict fishery visitation and benefits. Several investigators have developed multiple-site models to evaluate benefits associated with aquatic resources in a state or region of the country. If the site being evaluated is unique, substitute sites are not a factor. Additional research should be directed at the comparability of a variety of sites for recreational fishing.

Choice of the functional form is critical when estimating recreational demand and calculating consumer surplus in both types of models. Variability in estimates is dramatic based on the model selected. Most recent studies include models based on the semilog form. Frequently, samples need to be truncated. Ozuna et al. (1993) use and describe a truncated general Box-Cox regression model to select an appropriate functional form using truncated data to estimate consumer surplus.

Many potential problems have been identified in regard to the questioning procedure or design of contingent valuation surveys (Hoehn 1987). If respondents cannot relate the questions to real world situations, the survey could be biased. Strategic bias could be another problem. Survey results could be affected if people think they might benefit by giving a particular answer, but evidence of this kind of strategic behavior usually has not been found. Another concern is that a hypothetical change in fees in the survey might be objectionable to a segment of the respondents. Care must also be taken not to infer that an actual fee for entry could change. In bidding-game questions to determine willingness to pay, a bias can be introduced by selecting a given starting amount (in dollars) that an angler is willing to spend to participate. Thayer (1981) developed a test and adjustment to eliminate this problem (starting-point bias). Horowitz (1993) addressed the problems of nesting and sequencing, which deal with the scale or scope and ordering of outcomes of interest, in contingent valuation analyses.

8.3.2.3 Total Economic Valuation

Total economic valuation expands on economic value assessment by incorporating indirect, option, and existence values (described in 8.2.1), in addition to consumptive and nonconsumptive use values. Option prices or existence values of a variety of natural resources or sites for recreation have been considered. These values are accrued by nonparticipants solely from knowing a particular opportunity exists. For example, Greenley et al. (1981) measured the option value of maintaining the water quality of the South Platte River basin in Colorado, and Brookshire and Randall (1978) estimated consumptive use, option, and existence values for hunting and preserving wildlife resources in Wyoming.

The traditional contingent valuation and travel cost approaches can each measure portions of total economic value. Contingent valuation is the only technique, however, that currently can be adapted to measure the entire total economic value including indirect-use values and nonuser values (Randall 1987). The approach of estimating total economic value (all assigned values) is particularly useful for placing a value on endangered species or other nonexploited resources (Bishop et al. 1987).

8.3.3 Total Value Assessment

Total value assessment is a comprehensive concept of valuation in which all values—social, anthropological, political, philosophical, and economic—are considered concurrently (Talhelm and Libby 1987; Table 8.2). The problem is that no single valuation method addresses all areas of interest, so they must be addressed individually. Total value assessment considers all different kinds of value, including held and assigned values, in resolving fisheries conflicts. In particular, the use of social sciences must be emphasized to aid in management of fisheries resources (Talhelm 1987). Multidisciplinary research should expedite the integration of social sciences with traditional approaches to evaluating fisheries.

8.4   MANAGEMENT IMPLICATIONS OF SOCIOECONOMIC DATA

Socioeconomic data are essential for effective fisheries management because fisheries are managed for people as well as fishes. People includes the person who goes fishing with the family on vacation, the dedicated angler who fishes 50 d a year, commercial and Native American fishers who make a living from the resource, and the person who owns the bait-and-tackle shop. Fisheries biologists state that their intent is to manage the fishes to provide high-quality fishing. Criteria for making fishing better need to be defined to lay the groundwork for these management efforts. Fisheries managers will be successful if they (1) carefully consider the entire array of benefits fisheries offer, (2) make the effort to determine what angler and other groups want, and (3) develop a process to explain to people the differences between their expectations and a manager's ability to effect change in a particular ecosystem.

Fisheries management is the process of working with a given aquatic habitat and assemblage of organisms for the benefit of people in a recreational or commercial setting. Goals and objectives are generally set to approach OSY (see Chapter 2). Important benefits have been identified (section 8.2) and methods have been recommended to assess those benefits (section 8.3). This section documents the importance of socioeconomic data in the decision-making process and explains use of such data. Four general fisheries-related examples are discussed to demonstrate the utility of socioeconomic data.

8.4.1 Agency Planning

Planning is essential for the efficient, long-term operation of a natural resource agency. Typically, such an agency has the constitutional responsibility to preserve, protect, restore, and manage fisheries resources. The agency manages the resources for the people (users and nonusers alike), but administrators must be cognizant of the potential for socioeconomic impacts on regional economies as well. Socioeconomic data can provide identification of the clientele and their expectations from the fisheries under management. Entire programs can be evaluated on the basis of projected needs and desires.

One of the most important resource agency functions is asset allocation–selecting purposes and priorities for spending money. Budgets are prepared based on a program review and prioritization. From an agency standpoint, a benefit-cost analysis is appealing. Costs can be calculated accurately, but socioeconomic data are needed to estimate benefits. For example, the choices of building a fish hatchery, buying access sites on a river, and building community lakes could be evaluated rationally. Often the choice is not either-or but rather identifying the best mix of expenditures. Another example of the use of socioeconomic data would be justification for an endangered species program. An analysis of nonuser economic values would allow direct comparisons with user-oriented programs designed to enhance recreational fishing.

8.4.2 Recreational Fisheries Management

One objective that frequently appears in recreational fisheries management plans is to improve the quality of fishing (i.e., increase catch or harvest rates and the size of fish caught or increase angler use or satisfaction). Monitoring fishing quality is an important part of fisheries management. Examination of angler attitudes reveals that almost all anglers rate their fishing trips as excellent or good. If those same anglers are asked about their fishing success, however, a majority often rates it as fair or poor. Even without any management efforts, anglers can apparently have a good time given the opportunity to fish. Fisheries managers have the opportunity, though, to improve fishing success and increase the number of memorable fishing trips by considering angler preferences in management plans.

We must delve deeper into anglers' minds to maximize the benefits they derive from recreational fisheries, specifically concentrating on fishing success. Opportunities exist at most recreational fisheries to make improvements through species introductions, supplemental stocking (see Chapter 14), or fish habitat manipulation (see Chapters 9–12). Selection of the best approaches from a list of those that are biologically satisfactory should be based on calculations of socioeconomic values. As an example, assume an additional predator is needed to balance a fish community. If the choice is between two of equal biological value, which one should be selected? In another instance, is it desirable to double the catch rate of a species if harvest is reduced by 90%? Or, should a supplemental stocking program be dropped because the overall agency budget has been reduced by 10%? These questions cannot be answered without more information, preferably socioeconomic data related to people who use or could be affected by the selected policy.

8.4.3 Resolution of Environmental Problems

Knowing the complete array of a fishery's benefits, including socioeconomic values and impacts, is essential for protecting aquatic resources and giving decision makers comparable data to evaluate competing projects. Destruction of aquatic resources jeopardizes every value discussed in section 8.2. A catastrophic event could eliminate current recreational fishing (consumptive use), degrade water quality (nonconsumptive and indirect uses), and even eliminate nonuser values if the damage is irreparable (Figure 8.5). Socioeconomic impacts would also undoubtedly be significant because the reduction in recreational use would affect the local and regional economies.

Consider three examples of environmental problems for analysis: (1) an extensive oil spill into a river, (2) a dam failing to meet state water quality standards for tailwater releases, and (3) withdrawal of water from the Missouri River for nonfishery uses. If an oil spill results in an extensive fish kill, recovery of the replacement cost for the

Figure 8.5 The urgency for environmental recovery efforts (oil spill shown here) is predicated on socioeconomic concerns. (Photograph provided courtesy of Missouri Department of Conservation.)

fishes would be insufficient to cover the benefits foregone. Anglers would be affected because fishing quality would be diminished, and eventually businesses would be affected when anglers stopped returning to the area. Extirpation or threats to rare or endangered species would be another concern. Failure of discharges from Table Rock Dam to meet state standards of 6 mg/L dissolved oxygen affected the rainbow trout fishery in downstream Lake Taneycomo, Missouri. The results were reduced angler success and visitation to the area in the fall and a variety of socioeconomic impacts (Weithman and Haas 1982). Proposals have been made to divert water from the Missouri River for irrigation and industrial purposes. This water could yield measurable benefits for irrigation and industrial purposes or for recreational and commercial fisheries. The best way to resolve the conflict would be to conduct a complete study of the socioeconomic values and impacts associated with both uses.

8.4.4 Conflict Resolution

Conflicts between groups of fishers who simultaneously exploit a fishery resource are becoming more common. The problem is that certain freshwater fish stocks, especially those that are harvested commercially, run the risk of being overfished and excessively harvested to the point of depletion of those stocks well below MSY. Tensions mount, especially when the species of interest periodically has weak year-classes.

In general, from a value standpoint, recreational use is more efficient than commercial use of aquatic resources. The socioeconomic impacts resulting from changes in commercial fisheries could be significant, though, and therefore must be considered. What effects will occur in terms of employment and income with the closure of a commercial fishery? What reaction will people have when they lose their livelihood? A recent trend has been for states to buy out commercial fishing interests and thereby eliminate commercial harvest when conflicts with recreational anglers are perceived. Sutinen (1993) discusses optimal allocation of a fishery resource between recreational and commercial fisheries and makes the point that enforcement costs are a key factor in the determination.

An example of a potential conflict between fishing interests is the striped bass fishery in the Hudson River. An important recreational fishery has developed there even during the closure of commercial fishing due to PCB contamination. Should commercial fishing be permitted after the water quality problems are solved? A total value assessment would be in order to make this determination.

Native American fishing rights are another issue (see Chapter 4). Socioeconomic data could be used to compare benefits provided by recreational fishing versus a Native American fishery. The economic values and impacts of a recreational fishery might be very important. Likewise, the social values and impacts of a Native American fishery might be substantial. In this instance, legalities and politics might take precedence, but a socioeconomic valuation is still important in making all interests aware of the trade-offs involved in each potential decision.

8.5   CONCLUSION

Fisheries benefits are real and important (Figure 8.6) but measuring them is an art, not a science. No single approach to estimating socioeconomic benefits is perfect for all situations. Selection of an appropriate method depends on knowledge of the fishery in question, a list of the benefits of interest, the importance of the decision being made, and fiscal constraints.

In any fishery assessment it is critical that one distinguishes between an evaluation of values versus impacts. The purpose of the study will dictate which of the two categories is important. From an angler's viewpoint, values are important, and methods are available to place a dollar value on the experience from this perspective. From an economic standpoint, impacts are important, and methods are available to determine the contribution fisheries have on economics. Occasionally, a comprehensive study might require an evaluation of both categories of benefits. Care must be taken in any economic evaluation not to measure dollars more than once, realizing that at times the same benefit can be estimated in two different ways.

People are the ultimate beneficiaries of recreational and commercial fisheries management. Fisheries professionals need to know the social and economic benefits of their management efforts. Knowledge of socioeconomic benefits is useful in program development, program evaluation, and crisis management. Fisheries professionals are learning that the essence of fisheries management is establishing priorities, and this process involves not only biological data but fiscal constraints, legislative influence, management goals of an agency or its commission, public opinion, and a thorough analysis of socioeconomic benefits provided by fisheries.

Figure 8.6 Fisheries benefits are real and important! (Photograph provided courtesy of Missouri Department of Conservation.)

8.6 REFERENCES

Adamowicz, W. L. 1994. Habitat formation and variety seeking in a discrete choice model of recreation demand. Journal of Agricultural and Resource Economics 19:19–31.

Adamowicz, W. L., T. Graham-Tomasi, and J. J. Fletcher. 1989. Inequality constrained estimation of consumer surplus. Canadian Journal of Agricultural Economics 37:407–420.

Anderson, L. G. 1991. A note on market power in ITQ fisheries. Journal of Environmental Economics and Management 21:291–296.

Binkley, C. S., and W. M. Hanemann. 1976. The recreation benefits of water quality improvements: analysis of day trips in an urban setting. U.S. Environmental Protection Agency, NTIS (National Technical Information Service) PB-257719, Springfield, Virginia.

Bishop, R. C. 1987. Economic values defined. Pages 24–33 in Decker and Goff (1987).

Bishop, R. C., K. J. Boyle, and M. P. Welsh. 1987. Toward total economic valuation of Great Lakes fishery resources. Transactions of the American Fisheries Society 116:339–345.

Bockstael, N. E., I. E. Strand, K. E. McConnell, and F. Arsanjani. 1990. Sample estimation bias in the estimation of recreation demand functions: an application to sportfishing. Land Economics 66:40–49.

Brookshire, D. S., and A. Randall. 1978. Public policy alternatives, public goods, and contingent valuation mechanisms. Proceedings of the Western Economic Association, Honolulu, Hawaii, 53:20–26.

Brown, P. J., and M. J. Manfredo. 1987. Social values defined. Pages 12–23 in Decker and Goff (1987).

Brown, T. L. 1976. The 1973-75 salmon runs: New York's Salmon River sport fishery, angler activity, and economic impact. Cornell University, New York Sea Grant Institute NYSSGP-RS-76–025, Ithaca.

Brown, W. G., and F. Nawas. 1973. Impact of aggregation on the estimation of outdoor recreation demand functions. American Journal of Agricultural Economics 55:246–249.

Burt, O. R., and D. Brewer. 1974. Evaluation of recreational benefits associated with the Pattonsburg Dam Reservoir. University of Missouri, Office of Industrial Development Studies, Columbia.

Cesario, F. J. 1969. Operations research in outdoor recreation. Journal of Leisure Research 1:33–52.

Cesario, F. J. 1976. Value of time in recreation benefit studies. Land Economics 52:32–41.

Cohen, M. J. 1995. Technological disasters and natural resource damage assessment: an evaluation of the *Exxon Valdez* oil spill. Land Economics 77:65–82.

Connelly, N. A., and T. L. Brown. 1991. Net economic value of the freshwater recreational fisheries of New York. Transactions of the American Fisheries Society 120:770–775.

Decker, D. J., and G. R. Goff, editors. 1987. Valuing wildlife: economic and social perspectives. Westview Press, Boulder, Colorado.

Dwyer, J. F., J. R. Kelly, and M. D. Bowes. 1977. Improved procedures for valuation of the contribution of recreation to national economic development. University of Illinois, Water Resources Center, Research Report 128, Champaign-Urbana.

Englin, J., and R. Mendelsohn. 1991. A hedonic travel cost analysis for valuation of multiple components of site quality: the recreation value of forest management. Journal of Environmental Economics and Management 21:275–290.

Gan, C., and E. J. Luzar. 1993. A conjoint analysis of waterfowl hunting in Louisiana. Journal of Agricultural and Applied Economics 25:36–45.

Greenley, D. A., R. B. Walsh, and R. A. Young. 1981. Option value: empirical evidence from a case study of recreation and water quality. Quarterly Journal of Economics 96:657–674 (Wiley, New York).

Gregory, R. 1987. Nonmonetary measures of nonmarket fishery resource benefits. Transactions of the American Fisheries Society 116:374–380.

Haab, T. C., and K. E. McConnell. 1996. Count data models and the problems of zeros in recreation demand analysis. American Journal of Agricultural Economics 78:89–102.

Hanemann, W. M., and I. E. Strand. 1993. Natural resource damage assessment: economic implications for fisheries management. American Journal of Agricultural Economics 75:1188–1193.

Hansen, W. J. 1986. Valuating the recreational use of fishery resource sites with the travel cost method. Sport Fishing Institute, Technical Report V, Washington, D.C.

Hellerstein, D. M. 1991. Using count data models in travel cost analysis with aggregate data. American Journal of Agricultural Economics 73:860–866.

Hoehn, J. P. 1987. Contingent valuation in fisheries management: the design of satisfactory contingent valuation formats. Transactions of the American Fisheries Society 116:412–419.

Horowitz, J. K. 1993. A new model of contingent valuation. American Journal of Agricultural Economics 75:1268–1272.

Huppert, D. D., and R. D. Fight. 1991. Economic considerations in managing salmonid habitats. Pages 559–585 *in* W. R. Meehan, editor. Influences of forest and rangeland management on salmonid fishes and their habitats. American Fisheries Society, Special Publication 19, Bethesda, Maryland.

Hushak, L. J. 1987. Use of input-output analysis in fisheries assessment. Transactions of the American Fisheries Society 116:441–449.

Knetsch, J. L., R. E. Brown, and W. J. Hansen. 1976. Estimating expected use and value of recreation sites. Pages 103–115 *in* C. E. Gearing, W. W. Swart, and T. Var, editors. Planning for tourism development: quantitative approaches. Praeger, New York.

Lichatowich, J. 1992. Managing for sustainable fisheries: some social, economic, and ethical considerations. Pages 11–17 *in* G. H. Reeves, D. L. Bottom, and M. H. Brookes, technical coordinators. Ethical questions for resource managers. U.S. Department of Agriculture General Technical Report PNW-GTR-288.

Lichtkoppler, F. R. 1997. Ohio's Lake Erie charter fishing industry: 1985–1994. Fisheries 22(1):14–21.

Loomis, J. B. 1989. Test-retest reliability of the contingent valuation method: a comparison of general population and visitor responses. American Journal of Agricultural Economics 79:76–84.

Loomis, J. B., and D. S. White. 1996. Economic values of increasingly rare and endangered fish. Fisheries 21(11):6–11.

Malvestuto, S. P. 1983. Sampling the recreational fishery. Pages 397–419 *in* L. A. Nielsen and D. L. Johnson, editors. Fisheries techniques. American Fisheries Society, Bethesda, Maryland.

Malvestuto, S. P., and M. D. Hudgins. 1996. Optimum yield for recreational fisheries management. Fisheries 21(6):6–17.

McConnell, K. E., and I. Strand. 1981. Measuring the cost of time in recreation demand analysis: an application to sport fishing. American Journal of Agricultural Economics 63:153–156.

Mendelsohn, R., J. Hof, G. Peterson, and R. Johnson. 1992. Measuring recreation values with multiple destination trips. American Journal of Agricultural Economics 74:926–933.

Morey, E. R., R. D. Rowe, and M. Watson. 1993. A repeated nested-logit model of Atlantic salmon fishing. American Journal of Agricultural Economics 75:578–592.

Oster, S. 1977. Survey results on the benefit of water pollution abatement in the Merrimack River basin. Water Resources Research 13:882–884.

Ozuna, T., L. L. Jones, and O. Capps. 1993. Functional form and welfare measures in truncated recreation demand models. American Journal of Agricultural Economics 75:1030–1035.

Randall, A. 1987. Total economic value as a basis for policy. Transactions of the American Fisheries Society 116:325–335.

Rockland, D. B. 1985. The economic benefits of a fishery resource: a practical guide. Sport Fishing Institute, Technical Report I, Washington, D.C.

Rockland, D. B. 1986. Economic models: black boxes or helpful tools? A reference guide. Sport Fishing Institute, Technical Report II, Washington, D.C.

Smith, V. K., and W. H. Desvousges. 1985. The generalized travel cost model and water quality benefits: a reconsideration. Southern Economic Journal 52:371–381.

Steinhoff, H. W., R. G. Walsh, T. J. Peterle, and J. M. Petulla. 1987. Evolution of the valuation of wildlife. Pages 34–48 in Decker and Goff (1987).

Stevens, J. B. 1966. Recreation benefits from water pollution control. Water Resources Research 2:167–181.

Sutinen, J. G. 1993. Recreational and commercial fisheries allocation with costly enforcement. American Journal of Agricultural Economics 75:1183–1187.

Talhelm, D. R. 1987. Recommendations from the SAFR symposium. Transactions of the American Fisheries Society 116:537–540.

Talhelm, D. R., and L. W. Libby. 1987. In search of a total value assessment framework: SAFR symposium overview and synthesis. Transactions of the American Fisheries Society 116:293–301.

Teasley, R. J., J. C. Bergstrom, and H. K. Cordell. 1994. Estimating revenue-capture potential associated with public area recreation. Journal of Agricultural and Resource Economics 19:89–101.

Thayer, M. A. 1981. Contingent valuation techniques for assessing environmental impacts: further evidence. Journal of Environmental Economics and Management 8:27–44.

Vanderpool, C. K. 1987. Social impact assessment and fisheries. Transactions of the American Fisheries Society 116:479–485.

Walker, K., B. Rettig, and R. Hilborn. 1983. Analysis of multiple objectives in Oregon coho salmon policy. Canadian Journal of Fisheries and Aquatic Sciences 40:580–587.

Walsh, G., D. A. Greenley, R. A. Young, J. R. McKean, and A. A. Prato. 1978. Option values, preservation values, and recreational benefits of improved water quality. U.S. Environmental Protection Agency EPA-600/5-78-001, Washington, D.C.

Weithman, A. S., and R. O. Anderson. 1978a. An analysis of memorable fishing trips by Missouri anglers. Fisheries 3(1):19–20.

Weithman, A. S., and R. O. Anderson. 1978b. A method of evaluating fishing quality. Fisheries 3(3):6–10.

Weithman, A. S., and M. A. Haas. 1982. Socioeconomic value of the trout fishery in Lake Taneycomo, Missouri. Transactions of the American Fisheries Society 111:223–230.

Weithman, A. S., and S. K. Katti. 1979. Testing of fishing quality indices. Transactions of the American Fisheries Society 108:320–325.

Wilman, E. A. 1980. The value of time in recreation benefit studies. Journal of Environmental Economics and Management 7:272–286.

HABITAT MANIPULATIONS

Chapter 9

Watershed Management and Land Use Practices

THOMAS A. WESCHE AND DANIEL J. ISAAK

9.1 INTRODUCTION

> A hydrologist may be better able than a biologist to evaluate the dynamic fisheries production potential of a drainage basin. Certainly the two disciplines working together will come much nearer to accurately assessing the potential than will either working alone. (G. L. Ziemer 1971)

This was sound advice nearly 3 decades ago and it remains so today as fisheries managers find themselves faced with the challenge of evolving from just fish managers to water managers, land managers, and habitat managers. Declines of many important game fish populations, increased realization of broadscale habitat requirements, better awareness of the effects of human activities on aquatic systems, and the multijurisdictional nature of problems facing fisheries managers are forcing fisheries managers to approach their work from a much broader perspective than has been taken traditionally. Much as a stream integrates its surroundings, today's fisheries managers must be able to integrate their knowledge of management (Chapter 2), communication skills (Chapter 3), and fish biology (Chapter 6) with knowledge of watershed function. Therefore, in this chapter we present an overview of watershed structure, process, and function and stress the interdependence of fisheries resources upon fish biology and the geomorphology, hydrology, and land use of a watershed.

9.2 WATERSHED CHARACTERIZATION

A watershed is any sloping land surface that sheds water. A more functional definition is "all land enclosed by a continuous hydrologic drainage divide and lying upslope from a specified point on a stream." Typically, watershed is synonymous with drainage basin or catchment. A drainage basin is a watershed that collects and discharges its surface streamflow through one outlet or mouth, whereas a catchment is generally considered to be a small drainage basin (Hewlett and Nutter 1969).

What is implied when we use the term watershed management? The Society of American Foresters defined watershed management in 1944 as

> The management of the natural resources of a drainage basin primarily for the production and protection of water supplies and water-based resources, including the control of erosion and floods, and the protection of esthetic values associated with water.

The key words here are "production and protection of water supplies *and water-based resources*" which includes fishes.

One of the important functions of a watershed is to produce water. From a water supply standpoint, we must be concerned not only with the total quantity of water yielded but also with the timing of that yield (flow regimen) and its quality. Numerous variables interact within a watershed to control streamflow and the nature of the stream channels that convey the water. In general, these variables can be categorized as climatic, topographic, geologic, and vegetative. One must be aware of the strong interrelations among these controlling factors, their dominant influence on the character of watersheds, and ultimately, the role they play in regard to the zoogeography of fish species, population attributes, and fishery quality.

Climate can be defined as the long-term aggregate atmospheric condition produced by daily weather. The climate of an area has direct bearing on both the quantity of water yield and the flow regimen. Descriptors typically used to characterize a climate include

- air temperature (mean, maximum, and minimum),
- freeze dates (frost-free period),
- soil temperature,
- precipitation (quantity, type, and temporal and spatial distribution),
- humidity (quantity of water vapor in the air),
- evaporation (process by which water changes phase from liquid to vapor),
- wind (speed and direction),
- solar radiation, and
- atmospheric pressure.

The climate of an area is governed by latitude, elevation, proximity to oceans, and local topographic features. Together, these factors determine or modify airflow patterns and weather systems that dictate the ranges of precipitation, temperature, humidity, wind, and evaporation that a watershed experiences.

Geomorphology, the science that deals with relief features of the earth's surface, has been the discipline most responsible for developing quantitative techniques to measure topographic characteristics of drainage basins and for attempting to explain the influence of topographic characteristics in terms of watershed functions and processes (Brown 1970). Leopold and Langbein (1963) summarized the relation of these two objectives when they wrote, "In geomorphologic systems, the ability to measure may always exceed ability to forecast or explain."

Over the century that geomorphology has been recognized as a science, numerous topographic measures have been devised to describe watersheds. These measures characterize the drainage basin as a whole, the total channel system or network, reaches

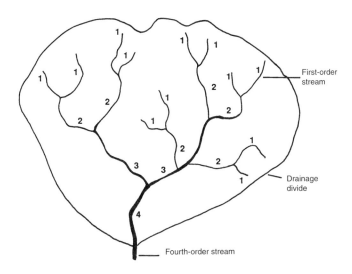

Figure 9.1 Stream ordering within a fourth-order watershed based on the Strahler (1957) method. Stream order increases by one only where two streams of equal order meet.

within the total channel network, and cross sections within a channel reach. For each of these components, measures of area, length, shape, and relief have been developed. In general, the area of the watershed dictates the quantity of water yield, whereas length, shape, and relief control the streamflow regimen and the rate of sediment yield.

Commonly measured topographic attributes of watersheds are listed in Table 9.1 along with their definition or derivation and a literature source for locating additional information. Figure 9.1 depicts the popular stream-ordering method developed by Strahler (1957), and Figure 9.2 depicts three common stream channel patterns. Stream cross sections can be described as shown in Figure 9.3.

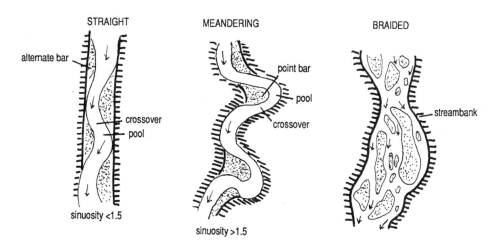

Figure 9.2 Patterns of stream channels. Figures adapted from Simons and Senturk (1977).

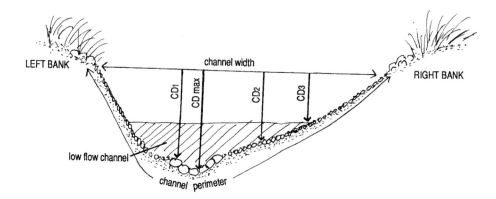

Figure 9.3   Cross section of a stream channel (looking downstream). Channel depth is denoted by CD.

Watershed descriptions should include reference to the rocks and sediments that underlie the drainage basin because geologic characteristics determine the nature and extent of groundwater storage, the type of material available for erosion and transport, and the chemical quality of the water. Groundwater-bearing formations sufficiently permeable to transmit and yield water are termed aquifers. The most common aquifer materials are unconsolidated sands and gravels that occur in stream valleys, old streambeds, coastal plains, dunes, and glacial deposits. Sandstone, a sedimentary rock, also serves as aquifer material; other sedimentary rocks such as shale and solid limestone do not. Where these rocks and others, such as igneous granites and metamorphic gneiss, are highly fractured, small water yields may be possible. Volcanic materials such as basalt and lavas can form aquifers if they are highly fractured or porous (Bouwer 1978).

Weathering is the physicochemical process whereby the minerals in rocks at or near the earth's surface are altered under normal temperatures and pressures. Weathering proceeds at varying rates depending on the type of rocks, climate, and vegetation. Topography in large measure determines the rate at which the product of weathering— sediment—is removed. The weathered products that remain in place are the basic materials from which soil is developed. Sediments, which range in size from large boulders to fine particles, are termed fluvial sediments when they are moved by water. Movement may occur as sheet erosion from land surfaces (particularly for the fine particles) or as channel erosion once runoff accumulates in streams. Fluvial sediments are transported within streams either as bed load, where the particles tumble over the streambed, or as suspended sediment in the turbulent flow.

Weathering of rock material is also an important determinant of stream water chemistries. Studies at Hubbard Brook, a northeastern hardwood experimental forest, revealed that while precipitation and atmospheric inputs were major sources of sulfur, nitrogen, chloride, and phosphorous, weathering was the primary source of calcium, magnesium, potassium, and sodium in the ecosystem (Likens et al. 1970).

The vegetative character of a watershed is a function of climate, soils, topography, and land use. In turn, the vegetation that covers the land surface plays a dominant role in controlling the timing and amount of water yield from a drainage basin, as well

Table 9.1 Commonly measured topographic attributes of watersheds.

Attribute	Symbol	Derivation of definition	Units	Source
Basin area	BA	For a specified stream location, that area, measured in a horizontal plane, enclosed by a drainage divide	km²	Horton (1945)
Basin length	BL	Length of the line, parallel to the main drainage line, from the headwater divide to a specified stream location	km	Potter (1961)
Basin perimeter	BP	Length of the boundary line along a topographic ridge that separates two adjacent drainage basins	km	Smith (1950)
Basin relief	BR	Highest elevation on the headwater divide minus the elevation at a specified stream location	m	Schumm (1956)
Channel slope	CS	Elevation at 85% of stream length minus elevation at 10% of stream length divided by stream length between these two points	m/m	Craig and Rankl (1978)
Drainage density	DD	TSL/BA	km/km²	Horton (1945)
Mean basin elevation	MBE	Sum of the products of the areas between contour lines and the average elevation between contour lines divided by the total area of the basin	m	Burton and Wesche (1974)
Sinuosity	S	Ratio of stream length to valley length		Knighton (1984)
Stream order	SO	Method of classifying streams as part of a drainage basin network (see Figure 9.1)		Strahler (1957)
Total stream length	TSL	Sum of the lengths of all streams within a drainage basin	km	Horton (1945)

as the quantity of soil available for transport to the stream channel as fluvial sediment. Hence, when soils have been denuded by land use activities, erosion control is focused primarily on rapid re-establishment of the plant community. Vegetation also contributes to the aesthetic qualities of a drainage basin, and watersheds are commonly characterized by the composition of their vegetation (e.g., coniferous forest, deciduous woodland, grassland, or shrubland).

9.2.1 Hydrologic Processes

The circulation of water from the oceans to the atmosphere, back to the land, and eventually back to the oceans is referred to as the hydrologic cycle. Anderson et al. (1976) considered the hydrologic cycle as a system comprising water storage compartments and the flow of water within and between them (Figure 9.4). Water is stored in the atmosphere and the oceans as well as in the soil, stream channels, lakes, ponds, groundwater, and vegetation. We call these water reservoirs individual storage compartments and the flows between them hydrologic processes. Water moves between storage areas in either the solid, liquid, or gaseous state. Definitions of the various hydrologic processes of concern to watershed managers are presented in Table 9.2.

Another way to consider the hydrologic cycle is in terms of the disposition of precipitation. This relationship can be expressed in a water balance equation,

$$P = R + E_T + S \tag{9.1}$$

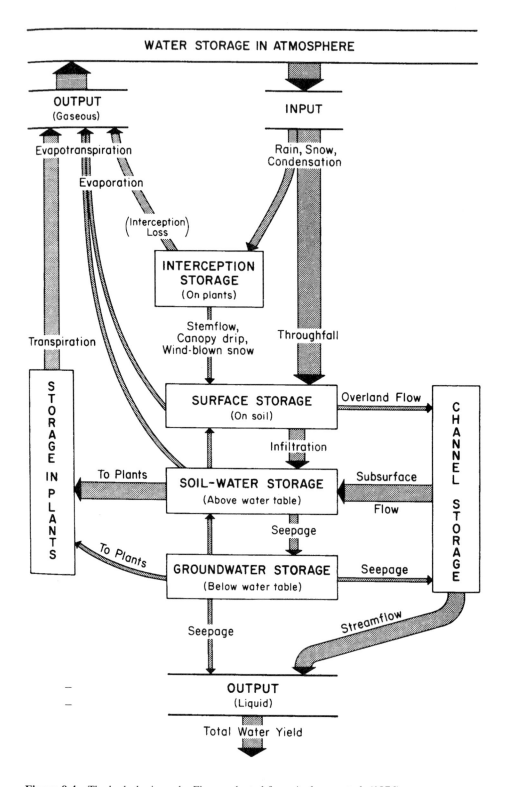

Figure 9.4   The hydrologic cycle. Figure adapted from Anderson et al. (1976).

Table 9.2 Definitions of hydrologic processes.

Process	Definition
Evaporation	The process by which water is transformed from the liquid or solid state into the vapor state.
Evapotranspiration	The process by which water is lost from soil and plant surfaces through evaporation and transpiration and the volume of this loss per unit land area.
Infiltration	The flow of liquid water into the soil above the water table.
Interception loss	The portion of the precipitation caught and held by vegetation and lost by evaporation, thus never reaching the ground.
Overland flow	The flow of rainwater or snowmelt over the land surface toward stream channels. Upon entering a stream, it becomes runoff.
Precipitation	The discharge of water, in liquid or solid state, from the atmosphere onto a land or water surface.
Seepage	The slow movement of water from soil-water storage to groundwater storage.
Streamflow	The discharge that occurs in a natural channel.
Subsurface flow	Water that infiltrates the soil surface and moves laterally through the upper soil layers until it enters a channel.
Throughfall	In a vegetated area, the precipitation that falls directly to the ground, including streamflow, canopy drip, and windblown snow.
Transpiration	The process by which water vapor escapes from a living plant, principally through the leaves, and enters the atmosphere.
Water yield	The total runoff from a drainage basin through surface channels and aquifers.

where P = precipitation,
       R = streamflow or total water yield,
      E_T = water lost by evapotranspiration, and
      S = change in storage or that portion of the precipitation that is retained or lost from storage in the earth's mantle.

Rearranging equation 9.1, the variables that a watershed manager can work with to alter water yields can be identified,

$$R = P - E_T - S. \tag{9.2}$$

These variables are typically expressed in terms of inches of water. This unit of measure contrasts somewhat with the measures of water normally applied by water users (acre-feet) and stream habitat biologists (cubic feet per second or ft^3/s). Perhaps an example is the simplest way to relate these units. If during a year, 48 in of precipitation falls on a 1,000-acre watershed and 50% of this total is realized as streamflow, the water yield is 24 in (48 in $\times$ 0.50). The total volume of water yielded during the year would be equal to 2,000 acre-feet (24 in $\times$ 1,000 acres $\times$ [1 ft/12 in]). The average discharge for the stream draining this watershed over the course of the year would be 2.76 ft^3/s (2,000 acre-feet/year $\times$ 43,560 ft^2/acre-foot $\times$ [1 year/365 d] $\times$ [1 d/86,400 s]).

Average annual precipitation over the contiguous United States is 30 in, the equivalent of 4.75 billion acre-feet of water. Of this total, 3.4 billion acre-feet (72%) are lost due to evapotranspiration from nonirrigated lands, and an additional 3% is consumptively used by irrigators, industry, and municipalities (Hewlett and Nutter 1969). Streamflow accounts for the remaining 25%.

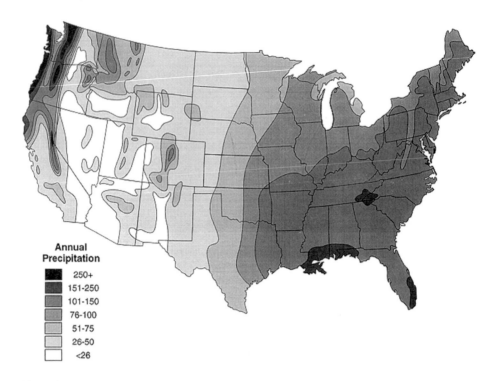

Figure 9.5 Average annual precipitation (in centimeters) across the contiguous United States. Map is from Satterlund (1972).

Precipitation is highly variable across the United States (Figure 9.5). Extremes range from more than 500 cm annually on the Olympic Peninsula of Washington to less than 13 cm across much of the desert southwest (Satterlund 1972). Such variation is due to storm paths, the distance and direction from large water bodies, location with respect to barriers to atmospheric flow, and elevation. For the most part, alpine and forested zones of the United States receive at least 50 cm of precipitation annually, whereas brush and grassland receive 25–50 cm of precipitation in semiarid zones and less than 25 cm in arid zones.

The proportion of precipitation that results in streamflow varies between geographic regions and vegetation zones. Table 9.3 summarizes precipitation–water yield relationships from several locations and vegetation zones across the country. In general, the percent of precipitation translating into streamflow increases with increasing precipitation. This is because evapotranspiration is the most constant variable in the water balance equation. Once this requirement is met, remaining precipitation is available for runoff. Within a given watershed, the percent of precipitation yielded as runoff will typically increase with elevation.

Often, water supply difficulties are not problems of water yield but rather are related to the timing of runoff. As anyone who has ever watched a stream for any length of time will attest, streamflow can be highly variable, not only from one year to the next but, in some cases, from one minute to the next. The hydrograph is a basic tool used by watershed managers to describe the streamflow regimen of a stream and the

Table 9.3 Precipitation–water yield relationships for different types of watersheds. Wyoming data are from Wesche (1982); all other data are from Anderson et al. (1976).

Location	Vegetation type	Midarea elevation (m)	Precipitation (cm)	Streamflow (cm)	% Water yield
New Hampshire	Northern hardwoods	625	122	69	56
West Virginia	Central Appalachian hardwoods	763	147	61	41
North Carolina	Southern Appalachian hardwoods	824	183	91	50
Minnesota	Bog black spruce	422	79	18	23
Oregon	Douglas fir	763	239	155	65
Arizona	Chaparral	1,296	69	5	7
Arizona	Ponderosa pine, white fir	2,181	81	8	9
Colorado	Lodgepole pine, spruce–fir	3,172	56	30	54
Wyoming	Alpine	3,355	127	97	76

volume of runoff from a drainage basin. A hydrograph is a plot showing some measure of discharge with respect to time for a given stream location. The components of a simple hydrograph are shown in Figure 9.6.

Hydrographs, which vary within and among watersheds, are dependent upon the watershed factors already discussed and weather factors including amount, intensity, duration, and distribution of the precipitation, storm direction, and temperature. Figure 9.7 depicts annual hydrographs from two Midwest streams and a mountain stream from the central Rocky Mountains. The midwestern river A shows flashy responses to both precipitation and snowmelt runoff, which is indicative of poor storage due to impermeable or shallow soil layers. The midwestern river B hydrograph shows little variation throughout the year, although river B had a precipitation regime similar to that of river A. The river B watershed, however, is characterized by deep, permeable, sandy soils with good infiltration and storage capability. The mountain stream hydrograph indicates the dominance of snowmelt runoff on the hydrology of the region. Approximately 75–80% of the annual water yield results from snowpack that accumulates during the winter months. Base flow during the remainder of the year is generally stable.

9.2.2 Sediment Transport

Sediment is a major product of watersheds and is produced in part by the hydrologic cycle. Sediment yield can be defined as the sediment transported out of the drainage basin from all fluvial erosion processes. It is typically measured in tons per unit land area per unit time. From a watershed management perspective, the sediment yield from a drainage basin is important because (1) an estimated 80% of water quality degradation results from erosion, (2) sediment interacts strongly with other water quality components, and (3) sediment yield is directly affected by land use activities (Anderson et al. 1976).

Erosion and sedimentation are the two phases of the process of detaching particulate material from one location and depositing it in another location. Erosion refers to the removal of the material whereas sedimentation refers to its deposition. Once transport has begun, the material is referred to as sediment (Satterlund 1972).

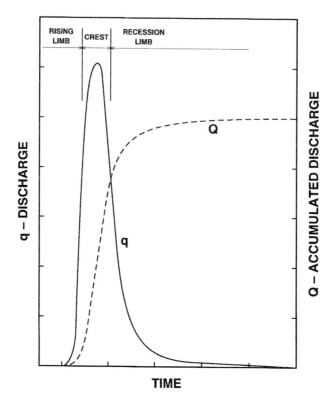

Figure 9.6 Component parts of a hydrograph. The cumulative volume of runoff relative to time is given by Q, and the discharge rate at a point in time is given by q.

There are three types of erosion within a watershed: surface, mass movement, and channel cutting, all of which can contribute significantly to the sediment yield from the drainage basin. Surface erosion is the detachment and removal of individual soil particles from the land surface and includes sheet erosion and rill and gully formation. Two hydrologic processes primarily involved with surface erosion are: raindrop splash and overland flow. Raindrops striking bare soil cause minor explosions that dislodge soil particles. These particles can be forced upward into suspension and moved downslope in overland flow (Brooks et al. 1991).

Mass movement of slopes is caused by gravity. It occurs when the cohesive and frictional strength of a land mass fails across an area of weakness. When the movement occurs at a rate perceptible to the eye, it is called a landslide. Creep occurs at a rate slower than the eye can perceive. Mass movement results from either an increase in shear stress, a decrease in friction, or both. Shear stress is the sum of all forces acting to displace the earth mass on the slope. Although the weight of the soil itself is generally the primary stress component, water, vegetation, wind, vibrations (e.g., earthquakes and explosions), erosive actions of streams, and soil creep on steep slopes may all contribute. Land use activities such as road building and timber harvest can also significantly influence the stress acting on a land mass.

Channel cutting refers to the detachment and movement of material from stream channels by flowing water. Although much of the suspended sediment load transported by a stream originates from erosion processes on the land surface of the watershed, the

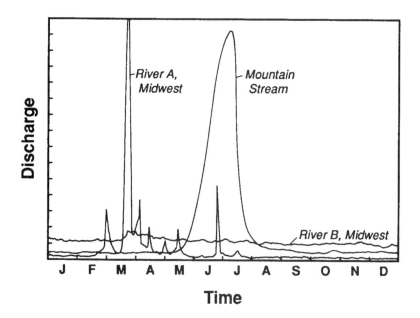

Figure 9.7 Annual hydrographs for two streams in the midwestern United States and a typical mountain stream from the central Rocky Mountains. Figure adapted from Hewlett and Nutter (1969).

bulk of the sediment transported as bed load is the result of channel erosion. Channel erosion, like streamflow, is highly variable but is most likely to occur at higher flows (Knighton 1984).

Sediment transport is a function of the availability of material to be eroded and the sediment-transporting ability of the streamflow. Each stream has a fixed sediment carrying capacity or competency, based upon its channel dimensions and water velocity (Leopold and Maddock 1953). A concept fundamental to sediment transport is stream power, the rate at which a stream does work. This expression of sediment transport capability is defined by the relationship,

$$\omega = \gamma \bar{D} \bar{V} S, \tag{9.3}$$

where ω = stream power (pounds per foot of channel width per second);
γ = specific weight of water (62.4 lbs/ft^3);
D = mean depth of the flow (ft);
V = mean velocity of the flow (ft/s); and
S = stream slope (ft/ft).

As discharge changes, both stream power and sediment carrying capacity change at a particular stream location. Natural stream channels are seldom uniform from one location to another. Therefore, even though discharge may be constant at two locations on a given stream, depth, velocity, and slope can differ, resulting in variations of stream power and transport capacity.

As a result of temporal and spatial variation in sediment transport ability, streams—even though they may appear stable—are characterized by a dynamic equilibrium. Sediment is eroded and transported primarily on the rising limb of the hydrograph

(Figure 9.6) and is deposited on the receding limb. Sediment is periodically moved from one channel location and deposited in another. Pools may be scoured of deposited sediment at high flows while riffles may be filled. The opposite occurs at lower discharges. The net result is that stream habitats for fishes are created and sediment tends to move downstream in a sort of leapfrog pattern, responding to the changing hydrograph and the hydraulic characteristics of the channel.

9.2.3 Riparian Zones

Riparian areas serve as transitional zones, or ecotones, between a watershed and its stream and have been defined as "zones of direct interaction between terrestrial and aquatic ecosystems" (Gregory et al. 1991). During the settlement of the western United States in the 1800s, riparian areas provided much of the food, shelter, and fuel necessary for pioneer survival. Today, riparian areas are recognized as being transportation corridors, high producers of timber and forage, key habitats for a diversity of wildlife, major components of quality fisheries habitat, prime recreation areas, and areas critical to the overall management of any watershed. Given this array of uses, it is not surprising that riparian areas have been the focus of much debate in recent years between user groups and management agencies.

Plant community composition in riparian zones generally differs from adjacent lands due to greater water availability and disturbance frequency (Hawk and Zobel 1974; Decamps et al. 1988). In unconstrained valleys, streams are flanked by the alluvial floodplains they have built. Over the course of decades and centuries, the stream wanders across the valley floor, reworking previously deposited alluvial sediments and leaving behind new sediments. The stream interacts extensively with its floodplain to create a diverse mosaic of patch types on the valley floor. Frequent disturbance and a shallow water table provide conditions favorable to a riparian plant community dominated by mesic, early successional species (e.g., alder, cottonwood, and willow) and differing markedly from upslope or adjacent plant communities (Gregory et al. 1991). Observation of a "river of green" during late summer along a lowland stream in a semiarid region highlights this difference and indicates the water dependence of riparian vegetation in wide valleys.

The lateral extent of fluvial processes is limited in constrained valleys by the imposition of valley walls. Floodplain development is restricted or nonexistent, and lands upslope of the stream are disturbed infrequently (Gregory et al. 1991). Further, the steepness of adjacent land slopes quickly increases the distance between the land surface and the water table. As a result, riparian plant species composition closely resembles species composition in upslope areas.

From watershed and fisheries management perspectives, riparian areas provide many important services. Streamside vegetation plays a role in controlling channel morphology. Not only do roots stabilize otherwise easily eroded streambanks, but pieces of large woody debris recruited into the stream from the riparian zone retain sediments that would otherwise be flushed from the stream (Speaker et al. 1984). Large woody debris in conjunction with fluvial processes also create a diversity of meso- and microhabitats important to stream fishes (Keller and Swanson 1979; Bisson et al. 1982).

Riparian areas serve to moderate environmental conditions experienced by stream biota. Riparian vegetation decreases variation in stream temperatures by shading streams from direct insolation and retarding heat loss. When overbank flooding occurs, aboveground portions of riparian plants increase floodplain roughness, thereby slowing water movement and promoting infiltration and recharge of the alluvial aquifer. Instantaneous water yield from a flood event is reduced, and water retained in the aquifer can later be released to improve base flow conditions. Riparian vegetation also acts to control nonpoint source pollution by filtering out sediments delivered from adjacent lands by overland flow. Because many plant nutrients (especially phosphorus and to some extent nitrogen) in surface runoff are attached to sediment particles, this filtering can help reduce nutrient loadings to the aquatic system (Karr and Schlosser 1978). In these ways, streambanks are built and fertile soils developed on alluvial areas.

As a result of moderated environmental conditions and maintenance of high-quality structural habitats, riparian zones strongly influence the integrity of the stream. Spawning beds for fishes and microhabitats for aquatic insects remain relatively free from damaging fine-sediment deposits. Filling of pools is reduced while water depths and structural diversity are maintained. Escape cover provided by undercut streambanks is maintained, and critical low-flow habitat is improved by augmented base flow levels. Water chemistries remain within tolerable levels for both fish and incubating eggs. In cold regions, narrower, deeper channels promote snow bridging, which can moderate winter conditions.

9.2.4 A Watershed Comparison

Consider two small watersheds, one located in the spruce–fir forest of the central Rocky Mountains, the other in the shortgrass prairie of the northern Great Plains. Although drainage basin size and stream order may be similar, they differ dramatically in the aquatic habitat provided and fish populations present. In the mountain watershed, basin elevation is high, precipitation is approximately 76 cm annually, and the hydrograph is dominated by spring snowmelt. Soils are thin with high infiltration rates. Overland flow seldom occurs, even though drainage density (see Table 9.1) is low. Water yield as a percent of precipitation is in the 40–50% range, and streamflow is perennial. Lands adjacent to the stream are steep and covered with conifers. The valley bottom is narrow and basin relief is high, resulting in a relatively steep stream slope. Within the channel itself, the thin soil mantle has been eroded, exposing the underlying geology of the drainage basin. The streambed is dominated by bedrock, boulders, and cobbles in the straight, steeper reaches; in the flatter sections, gravels and sands are deposited, and a meandering stream pattern may develop. Streambanks are poorly developed in the steeper reaches due to the shortage of building material and sediment and reflect the composition of the substrate. As the valley floor widens and stream slope decreases, sediments transported from the steep reaches can settle out along the channel boundaries, vegetation can become established, and bank development occurs. Sediment yield from the watershed is limited, which results in quite low suspended sediment transport and turbidity. Water temperatures remain cool and nutrient loadings low. Bed load transport of cobbles and gravels occurs primarily during the rising limb of the spring runoff hydrograph when available stream power is at or near its maximum. At this time, pocket pools are scoured behind the boulders that form much of the habitat in the steep stream sections while undercut banks and pools created by meander processes are scoured in the flats. On the receding limb of the hydrograph, as stream power lessens, these larger sediments can no longer be transported and are deposited along the alternate, point, and midchannel bars (see Figure 9.2). Through these integrated watershed processes, salmonid rearing and spawning habitats are created and maintained on an annual basis.

Consider now the same processes but under the different climatic, hydrologic, topographic, geologic, and vegetative conditions of the shortgrass prairie watershed.

Basin elevation is relatively low, and the climate is semiarid with an annual precipitation of approximately 36 cm. Over 40% of the annual precipitation occurs as rainfall during the summer months, primarily the product of thunderstorm activity. Soils are well developed and deep and have a high clay content and a low infiltration rate. The annual hydrograph is flashy. Streamflow is ephemeral; surface flow in the channel occurs only in response to precipitation, during which overland flow may occur even though the topography is gently rolling. Only a small percentage of annual precipitation leaves the watershed as streamflow due to the high evapotranspiration rate and the limited contribution of groundwater. Surface runoff is turbid due to the sparse vegetative ground cover on the uplands. In a small, low-relief watershed such as this, the low-gradient channel may show little development, often more closely resembling a grass-covered waterway. The only aquatic habitat present may be a series of unconnected "potholes" that support small populations of hardy species such as black bullhead or green sunfish. These potholes are formed at low spots in the channel where ephemeral standing water is present long enough to kill the grass cover. Subsequent high flows can then scour these areas to a depth sufficient to intersect the shallow groundwater aquifer and create a perennial pond environment.

Such a macroscopic comparison allows managers to make gross estimates of the fishery potential of diverse watersheds. Certainly even the most casual observer could discern the obvious differences between these watersheds. The important point to be made, however, is not that the watersheds are different but how and why they differ.

9.3 LINKAGES AMONG THE WATERSHED, STREAM, AND FISH POPULATIONS

The intricacies of the hydrologic cycle and the dynamics of sediment transport combined with underlying geology, topography, vegetation, and climate make it apparent that watersheds are complex systems. Because of their integrative nature, streams synthesize this complexity and provide intimate reflections of the watersheds they drain (Hynes 1975). Fish populations respond to the array of habitats provided by the stream, and predictable patterns in fish populations often occur over the length of the stream. A manager must be aware of the interactions between a watershed, a stream, and fish populations in the stream if a fishery is to be managed wisely.

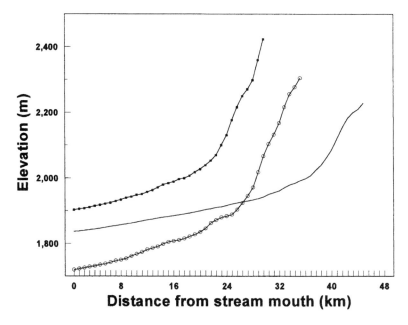

Figure 9.8 Stream slope profiles for three streams in the Salt River watershed of western Wyoming.

9.3.1 Watershed Influences on Environmental Gradients in Streams

Leopold and Maddock (1953) were some of the first to provide evidence that gradual, predictable changes in stream characteristics occur over the length of a stream. For example, the longitudinal profile of most streams is characterized by a concave shape (Figure 9.8), wherein upstream reaches of a stream are steeper than are downstream reaches. Because stream power is directly related to stream slope (equation 9.3), the ability of the stream to transport large sediment particles is greatest in upstream reaches and decreases in downstream reaches. As a result, average substrate size decreases over the length of a stream. Leopold and Maddock (1953) also described concomitant increases in stream dimensions and average water velocities—the former being an adjustment to the increasing size of the watershed contributing runoff to the stream and the latter due to decreased importance of streambed friction in slowing water flow. Fisheries biologists have since related changes in stream power, substrate size, water velocity, and stream size to the types and abundances of habitats that are important to fish (Grant et al. 1990; Hubert and Kozel 1993).

Many other environmental gradients related to the surrounding watershed have been documented over the course of a stream. Water chemistry is known to vary among watersheds; it also changes over the length of a stream depending on geology and land uses that occur upstream (Allen 1995). Average air temperatures decrease by 1°C for every 100 m increase in elevation (Brooks et al. 1991). Because water temperatures are strongly correlated with air temperatures, streams flowing at higher elevations tend to be colder than streams flowing at lower elevations. As a stream drops in elevation, not only is it exposed to warmer air temperatures, but increases in stream size limit the ability of adjacent vegetation to shade the stream. Therefore, insolation contributes increasingly to the warmer stream temperatures observed in downstream areas. Downstream reaches of streams also exhibit less temperature variability because of the thermal inertia associated with the greater volume of water (Ward 1985).

The role that pieces of large woody debris play in stream systems changes from upstream to downstream because of interactions with other stream gradients. In headwater areas, the width of the stream and its total power are often inadequate to transport pieces of large woody debris downstream. Instead, pieces of large woody debris remain where they are deposited within the stream channel and perform several functions dependent upon their final orientation (Richmond and Fausch 1995). Large woody debris orienting parallel to flow and adjacent to the bank can help shield the bank against the erosive action of the stream. These pieces of large woody debris often become integral parts of the bank where they help form overhead cover that can provide fishes low-velocity refuges from terrestrial predators. Pieces of large woody debris that come to rest perpendicular to the direction of flow may form dam and plunge pools, retain sediments that would otherwise be mobilized and lost at high flows, or contribute to the formation of debris jams that exhibit numerous combinations of cover, hydraulics, and substrates. As stream width and total power increase in downstream areas, however, fewer trees are of a size that bridges both banks, and an increasing proportion are transported downstream or oriented parallel to flow (Bilby and Ward 1989). Cover for fishes becomes less a function of large woody debris and increasingly a function of water depths, undercut banks, and overhanging vegetation.

Most environmental gradients associated with a stream are perceived to change gradually in a longitudinal fashion. However, this perception is not entirely accurate because a stream is also linked laterally to its surroundings, and abrupt changes in terrestrial conditions elicit rapid changes in the stream. For example, imagine a small stream flowing through a narrow canyon. This stream would be characterized by constraining valley walls, high stream slope and power, large average substrate size, and a channel pattern that is straight (Figure 9.2). Now imagine this stream exiting the canyon and flowing onto the wide, alluvial plain of a large river. The canyon walls no longer confine the stream, and the slope of the valley floor on the alluvial plain is less steep than is the valley floor slope in the canyon. Stream power decreases rapidly, and the stream loses the ability to transport the sediment load it has been carrying. As a result, a large amount of sediment is deposited at the mouth of the canyon, and a short stream reach with a braided channel pattern is formed. Once the sediment is deposited, the stream rapidly adjusts to its new surroundings and now exhibits a meandering stream pattern, little slope and power, and smaller substrate sizes. Similarly abrupt changes can occur where a stream passes through a geologic formation that is either more or less resilient than surrounding geologic materials (Hupp 1982). Stream character also changes abruptly at the confluence of two streams. In these instances, changes in water temperature, water chemistry, or substrate size are most probable, and the amount of change is directly related to the degree of difference between the watersheds being drained.

9.3.2 Streams as Templates

The hypothesis that habitat forms a template dictating the characteristics of the biotic community has been articulated by Southwood (1977, 1988). Vannote et al. (1980) applied this concept to stream systems in their description of the river con-

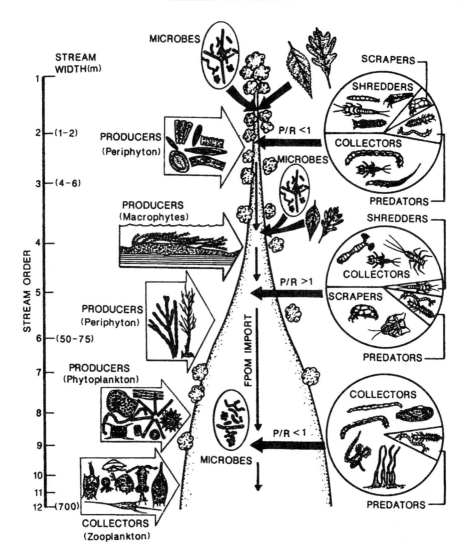

Figure 9.9 Trends in energy sources, ratios of autotrophic production to heterotrophic respiration (P/R), and aquatic insect functional groups along the river continuum. Fine particulate organic matter is denoted by FPOM. Figure is from Vannote et al. (1980).

tinuum concept. The basic premise of the river continuum concept is that the set of gradients occurring over the length of the stream determine how and in what form energy enters the stream, which in turn structures the aquatic insect community (Figure 9.9).

In headwaters reaches (orders 1–3), where stream size is small, shading from the riparian zone greatly limits the autotrophic production of organic matter within the stream. Instead, most energy is derived allochthonously in the form of leaf litter from the riparian zone. Aquatic insect communities are dominated by shredder and collector insect guilds that are capable of using the leaf litter as a food source. In downstream reaches (orders 4–6), the stream is wider, and the riparian zone does not completely shade the stream. Energy inputs are no longer dominated by the riparian zone but instead originate from autochthonous production within the reach and the import of organic materials from upstream. Aquatic insects from the collector guild still compose a significant portion of the insect community, but insects from the scraping guild that use autotrophic producers as a food source now supplant shredding insects. In the lowermost sections of streams (orders 7–12), the suspension of organic matter and sediments derived from upstream decreases water clarity and limits autotrophic production within the stream. Energy inputs are almost entirely derived from the import of fine particulate organic matter from upstream areas, and the aquatic insect community is now dominated by the collecting guild.

The river continuum concept was developed primarily to explain patterns in aquatic insect communities, but a similar conceptualization can be used to understand better game fish populations. Many investigators have related attributes of game fish populations to microhabitat characteristics (see Fausch et al. 1988 for a review). Given that microhabitat features vary relative to stream gradients (Hubert and Kozel 1993), it seems plausible that patterns should be manifest in game fish populations over the length of a stream. Evidence for this line of reasoning can be derived from numerous sources. Layher et al. (1987) attributed much of the variation in the biomass of spotted bass from Kansas streams to physicochemical variables that vary over the length of a stream. Similarly, Hubert et al. (1996) developed a model capable of predicting the potential maximum biomass of trout in Wyoming streams. From an initial pool of 20 habitat variables, 3 of the 4 variables entering the final model—elevation, stream slope, and stream width—could be classified as stream gradients.

Stream gradients also appear to influence length structures of game fish populations. Larscheid and Hubert (1992) noted increasing proportions of longer fish in populations of brook trout and brown trout as stream size increased and elevation and stream slope decreased. Similar trends in length structures of populations of Atlantic salmon and brown trout relative to stream slope were observed by Kennedy and Strange (1982). Fausch (1989) provided anecdotal evidence when he noted that large brook trout occurred in a stream reach with low stream slope and smaller brook trout were found immediately downstream in a reach with high stream slope.

A third population metric of interest to managers, relative abundance, has been related to stream gradients. Sowa and Rabeni (1995) noted that stream temperatures and water velocities could predict relative abundances of largemouth bass and smallmouth bass in Missouri streams. In montane settings, relative abundances of brown trout and rainbow trout are highest in the warmer waters found at lower elevations whereas brook trout and cutthroat trout are most abundant at higher elevations (Moore et al. 1986; Bozek and Hubert 1991). Where brook trout and cutthroat trout occur sympatrically, brook trout predominate in reaches with low to moderate stream slopes and cutthroat trout in reaches of higher stream slope (Fausch 1989). Work by Schlosser (1982) suggests that not only does the relative abundance of many game fishes increase in a downstream direction, but that life history strategies change relative to stream gradients. Fish species in environmentally variable headwater areas exhibit life histories characterized by short life spans, smaller body sizes, and earlier sexual maturity than do fish species in less variable downstream areas (Schlosser 1990). Another study linked differences in the allocation and use of energy by American shad to the length and steepness of the river ascended during spawning migrations (Glebe and

Leggett 1981). In steeper and longer streams, a smaller proportion of energy was allocated to reproductive products, more energy was used to reach spawning areas, and fewer adults exhibited semelparous life histories. Finally, Gresswell et al. (1997) linked variation in the average size of Yellowstone cutthroat trout and timing of its spawning migrations to watershed scale features such as basin area and mean aspect.

The many environmental gradients that occur over the length of streams result in an uneven distribution of habitats capable of supporting various life history stages of game fishes (Benda et al. 1992; Magee et al. 1996). Many fish populations adapt to these spatial heterogeneities by exhibiting mobile life histories and spatially structured populations. Bembo et al.'s (1993) study of brown trout from England's Usk River drainage provides an excellent example of a spatially structured game fish population. Brown trout were sampled from three tributary streams and the river main stem. Age distributions of brown trout from tributary streams were dominated by younger fish whereas the main stem had a larger proportion of adult fish. Further, juvenile fish were detected moving from the tributary streams to the main stem, and growth rates for adult fish were higher in the river main stem. Bembo et al.'s (1993) findings suggest tributary streams were used as spawning and nursery areas and the main-stem river for growth to maturity. These insights highlight the importance of the spatial juxtaposition of habitats and suggest that maintenance of connectivity within and across watersheds is an important management consideration for many game species.

9.3.3 Predictive Models

The linkages between a watershed and the characteristics of its fish populations have not gone unnoticed. Indeed, researchers and managers have shown increasing interest in discerning differences among watersheds and relating these differences to fisheries potential. If functional linkages can be established between drainage basin characteristics and stream habitat quality or fish carrying capacity, then statistical models can be developed to (1) prioritize streams for management purposes based upon their potential, (2) estimate the influence various types and intensities of land uses have had on fisheries, (3) better integrate fisheries into existing land classification systems, and (4) predict population attributes across broad spatial areas.

The best example of an effort to relate the attributes of a watershed to fish populations is the work by Lanka et al. (1987). Three sets of regression models capable of predicting salmonid biomass were developed. The first set of models was based exclusively on metrics that quantify the attributes of a drainage basin (e.g., Table 9.1). The second model set considered only instream habitat variables, and the third set of models was built from the entire pool of variables. Models based on basin metrics predicted salmonid biomass as well or better than did models based on instream habitat variables. Further, models built from the entire pool of variables were dominated by basin metrics. These results suggest that salmonids respond proximally to instream habitats but ultimately to watershed scale processes, at least in the small, montane watersheds studied. Lanka et al. (1987) provided a mechanistic explanation for their findings through additional analyses that revealed strong links between basin scale metrics and instream habitat variables.

9.4   LAND USE PRACTICES

The term land use encompasses many activities, including timber harvest, mining, agriculture, urbanization, and recreation. These activities can affect our stream
resources directly through the destruction of habitat by trampling, construction, and
channelization or indirectly through their influence on watershed processes that govern water yield, water regimen, and sediment production.

9.4.1   Land Use Effects on Water Yield and Regimen

A review of the hydrologic cycle (Figure 9.4) and the water balance equation
(equation 9.1) indicates the processes that can be influenced by land use to effect a
change in water yield. In general, those land use activities that produce augmented
water yields do so by reducing evapotranspirative water losses or by decreasing infiltration rates, thereby increasing overland flow to the channel.

The removal of vegetation or the conversion of one vegetative type to another are
traditional means for increasing water yields. Much of the work on vegetation modification has been conducted in the western United States, where periodic shortages of
useable water seriously affect the growth and economy of the region. Research conducted by the U.S. Forest Service at the Fraser Experimental Forest, located in the
forest snowpack zone in the central Rocky Mountains of Colorado, has provided insight regarding timber harvest–water yield relations. Here, snowfall is the key to water
yield, as it is throughout much of western North America and the northern latitudes.
Research results from the Fool Creek watershed, where timber was harvested in the
mid-1950s by clear-cutting alternate strips of varying width, indicate that over the past
30 years water yield has increased by almost 40%, the result of an extended spring
runoff. Peak flows and late-summer streamflow appear to be little affected. Analysis
indicates that the effect of the timber harvest treatment is diminished by about 0.1 cm
per year. At this rate, 70–80 years will be required for water yield to return to preharvest
levels (Alexander et al. 1985).

The water yield increase from the Fool Creek watershed can be attributed to several
factors. Evapotranspiration losses were reduced by the removal of timber from the alternating strips. Although the watershed realized only a small net increase in total snow accumulation, the snowpack was redistributed so that a larger percentage was deposited in the
cleared strips. The cleared strips had low soil moisture deficits and higher melt rates, thereby
causing meltwater to become available earlier in the season before evapotranspiration could
deplete it (Alexander et al. 1985). Timber harvest in forested watersheds outside the snowpack zone tends to produce even greater water yield increases (Satterlund 1972). In rainfall-
dominated areas, late-summer streamflow can be augmented but is variable dependent upon
the distribution and intensity of precipitation.

Removal of woody vegetation, such as chaparral and juniper, from semiarid watersheds has resulted in less-consistent water yield changes than observed in mesic,
forested conditions (Satterlund 1972). Most increases have been observed during years
of above-normal precipitation on moist lowland sites that have deep soils. Failures to
increase water yield generally occur during normal or dry years at upland sites that
have shallow soils.

Much interest has also been shown in the eradication of phreatophytes (plants obtaining their moisture primarily from the groundwater or capillary fringe just above the groundwater) from riparian areas to increase water yields. Because water seldom limits evapotranspiration losses from phreatophytes and because such communities often exhibit an "oasis effect" due to the dry, hot uplands that surround them, water losses per unit area are greater than in other vegetation types. Interest in managing phreatophytes has been high in the desert southwest. Heindl (1965) estimated that if all riparian vegetation was removed along major streams in Arizona, New Mexico, and western Texas, 3 million acre-feet of water could be salvaged annually. However, Campbell (1970) pointed out that predicted water yield increases following removal of riparian vegetation have not been realized. Based upon the limited research, he concluded (1) removal of riparian vegetation increases surface flow if sufficient plants previously existed on the site, (2) enhanced water yields are modest, most likely because of increased surface evaporation, and (3) periodic treatment is necessary, the cost of which must be included in the total cost of the water harvested. Certainly, given the immense value of riparian areas for multiple use, such a management strategy should proceed cautiously.

Agriculture is a major land use that has had a significant influence on water yield. Over much of the semiarid western United States, water withdrawal for irrigation of crop and meadow lands is a major water use. For example, in Wyoming over 90% of appropriated water is used for irrigation. The remaining 10% is used to meet the needs of industry, municipalities, and other uses within the state.

The storage and withdrawal of water for irrigation has played a major role in shaping the stream channels and riparian areas that exist today in the western United States. Depletion of streamflow during the spring runoff period reduces stream power available for transporting sediments. In the absence of these flushing flows, riparian vegetation encroaches, reducing the width and conveyance capacity of the active channel. A striking example of this is provided by the North Platte River in western Nebraska. Historically, the river possessed a wide, shallow, braided channel up to 750 m wide. Over the past 100 years, riparian vegetation has encroached, and the active channel now averages 100 m in width (Williams 1978).

Agriculture and water development are frequently viewed negatively with regards to their effects on aquatic and riparian habitats. Criticism may be justified in some cases, but these land uses can also benefit fisheries. Water removed from the channel and applied to the land for crop production may (1) encourage development of new riparian vegetation in portions of the former active channel, (2) reduce evaporative loss by storing water in the shallow, unconfined aquifer connected to the stream, (3) augment late-season base flows by means of return flows to the channel, (4) reduce the power of peak runoff flows, thereby preventing channel damage, and (5) encourage development of a narrower, deeper channel where aquatic habitat quantity may be reduced but quality is improved for many game species. These benefits are largely responsible for the development of many blue-ribbon trout fisheries across the western United States. Unfortunately, water development often has adverse consequences, and any benefits can be partially offset by the increased water temperature and salinities associated with return flows, the fragmentation of aquatic habitats, or alterations of hydraulic conditions necessary for life history completion in many species. Therefore, the labeling of all water development as either harmful or beneficial is unwise, and a better approach is to consider each project on its own merits.

Livestock grazing can also have an effect on water yield and regimen. Lusby (1970) compared runoff from ungrazed and grazed semiarid rangeland watersheds in western Colorado and found a 30% reduction in runoff from the ungrazed watershed. As pointed out by Holechek (1980), soils compacted by grazing and having little vegetative cover have greatly reduced infiltration rates. Water moves over the soil rather than into it, causing rapid delivery of runoff to stream channels. The resultant hydrograph is flashier than it would otherwise be, and groundwater storage is reduced, possibly leading to lower late-season base flows.

The need to provide rapid drainage of tillable agricultural lands and prevent flooding has led to the channelization of thousands of river miles across North America. Channelization involves removing channel roughness elements or forms (e.g., boulders, snags, and meander patterns) that impede flow. As a result, streamflow is carried more efficiently, flood waters pass more quickly, and the channel conveys greater flood peaks without overtopping the streambanks. Channelization, in association with improved field drainage, often result in higher peak flows than would be realized under natural watershed conditions.

Urbanization, including the development of roads, buildings, other municipal–industrial structures, parking lots, and such recreational developments as ski areas and second homes, can have significant effects on the hydrology of a watershed. Around metropolitan areas, asphalt and concrete cover soil, buildings replace trees and other native vegetation, and sewer systems replace streambeds. Several hydrologic processes are affected as a watershed becomes urbanized. These include infiltration and overland flow, soil moisture storage, evapotranspiration, runoff, and peak flows. Overall, the total water yield from an urbanized area can be expected to increase as a result of reduced infiltration, soil moisture storage, and evapotranspiration and increased overland flow. Peak flows increase as a result of augmented overland flows because the lag time between precipitation and runoff is reduced (Lazaro 1990).

Since about 1960, the growing American population has found increasing time for recreational pursuits. Much of this time is spent outdoors on forest and rangeland watersheds enjoying a variety of activities such as fishing, hunting, camping, sightseeing, hiking, and driving motorbes, four-wheel vehicles, and snowmobiles. From a hydrologic perspective, such activities have minimal effects when use remains low. However, as use increases and becomes concentrated, so do associated trampling, compaction, and trailing effects. Trail and road building that may accompany recreational activities often contribute more to soil compaction, enhancement of overland flow, and hydrograph alteration than do the uses for which they were built.

The land uses potentially influencing streams are not located randomly across a watershed. Instead, the attributes of the watershed dictate the spatial distribution of human activities just as the attributes of the watershed determine its fisheries potential. As a result, land uses occurring across a watershed often vary in a predictable manner, and the relative importance of their effects changes over the length of a stream. For example, many cities have been built next to large rivers because the river provides a convenient means of transporting goods and industry often needs large volumes of water for cooling. Additionally, the fertile soils and flat topographies of river flood-plains have proven amenable to the production of many agricultural crops. In headwater areas of many watersheds, topography and climate preclude the construction of large cities and agriculture, but forestry and recreation become common land uses because mountain forests attracts many recreationists and trees grow well in the mesic, cooler conditions found at higher elevations.

9.4.2   Land Use Effects on Sediment

Sediment is widely regarded as the greatest source of water pollution in the United States (USEPA 1973). Not only can excess sediment degrade the quality of aquatic habitat, it can also have numerous other effects, including

- reducing water storage volume in reservoirs, lakes, and ponds;
- reducing conveyance capacity of stream channels, thereby encouraging overbank flooding;
- creating turbidity, which reduces primary productivity and detracts from recreational use of water;
- degrading water quality for consumptive uses;
- increasing water treatment costs;
- damaging water distribution systems;
- serving as a carrier of other pollutants, such as heavy metals, herbicides, insecticides, and plant nutrients; and
- acting as a carrier of bacteria and viruses.

In 1966, the damage estimated to be caused annually by sediment in streams was placed at US$262 million. In 1988, the cost of sediment pollution exceeded the $1 billion mark.

Any land use activity that removes vegetative cover, disturbs the soil mantle, reduces infiltration rates, decreases soil moisture storage, or increases overland flow has the potential to increase sediment yield from a watershed. Mining, construction, agriculture, and silviculture are major sources of sediment pollution. Of these four, agriculture accounts for approximately 50% of the total contribution, but mining and construction have the highest erosion rates per unit of land area (Table 9.4).

If soil erosion is to be minimized, climate, topography, geology, hydrology, and vegetation must be carefully considered before the land use activity begins, as well as during and after the activity is completed. There are two general approaches for preventing soil erosion: runoff control and soil stabilization. Runoff control practices are designed to reduce the ability of runoff to cause erosion, whereas soil stabilization practices are intended to protect soil from the erosive action of precipitation, runoff, and wind. Specific sediment control practices can be further broken down into four categories: biological, structural, mechanical, and procedural control. For any particular land use activity, the manager may apply techniques from any or all of these categories.

Biological control through the re-establishment of vegetation on disturbed soil is critical for both runoff control and soil stabilization. Different plant species will have different erosion control values for particular applications, dependent upon climate, soil type, moisture conditions, land surface slopes, degree of cultivation, and specific

Table 9.4 Representative rates of erosion for various land uses (USEPA 1973).

Land use	Tons per square mile per year	Relative to forest = 1
Forest	24	1
Grassland	240	10
Abandoned surface mines	2,400	100
Cropland	4,800	200
Harvested forest	12,000	500
Active surface mines	48,000	2,000
Construction	48,000	2,000

biological and ecological characteristics of the plants themselves. Many reports have been written regarding the selection and application of plant species for different reclamation uses in different areas of the country (USEPA 1976). Another example of biological sediment control would be the introduction of beaver into degraded streams. The dam-building activities of beavers serve to trap sediments, store water, provide grade control for the channel, and encourage recovery of the riparian zone (Olson and Hubert 1994).

Numerous structural methods of sediment control have been developed. These may include, but certainly are not limited to, the proper installation of culverts, ponds, drainage pipes, and diversion channels to control runoff and the placement and construction of rock riprap, gabions, revetments, and grade-control structures (small dams) to stabilize eroding stream channels and watercourses (see Chapters 10, 18, and 19).

Mechanical sediment control measures deal primarily with the grading, shaping, and conditioning of the soil surface, often in association with revegetation efforts. These measures may include such treatments as manipulation of slope gradient and length, scarification of the soil surface by chiseling and gouging, and development of contour terraces or furrows to disrupt and slow surface runoff.

The fourth category of sediment control, procedural control, is equally important. By procedural we refer to the careful planning and timing of activities associated with a land use to prevent or minimize soil erosion. Such measures might include closure of a steep, unpaved forest road to vehicular traffic during wet seasons, the proper design of a highway through a region with unstable soils, or the development of an operations plan for a surface mine. Also included as procedural would be the selection of methods to remove harvested timber from a watershed.

9.4.3 Land Use Effects on Stream Habitats and Fish Populations

Despite efforts to limit the negative effects of land uses, streams are often adversely affected by human activities (Meehan 1991). These impacts are readily apparent when they occur rapidly in large, catastrophic events, such as a landslide that is triggered by land use activities on a steep slope or a mud slurry that is transported from a construction site to an adjacent stream by a thunderstorm. However, most land use effects are subtle in nature and result from a diffuse set of nonpoint sources that affect streams simultaneously and interact across time or space (Frissell et al. 1986; Frissell 1992). Detecting or directly linking a particular land use to a stream impact is difficult because of a general lack of unimpacted control sites and the integrative nature of streams. From the perspective of the stream, however, most land uses are similar in that they either alter the amount of water or the amount of sediment delivered to and through the stream channel. By understanding how the stream adjusts to altered sediment or water regimes, insights can be gained into the effects that land uses will have on stream habitats.

A stream is said to be graded when a balance exists between the inputs and outputs of sediment and water through the stream channel (Knighton 1984). Over the long term, the size of the channel is just large enough to convey materials from a watershed efficiently, and the channel is neither aggrading (filling with sediment) or degrading (eroding). The result of this equilibrium is a diverse array of high-quality habitats. Land uses, through their effects on sediment or water regimes, frequently move the stream from a graded state. For example, many land uses increase the amount of sediment transported to a stream. Without a concurrent increase in water yield, the stream lacks the competency to transport the additional sediment, and the stream channel begins to aggrade. Pool volume and substrate diversity decrease as fine sediments settle in the pool or fill spaces between coarser substrate particles. Conversely, land uses that result in greater water yields increase stream competency. The stream channel responds by degrading, or eroding, its banks and bed. Figure 9.10 depicts Lane's balance, a useful conceptual tool for understanding the interactions between land uses, changes in stream competency, and the alteration of stream channels.

Variation in the spatial distribution of human activities combined with gradients in stream size and power result in differences between the way headwater and lowland streams are affected by land uses. Headwater streams in montane areas are typically small but have high power per unit of streambed due to their high stream slopes. Land uses typically occurring in montane watersheds, such as silviculture, road building, or recreational activities, often lead to flashier hydrographs, increased water yields, and greater sediment loads. If riparian vegetation is removed in conjunction with land uses, the thermal stability of the stream and its nutrient base will also be affected (Barton et al. 1985). The net effect of land uses on steep headwater streams is to increase the variability of seasonally dynamic physicochemical conditions (Schlosser 1991). Structural habitat diversity is less affected because steep streams are usually sediment limited and have a power surplus that can transport increased sediment loads (Coats et al. 1985). If a small stream were located in a wide valley where it had low stream slope and power, not only could land uses lead to greater variability in physicochemical conditions, but the stream would lack the ability to transport an increased sediment load. Sediments would be deposited, the channel would aggrade, and the diversity of physical habitats would decrease (Rabeni and Smale 1995).

Large rivers, because of the volumes of water they transport, are buffered against changes in variation of seasonally dynamic physicochemical conditions, but their low power per unit of streambed makes them susceptible to deposition of sediments transported from upstream or derived from adjacent lands. Further, many land uses occurring next to large rivers involve activities such as snagging, channelization, and diking that serve to separate the river from its floodplain. The net effect of these activities is to homogenize a previously complex system and greatly decrease the diversity of habitats available to fishes. Based on historic surveying and snagging records dating back to the time of European settlement, Sedell and Froggatt (1984) described the separation of Oregon's Willamette River from its floodplain. Their results suggest that be-

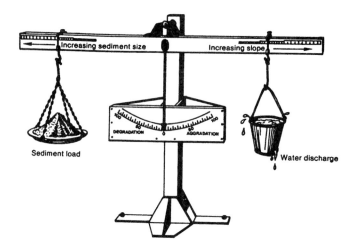

Figure 9.10 Lane's balance, a conceptual tool for understanding how changes in sediment or water yield affect stream channels. Figure is from Lane (1955).

tween 1854 and 1967, the Willamette River was changed from a complex series of multiple channels, sloughs, and backwaters containing large quantities of woody debris to a single main channel with few downed trees and over a fourfold decrease in the amount of land–water interface. Similar habitat simplification has taken place on most large rivers in developed countries, and the process is currently being repeated in developing countries across the globe. However, there are signs that this trend is partially reversing in developed countries where restoration strategies for large rivers are now being developed (Van Dijk et al. 1995), a turn of events stemming as least as much from periodic flooding and losses of human property as from an increased awareness of linkages between a large river and its floodplain (Bayley 1995).

Differences between the ways in which land uses affect small streams versus lowland areas result in differential responses in fish populations. Summarizing the work of others, Schlosser (1991) characterized the responses of fish populations in small streams as follows:

- higher growth rates of juvenile fishes and increased abundance of nongame fish species due to increased primary production during summer months; and
- greater temporal variation in fish abundance due to greater variability in dynamic physicochemical conditions.

If habitat complexity was also lost, fish population responses could include

- decreased diversity of adult and juvenile fishes as a result of lower habitat heterogeneity;
- fewer large fish due to decreased abundance of pools;
- more small fish because of increases in the amount of shallow-water habitats; and
- greater temporal variation in fish abundance due to the absence of large, structurally complex pools that serve as refugia during stressful periods.

Hicks et al. (1991) provide a more detailed review of the effects that land uses have on salmonids. Despite their work being geared toward salmonids in western streams, many of Hicks et al.'s insights are directly generalizable to other species given similarities between the effects that many land uses have on streams and the habitat requirements of most game fishes.

The difficulty of sampling large rivers, combined with a discrepancy between the distributions of the world's fisheries biologists and the large, relatively pristine river systems that remain, currently limit our understanding of fish population responses to land uses in large rivers. However, existing evidence suggests that the abundance and diversity of juvenile fishes decrease as a result of decreases in habitat diversity associated with the separation of rivers from their floodplains (Schiemer and Spindler 1989).

9.4.4   An Example of the Effects of Land Use History

An important question a fisheries manager must address is, why is the stream habitat I am trying to manage in the condition that it is? As should now be apparent, the answer will not be found just between the banks of a stream. Examine the photographs presented in Figure 9.11. From these, what can be perceived about historic land use in Wyoming's Douglas Creek watershed and the long-term trend of the aquatic habitat?

Douglas Creek is a moderate-size stream located high in the forest snowpack zone of the central Rocky Mountains. Average discharge is 30–40 ft^3/s. Several aspects of the channel evident from the photographs are its width in relation to its depth, the lack of well-developed streambanks even in less steep sections, the absence of woody debris, the establishment of vegetation on bars located on the inside of meanders, the lack of erosion on the banks opposite these bars, and the establishment of young lodgepole pine on the floodplain. As a result, habitat quality is marginal for the brown trout-dominated population.

The Douglas Creek channel is the product of a series of different land uses dating back to the late Nineteenth Century. Timber was first harvested across the watershed to provide ties for the railroad. In the absence of roads, each spring during snowmelt runoff logs were transported via massive tie drives down Douglas Creek. To facilitate these drives, the channel was cleared of woody debris and boulders, thereby reducing channel roughness and increasing water velocities. Gold mining became popular on Douglas Creek during the 1920s and 1930s. Evidence of the activity still remains in the form of long, partially vegetated dredge spoil piles that line the channelized streambanks in many areas. During the early 1960s, Rob Roy Dam was built farther upstream to provide municipal water. The vegetative encroachment present today, both in the active channel and on the floodplain, is a result of flow regulation. Rob Roy Dam has stabilized the natural flow regime, especially through the reduction of peak spring flows. As a result, vegetation has become established, streambanks are being rebuilt, and the oversized channel is gradually narrowing. Although this process is a slow one due to the short growing season and the limited sediment supply at 2,745 m, the trend for the stream and its fish population is in a positive direction.

Figure 9.11   Douglas Creek, Medicine Bow National Forest, Wyoming.

9.5 CONCLUSION

As with Douglas Creek, historic and present land use activities interact with the natural attributes of a watershed to leave their mark, however subtle, on all stream environments. Because the fish populations inhabiting streams are reflections of the quality, quantity, and juxtaposition of stream habitats arising from these interactions, the fisheries manager must understand the complexities of watershed function. When combined with the fact that most problems associated with a watershed cross political boundaries and must be resolved in an equally complex sociopolitical arena, watershed management becomes a very challenging endeavor. In light of this, fisheries managers of the future must become increasingly interdisciplinary and develop abilities to work productively with professionals from a variety of fields. Only by doing so can desired responses in fish populations be achieved and impacts to aquatic habitats minimized in the face of increasingly intensive land use.

9.6 REFERENCES

Alexander, R. R., and five coauthors. 1985. The Fraser Experimental Forest, Colorado: research program and published research 1937–1985. U.S. Forest Service General Technical Report RM-118.

Allen, J. D. 1995. Stream ecology: structure and function of running waters. Chapman and Hall, New York.

Anderson, H. W., M. D. Hoover, and K. G. Reinhart. 1976. Forests and water: effects of forest management on floods, sedimentation, and water supply. U.S. Forest Service General Technical Report PSW-18/1976.

Barton, D. R., W. D. Taylor, and R. M. Biette. 1985. Dimensions of riparian buffer strips required to maintain trout habitat in southern Ontario streams. North American Journal of Fisheries Management 5:364–378.

Bayley, P. B. 1995. Understanding large river–floodplain ecosystems. BioScience 45:153–158.

Bembo, D. G., R. J. H. Beverton, A. J. Weightman, and R. C. Cresswell. 1993. Distribution, growth and movement of River Usk brown trout. Journal of Fish Biology (Supplement A)43:45–52.

Benda, L., T. J. Beechie, R. C. Wissmar, and A. Johnson. 1992. Morphology and evolution of salmonid habitats in a recently deglaciated river basin, Washington State, USA. Canadian Journal of Fisheries and Aquatic Sciences 49:1246–1256.

Bilby, R. E., and J. W. Ward. 1989. Changes in characteristics and function of woody debris with increasing size of streams in western Washington. Transactions of the American Fisheries Society 118:368–378.

Bisson, P. A., J. L. Nielsen, R. A. Palmason, and L. E. Grove. 1982. A system of naming habitat types in small streams, with examples of habitat utilization by salmonids during low streamflow. Pages 62–73 in N. B. Armantrout, editor. Acquisition and utilization of aquatic habitat inventory information. American Fisheries Society, Western Division, Bethesda, Maryland.

Bouwer, H. 1978. Groundwater hydrology. McGraw-Hill, New York.

Bozek, M. A., and W. A. Hubert. 1991. Segregation of resident trout in streams as predicted by three habitat dimensions. Canadian Journal of Zoology 70:886–890.

Brooks, K. N., P. F. Ffolliott, H. M. Gregersen, and J. L. Thames. 1991. Hydrology and the management of watersheds. Iowa State University Press, Ames.

Brown, E. H. 1970. Man shapes the earth. Journal of Geography 136:74–84.

Burton, R. A., and T. A. Wesche. 1974. Relationship of duration of flows and selected watershed parameters to the standing crop estimates of trout populations. University of Wyoming, Water Resources Research Institute Series Publication 52, Laramie.

Campbell, C. J. 1970. Ecological implications of riparian vegetation management. Journal of Soil and Water Conservation 2:49–52.

Coats, R., L. Collins, J. Florsheim, and D. Kaufman. 1985. Channel change, sediment transport, and fish habitat in a coastal stream: effects of an extreme event. Environmental Management 9:35–48.

Craig, G. S., Jr., and J. G. Rankl. 1978. Analysis of runoff from small drainage basins in Wyoming. U.S. Geological Survey, Water-Supply Paper 2056.

Decamps, H., M. Fortune, F. Gazelle, and G. Pautou. 1988. Historical influence of man on the riparian dynamics of a fluvial landscape. Landscape Ecology 1:163–173.

Fausch, K. D. 1989. Do gradient and temperature affect distributions of, and interactions between, brook charr and other resident salmonids in streams? Physiology and Ecology Japan (Special Volume 1):303–322.

Fausch, K. D., C. L. Hawkes, and M. G. Parsons. 1988. Models that predict standing crop of stream fish from habitat variables: 1950–1985. U.S. Forest Service General Technical Report PNW-GTR-213.

Frissell, C. A. 1992. Cumulative effects of land use on salmon habitat in southwest Oregon coastal streams. Doctoral dissertation. Oregon State University, Corvallis.

Frissell, C. A., W. J. Liss, C. E. Warren, and M. D. Hurley. 1986. A hierarchical framework for stream habitat classification: viewing streams in a watershed context. Environmental Management 10:199–214.

Glebe, B. D., and W. C. Leggett. 1981. Latitudinal differences in energy allocation and use during the freshwater migrations of American shad and their life history consequences. Canadian Journal of Fisheries and Aquatic Sciences 38:806–820.

Grant, G. E., F. J. Swanson, and M. G. Wolman. 1990. Patterns and origin of stepped-bed morphology in high-gradient streams, Western Cascades, Oregon. Geological Society of America Bulletin 102:340–352.

Gregory, S. V., F. J. Swanson, W. A. McKee, and K. W. Cummins. 1991. An ecosystem perspective of riparian zones. BioScience 41:540–551.

Gresswell, R. E., W. L. Liss, G. L. Larson, and P. J. Bartlein. 1997. Influence of basin-scale physical variables on life history characteristics of cutthroat trout in Yellowstone Lake. North American Journal of Fisheries Management 17:1046–1064.

Hawk, G. M., and D. B. Zobel. 1974. Forest succession on alluvial landforms of the McKenzie River valley, Oregon. Northwest Science 48:245–265.

Heindl, L. A. 1965. Groundwater in the Southwest—a perspective. In J. E. Fletcher and G. L. Bender, editors. Ecology of groundwater in the southwestern United States. Arizona State University, Tempe.

Hewlett, J. D., and W. L. Nutter. 1969. An outline of forest hydrology. University of Georgia Press, Athens.

Hicks, B. J., J. D. Hall, P. A. Bisson, and J. R. Sedell. 1991. Responses of salmonid populations to habitat changes caused by timber harvests. Pages 483–518 in W. R. Meehan, editor. Influence of forest and rangeland management on salmonid fishes and their habitat. American Fisheries Society, Special Publication 19, Bethesda, Maryland.

Holechek, J. 1980. Livestock grazing impacts on rangeland ecosystems. Journal of Soil and Water Conservation 4:162–164.

Horton, R. E. 1945. Erosional development of streams and their drainage basins, hydrophysical approach to quantitative morphology. Geological Society of America Bulletin 56:275–370.

Hubert, W. A., and S. J. Kozel. 1993. Quantitative relations of physical habitat features to channel slope and discharge in unaltered mountain streams. Journal of Freshwater Ecology 8:177–183.

Hubert, W. A., T. D. Marwitz, K. G. Gerow, N. A. Binns, and R. W. Wiley. 1996. Estimation of potential maximum biomass of trout in Wyoming streams to assist management decisions. North American Journal of Fisheries Management 16:821–829.

Hupp, C. R. 1982. Stream-grade variation and riparian-forest ecology along Passage Creek, Virginia. Bulletin of the Torrey Botanical Club 109:488–499.

Hynes, H. B. 1975. A stream and its valley. International Association of Theoretical and Applied Limnology 19:1–15.

Karr, J. R., and I. J. Schlosser. 1978. Water resources and the land–water interface. Science 201:229–234.

Keller, E. A., and F. J. Swanson. 1979. Effects of large organic material on channel form and fluvial processes. Earth Surface Processes 4:361–380.

Kennedy, G. J. A., and C. D. Strange. 1982. The distribution of salmonids in upland streams in relation to depth and gradient. Journal of Fish Biology 20:579–591.

Knighton, D. 1984. Fluvial forms and processes. Edward Arnold, London.

Lane, E. W. 1955. The importance of fluvial geomorphology in hydraulic engineering. Proceedings of the American Society of Civil Engineers 81:1–17.

Lanka, R. P., W. A. Hubert, and T. A. Wesche. 1987. Relations of geomorphology to instream habitat and trout standing stock in small Wyoming streams. Transactions of the American Fisheries Society 116:21–28.

Larscheid, J. G., and W. A. Hubert. 1992. Factors influencing the size structure of brook trout and brown trout in southeastern Wyoming mountain streams. North American Journal of Fisheries Management 12:109–117.

Layher, W. G., O. E. Maughan, and W. D. Warde. 1987. Spotted bass habitat suitability related to fish occurrence and biomass and measurements of physiochemical variables. North American Journal of Fisheries Management 7:238–251.

Lazaro, T. R. 1990. Urban hydrology: a multidisciplinary perspective. Technomic Publishing, Lancaster, Pennsylvania.

Leopold, L. B., and T. Maddock. 1953. The hydraulic geometry of stream channels and some physiographic implications. U.S. Geological Survey Professional Papers 252.

Leopold, L. B., and W. B. Langbein. 1963. Association and indeterminacy in geomorphology. *In* C. C. Albrilton, editor. The fabric of geology. New York.

Likens, G. E., F. H. Bormann, N. M. Johnson, D. W. Fisher, and R. S. Pierce. 1970. Effects of forest cutting and herbicide treatment on nutrient budgets in the Hubbard Brook watershed-ecosystem. Ecological Monographs 40:23–47.

Lusby, G. C. 1970. Hydrologic and biotic effects of grazing versus nongrazing near Grand Junction, Colorado. U.S. Geological Survey Professional Papers 700B.

Magee, J. P., T. E. McMahon, and R. F. Thurow. 1996. Spatial variation in spawning habitat of cutthroat trout in a sediment-rich stream basin. Transactions of the American Fisheries Society 125:768–779.

Meehan, W. R. editor. 1991. Influences of forest and rangeland management on salmonid fishes and their habitats. American Fisheries Society, Special Publication 19, Bethesda, Maryland.

Moore, S. E., G. L. Larson, and B. Ridley. 1986. Population control of exotic rainbow trout in streams of a natural area park. Environmental Management 10:215–219.

Olson, R., and W. A. Hubert. 1994. Beaver: water resources and riparian habitat manager. University of Wyoming, Department of Renewable Resources, Laramie.

Platts, W. S. 1979. Relationships among stream order, fish populations, and aquatic geomorphology in an Idaho river drainage. Fisheries 4(2):5–9.

Potter, W. D. 1961. Peak rates of runoff from small watersheds. U.S. Department of Commerce, Bureau of Public Roads, Hydraulic Design Series 2, Washington, D.C.

Rabeni, C. F., and M. A. Smale. 1995. Effects of siltation on stream fishes and the potential mitigating role of the riparian buffering zone. Hydrobiologia 303:211–219.

Richmond, A. D., and K. D. Fausch. 1995. Characteristics and function of large woody debris in subalpine Rocky Mountain streams in northern Colorado. Canadian Journal of Fisheries and Aquatic Sciences 52:1789–1802.

Satterlund, D. R. 1972. Wildland watershed management. Wiley, New York.

Schiemer, F., and T. Spindler. 1989. Endangered fish species of the Danube River in Austria. Regulated Rivers Research & Management 4:397–407.

Schlosser, I. J. 1982. Fish community structure and function along two habitat gradients in a headwater stream. Ecological Monographs 52:395–414.

Schlosser, I. J. 1990. Environmental variation, life history attributes, and community structure in stream fishes: implications for environmental management and assessment. Environmental Management 14:621–628.

Schlosser, I. J. 1991. Stream fish ecology: a landscape perspective. BioScience 41:704–711.

Schumm, S. A. 1956. The evolution of drainage systems and slopes in badlands at Perth Amboy, New Jersey. Geological Society of America Bulletin 67:597–646.

Sedell, J. R., and J. L. Froggatt. 1984. Importance of streamside forests to large rivers: the isolation of the Willamette River, Oregon, U.S.A., from its floodplain by snagging and streamside forest removal. International Association of Theoretical and Applied Limnology 22:1828–1834.

Simons, D. B., and F. Senturk. 1977. Sediment transport technology. Water Resources Publications, Fort Collins, Colorado.

Smith, K. G. 1950. Standards for grading texture of erosional topography. American Journal of Science 248:655–668.

Southwood, T. R. E. 1977. Habitat, the template for ecological strategies? Journal of Animal Ecology 46:337–367.

Southwood, T. R. E. 1988. Tactics, strategies, and templates. Oikos 52:3–18.

Sowa, S. P., and C. F. Rabeni. 1995. Regional evaluation of the relation of habitat to distribution and abundance of smallmouth bass and largemouth bass in Missouri streams. Transactions of the American Fisheries Society 124:240–251.

Speaker, R., K. Moore, and S. V. Gregory. 1984. Analysis of the process of retention of organic matter in stream ecosystems. International Association of Theoretical and Applied Limnology 22:1835–1841.

Strahler, A. N. 1957. Quantitative analysis of watershed geomorphography. Transactions of the American Geophysical Union 38:913–920.

USEPA (U.S. Environmental Protection Agency). 1973. Methods for identifying and evaluating the nature and extent of nonpoint sources of pollutants. USEPA-4030/9-73-014, Washington, D.C.

USEPA (U.S. Environmental Protection Agency). 1976. Erosion and sediment control, surface mining in the eastern U.S., planning, volume 1. USEPA-625/3-76-006, Washington, D.C.

Van Dijk, G. M., E. C. L. Marteijn, and A. Schulte-Wulwer-Leidig. 1995. Ecological rehabilitation of the river Rhine: plans, progress and perspectives. Regulated Rivers Research & Management 11:377–388.

Vannote, R. L., G. W. Minshall, K. W. Cummins, J. R. Sedell, and C. E. Cushing. 1980. The river continuum concept. Canadian Journal of Fisheries and Aquatic Sciences 37:130–137.

Ward, J. V. 1985. Thermal characteristics of running waters. Hydrobiologia 125:31–46.

Wesche, T. A. 1982. The Snowy Range Observatory: an update. University of Wyoming, Water Resources Research Institute Series Publication 81, Laramie.

Williams, G. P. 1978. The use of shrinking river channels—the North Platte and Platte rivers in Nebraska. U.S. Geological Survey Circular 781.

Ziemer, G. L. 1971. Quantitative geomorphology of drainage basins related to fish production. Alaska Department of Fish and Game, Information Leaflet 162, Juneau.

Chapter 10

Stream Habitat Management

DONALD J. ORTH AND RAY J. WHITE

10.1 INTRODUCTION

Habitat for fishes is a place—or for migratory fishes, a set of places—in which a fish, a fish population, or a fish assemblage can find the physical and chemical features needed for life, such as suitable water quality, migration routes, spawning grounds, feeding sites, resting sites, and shelter from enemies and adverse weather. Although food, predators, and competitors *are not* habitat, proper places in which to seek food, escape predators, and contend with competitors *are* part of habitat, and a suitable ecosystem for a fish includes habitat for these other organisms as well.

Characteristics of the habitat play a large role in determining the numbers, sizes, and species of fishes that can be sustained. Habitat management is a vital aspect of stream fisheries management because it affects stream communities. Habitat management should, therefore, be coordinated with population assessment (Chapters 6 and 7) because characteristics of fish assemblages and harvest can be used to monitor stream habitat quality (Fausch et al. 1990; Simon 1999).

Stream habitat science and management are in a rapid stage of development. Actions, such as instream habitat treatments, must be coordinated with regulation of land use (Williams et al. 1997; Chapter 9) because streambanks, their vegetation, and watershed conditions ultimately affect habitat quality for fishes. There is a close interaction between riparian (streamside) vegetation and stream habitat in small as well as large streams. Consequently, an effective management program should treat the riparian-stream ecosystem as a single unit. Modern stream habitat managers must have a thorough knowledge of the influence of stream habitat on the many benefits that streams can provide to society (Meehan 1991; Stouder et al. 1996).

10.1.1 Goals of Stream Habitat Management

The goals of stream habitat management are to protect, restore, or improve stream habitat so that diverse natural habitats, coadapted populations, critical refugia, and natural ecosystem processes are maintained in the face of human activities. Although there may be target fish populations, these are not the sole goal of an effective habitat management program. Habitat protection involves preventing or reducing human-generated changes that adversely affect habitat. Habitat improvement can either restore degraded habitat (change it toward predisturbance

---

Box 10.1   Common Habitat Deficiencies

Unsuitable flow or temperature regime

Unstable banks and severe erosion

Too many or too few of nature's engineers (i.e., beavers)

Homogenous channel form

Elevated sediment input

Water quality degradation

Inadequate holding areas for fishes

Inadequate reproductive habitat for fishes

Reduced overwinter habitat for fishes

Blocked migration routes

Excessive or inadequate shading

---

condition) or enhance it (improve it beyond some less-than-optimal condition). Often the same methods are appropriate for restoring or enhancing streams with common habitat deficiencies (Box 10.1). Often these deficiencies derive from unfavorable watershed conditions (e.g., soil erosion or decreased surface perviousness) and cannot be corrected or offset by instream habitat treatments.

10.1.2 Cautions for Habitat Enhancement and Restoration

It is paramount to have sound fish- and fishing-oriented purposes for habitat enhancement and to let the design of projects follow from these. A common mistake is to regard stream habitat improvement as a collection of techniques, a matter of simply finding ideas in a how-to book and installing structures. Cookbook habitat work characterized by infatuation with techniques fails again and again to achieve proper effects. How to build a habitat structure is important but not as important as why—and, deriving from the why, selecting the right structure for a valid ecological purpose and placing it properly in relation to channel form. Guiding principles for stream habitat management are summarized in Box 10.2.

The trend in stream habitat management is toward ecosystem management (Chapter 5)—considering the whole stream ecosystem and its drainage basin and operating on a landscape scale (Beschta 1996; Bisson et al. 1996; Frissell et al. 1996; Sedell et al. 1996; White 1996). Managers should avoid radically altering streams in ways that, while enhancing habitat for the target organism(s), degrade it for other desirable organisms. Fish and wildlife managers have been criticized for narrowly focusing on habitat for game and food species and disregarding broader ecosystem concerns such as habitat diversity and species diversity. The danger of disregarding broader ecosystem concerns may be reduced if stream habitat enhancement simulates natural features of channel form and flow and incorporates measures to protect and restore natural riparian and watershed processes.

> **Box 10.2   Guiding Principles for Stream Habitat Management**
>
>   These principles have been adapted from Hunt (1993).
>
> Learn from nature. What factors make locally "good" streams good?
>
> Work with, not against, the inherent capacity of streams and watersheds to restore their biotic health.
>
> Focus on identifying limiting factors at work and try to eliminate or ameliorate those factors.
>
> Consider species-specific, age-specific, and season-specific requirements of fishes present, including environmental suitability and species interactions.
>
> Tailor management activities to the individual stream. Do not borrow successful designs from other dissimilar streams.
>
> Disguise artificiality of human structures or modifications to channel shape.
>
> Encourage the right kinds of streambank vegetation to become dominant, depending on the character of the stream and riparian zone.
>
> Preserve, restore, and accentuate the two most common natural characteristics of streams— the meandered channel and the riffle and pool sequence.
>
> Make the streamflow work beneficially. Bring main threads of flow close to resting and security cover.
>
> Whenever possible, maintain or enhance base flow. Management of the riparian zone and entire watershed may be needed to achieve greater and more stable base flow.
>
> Integrate habitat management of the stream, its riparian zone, and its watershed with other fisheries management techniques, especially angling regulations, to achieve synergistic benefits.
>
> Follow a logical sequence of habitat management steps (see section 10.3).

In improving habitat for sport fishes, the manager should avoid structures that impart an unnatural appearance. Most anglers value naturalness of the setting in which they conduct their sport (Chapter 8). The physical aspects valued by anglers who have long fished the project stream should be preserved if possible. In structural work, use of artificial materials (e.g., steel stakes, pins and wire, plastic fabrics and meshes) even for concealed components should be held to a minimum. Despite maintenance, everything built eventually disintegrates. Therefore, artificial materials can become an unsightly clutter of metal and plastic objects. In contrast, when rot or extreme flow breaks up structures made of materials such as logs and native rock, the result will look natural.

Periodic postconstruction maintenance can greatly extend the effective life of many types of habitat work. Just about anything that is built (a house, highway, or motor vehicle) must be inspected and maintained from time to time if it is to last. Maintenance is a frequently neglected aspect of habitat projects. Sometimes it is not considered in the overall project plan. Also, maintenance is a low-visibility function that is easy to drop from budgets during financial crunches.

A widespread tendency to overstabilize channels stems from fear of channel erosion. Moderate channel erosion is natural and creates fish habitat. People often wish to have streams locked into unchanging courses to facilitate navigation; to protect roads, buildings, cropland, and other property in floodplains; or just to keep things as they are. The result has been an emphasis on "hard stabilization" in channel engineering, and this emphasis tends to carry over into fish habitat work, sometimes overwhelming the fish habitat aspects.

Fish and other organisms may benefit by gradual shifts of stream course, but these shifts may destroy installed habitat structures. Solving this dilemma may often lie in relying less on permanent structures, placing more emphasis on stream-side vegetation and its "flexible stabilization," and restoring natural flow regimens. The thatch and binding of streamside vegetation can, through flexibility and regrowth, change interactively as erosion and deposition take place in the channel.

Severe channel erosion results from such landscape changes as clearing of forest, cropping or overgrazing land, and creating hard land surfaces associated with urbanization and highways. Consequent increased imperviousness of the land reduces the amount of rain and snowmelt water that enters the soil to recharge groundwater and maintain springs. Instead, more water runs off the land, causing high streamflow to occur more often, to reach higher peaks, and to last longer. As a result, streams in urbanizing landscapes tend to develop enlarged channels, which have even less water than before during low flow periods, and have water temperatures that are colder in winter and warmer in summer. Where this happens, people often pressure governments to reinforce streambanks. In contrast, the sound approach for fish habitat management is to change the human land uses that are causing the problem.

Streams exist in a state of dynamic equilibrium (Leopold et al. 1964; Rosgen 1996). Habitat managers should be cautious about unfavorably reducing the natural dynamism of streams. Toward this end, it will be helpful to let floodplains serve their natural function as areas to be flooded.

10.2 STREAM HABITAT FACTORS

Stream fisheries management requires an understanding of the interaction of habitat factors and fish populations. Five classes of habitat factors affect the distribution and abundance of stream fishes (Figure 10.1): streamflow, water quality, energy source, physical habitat structure, and biotic interactions (Karr et al. 1986). These habitat features interact to determine the characteristics of the fish community in a particular stream. Characteristics of the riparian zone and the watershed influence these five factors.

10.2.1 Riparian and Watershed Influence

Riparian zone and watershed characteristics intimately affect stream habitat. Riparian plants filter sediments and nutrients, provide shade, stabilize streambanks, provide cover in the form of large and small woody debris, and produce leaf litter energy inputs (Gregory et al. 1991). Altered riparian zones or watershed land use can detrimentally affect stream habitat (Meehan 1991). It is essential that riparian zones be protected and managed with an awareness of their influence on stream habitats.

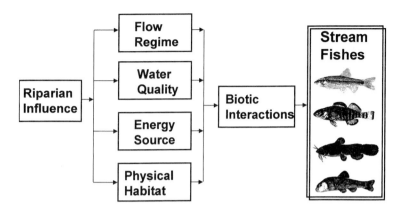

Figure 10.1 Stream habitat factors affecting distribution and abundance of stream fishes. Figure modified from Karr et al. (1986).

10.2.2 Streamflow Discharge

Streamflow discharge, the volume of water flowing past a cross section of channel per unit time, can have substantial short-term variability. In most streams, however, the seasonal pattern of discharge (flow regime) is often predictable within a region. The stream biota is adapted to the seasonal cycle of flooding and low flows, but flow can limit the biomass of fishes. For example, late-summer flow and the magnitude of seasonal flow variation significantly affect trout biomass in Wyoming streams (Binns and Eiserman 1979). Flow also affects fish populations indirectly because it affects the other four classes of habitat factors. For example, water temperatures and water quality will generally change more rapidly and become more extreme as streamflows decline.

Streamflow regime refers to the permanence and seasonal patterns of flow. Some streams have stable flow due to ground or glacial water sources, whereas others fluctuate erratically or are intermittent because they are fed primarily by overland runoff. In fact, there are 10 recognizable stream types based on hydrological patterns (Poff 1996). Although stable flow benefits some species of fishes, it is not always necessary or even desirable for all of them. High flows form important channel features by scouring pools and sorting streambed materials. High flows also serve as cues for timing of migration and spawning. Extreme low flows can limit fish production in many ways. Alteration of a stream's natural flow regime can adversely modify normal seasonal high and low flows to which the biota is adapted (Poff et al. 1997).

10.2.3 Water Quality

Factors that determine water quality, such as temperature, pH, turbidity, dissolved oxygen, alkalinity, dissolved nutrients, and presence or absence of anthropogenic toxicants, can directly affect stream fishes or can indirectly affect them through effects on food production. Maintaining regional water quality standards for point source discharges of harmful substances (Alabaster and Lloyd 1984) protects stream water quality, as does reducing nonpoint source pollutants by encouraging less abusive human activities on the land.

10.2.4 Energy Sources

Energy sources that support aquatic organisms are both autochthonous (produced within the stream) and allochthonous (entering from outside the aquatic system). Autochthonous energy sources include rooted aquatic plants and the aufwuchs community of algae, fungi, bacteria, and invertebrates on streambed materials. Important among allochthonous sources are such materials as leaves, needles, twigs, and pollen that fall into streams from terrestrial vegetation. Composition of the benthic invertebrate and fish communities reflects the dominant energy sources. Streamflow, water quality, and vegetation changes the amount of energy from allochthonous and autochthonous energy. Physical habitat structure can modify the efficiency with which energy sources are consumed and incorporated into animal tissue.

10.2.5 Physical Habitat Structure

Physical habitat structure refers to the channel and adjacent riparian characteristics of bed materials, water depth, current velocity, bank slope, and cover that determine the amount of suitable living space. Physical habitat requirements vary among fish species and life history stages. Many stream fishes spawn over or in rocky material of special sizes (Shirvell and Dungey 1983). Stream fishes tend to position themselves in and near foraging locations (determined largely by water depth and velocity) and escape cover. Some fishes, particularly small ones, occupy streambed crevices much of the time—to escape winter's very cold water, when they may become lethargic. The complexity of physical habitats is often correlated with diversity of fishes and resilience of fish assemblages (Pearsons et al. 1992). Also, in streams in which water quantity or quality are not limiting, cover is often positively correlated with fish density or biomass (Fausch et al. 1988), and management techniques that increase the amount of cover in a stream can substantially increase fish abundance (Hunt 1988).

10.2.6 Biotic Interactions

Biological interactions, particularly predation and competition, can affect a fish's habitat selection and use as well as its survival in a given habitat. The abundance of a given fish species that can be sustained in a stream may depend on the outcome of competition and predation. Among sites of comparable habitat quality, for example, those with piscivorous fishes may have smaller biomasses of nonpiscivorous fishes than do sites lacking piscivores (Bowlby and Roff 1986). Therefore, species introductions (Chapter 13) can result in changes in the abundance of fishes even if habitat quality or exploitation does not change.

10.3 STREAM HABITAT PROJECTS: ANALYSIS
AND MANAGEMENT

A stream habitat project is conducted as a sequence of steps, a process typically lasting several years and requiring an interdisciplinary team to help the fisheries biologist to classify and diagnose the stresses correctly (Figure 10.2). Stream hydrologists and morphologists are usually essential (Gordon et al. 1992). In analyzing stream habitats the manager must understand that river channels are dynamic and that only when sediment inputs are balanced with outputs will there be a predictable, stable channel form that fits the valley and terrain. If the channel is unstable it may evolve through a series of channel forms before reaching a new stable form. Services from such fields as geology, engineering, entomology, plant ecology, and livestock grazing are often needed during the project. Some government agencies have the requisite professionals on staff and have well-established procedures for following the appropriate steps. In other situations, a management team may have to be developed.

At a minimum, the following 12 steps should be included in a stream habitat project. The first seven form a planning phase. Manipulating habitat without proper planning is a common mistake. For complex projects it can be impractical to plan the whole project at one time. It may be better to refine project methods with the actual experience gained from pilot studies or from conducting a small part of the project initially. Some projects are repeatedly cycled through such adaptive refinement. At all steps, all involved parties should be given progress reports and their approval should be obtained at each step.

1. Objective setting. Determine the overall goal. For many streams, the goal will be to maximize resident game fish populations. For others, such as in anadromous or adfluvial fisheries, the goal may be to supply more young to an ocean, lake, or river for commercial fisheries or sportfishing. For still other streams, it may be to preserve or restore endangered fishes. Specific objectives are set later (White 1991a; Chapter 2).

2. Examination. Measure and analyze the stream's biological, physical, and chemical characteristics. It is fundamental to analyze the stream's hydrology and morphology—sources and flow regimes, channel form and mechanics, and underlying processes of the drainage basin. The objective is to determine if basin processes—erosion, sedimentation, water transport, and channel development among them—are in dynamic equilibrium and, if they are not, to estimate how and when equilibrium might be reached (Heede and Rinne 1990; Rosgen 1996). Inspecting the drainage basin's vegetation, soils, geology, and human uses and analyzing water chemistry data can help managers assess production potential and pollution hazards. Stream temperature should be surveyed during extreme hot and cold weather, and the progression of ice conditions—a major influence on fishes and habitat—should be observed in winter and during spring breakup. Fish habitat is inventoried and evaluated by various procedures (Platts et al. 1987; McMahon et al. 1996; Bain et al. 1999). A fish population inventory can reveal habitat deficiencies and perhaps rule out the need for certain other data. For example, an abundance of fast-growing young in a population that has few trophy size trout, despite being lightly fished, may indicate little likelihood of problems with water quality, food supply, or reproductive habitat; further investigation might concentrate on structural habitat. It is always advisable to survey aquatic and riparian plants, beaver workings, and major predatory wildlife. In special cases, it can be useful to analyze the benthic fauna.

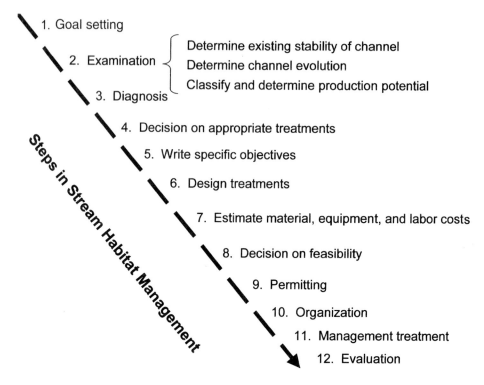

Figure 10.2   Steps involved in the process of stream habitat management.

3. Diagnosis. Based on the information gained from examining the resource, identify the stream's fish production potential and habitat problems. Some of the problems revealed (e.g., fish population deficiencies due to overfishing) may point to needs for management other than—or integrated with—habitat work. Resource examination and diagnosis is often a repeated cyclic process done in several steps appropriate to local conditions. Box 10.1 contains several common habitat deficiencies that can be corrected via habitat work.

4. Decision on appropriateness. If no significant habitat problem exists, or if the stream potential does not coincide with the broad objective, cancel the project. If potentials reasonably match objectives and there is a significant habitat problem to address, proceed to step 5.

5. Specific objective setting. Write definite, quantified results to be obtained by certain times (White 1991a; Chapter 2). Typically, a hierarchy of increasingly specific objectives is stated. At one level, the objective might be to increase the population of age-1 cutthroat trout by 80% within 5 years; appropriate subobjectives might include increasing spawning and juvenile rearing habitat by 80% within 2 years. Even more detailed objectives would state the amounts of riffle (containing specific sizes of gravel) to be created for spawning and the amounts of pools and cover of various types to be created for rearing.

6. Design treatments. Methods of habitat treatment are selected and developed to meet the specific objectives. The methods chosen are based on the examination and diagnosis steps (often with further mapping and detailed elevational

surveying) and on the management team's knowledge about stream mechanics and the ecology of the species involved. Several common designs are appropriate to only a limited range of channel types (see Hunt 1993 and Rosgen 1996 for specific guidance). The locations of riparian and in-channel treatments are designated on stream maps, detailed drawings of typical structures are made, and specifications and procedures are written.

7. Estimation of material, equipment, and labor costs. Based on the design and such factors as travel distances and physical accessibility of the project area, lists of the kinds and amounts of materials, equipment, and labor are drawn up. Prices are obtained from a variety of sources and a cost estimate is compiled. Needs for added funding may be seen at this step. Complete budgeting of costs should also include labor and materials for periodic examination and maintenance.

8. Decision on feasibility. If funding is not sufficient to cover costs, alter or cancel the project. If funding is adequate, proceed to step 9.

9. Permitting. Submit the project plan for requisite approval of government agencies. If the needed permits are denied, revise plans or cancel the project. If the permits are granted, proceed to step 10.

10. Organization. Arrange for materials, equipment, and labor. Draw up a work schedule. Train personnel.

11. Management treatment. Conduct the restoration or enhancement. Monitor funding. For quality control and correct interpretation of project design, it is especially important that members of the design team help direct the habitat treatments.

12. Evaluation of results. Periodically remeasure physical attributes of the stream and reinventory fish populations to see if results meet the stated objectives and to refine methods for better results in the future. Because most physical responses to stream management techniques develop for months or years, and fish population responses lag behind these, biological evaluation must span five or more years (Hunt 1976).

Special conditions of the stream, species present, management objectives, and sociopolitical setting may require other steps. For example, negotiation for access to the stream or surrounding property may be needed. Maintenance of the restorations or enhancements involved in many kinds of habitat projects is an ongoing process that must be considered in planning and budgeting.

10.4 HABITAT PROTECTION: PREVENTING AND REMEDYING ADVERSE HABITAT MODIFICATIONS

Fisheries managers are routinely involved in the review and comment process for a wide array of construction and operation permits. Through the permit review process essential stream habitats are protected from negative influences. Habitat protection requires sound policies and guidelines for dealing with all construction activities within the stream corridor, including channelization, stream cleaning (removal of large woody debris), and inadvertent installation of barriers to fish movement.

10.4.1 Stream Corridor Management

Effective management of stream habitats requires a broad view of the functioning ecosystem. Policies such as zoning and permitting in the stream corridor are critically important but often a neglected part of stream habitat management. The stream corridor contains three major components:

- the stream channel, which contains flowing water part of the year;
- the floodplain, which is inundated by floodwaters at some rare or infrequent interval; and
- the transitional upland fringe, which is the edge between the floodplain and surrounding terrain.

Therefore, the corridor includes the riparian forest, grassland and herbaceous cover, wetlands, islands, feeder streams, and floodplains. The goal of stream corridor management is to have a self-sustaining, ecologically sound, functioning floodplain. Usually this means restoring native grasses, shrubs, and trees to reduce sediment and nutrient inputs to streams. Streamside forest corridors serve several important functions in maintaining stream ecosystems. They

- filter sediment and suspended solids from runoff and incorporate it into the forest soil;
- transform nutrients and organic matter into mineral forms, which are synthesized into proteins by plants or bacteria or converted to gasses by denitrifying bacteria;
- serve as a long-term storage for nutrients as they are incorporated into plant and animal tissues; and
- serve as an energy source providing as much as 75% of the organic food base in wooded streams.

Stream corridor management involves encouraging and controlling vegetation in distinct zones, is not appropriate in wilderness situations, and can take various forms depending on objectives. For example, under one general scheme (Figure 10.3), for which the objective is nutrient management and sediment and erosion control, vegetation in the zone closest to the channel is left undisturbed, the next zone has only periodic tree cutting, and the third zone is kept in grasses that are mowed, as needed, to remove sequestered nutrients and encourage dense growth and soil stabilization (Welsch 1991; FISRWG 1998). Some efforts to protect streams with vegetated buffer strips have failed because the strips were too narrow and poorly maintained (Dillaha et al. 1986).

Success in stream corridor management often depends on the stream manager's ability to interact with landowners and decision-making bodies, such as town councils, county boards of supervisors, and planning district commissions (Riley

1998). The most effective form of communication is personal contact between private interests and public agencies (Turner 1997). As a poorly conceived land use plan in or near floodplains progress, it becomes increasingly difficult to alter the thinking behind the plan, and once a "development" is in place, correcting the damage may be impossible—or can be very expensive and consume many years. In dealing with landowners and planners, the manager should listen and try to understand their goals first (Chapter 3), then work with them to find ways to encourage proper actions.

One of the most widespread needs in stream corridor management is to exclude livestock from stream corridors. Alternatively, managers can prescribe special grazing plans to disrupt riparian vegetation minimally in large pastures of the West. Installing and maintaining fences along streams is expensive. Where streamside fencing is to be done, it is usually needed on both sides of the protected corridor, and special places for the livestock to cross the stream or to drink must be built (see Figure 10.4). It is sometimes cheaper in the long run to purchase whole pastures and convert them to fish and wildlife purposes.

10.4.2 Channelization

Channelization creates unfavorable stream habitat; the resulting uniform channel lacks pools, riffles, and boulders or log jams that are essential for sustaining fish abundance. Channelization typically involves some combination of straightening, widening, or deepening. In addition, streambank vegetation and instream obstructions (e.g., snags) are removed to increase channel capacity. Artificial reinforcement of streambanks is often required because stream power (see Chapter 9) has been increased and stabilizing bank vegetation has been removed. Channelization is conducted to

- drain wetlands and increase the amount of land useable for agriculture;
- reduce flooding in localized areas by increasing channel capacity;
- relocate channels for highway and other construction projects;
- permit navigation; and
- eliminate streambank erosion by eliminating bends in the river (a misconception).

Stream straightening results in a loss of important fish habitat features associated with natural meandering and pool–riffle patterns, in addition to decreasing stream length and increasing stream slope (see Chapter 9) and power. As a consequence, habitat diversity is reduced, and the altered channel has higher and more uniform velocity. Abundance of sport fishes can be 8–10 times greater in natural channels than in channelized parts of the same stream (Gebhards 1970).

Although stream channels may eventually recover from minor alterations through natural fluvial and biotic processes, recovery may require decades, perhaps centuries. Artificial riffles and pools, spaced approximately 5–7 channel widths apart (as in some natural streams; Leopold et al. 1964), have been successful in restoring a more natural fish and invertebrate fauna than typically exists in channelized streams without such mitigation (Edwards et al. 1984).

Figure 10.3 Schematic showing one approach toward the management of vegetation in the stream corridor for nutrient management and sediment and erosion control (Welsch 1991).

The following design principles can be used to reduce the adverse effects of channel modifications (Keller 1978; Nunnally 1978).

1. Where possible, avoid straightening and steepening channels.

2. Preserve and promote bank stability by leaving trees and shrubs in place, minimizing channel alteration, restoring vegetation in disturbed areas, and judiciously placing riprap or tree revetment (see sections 10.5.1 and 10.6.2).

3. Emulate nature in designing channel form by creating appropriately spaced pools, riffles, and meanders.

4. Clear snags and debris by hand rather than with heavy equipment (appropriate in small streams).

Instead of making severe channel modifications, undertake alternatives such as building levees far from the channel, constructing floodways, removing buildings and other structures from the floodplain, instituting zoning to prohibit building in flood-prone areas, and converting land uses in the floodplain to those that flooding will not harm. Many hard lessons have been learned from repeated attempts to channelize streams rather than accommodate their dynamic nature; therefore, many costly restoration projects are needed (Brookes 1988).

Figure 10.4 A cattle guard restricts cattle to designated water crossings without the maintenance and repair problems associated with cross-stream fencing, such as removing debris (Hunt 1993; reproduced courtesy of University of Wisconsin Press).

10.4.3 Removal of Woody Debris

Woody debris accumulates naturally in streams and plays important roles in stream mechanics and fish habitat; its clearance has often reduced stream carrying capacity for fishes. Woody material creates pools, increases structural complexity, provides fish cover, harbors invertebrates, traps gravel for spawning and invertebrate production, holds other organic matter, and increases channel stability. Logs and smaller woody matter enter channels by windthrow, the toppling effects of snow and ice, bank erosion, avalanches, and beaver activity. Riparian logging may increase woody debris in streams immediately but reduces long-term input. Log jams and other woody debris accumulations are often removed to restore channel capacity and reduce impoundment, eliminate barriers to fish migration and human navigation, and return streams to prelogging condition (Chamberlin et al. 1991; Hicks et al. 1991). However, natural log jams (even major ones) and beaver dams create valuable fish habitat, especially in steep streams (>1% bed slope), and anadromous fishes are adept at surmounting them (Sedell et al. 1988). In streams with slopes of about 0.5% or less, beaver dams can cause sedimentation and warming that severely harm coldwater fish production.

10.4.4 Barriers to Fish Migrations

Migration barriers have harmed valuable anadromous and adfluvial fish stocks (Chapter 24). Roadway culverts often hinder passage unless they are installed to meet special criteria for the fish involved (see section 10.5.2). Constructed dams are common barriers. Fishways are often needed in dams to allow adult fish passage to upstream spawning grounds. However, downstream passage through turbines, spillways, and conduits can kill many fish, so devices to guide fishes to other outlets may be needed. Changing the flow of rivers into a series of large, slow reservoirs can lethally delay migration of juvenile anadromous fish to the sea; this is increasingly recognized as a reason for declining fish populations.

10.5 HABITAT RESTORATION: REMEDYING STREAM ABUSES

Examination of stream habitats leads to diagnosis of habitat deficiencies (Box 10.1) that can be traced to specific disturbances and can be corrected to return habitats to predisturbance conditions. These corrective actions are habitat restoration and include corrections for excessive sediment input, blockage to fish passage, water withdrawal, and unsuitable flow regime.

10.5.1 Streambank and Grade Stabilization

Many streams are diagnosed with unstable banks, severe erosion, and excessive sediment inputs. Changes in vegetation, exposed soil layers, geomorphic conditions, hydrologic events, instream hydraulics, or some combination may be responsible. Although the first reasonable design solution is to reinforce the streambanks, the stream habitat manager must first determine whether the stream grade (longitudinal profile) is stable. If the stream grade is significantly changing, bank projects will fail no matter what materials or designs are used.

Effective techniques for stabilizing eroding streambanks often include revegetation, installation of revetments made of trees and brush, burial of rootwads into the bank, and stabilization of the bank with riprap. In designing streambank protection consider the basic erosional processes and conditions in streams (Henderson 1986) as well as the needs of fishes. Riprap is often used along the lower parts of banks. Large, dense, angular rock that is resistant to freeze shattering provides the best habitat for fishes. The rock should be loosely laid to create refugia and turbulence. The minimum size of rock to be used depends on the density of the rock and the stream power it must withstand. Rootwads must be buried trunk first in banks subjected to lateral scour, and the rootwad must be placed to divert currents at bankful stage (Rosgen 1996). Tree and brush revetments can be installed along with riprap to provide even more cover for fishes. Trees should be anchored parallel to banks and overlap each other by onethird to one-half. Section 10.6.2 describes these improvements in greater detail. Grasses, shrubs, and trees can be planted on the upper slopes and tops of the banks. The use of cedar tree revetments have proven to be cost-effective because they allow eroding materials to be trapped and seeded, creating a flow-resistant bank (Figure 10.5). Costs for tree revetments are about one-third those for stabilization of the bank with riprap (Roseboom and White 1990).

10.5.2 Fishway and Culvert Design

To design structures that will not impede migrating fish, biologists must understand the behavior and swimming capacity of fishes and work with hydraulic engineers who know how to apply principles of fluid mechanics. Swimming speeds of fishes vary according to species, body size, and environmental factors, especially temperature (Beamish 1978). The critical swimming speeds and leaping abilities of the species and the sizes of fishes in a particular stream must be known to design effective fishways, culverts, and guiding devices (Bender et al. 1992; Clay 1995).

The basic principles of design are that fishways should be readily passable by all migratory species in the stream, operate at all water levels in the forebay and tailrace of the barrier, be navigable by fishes without injury or undue stress, and have entrances that fishes can find, enter, and pass without delay. Fishways can be categorized based on whether the design incorporates (1) a series of steps (ladders), (2) baffles to reduce velocities (chutes or vertical-slot fishways), or (3) lift systems. The earliest fishway designs had a series of steps created by alternating weirs and pools (Orsborn 1987), hence the name fish ladder (Figure 10.6). Weir and pool fishways were modified with submerged orifices or vertical slots to permit passage through the fishway at a wider range of flows. A chute type fishway consists of a narrow channel with many closely spaced baffles to reduce water velocity. Designs that mechanically lift fish over barriers include various elevators and locks, the effectiveness of which may be limited by intermittent operation and need for frequent maintenance. Designs of the step and chute fishways must accommodate swimming capabilities of fishes. For step, or weir and pool, fishways, a maximum drop of 0.3 m is often recommended to permit rapid and easy migration, but this criterion is based on gross underestimates of leaping abilities of some species. New designs that attempt to match more closely fishway design with leaping and swimming capabilities will likely lead to more efficient and cost-effective fishways (Orsborn 1987).

The most common problems associated with culverts are (1) excessive drop at the downstream end, (2) velocities exceeding critical swimming speeds, (3) shallow flow, (4) lack of resting pools at inlets and outlets, and (5) upstream blockage due to debris. Each of these can be avoided by proper design.

A culvert should be designed so that the mean velocity of water passing through it is significantly less than the critical prolonged swimming speed for the species and size of fish involved. The critical prolonged swimming speed is the maximum velocity a fish can maintain for a given time, and designs should permit passage by fish in 5 min or less. Time required for passage is obtained by dividing the length of culvert by net velocity of the fish (swimming speed minus mean current velocity in culvert). Therefore, longer culverts should be designed with lower average velocities (perhaps including resting areas such as baffles) to avoid fatiguing fishes (Figure 10.7). In existing culverts, where velocity or shallowness limits passage, baffles or other devices that impart roughness can be installed to create more suitable water velocities and depths if the capacity of the culvert is sufficient. Resting areas should be available at the upstream and downstream ends of the culverts.

10.5.3 Screens and Guiding Devices

Alterations of stream habitat caused by dams and water withdrawals require screens or guiding devices to prevent fishes from entering certain areas, such as cooling water intakes, turbines, or irrigation systems. Many designs are available depending on the application (Colt and White 1991).

Figure 10.5 Streambank stabilization with cedar tree revetments before (top), during (middle), and after (bottom) construction on fourth-order Otter Creek, Cooper County, Missouri. (Photographs provided courtesy of Missouri Department of Conservation.)

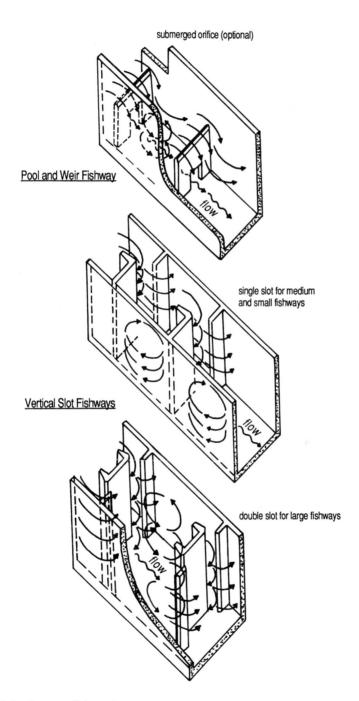

Figure 10.6 Common fishway types.

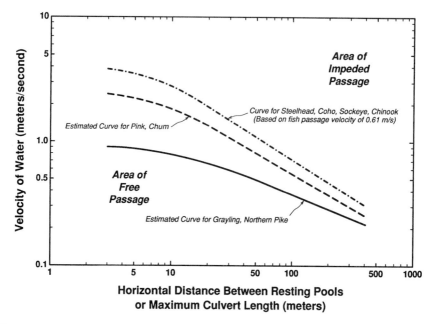

Figure 10.7 Maximum swimming capabilities of several fish species as they relate to water velocity and culvert length. These curves can be used to determine suitable water velocities, culvert lengths, and distances between resting pools for fishways and culverts. Figure modified from Slaney et al. (1980).

10.6 HABITAT ENHANCEMENT: IMPROVING INSTREAM HABITAT FOR FISHES

Various technical guides to stream habitat improvement for fishes—primarily for salmonids—have appeared since the 1960s (White and Brynildson 1967; Slaney et al. 1980; Hall and Baker 1982; Reeves and Roelofs 1982; Wesche 1985; Hunter 1991; Hunt 1993). In streams containing nonmigratory trout, most habitat work centers on cover and otherwise creating proper spaces for survival and growth of the large fish that anglers seek (Hunt [1988] reviewed 45 evaluations of such work in Wisconsin); improving reproductive and juvenile rearing habitat often are not major objectives because they are not in short supply or are achieved concomitantly. For migratory fishes that use streams primarily as reproductive areas, the principle objective is to create conditions for producing more out-migrating young, so work is commonly focused on enhancing spawning grounds and rearing habitat for juveniles.

10.6.1 Channel Form and Current Pattern

Stream-dwelling fishes are adapted to the channel forms that occur most commonly in nature. Therefore, simulating those forms is a sound principle to follow when improving existing channels or when relocating channels. Channel patterns are broadly classified as

1. steep, erosive channels, in which sediments erode away faster than they are replaced and in which water follows an irregular mix of rather straight and jagged courses over bedrock or over and through boulders;

2. meandering channels of moderate to low slope with pronounced pools at the outsides of bends and shallower riffles or silt bars between bends (these have sediment entering at about the same rate as it is carried away downstream and are for this and other reasons often considered the most stable of the forms; Leopold and Langbein 1966); and

3. braided channels, choked with sediment that is being fed in faster than it can be washed away.

Other, less common patterns exist that may be considered intergrades of the three main ones (Rosgen 1996).

Two major, large-scale aspects of channel structure prevail: the meandering form itself and, in steeper streams, the stair-stepped form. The remarkably regular, graceful winding pattern of meanders is the tendency of streams in traversing gentle slopes in various media, notably in soils (including gravels and cobble) that are easily eroded and transported but cohesive enough to form firm banks (Leopold and Langbein 1966). Therefore, the features of meandering predominate in many sections of most alluvial streams.

Natural stair-stepped formations are caused by log jams, accumulations of other woody debris, and beaver dams, as well as by bedrock ledges and groups of boulders. Some of these obstacles also occur in low-gradient meandering streams and may tend to obscure the meander pattern and interrupt its regularity but do not destroy the tendency.

Within-channel habitat forms are broadly classified as pools, riffles, glides (or runs), and falls. Bisson et al. (1982) identified three types of riffles in small, steep streams of western Washington as

1. low-gradient riffles, having a slope of less than 4% and a current of 20–50 cm/s;

2. rapids, occurring over large boulders and having a rather even slope greater than 4% and a current greater than 50 cm/s; or

3. cascades, also having a slope steeper than 4%, but the slope is uneven, and the cascade consists of series of small steps and pocket pools.

Additionally, Bisson et al. recognized six types of pools in these streams:

1. lateral-scour pools, by far the most prevalent type, occurring where, for various reasons, flow veers against a channel bank;

2. trench pools, which are longitudinal grooves having bedrock sides;

3. plunge pools, formed by water dropping vertically from a complete or nearly complete channel obstruction;

4. dammed pools, occurring upstream from channel obstructions;

5. backwater pools, formed along channel margins by major highwater eddies behind or beside large partial obstructions such as rootwads or boulders; and

6. secondary channel pools, which become isolated or dry up during normal low flow.

The habitat features of meandering channels are especially important because the meandering form is so prevalent. Pool and riffle undulation of the streambed results from the same processes that cause meandering and is often a feature to enhance. Long, lateral-scour pools develop along the current-bearing banks of meander bends and offer advantageous habitat, particularly for large fish. Riffles provide spawning and nursery habitat for many fishes and tend to form in the crossover areas between bends. The distance from pool to pool or riffle to riffle along the down-valley line (not along the channel, unless it is straight) is the same as the bend to bend spacing of the meander system—about five to seven times channel width. This interval can be a guideline for spacing of habitat improvement structures.

10.6.2 Direct Improvement of Channel Form and Current Pattern

The lateral-scour pools of meander bends can be enhanced by using vegetation (often restoring it) to increase cohesiveness of the current-bearing banks. Such enhancement leads to decreased lateral erosion of the channel and greater downcutting (active lowering of streambed elevation) and undercutting. Curved artificial reinforcements of banks, such as certain forms of riprap, log, or whole-tree revetment can be effective in achieving this (Figures 10.8, 10.9)—often in combination with promoting healthier bank vegetation. Properly using current deflectors in conjunction with the bank revetments (Figure 10.10) further develops the lateral-scour effect. This combination of current deflectors and revetments accentuates the meander form by making bends slightly tighter, that is, the radius of the meander curvature is shortened. Another way to guide current into creating or enhancing lateral scour is by installing low oblique sills of large logs. To fit in with the meandering form, these should be placed as shown in Figure 10.11.

In small, steep streams, the steps formed by log jams, beaver dams, and accumulations of large rock are important in stabilizing the channel, creating areas of slow current, trapping sediments such as gravel, and causing development of plunge pools and lateral-scour pools. Particularly in old-growth forest streams, such elements may be abundant. In contrast to the rather regular spacing of habitat features associated with meandering channels, the spacing of stair steps and associated features tend to be irregular. Steep streams may sometimes meander, but they are often highly irregular in course pattern, having relatively straight reaches interrupted by sharp bends. If the obstacles that create steps in steep streams are removed, the relatively straight parts of the channel develop very swift current during high water, which erodes the channel until large rough elements are exposed or until the streambed downcuts to low gradient. Improving fish habitat in

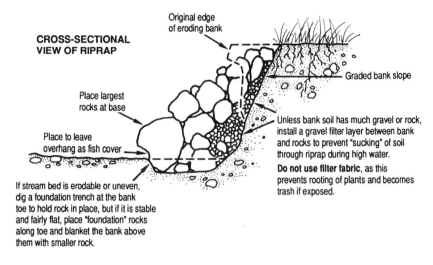

CROSS-SECTIONAL VIEW OF RIPRAP

Original edge of eroding bank

Graded bank slope

Place largest rocks at base

Place to leave overhang as fish cover

Unless bank soil has much gravel or rock, install a gravel filter layer between bank and rocks to prevent "sucking" of soil through riprap during high water.

Do not use filter fabric, as this prevents rooting of plants and becomes trash if exposed.

If stream bed is erodable or uneven, dig a foundation trench at the bank toe to hold rock in place, but if it is stable and fairly flat, place "foundation" rocks along toe and blanket the bank above them with smaller rock.

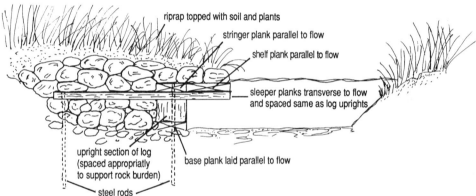

riprap topped with soil and plants

stringer plank parallel to flow

shelf plank parallel to flow

sleeper planks transverse to flow and spaced same as log uprights

upright section of log (spaced appropriatly to support rock burden)

base plank laid parallel to flow

steel rods

Figure 10.8 Rock structures for streambank revetment. Top figure is a cross-sectional view of riprap. Bottom figure shows a lunker structure with riprap above a cantilevered ledge to provide for fish along current-bearing banks.

steep channels largely involves enhancing or restoring the stepped form. This can be done by installing sills (check dams, drop structures, or overpours) made of boulders or logs. Better effects for fish habitat are often more possible with logs than with boulders. Promoting colonization by dam-building beavers can also create useful step formations.

A suitable holding area for a stream fish is a space having proper water depth, volume, and currents, as well as having ample cover, for the activities of a fish of a particular size and species. Particularly in shallow streams, cover for hiding is crucial. Cover consists of objects or channel formations that offer fish concealment from predators or visual isolation from competitors. General categories of security cover are (1) overhead bank cover (overhead referring to the fish's head, not ours), such as undercuts and objects associated with the streambank that are submerged coverts, under which fish can hide; (2) water depth, in small streams especially that afforded by deep pools; (3) midstream objects, such as boulders and aquatic plants; and (4) hydraulic features, such as areas of broken or turbu-

Figure 10.9   Using trees for streambank revetment is a low-cost, effective technique to solve eroding streambank problems. Width of trees should be at least two-thirds the maximum bank height.

lent water surface and masses of bubbles or foam. Cover associated with streambanks is usually the most prevalent kind of cover in meandering streams. Pools are made much more beneficial for fish if they adjoin bank cover or have cover objects within them. Various methods for making cover also create spaces with more favorable shape and current.

Generally, the most effective way to enhance cover is to create physical niches in, or objects jutting out from, the streambank, under which fish can position themselves when fleeing from attack or when resting. Trees, bushes, and bank ledges that overhang the water surface cast shadows in which trout lie during sunlit times (Fausch and White 1981), and such objects may, to varying degrees, directly obstruct the view of aerial predators. The closer the overhanging object is to the streambed, the darker and more protective its shadow will be. For daytime hiding, trout prefer submerged overhead covers, particularly those close to the bed. Abundance of trout, particularly large trout, is often closely correlated with the amount of overhead bank cover in streams.

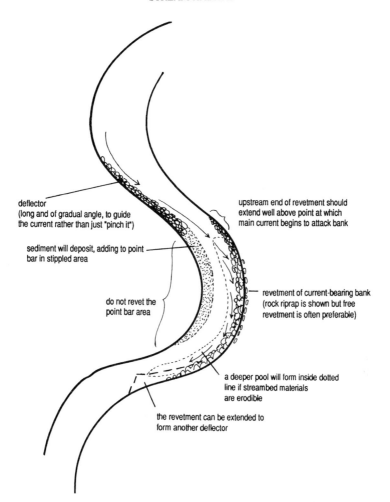

Figure 10.10   Current deflector to direct and speed currents to feeding and shelter sites increases habitat complexity in conjunction with lunker structures (Figure 10.11) for bank protection.

Installing a rock revetment, also called riprap (Figure 10.8), is the most common method for artificially protecting streambanks. If properly placed along current-bearing banks (e.g., concave banks of meander bends), riprap can benefit fish. Riprapping inner (convex or point-bar) banks of meanders is generally a waste of money and may harm the riparian ecosystem.

Tree revetment (Figures 10.5, 10.9), although less permanent, may be relatively inexpensive and much more beneficial to fishes than is riprap. Series of whole trees can be fastened in place with earth anchors and cables. When installed as a dense, overlapping "thatch," such revetment can simulate beneficial aspects of trees that naturally lodge along current-bearing banks. In some situations, as in restoring habitat along steep, forested rivers that would naturally have large masses of haphazard log jams, it may be appropriate to install trees in a jumbled fashion, with some trunks and root masses jutting out in various directions. Tree revetment immediately creates cover and—by trapping fine sediment—

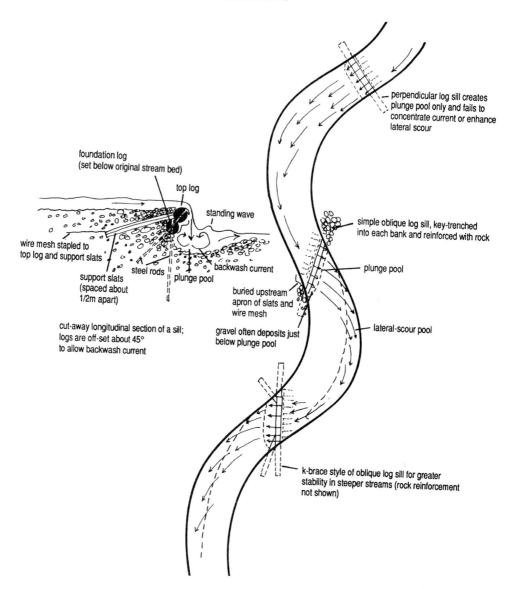

foundation log
(set below original stream bed)

top log

standing wave

perpendicular log sill creates
plunge pool only and fails to
concentrate current or enhance
lateral scour

wire mesh stapled to
top log and support slats

steel rods

backwash current

plunge pool

support slats
(spaced about
1/2m apart)

simple oblique log sill, key-trenched
into each bank and reinforced with rock

plunge pool

buried upstream
apron of slats and
wire mesh

cut-away longitudinal section of a sill;
logs are off-set about 45°
to allow backwash current

gravel often deposits just
below plunge pool

lateral-scour pool

k-brace style of oblique log sill for greater
stability in steeper streams (rock reinforcement
not shown)

Figure 10.11   Oblique log sills are designed and placed to create a downstream plunge pool and a lateral-scour pool.

speeds the development of riparian vegetation, making it a particularly suitable technique for restoring habitat in streams damaged by overgrazing and other destabilizing activities. The damaging activity should be halted before revetment is undertaken.

Stream pools and glides (runs) can be formed either by slightly raising the water level (damming, but not enough to reduce water velocity to pondlike conditions), by excavating, or by causing current to scour down the bed. Scour is by far the more important process in stream fish habitat improvement, and there are several general methods for achieving it.

The roughening of current-bearing surfaces, such as the leading edges of deflectors or banks lined with riprap, tree revetment, or natural vegetation, can create a current-pulling or velocity-concentrating effect that can cause lateral-scour pools to form. Often restoring or enhancing streambank vegetation induces considerable lateral-scour pool formation. A lateral-scour pool can also be formed by installing a sill oblique to channel alignment (White 1991b) so as to redirect current against one bank (Figure 10.11). The oblique sill can be built very low—a mere linear bump on the streambed—and still have the current-bending effect. Alternatively, a higher oblique drop structure can be built (bed slope permitting) to create a plunge pool, as well as a lateral-scour pool (Figure 10.11). Such structures simulate some of the important effects of large woody debris in streams.

In low-gradient streams (perhaps a slope of less than 0.5%), relatively long, gradually protruding wing deflectors can guide the current to flow more strongly against an area of the opposite bank that is downstream from the deflector. There is no standard angle for deflectors to jut from the bank. The receiving bank will, if cohesive and rough, resist lateral erosion, and scouring will occur in the bed along it. If the current-bearing edge of the deflector is also rough and tough, then it will have the current-pulling effect. Thus, a lateral-scour pool (or glide) will form along the structure and downstream from it on the opposite side of the stream (Figure 10.10).

Improving fish reproductive success can also be an objective in habitat management. This may involve creating better quantity or quality of spawning or nursery (juvenile rearing) areas. For fishes such as salmonids that spawn in gravel, certain channel modifications can increase the amount of gravel streambed available to them. General approaches are (1) removal of fine streambed sediments that overlie gravel beds, (2) removal of interstitial deposits of fine sediments within gravel beds, and (3) modification of streambed profile to increase deposition of gravel at key locations in streambeds. The first and second approaches may involve reducing input of fine sediment from the riparian corridor, from streambanks, or from upstream or installing deflectors to scour sand off gravels. In some situations, sediment can be removed by digging streambed pits (e.g., 1.5 m deep and 30–50 m long), then periodically re-excavating the sand and silt that drift into such traps (Hansen et al. 1983).

In temperate and colder climates, winter is generally a time of severe hardship and high mortality for stream-dwelling fishes. Not only does the water become unfavorably cold—near 0°C in many streams—but various types of ice, particularly anchor ice and frazil ice, form, and sudden movements of thick ice slabs can mechanically injure fish or cause other threatening conditions. Other hazards are the extreme winter low flows that occur in some regions and near 0°C snowmelt floods, which are much more dangerous than warmer floods. Stream fish move to special winter habitats that ameliorate the hardships, and it is important in habitat management to provide such refuges. The manager should know the special winter habitat requirements of the target species and its life stages. Various smaller fishes take refuge from predators and water current in the interstices of wood debris or under streambed rocks. Those too large to fit in such spaces may swim to lakes or to the quiet, protective depths of deep pools in larger streams. The deep habitat must be available and the routes to it passable.

10.7 STREAMFLOW MANAGEMENT

Dams and diversions artificially regulate discharge in many streams, often disrupting the natural flow regimes to which fishes are adapted (Collier et al. 1996). Sometimes the artificial regulation of flow can be done in ways that benefit fisheries. In arid and semiarid regions, water withdrawals for out-of-stream uses have often left large segments of stream channel partially or completely dewatered. In these regions, management agencies and concerned public groups have worked to reserve minimum instream flows for fish habitat. In water-rich regions, dewatering of stream channels is less common, but other aspects of flow regulation have significant effects on stream habitat and biota. To avoid or reduce adverse effects, fisheries managers should understand the effects of various flow modifications and be able to use a variety of analyses in determining impacts and recommending suitable flow regimes. In some cases, dam removal will be in the best public interests because dams were designed for a finite time span. Recent precedents include the Edwards Dam on Kennebec River, Maine, and Woolen Mills Dam on Milwaukee River, Wisconsin.

Responses of fishes and invertebrates to flow regulation vary (Gore and Petts 1989). In a review of case histories of regulated streams in the Pacific Northwest, Burt and Mundie (1986) found that following flow regulation salmonid populations declined in 76% of cases and increased or showed no change in 24% of cases. Most of the positive or benign effects occurred in cases in which flows had been little changed, had been increased, or had been decreased in high-flow months. Reduced stocks were commonly associated with flow decreases that reduced habitat. Other reasons for reduced stocks included barriers, sedimentation, fluctuating flows, impoundment, and altered temperature or water quality. The effects of flow modification on fish stocks are complex and difficult to predict.

Modifying a streamflow regime often requires a permit or environmental impact statement to comply with state, provincial, or federal statutes. Site-specific evaluations should be conducted, where possible. The hierarchy developed by Petts (1984) serves to organize the many possible effects according to time scale (Figure 10.12). First-order impacts are those occurring soon after dam construction, such as blocked migration and altered flow, reduced sediment load, altered water quality, and a new plankton composition and abundance. Second-order impacts occur later, typically as a result of first-order impacts. For example, major reduction in sediment loads and peak flows can result in gradual change in channel shape and bed materials. Third-order impacts are the cumulative effects on fishes and invertebrates stemming from first- or second-order impacts.

A major first-order impact of flood control impoundments is that maximum seasonal flow is reduced to an extent that leads to significant channel changes (second order). In addition, downstream water quality and temperature patterns can change, depending on reservoir stratification, outlet depth, and flow releases. Releases of hypolimnetic water are often cold and anoxic and have elevated nutrient, iron, manganese, and hydrogen sulfide concentrations. The downstream result is often conversion from warmwater to coldwater fisheries and to a benthos composed of only a few tolerant taxa.

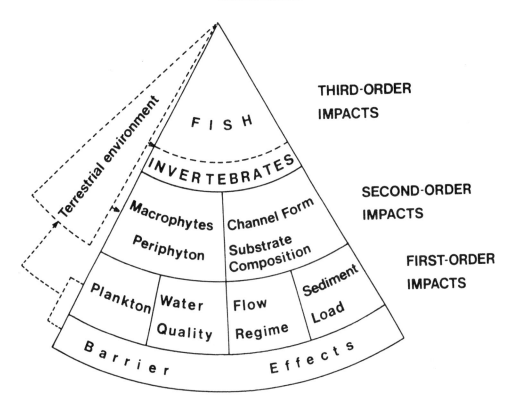

Figure 10.12 Hierarchical framework for examining the impacts of river impoundment. Figure modified from Petts (1984).

Water diversions without large impoundments usually significantly modify flow during the low-flow season only, and the effects of these diversions tend to be most severe during drought years. Where low flows are made more extreme, the reduced living space, reduced cover availability, and elevated temperature can significantly reduce fish populations. Although run-of-river (low-head) hydropower facilities may have minimal effects on flows, hydropower operations causing large daily fluctuations in flow can have dramatic effects. In hydropower "peaking" operations, water is stored in a reservoir, often with no minimum release, and run through turbines to satisfy peak demands, typically with sudden starts and stops. Rapidly fluctuating flow changes habitat faster than some fishes and invertebrates can endure. Periodic dewatering of the stream margins reduces insect abundance and diversity. Stranding of stream organisms may occur, depending on stream channel shape and the rate of change in discharge or depth. Fishes adapted to the shallow, slower-current areas along stream margins are less abundant in sites subjected to daily flow fluctuations (Bain et al. 1988).

10.7.1 Methods to Determine Acceptable Flow Regimes

Methods to determine acceptable flow regimes can be categorized as (1) discharge methods, (2) hydraulic rating, (3) habitat rating, and (4) biological response (Stalnaker 1995). These are ordered according to increasing resolution, data needs, and costs. The appropriate type of analysis depends on the stage of planning of the project and amount of controversy it involves. Instream flow study in fisheries first centered on finding the minimum flow required to maintain fishes in streams, which led to simplistic answers. The minimum flow concept is a myth and should be discarded (Stalnaker 1990). Today we realize that a wide variation of flows is required for different seasons and purposes (Hill et al. 1991). Furthermore, the administrative framework for decision making on flow allocations always involves trade-offs and compromises. See Gillilan and Brown (1997) for further background on legal and administrative strategies for protecting instream flows. Examples of the first three types of methods for determining acceptable instream flow regimes, or providing information for trade-off analysis, are described in the following paragraphs.

Discharge methods require no fieldwork and rely on streamflow statistics or drainage basin variables. The first step, where long-term data permit, is to reconstruct the natural flow regime. Restoring the integrity of river habitats will require the use of the natural flow regime as a guide when attempting to restore native species and ecosystems. There are five critical components of the flow regime that regulate ecosystem processes: magnitude, frequency, duration, timing, and rate of change.

Many dams and diversions were constructed long before adequate long-term streamflow gauging data were available to characterize a natural flow regime. In many cases, records of diversions will have to be added to gauging records, which are available for many United States' gauges on the worldwide web (www.usgs.gov). Box 10.3 outlines an approach of how to use historical records to develop management targets for streamflow management (Richter et al. 1996, 1997; Poff et al. 1997). For example, the Smith River, Virginia, was impounded in 1950 by Philpott Dam to provide flood control and hydropower. The present flow regime (Figure 10.13 top) is designed to generate hydropower during peak demand times (except weekends for safety reasons). The 32 parameters of the index of hydrologic alteration (Box 10.3) were compared for pre- and post-impact periods, and 3 were significantly altered: number of hydrograph reversals increased, low-flow pulses increased, and the 1-d minimum streamflow decreased (Figure 10.13 bottom). The range of variability approach (Box 10.3) would allow agencies to develop flow regimes so that these three parameters would fall within the target or tolerance range.

The Tennant (1976) method is also a commonly used discharge assessment technique. Because many streams show a similar pattern in the relationships between depth, velocity, and top width and discharge, expressed as a percentage of the mean annual flow, the Tennant method prescribes a fixed percentage of the unregulated mean annual flow. For the streams Tennant (1976) studied, flows below 10% of the mean annual flow result in degraded habitat conditions due to dramatic reductions in depth, velocity, and width and are considered minimum acceptable flows; flows of 30 to 60% result in more satisfactory conditions for fishes, wildlife, and related recreational values. This method has obvious shortcomings in that it assumes fixed percentages and similar channel types, and it ignores flow dynamics and extreme variation. The method should be used only to provide rough preliminary guidelines, which should be evaluated with field measurements (Orth and Leonard 1990).

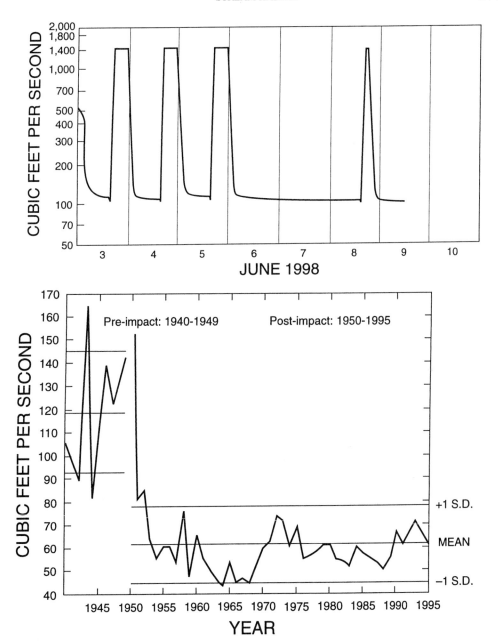

Figure 10.13 Current daily flow fluctuation (top) and comparison of minimum 1-d low flow before and after regulation by U.S. Army Corps of Engineers Philpott Dam (bottom) on Smith River at Bassett, Virginia.

Hydraulic rating methods require field measurements at different flows in order to relate hydraulic geometry variables to discharge. The most common hydraulic rating method uses the relationship between wetted perimeter and discharge to recommend flows to preserve wetted habitat. Wetted perimeter, the length of wetted streambed along a transect perpendicular to the direction

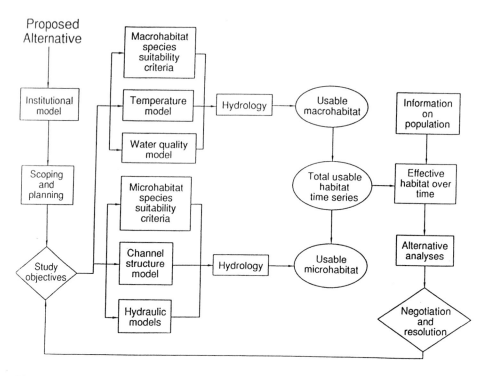

Figure 10.14   Overview of the instream flow incremental methodology.

of flow, obviously is related to wetted surface area of a streambed. The rating of habitats by hydraulic methods cannot be determined without additional measurements, such as current velocity, bed materials, or cover (Orth and Leonard 1990). The riverine community habitat assessment and restoration concept is another approach that compares present hydraulic characteristics to preregulation characteristics.

Habitat rating methods relate quantity and quality of habitat to streamflow. These methods usually address the specific requirements of selected aquatic organisms. The instream flow incremental methodology (IFIM), a common habitat rating method, involves a variety of models and information sources to describe a time series of usable habitat for various life stages of a given fish species under natural conditions and proposed project conditions (Bovee 1997). Figure 10.14 describes the major components involved in determining total usable habitat. Microhabitat suitability criteria, channel structure data, and hydraulic simulations are used to calculate amount of usable microhabitat. Weighted usable area is the measure of microhabitat area and is calculated by weighting the total surface area relative to optimum habitat. The physical habitat simulation system consists of computer software that determines relationships between weighted usable area and streamflow, information that can be used to negotiate acceptable flows. Other components of IFIM consist of water quality and temperature models and species suitability criteria for these factors (Figure 10.14).

Box 10.3 Characterization of the Natural Flow Regime

The natural flow paradigm maintains that the full range of natural intra- and interannual variation of hydrological regimes is necessary to sustain the native biodiversity and evolutionary potential of aquatic, riparian, and wetland ecosystems. The 32 biologically relevant parameters in the index of hydrological alteration have been proposed as a minimum set of descriptors for evaluating the degree of hydrologic alteration on proposed and existing dams. The range of variability approach then sets streamflow-based management targets (± 1 standard deviation from means). This approach permits managers to adopt interim management targets before conclusive long-term ecological research results are available to decision makers.

Flow regime characteristics	*Hydrological parameters*
Magnitude timing	Mean value for each month [a]
Magnitude duration	Annual minima 1-d means
	Annual maxima 1-d means
	Annual minima 3-d means
	Annual maxima 3-d means
	Annual minima 7-d means
	Annual maxima 7-d means
	Annual minima 30-d means
	Annual maxima 30-d means
	Annual minima 90-d means
	Annual maxima 90-d means
Timing	Julian date of each annual 1-d maximum
	Julian date of each annual 1-d minimum
Frequency duration	Number of high pulses each year
	Number of low pulses each year
	Mean duration of high pulses within each year (d)
	Mean duration of low pulses within each year (d)
Rates of change frequency	Means of all positive differences between consecutive daily values
	Means of all negative differences between consecutive daily values
	Number of rises
	Number of falls

[a] Each mean is considered 1 of 32 biological parameters.

Because the temporal dynamics of habitat are a major influence on fish populations in streams, any habitat rating methods should analyze the time series of habitat conditions to identify limiting events in the life cycles of key organisms. The purpose of such time series analyses is to relieve or avoid exacerbating these habitat bottlenecks induced by flow regulation (Stalnaker et al. 1996). A simplistic analysis of the relationship between usable habitat and streamflow is an intermediate step and *not* the final analysis.

Parts of IFIM have been widely applied in the western United States, Canada, France, and New Zealand. However, controversy exists over its general applicability (Barinaga 1996; Castleberry et al. 1996; Van Winkle et al. 1997), as is to be expected anytime a scarce resource is allocated. Major problems are that the IFIM does not take into account associated fish species (competitors, predators, or prey) or ecosystem processes. Furthermore, most applications attempt to get by with minimal field sampling to develop reliable results (Williams 1996). Stream habitat managers should be cautious in developing streamflow regimes to protect stream fisheries because our ability to predict biological responses to streamflow change is poor (Orth 1995). As a reality check the manager should always compare results to the natural flow statistics (Box 10.3).

10.8 CONCLUSION

Successfully managing habitat for fishes (and fishing) in streams requires thorough knowledge of and concern for streams, stream processes, watershed processes, and the kinds of fishes and other organisms involved. It requires knowing and caring about the uses and values that people have for the resource, such as angling, commercial fisheries, species protection, and aesthetics. It requires knowing when not to construct habitat improvement structures and to leave streams as they are. Increased abundance of fishes, better fishing, and ecosystem health are basic objectives. The modern trend is toward more professional management, toward more attention to design and planning, and toward managing in ways that derive from and are increasingly attuned to natural processes in streams and their surroundings—the processes to which the fishes are adapted. This trend—in contrast to the artificiality and concern for tidiness that characterized some past work—involves increased focus on the drainage basin, on riparian grazing and logging practices, on the roles of streambank vegetation, woody debris, and beaver, and on structural complexity within the channel. Many of the streams on our continent are not meeting their potential. Changing this awful situation will require the cooperation of private interests, professionals, and volunteers working in concert to identify habitat problems and seek innovative solutions. The volunteer movement is growing and should be encouraged by active involvement of fisheries professionals who are aware and supportive of programs such as Streams for the Future (Missouri), Stream Teams (Arkansas), Izaak Walton League of America Save our Streams, the Adopt-A-Stream Foundation's Streamkeepers, and Riverkeepers, Inc. (Murdoch et al. 1996; Cronin and Kennedy 1997; USEPA 1997; Riley 1998). Healthy streams are the signature of a vibrant society.

10.9   REFERENCES

Alabaster, J. S., and R. Lloyd, editors. 1984. Water quality criteria for freshwater fish, 2nd edition. Butterworth, London.

Bain, M. B., J. T. Finn, and H. E. Booke. 1988. Streamflow regulation and fish community structure. Ecology 69:382–392.

Bain, M. B., T. C. Hughes, and K. K. Arend. 1999. Trends in methods for assessing freshwater habitats. Fisheries 24(4):16-21.

Barinaga, M. 1996. A recipe for river recovery? Science 273:1648–1650.

Beamish, F. W. H. 1978. Swimming capacity. Pages 101–187 *in* W. S. Hoar and D. J. Randall, editors. Fish physiology, volume 7. Academic Press, New York.

Bender, M. J., C. Katopodis, and S. P. Simonovic. 1992. A prototype expert system for fishway design. Environmental Monitoring and Assessment 23:115–127.

Beschta, R. L. 1996. Restoration of riparian and aquatic systems for improved fisheries habitat in the Upper Columbia Basin. Pages 475–491 *in* D. J. Stouder, P. A. Bisson, and R. J. Naiman, editors. Pacific salmon and their ecosystems: status and future options. Chapman and Hall, New York.

Binns, N. A., and F. M. Eiserman. 1979. Quantification of fluvial trout habitat in Wyoming. Transactions of the American Fisheries Society 108:215–228.

Bisson, P. A., J. L. Nielsen, R. A. Palmason, and L. E. Grove. 1982. A system of naming habitat types in small streams, with examples of habitat utilization by salmonids during low streamflow. Pages 62–73 *in* N. B. Armantrout, editor. Acquisition and utilization of aquatic habitat inventory information. American Fisheries Society, Western Division, Bethesda, Maryland.

Bisson, P. A., G. H. Reeves, R. E. Bilby, and R. J. Naiman. 1996. Watershed management and Pacific salmon: desired future conditions. Pages 447–474 *in* D. J. Stouder, P. A. Bisson, and R. J. Naiman, editors. Pacific salmon and their ecosystems: status and future options. Chapman and Hall, New York.

Bovee, K. D., editor. 1997. The compleat IFIM: a coursebook for IF 250. National Biological Service, Fort Collins, Colorado.

Bowlby, J. N., and J. C. Roff. 1986. Trophic structure in southern Ontario streams. Ecology 67:1670–1679.

Brookes, A. 1988. Channelized rivers: perspectives for environmental management. Wiley, New York.

Burt, D. W., and J. H. Mundie. 1986. Case histories of regulated stream flow and its effects on salmonid populations. Canadian Technical Report of Fisheries and Aquatic Sciences 1477.

Castleberry, D. T., and 11 coauthors. 1996. Uncertainty and instream flow standards. Fisheries 21(8):20–21.

Chamberlin, T. W., R. D. Harr, and F. H. Everest. 1991. Timber harvesting, silviculture, and watershed processes. Pages 181–205 *in* W. R. Meehan, editor. Influences of forest and rangeland management on salmonid fishes and their habitats. American Fisheries Society, Special Publication 19, Bethesda, Maryland.

Clay, C. H. 1995. Design of fishways and other fish facilities, 2nd edition. Lewis Publishers, Boca Raton, Florida.

Collier, M., R. H. Webb, and J. C. Schmidt. 1996. Dams and rivers: primer on the downstream effects of dams. U.S. Geological Survey Circular 1126, Reston, Virginia.

Colt, J., and R. J. White, editors. 1991. Fisheries bioengineering symposium. American Fisheries Society, Symposium 10, Bethesda, Maryland.

Cronin, J., and R. F. Kennedy, Jr. 1997. The riverkeepers: two activists fight to reclaim our environment as a basic human right. Scribner, New York.

Dillaha, T. A., J. H. Sherrard, and D. Lee. 1986. Long-term effectiveness and maintenance of vegetative filter strips. Virginia Polytechnic Institute and State University, VPI-VWRRC-BULL 153, Blacksburg.

Edwards, C. J., B. L. Griswold, R. A. Tubb, E. C. Weber, and L. C. Woods. 1984. Mitigating effects of artificial riffles and pools on the fauna of a channelized warmwater stream. North American Journal of Fisheries Management 4:194–203.

Fausch, K. D., C. L. Hawkes, and M. G. Parsons. 1988. A review of models that predict standing crop of stream fish from habitat variables: 1950–1985. U.S. Forest Service General Technical Report PNW-GTR-213.

Fausch, K. D., J. Lyons, J. R. Karr, and P. L. Angermeier. 1990. Fish communities as indicators of environmental degradation. Pages 123–144 *in* S. M. Adams, editor. Biological indicators of stress in fish. American Fisheries Society, Symposium 8, Bethesda, Maryland.

Fausch, K. D., and R. J. White. 1981. Competition between brook trout *(Salvelinus fontinalis)* and brown trout *(Salmo trutta)* for positions in a Michigan stream. Canadian Journal of Fisheries and Aquatic Sciences 38:1220–1227.

FISRWG (Federal Interagency Stream Restoration Working Group). 1998. Stream corridor restoration: principles, processes, and practices. National Technical Information Service, Springfield, Virginia (draft version at http://www.usda.gov/stream_restoration/).

Frissell, C. A., W. J. Liss, R. E. Gresswell, R. K. Nawa, and J. L. Ebersole. 1996. A resource in crisis: changing the measure of salmon management. Pages 411–444 *in* D. J. Stouder, P. A. Bisson, and R. J. Naiman, editors. Pacific salmon and their ecosystems: status and future options. Chapman and Hall, New York.

Gebhards, S. 1970. The vanishing stream. Idaho Wildlife Review 22(5): 3–8.

Gillilan, D. M., and T. C. Brown. 1997. Instream flow protection: seeking a balance in western water use. Island Press, Washington, D.C.

Gordon, N. D., T. A. McMahon, and B. L. Finlayson. 1992. Stream hydrology. An introduction for ecologists. Wiley, New York.

Gore, J. A., and G. E. Petts, editors. 1989. Alternatives in regulated river management. CRC Press, Boca Raton, Florida.

Gregory, S. V., F. J. Swanson, W. A. McKee, and K. W. Cummins. 1991. An ecosystem perspective of riparian zones. BioScience 41:540–551.

Hall, J. D., and C. O. Baker. 1982. Rehabilitating and enhancing stream habitat: 1. Review and evaluation. U.S. Forest Service General Technical Report PNW-138.

Hansen, E. A., G. R. Alexander, and W. H. Dunn. 1983. Sand sediment in a Michigan trout stream, part I. A technique for removing sand bedload from streams. North American Journal of Fisheries Management 3:355–364.

Heede, B. H., and J. N. Rinne. 1990. Hydrodynamic and fluvial morphologic processes: implications for fisheries management and research. North American Journal of Fisheries Management 10:249–268.

Henderson, J. E. 1986. Environmental designs for streambank protection projects. Water Resources Bulletin 22:549–558.

Hicks, B. J., J. D. Hall, P. A. Bisson, and J. R. Sedell. 1991. Responses of salmonids to habitat changes. Pages 483–518 *in* W. R. Meehan, editor. Influences of forest and rangeland management on salmonid fishes and their habitats. American Fisheries Society, Special Publication 19, Bethesda, Maryland.

Hill, M. T., W. S. Platts, and R. L. Beschta. 1991. Ecological and geomorphological concepts for instream and out-of-channel flow requirements. Rivers: Studies in the Science, Environmental Policy, and Law of Instream Flow 2:198–210.

Hunt, R. L. 1976. A long-term evaluation of trout habitat development and its relation to improving management-related research. Transactions of the American Fisheries Society 105:361–364.

Hunt, R. L. 1988. A compendium of 45 trout stream habitat development evaluations in Wisconsin during 1953–1985. Wisconsin Department of Natural Resources, Technical Bulletin 162, Madison.

Hunt, R. L. 1993. Trout stream therapy. The University of Wisconsin Press, Madison.

Hunter, C. J. 1991. Better trout habitat: a guide to stream restoration and management. Island Press, Washington, D.C.

Karr, J. R., K. D. Fausch, P. L. Angermeier, P. R. Yant, and I. J. Schlosser. 1986. Assessing biological integrity in running waters: a method and its rationale. Illinois Natural History Survey, Special Publication 5, Champaign.

Keller, E. A. 1978. Pools, riffles, and channelization. Environmental Geology and Water Sciences 2:119–127.

Leopold, L. B., and W. B. Langbein. 1966. River meanders. Scientific American 214(6): 60–70.

Leopold, L. B., M. G. Wolman, and J. P. Miller. 1964. Fluvial processes in fluvial geomorphology. Freeman, San Francisco.

McMahon, T. E., A. V. Zale, and D. J. Orth. 1996. Aquatic habitat measurements. Pages 83–120 *in* B. R. Murphy and D. W. Willis, editors. Fisheries techniques, 2nd edition. American Fisheries Society, Bethesda, Maryland.

Meehan, W. R., editor. 1991. Influences of forest and rangeland management on salmonid fishes and their habitats. American Fisheries Society Special Publication 19, Bethesda, Maryland.

Murdoch, T., M. Cheo, and K. O'Laughlin. 1996. The streamkeeper's field guide: watershed inventory and monitoring methods. The Adopt-A-Stream Foundation, Everett, Washington.

Nunnally, N. R. 1978. Stream renovation: an alternative to channelization. Environmental Management 2:403–411.

Orsborn, J. F. 1987. Fishway design practices. Pages 122–130 *in* M. J. Dadswell and five coeditors. Common strategies of anadromous and catadromous fishes. American Fisheries Society, Symposium 1, Bethesda, Maryland.

Orth, D. J. 1995. Food web influences on fish population responses to instream flow. Bulletin Francais de la Peche et de la Pisciculture 327/328/329:317–328.

Orth, D. J., and P. M. Leonard. 1990. Comparison of discharge and optimizing methods for recommending instream flows. Regulated Rivers: Research & Management 5:129–138.

Pearsons, T. N., H. W. Li, and G. A. Lamberti. 1992. Influence of habitat complexity on resistance to flooding and resilience of stream fish assemblages. Transactions of the American Fisheries Society 121:427–436.

Petts, G. E. 1984. Impounded rivers. Wiley, New York.

Platts, W. S., and 12 coauthors. 1987. Methods for evaluating riparian habitats with applications to management. U.S. Forest Service General Technical Report INT-221.

Poff, N. L. 1996. A hydrogeography of unregulated streams in the United States and an examination of scale-dependence in some hydrological descriptors. Freshwater Biology 36:71–91.

Poff, N. L., and seven coauthors. 1997. The natural flow regime. BioScience 47:769–784.

Reeves, G. H., and T. D. Roelofs. 1982. Rehabilitating and enhancing stream habitat: 2. Field application. U.S. Forest Service General Technical Report PNW-140.

Richter, B. D., J. V. Baumgartner, J. Powell, and D. P. Braun. 1996. A method for assessing hydrologic alteration within ecosystems. Conservation Biology 10:1163–1174.

Richter, B. D., J. V. Baumgartner, R. Wigington, and D. P. Braun. 1997. How much water does a river need? Freshwater Biology 37:231–249.

Riley, A. L. 1998. Restoring streams in cities: a guide for planners, policymakers, and citizens. Island Press, Washington D.C.

Roseboom, D. P., and B. White. 1990. The Court Creek restoration project. Pages 27–39 *in* Erosion control: technology in transition. Proceedings of conference XXI. International Erosion Control Association, Washington, D.C.

Rosgen, D. 1996. Applied river morphology. Wildland Hydrology, Pagosa Springs, Colorado.

Sedell, J. R., P. A. Bisson, F. J. Swanson, and S. V. Gregory. 1988. What we know about large trees that fall into streams and rivers. Pages 47–81 *in* C. Maser, R. F. Tarrant, J. M. Trappe, and J. F. Franklin, editors. From the forest to the sea: a story of fallen trees. U.S. Forest Service, General Technical Report PNW-GTR-229.

Sedell, J. R., G. H. Reeves, and P. A. Bisson. 1996. Habitat policy for salmon in the Pacific Northwest. Pages 375–387 *in* D. J. Stouder, P. A. Bisson, and R. J. Naiman, editors. Pacific salmon and their ecosystems: status and future options. Chapman and Hall, New York.

Shirvell, C. S., and R. G. Dungey. 1983. Microhabitats chosen by brown trout for feeding and spawning in rivers. Transactions of the American Fisheries Society 112:355–367.

Simon, T. P., editor. 1999. Assessing the sustainability and biological integrity of water resources using fish communities. CRC Press, Boca Raton, Florida.

Slaney, P. A., R. J. Finnigan, D. E. Marshall, J. H. Mundie, and G. D. Taylor. 1980. Stream enhancement guide. Canada Department of Fisheries and Oceans and British Columbia Ministry of Environment, Stream Enhancement Research Committee, Vancouver.

Stalnaker, C. B. 1990. Minimum flow is a myth. U.S. Fish and Wildlife Service Biological Report 90(5):31–33.

Stalnaker, C. B. 1995. Fish habitat evaluation models in environmental assessments. Pages 140–162 *in* S. G. Hilderbrand and J. B. Cannon, editors. Environmental analysis: the NEPA experience. CRC Press, Boca Raton, Florida.

Stalnaker, C. B., K. D. Bovee, and T. J. Waddle. 1996. Importance of the temporal aspects of habitat hydraulics to fish population studies. Regulated Rivers: Research & Management 12:145–153.

Stouder, D. J., P. A. Bisson, and R. J. Naiman, editors. 1996. Pacific salmon and their ecosystems: status and future options. Chapman and Hall, New York.

Tennant, D. L. 1976. Instream flow regimens for fish, wildlife, recreation and related environmental resources. Fisheries 1(4):6–10.

Turner, W. M. 1997. Achieving private sector involvement and its implications for resource professionals. Pages 158–176 *in* J. E. Williams, C. A. Wood, and M. P. Dombeck, editors. Watershed restoration: principles and practices. American Fisheries Society, Bethesda, Maryland.

USEPA (U.S. Environmental Protection Agency). 1997. Volunteer stream monitoring: a methods manual. USEPA 841-B-97–003.

Van Winkle, W., and eight coauthors. 1997. Uncertainty and instream flow standards: perspectives based on hydropower research and assessment. Fisheries 22(7):21–22.

Welsch, D. J. 1991. Riparian forest buffers: function and design for protection and enhancement of water resources. U.S. Forest Service Report NA-PR-07–91.

Wesche, T. A. 1985. Stream channel modifications and reclamation structures to enhance fish habitat. Pages 103–163 *in* J. A. Gore, editor. The restoration of rivers and streams. Butterworth, Stoneham, Massachusetts.

White, R. J. 1991a. Objectives should dictate methods in managing stream habitat for fish. Pages 44–52 *in* J. Colt and R. J. White, editors. Fisheries bioengineering symposium. American Fisheries Society, Symposium 10, Bethesda, Maryland.

White, R. J. 1991b. Resisted lateral scour in streams–its special importance to salmonid habitat and management. Pages 200–203 *in* J. Colt and R. J. White, editors. Fisheries bioengineering symposium. American Fisheries Society, Symposium 10, Bethesda, Maryland.

White, R. J. 1996. Growth and development of North American stream habitat management for fish. Canadian Journal of Fisheries and Aquatic Sciences 53(Supplement 1):342–363.

White, R. J., and O. M. Brynildson. 1967. Guidelines for management of trout stream habitat in Wisconsin. Wisconsin Department of Natural Resources Technical Bulletin 9.

Williams, J. G. 1996. Lost in space: minimum confidence intervals for idealized PHABSIM studies. Transactions of the American Fisheries Society 125:458–465.

Williams, J. E., C. A. Wood, and M. P. Dombeck, editors. 1997. Watershed restoration: principles and practices. American Fisheries Society, Bethesda, Maryland.

Chapter 11

Lake and Reservoir Habitat Management

ROBERT C. SUMMERFELT

11.1 INTRODUCTION

Management of a fishery is everything that is done to sustain or enhance a fishery, including regulation of harvest, stocking, and habitat protection and enhancement. The manager manages by attempting to influence some or all parts of the fishery to achieve a desired outcome. So, what is a fishery? A fishery is a system of interacting, interdependent, and interrelated components; these components are the fishes, the space in which fish live (i.e., the environment or habitat), other biota, and the anglers and other people who have influence on or are affected by the use of the fishery (Nielsen and Lackey 1980). Obviously, an ecosystem perspective requires inclusion of much more than the fish; the multidimensional space of water that fills the basin, that is, the lake, the tributary streams, and the physicochemical (abiotic) and biotic components of the lake and its basin, the watershed, and the manifold aspects of the human dimension are all part of the ecosystem (Miranda and DeVries 1996).

Acceptance of a broader ecosystem management perspective (Chapter 5) forces the fishery manager to "get out of the water" and into the watershed (Chapter 9) to protect or enhance habitat and also to communicate with anglers and others with vested interest in the fishery (Chapter 3). This chapter attempts to present an ecosystem perspective of habitat management for standing bodies of water (lacustrine habitats) that are natural or artificial in origin and larger than ponds (Chapter 21) but smaller than the Great Lakes (Chapter 23).

The manager taking an ecosystem approach attempts to manage (i.e., to manipulate, control, or direct) or restore (i.e., to bring back to some previous condition) some aspect of the ecosystem by strategies that will preserve ecosystem integrity and produce a specific outcome of benefit to the fishery. Given the complexity of large lake ecosystems, as well as the cost of carrying out meaningful changes that have substantive effects on the fishery, ecosystem management of lake and reservoir habitat is difficult, and unpredictable responses are commonplace. Also, because a fishery includes people—the anglers who use the fishery resource as well as other people who may be affected by or who can influence a management decision (e.g., water level manipulation)—their preferences and needs must be considered in management of habitat.

A new paradigm for habitat manipulation of large aquatic ecosystems is the use of strategies and tactics of adaptive environmental assessment and management (Halbert 1993). Adaptive management means carrying out management as a field experiment, with testable hypotheses that are evaluated in terms of effectiveness in achieving goals and objectives that are compatible with ecological, economic, and social attributes of the system. Adaptive management provides a link between fisheries science and management, whereby results of well-designed management experiments provide feedback to adjust management strategies if necessary and contribute to the body of fisheries science knowledge.

11.1.1 Lakes and Reservoirs

Traditionally, large artificial lakes (impoundments or reservoirs) have been considered bodies of water with a surface area of 200 ha or more. Natural or artificial bodies of water that range from 10 to 200 ha are considered intermediate, and ponds are less than 10 ha. Cooke et al. (1993) called natural lakes, lakes, and artificial lakes, reservoirs. In Wisconsin and some other places, the term flowage (the state of being flooded) is used for both an artificial and enlarged natural lakes. Some natural lakes are classified as reservoirs after their water surface elevation is artificially raised and the original area or volume more than doubled (i.e., flowage). In this chapter, lake refers to a body of water of natural origin and reservoir refers to an artificial body of water 200 ha or more.

The majority of lakes in North America are of glacial origin (8,000–10,000 years old). They are important fish habitats and conspicuous parts of the landscape of Canada and the northern United States from Minnesota to New England, which are areas overrun by the last continental glaciation (Frey 1963). The origin of lakes, however, is diverse. Hutchinson's (1957) monograph is the classic source for descriptions of the origins of natural lakes, but the information is summarized to a greater or lesser extent in most textbooks of limnology (e.g., Wetzel 1983).

Large, main-stem reservoirs drain basins with diverse land uses. The quantity and quality of water can change greatly during the year. Reservoir water levels can fluctuate widely as reservoirs are filled and drawn down for flood control, irrigation, or power production. Because of their variable ecological conditions, reservoirs pose challenging management problems. For the fishery manager and the angler, both of whom should have keen interest in fish habitat, it is important to note that there are major differences in habitat features between lakes and reservoirs, and both are affected by climate and human use.

In addition to being used for angling, reservoirs are used for hydropower generation, flood control, navigation, irrigation, power plant cooling, municipal and industrial water supply, and streamflow augmentation. There are few large, single-purpose reservoirs, and most reservoirs are storage elements of a local or regional water resource system. Multiple use often results in competitive and conflicting demands on the reservoir management plan, which places constraints on fisheries management options. The operation plan, which is a statement of the principal social aim that dominates reservoir function, involves basinwide considerations (Peters 1986). Socioeconomic issues and human dimensions of reservoir management are facets of the multidimensional aspects of reservoir management.

Thornton et al. (1990) and Cooke et al. (1993) itemized detailed differences between natural and artificial lakes. They differ in shape, age, hydraulic residence time, depth of the outlet, location of deep water, water level fluctuations, drainage area, watershed area to lake surface area ratio, and nutrient (phosphorus and nitrogen) load-ing. For example, the average lake has a watershed area to lake surface area ratio of 33, compared with a ratio of 93 for reservoirs; phosphorus loading in reservoirs can be 195% greater than in natural lakes (Cooke et al. 1993). As a consequence of just these two differences, the aging process of reservoirs is faster than it is in natural lakes because of more rapid basin filling (Kimmel and Groeger 1986).

Pump storage and cooling impoundments are special categories of reservoirs. Pump storage reservoirs are used for hydropower generation; water is pumped to a higher-level impoundment during periods of low power demand so that the water can drive turbines by gravity during periods of peak demand. Cooling impoundments are constructed primarily for the dissipation of waste heat from power plants (Olmsted and Clugston 1986).

Flooding riverine habitats to make reservoirs has formed aquatic habitat in many localities where few or no lakes previously existed, and these lakes have produced considerable environmental alteration. Large reservoirs have inundated substantial portions of major rivers in North America, thereby limiting the stretches of free-flowing rivers. In the Midwest, reservoirs have eliminated natural spawning sites for paddle-fish (Pfleiger 1975); the many dams and diversions of the Columbia River have ad-versely affected stocks of anadromous salmon (Collins 1976).

The upper Mississippi River, the Ohio River, and many other modified riverine systems have run-of-the-river navigation pools that are formed by locks and dams. The navigation pools are not classified as reservoirs because they have storage ratios (the ratio of the reservoir water volume at the listed surface area to the annual discharge volume) of less than 0.01, which means that the annual flow rate is 100 times greater than the average annual pool volume (Jenkins and Morais 1971). The short turnover time (also called residence time) of these environments makes them more riverine than lacustrine. In comparison, Dale Hollow Lake, Tennessee, has a storage ratio of 1.16; Fort Peck, Montana, 2.36; and Grandby Reservoir, Colorado, 3.00 (Jenkins and Morais 1971).

Reservoirs have a tailwater, which is the river area immediately downstream from the dam that is strongly influenced by the fluctuations in reservoir discharge. Fish abundance in reservoir tailwaters is often substantial (Moser and Hicks 1970), and angler harvest in the tailwater may rival that in the reservoir. Obviously, tailwater fish-eries are affected by temperature, dissolved oxygen, and gas pressure in the discharge, the quantity and timing of reservoir releases (Peters 1986), and the quality of angler access. Tailwater fishes originate from both the river and the reservoir. The dam blocks upstream movement of fishes, thereby concentrating them in the tailwater, and the tailwater also provides an abundance of food for predaceous fishes when the current and water temperature are favorable for feeding and spawning activity. Although fish-ing may be better in the tailwater than in the reservoir, fluctuations in abundance of fish stocks is similar to the fluctuations in the reservoir (Walburg 1971, Walburg et al. 1971). For the most part, management of tailwater fish habitat must be directed at modifications of the quality and quantities of reservoir releases. This includes using regulated penstock discharges to control water temperature and oxygen content and making structural modifications to avoid deep plunge pools that cause gas supersatura-tion in fishes. The penstock is the vertical structure in a reservoir that is used to direct water flow to a generator or that functions in downstream discharge.

11.1.2 Habitat Classification

Lakes and reservoirs are ecosystems in which an assemblage of species (community) lives, and it is also a place for a fishery. A fishery may be named for the habitat, or a fishery may be named for a species or species complex, such as a crappie fishery or a largemouth bass–bluegill fishery. Traditionally, management has been directed toward single-species populations (e.g., largemouth bass), but an ecosystem approach requires consideration of the influence of management efforts on community structure and ecosystem function. Habitat classification can contribute to understanding both structure and function of lake ecosystems. Lake and reservoir classification is typically based on abiotic characteristics of the water (e.g., total dissolved solids), lake basin morphometry (e.g., mean depth), nutrient loading, and water residence time. From a landscape perspective, latitude and climate have dominant effects on lake biota.

Schemes for lake habitat classification are plentiful (Busch and Sly 1992). Leach and Herron (1992) reviewed 32 categories of classification systems or indices (schemes and typologies), and grouped them into five major categories: (1) origin, shape, and location; (2) physical properties (optical and thermal mixing); (3) chemical properties (water quality); (4) trophic status; and (5) assemblages of fish species and fish habitat. Hutchinson (1957) distinguished 76 lake types grouped under 11 processes or events responsible for the origin of the lake basin. The most familiar physical classification schemes are based on the nature of thermal stratification and frequency of mixing; there are five to eight classes based on the frequency of mixing and thermal conditions (Leach and Herron 1992). In north temperate lakes and reservoirs the dimictic situation is most common, whereby a lake or reservoir is stratified in the summer and undergoes mixing in the spring and fall.

Because of widespread concern over cultural eutrophication (nutrient enrichment from human sources), indices of trophic status are abundant. These indices include the familiar oligotrophic–mesotrophic–eutrophic classification, which has many definitions based on hypolimnetic oxygen deficit, primary production, total phosphorus, total nitrogen, chlorophyll a, Secchi disk transparency, and organic matter content of lake sediments. There are several multiple-parameter indices of lake productivity, such as the morphoedaphic index (MEI), trophic state index, and others (Leach and Herron 1992). Some of these indices are attempts to predict fish standing stock or yield (Leach et al. 1987). For example, Jenkins and Morais (1971) developed multiregression models to predict sport fish harvest from environmental variables, and Ryder (1965) developed the MEI in an attempt to predict fish yield in north-temperate lakes based on the ratio between total dissolved solids and mean depth. Applications of the MEI include prediction of (1) angling yield, (2) commercial fish yield from a new reservoir, and (3) responses to cultural eutrophication (Ryder et al. 1974). An analysis by Downing et al. (1990) of fish production in lakes covering a wide range of geographic areas and trophic status indicated that fish production was not correlated with MEI. However, fish production was correlated with annual phytoplankton production and mean total phosphorus concentration.

Lake classification schemes based on fish species and fish habitat have been developed for Ontario, Wisconsin, Michigan, and other regions (Leach and Herron 1992). Lake size and depth, as well as many physical and chemical habitat characteristics, have been included in such classification schemes. Surface area and either maximum or mean depth have been the most significant habitat characteristics determining fish species associations, but other important variables have been Secchi disk transparency, total dissolved solids, and pH in acid rain-stressed lakes. Schneider (1981) found that the better fishing lakes in Michigan were the deeper lakes that had high transparency and thermal stratification in the summer with a layer of well-oxygenated water but did not have excessive density of vascular aquatic plants.

Comprehensive lake classification schemes provide general perspective on lake types and community approaches to management, as well as providing indices of fish production and yield (Leach and Herron 1992). These schemes, however, contribute little to identification of critical habitat needs and development of habitat management strategies within a lake for a single species or a community; for example, spawning or nursery habitat requirements for specific life stages of largemouth bass. Rather than needing whole lake models, managers need an understanding of relationships between specific habitat features and needs of single species and fish communities. Sly and Busch (1992) cited Aggus and Bivin (1982) as an example of habitat suitability models for warm- and coolwater fish communities and McMahon et al. (1984) as an example of a single-species model.

11.1.3 Lake and Reservoir Surveys

A substantial part of a fishery manager's activities can be characterized as surveying and monitoring. Aquatic habitat surveys require careful consideration of variable selection and study design, capture methods for fishes and other biota, and care and handling of sampled organisms (Murphy and Willis 1996). One of the major reasons for a survey is to check on the health and well-being of the lake or reservoir and to identify problems (symptoms) and their causes. A problem has been defined as a limitation on a desired use of the lake (Olem and Flock 1990). The manager's basic tool for identifying problems is to survey (1) the fish community, (2) the environment (physicochemical and biological components), and (3) the people (creel) who use the resource. A survey can be regarded as a formal inspection based on a predefined plan that uses standardized sampling methods with statistically acceptable rationale. A fishery survey is an essential first step in acquiring the information needed to understand the cause of a problem, to provide options to resolve the problem, and, if appropriate, to develop a lake management plan. Such problems are summarized in Box. 11.1.

Clearly, a precise statement of the problem is needed to focus on causes and remedial action. Corrective action, as with other aspects of fisheries management, is usually complex and must be undertaken with an understanding of each lake's environment (Weithman and Haas 1982).

Box 11.1   Types of Problems Requiring Survey Information

Fish	Environment	People
missing year-class	acidification	angler dissatisfaction
fish kills	hypolimnetic anoxia	reduced angling effort
poor size structure	eutrophication	poor catch quality
poor growth	problem animals (e.g., carp)	user conflicts
poor condition index	excessive plant growth	overcrowding of anglers

11.2   MANAGEMENT OF ENVIRONMENTAL QUALITY

Habitat loss and degradation is a threat to the integrity of aquatic environments and the sustainability of fisheries resources. In this section, several categories of threats to habitat quantity and quality are considered.

11.2.1 Watershed Considerations

Obviously, land–water interactions control and modify the physicochemical environment of lakes and reservoirs. Many in-lake problems are the result of the transport of eroded soil, nutrients, and contaminants from the watershed to the lake. Thus, it is folly to consider lake habitat management without appreciating the linkage between in-lake environmental problems of turbidity, decreasing depth, and algal blooms and point and nonpoint source pollution. In the long term, holistic approaches (i.e., watershed scale) are needed to maintain or restore biotic diversity and community metabolism. Soil erosion and nutrients must be controlled at their sources because a quality fishery cannot be developed or sustained when there are chronic pollution problems and periodic fish kills.

Nonpoint source pollution, in contrast to a point source such as the effluent pipe of a paper mill, originates from diffuse sites throughout a drainage basin (i.e., watershed or catchment). The major kinds of nonpoint source pollutants are soil, nutrients, particulate and dissolved organic matter, and pesticides. Sediment originates from croplands, pastures, forests, and construction sites and is carried by runoff; organic matter and inorganic nutrients (nitrogen and phosphorus) originate from barnyards, feedlots, manured fields, and row crops; and pesticides originate from orchards, croplands, and forests. Some pesticides and nutrients are water soluble, and others are transported from field to stream attached to soil; for example, phosphorus is attached to soil, but various forms of nitrogen are water soluble. Development of effective watershed management strategies requires an understanding of hydrological processes and land use (Brooks et al. 1997; Chapter 9).

Cultural eutrophication and sedimentation from human activities have severely deteriorated fisheries and limited the recreational opportunities of lakes and older impoundments (Likens 1972; Born et al. 1973; Henderson-Sellers and Markland 1987). Although it is difficult to reduce inputs of nutrients and sediments from diffuse agricultural and residential lands in large watersheds, interagency efforts through the U.S. Environmental Protection Agency (EPA) combined with effective public hearings and other public relations efforts have demonstrated watershed improvements can protect and restore lakes, at least in modest-size basins (Born et al. 1973). Basically, sediment and excessive nutrients have been kept out of lakes by best management practices on agricultural and forest lands, and by effective control of municipal and industrial wastes. If nutrient inputs cannot be reduced, the only course of action is to manage around the problem with in-lake restoration and management strategies (Olem and Flock 1990; section 11.2.2).

Generally, enforcement of pollution laws and even the basic monitoring of pollutant levels in the environment and in fishes are not the legal responsibility of most fisheries agencies. The fisheries biologist, however, plays a key role in directing violations to the attention of appropriate agency officials. Close cooperation between fisheries and environmental (pollution) control agencies is essential for environmental protection.

In 1996 the EPA initiated the National Watershed Assessment Program "to collect, organize, and evaluate multiple sources of environmental information at the watershed level" (Lehman 1997). To date, two phases of the plan have been initiated. In phase one, the EPA has been aggregating diverse water quality data on 2,150 watersheds in the United States. In phase two, the database and site-specific monitoring studies will be used to decide on sites for pollution prevention and remediation (Lehman 1997).

11.2.1.1 Sediment

Sediment deposition has degraded and eliminated valuable fish habitat in rivers and streams (Waters 1995). Small reservoirs can be easily overwhelmed by sediment deposition, and it is not surprising that the filling of large, run-of-the-river reservoirs with sediment is the dominant aging process (Kimmel and Groeger 1986). The deposition of solids decreases lake and reservoir storage capacity, smothers fish spawning sites (e.g., lake trout and walleye), reduces the diversity and abundance of many kinds of aquatic life, and encourages the invasion of macrophytes. In northern lakes and reservoirs, reduction in water depth increases the likelihood of winterkill (see section 11.2.2.8).

Sediment originates from erosion processes within the lake's drainage area, including the river channels, and along the shoreline. In terms of absolute volume, the watershed is the major contributor of sediment. Substantial basin filling and reduced water depth changes the entire ecology of lentic systems (Brugam 1978; Thornton et al. 1990). Throughout the Corn Belt and in other areas with watersheds dominated by row crops, the greatest threat to water quality and lake aging is soil erosion. An acceptable annual soil loss is 11.2 metric tons per hectare; in the worst cases, annual soil losses may be six times greater. Lake volume may be totally lost to siltation (with sediment at the lip of the spillway), as has happened at the reservoir Lake Ballenger, Texas, and Mono Reservoir, California (Owen 1980).

In addition to reducing the water storage capacity of reservoirs, suspended solids contribute adsorbed nutrients and turbidity. Turbidity reduces the depth to which light can penetrate, thereby limiting primary production. Reservoirs located where colloidal clays are abundant may have persistent muddy water conditions long after a given runoff event. Suspended solids, except for colloidal clay, settle when they reach the reservoir. The larger and denser sand particles settle first, forming deltas at the headwaters of reservoirs. The lighter silt and clays are dispersed throughout the reservoir. The distribution of sediment forms a horizontal gradient in reservoirs that strongly influences the spatial distribution of benthic invertebrates, fishes that prey on them, and some piscivorous fishes, such as flathead catfish (Summerfelt 1971).

Agricultural best management practices to reduce the loads of sediment and nutrients entering lakes and streams include crop rotation, vegetative cover, minimum tillage, nutrient management, riparian buffer strips, and use of structural devices such as grassed waterways (filter strips), terraces, sediment retention basins, and erosion control weirs (Johengen et al. 1989). Since the 1930s, many programs of the U.S. Department of Agriculture have provided guidance and cost-share assistance (economic incentives) for terrace and pond construction and other practices to reduce soil erosion.

Sediment dikes are low-head, rock-covered dams located at the upper ends of major arms of modest-size reservoirs. As inflow reaches the sediment dike, the current velocity of the sediment-ladened inflowing water is slowed enough to deposit much of the sediment above the sediment dike (McGhee 1990).

If a lake basin has been made too shallow from sedimentation, hydraulic dredging seems to be the only method to remove the materials. Hydraulic dredging is very expensive, and it is often difficult to find sites for dredge disposal (Cooke and Olem 1990). In small lakes and reservoirs, overwinter drawdown may consolidate a flocculent sediment and slightly deepen the lake (Beard 1973).

11.2.1.2 Organic Matter

Inputs of organic matter into lakes can be harmful or beneficial, depending on the relative amounts and the inherent fertility of the lake. Detritus, the particulate, unrecognizable organic matter of plant and animal origin, is the basis for food chains of turbid reservoirs because primary productivity is often light limited rather than nutrient limited (Marzolf and Osborne 1971). Because the major source of organic matter for fish production in many turbid main-stem impoundments is derived from the watershed (allochthonous) and not from primary productivity (autochthonous), the microbial–detritus food web is an important pathway influencing the dominance of bottom-feeding fishes in reservoirs that have a high water exchange rate (Jenkins 1974).

The biochemical oxygen demand (BOD) for decomposition of organic matter may remove oxygen faster than it can be produced by algae or entrained at the surface. High BOD levels result in hypolimnetic oxygen depletion and increased incidence of fish kills. Small- and intermediate-size lakes and reservoirs can be quickly overwhelmed by excessive inputs of organic matter, but even large lakes such as Lake Erie have been adversely affected by chronic nutrient-induced dissolved oxygen depletion of the hypolimnion (Chapter 23).

11.2.1.3 Nutrient Enrichment

Phosphorus additions from external sources and regeneration of phosphorus from lake sediments are the major forces driving primary productivity of lakes and reservoirs. The trophic status of lakes, commonly designated oligotrophic, mesotrophic and eutrophic, may be based on one or more interrelated variables: total phosphorus, total nitrogen, chlorophyll a, transparency, or the rate of primary production (Leach and Herron 1992). Heiskary (1985) classified 65% of Minnesota's 12,034 lakes as eutrophic or hypereutrophic based on total phosphorus concentrations. Bachmann (1980) regarded lakes with greater than 23 mg/m^3 total phosphorus as eutrophic, but there is not agreement on a single concentration value for defining trophic state.

Nutrient enrichment from human sources (cultural eutrophication) is the major cause of eutrophication in North America (Edmondson 1969). Cultural eutrophication is the major and most persistent water quality problem throughout the United States. Phosphorus inputs usually come from agricultural sources and municipal effluents. The consequence of eutrophication is manifested by frequent occurrence of algal blooms, decreased water transparency, scums and mats of blue-green algae, and dense littoral zone beds of submergent and emergent vegetation.

Dense blooms of algae are deleterious to almost all uses of water, making a lake less desirable as a place for fishing, swimming, and boating as well as causing taste and odor problems in drinking water that are costly to treat. Algae may shade out, or limit, light for macrophytes. In moderate abundance, macrophytes are desirable; they provide structure for the epiphytic macroinvertebrate (phytomacrofauna) prey of fishes. Excessive macrophyte abundance can inhibit fishing and boating and adversely affect predator–prey balance by affording too much cover for the prey. Rooted macrophytes may pump nutrients from the sediment into their plant tissue, which puts the nutrients back into circulation when the plants decompose. The survival of the entire fish community may be jeopardized when a dense plant growth decomposes, resulting in oxygen depletion.

Aside from situations in which light necessary for algal growth is limited by the presence of inorganic materials, particularly suspended colloidal clay, transparency varies in relationship to the mass of algae, which, in turn, varies in proportion to the amount of phosphorus in the water (Edmondson 1969; Jones and Bachmann 1974). Quantification of nutrient loading rates can be used to predict lake trophic state (Carlson 1977). A trophic status index, which refers to the nutritional status of a lake, can be used as a relative assessment of the degree to which nutrients and algal biomass are present in a lake (Carlson 1977). Carlson proposed indices based on Secchi disk depth, chlorophyll *a*, and total phosphorus.

Reducing phosphorus inputs at the source is the obvious solution to cultural eutrophication, but when this is not practical, the manager has several options for in-lake treatment: dilution and flushing, phosphorus precipitation and inactivation, sediment removal (hydraulic dredging), water level drawdown, whole lake aeration (destratification), hypolimnetic aeration, and mechanical or biological harvesting of macrophytes (Dunst et al. 1974; Cooke et al. 1986; Henderson-Sellers and Markland 1987; Cooke and Olem 1990). Some of these options are discussed in section 11.2.2.

11.2.1.4 Acidification

Acidification of aquatic ecosystems is one of the most serious types of environmental pollution. The acidification of surface waters is evident in North America (Figure 11.1) and Europe (van Breeman et al. 1984). Acidification of aquatic ecosystems requires the combination of two factors: (1) poor buffering capacity (i.e., low acid-neutralizing capacity) and (2) acid deposition (i.e., low average pH of precipitation). Acid deposition includes various forms of precipitation containing sulfuric acid (H_2SO_4) and nitric acid (HNO_3) and deposition of dry particles containing sulfate and nitrate salts. Thus, sulfuric and nitric acids are secondary pollutants from acid precursors (SO_2 and NO_x), the primary pollutants released from tall smokestacks of coal-burning power plants, smelters, and steel mills and emitted from trucks and automobiles. The sulfur in the emission originates from the sulfur in the coal; the nitrogen, however, originates from the air; and the NO_x is formed from combustion at high temperatures.

Tall stacks, which disperse a plume of the acid precursors, "air mail" the pollution over large geographical areas. The dispersion effectively converts a point source pollutant into a nonpoint source. In the United States, the 1990 Clean Air Act mandated substantial reductions in sulfur dioxide emission. Although reductions are being achieved, lake acidification has not been reversed because of an increased supply of nitric acid originating from all forms of combustion but particularly from automobiles and trucks (Simonin 1998).

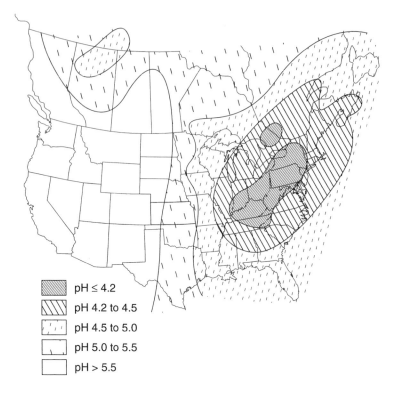

Figure 11.1   Acidity (pH) of precipitation across Canada and the United States.

Legend:
- pH ≤ 4.2
- pH 4.2 to 4.5
- pH 4.5 to 5.0
- pH 5.0 to 5.5
- pH > 5.5

Acid precipitation is a major threat to fisheries and all other forms of aquatic life in acid-sensitive lakes in the northeastern United States and eastern Canada (Beamish 1976; Schofield 1976; Johnson 1982). Effects of acidification on fishes follow from direct effects of low pH and solubilization of an aluminum ion that is toxic to fish embryos and larvae (Baker 1982). The pH level at which there is no effect on reproduction in fishes is around 6.5. Low pH upsets calcium metabolism of fishes and protein deposition in developing eggs. Acid stress may also coagulate gill mucous, damage gill epithelia, and upset electrolyte homeostasis (Fromm 1980). Acidification is especially serious for glacial lakes of Scandinavia, nearly all of Canada east of Lake Winnipeg, the Adirondack, and Appalachian mountains in the eastern United States, the upper midwestern United States, and high-elevation lakes of mountain regions of western United States.

Lakes acidified by acidic deposition may be limed to counteract acidification, as has been done on about 5,000 lakes in Sweden since 1997 (Olem and Flock 1990). Lime ($CaCO_3$) increases the acid-neutralizing capacity (equation 11.1) by increasing the concentration of bicarbonate ion (HCO_3^-):

$$CaCO_3 + H_2O + CO_2 <\!\!-\!\!> 2HCO_3^- + Ca^+. \qquad\qquad (11.1)$$

Bicarbonate reacts with acids (H^+ ions) to form CO_2 and H_2O, thereby preventing a major change in pH (equation 11.2) until the bicarbonate alkalinity (acid-neutralizing capacity) is exhausted:

$$2HCO_3^- + 2H^+ <—> 2H_2O + 2CO_2. \qquad (11.2)$$

It is not economically feasible, however, to lime lakes that have less than 6 months water retention time (Olem and Flock 1990), and it is certainly not economically feasible to carry out aerial application of lime for many lakes in remote, roadless areas. In any case, without control over acid deposition, liming must be repeated to prevent reacidification (Driscoll et al. 1987).

11.2.1.5 Contaminants

Environmental contaminants are a major problem in many aquatic environments (Cairns et al. 1984). Contaminants have both direct and indirect effects on aquatic life (Cairns et al. 1984; Nriagu and Simmons 1984); they may kill selectively or they may kill all of the biota. Contaminants emanating from point sources (e.g., mines and petrochemical, municipal and industrial sites) may be chemical or physical in nature. Physical pollutants include solids (e.g., mine tailings and fiber from paper mills) and waste heat (thermal effluents). Chemical contaminants include a large variety of toxic substances, including substances that have been associated with elevated tumor incidence in fishes (Black 1983). Some substances from paper mills and other sources produce taste and odor problems that render the fish unsuitable for human consumption. Insecticides are often toxic to fish in concentrations of micrograms per liter. Moreover, accumulation of contaminants in fish flesh may require restrictions on marketing the fish, hazard warnings to the public (fish consumption advisories) about eating the fish, and even closure of the fishery. However, fish kills are caused by a diversity of factors, not just toxic substances (Meyer and Barclay 1990).

11.2.1.6 Heated Effluents

Thermal pollution is a physical form of point source pollution that occurs when lakes and rivers are used as a water source for cooling condensers of power plants. Heated water effluents of power plants may affect fish spawning and growth and cause gas bubble disease. Sudden shutdowns may result in fish kills from temperature shock (Olmsted and Clugston 1986). Additionally, pumping large volumes of water for condenser cooling causes fish entrainment and impingement problems at water intake areas (Figure 11.2). Eggs, larval and fingerling fishes, and invertebrates that are pulled by the current through the mesh of barrier screens are said to be entrained; larger fishes that are held against the barrier screens are said to be impinged (Fletcher 1990). In the United States, once-through cooling is used by 60% of the fossil fuel power plants and 74% of the nuclear-fueled power plants to condense exhaust steam from the turbines (Schubel and Marcy 1978). An entrained organism going through a once-through cooling system will experience a variety of lethal physical stressors including shear force, impact, abrasion, and a temperature rise of 11°C when passing through the condensers.

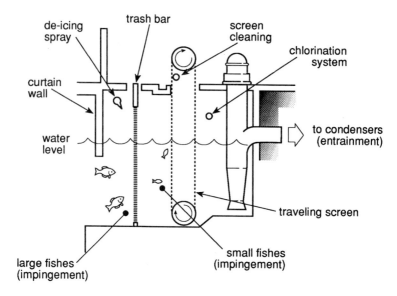

Figure 11.2   Impingement and entrainment problems at a steam-generating power plant. Large fishes are killed by impingement on the trash bars or the traveling screen (2 cm mesh). Larval and juvenile fishes, zooplankton, and other organisms small enough to pass through the trash bar and traveling screen are entrained with the flow and killed by high temperatures, pressure changes, and mechanical damage from passage through the condensers and plumbing of the power plant. Figure is redrawn from Zweiacker and Bowles (1976).

11.2.2 Lake Management

Habitat management considerations include lake treatments to protect and en-hance the environment for the benefit of the aquatic community and the quality of the angling experience.

11.2.2.1 In-Lake Erosion Control

Waves generated by a long fetch (distance over the water surface that wind will blow) or the wake of power boats may cause damage to earthen dams and exposed shorelines. Suspended solids and sediment derived from shoreline erosion may de-grade spawning habitat, increase lake turbidity, and have other effects similar to nonpoint source sediment inputs. Barren, windswept shorelines are poor food producers, unsuit-able habitat for spawning of nest-building fishes, and poor nursery habitat for young of both game and nongame fishes (Summerfelt 1975).

Management options to protect shorelines from erosion include (1) grading steep slopes and terracing above the water line to prevent erosion and slumping; (2) con-structing a retaining wall (bulkhead) along the waterfront or riprapping the land–water interface with rocks, gabions, or beach stabilization mats; (3) reducing the energy of incoming waves by use of breakwaters (armored earthen structures or floating-tire structures); and (4) instituting and enforcing wake limits for power boats in nearshore areas with heavy boat traffic.

Figure 11.3 A shoreline that has been riprapped (armored) with crushed rock to reduce in-lake source of sediment derived from shoreline erosion. (Photograph is courtesy of Mike McGhee, Iowa Department of Natural Resources.)

Steep slopes may need to be reshaped by terracing above the water line and riprapping the foot of the slope. Given the lay of the lake in relation to prevailing winds, selected segments of shoreline can be protected by wave dissipation revetments, erosion control dikes, and riprapping (Figure 11.3).

Riprapping the face of earthen dams is usually done at the time of lake construction to protect the dam, and that cost is born by the agency responsible for reservoir construction (e.g., U.S. Army Corps of Engineers). Riprap on the face of the dam has been found to be a valuable spawning site for walleye (Grinstead 1971) and a probable spawning site for cavity-spawning fishes such as channel catfish and flathead catfish. Riprapping the shoreline is also called armoring of the shore. The most sensitive shoreline sites are those exposed to a long fetch or wakes produced by boat traffic. When extensive, armoring of the shoreline is controversial because of its high cost as well as for aesthetic and ecological considerations. The expense of rock riprap will vary depending on the area to be covered, trucking distance to a source of rock, and access to the lake shore. To reduce costs, structures have been made of used automobile tires (erosion control mats) (Candle 1985), but many people regard tire structures as aesthetically unacceptable.

A naturally vegetated littoral zone dampens the erosive effects of wave action against the shoreline. Thus, a stable and well-vegetated land–water interface is considered by many to be a biological means to protect and stabilize shorelines. Vegetation provides many benefits other than erosion control. Macrophytes or flooded terrestrial vegetation provide habitat for vegetation spawners (e.g., northern pike and yellow perch) and nursery areas for young-of-the-year fish, such as largemouth bass (see section 11.2.2.5).

At average pool level, macrophytes are generally absent from the littoral zone of flood control and irrigation reservoirs because of rising and falling water levels. The nearshore portion of the littoral zone may be dewatered on an annual cycle or for several years during a drought; this area is often called the fluctuating zone of the reservoir. The annual cycles of dewatering eliminates aquatic macrophytes, and episodic, high water levels reduce encroachment of grasses or mesophytic plants that cannot withstand prolonged inundation. In flood control reservoirs, plants above the normal pool level are sometimes inundated for several weeks as the reservoir meets its obligation for storage of floodwaters. Thus, the margin of a fluctuating reservoir is an unstable ecotone and often lacking in permanent colonization of aquatic, wetland, or terrestrial plants.

Effective management of the littoral zone of reservoirs is a major environmental challenge and a research topic that has not received the attention it deserves. Among the prairie grasses of the Great Plains, reed canarygrass and western wheatgrass tolerate long periods of inundation (McKenzie 1951). Gamble and Rhoades (1964) reported that duration and depth of submergence, season, and frequency of flooding are important in determining survival of prairie grasses from flooding; Bermuda grass, buffalo grass and knotgrass tolerated up to 20 d of submergence, whereas weeping lovegrass and most bluestem species survived up to 10 d of flooding. Young (1973) reports that weeping lovegrass and Sericea lepedeza (*Lespedeza cuneata*) are usually beneficial in producing a quick cover on bare soil surfaces after construction of a dam or following drawdown. Maiden cane, a native perennial grass occurring in freshwater marshes and wetlands, has shown promise for reducing wave action damage at the waterline because it can grow under a variety of site and soil conditions over the widest geographical area of the United States (Young 1973). Fraisse et al. (1997) examined seed germination in a greenhouse to evaluate the effects of drought and immersion on plant growth to determine the suitability of plant species for vegetation of reservoir margins.

11.2.2.2 Fertilization

Lake fertility or lake trophic state is indicated by one or a combination of parameters including nutrient concentration (total phosphorus), chlorophyll *a* (a surrogate measurement of algal biomass), and measurements of primary productivity. Although there is a positive relationship between sport fish harvest and concentration of chlorophyll *a* in lakes and reservoirs (Jones and Hoyer 1982), it is usually undesirable and impractical to consider fertilization of a large lake or reservoir (>200 ha). A few experimental attempts have been made at fertilizing coves of large reservoirs, the purpose being to enhance primary productivity and zooplankton populations to benefit survival of larval or fingerling fishes. Generally, rapid exchange rates between water in the coves and the main reservoir will dilute the application. Even for oligotrophic reservoirs, it is unrealistic to obtain a favorable cost-benefit ratio for fertilization of a large reservoir on a sustained basis given the rapid flushing rate.

Nutrient enrichment increases the hypolimnetic oxygen deficit, encourages algae blooms and macrophyte problems, and increases the incidence of summer and winter fish kills. Shortly after Swingle's early studies on pond fertilization in Alabama (Swingle and Smith 1947), problems related to fertilization quickly surfaced (Hasler 1947; Ball

1952). It is now obvious that enhancement of biological production of lakes by application of inorganic fertilizers (phosphorus compounds) is undertaken with great risks. Given some control of lake levels, the "trophic upsurge" phenomenon (that is, high secondary productivity) of a new reservoir can be recreated without artificial fertilization (Ploskey 1986; section 11.2.2.10).

An inverse relationship often occurs between macrophyte abundance and chlorophyll *a* concentration in lakes; that is, a large influx of nutrients from surface runoff often promotes an algal bloom, which reduces light penetration and abundance of macrophytes (Colle et al. 1987). The application of fertilizers to ponds with the objective of stimulating phytoplankton blooms to shade out extensive macrophytes may create further macrophyte problems in the same or subsequent years because of the ability of rooted macrophytes to recycle nutrients from the sediments.

11.2.2.3 Vegetation Control

Dense beds of emergent and submergent macrophytes may reduce aesthetic quality and recreational value of lakes and reservoirs, and they may retard growth of both predator and prey species (Engel 1995; Olson et al. 1998). For these reasons, more attention is given to controlling vegetation than to its enhancement (but see section 11.2.2.5 on waterscaping). Infestations of invasive, nonnative aquatic plants such as hydrilla and Eurasian water milfoil have received major attention because these plants may form a dense growth that interferes with fishing and boating and displaces or replaces native species of aquatic plants (Nichols 1994). In Orange Lake, Florida, excessive macrophyte growth caused a reduction in angling, which, in turn, caused a 90% loss in revenue from the sport fishery (Colle et al. 1987). Typically, the economic value of an unimpaired sport fishery will far exceed the cost of aquatic plant control.

Control of aquatic vegetation may be undertaken by mechanical, chemical, and biological methods. Mechanical methods include cutting, dredging to deepen the water, covering bottoms with plastic film or other synthetic materials, and drawing down the water level to expose the shallow water to drying and freezing (Dunst and Nichols 1979). Mechanical harvesting (cutting) has improved in cost effectiveness with innovations in equipment, but it is still expensive, labor intensive, and a never-ending process akin to mowing the lawn. However, mechanical harvesting of heavy growths of macrophytes in high-public-use areas avoids the hazards of chemicals and use of exotic organisms. Cutting and removing aquatic vegetation, along with the nutrient content in the vegetation, is regarded as an in-lake tool for water quality management.

Chemical methods involve use of registered herbicides. These chemicals often have specified withdrawal intervals for different uses; that is, the period of time that must pass after chemical treatment before the water can be used for drinking, swimming, irrigation, or livestock watering. The label on many herbicides contains precautionary statements regarding hazards to humans and domestic animals, such as "Do not use water for any purpose for 7 days after treatments." Because most lakes have multiple uses, such regulatory restrictions greatly reduce the choice of registered herbicides. Moreover, herbicides are costly, and their effectiveness is usually limited to the year of application.

Biological control of aquatic macrophytes with crayfishes or herbivorous fishes (e.g., grass carp or tilapia) has been effective in some lakes, but biological control is controversial. There is a widespread fear that grass carp may become a problem species similar to the common carp. These concerns have resulted in restrictions and prohibitions on using any kind of nonnative fish (see Chapter 13). If a herbivorous fish escapes into a river system, it might compete directly with waterfowl for rhizomes or tubers of aquatic plants such as arrowhead and pondweed. Some state agencies allow the use of triploid grass carp, which are presumed to be sterile.

11.2.2.4 Landscaping

Management of terrestrial vegetation can enhance environmental quality for fishes and anglers, as well as improve aesthetics. In low-relief topography, a windbreak of trees that is aligned perpendicular to the prevailing winds can reduce low-level wind velocities and, in turn, the fetch and wave height impacting a shoreline (Young 1973). Reduction of wind can provide protection for fishes spawning in shallow water as well as create quiet-water areas for angling. In the southeastern United States, a mowed grassy area is preferred around the periphery of ponds and intermediate-size reservoirs for aesthetic reasons and to eliminate habitat for ticks, chiggers, and poisonous snakes. Downed trees along the shoreline add natural brushy spots for spawning, and they may serve to concentrate crappies or sunfishes (*Lepomis* spp.) for the benefit of anglers.

11.2.2.5 Waterscaping

Management of vegetation for production of desirable food and cover is a major theme of terrestrial wildlife management, yet vegetation enhancement and selective control is a neglected topic in aquatic ecosystem management. Certainly, flowering macrophytes enhance the aesthetic appearance of a lake, and the tubers of many kinds of macrophytes provide food for waterfowl, muskrats, and other wildlife. Macrophytes also serve as cover for fishes and they function as attachment surfaces for periphyton and the many kinds of phytomacrofauna that are important food for bluegill and other fishes (Schramm and Jirka 1989). Engle (1989) proposed use of tree shade, cruising lanes, and dense growth of submerged macrophytes as ways to create habitat diversity (Figure 11.4).

Although most aquatic biologists have regarded Eurasian water milfoil as a serious scourge because of its rank growth, Engle (1995) pointed out that in water too turbid to support native plant growth Eurasian water milfoil can improve fish production. When it is managed by an integrated control program that limits its growth to 10–40% of the lake surface, it can be an asset to expanding the food base and protecting emerging year-classes from piscivores.

Olson et al. (1998) suggested harvesting macrophytes with mechanical weed cutters to improve fish growth and population size structure in lakes with high densities of submergent macrophytes and stunted fish populations. Removal of 20% of the macrophytes from the littoral zone by cutting a series of evenly spaced, deep channels resulted in substantially increased growth rates of some age-classes of both bluegill and largemouth bass. Although fewer than 25% of the channels remained open for more than 1 year, they had a permanent effect on the affected year-classes.

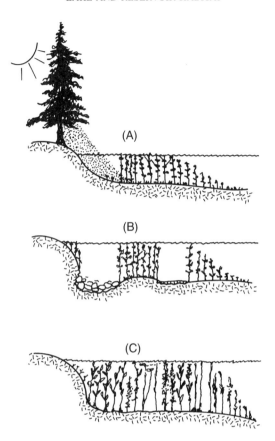

Figure 11.4 (**A**) A row of trees planted on a south shoreline will shade the water and retard near-shore growth of aquatic plants. (**B**) Cruising lanes for boats and predator fishes can be developed through heavy vegetation by covering the bottom with fiberglass screens topped with rock. (**C**) Waterscaping may be done with broad-leaved pond weeds and water lilies to provide vertical structure for attachment of the phytomacrofauna, which are important food for fishes and waterfowl; shelter for prey species; and spawning habitat. Figure is redrawn from Engle (1989).

Crowder and Cooper (1979) found that fish growth was greater at a macrophyte density sufficient to provide cover for invertebrates, so that the fishes did not overexploit their prey, yet not a density too great to prevent effective foraging. The idea of an optimal macrophyte density implies a vegetation enhancement or control program that considers effects on fish–invertebrate relationships and piscivory. A patterned macrophyte harvesting procedure has been recommended to obtain edge effects (Loucks et al. 1979; Olson et al. 1998).

Survey data from Texas reservoirs showed a strong positive linear relationship between submerged vegetation (up to 20% coverage) and largemouth bass recruitment and standing crop (Durocher et al. 1984).

Enhancement of aquatic vegetation may include direct planting of tubers and fertilization. In reservoirs with a barren fluctuating zone that macrophytes are unable to colonize, the dewatered area may be planted with an annual rye grass or other cover crop by aerial seeding. The cover crop provides benefits similar to macrophytes, but it decomposes quickly after flooding.

11.2.2.6 Stratification and Destratification

Many North American lakes develop stratification both in summer and winter. Summer stratification typically results in three layers (or, etymologically speaking, three lakes): (1) a warm, well-mixed upper layer, or lake, (epilimnion) with oxygen levels near saturation for the temperature and atmospheric pressure; (2) a middle layer (metalimnion) with rapidly changing temperature (thermocline); and (3) a lower layer of cold water (hypolimnion). In lakes that undergo winter stratification, the coldest water (0°C) is nearest the surface or adjacent to the ice, and the warmest, heaviest water (4°C) is near the bottom (a fact that accounts for fish feeding through the winter).

By midsummer, many stratified lakes have an oxygen-depleted hypolimnion, typically with oxygen concentrations less than 1 mg/L. Because oxygen concentrations need to be more than 2 mg/L for a water stratum to be continuously occupied by even the most low-oxygen-tolerant fishes, the hypolimnion of eutrophic lakes is generally devoid of fishes. The anoxic hypolimnion is also impoverished of many desirable kinds of benthic invertebrates that are important fish foods. Fishes that cannot tolerate the warm temperatures of the epilimnion may be "sandwiched" between thermally lethal epilimnetic water above and cooler but anoxic water below. The environmental deterioration that accompanies stratified conditions has caused summerkill of some coldwater species.

The inhabitants of the epilimnion are also adversely affected by pronounced stratification. Warmwater fishes may cease growing because of high epilimnetic temperatures. Summer stratification may also be an important contributing factor to the occurrence of winterkill because accumulated hypolimnetic BOD is not metabolized during the relatively short fall turnover, and carryover into the winter exacerbates wintertime oxygen demand.

Since the early 1960s, destratification has been used to improve the quality of water in municipal water supply lakes and released from reservoirs (Toetz et al. 1972; ASCE 1979; Burns and Powling 1981). Also, destratification of a reservoir can be used to improve a tailwater fishery. In lakes where it is not necessary to maintain a cold hypolimnion for trout or other fishes, the lake may be destratified or hypolimnetic water may be discharged. To enhance water quality in the tailwater, multilevel outlets make selection of the depth of water discharge possible (Moen and Dewey 1978).

Potential benefits of artificial destratification include expanded fish habitat, amelioration of symptoms of eutrophication (changes in algae composition dominance from blue-green to green algae or diatoms), an increase in the diversity and average size of the benthos, and avoidance of summerkill and winterkill (Figure 11.5). For example, destratification of a 9.7-ha lake, where as much as 82% of the lake bottom was below the thermocline when the lake was stratified, increased channel catfish harvest by 227% (Mosher 1983). The cause of the increase in angler harvest of catfish is not clear, but it is known that anglers often fish in the hypolimnion of stratified lakes, where there are no fish.

Two major destratification systems are air bubble and mechanical pumps (Toetz et al. 1972). Air bubble systems use compressors or air blowers to force air through diffusers or Helixors (pipes constructed with an internal spiral that lengthens the travel distance of the water for the purpose of increasing oxygen transfer to the water). By entraining an envelope

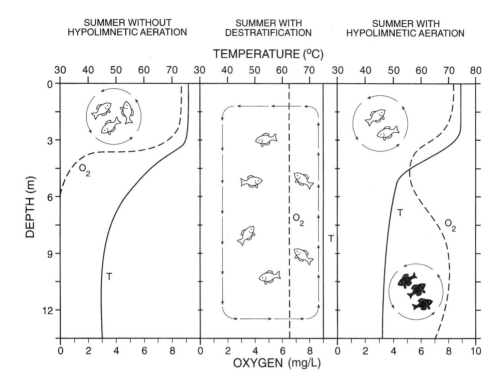

Figure 11.5 Influence of summer hypolimnetic aeration on stratification and vertical distributions of fishes. Figure is redrawn from Fast (1981).

of cold water, the rising air bubbles create an upflow of cold water to the lake surface. The upward movement of air bubbles develops a mixing cell and pulls in surrounding water. Destratification by air injection is an effective reservoir management technique to expand fish habitat (Summerfelt 1981; Fast and Hulguist 1989).

Mechanical pumping is done most efficiently by use of a high-volume axial flow pump (Garton et al. 1978). In the summer, the pump is operated in a down-flow direction to force warm surface water to the bottom of the lake to destratify the lake. In cold climates, where ice cover and winterkill occur, the pump is operated in an upwelling mode to force warmer water to the surface in order to melt an opening in the ice (Summerfelt et al. 1981).

11.2.2.7 Hypolimnetic Aeration

Hypolimnetic aeration is a method to increase the oxygen content of the hypolimnion without disrupting thermal stratification (Wirth 1981). Hypolimnetic water is air-lifted from near the lake bottom by means of positive displacement compressors that are powered by electric motors; these compressors may include aeration or oxygen injections (Ashley 1985). Water is pumped upward within a pipe to an aeration box on the lake surface where it is aerated before it is returned to the hypolimnion through a separate pipe. Usually aeration is accomplished without appreciable warming; coldwater habitat can be preserved for trout or other coldwater fishes.

Hypolimnetic aeration is becoming increasingly important as a fisheries management and water quality improvement technique. Ashley (1983) found that hypolimnetic aeration of a eutrophic lake enhanced sediment and water column oxidation of organic material, reduced ammonia concentrations by enhancing nitrification, and vented carbon dioxide. It also has potential applications for developing two-story fisheries or enhancing habitat quality.

11.2.2.8 Winterkill Lakes

It has long been evident that winterkill of a fish population in shallow, ice-covered lakes is the result of an oxygen reduction below a concentration at which fishes can survive (Greenbank 1945). The oxygen decline begins when ice cover seals the water surface from direct contact with the air; snow-covered ice reduces light penetration needed for algal photosynthesis. A partial or total winterkill of a lake's population is a serious lake management problem in northern latitudes (Schneberger 1971). Shallow and fertile lakes with a thick organic substrate have a high metabolic demand for oxygen (Cross and Summerfelt 1987). Shallow lakes also have a small storage capacity for oxygen, and the close association of the water mass with the substrate may quickly exhaust the oxygen when snow cover is heavy or persists for long periods.

Fishes vary in the minimum amount of oxygen they require for survival. As a consequence, winterkills rarely kill the entire fish population. Usually the game fish are more sensitive to low oxygen levels than are common carp, buffalos, and bullheads. Following a winterkill, the surviving common carp and bullheads spawn successfully and develop a strong year-class because of the absence of predatory fishes. Common carp and bullheads then become ecologically dominant and strongly influence environmental quality by making the water extremely turbid. The turbidity limits or prevents development of aquatic algae and macrophytes. Consequently, a winterkill often requires whole lake renovation (poisoning) prior to restocking game species.

Many kinds of aeration devices are available (bubbles or mechanical pumps) to melt an opening in the ice or to keep a section of lake from freezing (Johnson and Skrypek 1975; Smith et al. 1975). Pumping water melts the ice by expending the heat stored in the lake. The opening in the ice allows direct contact between water and air and also allows light penetration. Actual oxygen input from the aeration apparatus itself is a small proportion of the total oxygen input that occurs through the agitated lake surface in ice-free areas. An opening in the ice, or a thinning of the ice near the opening, is a hazard. The affected area must be cordoned off with snow fencing, signs must be posted, and educational programs must be carried out to make the public aware of the danger.

11.2.2.9 Dewatering

Reservoirs in semiarid regions are vulnerable to periodic, long-term dewatering from droughts. Intentional dewatering is called drawdown. Deep drawdowns of reservoir water level may occur on an annual cycle in reservoirs used for irrigation, navigation, power production, or flood control. In flood control reservoirs, purposeful drawdowns anticipate high spring floods; water levels then rise rapidly when river flows are impounded to prevent downstream flooding. Daily drawdowns of considerable magnitude occur in pump storage reservoirs where water is used to meet the morning and evening peak demand for power. Both short-term and chronic reductions in lake levels are detrimental to recreational facilities, development of littoral zone plant communities, and fish spawning. Common carp spawning has been reduced by exposure of spawning areas immediately following spawning.

Purposeful dewatering has been used as a fishery management manipulation to ameliorate effects of eutrophication in older, intermediate-size reservoirs (Born et al. 1973). During drawdown, flocculent can consolidate and an opportunity may exist to excavate nutrient-rich sediments and deepen areas prone to excessive macrophytes. In-lake habitat may be added during a drawdown.

Drawdown may be used to increase vulnerability of prey to predation. Largemouth bass predation on small bluegills has been enhanced by lowering water levels to force the bluegills out of the shelter of littoral vegetation. Bennett et al. (1969) reduced the bluegill population 60% by means of a fall drawdown at Ridge Lake, Illinois, which decreased lake surface area by 35%. A summer drawdown that reduced lake surface area by 42% and volume by 58% on Little Dixie Lake, Missouri, reduced the percentage of largemouth bass with an empty stomach and accelerated their growth (Heman et al. 1969). The density of fry and intermediate-size bluegills was reduced by the process of stranding the fish in vegetation and through increased predation of small fish by largemouth bass. Reduction in the abundance of bluegill may result in improved survival of largemouth bass fry the following spawning season. On the other hand, the vulnerability of a pelagic (open water) prey species such as a gizzard shad may be unaffected by drawdown (Wood and Pfitzer 1960; Lantz et al. 1967).

11.2.2.10 Water Level Fluctuation

Reservoir littoral zones are highly unstable because fluctuations in water level in reservoirs are more extreme on an annual basis than those observed in natural lakes. Water level fluctuations (daily, seasonal, and annual) are often found to be critically important environmental variables. It has been inferred that reservoirs with large annual changes in water level would likely have less stable fish assemblages (Carline 1986). Because recruitment and growth of many fishes are responsive to the hydrological regime of reservoirs, water level management plans may be used to meet management objectives for different species of fishes (Willis 1986). Spring surface area, summer surface area, annual change in area, and spring flooding are important factors affecting August fish standing crop (Ploskey 1986). An increase in year-class strength occurs in many fishes after spring flooding, but this is nullified when summer drawdown is extensive. Maintaining above-average water levels for as long as possible after successful spawning is necessary to protect young of the year of some fishes (Mitzner 1981).

Given some control of lake levels, the trophic upsurge phenomenon of a new reservoir can be recreated (Ploskey 1986). Trophic upsurge refers to the 5–10 years of highly successful sport fish production that follows initial reservoir filling. A period of 3–4 years of low water levels, when growth of terrestrial vegetation takes place in exposed areas, followed by flooding of substantial areas of vegetation, can simulate trophic upsurge, increase productivity, and bring on strong year-classes of certain fishes (Ploskey 1986). Water level management has been an effective fisheries management strategy for walleye, white bass, and white crappie in Kansas reservoirs (Willis 1986).

Daily fluctuations in lake levels are pronounced in pump storage reservoirs used for meeting peak power demand, and these fluctuations are basically detrimental to fishes and fishing. At this time, it is unlikely that a quality fishery will be found in a reservoir managed for pump storage that produces more than a 3-m diurnal flux in the lake level.

11.3 MANAGEMENT OF FISH POPULATIONS

11.3.1 Management of Spawning Habitat

Manipulations of habitat can be used to control undesirable fishes. In lakes with inlet waterways (i.e., drainage lakes in contrast to seepage lakes) and in nearly all reservoirs, common carp make upstream spawning migrations to marshes and flood-plains inundated by spring runoff. Weirs, suitably constructed across streams or the mouth of marshes, can block the upstream migration of common carp to spawning sites. Weirs are used in Iowa lakes to prevent common carp from migrating into connecting marshes. Unfortunately, weirs are not discriminating, and they may block migration of desirable species as well as blocking boat traffic. Weirs are also vulnerable to destruction by floods.

Sharp reductions in lake level may reduce sites for spawning common carp or expose eggs to drying. In Fort Randall Reservoir, South Dakota, a drawdown of 0.5–0.6 m immediately after major episodes of common carp spawning in shallow (0.3 m) flats killed the exposed eggs; the drawdowns were considered responsible for poor year-class strength of common carp in the 3 years that drawdowns were conducted (Shields 1958). Drawdowns, however, may adversely affect the reproductive success of other fishes that spawn in the littoral zone.

Management of spawning habitat should be considered when recruitment of desirable fish species is inconsistent, inadequate, or nonexistent. Spawning of white bass or walleye may be assisted by removal of barriers to upstream passage. When spawning habitat is suitable but larval fish survival is poor, other approaches are needed. Larval fish survival, and thus year-class strength, is influenced by many density-independent, abiotic factors (e.g., wind, temperature, and water level fluctuations) and density-dependent, biotic factors (e.g., abundance of prey and predators of larval fish). Maintenance stocking may be a more practical and effective alternative to habitat manipulation because the spawning requirements of some fishes (e.g., striped bass or rainbow trout) are difficult to accommodate. In many lakes, walleye populations are routinely maintained by stocking fry or fingerling fish produced in hatcheries (Conover 1986). Of course, hatchery production is necessary to provide hybrids, such as tiger muskellunge, hybrid striped bass, and saugeye (see Chapter 14).

Management manipulations to enhance spawning habitat include flooding of terrestrial vegetation and use of artificially constructed spawning sites. Fishes that spawn in lakes include gravel spawners (e.g., lake trout and walleye), vegetation spawners (e.g., northern pike and yellow perch), nest builders (e.g., largemouth bass, bluegill, and other fishes), and cavity spawners (e.g., channel and flathead catfishes). Constructing artificial spawning reefs of gravel, connecting marshes, and shoreline protection are useful techniques to enhance spawning of these fishes.

In lakes, walleye generally spawn at depths of 0.3–3.0 m on gravel and rubble shoals. In reservoirs, walleye generally spawn in water less than 1.5 m deep over gravel and rubble substrates, commonly on the rock riprap of the dam face. Egg survival from walleye spawning on other bottom types is poor (Prentice and Clark 1978). Newberg (1975) has described criteria for construction and placement of walleye spawning shoals.

Although it has been suggested that spawning devices such as drain tiles, milk cans, and grease cans be added to lakes to provide spawning sites for channel catfish (Nelson et al. 1978), doing so will probably be of little value in large reservoirs because of access to riverine spawning sites. In small and intermediate-size lakes that also contain largemouth bass, it is not the lack of spawning habitat that causes poor recruitment of channel catfish, but rather it is the exceptional vulnerability of channel catfish fry to predation by largemouth bass. Maintenance stocking of channel catfish in ponds and small lakes requires a nonvulnerable-size fish (>17.5 cm).

Recruitment of largemouth bass is inadequate in many lakes and reservoirs because of density-independent environmental factors. A sudden decline in water temperature may cause male largemouth bass to abandon the nest and stop spawning (Kramer and Smith 1962). Sometimes the fish will resume spawning when the water warms. On the other hand, shoreline areas exposed to a long fetch will receive continuous exposure to wind-driven waves, and such environments are generally inhospitable habitats for spawning of nest-building fishes such as largemouth bass (Summerfelt 1975) and crappies (Mitzner 1991) or vegetation spawners such as yellow perch (Clady 1976). A reservoir dominated by great stretches of barren, windswept shoreline will probably have poor recruitment of these fishes. A reservoir with an irregular shape will have a greater proportion of protected spawning habitat (Figure 11.6).

After young largemouth bass have dispersed from the nest, predation becomes a factor in survival. At that time, survival is enhanced by the presence of macrophytes or flooded terrestrial vegetation present from when the reservoir first filled or from growth following one or more years of low water levels. A high correlation between the amount and duration of flooded terrestrial vegetation and abundance of young-of-the-year largemouth bass in a given year indicates that flooded terrestrial vegetation is essential for providing food and protection (Aggus and Elliott 1975; Shirley and Andrews 1977). It has been recommended that fisheries management effort be directed to produce a high water level through most of a growing season every 3 or 4 years to flood terrestrial vegetation. An annual inundation will not produce a strong year-class each year because the terrestrial vegetation will not be available unless the shoreline is exposed for two or three successive years.

There are pros and cons to clearing trees from reservoir basins (Ploskey 1986). Inundated timber increases vertical habitat for attached organisms, food abundance, and spawning habitat as well as providing cover (Figure 11.7). Studies on distribution of larval gizzard and threadfin shads showed they were more abundant in offshore inundated timber than in open-water areas; this distribution reflected both extensive use of the flooded trees as spawning sites and increased availability of food or shelter (Van Den Avyle and Petering 1988).

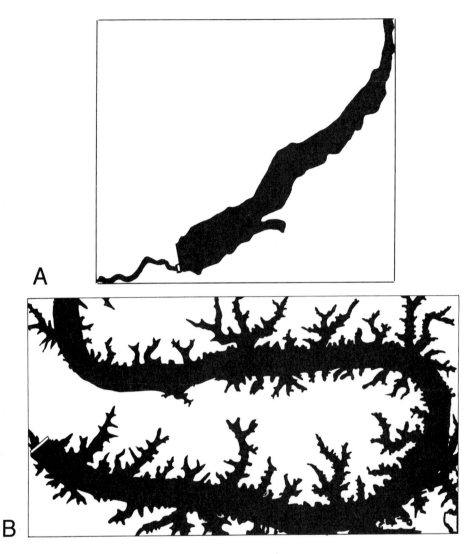

Figure 11.6   Shoreline development is contrasted in reservoirs with (**A**) wedge and (**B**) dragon shapes. The wedge-shaped reservoir has few arms, coves, and other shoreline irregularities; its littoral habitat is generally void of macrophytes and often dominated by windswept beaches. The wedge-shaped reservoir often lacks suitable spawning and nursery habitat for fishes such as largemouth bass. The dragon-shaped reservoir has a large shoreline development index and an abundance of wind-sheltered coves, which often contain submerged standing timber.

Submerged tire reefs may be used for spawning sites by some fishes, but given the small size of the reefs it is unlikely that an artificial reef will substantially influence the year-class strength of fishes in large reservoirs. Research has shown that submerged structures of either natural or artificial materials are soon colonized by an abundance of aquatic invertebrates and a surface film of periphyton. Artificial structures contain such small surface areas relative to other objects in the lake, such as rock riprap or standing timber, that they cannot be expected to have an substantial effect on primary production or invertebrate biomass of an entire reservoir. Large areas of standing timber in the lake basin will more likely provide meaningful habitat than postconstruction additions of artificial reefs.

Figure 11.7   Retaining standing timber at the time of reservoir construction provides vertical habitat for attachment of invertebrates and cover for many kinds of fishes. Figure reproduced from Linder et al. (1990).

11.3.2 Use of Subimpoundments as Nursery Areas

Arms or coves or a segment of a reservoir partitioned from the larger body may be stocked with fry and used for rearing northern pike, largemouth bass, walleye, striped bass, and other species to a nonvulnerable size for stocking. A drainable pond located to allow direct draining into the adjacent reservoir is best. In Arkansas, a nursery pond was used to rear fry of walleye, channel catfish, largemouth bass, and striped bass (Keith 1970; Chapter 14). It was particularly successful in developing a walleye fishery in Norfork and Bull Shoals reservoirs from fingerlings released from the nursery ponds.

11.3.3 Management of Fluctuating Zone Vegetation as Nursery Areas

Provision for protected spawning and nursery areas during reservoir planning can prevent negative effects of the exposed fluctuation zone. As an operational procedure, however, it is difficult to maintain vegetation in the fluctuation zone of large reservoirs. Nelson and Walburg (1977) found that strong yellow perch year-classes were produced in upper Missouri River impoundments when spring water flooded shorelines that had developed terrestrial (mesophytic) plants as a result of being out of the water for one or more years.

Fluctuation zone seeding has been used on Tennessee Valley Authority and Kansas reservoirs (Groen and Schroeder 1978) with generally favorable results on the fish populations. Strange et al. (1982) found that fluctuation zone seeding (rye grass) had a positive effect on the abundance and growth of young-of-the-year largemouth bass in early summer. Numbers of aquatic insects and small sunfishes also increased.

11.3.4 Management of Water Levels to Enhance Recruitment

Lakes with seasonally stable water levels usually develop a variety of macrophyte species in the littoral zone. However, most large multipurpose reservoirs are characterized by extreme seasonal fluctuations in water level that eliminate macrophytes from the fluctuating zone. The lack of vegetative cover in the littoral zone and the coincident harsh physical environment results in poor spawning habitat for nest-building fishes or excessive cannibalism of the year-class. The absence of cover (i.e., suitable nursery habitat) eliminates protection needed by young-of-the-year fishes. The consequence of this type of habitat problem for largemouth bass is weak or inconsistent year-class strength in many large reservoirs. Variation in young-of-the-year abundance in fall samples is related to early-season conditions that affect spawning and survival to the swim-up stage or factors that affect growth and survival of the fingerlings from early to late summer. Predation, including cannibalism, and starvation are the most important factors affecting survival in the postspawning period.

Flood control, power generation, and navigation have priority when government management agencies (e.g., U.S. Army Corps of Engineers, Tennessee Valley Authority, U.S. Bureau of Reclamation) and public utilities establish water level schedules; thus, adjustments to a reservoir plan for fisheries are usually limited, but adjustments have been negotiated. In Kansas, Groen and Schroeder (1978) developed a water level management plan for walleye, and Beam (1983) modified the plan for white crappie. A careful analysis of the relationship between water storage and year-class strength can be a useful predictor of potential harvest of crappies (Mitzner 1981). In natural lakes, lake levels have substantial effects on the quality and quantity of spawning habitat and year-class strength (Kallemeyn 1987).

11.3.5 Loss of Fishes Via Lake and Reservoir Discharges

Large discharges of water following the spawning season may result in loss of large numbers of planktonic fish larvae (Walburg 1971). Water discharge rates may be too swift for many young-of-the-year fishes to avoid entrainment in the discharge. Fishes (larvae to adults) may be lost from an artificial lake through subsurface and surface discharges. The loss of walleye fry in subsurface discharges substantially affected year-class strength in Lewis and Clark Lake, a main-stem impoundment on the Missouri River, South Dakota (Walburg 1971). Loss from natural lakes will be limited to surface discharges. Walleye fry survival in Kansas reservoirs has been related to the storage ratio (reservoirs with low storage ratios have high discharges); reservoirs with low storage ratios had lower densities of walleye (Willis and Stephen 1987).

Loss of adult fishes from a reservoir may occur through subsurface penstock discharges, but more often fishes are lost from surface discharges over the spillway and through roller and taintor gates (sluices used to control flow of water over the dam). Substantial losses of fishes from reservoir discharges have been reported; for example, Moen and Dewey (1978) estimated 83.3 million fish larvae were lost in one year and 122.4 million in a second year. Clark (1942) estimated a loss of 47 fish per hectare over 68 d in a surface discharge from a 607-ha Ohio reservoir. Louder (1958) reported fish losses over the spillway from two reservoirs in southern Illinois. Losses were as high as 116 fish per hectare from Lake Murphysboro (64.8 ha) during 1,152 h of overflow and 1.5 fish per hectare from Little Grassy Lake (486 ha) in 885 h. Fish loss is more common over modern concrete spillways with laminar flow at the tip than over a drop-board spillway with turbulent, noisy flow (the noise may repel the fish).

There is no practical way to prevent loss of larvae or adult fishes through discharges of large reservoirs. However, loss of pelagic fish larvae, such as walleye, from a large reservoir can be reduced if a reservoir water level management plan is in effect that stores as much water as possible at the time the fish are hatched (Kallemeyn 1987).

On smaller lakes that have a small spillway overflow for a short spring interval, it is possible to construct simple fish barriers that are effective. Powell and Spencer (1979) devised a horizontal parallel-bar barrier to prevent fish loss over spillways (see Figure 21.6). The barrier allowed small debris to slip through the bars and reduced maintenance and risk of dam failure. Bar spacing can be designed to retain fishes of different size, but these barriers cannot prevent the passage of fry and fingerlings.

11.4 THE HUMAN DIMENSION OF THE FISHERY

In the United States, and probably in Canada as well, lakes and reservoirs are the major habitats fished by anglers. In 1996, there were an estimated 28.9 million freshwater anglers (excluding the Great Lakes) 16 years old and older. Of these, 86% fished in lakes, reservoirs, and ponds; the balance in rivers and streams (USDI 1997). The popularity of lakes and reservoirs to anglers is the driving force for the socioeconomic value of these habitats. Broadly interpreted, of course, the human dimension of a fishery is more than the anglers. It includes the sum total of the socioeconomic aspects of the fishery, for example, the nonangler uses such as boating, swimming, hunting, and wildlife viewing. Because this chapter is focused on habitat of lakes and reservoirs, this section considers the habitat components that anglers desire; that is, those habitat features (access, amenities, and quality of surroundings) that enhance the quality of the angling experience and fishing success. Amenities include all-weather roads, docks, fish-cleaning stations, restrooms, drinking water, lighting, picnic facilities, and camping facilities.

11.4.1 Access Roads, Ramps, and Jetties

Most anglers will not walk far from a parking lot to fish on the bank, so good road access for bank anglers, as well as boat access for boat anglers, are generally essential. Fishing jetties and earthen dikes that are riprapped to reduce erosion, are designed to provide bank anglers with access to deeper-water fish habitat.

11.4.2 Fish Attractors

Artificial structures placed in lakes to attract and concentrate fishes for the purpose of improving angling catch rates may be called artificial reefs or fish attractors. These include brush piles, structures made with tires, and stake beds. Artificial structures are valuable additions to habitat when lake basins are barren and the average angler has difficulty locating fish. When properly located and marked with buoys or identified on lake maps, artificial structures help anglers locate fish in lakes without flooded timber or natural features that concentrate fishes. Phillips (1990) developed a useful pocket guide to the construction of artificial reefs that illustrates a variety of traditional and new types of reefs. Performance and longevity of artificial reefs are determined by materials, design (configuration), and construction (Brown 1986; Seaman and Sprague 1991).

There are also legal, economic, and regulatory issues related to siting and construction of reefs that must be considered (D'Itri 1986). Natural and artificial structures that concentrate fishes and improve angler catch rates (e.g., see Prince and Brouha 1974; Wilbur and Crumpton 1974) are attractive to sunfishes, crappies, largemouth bass, and catfishes (*Ictalurus* spp.). Crappies are most likely to be the species caught in greatest numbers near reefs. Several studies have shown that catch rate of sport fishes is greater near reefs than in otherwise comparable areas, but there is also evidence that artificial structures have increased the potential for overharvest of largemouth bass, defined as the number of days it takes anglers to catch 40% of the largemouth bass. Of course, overharvest is less problematic when appropriate size limits are in place and angler compliance is the rule.

Natural and artificial structures concentrate fishes by providing cover, food, and spawning habitat (Prince et al. 1975). Structures will be colonized by periphyton and will provide invertebrate food for fishes (Pardue 1973; Prince et al. 1977). However, it is unlikely that even a hectare of artificial structure in a large, multipurpose reservoir will be sufficient to affect significantly population rate functions (recruitment, growth, and mortality) of even a single species for the reservoir as a whole. In most cases, is it more accurate to consider that the function of artificial structures is solely to attract fishes and facilitate angler catch rate.

11.5 DESIGN OF LAKES FOR FISHERIES MANAGEMENT

Construction of new fishing lakes provides the opportunity to develop habitat diversity and to add conveniences for anglers and other resource users. It is more practical to include access, boat-launching facilities, fishing jetties, and submerged fish attractors in the planning stages and in the overall project budget than to fund them

Figure 11.8 A fishing jetty (background) and an associated stake bed were installed during lake renovation when the lake level was reduced. (Photograph is courtesy of Mike McGhee, Iowa Department of Natural Resources.)

from the annual operating budget of the fisheries management agency. Also, it costs less to construct structures before flooding a new impoundment than to add them after the lake basin is filled with water.

It is desirable to leave trees in coves to provide surface for periphyton coloniza-tion and invertebrate foods for fishes, to provide spawning habitat, and to attract fishes to sites convenient to anglers or out of the way of boat traffic. Because several sport fishes concentrate in coves with standing timber (Willis and Jones 1986), angler har-vest is higher from timbered coves than nonwooded ones. If the addition of artificial structures (e.g., brush piles or tire reefs) is desired, then it is even more important to leave a substantial area of natural woody vegetation standing in small coves, where the structure will not be a hazard for boaters and water skiers. The preimpoundment phase is a good time to construct fish attractors in proximity to roads and boat access (Figure 11.8).

Design of a reservoir outlet structure is one of the most important considerations in reservoir planning for water quality management. Penstocks with multiple discharge points provide options to regulate the water quality of the discharge. The discharge level may be above or below the thermocline as determined by water quality within each strata and by biological needs within the reservoir and tailwater (Figure 11.9).

Depth and slope of the basin are important morphometric factors determining environmental quality and rate functions of fish populations (Hill 1986). Extremely deep lakes are generally unproductive, but very shallow lakes are often turbid or they may have excessive growths of algae or macrophytes. Sediment basins, essentially subimpoundments, can be used to retain settleable solids in small upstream areas. Site selection, dam height, and construction initiatives to maximize shoreline development will increase the land–water interface, enhancing spawning habitat and the productive potential of the littoral zone (Ryder 1978; Olmsted and Clugston 1986). Barriers should be considered to prevent common carp from reaching their spawning habitats.

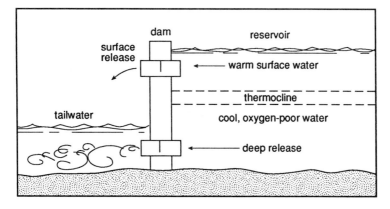

Figure 11.9 Biological, physical, and chemical characteristics of the tailwaters of reservoirs are strongly influence by volume and level of discharge (surface, bottom, or multiple) from the reservoir. In the summer, surface discharges are warm and deep, near-bottom releases are cold and may be anoxic.

Gravel shoals may benefit spawning of walleye and smallmouth bass, shallow vegetated bays are useful for yellow perch and northern pike, and reducing the extent of fetch and barren windswept beaches improve recruitment of largemouth bass and crappies. Selected areas of shoreline should be considered for riprapping when erosion would adversely affect the quality of either the aquatic or surrounding terrestrial environment.

11.6   CONCLUSIONS

Lakes and reservoirs are dynamic ecosystems with seasonal and annual cycles that are superimposed on long-term trends in weather and ecological succession. Surveys and monitoring are needed to identify problems within these ecosystems and to develop sound management plans. Problem solving and the enhancement of the quantity and quality of reservoir and lake habitat may be accomplished by long-term commitment to strategies and tactics of adaptive management, whereby management efforts are continually evaluated and adjusted as deemed necessary. Ecological studies of fish and lacustrine environments have provided a basic understanding of trophic relationships and environmental factors that control fish production and harvest. Fishery scientists, however, still need more information on the habitat requirements of individual species and species interactions in complex communities (Carpenter 1988).

11.7   REFERENCES

Aggus, L. R., and G. V. Elliott. 1975. Efforts of cover and food on year class strength of largemouth bass. Pages 317–322 *in* H. Clepper, editor. Black bass biology and management. Sport Fishing Institute, Washington, D.C.

Aggus, L. R., and W. M. Bivin. 1982. Habitat suitability index models: regression models based on harvest of coolwater and coldwater fishes in reservoirs. U.S. Fish and Wildlife Service, FWS/OBS-82./10.25, Washington, D.C.

ASCE (American Society of Civil Engineers). 1979. Symposium on re-aeration research. ASCE, Hydraulics Division, New York.

Ashley, K. I. 1983. Hypolimnetic aeration of a naturally eutrophic lake: physical and chemical effects. Canadian Journal of Fisheries and Aquatic Sciences 40:1343–1359.

Ashley, K. I. 1985. Hypolimnetic aeration: practical design and application. Water Research 19:735–740.

Bachmann, R. W. 1980. The role of agricultural sediments and chemicals in eutrophication. Journal of Water Pollution Control Federation 52:2425–2432.

Baker, J. P. 1982. Effects on fish of metals associated with acidification. Pages 165–167 *in* R. E. Johnson, editor. Acid rain/fisheries. American Fisheries Society, Northeastern Division, Bethesda, Maryland.

Ball, R. C. 1952. Farm pond management in Michigan. Journal of Wildlife Management 16:269.

Beam, J. H. 1983. The effect of annual water level management of population trends of white crappie in Elk City Reservoir, Kansas. North American Journal of Fisheries Management 3:34–40.

Beamish, R. J. 1976. Acidification of lakes in Canada by acid precipitation and the resulting effects on fishes. Water, Air, and Soil Pollution 6:501–514.

Beard, T. D. 1973. Overwinter drawdown: impact on the aquatic vegetation in Murphy Flowage, Wisconsin. Wisconsin Department of Natural Resources Technical Bulletin 61.

Bennett, G. W., H. W. Adkins, and W. F. Childers. 1969. Largemouth bass and other fishes in Ridge Lake, Illinois, 1941–1963. Illinois Natural History Survey Bulletin 30:1–67.

Black, J. J. 1983. Field and laboratory studies of environmental carcinogens in Niagara River fish. Journal of Great Lakes Research 9:326–334.

Born, S. W., T. L. Wirth, J. O. Peterson, J. P. Wall, and D. A. Stephenson. 1973. Dilutional pumping at Snake Lake, Wisconsin—a potential renewal technique for small eutrophic lakes. Wisconsin Department of Natural Resources Technical Bulletin 66.

Breck, J. E., R. T. Prentki, and O. L. Loucks, editors. 1979. Aquatic plants, lake management, and ecosystem consequences of lake harvesting. University of Wisconsin, Institute for Environmental Studies, Madison.

Brooks, K. N., P. F. Ffolliott, H. M. Gregersen, and L. F. DeBano. 1997. Hydrology and the management of watersheds, 2nd edition. Iowa State University Press, Ames.

Brown, A. M. 1986. Modifying reservoir fish habitat with artificial structures. Pages 98–102 in Hall and Van Den Avyle (1986).

Brugam, R. B. 1978. Human disturbance and the historical development of Linsely Pond. Ecology 59:19–36.

Burns, F. L., and I. J. Powling, editors. 1981. Destratification of lakes and reservoirs to improve water quality. Australian Government Publishing Service, Canberra.

Busch, W.-D. N., and P. G. Sly. 1992. The development of an aquatic habitat classification system for lakes. CRC Press, Boca Raton, Florida.

Cairns, V. W., P. V. Hodson, and J. O. Nriagu. 1984. Contaminant effects on fisheries. Wiley, New York.

Candle, R. D. 1985. Scrap tires as artificial reefs. Pages 293–302 in F. M. D'Itri, editor. Artificial reefs: marine and freshwater applications. Lewis Publishers, Chelsea, Michigan.

Carline, R. F. 1986. Assessment of fish populations and measurements of angler harvest. Pages 46–56 in Hall and Van Den Avyle (1986).

Carlson, R. E. 1977. A trophic state index for lakes. Limnology and Oceanography 22:361–369.

Carpenter, S. R., editor. 1988. Complex interactions in lake communities. SpringerVerlag, New York.

Clady, M. D. 1976. Influence of temperature and wind on the survival of early stages of yellow perch, Perca flavescens. Journal of the Fisheries Research Board of Canada 33:1887–1893.

Clark, C. F. 1942. A study of the loss of fish from an artificial lake over a wasteweir, Lake Loramie, Ohio. Transactions of the North American Wildlife Conference 7:250–256.

Colle, D. E., J. V. Shireman, W. T. Haller, J. C. Joyce, and D. E. Canfield, Jr. 1987. Influence of hydrilla on harvestable sport-fish populations, angler use, and angler expenditures at Orange Lake, Florida. North American Journal of Fisheries Management 7:410–417.

Collins, G. G. 1976. Effects of dams on Pacific salmon and steelhead trout. U.S. National Marine Fisheries Service Marine Fisheries Review 38(1):39–46.

Conover, M. C. 1986. Stocking cool-water species to meet management needs. Pages 31–39 in R. H. Stroud, editor. Fish culture in fisheries management. American Fisheries Society, Fish Culture Section and Fisheries Management Section, Bethesda, Maryland.

Cooke, D., and H. Olem. 1990. Lake and reservoir restoration and management techniques. Pages 117–160 in H. Olem and G. Flock, editors. Lake and reservoir restoration guidance manual, 2nd edition. U.S. Environmental Protection Agency, EPA 440/4-90-006, Washington, D.C.

Cooke, G. D., E. B. Welch, S. A. Peterson, and P. R. Newroth. 1986. Lake and reservoir restoration, 2nd edition. Butterworth, Stoneham, Massachusetts.

Cooke, G. D., E. B. Welch, S. A. Peterson, and P. R. Newroth. 1993. Restoration and management of lakes and reservoirs, 2nd edition. Lewis Publishers, Boca Raton, Florida.

Cross, T. K., and R. C. Summerfelt. 1987. Oxygen demand of lakes: sediment and water column BOD. Lake and Reservoir Management 3:109–116.

Crowder, L. B., and W. E. Cooper. 1979. The effects of macrophyte removal on the feeding efficiency and growth of sunfishes: evidence from pond studies. Pages 251–268 in Breck et al. (1979).

D'Itri, F. M. 1986. Artificial reefs: marine and freshwater applications. Lewis Publishers, Chelsea, Michigan.

Downing, J. A., C. Plante, and S. Lalonde. 1990. Fish production correlated with primary productivity, not the morphoedaphic index. Canadian Journal of Fisheries and Aquatic Sciences 47:1929–1936.

Driscoll, C. T., G. F. Fordham, W. A. Ayling, and L. M. Oliver. 1987. The chemical response of acidic lakes to calcium carbonate treatment. Lake and Reservoir Management 3:404–411.

Dunst, R. C., and nine coauthors. 1974. Survey of lake rehabilitation techniques and experiences. Wisconsin Department of Natural Resources Technical Bulletin 75.

Dunst, R., and S. Nichols. 1979. Macrophyte control in a lake management program. Pages 411–418 *in* Breck et al. (1979).

Durocher, P. P., W. C. Provine, and J. E. Kraai. 1984. Relationship between abundance of largemouth bass and submerged vegetation in Texas reservoirs. North American Journal of Fisheries Management 4:84–88.

Edmondson, W. T. 1969. Eutrophication in North America. Pages 124–149 *in* Eutrophication: causes, consequences, correctives. National Academy of Sciences, Washington, D.C.

Engle, S. 1989. Lake use planning in local efforts to manage lakes. Pages 101–105 *in* Proceedings of a national conference on enhancing states' lake management programs. Northeastern Illinois Planning Commission, Chicago.

Engel, S. 1995. Eurasian watermilfoil as a fishery management tool. Fisheries 20(3):20–27.

Fast, A. 1981. Hypolimnetic aeration. Pages 201–218 *in* F. L. Burns and I. J. Powling, editors. Destratification of lakes and reservoirs to improve water quality. Australian Government Publishing Service, Canberra.

Fast, A. W., and R. G. Hulguist. 1989. Oxygen and temperature relationships in nine artificially aerated California reservoirs. California Fish and Game 75:213–217.

Fletcher, R. I. 1990. Flow dynamics and fish recovery experiments: water intake systems. Transactions of the American Fisheries Society 119:393–415.

Fraisse, T., E. Muller, and O. Decamps. 1997. Evaluation of spontaneous species for the revegetation of reservoir margins. Ambio 26:375–381.

Frey, D. G., editor. 1963. Limnology in North America. University of Wisconsin Press, Madison.

Fromm, P. O. 1980. A review of some physiological and toxicological responses of freshwater fish to acid stress. Environmental Biology of Fishes 5:79–93.

Gamble, M. D., and E. D. Rhoades. 1964. Effects of shoreline fluctuation on grasses associated with upstream flood prevention and watershed protection. Agronomy Journal 56:21–23.

Garton, J. E., R. G. Strecker, and R. C. Summerfelt. 1978. Performance of an axial-flow pump for lake destratification. Proceedings of the Annual Conference Southeastern Association of Fish and Wildlife Agencies 30(1976):336–347.

Greenbank, J. 1945. Limnological conditions in ice-covered lakes, especially as related to winterkill of fish. Ecological Monographs 15:343–392.

Grinstead, B. G. 1971. Reproduction and some aspects of the early life history of walleye, *Stizostedion vitreum* (Mitchill) in Canton Reservoir, Oklahoma. Pages 41–51 *in* G. E. Hall, editor. Reservoir fisheries and limnology. American Fisheries Society, Special Publication 8, Bethesda, Maryland.

Groen, C. L., and T. A. Schroeder. 1978. Effects of water level management on walleye and other coolwater fishes in Kansas reservoirs. Pages 278–283 *in* R. L. Kendall, editor. Selected coolwater fishes of North America. American Fisheries Society, Special Publication 11, Bethesda, Maryland.

Halbert, C. L. 1993. How adaptive is adaptive management? Implementing adaptive management in Washington state and British Columbia. Reviews in Fisheries Science 1:261–283.

Hall, G. E., and M. J. Van Den Avyle. 1986. Reservoir fisheries management: strategies for the 80's. American Fisheries Society, Southern Division, Reservoir Committee, Bethesda, Maryland.

Hasler, A. D. 1947. Eutrophication of lakes by domestic drainage. Ecology 28:383–395.

Heiskary, S. A. 1985. Trophic status of Minnesota lakes. Minnesota Pollution Control Agency, Division of Water Quality, Monitoring and Analysis, Minneapolis.

Heman, M. L., R. S. Campbell, and L. C. Redmond. 1969. Manipulation of fish populations through reservoir drawdown. Transactions of the American Fisheries Society 98:293–304.

Henderson-Sellers, B., and H. R. Markland. 1987. Decaying lakes: the origin and control of cultural eutrophication. Wiley, Chichester, UK.

Hill, K. R. 1986. Classification of Iowa lakes and their standing stocks. Lake and Reservoir Management 2:105–109.

Hutchinson, G. E. 1957. A treatise on limnology. I. Geography, physics, and chemistry. Wiley, New York.

Jenkins, R. M. 1974. Prediction of fish production in Oklahoma reservoirs on the basis of environmental variables. Pages 11–20 *in* L. G. Hill and R. C. Summerfelt editors. Annals of the Oklahoma Academy of Science Publication 5, Oklahoma Geological Survey, Norman.

Jenkins, R. M., and D. I. Morais. 1971. Reservoir sport fishing effort and harvest in relation to environmental variables. Pages 371–381 *in* G. E. Hall, editor. Reservoir fisheries and limnology. American Fisheries Society, Special Publication 8, Bethesda, Maryland.

Johengen, T. H., A. M. Beeton, and D. W. Rice. 1989. Evaluating the effectiveness of best management practices to reduce agricultural non-point source pollution. Lake and Reservoir Management 5:63–70.

Johnson, R. E., editor. 1982. Acid rain/fisheries. American Fisheries Society, Northeastern Division, Bethesda, Maryland.

Johnson, R., and J. Skrypek. 1975. Prevention of winterkill of fish in a southern Minnesota lake through use of a helixor aeration and mixing system. Minnesota Department of Natural Resources, Division of Fish and Wildlife, Investigational Report 336, St. Paul.

Jones, J. R., and R. W. Bachmann. 1974. Prediction of phosphorus and chlorophyll levels in lakes. Journal of the Water Pollution Control Federation 48:2176–2182.

Jones, J. R., and M. V. Hoyer. 1982. Sportfish harvest predicted by summer chlorophyll a concentration in midwestern lakes and reservoirs. Transactions of the American Fisheries Society 111:176–179.

Kallemeyn, L. W. 1987. Correlations of regulated lake levels and climatic factors with abundance of young-of-the-year walleye and yellow perch in four lakes in Voyageurs National Park. North American Journal of Fisheries Management 7:513–521.

Keith, W. E. 1970. Preliminary results in the use of a nursery pond as a tool in fishery management. Proceedings of the Annual Conference Southeastern Association of Game and Fish Commissioners 23(1969):501–511.

Kimmel, B. L., and A. W. Groeger. 1986. Limnological and ecological changes associated with reservoir aging. Pages 103–109 *in* Hall and Van Den Avyle (1986).

Kramer, R. H., and L. L. Smith, Jr. 1962. Formation of year classes in largemouth bass. Transactions of the American Fisheries Society 91:29–41.

Lantz, K. E., J. T. Davis, J. S. Hughes, and H. E. Schafer, Jr. 1967. Water level fluctuation—its effect on vegetation control and fish population management. Proceedings of the Annual Conference Southeastern Association of Game and Fish Commissioners 18(1964):482–492.

Leach, J. H., and five coauthors. 1987. A review of methods for prediction of potential fish production with application to the Great Lakes and Lake Winnipeg. Canadian Journal of Fisheries and Aquatic Sciences 44(Supplement 2):471–485.

Leach, J. H., and R. C. Herron. 1992. A review of lake habitat classification. Pages 27–57 *in* W.-D. N. Busch and P. G. Sly, editors. The development of an aquatic habitat classification system for lakes. CRC Press, Boca Raton, Florida.

Lehman, S. 1997. The national watershed assessment program. Fisheries 22(5):25.

Likens, G. E. 1972. Eutrophication and aquatic ecosystems. Pages 14–348 *in* F. G. Howell, J. B. Gentry, and M. H. Smith, editors. Mineral cycling in southeastern ecosystems. National Technical Information Service, CONF-740513, Springfield, Virginia.

Linder, A., and eight coauthors. 1990. Largemouth bass in the 1990s. In-Fisherman, Brainerd, Minnesota.

Loucks, O. L., and seven coauthors. 1979. Conference findings: an overview. Pages 421–434 *in* Breck et al. (1979).

Louder, D. 1958. Escape of fish over spillways. Progressive Fish-Culturist 20:38–40.

Marzolf, G. R., and J. A. Osborne. 1971. Primary production in a Great Plains reservoir. Internationale Vereinigung fur theoretische und angewandte Limnologie Verhandlungen 18:126–133.

McGhee, M. 1990. Prescription for an aging lake. Iowa Conservationist 49(10):28–31.

McKenzie, R. E. 1951. The ability of forage plants to survive early spring flooding. Scientific Agriculture 31:358–367.

McMahon, T. E., J. W. Terrell, and P. C. Nelson. 1984. Habitat suitability information: walleye. U.S. Fish and Wildlife Service, FWS/OBS-82/10.56, Washington, D.C.

Meyer, F. P., and L. A. Barclay. 1990. Field manual for the investigation of fish kills. U.S. Fish and Wildlife Service Resource Publication 177.

Miranda, L. E., and D. R. DeVries, editors. 1996. Multidimensional approaches to reservoir fisheries management. American Fisheries Society, Symposium 16, Bethesda, Maryland.

Mitzner, L. 1981. Influence of floodwater storage on abundance of juvenile crappie and subsequent harvest at Lake Rathbun, Iowa. North American Journal of Fisheries Management 1:46–50.

Mitzner, L. 1991. Effect of environmental variables upon crappie young, year-class strength, and the sport fishery. North American Journal of Fisheries Management 11:534–542.

Moen, T. E., and M. R. Dewey. 1978. Loss of larval fish by epilimnial discharge from DeGray Lake, Arkansas. Arkansas Academy of Science Proceedings 32:65–67.

Moser, B. B., and D. Hicks. 1970. Fish production of the stilling basin below Canton Reservoir. Proceedings of the Oklahoma Academy of Science 50:69–74.

Mosher, T. D. 1983. Effects of artificial circulation on fish distribution and angling success for channel catfish in a small prairie lake. North American Journal of Fisheries Management 3:403–409.

Murphy, B. R., and D. W. Willis, editors. 1996. Fisheries techniques, 2nd edition. American Fisheries Society, Bethesda, Maryland.

Nelson, R. W., G. C. Horak, and J. E. Olson. 1978. Western reservoir and stream habitat improvements handbook. U.S. Fish and Wildlife Service FWS/OBS-76/56.

Nelson, R. W., and C. H. Walburg. 1977. Population dynamics of yellow perch *(Perca flavescens)* sauger *(Stizostedion canadense)* and walleye (S. *vitreum vitreum)* in four main stem Missouri River reservoirs. Journal of the Fisheries Research Board of Canada 34:1748–1763.

Newberg, H. J. 1975. Evaluation of an improved walleye *(Stizostedion vitreum)* spawning shoal with criteria for design and placement. Minnesota Department of Natural Resources Section of Fisheries Investigational Report 340.

Nichols, S. A. 1994. Evaluation of invasions and declines of submersed macrophytes for the Upper Great Lakes region. Lake and Reservoir Management 10:29–33.

Nielsen, L. A., and R. T. Lackey. 1980. Introduction. Pages 3–14 *in* R. T. Lackey and L. A. Nielsen, editors. Fisheries management. Wiley, New York.

Nriagu, J. O., and M. S. Simmons. 1984. Toxic contaminants in the Great Lakes. Wiley, New York.

Olem, H., and G. Flock, editors. 1990. Lake and reservoir restoration guidance manual, 2nd edition. U.S. Environmental Protection Agency, EPA 440/4-90-006, Washington, D.C.

Olmsted, L. L., and J. P. Clugston. 1986. Fishery management in cooling impoundments. Pages 227–237 *in* Hall and Van Den Avyle (1986).

Olson, M. H., and nine coauthors. 1998. Managing macrophytes to improve fish growth: a multi-lake experiment. Fisheries 23(2):6–12.

Owen, O. S. 1980. Natural resource conservation. Macmillan, New York.

Pardue, G. B. 1973. Production response of the bluegill sunfish, *(Lepomis macrochirus)* Rafinesque, to added attachment surface for fish food organisms. Transactions of the American Fisheries Society 102:622–626.

Peters, J. C. 1986. Enhancing tailwater fisheries. Pages 278–285 *in* Hall and Van Den Avyle (1986).

Pfleiger, W. L. 1975. The fishes of Missouri. Missouri Department of Conservation, Jefferson City.

Phillips, S. H. 1990. A guide to the construction of freshwater artificial reefs. Sport Fishing Institute, Washington, D.C.

Ploskey, G. R. 1986. Effects of water-level changes on reservoir ecosystems with implications for fisheries management. Pages 86–97 *in* Hall and Van Den Avyle (1986).

Powell, D. H., and S. L. Spencer. 1979. Parallel-bar barrier prevents fish loss over spillways. Progressive Fish-Culturist 41:174–175.

Prentice, J. A., and R. C. Clark, Jr. 1978. Walleye fishery management program in Texas—systems approach. Pages 408–416 *in* R. L. Kendall, editor. Selected coolwater fishes of North America. American Fisheries Society, Special Publication 11, Bethesda, Maryland.

Prince, E. D., and P. Brouha. 1974. Progress of the Smith Mountain Reservoir artificial reef project. Texas A&M University Sea Grant College TAMU-SG-74-103:68-72.

Prince, E. D., O. E. Maughan, P. Brouha. 1977. How to build a freshwater artificial reef. Virginia Polytechnic Institute and State University, Sea Grant Extension Publication VPI-SG-77-2, 2nd edition, Blacksburg.

Prince, E. D., R. F. Raleigh, and R. V. Corning. 1975. Artificial reefs and centrarchid basses. Pages 498–505 *in* H. Clepper, editor. Black bass biology and management. Sport Fishing Institute, Washington, D.C.

Ryder, R. H. 1965. A method for estimating the potential fish production of north-temperature lakes. Transactions of the American Fisheries Society 94:214–218.

Ryder, R. H. 1978. Fish yield assessment of large lakes and reservoirs—a prelude to management. Pages 403–423 in S. D. Gerking, editor. Ecological freshwater fish production. Blackwell Scientific Publications, Oxford, UK.

Ryder, R. H., S. R. Kerr, K. H. Loftus, and H. A. Regier. 1974. The morphoedaphic index, a fish yield estimator—review and evaluation. Journal of the Fisheries Research Board of Canada 31:663–688.

Schneberger, E., editor. 1971. A symposium on the management of midwestern winterkill lakes. American Fisheries Society, North Central Division, Special Publication 1, Bethesda, Maryland.

Schneider, J. C. 1981. Fish communities in warmwater lakes. Michigan Department of Natural Resources, Fisheries Research Report 1980, Lansing.

Schramm, H. L., Jr., and K. J. Jirka. 1989. Epiphytic macroinvertebrates as a food resource for bluegills in Florida lakes. Transactions of the American Fisheries Society 118:416–426.

Schofield, C. L. 1976. Acid precipitation: effects on fish. Ambio 5:228–230.

Schubel, J. R., and B. C. Marcy, Jr. 1978. Power plant entrainment: a biological assessment. Academic Press, New York.

Seaman, W., Jr., and L. M. Sprague. 1991. Artificial habitats for marine and freshwater fisheries. Academic Press, San Diego, California.

Shields, J. T. 1958. Experimental control of carp reproduction through water drawdowns in Fort Randall Reservoir, South Dakota. Transactions of the American Fisheries Society 87:23–33.

Shirley, K. E., and A. K. Andrews. 1977. Growth, production, and mortality of largemouth bass during the first year of life in Lake Carl Blackwell, Oklahoma. Transactions of the American Fisheries Society 106:590–595.

Simonin, H. 1998. The continuing saga of acid rain. New York Conservationist 52(5):4–5.

Sly, P. G., and W.-D. N. Busch. 1992. Introduction to the process, procedure, and concepts used in the development of an aquatic habitat classification system for lakes. Pages 1–13 in W.-D. N. Busch and P. G. Sly, editors. The development of an aquatic habitat classification system for lakes. CRC Press, Boca Raton, Florida.

Smith, S. A., D. R. Knauer, and T. L. Wirth. 1975. Aeration as a lake management technique. Wisconsin Department of Natural Resources Technical Bulletin 87.

Strange, R. J., W. B. Kittrell, and T. D. Broadbent. 1982. Effects of seeding reservoir fluctuations zones on young-of-the-year black bass and associated species. North American Journal of Fisheries Management 2:307–315.

Summerfelt, R. C. 1971. Factors influencing the horizontal distribution of several fishes in an Oklahoma reservoir. Pages 425–439 in G. E. Hall, editor. Reservoir fisheries and limnology. American Fisheries Society, Special Publication 8, Bethesda, Maryland.

Summerfelt, R. C. 1975. Relationship between weather and year-class strength of largemouth bass. Pages 166–174 in H. Clepper, editor, Black bass biology and management. Sport Fishing Institute, Washington, D.C.

Summerfelt, R. C. 1981. Fishery benefits of lake aeration: a review. Pages 419–445 in F. L. Burns and I. J. Powling, editors. Destratification of lakes and reservoirs to improve water quality. Proceedings of joint United States/Australia seminar and workshop. Australian Government Publishing Service, Canberra.

Summerfelt, R. C., B. R. Holt, and A. K. McAlexander. 1981. Fall and winter aeration of lakes by mechanical pumping to prevent fish kills. Pages 933–942 in H. G. Stefan, editor. Proceedings of symposium on surface water impoundments. American Society of Civil Engineers, Hydraulics Division, New York.

Swingle, H. S., and E. V. Smith. 1947. Management of farm fish ponds. Alabama Polytechnic Institute, Agriculture Experiment Station Bulletin 254.

Thornton, K. W., B. L. Kimmel, and F. E. Payne. 1990. Reservoir limnology. Wiley, New York.

Toetz, D. W., R. C. Summerfelt, and J. Wilhm. 1972. Biological effects of artificial destratification in lakes and reservoirs—analysis and bibliography. U.S. Department of the Interior, Bureau of Reclamation, Report RED-ERC-72-33, Washington, D.C.

USDI (U.S. Department of the Interior, Fish and Wildlife Service, and U.S. Department of Commerce, Bureau of Census). 1997. 1996 national survey of fishing, hunting, and wildlife associated recreation. U.S. Fish and Wildlife Service, Publication FHW/96 NAT, Washington, D.C.

van Breeman, N., C. T. Driscoll, and J. Mulder. 1984. Acidic deposition and internal proton sources in acidification of soil and waters. Nature (London) 307:599–604.

Van Den Avyle, J. J., and R. W. Petering. 1988. Inundated timber as nursery habitat for larval gizzard shad and threadfin shad in a new pumped storage reservoir. Transactions of the American Fisheries Society 117:84–89.

Walburg, C. H. 1971. Loss of young fish in reservoir discharge and year-class survival, Lewis and Clark Lake, Missouri River. Pages 441–448 in G. E. Hall, editor. Reservoir fisheries and limnology. American Fisheries Society, Special Publication 8, Bethesda, Maryland.

Walburg, C. H., G. L. Kaiser, and P. L. Hudson. 1971. Lewis and Clark Lake tailwater biota and some relations of the tailwater and reservoir fish populations. Pages 449–467 in G. E. Hall, editor. Reservoir fisheries and limnology. American Fisheries Society, Special Publication 8, Bethesda, Maryland.

Waters, T. F. 1995. Sediment in streams: sources, biological effects, and control. American Fisheries Society, Monograph 7, Bethesda, Maryland.

Weithman, A. S., and M. A. Haas. 1982. Socioeconomic value of the trout fishery in Lake Taneycomo, Missouri. Transactions of the American Fisheries Society 111:223–230.

Wetzel, R. G. 1983. Limnology, 2nd edition. Saunders, Philadelphia.

Wilbur, R. L., and J. E. Crumpton. 1974. Florida's fish attractor program. Texas A&M University Sea Grant College TAMU-SG-74-103:39-46.

Willis, D. W. 1986. Review of water level management of Kansas reservoirs. Pages 110–114 in Hall and Van Den Avyle (1986).

Willis, D. W., and L. D. Jones. 1986. Fish standing crops in wooded and non-wooded coves of Kansas reservoirs. North American Journal of Fisheries Management 6:105–108.

Willis, D. W., and J. L. Stephen. 1987. Relationship between storage ratio and population density, natural recruitment, and stocking success of walleye in Kansas reservoirs. North American Journal of Fisheries Management 7:279–282.

Wirth, T. L. 1981. Experiences with hypolimnetic aeration in small reservoirs in Wisconsin. Pages 457–467 in F. L. Burns and I. J. Powling, editors. Destratification of lakes and reservoirs to improve water quality. Australian Government Publishing Service, Canberra.

Wood, R., and D. W. Pfitzer. 1960. Some effects of water-level fluctuations on the fisheries of large impoundments. Pages 118–138 in Soil and water conservation volume 4. International Union for the Conservation of Nature and Natural Resources, Brussels, Belgium.

Young, W. C. 1973. Plants for shoreline erosion control in southern areas of the United States. Geophysical Monograph 17:798–803.

Zweiacker, P. L., and L. G. Bowles. 1976. Aquatic effects of steam-electric generating plants. Annals of the Oklahoma Academy of Science 5:124–135.

Chapter 12

Maintenance of the Estuarine Environment

WILLIAM H. HERKE AND BARTON D. ROGERS

12.1 INTRODUCTION

Estuarine zones are extremely complex transition areas between freshwater and marine habitats. They are the permanent home for numerous species and the temporary home for many others. Without estuarine zones, our saltwater commercial and recreational fisheries would be severely impaired. Even so, human actions continue to degrade estuaries, threatening these fisheries.

12.1.1 Habitat Description

The classic definition of an estuary is a semienclosed coastal body of water having a free connection with the open sea and within which the seawater is measurably diluted with freshwater deriving from land drainage (Cameron and Pritchard 1963). Pritchard (1967) categorized estuaries into four basic types: drowned river valley, bar built, fjord, and tectonically produced. Drowned river valleys form the typical estuaries along the eastern coast of North America (e.g., Chesapeake Bay). Bar-built estuaries occur when offshore sand bars build above sea level and create a sound; the Outer Banks of North Carolina form barrier islands to make Albemarle and Pamlico sounds. On the West Coast, glacial gouging, such as along the coast of British Columbia and southeastern Alaska, has created fjord-type estuaries; tectonic processes, such as faulting and subsidence, formed others (e.g., San Francisco Bay). Estuaries provide numerous gradients from mouth to head, surface to bottom, and side to side, resulting in many diverse habitats and complex natural interactions within a relatively small area.

Lagoons also occur in the transition zone between the land and the sea; they bear strong similarities to bar-built estuaries.

> Coastal lagoons are usually differentiated from estuaries on a geomorphological basis; an estuary is commonly considered as the mouth of a river while a coastal lagoon is an embayment separated from the coastal ocean by barrier islands....
>
> From an ecological point of view, however, coastal lagoons and estuaries constitute a similar type of ecosystem and we can speak of a lagoon–estuarine environment....
>
> The lagoon–estuarine environment is characterized as a coastal ecotone, connected to the sea in a permanent or ephemeral manner. They are shallow bodies of water, semi-enclosed of variable volumes depending on local climatic and hydrologic con-

ditions. They have variable temperatures and salinities, predominantly muddy bottoms, high turbidity, and irregular topographic and surface characteristics. The flora and fauna have a high level of evolutionary adaptation to stress conditions, and have originated from marine, freshwater and terrestrial sources. The biota of these coastal ecosystems include a varied flora and fauna; this biota is directly important to man for what it yields, and to many marine and fresh water organisms which use estuaries. In this natural conditions [sic], the ecosystem incorporates a balanced network of biotic interrelationships. The natural balance is all too easily upset by human impacts. (Day and Yáñez-Arancibia 1985:17–18)

The above description of the lagoon–estuarine environment differs from that of Cameron and Pritchard (1963) and Pritchard (1967) by requiring that the water bodies be shallow and by not requiring a free connection to the open sea. Moreover, neither Cameron and Pritchard's nor Day and Yáñez-Arancibia's description specifically includes adjoining transition zones, such as coastal marshes, tidally influenced freshwater habitats, or areas offshore of estuaries, particularly where large rivers empty into the ocean. In this chapter, an estuarine zone is defined as an environmental system consisting of the estuary (including lagoons) and those transitional areas consistently influenced or affected by water from the estuary.

In the estuarine zone's natural state, sedimentation and sea level variations cause changes on a geologic time scale, whereas climatological (such as El Niño changes, annual sea level changes, droughts, and hurricanes) or tectonic events cause shorter-term effects (Wolfe and Kjerfve 1986). Compared with most ecosystems, many scientists consider the estuarine zone to be a harsh environment because of its rapid and sometimes extreme short- and long-term changes in physicochemical factors.

Wind-driven currents and water level fluctuations are the dominant physical forces in certain bar-built estuaries and shallow coastal marshes. However, in most estuaries, tides and their associated tidal currents exert the primary short-term control of many of the physical processes. Generally there are three types of tides: semidiurnal, diurnal, and mixed. Semidiurnal tides, such as those on the Atlantic coast, produce two high-water and two low-water periods during an approximate 24-h period. Diurnal tides occur on the northern Gulf of Mexico coast and generally produce one high and one low tide per day. Mixed tides oscillate between semidiurnal and diurnal and are characteristic of Pacific coast estuaries and some areas of the Gulf of Mexico. Tidal amplitudes vary regionally; they range from less than 0.3 m in the northern Gulf of Mexico to 2–2.5 m along Georgia and South Carolina to over 16 m in the Bay of Fundy. They also vary monthly and seasonally in each region. The physicochemical effects of the tides on the resident biota are extensive and highly variable; the abiotic variability that characterizes coastal environments, compared with more static environments such as inland lakes, increases the complexity of biotic–abiotic interactions. The subject of estuarine physical processes is much more complex than presented here (for more see Kjerfve 1989).

Most physicochemical factors vary during the hours of a tidal cycle, seasonally, and from year to year. As the tidal current ebbs and floods, the physicochemical factors at any particular point will vary according to the characteristics of the water passing that point at that moment. Increased riverine input on a seasonal basis (e.g., spring thaw or rainy season) can reduce salinities throughout the estuary for long periods. Shallow estuaries that tend to be dominated by wind forces may show tremendous variability in water levels, temperature, and salinity on an irregular basis. For the shallow marshes of southwestern Louisiana, short-term changes in some physicochemical factors are caused by passages of strong atmospheric cold fronts (Figure 12.1). Climatic variations of longer duration also result in long-term variability.

Chemical processes are no less complex than are physical processes, and salinity is the dominant factor. Salinities of open oceans generally range from 33 to 37‰ (Sverdrup et al. 1942), whereas in the estuarine zone they usually vary from

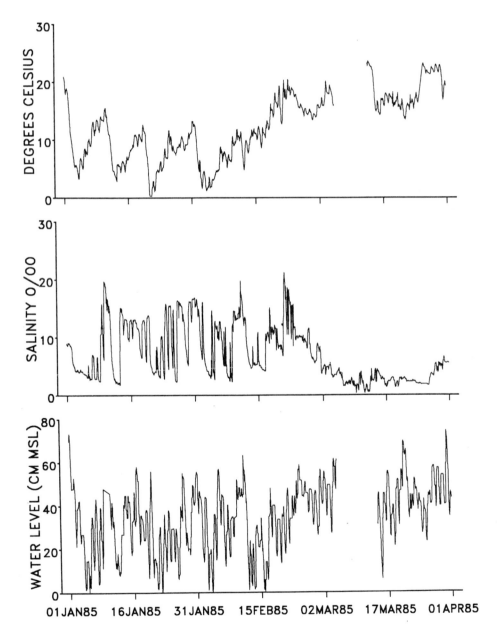

Figure 12.1   Hourly values for temperature, salinity, and water level (mean sea level, MSL) in Grand Bayou in southwestern Louisiana. The large, abrupt changes in January and early February were caused by the passage of strong cold fronts. Data gaps were due to data logger malfunction.

nondetectable to about 35‰. Hypersaline lagoons, such as the Laguna Madre on the southeastern Texas coast, may have considerably higher salinities—as high as 290‰ (Copeland 1967).

Water and soil salinities, along with physical factors that control these salinities (e.g., periodicity of flooding or desiccation, subsurface circulation, drainage, and porosity), determine the types of vegetation that occur in the estuarine zone. Lower variations in salinity and water levels characterize the landward zones, and if salinities are generally less than 5‰, freshwater vegetation predominates (Mitsch and Gosselink 1986). As salinity and tidal amplitude increase in the seaward zones, the vegetation changes to plants tolerant of brackish and saline water, which frequently results in less diverse or essentially monotypic stands of marsh plants, mangroves, or one of the submerged sea grasses. Many studies have shown these marsh and mangrove areas and sea grass beds to exhibit some of the highest primary productivity in the world. Their primary production, and the nutrients from both land and sea that are periodically or constantly supplied to or regenerated within the estuarine zone, make it extremely productive for fisheries. The complexity and productivity of the estuarine zone, exhibiting many alternate pathways, give nekton communities some year-to-year stability in a variable physical ecosystem.

12.1.2   Life Cycles of Estuarine Zone Organisms

Biological adaptations to an ever-changing environment are the key factors in the life histories of coastal fisheries species. Yáñez-Arancibia et al. (1985) discussed the hypothesis that the life histories of the dominant estuarine species are coupled with the physical processes in the ecosystem as evolutionary adaptations to stress conditions. Life histories of these species include migration patterns from marine, freshwater, and truly estuarine sources. Many of these organisms are euryhaline and eurythermal; that is, they can tolerate a wide range of salinities or temperatures, respectively. Some do this at all stages of their life cycle, whereas others accommodate wide ranges of physical factors by varying their location along the numerous estuarine gradients as their tolerance changes during their successive life states.

Although several different categorical schemes have been applied to fisheries organisms in the estuarine zone, two basic categories will be discussed here: residents and transients. (Anadromous fish species are discussed in Chapter 24.)

Residents are organisms that remain in the estuarine zone and accordingly must be able to tolerate varying physical and chemical conditions throughout their life. Except for oysters, crabs, and a few others, most of these species are not of direct commercial or recreational importance. However, some have great ecological significance in that they provide important links in the food web for commercial and recreational species. They may also be important in the process of decomposing the remains of vascular plants to detritus (which forms the base of some food webs) or in recycling or resuspending various nutrients into the water column.

Many of the transient organisms are important recreationally and commercially. Some transient species occur in the estuarine zone only occasionally and do not require its use to complete their life cycle. Others must use the estuarine zone for some portion of their life cycle if appreciable numbers are to survive; these are classified as estuarine-dependent species. (Some scientists prefer the term estuarine–marine.) Excluding anadromous and catadromous species, estuarine-dependent organisms of the

Atlantic and northern Gulf coasts typically spawn offshore, nearshore, or in lower bays. The larvae or juveniles, or both, migrate into the estuarine zone, often proceeding into the lower-salinity waters. In Louisiana, shallow-water marsh areas tend to have considerably higher nekton densities than do the adjacent, large open-water areas (Herke 1971; Day et al. 1982). The young organisms grow rapidly for a few weeks to a few months in this nursery area and then begin their return to the open sea to complete their life cycle. The life cycle of penaeid shrimp (Figure 12.2) serves as a good example of an estuarine-dependent species.

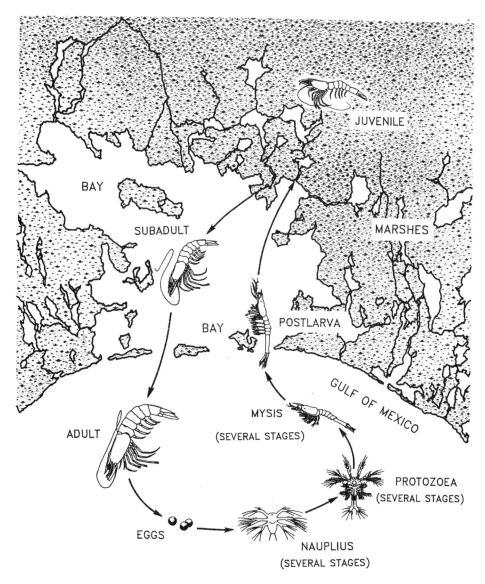

LIFE CYCLE OF PENAEID SHRIMPS

Figure 12.2 Life cycle of penaeid shrimps of the U.S. Atlantic coast and northern Gulf of Mexico. Although fish undergo fewer morphological changes, these life cycle movements are characteristic of most economically important estuarine-dependent species in this area. Drawing by Michelle Lagory.

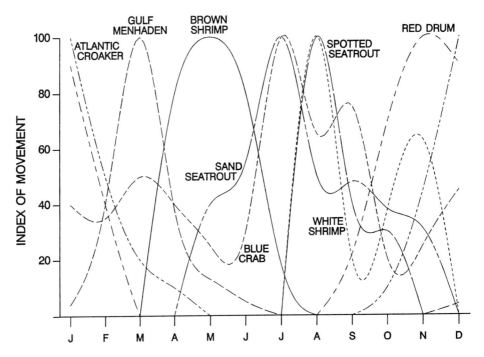

Figure 12.3 Index of movement (immigration and emigration combined), averaged over 1980–1982, for eight important species as measured by over 6,000 passive trap samples in southwestern Louisiana. Index is the average catch per trap each month divided by the highest monthly average and multiplied by 100; thus, the index for the month of highest catch is 100. At least 120 additional species migrated through the same channels during the year. Figure is from Rogers and Herke (1985).

Most estuarine-dependent species have annual peak migrations in and out of the nursery areas; timing of these peaks varies among species. At any given time during the year several species are using a nursery area; evolutionary processes have partitioned out the load so that there is efficient use year-round. Figure 12.3 depicts the timing of migrations in the Louisiana coastal marsh of eight commercially or recreationally important species.

There are fewer estuaries on the West Coast, and open-coast species predominate. Even so, there are a number of estuarine-dependent species (for example, several species of salmon and one or more species of flounders and crabs), but their life cycles do not necessarily follow the pattern discussed above.

Tropical estuarine-zone populations are characterized by greater numbers of species, but members of these species occur in less abundance than do members of the dominant species of cooler regions. Juvenile fishes and shrimps in temperate estuarine areas appear to respond, by emigration, to the interaction of seemingly minor changes in environmental factors (Herke 1971, 1977). There is less variation in physical factors in the tropics, especially in temperature and salinity. Nevertheless, it has been proposed that even in the tropics small variations in such factors as salinity, temperature, turbidity, sediment type, hydrography, and circulation patterns may be very important elements in controlling species biology and production (see Day and Yáñez-Arancibia 1985).

12.1.3   Value of Estuarine-Dependent Fisheries

Fisheries of the estuarine zone are of great economic importance. In many areas, coastal fisheries dominate the entire economy, from docks and processing plants to marinas and boat sales to restaurants and sometimes even the tourist industry.

An estimated two-thirds of the commercial fisheries catch in the Atlantic and associated waters is composed of estuarine-dependent species. In the Pacific Coast, marine fisheries organisms are more important in the commercial catch, but estuarine-dependent species still account for approximately 46% in landings value (McHugh 1966). In 1997, U.S. commercial fisheries landings equaled 4.4 billion kilograms with a US$3.5 billion dockside value (NMFS 1998). These amounts are conservative because an unknown proportion of the catch is unreported or underreported. Although these values include species that are not considered estuarine dependent, menhaden constituted 20% (900 million kilograms) of the landings, and penaeid shrimp accounted for 16% ($544 million) of the value. These two groups are classically estuarine dependent. Pacific salmon, which are also estuarine dependent, were fourth in both landings and value (258 million kilograms valued at $270 million). These three groups alone accounted for about 40% by weight and 25% by value of the entire U.S. commercial catch. Consider also that these are dockside values only; the total economic value is considerably greater (see Chapter 8). From the commercial viewpoint, the estuarine zone also attracts many tourists, and part of its attraction is the prevalence of tasty fresh seafood such as Maryland blue crab, Pacific salmon, and Louisiana shrimp creole.

Determining the recreational and aesthetic value of estuarine fisheries is more difficult but still very important (see Chapter 8). The National Marine Fisheries Service (NMFS 1998) estimated U.S. marine recreational finfish catch in 1997 was 336 million fishes. Estuarine fisheries are obviously of great value both commercially and recreationally.

12.2   ESTUARINE HABITAT LOSS AND CHANGE

Habitat loss and change in the estuarine zone is constantly occurring. This loss and change can be categorized as either natural or human induced, although often the effects of the latter compound (or are at least additive to) and are more rapid than natural effects.

12.2.1   Natural Sources of Change

A major natural phenomenon is subsidence, which results when geologic processes cause a general lowering of the land area. Shallow-water areas become deeper, and areas that were upland become submerged. Many of the plants die and are replaced by other species, or no vascular plants survive and eventually open water occurs. These habitat changes cause a shift in the food web and a subsequent change in the macrofaunal community structure.

Sediment deposition causes essentially the opposite situation. Open-water areas fill in, sometimes resulting in the smothering of productive shellfish beds or submerged sea grass flats, and mud flats or emergent marsh appear.

Natural changes in apparent sea level (land subsidence plus change in eustatic [worldwide] sea level) have resulted in estuarine zone changes throughout both geologic and recent history. Stevenson et al. (1986) reviewed 15 areas dated with lead 210 or cesium 137 and found that substrate accretion at 4 areas clearly has not kept pace with apparent sea level rise; moreover, these 4 areas were from regions that compose approximately half the tidal marsh acreage in the United States.

Droughts may cause normally brackish areas to become hypersaline, as happens along the Texas coast. Conversely, abnormally high rainfall sometimes causes extreme drops in salinity, as when much of Chesapeake Bay became fresh in 1973 following a hurricane. Hurricanes may also increase salinities and turbidities and cause large-scale physical changes in the estuarine zone.

Tidal passes between barrier islands open and close, causing hydrographic changes, and the barrier islands themselves may migrate to occupy different positions and configurations.

Naturally occurring estuarine habitat losses or changes are usually not too serious, at least in the long term. They may be minor in scope, offset by compensating changes elsewhere; they may be short term; or they may occur slowly enough to allow the flora and fauna to adjust to the changes. Estuarine organisms evolved in a highly variable environment and are generally resilient to natural changes. Populations decimated by natural catastrophes have the ability to rebuild once conditions return to "normal." Unfortunately, the same is not necessarily true of populations diminished by human-induced habitat loss and change because conditions may never return to normal.

12.2.2   Human-Induced Sources of Change

Construction of ports and marinas produces one of the most obvious human-induced effects on estuaries because such construction results in a direct loss of habitat. Construction also causes altered flow patterns and increased pollution. Channelization of existing waterways, or construction of new ones, also results in observable direct loss of habitat; however, it is the subsequent alteration of flow patterns that frequently causes more serious harm. For example, greater and more rapid tidal exchanges may (1) cause sediment deposition over productive areas, (2) change the faunal community composition, or (3) change the salinity regime sufficiently to kill existing vegetation and prevent establishment of alternate species, thereby reducing the detrital food base and also causing substrate erosion. The myriad canals and associated levees dug through marshes for oil field access and navigation have been indicated as a major cause of the annual loss to coastal erosion of about 13,000 ha in Louisiana (Turner 1987).

Human population density in coastal areas is increasing rapidly. For example, the U.S. population per square mile in 1950 and 1985, respectively, was 283.3 and 434.7 along the Atlantic coast; 59.8 and 151.9 along the Gulf of Mexico; and 97.6 and 237.6 along the Pacific coast (U.S. Bureau of the Census 1986). Acceding to the tremendous desire to live near the coast results in a high environmental cost. The need for such amenities as housing, roads, bridges, and waste disposal increases as population increases. Beachfront construction destabilizes the natural responses of beaches and sand dunes to storms and erosion.

Increases in the rate of sea level rise may result from human activities (i.e., global climate change). Combustion of fossil fuels, deforestation, and cement manufacture have caused the atmospheric concentration of carbon dioxide to increase by 20% (Titus 1986). Other trace gases that also trap heat (the greenhouse effect), such as chlorofluorocarbons, have also been increasing as a result of human activities (Ramanathan et al. 1985). The increase in the earth's temperature will cause expansion of seawater and increased seaward drainage from the melting of glacial ice. Estimates of sea level rise range from 10 to 21 cm by 2025 and 57 to 368 cm by 2100 (Titus 1986).

The effects of sea level rise on the estuarine zone will be wide ranging; the most obvious will be increased inundation and erosion. Salt water will intrude farther inland, changing the flora and fauna and even threatening sources of drinking water. Narrow marsh zones on high-relief coastlines (e.g., northwestern coast of North America) may move farther inland or may disappear. Wide marsh zones, normally associated with low-relief coast lines (e.g., southeastern and Gulf coasts of the United States) may migrate inland, but large areas will be lost if the habitats cannot migrate inland or adjust as quickly as the sea level rises. At present it is difficult to predict specific marsh responses to sea level rise.

The effects of sea level rise on shoreline developments could be tremendous. The choice would be either to retreat or to build levees and install pumps to hold back the water (Titus 1986). Either option will be expensive, and the latter will prevent marshes from migrating landward. Another problem associated with sea level rise is that areas that are now upland, and practically unregulated by governing agencies, will eventually become wetlands. Will these areas then become regulated? Legal battles will probably arise from this issue. We have barely touched on the complex, and still largely conjectural, subject of the effects of global climate change on estuarine-dependent fisheries (see Kennedy 1990).

Water control structures, impoundments, ditches, and levees or dikes are placed in estuaries and associated wetlands for various reasons. Usually, the main purpose of these structures is to maintain a certain hydrological regime that favors agriculture or particular animal groups such as ducks. Ditching or diking has been widely used in the past for mosquito control and still is sometimes. Impoundments are constructed and managed in Florida to reduce production of salt-marsh mosquitoes. Initially, these impoundments were managed strictly for mosquito control; however, with recent scientific information, management procedures have been incorporating measures to mitigate damages of these impoundments to fish and wildlife habitat. Elsewhere, levees are sometimes placed somewhat incidentally, as in the construction of causeways; other alterations may be caused by the piling of spoil along the side of a dredged channel. Such levees and spoil piles can result in unintentional areas of impoundment and soil waterlogging and result in eventual dieback of the marsh vegetation (Mendelssohn et al. 1981). In some areas, notably coastal Louisiana, levees and water control structures are also employed in an attempt to reduce saltwater intrusion and to provide seasonal drawdowns for reestablishment of emergent vegetation. Because many estuarine-dependent organisms migrate into the estuarine zone as larvae or juveniles (or both), they generally migrate with tidal flow. If tidal flow is substantially altered or prevented, migration pathways to nursery areas are altered or blocked.

Until recently, scientific documentation of the effects of coastal impoundments on fisheries has been lacking. In South Carolina, a multifaceted study was conducted in which the objectives were to evaluate the effects of coastal impoundments on sedimentation, hydraulic cycles, nutrient exchange, primary productivity, and the plankton, benthos, nekton, and wildlife communities (DeVoe and Baughman 1987). Total annual tidal exchange of dissolved organic carbon, particulate organic carbon, and phytoplankton biomass was similar between the impoundments and an adjoining natural area, but the timing of this exchange varied because exchange from the impoundments only could occur when their water control structures were open. Artificial timing of water exchange between impoundments and the natural system interfered with the normal life cycles of many nektonic species. On the other hand, management of impoundments for certain wildlife food plants created improved habitat for ducks. Thus, managed coastal impoundments may provide good habitat for target species (ducks in this case) but create barriers to migrations of many estuarine-dependent organisms. The effect of these barriers can be at least partially mitigated by operation of the water control structures to allow passage of estuarine-dependent fisheries organisms during their periods of peak movement.

Solid, low-level dams are used in Louisiana to create a similar, but much larger, form of semi-impoundment. The dams (called weirs in Louisiana) are placed in tidal bayous or canals; their crests are set about 15 cm below average marsh soil level. Water flows into the marsh when the tide rises above the level of the weir's crest and the water inside; as tide level falls, water flows out only until the water level reaches crest level. Semi-impoundments often provide, at least temporarily, improved duck habitat by increasing growth of submergent vegetation, such as widgeongrass. However, studies indicate semi-impoundments may result in the loss of emergent marsh plants, thus increasing marsh erosion. Moreover, these weirs substantially reduce the export of estuarine-dependent organisms from the impounded areas. When the total export of fisheries organisms from two matched 35-ha ponds was compared over two complete annual cycles, 107 species were caught as they exited the pond with no weir, whereas only 83 species were caught as they exited the pond with the weir (Herke et al. 1992). For both years, export in terms of abundance and biomass of nearly all estuarine-dependent species was substantially reduced from the pond with the weir.

Pollution is possibly the most serious of all threats to the estuarine zone. Pollution can come from point sources such as sewage and industrial outfalls and from nonpoint sources such as farm and urban runoff. Pollution can originate far up the watershed or even originate from outside the watershed via intentional transport (e.g., barge, truck, or rail) or atmospheric deposition. Acid rain may be a less serious problem in estuarine than in fresh waters because of the buffering effect of sea salts. However, Fisher et al. (1988) estimated that in 1984, 39% of the nitrogen input to Chesapeake Bay was from atmospheric deposition.

The effects of pollution on estuarine organisms can be subtle, hard to measure, and difficult to understand, such as reduced growth rates and increased susceptibility to predators, disease, and environmental stresses. The effects may be somewhat more apparent, such as increased internal lesions and cancerous growths in fishes correlated with increased levels of toxic materials (e.g., Malins et al. 1982) or externally visible anomalies and disease conditions statistically associated with pollutants (e.g., Ziskowski et al. 1987). Still, the effects of pollution may be obvious and catastrophic, such as the occurrence of massive fish kills. Types of pollution vary from location to location, and the means of reducing it must be tailored to individual sites and pollutants.

Human-induced changes in freshwater inflow to estuaries may have poorly understood, subtle, yet far-reaching effects. Inflow changes may result from nearby stream diversion or from actions taken hundreds of kilometers up into watersheds, such as irrigation withdrawal or dam construction and subsequent changes in the freshwater discharge cycle. Dams also block migrations of some species that use estuaries and trap sediment that otherwise would have been transported to the estuarine zone to form substrate for marsh building.

Changes in river courses cause a decrease in sediment input to some areas and a subsequent increase in other areas; changes in the Santee and Cooper river deltas in South Carolina furnish an example. Some years ago, dams were constructed that created two large reservoirs and diverted some of the Santee River flow down the Cooper River. Siltation in the Cooper River delta increased, and brackish marsh crept up the Santee River. Now, water is being redirected to increase flow in the Santee River and reduce siltation in Charleston harbor at the mouth of the Cooper River. Increased flow down the Santee River will move the interface of fresh and brackish water seaward and increase siltation in the Santee delta.

Changes in freshwater inflow affect the composition of phytoplankton species and the factors responsible for phytoplankton growth, such as light, nutrients, and the vertical stability of the water column. Freshwater inflow effects are less well defined for zooplankton, and elucidation of the effects on larval fishes and the resulting adult populations is hampered by a lack of understanding of the factors governing larval survival and year-class strength. Much work remains before a comprehensive understanding of the role of freshwater inflow to estuaries is achieved (Drinkwater 1986). However, some direct effects of inflow changes can be deduced. For example, in some estuaries, larvae are retained in the estuary by the counterflowing tidal and riverine currents (e.g., Graham 1972), and changes in freshwater inflow undoubtedly affect larval retention and year-class strength.

There is an abundance of literature describing the actual, or potential, benefits of mariculture—the farming of saltwater aquatic organisms. However, it must be recognized that such endeavors may adversely affect the estuarine environment. Loss of valuable natural nursery grounds, accelerated eutrophication caused by mariculture effluents, and interference with natural water movements, as well as the host of problems associated with introduced species and stocks (see Chapter 13), are just a few of the potential impacts culture ventures can pose to the estuary. Such risks need to be considered in the benefit-cost ratio of existing and potential culture operations that are in proximity to estuaries.

12.3 THE MANAGEMENT PROCESS

Up to this point we have discussed some natural processes, fisheries values, and human-induced perturbations in the estuarine zone. Responsibility for maintenance of the physical and biological integrity of the estuarine zone lies with individual, private group, local jurisdiction, state, regional, national, and international interests. How some of these interests attempt to maintain the estuarine zone will now be summarized.

12.3.1   Federal Agency Responsibilities

There are many federal laws and regulations in North America that are intended to help maintain the physical and biological integrity of the estuarine zone (many of these are discussed in Chapter 4). Sometimes there are overlaps and sometimes gaps of responsibility. Box 12.1 presents a partial summary of the regulatory and management functions for some of the major U.S. federal agencies that attempt to protect the estuarine zone. The following section elaborates on only a few points.

In the United States, the Water Quality Act of 1987 authorizes the Environmental Protection Agency to work with states and other federal agencies to conduct a National Estuary Program, a comprehensive study of the management needs of selected estuaries of national significance. As of January 1997, the National Estuary Program was addressing 28 estuaries—14 on the East Coast, 7 on the Gulf Coast, 6 on the West Coast, and 1 in Puerto Rico. Studies of other estuaries may be added in the future. The product of the study of each estuarine system will be a comprehensive conservation and management plan recommending specific pollution abatement and living resource management actions that should be undertaken to reach specific environmental goals for that estuary. Each study involves environmental assessment projects to characterize the status and trends in the estuary's health and resources; pollution control feasibility studies; public involvement and information activities; and the development of data management systems to support both the initial study and follow-up monitoring to assess success of any cleanup actions taken. A major objective of the National Estuary Program is to transfer information gained in its studies to a wide audience of environmental scientists, government agencies, and other interested groups concerned with estuarine environmental protection.

Under the U.S. Coastal Zone Management Act of 1972 (see Office of Ocean and Coastal Resource Management, Box 12.1), federal agency activities directly affecting the coastal zone must be consistent with programs of affected coastal states to the maximum extent possible (Archer and Knecht 1987). As of January 1997, 30 of 35 eligible coastal states and territories had federally approved coastal management programs. In addition, the 1972 act established a national system of estuarine research reserves to be managed jointly by the state and federal governments. These reserves represent various biogeographical regions and estuarine types in the United States. They are established to provide opportunities for long-term research, education, and interpretation. As of January 1997, 21 reserves had been established.

In Canada, the federal Department of Fisheries and Oceans' long-term policy objective is the achievement of an overall net gain of the productive capacity of fisheries habitats. This is to be achieved through restoration of degraded fisheries habitat, development of new habitat, and application of the no-net-loss principle in regard to habitat conservation. Under this principle, the Department of Fisheries and Oceans will strive to balance unavoidable habitat losses with habitat replacement on a project-by-project basis. For most estuaries where a number of major projects are planned, an estuary management plan is put together. Conflicting goals that could affect fish habitat are then evaluated. Although the results may be mixed, this comprehensive evaluation process works better than examining each development in isolation. Evaluating projects individually can lead to insidious, cumulative harm from numerous projects each causing at least a small amount of damage to the fisheries habitat.

Box 12.1   Partial Summary of Regulatory and Management Functions of Some Federal Agencies of the United States, Canada, and Mexico in Regard to Estuaries [1]

UNITED STATES [2]

Army Corps of Engineers (ACE)—Administers Section 10 of the River and Harbor Act of 1899 and the permit program of Section 404 of the Clean Water Act. Processes permit applications by soliciting comments from the public and concerned agencies via a comment period and conducts a full public interest review of probable impacts, including cumulative impacts, of the proposed activity.

Environmental Protection Agency (EPA)—Comments to ACE on Section 10 and Section 404 permit applications. Section 404(c) of the Clean Water Act provides the EPA with authority to prevent discharge of dredged or fill material into a defined area of waters whenever such discharge will have an adverse effect on municipal water supplies, shellfish beds and fishery areas, wildlife, or recreational areas.

Fish and Wildlife Service (FWS)—Under authority of the Fish and Wildlife Coordination Act, reviews and comments on Section 10 and Section 404 permit applications and federal water resource project proposals and provides technical assistance to protect fish and wildlife resources and mitigate project impacts. Other FWS activities in estuarine habitats include research on estuarine systems, management of anadromous fisheries, protection and management of endangered species, and wetland acquisition and habitat management on wildlife refuges.

National Oceanic and Atmospheric Administration (NOAA)

National Marine Fisheries Service—Evaluates and provides recommendations on federal regulatory and construction activities to protect living marine resources and mitigate adverse effects. Conducts research on life history of living marine resources and on factors affecting these resources and their habitat.

Estuarine Programs Office—Coordinates estuarine research policy for NOAA under the mandate of Public Law 99-659.

Office of Ocean and Coastal Resource Management—Administers the Coastal Zone Management Act of 1972 through which the federal government offers states bordering the coasts or Great Lakes financial help to develop state coastal zone management plans that meet federal standards.

CANADA [3]

Environmental Protection—Conducts sanitary and bacteriological surveys on the quality of shellfish-growing waters and determines if sale is safe for human consumption; monitors effluent quality and compliance with regulations; monitors nonpoint source pollution; and handles all ocean dredging and dumping activities.

Inland Waters and Lands—Develops and implements water management agreements between the federal and provincial governments; maintains data on water and land use; conducts studies relative to water and land management issues; and monitors streamflow and collects chemical, biological and bacteriological data on water, sediment, and biota.

Canadian Wildlife Service—Manages migratory birds and conservation of their habitat; acquires and manages wetlands; manages rare and endangered species; and studies effects of pollutants on migratory birds and their habitat.

Atmospheric Environment Service—Provides climatological information on weather and conducts atmospheric research.

Department of Fisheries and Oceans—Manages fisheries and fish habitat; reviews all projects that may affect fisheries or fish habitat; provides information on fish stocks, their locations, and catch regulations; and regulates commercial fisheries and enforces the Fisheries Act.

Energy, Mines and Resources—Monitors temporal changes in coastal and estuarine morphology and sedimentation rates.

MEXICO [4]

Mexico has almost 150 well-defined lagoon ecosystems. These represent more than 1.5 million hectares of lagoon-estuarine surface. One-third of the Mexican littoral region is composed of lagoon-estuarine ecosystems. The main institutions involved with the regulation of the coastal environment are Secretaria de Pesca (Instituto Nacional de la Pesca), Secretaria de Marina, Secretaria de Desarrollo Urbano y Ecología, Secretaria de Agricultura y Recursos Hidráulicos, Petróleos Mexicanos, Secretaria de Industria y Comercio, and the state governments. See Chapter 4 for further details.

---

[1] These governmental functions evolve over time; check with agency headquarters or regional offices for latest developments.

[2] Portions of the United States section were excerpted from an unpublished draft manuscript by Steve Gilbert and Roger Banks, U.S. Fish and Wildlife Service.

[3] Information for Canada was taken from a report by Environment Canada entitled "A Profile of Important Estuaries in Atlantic Canada."

[4] Information for Mexico was taken from a 1988 letter from Dr. Alejandro Yáñez-Arancibia, Laboratorio de Ictiologia y Ecologia Estuarina, Instituto de Ciencias del March y Limnologia, Universidad Nacional Autonoma de Mexico.

The General Law of Ecological Equilibrium and Protection of the Environment (Chapter 4) of Mexico includes environmental policy regarding regulation, protection, management, and exploitation of aquatic ecosystems and living resources, as well as directing educational efforts regarding the rational use of water. Estuaries, along with all other water bodies, are within the purview of this all-inclusive environmental legislation.

12.3.2 State and Local Jurisdiction Responsibilities

All states in the United States have some means of regulating activities within their boundaries. Most coastal states have met the federal Coastal Zone Management Act requirements and thus receive federal dollars to support a coastal zone management agency. However, the names, powers, and administrative procedures of these agencies vary from state to state. If a state has a procedure for issuing permits before certain actions may take place in the estuarine zone, it is generally administered by the coastal zone management agency. Many local jurisdictions have similar permitting requirements. If the actions contemplated are sufficiently local in nature, a state may give automatic approval to permits issued by local authorities.

Besides any coastal zone management agency that may exist, all states have at least one fish and wildlife agency, and sometimes other agencies, that may review and comment on Section 10 and Section 404 permit requests (see Box 12.1 and Chapter 4) being processed by the U.S. Army Corps of Engineers. Some local jurisdictions also review and comment on these permit requests.

In Canada and Mexico, maintaining the biological integrity of the estuarine zone is primarily a responsibility of the federal government; provincial, state, and local governments play relatively minor roles.

12.3.3   Private Organization Responsibilities

Many private organizations work to protect the estuarine zone, of which we have space to name only a few. In the United States two of the larger organizations that buy title to land or obtain conservation easements on estuarine property are The Nature Conservancy and the National Audubon Society. The Environmental Defense Fund and the Natural Resources Defense Council initiate legal action to force compliance with environmental laws, including those pertaining to the estuarine zone; depending on circumstances, these organizations sue lawbreakers, governmental agencies failing to enforce the law, or both. The National Wildlife Federation carries on an extensive public education program on the need to preserve wildlife habitat, with considerable emphasis on wetlands. There are also many other smaller private conservation groups that seek to protect the estuarine zone (see Box 12.2). For example, the San Juan Preservation Trust provides a low-cost, effective program to encourage coastline conservation on the San Juan Islands in Puget Sound, Washington (Myhr 1987).

In Canada, the Pacific Estuary Conservation Program, headed by the Nature Trust of British Columbia, coordinates two other private groups: Wildlife Habitat Canada and Ducks Unlimited Canada. The main goal of this program is to promote conservation and enhancement of coastal wetland habitats in British Columbia that have high fish and wildlife values. This program has secured lands in the Cowichan, Nanaimo, and Fraser estuaries, as well as the Baynes Sound. The British Columbia Ministry of Environment and Parks and Ministry of Forests and Lands assist in managing acquired lands and rehabilitation projects as well as providing technical assistance. The Canadian Wildlife Service and the Department of Fisheries and Oceans have served as advisors to the Pacific Estuary Conservation Program.

Organizations similar to those in the United States and Canada also exist in Mexico, but they are fewer in number.

12.4   MANAGEMENT NEEDS

With the multiplicity of laws and agencies dedicated to maintenance of the estuarine zone, one might think its maintenance would be assured. Unfortunately, such is not the case. Chapter 4 pointed out many of the legal complexities and problems involved, but there are numerous others. One is loopholes in existing laws. For instance,

---

Box 12.2   Information Sources

1. Information about land conservation through conservation easement programs can be obtained from The Land Trust Alliance, 1319 F Street, NW, Suite 501, Washington, D.C. 20004. This is the national organization serving hundreds of local and regional land conservation groups in the United States.

2. Save the Bay, Inc., 434 Smith Street, Providence, Rhode Island 02908–3770. This organization has been involved for many years in the protection of Narragansett Bay and its watershed and can furnish advice based on that long experience. It has published an illustrated paperback book entitled *The Uncommon Guide to Common Life of Narragansett Bay*. Although aimed particularly at Narragansett Bay, this book is informative about estuaries in general. There are chapters on plants, invertebrates, fishes, birds, and mammals.

3. Some consistent sources of information on wetlands restoration are the publications of Hillsborough Community College, Institute of Florida Studies, Plant City; Society of Wetland Scientists, Lawrence, Kansas; and the Office of Conference Services, Colorado State University, Fort Collins.

---

although Section 404 of the U.S. Clean Water Act forbids a developer to discharge dredge or fill material into waters of the United States without an Army Corps of Engineers permit, the developer needs no permit so long as the dredged material is not deposited in the water. Furthermore, it is nearly impossible to draft laws or agency regulations that cover all situations. Therefore, additional means for maintaining the estuarine zone are needed.

Although preservation of the estuarine zone in its pristine condition would be the best means of maintaining it, doing so is impossible because of actions that occur outside the estuarine zone. Moreover, preservation usually has to be balanced with economic development. Additionally, in many areas, a passive preservationist approach will not address prior human alteration of the estuarine zone that continues to cause deterioration. Thus, management is required.

Humans are continually proposing projects that will affect estuaries. Understanding how ecosystems function is a necessity before a project can be evaluated for its potential effects; thus, a basic need of management is the availability of sound scientific information. In many cases, this information is readily available but not known by those who need it. Improved transfer of scientific knowledge would provide for more effective fisheries management (Loftus 1987). Computerized literature searches are highly useful in this regard; most major university libraries can perform this service. In addition, much pertinent information exchange occurs over the Internet. Also, many general answers can be obtained from university scientists or from state fish and wildlife agencies. More specific answers may require new scientific study, which may take several years to authorize, fund, and conduct. Research should start early enough in the planning process so that results will be available when a management decision is needed.

12.4.1 Informed Use

Informed use is probably the most realistic approach to the maintenance of the estuarine zone, and the most important aspect of informed use is public awareness and involvement. The public should be told of potential habitat losses (and what these would mean in terms of fisheries losses) that will likely occur with a specific project. Most agencies have a mechanism to hear public comment (Chapters 2 and 3), and sincere efforts should be made to obtain timely public comment and involvement. Sometimes the public comment is all too late and comes in the form, where are all the fish? The public should be made aware of just how important a marsh or estuary is to recreational and commercial fishing and to the enjoyment of seafood and the economy of the area. If the public is aware of potential losses, and becomes actively involved, it can demand a revised plan that will at least reduce, if not eliminate, effects that would be detrimental to fisheries (see Box 12.2 and Chapter 3.)

One informed-use approach to a sound management decision is to bring together all concerned interests, furnish them all available information, and get them working toward a mutually acceptable common goal. For example, in Washington State, the Hood Canal Coordinating Council was established in 1985. It is charged with preparing and implementing a comprehensive management plan for environmental resource protection and enhancement in the region, an arm of Puget Sound. Membership on the council is composed of one county commissioner from each of the three bordering counties, one representative from each of the two affected Native American tribes, and one representative from the state Ecological Commission (Simmons 1987). Such cooperative approaches to environmental protection are potentially much more effective than the uncoordinated, piecemeal approach so often used elsewhere. On the other hand, special management groups are sometimes so large and entrenched in local political issues that meaningful management decisions are grossly impeded.

12.4.2 Long-Range Planning

This chapter has discussed project-specific management, but a preferable management regime is to set a long-term policy for development in an area. The stewards of a community should determine the amount and type of development that will be allowed in the coastal zone. This will involve a much higher degree of governmental land use control than the public historically has been willing to accept. However, as the human population continues to expand, the public must eventually abandon its frontier ethic. To continue to resist land use controls will result in intolerable damages to all environments, including the estuarine zone.

Recognizing the long-range need for land use controls, in 1984 the Maryland legislature passed the Chesapeake Bay Critical Area Law. This farsighted action resulted in land use restrictions much more stringent and widespread than generally exist elsewhere in North America. The law designated a critical area consisting of all waters of the Chesapeake Bay and extending 305 m beyond landward boundaries of state or private wetlands and heads of tides. The law also affects the streamside use of all the bay's perennial tributaries. Local jurisdictions are required to classify and delineate lands within their portion of the critical area into one of three categories as defined by criteria set out in the law: intense development, limited development, and resource conservation areas. Limits are set on how much of a local jurisdiction can be in each category. The law specifies many things that must, or must not, be done in the critical area. For example, no commercial timber harvesting is allowed within 15.25 m of mean high water of the bay or its perennial tributaries; all farms were required to have soil and water management plans within 5 years; and a 30.5-m-minimum vegetated buffer along tidal waters and streams is required for all new development. Further details can be found in Salin (1987).

Oregon passed a law designed to maintain the functional characteristics and processes of an estuary—such as its natural biological productivity, habitats and species diversity, unique features, and water quality—when intertidal or tidal marsh resources are destroyed by removal or fill activities. The procedures for enforcing the law, and the numerical habitat rating system involved, are described in Hamilton (1984).

12.4.3   Restoration

It is possible to restore estuarine habitat, but usually it is very expensive. The Chesapeake Bay again furnishes a good example. Because of the increased intensity of human activities within the Chesapeake Bay watershed, the health of the bay has declined rapidly since 1950 (Salin 1987). To restore the bay's health will require more than enforcement of the Chesapeake Bay Critical Area Law. Throughout the watershed it will require increased sewage treatment plant capacity, urban storm water projects, and construction of industrial pretreatment facilities to remove toxic materials. The total cost may run into billions of dollars. However, failure to undertake these actions will result in even greater losses, in terms of the economy, aesthetics, and health. (See section 5.3.4 for a summary of Chesapeake Bay restoration efforts.)

The most economical means of restoring marsh in the estuarine zone may be wise use of dredge spoil. Most ports and waterways maintain some level of dredging to keep ship channels open. With proper planning, spoil may be placed or directed such that it results in marsh building. However, this too may be expensive if the spoil must be transported very far.

Sometimes diversion of waterflow may be the most cost-effective means of restoring estuarine habitat. In Louisiana, diversions from the Mississippi River are being used to reestablish more favorable salinity regimes and build marsh through the additional input of sediment. Now in the planning stage are massive diversions to reintroduce some of the river water and sediment presently flowing out onto the shelf. Of course, in other situations this is sometimes a matter of "robbing Peter to pay Paul," as demonstrated by the previously mentioned case of the rediversion of the Santee–Cooper river system in South Carolina.

Where estuarine habitat is scarce and highly desired, small areas are sometimes restored or created by more energy-intensive methods, which can be very expensive. In northern California, a 12-ha parcel of lumber and plywood mills occupying original estuary and salt-marsh wetlands, was redeveloped into estuarine and freshwater wetlands at a cost of $280,000 (Allen and Hull 1987). In southern California there are ongoing plans to restore a 243-ha estuarine lagoon at an estimated cost of $15–25 million (Marcus 1987). It is readily apparent that it is much more economical to preserve already existent estuarine habitat than to create it.

Box 12.3   Sage Advice

The best and often the only way to deal with global problems is, paradoxically, to look for solutions peculiar to each locality....If we really want to do something for the health of our planet, the best place to start is in the streets, fields, roads, rivers, marshes, and coastlines of our communities.

Rene Dubos

Examine each question in terms of what is ethically and aesthetically right, as well as what is economically expedient. A thing is right when it tends to preserve the integrity, stability, and beauty of the biotic community. It is wrong when it tends otherwise.

Aldo Leopold

Unfortunately, although restored areas may become vegetated and appear the same as marsh formed naturally, there is question as to whether such human-constructed marshes function as well ecologically, or are as productive of fisheries organisms, as naturally occurring marshes (Josselyn et al. 1987; Minello et al. 1987). The techniques have not been used long enough for good evaluation (see Box 12.2).

12.5   CONCLUSION

Laws, agencies, and scientific information are the foundation for maintenance of the estuarine environment. However, as this chapter has described, numerous human-induced impacts can disrupt these "not quite fresh and not quite marine" ecosystems. In the final analysis, maintenance of the biological integrity of the estuarine zone depends on an interested, informed, and actively participating public that knows what is right for its own locality (Box 12.3).

12.6   REFERENCES

Allen, G. H., and D. Hull. 1987. Restoring of Butcher's Slough estuary–a case history. Pages 3674–3687 *in* Magoon et al. (1987).

Archer, J. H., and R. W. Knecht. 1987. The U.S. National Coastal Zone Management Program–problems and opportunities in the next phase. Coastal Zone Management Journal 15:103–120.

Cameron, W. M., and D. W. Pritchard. 1963. Estuaries, volume 2. Pages 306–324 *in* M. N. Hill, editor. The sea. Ideas and observations on the progress in the study of the seas. Wiley-Interscience, New York.

Copeland, B. J. 1967. Environmental characteristics of hypersaline lagoons. Contributions in Marine Science 12:207–218.

Day, J. W., Jr., C. S. Hopkinson, and W. H. Conner. 1982. An analysis of environmental factors regulating community metabolism and fisheries production in a Louisiana estuary. Pages 121–136 *in* V. S. Kennedy, editor. Estuarine comparisons. Academic Press, New York.

Day, J. W., Jr., and A. Yáñez-Arancibia. 1985. Coastal lagoons and estuaries as an environment for nekton. Pages 17–34 *in* A. Yáñez-Arancibia, editor. Fish community ecology in estuaries and coastal lagoons: towards an ecosystem integration. Universidad Nacional Autónoma de México Press, Ciudad Universitaria, Mexico, Distrito Federal.

DeVoe, M. R., and D. S. Baughman, editors. 1987. South Carolina coastal wetland impoundments: ecological characterization, management, status, and use, volume 1. Executive summary. South Carolina Sea Grant Consortium, Publication SC-SG-TR-86–1, Charleston.

Drinkwater, K. F. 1986. On the role of freshwater outflow in coastal marine ecosystems: a workshop summary. NATO ASI (Advanced Science Institutes) Series Series G7:429–438.

Fisher, D., J. Ceraso, T. Mathew, and M. Oppenheimer. 1988. Polluted coastal waters: the role of acid rain. Environmental Defense Fund, New York.

Graham, J. J. 1972. Retention of larval herring within the Sheepscot estuary of Maine. U.S. National Marine Fisheries Service Fishery Bulletin 70:299–305.

Hamilton, S. F. 1984. Estuarine mitigation: the Oregon process. Oregon Division of State Lands, Salem.

Herke, W. H. 1971. Use of natural, and semi-impounded, Louisiana tidal marshes as nurseries for fishes and crustaceans. Doctoral dissertation. Louisiana State University, Baton Rouge. (Also: University Microfilms, Order 71–29,372, Ann Arbor, Michigan.)

Herke, W. H. 1977. Life history concepts of motile estuarine-dependent species should be re-evaluated. Privately published. (Available from W. H. Herke, 555 Staring Lane, Baton Rouge, Louisiana 70810.)

Herke, W. H., E. E. Knudsen, P. A. Knudsen, and B. D. Rogers. 1992. Effects of semi-impoundment of Louisiana marsh on fish and crustacean nursery use and export. North American Journal of Fisheries Management 12:151–160.

Josselyn, M. N., J. Duffield, and M. Quammen. 1987. An evaluation of habitat use in natural and restored tidal marshes in San Francisco Bay, California. Pages 3085–3094 in Magoon et al. (1987).

Kennedy, V. S. 1990. Anticipated effects of climate change on estuarine and coastal fisheries. Fisheries 15(6):16–24.

Kjerfve, B. 1989. Estuarine geomorphology and physical oceanography. Pages 47–77 in J. W. Day, Jr., C. Hall, M. Kemp, and A. Yáñez-Arancibia. Estuarine ecology. Wiley-Interscience, New York.

Loftus, K. H. 1987. Inadequate science transfer: an issue basic to effective fisheries management. Transactions of the American Fisheries Society 116:314–319.

Magoon, O. T., and five coeditors. 1987. Coastal zone '87. American Society of Civil Engineers, New York.

Malins, D.C., and five coauthors. 1982. Chemical contaminants and abnormalities in fish and invertebrates from Puget Sound. National Technical Information Service, PB83-115188, Springfield, Virginia.

Marcus, L. 1987. Wetland restoration and port development: the Batiquitos Lagoon case. Pages 4152–4166 in Magoon et al. (1987).

McHugh, J. L. 1966. Management of estuarine fisheries. Pages 133–154 in R. F. Smith, A. H. Swartz, and W. H. Massmann, editors. A symposium on estuarine fisheries. American Fisheries Society, Special Publication 3, Bethesda, Maryland.

Mendelssohn, I. A., K. L. McKee, and W. H. Patrick, Jr. 1981. Oxygen deficiency in Spartina alterniflora roots: metabolic adaptation to anoxia. Science 214:439–441.

Minello, T. J., R. J. Zimmerman, and E. F. Klima. 1987. Creation of fishery habitat in estuaries. Pages 106–117 in M. C. Landin and H. K. Smith, editors. Beneficial uses of dredged material. U.S. Army Corps of Engineers, Waterways Experiment Station, Vicksburg, Mississippi.

Mitsch, W. J., and J. G. Gosselink. 1986. Wetlands. Van Nostrand Reinhold, New York.

Myhr, R. O. 1987. Private coastline conservation management: the land trust in the San Juan Islands, Washington. Pages 3266–3273 in Magoon et al. (1987).

NMFS (National Marine Fisheries Service). 1998. Fisheries of the United States, 1997. U.S. NMFS Current Fishery Statistics 9700.

Pritchard, D. W. 1967. What is an estuary: physical viewpoint. Pages 3–5 in G. H. Lauff, editor. Estuaries. American Association for the Advancement of Science Publication 83, Washington, D.C.

Ramanathan, V., R. J. Cicerone, H. B. Singh, and J. T. Kiehl. 1985. Trace gas trends and their potential role in climate change. Journal of Geophysical Research 90(D3):5547–5566.

Rogers, B. D., and W. H. Herke. 1985. Estuarine-dependent fish and crustacean movements and weir management. Pages 201–219 in C. F. Bryan, P. J. Zwank, and R. H. Chabreck, editors. Proceedings of the fourth coastal marsh and estuary management symposium. Louisiana Cooperative Fish and Wildlife Research Unit, Louisiana State University Agricultural Center, Baton Rouge.

Salin, S. L. 1987. Maryland's critical area program: saving the bay. Pages 208–221 in Magoon et al. (1987).

Simmons, D. M. 1987. A new approach to watershed planning. Pages 2726–2740 in Magoon et al. (1987).

Stevenson, J. C., L. G. Ward, and M. S. Kearney. 1986. Vertical accretion in marshes with varying rates of sea level rise. Pages 241–259 *in* D. A. Wolfe, editor. Estuarine variability. Academic Press, San Diego, California.

Sverdrup, H. U., M. W. Johnson, and R. H. Fleming. 1942. The oceans. Prentice-Hall, New York.

Titus, J. G. 1986. Greenhouse effect, sea level rise, and coastal zone management. Coastal Zone Management Journal 14:147–171.

Turner, R. E. 1987. Relationship between canal and levee density and coastal land loss in Louisiana. U.S. Fish and Wildlife Service Biological Report 85(14).

U.S. Bureau of the Census. 1986. Statistical abstract of the United States: 1987, 107th edition. U.S. Bureau of the Census, Washington, D.C.

Wolfe, D. A., and B. Kjerfve. 1986. Estuarine variability: an overview. Pages 3–17 *in* D. A. Wolfe, editor. Estuarine variability. Academic Press, San Diego, California.

Yáñez-Arancibia, A., and six coauthors. 1985. Ecología de poblaciones de peces dominantes en estuarios tropicales: factores ambientales que regulan las estrategias biológicas y la producción. Pages 311–365 *in* A. Yáñez-Arancibia, editor. Fish community ecology in estuaries and coastal lagoons: towards an ecosystem integration. Universidad Nacional Autónoma de México Press, Ciudad Universitaria, México, Distrito Federal.

Ziskowski, J. J., and five coauthors. 1987. Disease in commercially valuable fish stocks in the northwest Atlantic. Marine Pollution Bulletin 18:496–504.

COMMUNITY MANIPULATIONS

Chapter 13

Management of Introduced Fishes

HIRAM W. LI AND PETER B. MOYLE

13.1 INTRODUCTION

The introduction of fish species into waters outside their native ranges has occurred since the common carp was first moved around by the Chinese over 3,000 years
ago and by the Romans 2,000 years ago (Balon 1974). During the Middle Ages, common carp was spread throughout Europe by monastic orders, and Scandinavians were
busy stocking "barren" alpine lakes with salmonids. In the following centuries, the
great expansion of western civilization was accompanied by worldwide distribution of
European and American plants and animals, including fishes. This transplanting was a
reflection of western attitudes that natural systems could be improved by introducing
familiar, and therefore "superior," species (Crosby 1986). Introduced fishes are now
part of aquatic ecosystems throughout the world, and some are abundant enough in
their adopted countries to support important fisheries, such as trout and salmon in New
Zealand, Mozambique tilapia in Sri Lanka, common carp in Europe, and brown trout
in North America (Lever 1996). Only recently have such widely known "success"
stories been balanced by the realization that many introductions have done more harm
than good.

The damage done by introduced plants and animals to natural systems has caused
ecologists and resource managers much concern (Kornberg and Williamson 1986;
Mooney and Drake 1986; Holcik 1991; Nesler and Bergersen 1991; Williamson 1996).
The damage is especially severe in freshwater environments. Lassuy (1994) found that
species introduction was cited as a contributing factor in 68% of fish extinctions and
70% of fishes listed by the U.S. government as endangered or threatened. A similar
situation exists in Mexico (Contreras-Balderas and Lozano-Vilano 1994). In "Threatened Fishes of the World," a regular feature in the journal *Environmental Biology of
Fishes*, 49% of the 42 special accounts of threatened fishes list species introduction as
a contributing factor to the decline of these threatened species. These statistics reflect
a common observation that sustainable benefits have been sacrificed for short-term
gains (Baltz 1991; Ogutu-Ohwayo and Hecky 1991; Philipp 1991; Spencer et al. 1991).
The deliberate introduction of species of fishes and invertebrates to improve fisheries
is still a common management practice, especially in third world countries; however,
species spreading by other means is an even bigger, and growing, problem. In industrialized countries, aquatic organisms are introduced through canal systems, by the unauthorized planting of fishes by anglers, by hitchhiking on boats, through bait buckets,

and via other means. In all countries, the globalization of trade has led to frequent introductions of aquatic organisms through the dumping of ballast water, the release of aquarium fishes and plants, and the escape of fishes from aquaculture operations. Therefore, it is important to understand the effects of introduced species on native species and ecosystems. Without such understanding, well-intentioned management programs can create problems that actually subvert the original management intent. More importantly, fisheries managers must recognize that invasions of nonnative species rival habitat change and global warming as a threat to fisheries and aquatic ecosystems (Vitousek et al. 1996).

Here we provide (1) an overview of the reasons for introducing aquatic organisms and examples of successes and failures, (2) an introduction to ecological concepts important to understanding the effects of introductions, (3) some empirical rules regarding invading aquatic species, (4) management alternatives to introducing new species, and (5) guidelines for evaluating proposed introductions.

13.2   REASONS FOR FISH INTRODUCTIONS

Fishes and other aquatic organisms have been introduced for many and often multiple reasons: (1) to increase local food supplies, (2) to enhance sport and commercial fishing, (3) to manipulate aquatic systems (e.g., to control aquatic pests or reduce stunted fish populations), and (4) to improve aesthetics. Introductions have also often been a by-product of other activities.

13.2.1   Food Supply

The earliest introduced species were semidomesticated animals and plants that were moved about to create more reliable local food supplies. Domesticated plants and animals were keystones in the development of human culture, and their spread was a natural outcome of human expansion across the globe. As human populations grew and moved, commensal species such as rats and weedy plants also spread, and feral populations of domestic animals developed. Fishes were relatively late additions to the ranks of domestic and commensal species; most fish species were added after 1850. However, the worldwide spread of common carp and African tilapias began well before this time, and today they are a major source of protein in a hungry world. Fishes such as common carp and tilapia are hardy, so they can be easily transported, they establish populations quickly in a variety of new environments, and they grow rapidly, especially in ponds. These traits are characteristics of most animals used in aquaculture; they are also the characteristics of pest species.

In response to the increased human demand for fishes and static or declining wild fish populations, aquaculture today is a rapidly growing industry. Fish farmers tend to use only familiar species, which can lead to two problems: (1) native fishes well adapted to local conditions are often ignored for aquaculture (exceptions being salmonids and channel catfish) and (2) nonnative species escape into local waterways. The latter problem can result in the disruption of local wild fish populations, if not through direct interactions with the introduced fishes, then through exposure to new diseases and parasites.

13.2.2 Fisheries Enhancement

Izaak Walton was one of the first anglers to claim that an introduced fish was superior to native forms. In 1653, he pronounced the recently introduced common carp the "queen" of England's rivers because of its superior qualities as a sport fish. Local fish acclimatization societies formed in various countries, many of them predecessors of present fisheries management agencies. By the late 1800s, introducing species to solve management problems in North America had become a major activity of state and federal agencies. As a result, the dominant fishes in many lakes and rivers in North America are introduced sport fishes (Moyle 1986).

This pattern has been repeated throughout the world but particularly in the British Empire, where sportfishing was a favorite upper-class affectation, and local fishes were often considered to be unsuitable prey for the sophisticated angler. Favored sport fishes from Europe and North America were consequently brought to distant lands, often with great difficulty. Today species such as largemouth bass, rainbow trout, and brown trout enjoy virtual global distribution. Such fishes are still the backbone of sport fisheries in many areas, but there is a growing realization that they are often poor substitutes for native fishes, such as cutthroat trout of the interior basins of western North America. Similarly, the reduction in native cyprinodonts in Lake Titicaca (Peru–Bolivia), an important native subsistence fishery, resulted from the introduction of rainbow trout from North America and a predatory atherinid from elsewhere in South America.

Some of the most successful uses of introduced sport fishes illustrate the dictum of Giles (1978) that "importations are an admission of defeat in managing native populations to meet existing needs." The best example of this is found in Lake Michigan, which had its native fish communities severely disrupted by the invasions of a predator (sea lamprey), an efficient planktivore (alewife), and an effective predator on larval fish (rainbow smelt). Combined with severe overfishing of native fishes, the result was a nearly complete collapse of the native sport and commercial fisheries (see Chapter 23). Previous experience suggested that predatory Pacific salmonids could greatly reduce the numbers of alewives and rainbow smelt. In addition, these salmonids co-evolved with large lampreys, so they were less likely to be affected by lamprey predation. This proved to be the case, and a spectacular fishery for salmon soon developed. The reductions in rainbow smelt and alewives resulted in a marked increase in the numbers of some native planktivores through reduced competition for zooplankton and, perhaps, reduced predation on planktonic larvae. In a way, the introduction of Pacific salmonids worked too well. Salmon and steelhead populations exploded as they mined the huge biomass of alewives, creating expectations in anglers that the fabulous fishing would continue indefinitely. However, alewife populations are now being kept at moderate levels by salmonid predation, which, in turn, limits the number of salmon and trout the system can support (Eck and Brown 1985).

An additional factor helping to improve fisheries in the Great Lakes has been control programs for sea lampreys. This reduction in sea lampreys has lead to efforts to restore populations of the native salmonid predator, lake trout, through hatchery production. Stowe et al. (1995) have questioned this move to restore lake trout populations because lake trout accumulate more toxic contaminants in their flesh than do the nonnative salmonids, making them less suitable for harvesting. Either way, the new ecosystem has to be maintained with considerable effort to maintain production of fishes for human use.

A major problem with introducing predatory fishes to solve a management problem is that they usually spread, by natural or artificial means, from the water in which they were introduced. This is well illustrated by the introductions of sport fishes into reservoirs of the Columbia River basin in order to mitigate for losses of riverine salmon and trout fisheries. Unfortunately, the introduced species have invaded unimpounded reaches of the rivers and prey upon or compete with native fishes in both reservoirs and rivers. Today the native fishes of the Columbia and Snake rivers are now exposed to a greater intensity of piscivory than experienced during their evolutionary history because of the presence of introduced walleye, channel catfish, and smallmouth bass (Li et al. 1987; Tabor et al. 1993). Juvenile salmonids constitute a significant part of the diet of these fishes, and the establishment of these predators therefore counters the official policy of doubling salmonid escapement by the turn of the century.

There have been relatively few successful introductions of commercial fishes although attempts have been numerous, especially in marine environments. Some of the more successful introductions have been those made to benefit both sport and commercial fisheries, such as the introduction of striped bass and American shad in California in the 1870s and the spread of salmonids to coldwater lakes around the world. When angler demand for such fishes becomes high, the commercial fisheries are often banned. In North America, the main commercial fisheries of introduced fishes focus on species not favored by anglers, such as buffalo fishes in Arizona reservoirs, Sacramento blackfish in a Nevada reservoir, and common carp in rivers and reservoirs of the Midwest.

Often the poor performance of a sport fishery is attributed to the absence of an adequate prey base for predatory sport fishes. A common management practice is to introduce a suitable prey organism on the assumption that a new prey base will enhance growth and survival of sport fishes (see Chapter 14). Thus, species such as threadfin shad, gizzard shad, and golden shiner have been widely introduced into warmwater lakes and reservoirs as forage, and rainbow smelt and opossum shrimp *(Mysis relicta)* have been widely introduced into coldwater lakes and reservoirs. Such introductions have had mixed success. Growth rates of adult game fishes may accelerate following the introduction, and an initial period of outstanding fishing develops. However, it is common for the growth and survival rates of juveniles of the same game fishes to decrease because the forage species compete with them for food. Thus, the growth rates of juvenile white and black crappies in a California lake decreased considerably following the introduction of inland silverside while the growth rate of adults increased. The net result with time was that the large crappies preferred by anglers were the same size at a given age as they were before the introduction (Li et al. 1976). Not surprisingly, some of the most successful forage fish introductions have occurred in reservoirs where the game fish, usually trout, are planted at a size large enough to prey immediately on the forage fishes.

A better understanding of predator–prey relationships by fisheries biologists has led to much less frequent use of predator and forage introductions as a management tool. Unfortunately, segments of the angling public are not as well informed. The unauthorized introductions of sport and forage fishes is a major headache for managers in many areas today and is a legacy of the enthusiastic use of this tool by managers in the past. In Montana alone, over 200 illegal fish introductions have been recorded in recent years (Rahel 1997). In California, anglers have tried to establish northern pike in streams and reservoirs despite the danger it poses to native salmon fisheries and the high costs of eradicating established northern pike populations. "Johnny Appleseed" introductions of percids and centrarchids to reservoirs in Oregon are occurring with alarming frequency.

13.2.3 Manipulation of Aquatic Systems

The use of fishes as biological control agents for aquatic pests such as mosquitoes, disease-bearing snails, and aquatic weeds for control of populations of stunted fishes is a very appealing concept. If effective, a successful introduction for biological control can obviate the need for pesticides, be inexpensive, and have a long-lasting effect. The increased use of fishes for biological control in recent years has been spurred by the purported successes in the use of insects as control agents in agriculture and the development of biological control theory by applied entomologists. The cost of pesticides is also rapidly increasing, as is public concern about the effects of pesticides on nontarget organisms and human health. As a result, species with good track records for biological control are being spread worldwide, most prominently the eastern and western mosquitofishes and grass carp. Mosquitofishes are perhaps the most widely distributed fishes in the world today, found virtually everywhere where the climate is suitable. They are successful at controlling mosquitoes because they can live in stagnant water, reproduce and grow rapidly, are voracious insectivores, and are easy to raise in large numbers (Swanson et al. 1996). The distribution of the grass carp may someday rival that of the mosquitofishes because it is being widely introduced as a fish that is not only effective at controlling aquatic weeds (often introduced species themselves) but as one that converts these weeds to edible fish flesh. The grass carp is hardy, easy to culture, a voracious grazer, and relatively nonselective in its choice of plants to eat.

Even though there are many success stories in the use of mosquitofishes, grass carp, and other fishes for biological control, the use of such fishes also entails some risk. This is because fishes typically select a wide variety of prey species and will forage on pest species only when those species are abundant and easily available. A given species of fish is also longer lived than most of its prey, so it has relatively slow population responses to increases or decreases in prey abundance. As an example, if densities of mosquitofishes are too low, their presence may actually increase mosquito populations because mosquitofishes preferentially prey on the larger insects that are natural predators of mosquito larvae; they switch to mosquitoes only after the large insects are depleted. Mosquitofishes also can displace native fishes that may actually be better at mosquito control in some types of habitat (Miller 1961; Schoenherr 1981; Ahmed et al. 1988).

One reason introductions for biological control are often cited as success stories is that they are superficially less expensive than other means of control, such as pesticides, which require frequent application. In fact, however, long-term success using fishes for biological control is often labor intensive. Management of grass carp, for example, requires clear objectives concerning levels of water quality, vegetative cover, and the desired community structure (Cassani 1996). Evaluation of whether these objectives are being met requires constant monitoring of the aquatic community and grass carp production (Cassani 1996). Long-term public commitment may be critical because effective management may require acceptable levels of aquatic macrophytes rather than their elimination. Hence, use of grass carp may require ceaseless manipulation of size and age structure for acceptable levels of control. For this reason alone, grass carp as a biological control agent has been a failure in Oregon.

Piscivores have been introduced to control prey, but aside from the Great Lakes experience evidence of the success of this strategy in improving fisheries is largely equivocal (Noble 1981; Wydowski and Bennett 1981). The main reason for equivocal results is that predator–prey interactions are more complex than generally realized. For example, the tiger muskellunge, which is a sterile hybrid of the muskellunge and northern pike, has been proposed as a safe management tool to reduce populations of stunted sunfishes (*Lepomis* spp.) in lakes and ponds. However, tiger muskellunge are effective only when sunfish densities are high, vegetation is sparse, no alternative prey are available, and the tiger muskellunge themselves are too large to be eaten by largemouth bass (Tomcko et al. 1984)—a set of conditions not often met.

Limnologists have noted that water quality may be improved by manipulating the food chain (Shapiro and Wright 1984). One method is to introduce piscivores to reduce populations of small fishes that prey on zooplankton. Zooplankton populations then increase, and this results in increased grazing by the zooplankton on phytoplankton, causing an increase in water clarity. Such shifts have been noted by Scavia et al. (1986) in Lake Michigan, but it is unusual for predation to be so effective. In addition, the goals of water quality management and fisheries management may not be compatible; high harvest rates of predatory fishes may result in increases of planktivores and decreased water clarity. In general, the introduction of a piscivore into an aquatic system is likely to cause drastic changes (Moyle and Light 1996).

13.2.4   By-Product Introductions

Thanks to modern transportation and water delivery systems, aquatic ecosystems are being transformed all over the world by introductions that are the by-products of some other human activity. In the past, such introductions have been referred to as "accidental," which made it easier to excuse them and not to take action to control them. Today, we know these introductions are happening by a wide variety of means, and most are preventable. Consequently, they should be regarded as deliberate introductions.

The most severe type of by-product introduction is from ballast water of ships. The huge cargo ships that go back and forth across our seas carry millions of gallons of water that is dumped when a port is reached. This water contains hundreds of species of estuarine organisms, including fishes, that are changing estuarine and other ecosystems worldwide in what Carlton and Geller (1993) call "ecological roulette." The most famous example of a ballast water introduction is the zebra mussel, which has caused major ecological changes to the Great Lakes and billions of dollars in economic damage (Ludyanskiy et al. 1993). The zebra mussel is now spreading rapidly by other means throughout North America (Johnson and Padilla 1996). Further ecological changes to the Great Lakes are being signaled by invasions of other ballast water organisms, such as the ruffe, the round goby, and the spiny water flea *Bythotrephes cederstroemi*. Perhaps the best-documented estuary in terms of invasions is San Francisco Bay. At least 212 new species have become established, mostly though ballast water (Nichols et al. 1990; Cohen and Carlton 1995). Most of the common species of invertebrates now are introduced species, and they dominate the food webs of the estuary.

Another important type of by-product introduction is bait bucket introductions. Often these introductions are simply the result of anglers collecting baitfishes from one stream, fishing a stream in a different drainage, and then releasing their unused bait under the assumption they are providing forage for the game fishes. Such cryptic introductions are more common than is generally realized and can explain many anomalies in fish distribution, especially in the eastern United States. For example, brook stickleback, often captured with bait minnows but not used as bait, may have been widely distributed as a contaminant via bait buckets (Ludwig and Leitch 1996). In the West, major invasions of eastern minnows such as fathead minnow, red shiner, and golden shiner are largely the result of bait bucket introductions (Dill and Cordone 1997).

Fishes and invertebrates have also become established in new waters after being transported through canals, carried as hitchhikers in hatchery trucks, or released by aquarists or after escaping from fish farms. Zebra mussels and Eurasian water milfoil (an aquatic plant pest) have been spread by hitchhiking on boats. More insidious as hitchhikers are disease organisms, such as the myxosporidian that causes whirling disease. This disease is native to European salmonids but is now spreading throughout the United States with potentially devastating effects on nonresistant native salmonids. Increasingly, aquaculture is a source of disease organisms spreading to new areas; negative effects accrue to both the industry and wild populations of aquatic organisms (Hastein and Lindstad 1991; Stewart 1991).

The pet trade is another source of introductions, either because fishes escape from tropical fish farms or because people dump or flush unwanted charges in local streams and lakes. Such introductions are a major problem in tropical and subtropical areas such as Florida and Hawaii (Courtenay and Stauffer 1990). Even the spread of fishes, crayfishes, and frogs, normally attributed to anglers through bait buckets, may be partly the result of people releasing into local waterways unwanted pets collected from other waters, often quite distant.

13.3   INTRODUCTIONS AND ECOLOGICAL THEORY

A species introduction is a type of ecological perturbation that, if successful, will alter the biotic community into which the species has become part. Therefore, in this section we discuss the use of ecological theory to predict how a species proposed for introduction is likely to alter the receiving community, as well as the degree of community disturbance it is likely to cause. Relevant theory falls into two general headings: the niche concept and the concept of limiting similarity. These two ideas form the basis of the theory of island biogeography, which can be used as a general framework to explain how colonization and extinction processes shape communities of organ-

isms. Following the presentation of these concepts, the mechanisms that enable introduced species to invade and alter biotic communities will be discussed: (1) competition, (2) parasite–host interactions, (3) predation, (4) habitat modification, (5) indirect
interactions, and (6) hybridization. We end this section by integrating concepts from
an evolutionary perspective because the mechanisms affecting community structure
are dynamic and have an historical legacy.

13.3.1 Niche Concept

The most widely accepted description of the niche is that of Hutchinson (1958),
who developed the idea as a multidimensional attribute of a species or population that
contains many axes describing where and how the species lives. There are axes for
such dimensions as prey size, prey type, depth of water, and velocity of water. The
niche is therefore a characteristic of the organism, not the environment. The ecological
niche as defined by Hutchinson (1958) has two aspects: the fundamental niche and the
realized niche (Figure 13.1). The fundamental niche is the total capacity of the organism to perform activities over a wide range of environmental conditions; it is circum-
scribed by genetically determined physiological limits. The realized niche is the fraction of the fundamental niche that is expressed as an adaptation to local conditions. For
example, the presence of a competitor or predator of a species may prevent that species
from using food or space it would use in their absence and result in a rather narrow
realized niche (niche compression). If the constants imposed by other organisms are
removed, the realized niche becomes larger (niche expansion). Therefore, at any given
time and place a population of a species is using only a small part of its fundamental
niche. A classic example of how the realized niche of an introduced organism can
change is illustrated by the problems caused by the introductions of opossum shrimp
into lakes throughout the intermountain West. It was presumed that this small shrimp
would function elsewhere as it does in Kootenay Lake, British Columbia, Canada,
where it subsists largely on phytoplankton and detritus. Instead, in new environments
the opossum shrimp has proven to feed preferentially on zooplankton. As a result, it
has eliminated large zooplankton species that were often important foods for the very
game fishes whose populations it was supposed to enhance (Lasenby et al. 1986; Nesler
and Bergersen 1991).

The niche of a species changes as the species grows. Most fishes feed on quite
different and smaller prey while young and switch to larger prey as they mature. These
shifts in diets are often accompanied by shifts in habitat use. Most introductions are
made without regard to the distinct niches of the early life history stages of the fishes in
question, despite the fact that interspecific interactions for species change with life
stage (Werner 1986). For example, introduced forage fishes can compete with juvenile
piscivores for invertebrate prey or they can prey upon the young of their predators
(Crowder 1980; Kohler and Ney 1980).

Lack of understanding of the Hutchinsonian niche has led to the introduction of
species to fill "vacant," "unoccupied," or "empty" niches. The vacant-niche concept
confuses available, apparently unused, resources with the ecological function of the
organism (Herbold and Moyle 1986). Usually the vacant niche is identified by the
absence of a familiar link in a food web (e.g., planktivorous fish in a lake), the pres-

A. Bivariate Niche Dimensions

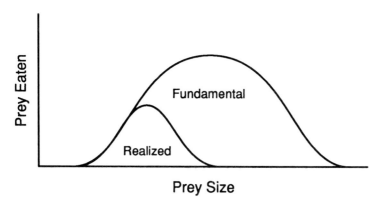

B. Multivariate Niche Dimensions

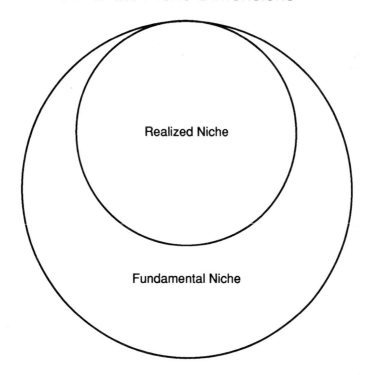

Figure 13.1 The concepts of the fundamental and realized niches (Hutchinson 1958) displayed as a single-niche axis (A) and in multiple-niche dimensions (B). Note that the realized niche in (A) may be limited by several factors, such as predator size distribution, size distribution of available prey, and available prey species.

ence of an abundant food resource located in a particular habitat, or both. Based upon observations of its success in other aquatic systems, a species is introduced to use that resource on the assumption that it will perform in the new system just as it did in the old. This assumption is often wrong because the niche expression of an organism in any particular ecological community is limited by local physicochemical and biological constraints. For example, early studies on opossum shrimp indicated that in their native lakes they were detritus feeders; consequently, they were widely introduced to fill the "vacant" detritivore niche. In new environments, the shrimp unexpectedly became major predators on zooplankton and have caused drastic changes to lake ecosystems (Lasenby et al. 1986; Nesler and Bergersen 1991). Williamson (1996) argues that a vacant niche reflects a missing functional role in the community, at least as we understand it. However, an invading species may occupy a so-called vacant niche and still compete for resources, although using these resources in a different way. As Williamson indicates, however, use of the term in this way is not particularly useful because it requires perfect knowledge of the system being invaded in order to predict effects of an introduction.

13.3.2   Limiting Similarity and Island Biogeography

When a new species invades an ecosystem, five outcomes are possible: (1) species addition with species extinction(s) (species replacement), (2) species addition without niche compression of similar species, (3) species addition with niche compression, (4) species extinction(s) because of alterations of food webs or the environment, and (5) failure of the invading species to become established. These are predictions based upon the theory of limiting similarity (MacArthur and Levins 1967) and the theory of island biogeography (MacArthur and Wilson 1967). One consequence of limiting similarity is that in a community of species that have all invaded at one time or another (which is true of most natural biological communities), the sequence of invading species makes a big difference as to which species ultimately become part of an assemblage. If species A and B are competitors, then if B arrives in a region before A it may prevent A from becoming established. This development is known as assembly theory, which has some potential in predicting the success or failure of introductions (Drake 1990; Moyle and Light 1996).

The theory of limiting similarity undergirds the theory of island biogeography. It is also known as the species packing hypothesis: as the number of species in an assemblage increases, greater partitioning of resources occurs, resulting in niche compression. At some level resources cannot be partitioned further because there is a limit to the degree of niche similarity among species.

The theory of island biogeography applies not only to actual islands but to isolated patches of habitats as well. Freshwater habitats are typically islands of water surrounded by a sea of land. The number of species supported by an island is a function of accessibility to colonists, as well as patch size and habitat diversity. These factors affect immigration and extinction rates, thus limiting the number of species that an island can support. One prediction from the theory of island biogeography is that if repeated introductions are made into an aquatic system, extinction rates will be high, and the fish community will become unstable and difficult to manage (Magnuson 1976).

The logical extension of these two theories is that a species-rich fauna will be more resistant to invasion than one that is species poor (Diamond and Case 1986), but this generality is far from universal. For instance, some depauperate native fish communities in undisturbed California streams have repeatedly resisted invasions by introduced species (Baltz and Moyle 1993), whereas the complex riverine faunas of the eastern United States have been readily invaded by common carp and other species. The reasons that the two theories have limited predictive value is that they presume steady state conditions and assume that competition among species is the most important factor structuring communities. These theories also do not take into account that moderately disturbed environments often have the highest diversity of species because random environmental fluctuations prevent the most efficient competitors from becoming dominant. Highly disturbed environments are most likely to contain only a few tolerant species, typically human symbionts like common carp, goldfish, fathead minnow, and mosquitofishes.

13.3.3   Mechanisms Affecting Community Structure

13.3.3.1   Competition

Competition occurs when two organisms require a given resource that is in limited supply. Competition is frequently cited as a major reason why introduced fishes replace native fishes, but most of the evidence is anecdotal or inferential and does not demonstrate conclusively that there is some limiting resource (Fausch 1988; Ross 1991). The most dramatic example of competition occurs among territorial salmonids where aggressive behavior leads to the dominance of one species over another (interference competition). For instance, Fausch and White (1986) found that introduced adult brown trout displaced native adult brook trout from the best habitats, making the brook trout more vulnerable to fishing and other forms of predation.

Exploitation competition has not been as well documented as has interference competition because it is not as conspicuous. However, it may be extremely common. It occurs when one species uses resources more quickly and more efficiently than the other. Thus, the redside shiner largely replaced juvenile rainbow trout in the littoral zone of a Canadian lake because it more efficiently exploited invertebrates associated with beds of aquatic macrophytes (Johannes and Larkin 1961). Declines of kokanee in lakes of western North America have been caused by the near elimination of large zooplankton species by introduced opossum shrimp (Lasenby et al. 1986), whereas declines and extinction of whitefishes in the Great Lakes have been at least partially caused by the removal of large zooplankton species by introduced alewife (Crowder and Binkowski 1983). Less characteristically, introduced creek chub forced brook trout in Quebec lakes to switch from feeding on benthos to feeding on zooplankton, apparently causing reduced growth rates (Mangan and Fitzgerald 1984).

13.3.3.2   Parasite–Host Interactions

Introduced species can be sources of introduced diseases that severely deplete native populations. The transfer of disease organisms (e.g., microparasitic bacteria and viruses and macroparasitic cestodes and nematodes) worldwide during the Twentieth Century has been without precedent (Ganzhorn et al. 1992). The European crayfish was virtually eliminated from northern Europe by a disease brought in by signal crayfish from North America. Ironically, the demand this decline created for imported cray-fishes in Scandinavian countries caused the development of fisheries for crayfishes in California. In the former Soviet Union, attempts to introduce a new species of sturgeon into the Caspian Sea failed, but the attempt did succeed in introducing a sturgeon parasite that devastated the populations of native sturgeons.

A more subtle effect of parasites is their role of giving advantage to an immune host over a susceptible competitor by introducing an energy drain and an additional source of mortality. Coexistence between native and introduced species may depend on relative degrees of immunity from reciprocal parasites. Thus, whitefishes are re-stricted to benthic prey in the presence of cisco, and they are further disadvantaged because of their increased susceptibility to parasites hosted by benthic invertebrates (Holmes 1979). One of the causes of the decline of the woundfin, an endangered spe-cies in Utah, is heavy infestations by an Asiatic tapeworm *Bothriocephalus acheilognathi*. This tapeworm accompanied the introduced red shiner, which seems to be replacing the woundfin, in part, because it is more resistant to the tapeworm. The red shiner, in turn, picked up the tapeworm in its own native range from introduced grass carp (Deacon 1988).

13.3.3.3   Predator–Prey Interactions

Predation is a powerful evolutionary force, and there are many studies that dem-onstrate that top carnivores can determine not only the kinds and numbers of potential prey species but also the kinds and numbers of species at lower trophic levels. How-ever, the most dramatic effects of predators are often on their prey species, and the introduction of a predator into a system containing prey species not evolved to counter its particular style of predation can lead to dramatic changes in the numbers and diver-sity of the prey assemblage (Li et al. 1987; Arthington 1991; Holcik 1991). There are many examples of how introduced predators have altered biotic communities of inland waters of North America, but in this section we will focus on the introduction of Nile perch into Lake Victoria in East Africa. We do this because it has been the most devas-tating introduction in modern times and because the introduction was based, in part, on the advice of western fisheries biologists.

Lake Victoria, like other large rift lakes in the region, supports an incredibly rich fauna of haplochromine cichlids—at least 300 species—which evolved within the lake from lim-ited ancestors. The introduction of the large, predatory Nile perch into Lake Victoria was proposed because, from a western perspective, the abundant but small cichlids did not provide an adequate fishery for the native peoples. However, an active artisanal fishery existed, and small cichlids were eaten locally. Gee (1965) considered the introduction de-sirable because it would control the haplochromine cichlids and convert them to more useful, larger fish. While the debate on whether or not to introduce the Nile perch was occurring, it mysteriously appeared in the lake, justifying further introductions. Until about 1975, the Nile perch populations remained small. Suddenly, the populations exploded and in the process extirpated or greatly reduced the populations of most of the cichlids and 40 species of noncichlids (Hughes 1986). This process was accelerated by the simultaneous eutrophication of the lake, perhaps stimulated by the Nile perch invasion. Eutrophication has interrupted the breeding behavior of cichlids through a reduction in water clarity (Seehausen et al. 1997). Today, three species dominate the fish fauna: Nile perch, an introduced cichlid, and a native zooplanktivore. The prey of the Nile perch has shifted from small cichlids to shrimp, and its own young, and the entire functioning of the lake ecosystem has changed (Goldschmidt et al. 1993).

Initially, Nile perch did not find favor with the local peoples because it was too large to preserve by traditional methods of sun drying and difficult to catch by traditional methods. However, people soon adjusted their gear to catch the Nile perch and learned to process the fish for oil, which they used for cooking it. The fishery has greatly expanded, and much of the Nile perch catch is now exported, giving at least some people an economic advantage. However, the need to rend oil from the fish has increased the demand for charcoal. As a result, greater local deforestation is occurring (Kaufman 1992). Because of the apparent success of the fishery, however, Nile perch may be planted in other rift lakes with similar devastating effects on local ecosystems.

The saga of the Nile perch is similar to what happened in the Laurentian Great Lakes following the invasion of the sea lamprey: drastic declines of fishes not adapted to lamprey predation. It is likely that the large native fishes would have disappeared from the Great Lakes altogether if a massive program of sea lamprey control had not been initiated and sustained at considerable cost. Predators need not be large in size, like Nile perch or sea lamprey, to have a major impact; predation on eggs and larvae of native fishes by introduced zooplanktivores can also cause population declines (Crowder 1980; Kohler and Ney 1980).

13.3.3.4   Complex and Indirect Effects

The effects of introductions are often more complicated than previously described. The effects may be noticeable only as a gradual, indirect restructuring of the recipient biotic community (Ross 1991). Two interrelated types of effects are most common: habitat modification and cascading trophic interactions.

The success of the common carp is at least partly the result of its ability to modify the shallow-water habitats it favors. It roots up the bottom and aquatic macrophytes, making ponds and shallow lakes more turbid. Increased turbidity decreases the abundance of visual-feeding predators and competitors. Similarly, declines of sport fisheries have been associated with the invasion of the rusty crayfish in lakes where it eliminates aquatic plant beds that are used for cover by juvenile fishes and as a source of invertebrates for larger fishes (Lodge et al. 1985).

Some introductions are made because the species can alter habitats. The best example of this is the grass carp, which can eliminate large beds of troublesome aquatic macrophytes. A study by Rowe (1984) on a New Zealand lake revealed some of the possible indirect effects of macrophyte removal. After the introduction of grass carp, the following sequence of events occurred: (1) removal of macrophytes; (2) increase in phytoplankton production; (3) increase in zooplankton production leading to an increase in growth rates of rainbow trout; (4) increase in predation on rainbow trout by cormorants because of lack of cover for the fish; and (5) shifts in feeding habits, relative density, and growth rates of other resident fishes.

When a top predator is introduced into a lake and severely reduces the populations of dominant planktivorous fishes, the effect cascades down the food chain and results in alterations of zooplankton and phytoplankton abundance and species composition. This, in turn, affects the abundance of other fishes, including benthic species. Such effects have been called cascading trophic interactions (Carpenter et al. 1985). Thus, Scavia et al. (1986) suggest that the introduction of predators coupled with some climatic changes have drastically altered the trophic structure of Lake Michigan. Similar cascades have also been noted in streams. Townsend (1996) has shown that the introduction of brown trout into New Zealand has drastically changed the stream ecosystems because the brown trout has greatly reduced the abundance of grazing insects, which has resulted in an increased abundance of algae. The likelihood that an introduced species will produce such cascading effects seems to be related to ecosystem productivity (Li and Moyle 1981). All other things being equal, more eutrophic (up to a point) environments are likely to support more species and be less susceptible to disruption by introductions. For example, large cladocerans (zooplankton) and introduced opossum shrimp coexist in mesotrophic lakes but not in oligotrophic lakes (Lasenby et al. 1986).

Occasionally, the effects of an introduction can cascade into terrestrial systems. For example, introduction of opossum shrimp into Flathead Lake, Montana, ultimately caused increases in the mortality of bald eagles (Spencer et al. 1991). The opossum shrimp competed with kokanee for zooplankton, which resulted in fewer salmon carcasses being available for bald eagles and grizzly bears to scavenge upon in spawning streams (Figure 13.2). Bald eagles then shifted their foraging efforts to scavenging road-killed animals, with unfortunate consequences to both motorists and birds (J. A. Stanford, University of Montana, personal communication).

13.3.3.5 Hybridization

Introduced fishes commonly hybridize with closely related native species, usually to the detriment of the native species (Krueger and May 1991). For example, hybridization has resulted in elimination of cutthroat trout by rainbow trout in much of the Great Basin (Behnke 1992). Hybridization with introduced species and subsequent introgression through backcrossing has been found to be extensive and pervasive, leading to the decline in the threatened Apache trout and the Pecos pupfish (Wilde and Echelle 1992; Carmichael et al. 1993). In recent years, another problem identified is hybridization between genetically distinct stocks of the same species. Thus indiscriminate introductions of Florida strain largemouth bass and northern strain largemouth bass may result in a loss of fitness of both strains in their native ranges (Philipp 1991). The same concern is shared by salmonid biologists. Various genetic stocks may be highly adapted to local conditions, so interbreeding with other stocks, especially hatchery strays, may reduce a stock's ability to respond to fluctuations in local environmental conditions (Utter 1981; Campton and Johnston 1985; Meffe 1992).

13.3.3.6 Adaptation and Evolutionary Processes

Adaptability to hydrologic conditions can affect interactions among native and nonnative fishes and determine whether or not nonnative species will become established in a new environment. Coexistence of the Gila topminnow with introduced western

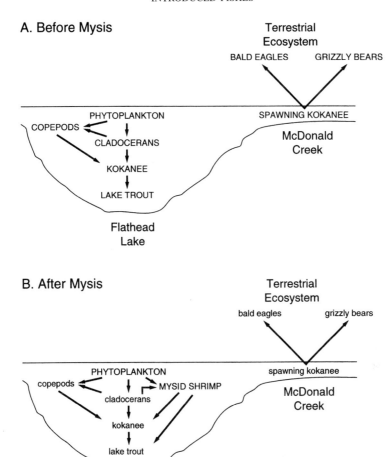

Figure 13.2 Effects of introductions of opossum shrimp (*Mysis*) to Montana's Flathead Lake trophic network. Arrows denote the direction of prey consumption. The lowercase lettering in (B) denotes the diminution of affected populations. (Figure modified from Spencer et al. 1991.)

mosquitofish in the desert streams of Arizona is predicated upon periodic flash floods (Meffe 1984). The topminnow has evolved behaviors to cope with sudden increases in discharge, but the nonnative western mosquitofish has not. In the absence of periodic flash flooding, predation by western mosquitofish would extirpate the Gila topminnow within a few years. Similarly, the fish community of a small mountain stream seems to shift between two equilibria, one dominated by native fishes and one dominated by introduced brown trout; the dominant species is determined by the recent hydrological history of the stream (Strange et al. 1992).

It is possible for introduced species to adapt to new environments through natural selection and spread beyond their expected range. Thus, intense selection pressure through periodic winter kills may select for cold-hardy strains of mosquitofishes and swordtails, enabling these fishes to invade areas where they are not wanted. Similarly, rainbow trout in western Australia now live at higher temperatures than other stocks (Arthington and Mitchell 1986). Natural selection may also reduce the impact of an invading species. Townsend (1996) found that mayflies in New Zealand streams switched from diurnal to nocturnal grazing in the presence of brown trout, reducing the predation pressure of trout on the insects. The change in behavior had a genetic basis. In short, evolutionary processes can both intensify problems brought about by invading species and allow the invaded systems to adjust to presence of the introduced species.

It might be presumed that due to their mere proximity interbasin transfers are safer introductions than are introductions of exotic species from more distant waters. This is not so. Brown and Moyle (1997) found that the interbasin transfer of Sacramento pikeminnow and California roach caused major changes to a large coastal river ecosystem, whereas species from more distant areas had relatively small impacts.

13.4   A THEORETICAL FRAMEWORK FOR EVALUATING SPECIES INTRODUCTIONS

Although various aspects of theory have given us some insights concerning the problems of nonnative invasions, an overarching theory enabling us to predict the effects of species introductions has not yet been fully realized (Kareiva 1996). This gap has been a stimulant to develop a broader and more cohesive framework. In the next two sections, we present two approaches as our contribution toward the development of a theoretical framework: mathematical modeling using signed digraphs (i.e., loop analysis) and a set of empirically derived rules.

13.4.1   Community Analyses and Signed Digraphs

Communities can be represented as mathematical models that provide metaphorical insights concerning the natural world. Most community models are composed of the following fundamental basic parts: (1) links that qualitatively describe the interactions between species (e.g., trophic, competitive, mutualistic, or parasitic), (2) qualitative measures of the strength of the interactions among organisms, (3) Malthusean-related parameters (e.g., production, births, and deaths), and (4) positive and negative feedback. Of interest to the community ecologist are the factors governing the membership and the structure of the community over time. This field is still in its infancy, but the current thinking can be summarized as follows.

1. Self-dampening properties in the form of negative feedback stabilize community structure.

2. Positive feedback destabilizes communities.

3. Contrary to May's (1974) assertion, community complexity does lead to stability and persistence (McCann et al. 1998). As the numbers of species increases, interaction strength must decrease (May 1974); weaker predator–prey interactions can dampen population oscillations (McCann et al. 1998).

The corollaries are that invasive species pose great risks when they increase positive feedback in the community and they create strong interspecific interactions within the community. A highly voracious predator or a very competitive species can change the structure of the native community.

From a certain sense, predicting the effects of an invading species would appear to be easy. There are established models to follow, and it would seem that all that is needed are the interaction values. The problem is that the data are scant. Another problem is that interaction effects may change under novel circumstances. How then might these models be useful in predicting impacts?

We suggest that an array of potential outcomes can be modeled using qualitative mathematical models called signed digraphs (Li et al., in press). This powerful technique has been used to evaluate ecological change and community stability (e.g., Li and Moyle 1981; DeAngelis et al. 1986; Lane 1986; Schmitz 1997). Signed digraphs are diagrams that show the direction and effect of one species upon the next, hence directed graph or digraph. Many ecologists have intuitively expressed community interactions in signed digraphs, for example, trophic webs, although they have not explored the relationships mathematically (e.g., Schoener 1989; Menge 1992; Hacker and Gaines 1997). These relationships can be translated mathematically into community or Jacobian matrices and explored using matrix algebra (Levins 1974; Puccia and Levins 1985). If quantitative data are available, the interactions are said to be specified. Qualitative analysis interactions are unspecified. The effect of one species on the next (feedback) is denoted as + (positive feedback), 0 (no feedback), or −(negative feedback). Feedback acts on a single population or among populations (Figure 13.3). Relationships between populations are often denoted as interaction pairs: $(+,-)$ can represent predator–prey or parasitic relationships; $(0,-)$ amensalism; $(+,0)$ commensalism; $(+,+)$ mutualism; and $(-,-$ or $-,0)$ interference competition, depending on the symmetry of the relationship. Modern software packages with symbolic processors are now widely available to alleviate the tedium and laborious task of hand calculation.

Qualitative analyses of signed digraphs have several advantages for analyses of the effects of invasive species: (1) results are mathematically rigorous and the results should be accurate, although not precise, as long as the depiction of interactions is correct; (2) "what-if" plausible scenarios can be generated by an intuitive natural historian when data are sparse; (3) predictions can be made concerning community stability, acute or pulse disturbances (Levine 1992), and chronic or press disturbances (Bender et al. 1984); (4) the influence of direct and indirect food web interactions can be examined through feedback networks; and (5) simulations can be run quickly on personal computers. Digraphs provide a powerful heuristic tool that can generate interesting hypotheses for managing communities.

Here is a hypothetical example. A nonnative species has been discovered in a small lake. From its morphology, the fish appears to be a small, benthic invertivore, and only three published papers can be found that discuss its life history. A useful strategy would be to simulate all feasible scenarios with respect to community interactions with this species and then to examine potential effects on target species and community stability. Let us say that 10 out of 15 possible scenarios suggest that the introduction will destabilize community structure. Therefore, the risk that the community will change is 66%. This risk factor is just a hypothesis, but working through the process will help identify ecological interactions of importance, place priorities concerning research, and determine the likelihood that control is possible.

Mathematical models of community interactions have generated interesting insights. Invasive species can destabilize communities by altering keystone interactions, which lead to simplified communities (Li et al., in press). Hierarchical spatial partitioning of community members tends to increase community stability (Tansky 1978). Indirect interactions (i.e., complementary feedback) are important for community persistence because they reflect food chain effects (Puccia and Levins 1985). Native predators can ameliorate the effects of invasive prey to a degree if the native species are generalist feeders and act as keystone predators (McCann et al. 1998; Castillo et al., in press). It is possible for nonnative organisms to coexist with native species depending upon the overall feedback strength of the system (Castillo et al., in press).

13.4.2 Empirical Rules

Because ecological theory is still a mixed blessing when it comes to predicting the success and effects of introduced species, Moyle and Light (1996) promulgated 12 empirical rules governing species introductions and invasions that seem to hold for most cases.

1. Most invaders fail to become established. Most failed introductions and invasions are not recorded, but there are many failures. Brown and Moyle (1997) found that in the Eel River, California, many introduced species failed to become established in the river despite annual invasions.

2. Most successful invaders become integrated into the communities being invaded, without extirpations of resident species. Although most attention is paid to the invasions that do have negative impacts, a surprising number of invading species seem to have been integrated into local assemblages containing mainly native species. This does not mean, of course, that the systems have not been changed as a result of the invasion. A caveat to this rule is that invaded communities have been studied for only very short timespans, so long-term effects may be different than observed effects. See also rule 6.

3. All aquatic systems are invasible, and invasibility is not related to diversity of the resident organisms. Introductions have been made into aquatic systems ranging from tropical reefs to desert springs, from mountain lakes to estuaries. Thus, diversity is not a good predictor of the success of an introduction.

4. Major community effects of invasions most often occur when the number of resident species is low. When there are few species in a system, an invading species is most likely to have noticeable effects, for example, the elimination of a pupfish from a desert spring by predatory largemouth bass. Even Lake Victoria may be an example of this rule in that the cichlid species driven to extinction by the Nile perch were all part of a species flock that evolved rapidly in the lake from a common ancestor, so all had an identical inability to avoid the predator. One corollary of this rule is that oligotrophic systems are more likely to be altered by a successful invader than are naturally eutrophic systems.

5. In systems that have been minimally altered by human activity, fishes most likely to be successful invaders are top predators and omnivore–detritivores. Both types of invaders are likely to find abundant resources in the system being invaded that allow for rapid expansion of their populations.

SIGNED DIGRAPHS

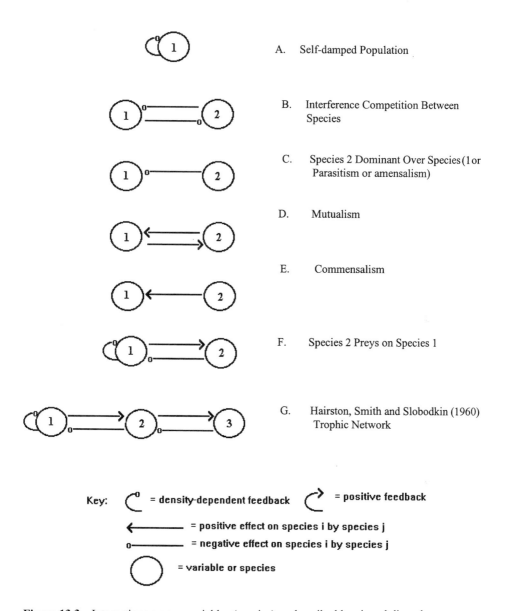

A. Self-damped Population

B. Interference Competition Between Species

C. Species 2 Dominant Over Species (1 or Parasitism or amensalism)

D. Mutualism

E. Commensalism

F. Species 2 Preys on Species 1

G. Hairston, Smith and Slobodkin (1960) Trophic Network

Key: C⁰ = density-dependent feedback   = positive feedback

 = positive effect on species i by species j

0 = negative effect on species i by species j

 = variable or species

Figure 13.3 Interactions among variables (species) as described by signed digraphs.

6. Piscivorous invaders are most likely to alter the fish assemblages they invade, whereas omnivores and detritivores are least likely to do so. This observation has been used to justify the introduction of three species of detritivorous fish into the Sepik River, New Guinea, to increase protein supplies for local people (Coates 1993) and also to devote considerable resources to attempts to eradicate predatory northern pike and white bass in California.

7. In aquatic systems with intermediate levels of human disturbance, any species with the right physiological and morphological characteristics can become established. The rule is a bit broad to be very useful, but it does indicate why such caution has to be taken with introductions.

8. In the long term, or in relatively undisturbed aquatic systems, success of an invader will depend on a close match between its physiological and life history requirements and the characteristics of the system being invaded. The importance of the match explains why species are highly likely to become successful invaders in nearby but isolated lakes and streams or above natural barriers. It also explains why many invasions fail after initial success (Lodge 1993; Brown and Moyle 1997).

9. Invaders into an aquatic system are much more likely to become established when native biotic assemblages have been temporarily disrupted or depleted. Most communities have some biotic resistance to invasions. For example, a local predator is likely to eat a new invader that has not yet adapted to local conditions. Thus fisheries that deplete populations of predatory fishes may make the system more vulnerable to invasion by a new species.

10. Integration of an invading species into an established assemblage is much more likely in an aquatic system permanently altered by human activity than in a lightly disturbed system. Estuaries, large rivers, and many lakes may be especially invasible by fishes and invertebrates because they have been heavily polluted or otherwise altered.

11. The invasibility of an aquatic system is related to interactions among environmental variability, predictability, and severity. In other words, invaders are less likely to be successful in extreme environments (e.g., highly alkaline lakes) or highly variable and unpredictable environments (e.g., some streams and coastal upwelling zones).

12. Invaders are most likely to extirpate native species in aquatic systems with extremely high or extremely low variability or severity. Some of the most disastrous invasions recorded have been in fairly constant environments such as tropical lakes or desert springs. Human-altered streams often have extremely high variability and severity of conditions, favoring tolerant invaders such as common carp, fathead minnow, and red shiner.

13.5 ALTERNATIVES TO INTRODUCTIONS

To paraphrase Aldo Leopold (1938), introductions can serve as the perfect alibi for postponing the practice of fisheries management. Although most fishes are introduced with the best intentions and may live up to the manager's expectations at the local level, at least for short periods of time, unexpected negative effects of introductions often outweigh the positive effects. This phenomenon was called the "Frankenstein effect" by Moyle et al. (1986) after the central figure of Mary Shelley's 1818 novel. Dr. Frankenstein thought he was creating an improved version of man but instead created a monster. There are many examples of the Frankenstein effect in North

America, including many not mentioned previously: common carp have reduced the ability of waterfowl refuges to support ducks by eliminating the aquatic macrophytes upon which the ducks feed; introduced cyprinids prey on the eggs and larvae of endangered fishes in the Colorado River, negating most efforts to restore the natives; and introduced trout eliminate frogs from mountain lakes.

Because the side effects of an introduction are difficult to predict, it is prudent to first consider alternative management strategies when an introduction is suggested as a solution to a management problem. Such strategies include (1) use of native fishes already present, (2) better water management, (3) habitat protection, and (4) use of sterile fishes.

13.5.1 Use of Native Fishes

Often a fish species is introduced simply because a fisheries manager is familiar with it. Native fishes may be underused because of ignorance. Often native species are superior to introduced species for aquatic pest control and for forage species because they are better adapted to local conditions. Ahmed et al. (1988) provided a system for rating the suitability of native fishes for aquatic pest control that could be modified for other purposes. Intensive investigation of native species should be conducted to see if they fit the management need before any introduction is made. For example, cisco and lake trout form a coadapted predator–prey complex in many of the Canadian boreal lakes. Therefore, cisco introductions to lakes with lake trout have long-lasting positive benefits; ciscoes enhance the production of lake trout (Matuszek et al. 1990) although they presumably alter the invertebrate communities in the lakes.

There are many situations in which increased fishing opportunities can be provided simply by educating the public about the value of underused species such as large cyprinids and suckers. Often these fishes already support small fisheries for people who know how to catch and prepare them. Expanding such fisheries can be accomplished by educating anglers on angling techniques and new ways of preparing the fishes to eat. For example, suckers are highly acceptable to North American palates if smoked or pickled.

13.5.2 Improved Water Management

Increasingly, North American waters are dammed, diverted, polluted, or otherwise made less suitable for fishes, especially high-value species like trout and salmon. Fisheries managers are given responsibility for managing the fish populations in such altered waters and are often blamed if the fishing is poor. Introductions of tolerant species or of hatchery-reared fish are then made as a way of providing better fishing because reversing environmental degradation is so difficult. Yet introductions into such altered waters are often undesirable because, although they may temporarily create better fishing, they may also prevent the pubic from realizing the seriousness of the water quality and water management problems. For example, the unique fishes of the Colorado River survived the environmental changes of numerous impoundments, but introduction of predatory fishes may have guaranteed their extinction or at least made them forever dependent on artificial propagation (Minckley 1991). If possible, attention should be focused on the environmental causes of lost fisheries, and the blame for poor fishing should be shifted onto the shoulders of water managers and polluters. They, in turn, should have the responsibility for providing the conditions necessary to support resident sport fishes. It is likely that such improvements would benefit not only fisheries but also entire communities of aquatic organisms of which the sport and commercial species are part.

A good example of the need for better water management is found in most reservoirs, which are often built using fisheries values as a partial justification. Reservoir water levels and outflows are rarely managed specifically for fishes even for short periods of time. For example, stabilizing water levels during the spawning period of centrarchids can greatly enhance their populations whereas releases from reservoirs to enhance streamflows can improve survival of migrating salmonids.

13.5.3   Habitat Protection

It is axiomatic that native fishes and most high-value sport fishes thrive in relatively undisturbed habitats or in habitats that most closely resemble those in which they evolved. Habitat protection and, if necessary, restoration, should therefore be cornerstones of fisheries management. These measures are often slow to produce results and are difficult to accomplish; however, they can result in more predictable and more manageable fisheries resources than are likely to occur through repeated species introductions.

Habitat is the template for community structure because geomorphic, climatic, energetic, and hydrologic factors set boundary conditions for community interactions (Steedman and Regier 1987). As such, the role of habitat should be carefully examined before introductions are made. The near extinction of lake whitefish after the introduction of walleye in Canadian lakes occurred because the role of habitat was not well understood. Mistakes such as this can be avoided in the future because the faunal composition of lakes can now be predicted from area, mean depth, water transparency, and nutrient availability; certain lakes are best suited for northern pike, others for lake trout (Marshall and Ryan 1987). The effects of rainbow smelt introductions were less severe in Lake Champlain than in other Laurentian Great Lakes because the lake's morphometry allowed for greater thermal, spatial, and temporal segregation of rainbow smelt from other fishes (Colby et al. 1987).

Ultimately, habitat management depends on managing the entire system of which the habitat is part: the watershed (see Chapter 9) and, beyond that, the bioregion (see Chapter 5). In the long run, watershed management is both practical and likely to produce much better results than the quick fixes from introductions.

13.5.4   Use of Sterile Fishes

There are some situations, mainly in highly disturbed or artificial habitats, in which special management problems (mosquitoes, weeds, or intensive fishing pressure) make the introduction of fish the best management strategy. If at all possible, the fish planted in such situations should not be able to escape and populate habitats where it is not wanted. The best way to ensure this is to plant a fish that cannot reproduce so that the introduction is a reversible phenomenon. The high demand for grass carp for weed control, coupled with the widespread fear of the ecological damage it might cause, has lead to the development of sterile, triploid stocks for planting in open waters. Another method that is being explored for grass carp, tilapia, and other fishes is to produce fish of only one sex (Shelton 1986). The biggest concern with planting fish incapable of reproduction is assuring that all fish are in fact sterile or of the same sex.

13.6   GUIDELINES FOR POTENTIAL INTRODUCTIONS

Despite the many problems created by introduced species, further introductions seem inevitable. For example, the newsletter of the Introduced Fish Section of the American Fisheries Society exists, in part, to inform its readers of new and proposed introductions in North America. Given the recent advances in ecological theory and our increased knowledge of the effects of introductions, no introduction today should be made without a detailed evaluation of its potential effect. Lasenby et al. (1986) pointed out that many of the problems created by opossum shrimp introductions would have been avoided if evaluation procedures such as those suggested by Li and Moyle (1981) had been followed. Kohler and Stanley (1984) and Kohler (1992) provide a detailed protocol, including a review and decision model, for evaluating potential introductions. Similar guidelines are provided by the International Council for the Exploration of the Sea (ICES 1995) and are recommended for implementation by the Aquatic Nuisance Species Task Force (1994), a special committee created by the U.S. Congress.

The following guidelines need to be considered when an introduction is proposed.

1. No introductions should be made into the few aquatic systems left that show little evidence of human disturbance. Oligotrophic, nutrient-poor, or open marine systems should, in general, be considered to be poor sites for introductions.

2. Introductions should be considered mainly for systems that have been so altered by human activity that original fish communities have been disrupted or eliminated.

3. Introductions should be considered mainly for bodies of water that are sufficiently isolated so that uncontrolled spread of the introduced species is unlikely. Because most problem waters are not isolated, the best alternative is to evaluate potential effects of the introduction on all connected waters, no matter how distant. Nearby unconnected waters also must be evaluated, as anglers are fond of moving fish around.

4. Any system being considered for an introduction should have its biota thoroughly inventoried, and a list of species that might be sensitive to the introduction should be developed. Special consideration should be given to rare species or species ecologically most similar to the species proposed for introduction.

5. From the inventory, species should be categorized according to functional groups by habitat and trophic position. Food webs should be constructed using whatever information is available, and the potential effect of the introduction on trophic structure evaluated. This would essentially provide an overview of the possible interactions among native and introduced species.

6. If major gaps in understanding emerge from the above exercise, then further research should be conducted on the system. Particularly recommended are experiments with the proposed introduction in isolated ponds or laboratory systems.

7. The potential of a proposed introduction for bringing new parasites or diseases into a recipient system should be thoroughly investigated. Ideally, a proposed species should be raised under quarantine conditions for several generations before being introduced.

8. The life history characteristics of the potential introduction should be thoroughly understood before the introduction is made. The ideal introduced species will be fairly specialized and have coevolved with many members of the assemblage to which it is being added. The broader the diet of the species, the more likely it will be to create unexpected problems. It should have low vagility in case it escapes from the original site of introduction so that its spread will be easier to control.

9. Each proposed introduction should be evaluated by an independent review panel of scientists familiar with ecological principles relating to aquatic systems. It is important not to be too hasty with an introduction, as most are irreversible.

13.7   CONCLUSIONS

It is clear that the problem of species invasions, especially from unauthorized and by-product introductions, will not be solved by voluntary actions on the part of citizens, corporations, or governments. Federal regulations to control species introductions already exist in both the United States and Canada (Stanley et al. 1991), but state and provincial regulations are highly variable. Dentler (1993) reviewed the laws available to control introductions in the United States at the federal and state (Washington) levels and concluded they are largely inadequate. For example, the federal Nonindigenous Aquatic Nuisance Prevention and Control Act of 1990, although a strong act in many respects, has been applied mainly to ballast water introductions in the Great Lakes. It does not solve the major problems of ballast water introductions elsewhere but relies on voluntary compliance of shipowners to eliminate or reduce the introductions. Although techniques are available to reduce the problem (Locke et al. 1993), there is little evidence of their widespread use outside the Great Lakes. One way to approach the problem of by-product introductions in particular is to treat them as pollution that must be halted under the Clean Water Act. Indeed, in January 1999, a coalition of environmental groups petitioned U.S. Environmental Protection Agency to regulate discharges of ballast water under the Clean Water Act in order to stop the continued introduction of new organisms. If unauthorized introductions cannot be halted under existing laws, then stronger, more effective laws are needed (Moyle 1991). New problems that require better regulation are also emerging, such as the development of transgenic fishes for aquaculture and other purposes (Kapuscinski and Hallerman 1990).

The management of natural resources ultimately is based on value systems. Political, social, economic, and aesthetic values are powerful engines that drive management policies. Human values have permitted and even encouraged introductions as a management tool. This is likely to be the case for some time to come, but the indiscriminate methods of the past are no longer acceptable. The American Fisheries Society has adopted a position statement (Kohler and Courtenay 1986) that advises caution and restraint with respect to introduced species.

We know that introduced species will alter the communities into which they are introduced, but we have a hard time predicting precisely what the changes will be in most cases. Therefore, extreme caution should be exercised in every proposed introduction to prevent irreversible damage to natural systems. Considerable effort needs to be made to prevent unauthorized and by-product introductions. The Frankenstein effect should always be kept in mind.

Fisheries managers also need to be cautious because values are changing, especially in the face of expanding human populations and declining natural habitats. Few fisheries managers in the 1950s or 1960s had an inkling of the present public concern for saving endangered species, preserving natural diversity, maintaining water quality, protecting wild lakes and streams, or even being able to angle for native fishes. Few also anticipated that the rapid expansion of international trade and efficient transportation systems would create many new problems with invading species. Accordingly, present-day managers need to keep as many options open as possible for future fisheries managers by critically examining their planned introductions and by seeking to enact regulations to avoid unplanned ones. As Rahel (1997:9) states, "Fisheries managers...need to consider their legacy to future generations: Will they be remembered as Johnny Appleseed or Dr. Frankenstein?"

13.8 REFERENCES

Ahmed, S. S., A. L. Linden, and J. J. Cech, Jr. 1988. A rating system and annotated bibliography for the selection of appropriate indigenous fish species for mosquito and weed control. Bulletin of the Society of Vector Ecologists 13:1–59.

Aquatic Nuisance Species Task Force. 1994. Findings, conclusions, and recommendations of the intentional introductions policy review. Report of the Aquatic Nuisance Species Task Force to U.S. Congress, Washington, D.C. (Available from the U.S. Fish and Wildlife Service.)

Arthington, A. H. 1991. Ecological and genetic impacts of introduced and translocated freshwater fishes in Australia. Canadian Journal of Fisheries and Aquatic Sciences 48(Supplement 1):33–43.

Arthington, A. H., and D. S. Mitchell. 1986. Aquatic invading species. Pages 34–53 in R. H. Groves and J. J. Burdon, editors. Ecology of biological invasions, and Australian perspective. Australian Academy of Science, Canberra.

Balon, E. K. 1974. Domestication of the carp, Cyprinus carpio L. Royal Ontario Museum, Miscellaneous Publications, Toronto.

Baltz, D. 1991. Introduced fishes in marine systems and seas. Biological Conservation 56:151–178.

Baltz, D. M., and P. B. Moyle. 1993. Invasion resistance to introduced species by a native assemblage of California stream fishes. Ecological Applications 3:246–255.

Behnke, R. J. 1992. Native trout of western North America. American Fisheries Society, Monograph 6, Bethesda, Maryland.

Bender, E. A., T. J. Case, and M. E. Gilpin. 1984. Perturbation experiments in community ecology: theory and practice. Ecology 65:1–13.

Brown, L. R., and P. B. Moyle 1997. Invading species in the Eel River, California: successes, failures, and relationships with resident species. Environmental Biology of Fishes 49:271–291.

Campton, D. E., and J. M. Johnston. 1985. Electrophoretic evidence for a genetic admixture of native and nonnative rainbow trout in the Yakima River, Washington. Transactions of the American Fisheries Society 114:782–793.

Carlton, J. T., and J. Geller. 1993. Ecological roulette: the global transport and invasion of nonindigenous marine organisms. Science 261:239–266.

Carmichael, G. J., J. N. Hanson, M. E. Schmidt, and D. C. Morizot. 1993. Introgression among Apache, cutthroat, rainbow trout in Arizona. Transactions of the American Fisheries Society 122:121–130.

Carpenter, S. R., J. R. Kitchell, and J. R. Hodgson. 1985. Cascading trophic interactions and lake ecosystem productivity. BioScience 35:635–639.

Cassani, J. R. 1996. Ponds and small impoundments. Pages 48–74 in J. R. Cassani, editor. Managing aquatic vegetation with grass carp: a guide for water resource managers. American Fisheries Society, Introduced Fish Section, Bethesda, Maryland.

Castillo, G. C., H. W. Li, J. Chapman, and P. A. Rossignol. In press. Absence of overall feedback in a benthic estuarine community: a system potentially buffered from impacts of biological invasions. Estuaries.

Coates, D. 1993. Fisheries ecology and management of a large tropical Australasian river basin, the Sepik-Ramu, New Guinea. Environmental Biology of Fishes 38:345–368.

Cohen, A. N., and J. T. Carlton. 1995. Nonindigenous aquatic species in a United States estuary: a case study of the biological invasions of the San Francisco Bay and delta. Report for U.S. Fish and Wildlife Service, San Francisco.

Colby, P. J., P. A. Ryan, D. H. Schupp, and S. L. Searns. 1987. Interactions in north-temperate lake fish communities. Canadian Journal of Fisheries and Aquatic Sciences 44(Supplement 2):104–128.

Contreras-Balderas, S., and M. Lozano-Vilano. 1994. Water, endangered fishes, and development perspectives in arid lands of Mexico. Conservation Biology 8:379–387.

Courtenay, W. R. Jr., and J. R. Stauffer, Jr. 1990. The introduced fish problem and the aquarium industry. Journal of the World Aquaculture Society 21:145–159.

Crosby, A. W. 1986. Ecological imperialism: the biological expansion of Europe, 900–1900. Cambridge University Press, Cambridge, UK.

Crowder, L. B. 1980. Alewife, rainbow smelt, and native fishes in Lake Michigan: competition or predation? Environmental Biology of Fishes 5:225–233.

Crowder, L. B., and F. P. Binkowski. 1983. Foraging behaviors and the interaction of alewife, *Alosa pseudoharengus*, and bloater, *Coregonus hoyi*. Environmental Biology of Fishes 8:105–113.

Deacon, J. E. 1988. The endangered woundfin and water management in the Virgin River, Utah, Arizona, Nevada. Fisheries 13(1):18–29.

DeAngelis, D. L., W. L. Post, and C. C. Travis. 1986. Positive feedback in natural systems. Biomathematics, v. 16. Spring-Verlag, Berlin.

Dentler, J. L. 1993. Noah's farce: the regulation and control of exotic fish and wildlife. University of Puget Sound Law Review 17:191–242.

Diamond, J., and T. Case. 1986. Overview: introductions, extinctions, exterminations, and invasions. Pages 65–79 in J. Diamond and T. Case, editors. Community ecology. Harper and Row, New York.

Dill, W. A., and A. J. Cordone. 1997. History and status of introduced fishes in California, 1871–1996. California Department of Fish and Game Fish Bulletin 178.

Drake, J. A. 1990. Communities as assembled structure: do rules govern pattern? Trends in Ecology & Evolution 5:159–164.

Eck, G. W., and E. H. Brown, Jr. 1985. Lake Michigan's capacity to support lake trout (*Salvelinus namaycush*) and other salmonines: an estimate based on the status of prey populations in the 1970s. Canadian Journal of Fisheries and Aquatic Sciences 42:449–454.

Fausch, K. D. 1988. Tests of competition between native and introduced salmonids in streams; what have we learned? Canadian Journal of Fisheries and Aquatic Sciences 45:2238–2246.

Fausch, K. D., and R. J. White. 1986. Competition among juveniles of coho salmon, brook trout, and brown trout in a laboratory stream, and implications for Great Lakes tributaries. Transactions of the American Fisheries Society 115:363–381.

Ganzhorn, J., J. S. Rohovech, and J. L. Fryer. 1992. Dissemination of microbial pathogens through introductions and transfers of finfish. Pages 175–192 in A. Rosenfield and R. Mann, editors. Dispersal of living organisms into aquatic ecosystems. University of Maryland, Maryland Sea Grant Program, College Park.

Gee, J. M. 1965. The spread of Nile perch (*Lates niloticus*) in East Africa with comparative biological notes. Journal of Applied Ecology 2:407–408.

Giles, R. H. 1978. Wildlife management. Freeman, San Francisco.

Goldschmidt, T., F. Witte, and J. Wanink. 1993. Cascading effects of the introduced Nile perch on the detritivorous/phytoplanktivorous species in the sublittoral areas of Lake Victoria. Conservation Biology 7:686–700.

Hacker, S. D., and S. D. Gaines. 1997. some implications of direct positive interactions form community species diversity. Ecology 78:1990–2003.

Hastein, T., and T. Lindstad. 1991. Diseases in wild and cultured salmon: possible interaction. Aquaculture 98:277–288.

Herbold, B., and P. B. Moyle. 1986. Introduced species and vacant niches. American Naturalist 128:751–760.

Holcik, J. 1991. Fish introductions in Europe with particular reference to its central and eastern part. Canadian Journal of Fisheries and Aquatic Sciences 48(Supplement 1):13–23.

Holmes, J. C. D. 1979. Parasite populations and host community structure. Pages 27–46 in B. B. Nicol, editor. Host-parasite interfaces. Academic Press, New York.

Hughes, N. R. A. 1986. Changes in the feeding biology of the Nile perch *Lates nilotica* (L.) (Pisces:Centropomidae) in Lake Victoria, East Africa since its introduction in 1960 and its impact on the native fish community of the Nyanza Gulf. Journal of Fish Biology 29:541–548.

Hutchinson, G. E. 1958. Concluding remarks. Cold Spring Harbor Symposia on Quantitative Biology 22:415–427.

ICES (International Council for the Exploration of the Sea). 1995. ICES code of practice on the introductions and transfers of marine organisms. ICES, Copenhagen.

Johannes, R. E., and P. A. Larkin. 1961. Competition for food between redside shiners (*Richardsonius balteatus*) and rainbow trout (*Salmo gairdneri*) in two British Columbia lakes. Journal of the Fisheries Research Board of Canada 18:203–220.

Johnson, L. E., and D. K. Padilla. 1996. Geographic spread of exotic species: ecological lessons and opportunities from the invasion of the zebra mussel *Dreissena polymorpha*. Biological Conservation 78:23–33.

Kapuscinski, A. R., and E. M. Hallerman. 1990. AFS position statement on transgenic fishes. Fisheries 15(4):2–5.

Kareiva, P. 1996. Developing a predictive ecology for nonindigenous species and ecological invasions. Ecology 77:1651–1652.

Kaufman, L. S. 1992. Catastrophic change in a species-rich freshwater ecosystem: lessons from Lake Victoria. BioScience 42:846–858.

Kohler, C. C. 1992. Environmental risk management of introduced aquatic organisms in aquaculture. ICES Marine Science Symposium 194:15–20.

Kohler, C. C., and W. R. Courtenay, Jr. 1986. American Fisheries Society position on introduction of aquatic species. Fisheries 11(2):39–42.

Kohler, C. C., and J. J. Ney. 1980. Piscivory in a land-locked alewife (*Alosa pseudoharengus*) population. Canadian Journal of Fisheries and Aquatic Sciences 37:1314–1317.

Kohler, C. C., and J. G. Stanley. 1984. A suggested protocol for evaluating proposed exotic fish introductions in the United States. Pages 387–406 in W. R. Courtenay, Jr., and J. R. Stauffer, editors. Distribution, biology and management of exotic fishes. Johns Hopkins University Press, Baltimore, Maryland.

Kornberg, H., and M. H. Williamson, editors. 1986. Quantitative aspects of the ecology of biological invasions. Philosophical Transactions of the Royal Society of London B 314:501–742.

Krueger, C. C., and B. May. 1991. Ecological and genetic effects of salmonid introductions in North America. Canadian Journal of Fisheries and Aquatic Sciences 48(Supplement 1):66–77.

Lane, P. A. 1986. Symmetry, change, perturbation, and observing mode in natural communities. Ecology 67:223–239.

Lasenby, D. C., T. G. Northcote, and M. Furst. 1986. Theory, practice and effects of *Mysis relicta* introductions to North American and Scandinavian lakes. Canadian Journal of Fisheries and Aquatic Sciences 43:1277–1284.

Lassuy, D. R. 1994. Aquatic nuisance organisms: setting national policy. Fisheries 19(4):14–17.

Leopold, A. S. 1938. Checkaremia. Outdoor America 3:3.

Lever, C. 1996. Naturalized fish of the world. Academic Press, San Diego, California.

Levine, R. L. 1992. An introduction to qualitative dynamics. Pages 267–331 *in* R. L. Levine and H. E. Fitzgerald, editors. Analysis of dynamic psychological systems, volume 1: basic approaches to general systems, dynamic systems, and cybernetics. Plenum Press, New York.

Levins, R. 1974. The qualitative analysis of partially-specified systems. Annals of the New York Academy of Science 231:123–138.

Li, H. W., and P. B. Moyle. 1981. Ecological analysis of species introductions in aquatic systems. Transactions of the American Fisheries Society 110:772–782.

Li, H. W., P. B. Moyle, and R. W. Garrett. 1976. Effect of the introduction of the Mississippi silverside (*Menidia audens*) on the growth of the black crappie (*Pomoxis nigromaculatus*) and white crappie (*Pomoxis annularis*) in Clear Lake, California. Transactions of the American Fisheries Society 105:404–408.

Li, H. W., P. A. Rossignol, and G. Castillo. In press. Risk analysis of species introduction: Insights from qualitative modeling. *In* R. Claudi and J. Leach, editors. Nonindigenous fresh water organisms in North America; vectors of introduction, biology and impact. CRC Press, Boca Raton, Florida.

Li, H. W., C. B. Schreck, C. E. Bond, and E. Rexstad. 1987. Factors influencing changes in fish assemblages of Pacific Northwest streams. Pages 193–202 *in* W. J. Matthews and D. C. Heins, editors. Community and evolutionary ecology of North American stream fishes. University of Oklahoma Press, Norman.

Locke, A., D. M. Reid, H. C. van Leeuwen, W. G. Sprules, and J. T. Carlton. 1993. Ballast water exchange as a means of controlling dispersal of freshwater organisms by ships. Canadian Journal of Fisheries and Aquatic Sciences 50:2086–2093.

Lodge, D. M. 1993. Biological invasions: lessons for ecology. Trends in Ecology & Evolution 8:133–137.

Lodge, D. M., A. L. Beckel, and J. J. Magnuson. 1985. Lake bottom tyrant. Natural History 94(8):32–37.

Ludwig, H. R., Jr., and J. A. Leitch. 1996. Interbasin transfer of aquatic biota via angler's bait buckets. Fisheries 21(7):14–18.

Ludyanskiy, M. L., D. McDonald, and D. MacNeill. 1993. Impact of the zebra mussel, a bivalve invader. BioScience 43:533–544.

MacArthur, R. H., and R. Levins. 1967. The limiting similarity, convergence and divergence of coexisting species. American Naturalist 101:377–385.

MacArthur, R. H., and E. O. Wilson. 1967. The theory of island biogeography. Princeton University Press, Princeton, New Jersey.

Magnuson, J. J. 1976. Managing with exotics-a game of chance. Transactions of the American Fisheries Society 105:1–10.

Mangan, P., and G. J. Fitzgerald. 1984. Mechanisms responsible for the niche shift of brook char, *Salvelinus fontinalis* Mitchill, when living sympatrically with creek chub, *Semotilus atromaculatus* Mitchill. Canadian Journal of Zoology 62:1543–1555.

Marshall, T. R., and P. A. Ryan. 1987. Abundance patterns and community attributes of fishes relative to environmental gradients. Canadian Journal of Fisheries and Aquatic Sciences 44(Supplement 2):198–215.

Matuszek, J. E., B. J. Shuter, and J. M. Casselman. 1990. Changes in lake trout growth and abundance after introduction of cisco into Lake Opeongo, Ontario. Transactions of the American Fisheries Society 119:718–729.

May, R. M. 1974. Stability and complexity in model ecosystems. Monographs in Population Biology. Princeton University Press, Princeton, New Jersey.

McCann, K., A. Hastings, and G. R. Huxel. 1998. Weak trophic interactions and the balance of nature. Nature 395:794–798.

Meffe, G. K. 1984. Effects of abiotic disturbance on coexistence of predator and prey fish species. Ecology 65:1525–1534.

Meffe, G. K. 1992. Techno-arrogance and halfway technologies: salmon hatcheries on the Pacific coast of North America. Conservation Biology 6:350–354.

Menge, B. A. 1992. Community regulation: under what conditions are bottom-up factors important on rocky shores? Ecology 73:755–765.

Miller, R. R. 1961. Man and the changing fish fauna of the American southwest. Papers of the Michigan Academy of Science Arts and Letters 46:365–404.

Minckley, W. L. 1991. Native fishes of the Grand Canyon region: an obituary? Pages 124–177 *in* Colorado River ecology and dam management. National Academy Press, Washington D.C.

Mooney, H. A., and J. A. Drake, editors. 1986. Ecology of biological invasions of North America and Hawaii. Springer-Verlag, New York.

Moyle, P. B. 1986. Fish introductions into North America: patterns and ecological impact. Pages 27–43 *in* H. A. Mooney and J. A. Drake, editors. Ecology of biological invasions of North America and Hawaii. Springer-Verlag, New York.

Moyle, P. B. 1991. Ballast water introductions. Fisheries 16(1):4–6.

Moyle, P. B., H. W. Li, and B. A. Barton. 1986. The Frankenstein effect: impact of introduced fishes on native fishes in North America. Pages 415–426 *in* R. H. Stroud, editor. Fish culture in fisheries management. American Fisheries Society, Fish Culture Section and Fisheries Management Section, Bethesda, Maryland.

Moyle P. B., and T. Light. 1996. Biological invasions of fresh water: empirical rules and assembly theory. Biological Conservation 78:149–162.

Nesler, T. P., and E. P. Bergersen, editors. 1991. Mysids in fisheries: hard lessons from headlong introductions. American Fisheries Society, Symposium 9, Bethesda, Maryland.

Nichols, F. H., J. K. Thompson, and L. Schemel. 1990. Remarkable invasion of San Francisco Bay (California, U.S.A.) by the Asian clam *Potamocorbula amurensis*. II. Displacement of a former community. Marine Ecology Progress Series 66:95–108.

Noble, R. L. 1981. Management of forage fishes in impoundments of the southern United States. Transactions of the American Fisheries Society 110:738–750.

Ogutu-Ohwayo, R., and R. E. Hecky. 1991. Fish introductions in Africa and some of their implications. Canadian Journal of Fisheries and Aquatic Sciences 48(Supplement 1):8–12.

Philipp, D. P. 1991. Genetic implications of introducing Florida largemouth bass, *Micropterus salmoides floridanus*. Canadian Journal of Fisheries and Aquatic Sciences 48(Supplement 1):58–65.

Puccia, C. J., and R. Levins. 1985. Qualitative modeling of complex systems. An introduction to loop analysis and time averaging. Harvard University Press, Cambridge, Massachusetts.

Rahel, F. J. 1997. From Johnny Appleseed to Dr. Frankenstein: changing values and the legacy of fisheries management. Fisheries 22(8):8–9.

Ross, S. T. 1991. Mechanisms structuring stream fish assemblages: are there lessons from introduced species? Environmental Biology of Fishes 30:359–368.

Rowe, D. K. 1984. Some effects of eutrophication and the removal of aquatic plants by grass carp (*Ctenopharyngodon idella*) on rainbow trout (*Salmo gairdneri*) in Lake Parkinson, New Zealand. New Zealand Journal of Marine and Freshwater Research 18:115–137.

Scavia, D., G. L. Fahnenstiel, M. S. Evans, D. J. Jude, and J. T. Lehman. 1986. Influence of salmonine predation and weather on long-term water quality in Lake Michigan. Canadian Journal of Fisheries and Aquatic Sciences 43:435–443.

Schmitz, O. J. 1997. Press perturbations and the predictability of ecological interactions in a food web. Ecology 78:55–69.

Schoener, T. W. 1989. Food webs from the small to the large. Ecology 70:1559–1589.

Schoenherr, A. A. 1981. The role of competition in the replacement of native fishes by introduced species. Pages 173–203 *in* R. J. Naiman and D. L. Soltz, editors. Fishes in North American deserts. Wiley, New York.

Seehausen, O., J. J. M. van Alphen, and F. Witte. 1997. Cichlid fish diversity threatened by eutrophication that curbs sexual selection. Science 277:1808–1811.

Shapiro, J., and D. I. Wright. 1984. Lake restoration by biomanipulation: Round Lake, Minnesota, the first two years. Freshwater Biology 14:371–383.

Shelton, W. L. 1986. Reproductive control of exotic fishes-a primary requisite for utilization in management. Pages 427–434 *in* R. H. Stroud, editor. Fish culture in fisheries management. American Fisheries Society, Fish Culture Section and Fisheries Management Section, Bethesda, Maryland.

Spencer, C. N., B. R. McClelland, and J. A. Stanford. 1991. Shrimp stocking, salmon collapse, and eagle displacement cascading interactions in the food web of a large aquatic ecosystem. BioScience 41:14–21.

Stanley, J. G., R. A. Peoples, Jr., and J. A. McCann. 1991. U.S. federal policies, legislation, and responsibilities related to importation of exotic fish and other aquatic organisms. Canadian Journal of Fisheries and Aquatic Sciences 48 (Supplement 1):162–166.

Steedman, R. J., and H. A. Regier. 1987. Ecosystem science for the Great Lakes: perspectives on degradative and rehabilitation transformations. Canadian Journal of Fisheries and Aquatic Sciences 44(Supplement 2):95–103.

Stewart, J. E. 1991. Introductions as factors in diseases of fish and aquatic invertebrates. Canadian Journal Fisheries Aquatic Sciences 48(Supplement 1):110–117.

Stowe, C. A., S. R. Carpenter, C. P. Madenjian, L. A. Eby, and L. J. Jackson. 1995. Fisheries management to reduce contaminant consumption. BioScience 45:752–758.

Strange, E. M., P. B. Moyle, and T. C. Foin. 1992. Interactions between stochastic and deterministic processes in stream fish community assembly. Environmental Biology of Fishes 36:1–15.

Swanson, C., J. J. Cech, Jr., and R. H. Piedrahita. 1996. Mosquitofish: biology, culture, and use in mosquito control. Mosquito and Vector Control Association of California, Elk Grove.

Tabor, R. A., R. S. Shively, and T. P. Poe. 1993. Predation on juvenile salmonids by smallmouth bass and northern squawfish in the Columbia River near Richland, Washington. North American Journal of Fisheries Management 13:831–838.

Tansky, M. 1978. Stability of multispecies predator–prey system. Memoirs of the Faculty of Science Kyoto University Series of Biology.

Tomcko, C. M., R. A. Stein, and R. F. Carline. 1984. Predation by tiger muskellunge on bluegill: effects of predator experience, vegetation, and prey density. Transactions of the American Fisheries Society 113:588–594.

Townsend, C. R. 1996. Invasion biology and ecological aspects of brown trout, *Salmo trutta,* in New Zealand. Biological Conservation 78:13–22.

Utter, F. M. 1981. Biological criteria for the definition of species and distinct intraspecific populations of anadromous salmonids under the U.S. Endangered Species Act of 1973. Canadian Journal of Fisheries and Aquatic Sciences 38:1625–1635.

Vitousek, P. M., C. M. D'Antonio, L. L. Loope, and R. Westbrooks. 1996. Biological invasions as global environmental change. American Scientist 84:468–478.

Werner, E. E. 1986. Species interactions in freshwater fish communities. Pages 344–358 *in* J. Diamond and T. J. Case, editors. Community ecology. Harper and Row, New York.

Wilde, G. R., and A. A. Echelle. 1992. Genetic status of Pecos pupfish populations after establishment of a hybrid swarm involving an introduced congener. Transactions of the American Fisheries Society 121:277–286.

Williamson. M. 1996. Biological invasions. Chapman and Hall, London.

Wydowski, R. S., and D. H. Bennett. 1981. Forage species in lake and reservoirs of the western United States. Transactions of the American Fisheries Society 110:764–771.

Chapter 14

Stocking for Sport Fisheries Enhancement

ROY C. HEIDINGER

14.1   INTRODUCTION

Fish are currently being raised for stocking in sport fisheries by federal and state agencies and by private fish culturists. Including private stocking, I estimate that approximately 2.5 billion sport fishes are stocked annually in the United States and Canada. To understand the history of fish stocking in the United States, one must become familiar with the major developments in fisheries management.

At the time the United States was settled, fishes were abundant, but as population centers developed certain stocks of fishes were reduced by overharvest and environmental degradation. Spawning runs of some fish stocks fluctuated widely from year to year. Because the commercial harvest of many of these stocks occurred during the spawning run, processing and shipping facilities were either underused or could not handle the number of fish caught. Such events eventually led to public concern for the resource.

Trout hatcheries were started in the mid-1800s, and there was a general belief that most of the problems associated with low catch rates were due to a failure in reproduction, which could be corrected by stocking. Theodatus Garlick, who is considered the father of fish culture in the United States, published his book on artificial propagation in 1857 (Thompson 1970). At that time fisheries management was dominated by fish culture. In 1864, Seth Green had opened a private trout hatchery, and by 1870 there were approximately 200 private individuals practicing fish culture (Thompson 1970). In fact, what we now know as the American Fisheries Society was founded in 1870 as the American Fish Culturists' Association (Clepper 1970); the name was changed to the American Fisheries Society in 1884. By 1871, 10 U.S. coastal states had established fisheries commissions (McHugh 1970), and in that same year the U.S. Congress authorized the formation of the U.S. Fish Commission, of which Spencer Fullerton Baird was named the first commissioner. A milestone was reached in 1872 when Congress appropriated US$15,000 for the production and introduction of American shad, salmon, and other valuable species (Clepper 1970). This was the beginning of federal support for fisheries management.

During the 1930s and 1940s some researchers began to question the merits of wholesale stocking. Swingle and Smith at Auburn University determined that the basic management problem in small ponds was not too few fish but rather too many small- and intermediate-size fish (Swingle 1970). From the late 1940s to the 1960s the failure of many stocking programs led to a reduction in the number of hatcheries and a corresponding reduction in stocking, especially for species in habitats in which populations were not self-sustaining.

Passed by Congress in 1950, the Federal Aid in Fish Restoration Act, commonly called the Dingell–Johnson Act, significantly increased the number of dollars that could be used by states for fisheries management. The Wallop–Breaux Aquatic Resources Trust Act of 1984 increased the funds from approximately $35 million to approximately $300 million in fiscal year 1997.

In the 1970s, the U.S. Congress reexamined the U.S. Fish and Wildlife Service's (FWS) responsibilities for fish stocking. Concern was expressed about the role of the FWS in providing fishes for stocking private farm ponds. In the late 1970s, FWS terminated this practice except in eight southeastern states (Chandler 1985). During the early 1980s, the FWS identified 11 species of fish or groups of fishes as being of national concern, so with a new emphasis on only nationally significant species, FWS closed 31 of their 104 hatcheries in fiscal year 1983 (Chandler 1985).

Similarly, from the 1970s to the present there has been a general reevaluation of state stocking programs. According to Keith (1986), fish stocking is a major form of fisheries management in some states but plays only a minor role in others. In most cases, agencies' philosophies regarding stocking have developed over a long period through routine monitoring and informal observation; political expediency has also played a major role. More recently, emphasis has been placed on deliberate, biologically justifiable stockings designed to optimize recreational fishing in many states (Smith and Reeves 1986). For additional information on the use and effects of cultured fishes on sport fisheries see Schramm and Piper (1995) and Stroud (1986).

14.1.1   Current Philosophies on the Use of Stocking

Even with the same information available to them, biologists do not always agree on the best use of a resource. For example, many biologists would argue that it was a mistake to stock salmon in the Great Lakes. At the extremes there are two broad approaches to fisheries management in North America: that of the purist and that of the pragmatist. The purist's philosophy places considerable emphasis on trying to maintain a near-pristine state, especially when dealing with natural bodies of water such as trout streams and natural lakes. Thus, the purist's goal is to maintain or return the fish community to the way it was before it was disturbed by human intervention. The pragmatist's philosophy takes the waters as they are—some of them humanly created and all of them humanly altered in some way—and seeks to establish a fish community that is suitable for the existing environment and for the desired purpose. In many cases, establishing a suitable fish community is done by stocking with either predator or prey species. The views of most biologists fall somewhere between these two philosophies.

One should remember that not every body of water lends itself to the same management protocol and, therefore, not every body of water should be managed the same way. Also, because the expectations of users differ, some bodies of water should be managed for different user groups. The idea of something for everyone makes sense from both a biological and public relations point of view.

14.1.2   Goals, Objectives, and Criteria for Stocking Programs

Noble (1986) discussed the relationship among goals, objectives, and criteria as these terms relate to stocking as a tool in fisheries management. Goals of stocking programs have tended to be amorphous, so it is difficult to determine if they have been

Table 14.1 Some of the more commonly stocked warmwater species.

Taxa	Number of states		Number stocked/ha [a]	
	Mid-1980s [a]	1995–1996 [b]	Fry	Fingerlings
Centrarchidae				
Largemouth bass	41	33	124–247	12–494
Smallmouth bass [c]	22	20		
Bluegill	29	30	618–3,706	Variable
Redear sunfish [d]	14	17		
Crappies	11	20		
Cyprinidae				
Grass carp	8	15		2–37
Ictaluridae				
Channel catfish	35	34		25–1,236
Flathead catfish	8	3		
Blue catfish	10	7		12–124
Percichthyidae				
Striped bass	19	17	2–124	2–124
Hybrid striped bass	26	23	Up to 247	12–49
Forage species [e]	8	7		

[a] Data are from Smith and Reeves 1986.
[b] Data are author's.
[c] Number stocked usually dependent upon hatchery allocation.
[d] Fifteen to 30% of all sunfishes stocked in the reporting states.
[e] Fathead minnow, threadfin shad, gizzard shad, golden shiner, blueback herring, creek chubsucker, and spottail shiner.

attained. The goal of stocking to enhance a fishery is difficult to quantify. Objectives, in contrast to goals, should be expressed in explicit terms, with achievement being in distinct, measurable form. For example, a quantifiable objective would be to maintain a catch rate of one fish per hour or an electrofishing rate of 100 largemouth bass per hour in a given body of water. Implementation of a stocking program implies a benefit, but this benefit may accrue at the population, community, fishery, agency, or societal level. It is not necessary that all levels will benefit (Noble 1986).

Potter and Barton (1986) reviewed a number of agencies' stocking goals and criteria for coldwater fisheries. Explicit, quantifiable objectives tended not to be included; instead, more general goals were used, such as to maintain quality fisheries, protect quality and diversity, maintain a healthy aquatic environment, meet the demand for angling, and provide a diversity of fish species.

14.2 CURRENT STOCKINGS IN NORTH AMERICA

The extent of fish stocking and the kinds of fishes being stocked, especially by the states and federal government in the United States, were reviewed in the mid-1980s. Stocking programs reviewed were subdivided into the following groups: (1) warmwater species (Smith and Reeves 1986); (2) coolwater species (Conover 1986); (3) coldwater species (Wydoski 1986); (4) anadromous species (Moring 1986); and (5) marine fisheries (Richards and Edwards 1986). Stocking records are difficult to summarize with respect to size of fish stocked, in part because of the terminology used. A partial list of terms for sizes stocked include sac-fry, fry, fingerling, smolt, subcatchable, catchable, broodstock, adult, subadult, yearling, trophy, fall fingerling, spring fingerling, and phases I, II, and III.

Table 14.2 Number (×1,000) of warmwater sport fishes stocked into inland waters of the United States and Canada during 1995 or 1996. Number of states stocking taxa is in parentheses.

Taxa	Total number stocked	Number stocked by life stage			Number stocked by jurisdiction	
		Fry	Fingerlings	>203mm	States	Federal
Acipenseridae						
Lake sturgeon	40 [d]		40		28 (2)	9
Shovelnose sturgeon	36		36		35 (1)	1
Centrarchidae						
Black crappie	3,187	1	3,167	19	2,658 (17)	529
Bluegill	16,238		16,232	6	12,794 (30)	3,444
Hybrid crappie	905		905		905 (2)	
Hybrid sunfishes[b]	2,607		2,607		2,607 (10)	
Largemouth bass	21,089	2,138	18,874	77	17,914 (33)	3,175
Redbreast sunfish	3,334		3,334	[a]	2,433 (2)	901
Redear sunfish	3,322		3,322		2,999 (17)	323
Rock bass	35		35		35 (1)	
Smallmouth bass	2,890 [c]		2,888	2	1,865 (20)	1,000
White crappie	617		589	28	543 (10)	74
Cyprinidae						
Grass carp	40	12	25	3	40 (14)	
Ictaluridae						
Black bullhead	112		111	1	92 (3)	20
Blue catfish	1,256	89	1,118	49	1,109 (7)	147
Brown bullhead	30		30		30 (1)	
Channel catfish	14,984	2,394	8,808	3,782	12,326 (34)	2,658
Flathead catfish	13		8	5	13 (3)	
Percichthyidae						
Hybrid striped bass	33,053	12,247	20,806		33,053 (23)	
Striped bass	33,789	5,924	27,865		24,886 (17)	8,903
White bass	1,919	1,759	152	8	1,919 (3)	
Polyodontidae						
Paddlefish	407	17	340	50	96 (5)	311
Sciaenidae						
Red drum	1,231		1,231		1,231 (1)	
Total	141,134	24,581	112,523	4,030	119,611	21,495
Percent		17	80	3	85	15

[a] Less than 1,000.
[b] *Lepomis* spp.
[c] 25,000 stocked by two Canadian provinces.
[d] 3,000 stocked by one Canadian province.

The number of warmwater species stocked by the states has decreased over the last 12–14 years. Forty-three of the 50 states surveyed by Smith and Reeves (1986) reported stocking warmwater species. In 1995–1996, 35 of 49 states responding stocked warmwater species; 1 of 9 Canadian provinces did so. Channel catfish was stocked by more states than any other taxa followed by largemouth bass and bluegill (Table 14.1). Almost half (47%) of the 141 million warmwater fishes stocked by government agencies were hybrid striped bass and striped bass (Table 14.2). Seventy-three percent of these were stocked as fingerlings.

In a 1984 survey of 47 states and 9 provinces, 48 reported stocking coolwater taxa such as walleye, sauger, yellow perch, muskellunge, tiger muskellunge, and saugeye. In 1995–1996, 70 federal hatcheries, 33 states, and 5 provinces reported stocking 1.269 billion individuals in 7 taxa (Table 14.3). This total is up from the 1,157 billion reported for the 1983–1984 season (Conover 1986). Approximately 92% of the individuals stocked in 1995–

Table 14.3 Number (×1,000) of coolwater sport fishes stocked into inland waters of the United States and Canada during 1995 or 1996. Number of states or provinces stocking taxa is in parenthesis.

Taxa	Total number stocked	Number stocked by life stage				Number stocked by jurisdiction		
		Eggs	Fry	Fingerlings	>203mm	States	Federal	Provinces
Esocidae								
Muskellunge	4,288		3,714	565	9	4,267 (17)		21 (1)
Northern pike	32,334		23,494	8,908	26	22,425 (16)	6,801	3,202 (2)
Tiger muskellunge	707		123	584		707 (15)		
Percidae								
Sauger	10,936		9,906	1,030		10,936 (6)		
Saugeye	35,334		26,065	9,269		35,310 (6)	24	
Walleye	1,172,025	4,502	1,120,789	46,696	38	977,389 (31)	30,822	163,814 (5)
Yellow perch	13,686		9,614	3,976	96	12,990 (12)	631	65 (2)
Total	1,269,404	4,502	1,193,705	71,028	169	1,064,024	38,278	167,102
Percent		<1	94	6	<1	84	3	13

1996 are walleye. Except for tiger muskellunge, most (94%) of the coolwater species are stocked as fry. The reason that tiger muskellunge are stocked as fingerlings instead of as fry reflects the stocking preferences of fisheries biologists in the few states (i.e., Michigan, New York, and Pennsylvania) that stock the majority of these hybrids. The tiger muskellunge accepts a prepared diet more readily than the muskellunge; thus, it is more practical to rear the tiger muskellunge to an advanced fingerling size.

Of the coldwater species stocked in inland waters, approximately 256 million trout and salmon were stocked by federal and state hatcheries in 1983 (Wydoski 1986). In 1995–1996, 70 federal hatcheries, 48 states, and 8 provinces stocked 276 million coldwater sport fishes (Table 14.4). Of these fish, 54% were stocked as fingerlings and 24% as catchable size.

Table 14.4 Number (×1,000) of coldwater sport fishes (Salmonidae) stocked into inland waters of the United States and Canada during 1995 or 1996. Number of states or provinces stocking taxa is in parenthesis.

Taxa	Total number stocked	Number stocked by life stage				Number stocked by jurisdiction		
		Eggs	Fry	Fingerlings	>203mm	States	Federal	Provinces
Arctic char	296			287	9	242 (3)		54 (2)
Arctic grayling	239		87	145	7	239 (6)		
Atlantic salmon	1,751		112	1,632	7	1,183 (5)	111	457 (4)
Brook trout	12,845	343	325	7,529	4,648	8,584 (24)	198	4,063 (9)
Brown trout	23,404	140	628	14,407	8,229	20,857 (38)	1,495	1,052 (7)
Chinook salmon	14,838		2,153	12,439	246	14,733 (11)		105 (1)
Coho salmon	7,479			6,987	492	2,687 (6)	4,792	
Cutthroat trout	13,699	605	342	11,736	1,016	11,066 (11)	2,179	454 (3)
Golden trout	91			18	73	91 (3)		<1 (1)
Kokanee	26,687	2,423	12,697	11,420	147	23,631 (10)	910	2,146 (1)
Lake trout	19,675	82		18,672	921	7,782 (16)	7,190	4,703 (4)
Lake whitefish	35,619	30,450	5,000	169		34 (1)		35,585 (2)
Rainbow trout	117,243	806	3,347	61,923	51,167	95,007 (42)	12,161	10,075 (8)
Splake	2,339	150		1,974	215	1,409 (10)		930 (3)
Tiger trout	36			25	11	36 (4)		<1 (1)
Total	276,241	34,999	24,691	149,363	67,188	187,581	29,036	59,624
Percent		13	9	54	24	68	10	22

Table 14.5 Numbers (×1,000) of anadromous salmon fry, fingerlings, and smolt stocked by federal hatcheries. Table based on data from Department of Commerce and Labor (1906), USFWS (1975, 1986, 1995).

Taxa	1905	1975	1986	1995
Atlantic salmon	1,017	421	5,597	8,718
Chinook salmon	21,625	54,644	94,318	52,933
Chum salmon		9,776	14,292	4,975
Coho salmon	10,634	10,064	21,268	4,792
Total	33,276	74,905	135,475	71,418

Large numbers of anadromous Pacific salmon and Atlantic salmon continue to be stocked on the east and west coasts by both federal (Table 14.5) and state agencies (Table 14.6). Other anadromous species that are stocked include striped bass and American shad. From 1992 to 1993, hatchery-stocked juvenile American shad have made up from 44 to 99% of the year-class above Conowingo Dam on the Susquehanna River (Hendricks 1995).

Due to declining fish stocks, interest in marine fish stocking is increasing in the United States and throughout the world (Blankenship and Leber 1995). From 1975 to 1982, 56 million red drum were stocked into bays along the coast of Texas; 15% were stocked as eggs, 80% as fry, and 5% as fingerlings (Richards and Edwards 1986). From 1983 to 1993 over 140 million fingerlings were released (McEachron et al. 1995). The state of Florida is stocking snook. Striped mullet have been stocked in Hawaii (Leber et al. 1995) and white seabass in California (Kent et al. 1995). As of 1978, Japan had stocked over 9 million marine fish of 12 different species (Richards and Edwards 1986).

Fish are being stocked by private fish culturists in at least 46 of the 50 states (Davis 1986). In over one-half of the states that Davis surveyed, the respondents replied that there was no record of the extent of the private fish culture industry in their states. However, based on the positive responses, private production of channel catfish, rainbow trout, largemouth bass, and sunfishes (*Lepomis* spp.) for stocking exceeds 660 million fish annually. Smaller numbers of many other species are also raised. The 1997

Table 14.6 Numbers (×1,000) of anadromous salmon stocking by states in 1995 or 1996.

State	Salmon taxa					
	Atlantic	Chinook	Chum	Coho	Pink	Sockeye
Alaska		4,211		2,876		
California		5,318		10		
Connecticut	3,162					
New Hampshire	1,246					
Oregon	7	38,199		11,774		21
Vermont	30					
Washington		105,461	42,654	44,585	4,491	7,184
Total	4,445	153,189	42,654	59,245	4,491	7,205

Aquaculture Magazine Buyer's Guide lists 87 private hatcheries that produce large-mouth bass, 89 that produce channel catfish, 67 that produce bluegill, and 27 that produce hybrid sunfishes.

14.3   INTRODUCTION AND ENHANCEMENT STOCKING PROGRAMS

There is a lack of uniform stocking terminology, which has management implications (Laarman 1978; Radonski and Martin 1986; Smith and Reeves 1986). For the purposes of this chapter, stocking programs are divided into two categories: introduction and enhancement; enhancement stocking is subdivided into maintenance and supplemental.

Introductory stocking of species into waters where they do not exist has been practiced for many years in North America. The classic combination of largemouth bass, bluegill, and channel catfish for stocking new or renovated farm ponds is just one example that is perceived as successful by the angling public (Chapter 21). Most of the freshwater sport fishes presently found in California were introduced, and Pacific Coast striped bass populations have resulted from stocking programs. Nonnative species have been introduced into many bodies of water (Chapter 13). In newly constructed lakes and reservoirs and in renovated bodies of water, introductions are usually successful in terms of relatively high survival of the stocked fishes, because of a high density of vulnerable food organisms and a low density of predators. Introduced fishes become self-sustaining only if all limiting environmental factors fall within the species' ranges of tolerance. For example, it is necessary that suitable habitat, food supply, and water temperatures exist for spawning. Even if habitat and environmental conditions are favorable and spawning takes place, predators can reduce or prevent recruitment.

When recruitment is limited, it may be necessary to augment the population through enhancement stocking (Figure 14. 1). If the stocking is done to enhance a weak year-class, this is called supplemental stocking. Maintenance stocking is used if recruitment fails completely, which in many cases is anticipated (e.g., when striped bass are stocked into a reservoir without a major headwater stream or when certain hybrids or triploid fish are stocked). Laarman (1978) evaluated walleye stocking in 125 bodies of water over a 100-year period. Approximately 48% of the introductory stockings and 32% of the maintenance stockings were successful, but only 5% of the supplemental stockings succeeded.

Keith (1986) reviewed introduction and enhancement stocking in reservoirs. In stocking coldwater reservoirs, some effort was made to exclude nonnative species; warmwater reservoirs were more intensively stocked with nonnative species. Keith concluded:

> Impacts on prey populations have varied from unnoticeable to severe depletion, and effects on native predator fishes have been judged largely insignificant. However, displacement of ecologically similar species, predation on desirable native species and production of intergrade populations have been documented. (Keith 1986:144)

Figure 14.1   Fingerling and adult fishes are transported in specially designed, insulated tanks supplied with oxygen.

The enhancement strategy of stocking subharvestable-size fish is called put, grow, and take. Under certain conditions, harvestable-size fish are stocked that are not expected to reproduce or even grow significantly before they are caught. These put-and-take fisheries are characteristically used where fishing pressure is extremely high, such as urban areas, or in bodies of water that will support fish for only a limited time during the year. For example, trout are stocked during the fall and spring in temporary streams or in lakes where they die in the summer because of low oxygen and high temperatures. Some stockings, such as rainbow trout in the White River below Bull Shoals Dam in Arkansas, initially created a put-grow-and-take fishery, but as the fishing pressure increased, it became a put-and-take fishery. Even though rainbow trout grow 2.5 cm per month in the White River, most are removed by fishing within a few weeks after stocking. In general, states can use Wallop–Breaux monies for put-grow-and-take stocking but not for put-and-take stockings. On average, fish have to be 110% of stocked size when harvested for states to use federal aid to support the program.

14.4   STOCKING TECHNIQUES

Why stock? Which taxa? What size? What number? When and where? What quality? These questions should be asked before a stocking program is implemented. It may not be possible to answer all of the questions adequately, but they still should be asked. In addition, the biologist should always keep in mind what is biologically feasible and desirable versus what is economically feasible and desirable.

14.4.1   Why Stock?

Normally, fishes are stocked to fulfill perceived needs. These needs range from a carefully conceived plan to stock a predator or prey species to frivolous stocking of fish because hatcheries happen to have a surplus on hand. Quite frequently, political or social issues override the biological ones. Ideally, when considering why to stock, the fisheries manager states goals and objectives and implements evaluation procedures. Some of the most common reasons for stocking are (1) to introduce a new sport or commercial species, (2) to introduce a new forage species, (3) to establish a biological control, (4) to start or maintain a put-and-take or put-grow-and-take fishery, (5) to establish a trophy fishery, (6) to supplement a year-class, (7) to satisfy public and political pressure to stock, (8) to increase angler satisfaction, (9) to mitigate an existing situation, (10) to reestablish a species that has been lost from an area (some endangered species plans call for stocking), and (11) to redistribute fishing pressure. A risk–benefit analysis is particularly important when different genetic stocks are used or a species is stocked outside its native range.

A major concern at this stage of program development is the possible negative effects of the stocking (see Chapter 13). If long-term community changes brought about by stocking are not considered, the costs of evaluating the results of a stocking program are much greater than those of stocking the fish in the first place. Even though evaluation costs are greater than stocking costs, these costs are not a compelling argument for no evaluation. Many agencies are currently using genetic or marking techniques, such as oxytetracycline, to evaluate relative survival of fish stocked at various sizes and the contributions stocked fish make to the year-class (Brooks et al. 1994). Few studies have tried to determine the compensatory effect that stocked fish have on naturally produced fish.

14.4.2   Which Taxa?

When considering an endangered species or supplementation of a year-class, the biologist has little choice of species to stock. Mitigation stocking is also usually targeted to a specific species. Choices may or may not be available when an agency attempts to satisfy public pressure for stocking. Often such pressure comes from organizations that are interested in a species. Normally, choices are available when the goal is to fill a niche by introducing new species that will be self-sustaining or to start put-grow-and-take and put-and-take fisheries. In many areas of North America there is a tremendous latent demand for fishing, which can be released by starting a put-and-take or put-grow-and-take fishery. However, once the fishery has become established, public pressure often demands continuation of the stocking program. This demand for stocking can develop into a public relations problem if the hatchery system cannot meet the demand for fish.

Another difficult situation can arise if a body of water is being used for collecting broodfish or eggs. For example, when a lake is being used as a source of walleye broodfish and a dense population of hybrid striped bass is established, the hybrids may become a tremendous nuisance in the gill nets that are used to collect walleye broodfish.

Table 14.7   Selected list of taxa that are stocked and some of the pertinent biological characteristics to consider when choosing a fish for stocking.

Taxa	Characteristics
Atherinidae	
Inland silverside	Forage fish; winterkills but can tolerate colder temperatures than threadfin shad; young of the year reproduce in Midwest
Centrarchidae	
Black crappie	Easier to handle and transport than is white crappie; tends to predominate over white crappie in northern and southern portions of United States; does not readily accept a prepared diet
Bluegill	Becomes stunted in small ponds and can limit largemouth bass recruitment; readily accepts prepared diets
Green sunfish	Very vulnerable to largemouth bass predation
Hybrid sunfishes	Grows faster than parentals; certain F_1s are predominately males; F_1s tend to be fertile
Largemouth bass	Sport fish; Florida subspecies cannot survive in cold water as well as northern subspecies
Redear sunfish	Harder to catch than bluegill; capable of eating mollusks; does not readily accept a prepared diet
Smallmouth bass	Grows well on insects and crayfishes as forage; grows well at warm temperatures but does not recruit in southern states in ponds with largemouth bass and sunfishes present; in the southern part of its range it recruits in streams
White crappie	Tends to overpopulate in small ponds and lakes or does not recruit; tends to dominate over black crappie in turbid water; does not readily accept a prepared diet
Clupeidae	
Gizzard shad	Very fecund forage species; not desired in small pond or lake when managing for sunfishes; spawns at 2 years
Threadfin shad	Very fecund forage species; young of the year spawn; winterkills at temperatures below 8°C
Cyprinidae	
Common carp	Commercial species; capable of eating infauna; fecund; long lived; wide temperature tolerance
Fathead minnow	Forage fish; so vulnerable that it tends to be eliminated by largemouth bass
Golden shiner	Has been stocked in small lakes and ponds as forage for largemouth bass; tends to be more successful in northern part of largemouth bass range; in Midwest may overpopulate and limit largemouth bass recruitment
Grass carp	Used as biological control of vegetation; stocked at 6 to 74 fish per hectare; triploids are available; not approved in all states; commonly reaches 14 kg; very vulnerable to largemouth bass predation below 20 cm
Esocidae	
Muskellunge	Trophy sport fish; fry are very vulnerable to fish predation
Tiger muskellunge	Accepted as trophy sport fish by most muskie anglers; easier to raise to advanced fingerling stage on prepared diet than parentals; sterile
Ictaluridae	
Black and yellow bullheads	Used in urban fisheries; tend to reproduce at small size (15 cm); dense populations capable of keeping a pond muddy in areas of colloidal clay
Channel catfish	Requires cavity in which to spawn; very vulnerable to largemouth bass below 15–20 cm; may not recruit in small ponds; will readily accept prepared diets
Percichthyidae	
Hybrid striped bass	Cross using female striped bass and male white bass grows larger than reciprocal; easier to train to take prepared diet than parentals; will backcross
Striped bass	Pelagic sport fish capable of eating large shads; floating eggs require large headwater stream to recruit; some populations are maintained by stocking fry
Percidae	
Saugeye	Currently being investigated; there is some indication it may produce a fishery where walleye stocking has failed; fertile and will backcross
Walleye	Sport fish; some populations can be maintained by fry stocking, others require fingerling stocking

Table 14.7 Continued.

Taxa	Characteristics
Salmonidae	
Atlantic salmon	Unlike Pacific salmon, all do not die after spawning; high sport fish quality; requires a forage fish to obtain large size
Brook trout	May outcompete other trout in cold headwater areas; very vulnerable to fishing; more acceptable to anglers at a small size
Brown trout	Harder to catch and longer lived than rainbow or brook trout; many strains available; provides a trophy component; more tolerant of warm water
Coho salmon	Requires approximately 18 months in hatchery to reach a size for imprinting; tends to school more densely than do chinook salmon; survives 6–12 months in Great Lakes fishery; many stocks on West Coast
Chinook salmon	Requires 6 months in hatchery to reach size for imprinting; survives 24–26 months in Great Lakes fishery; many unit stocks on West Coast
Kokanee	Small, freshwater form of sockeye salmon that feeds on zooplankton during its entire life; very susceptible to competition
Rainbow trout	Tolerates low oxygen better than most trout; very vulnerable to largemouth bass and striped bass predators; used frequently in two-story fish populations; highly vulnerable to sportfishing; can grow to a harvestable size on zooplankton

A public relations problem can also develop when fish populations, such as walleye or striped bass, are established in lakes and rivers that have a commercial fishery based on entanglement gear, such as gill nets or trammel nets. When anglers observe sport species trapped in entanglement gear, they often demand closure of the commercial fishery.

It is beyond the scope of this chapter to list all of the species that are being stocked in North America and all of the biological characteristics that should be considered when choosing a fish for stocking. Table 14.7 is presented to give the reader information on the range of biological characteristics that influence the choice of which fish to stock. Hybrids are often stocked because they are easier to raise in hatcheries than are the parental species and tend to have a higher return to the angler. Most freshwater hybrids are fertile; therefore, special consideration should be given to potential degradation of the parental species gene pool. In 1995 or 1996, 38 states stocked at least one hybrid; 18 states, two; 4 states, three; 3 states, four; and 1 state, five. The federal hatchery system and three Canadian provinces each stocked one hybrid, and one Canadian province stocked two hybrids.

14.4.3 What Size?

In general, there is a positive correlation between the size of fish stocked and their survival. However, it costs more to rear fingerlings than it does to rear fry, and normally a hatchery system has more fry than fingerlings available for stocking. Thus, the decrease in vulnerability and increase in relative survival of larger fish must be weighed against the higher cost of producing larger fish. Once the relative cost of production and relative survival of the different sizes of a fish have been determined for various environmental conditions, the optimal stocking size can be calculated. Unfortunately, little information has been published on this subject. One problem has been marking fry and fingerlings without inducing size-selective mortality so that relative survival rates can be calculated. With the advent of better marking techniques, such as coded wire tags and chemical and genetic markers, studies can now be conducted to determine the optimum size of fish to stock. An alternative to stocking different sizes of fish the same year is to stock different sizes of fish in alternate years and then evaluate year-class strength. The problem with this approach is that many factors other than size of fish stocked can contribute to year-class strength, and the effect of size cannot always be determined.

In general, when fry are stocked into a new or renovated body of water that is free of predator fishes, their chance of survival is much higher than when they are stocked into established fish populations. When predators are present, an abundant food supply for the fry must be present. Even if the small fish do not die of starvation, a low food supply reduces their growth rate and increases their period of vulnerability.

Eggs are stocked on only a limited basis and primarily in the coldwater group of sport fishes (Table 14.4). Red drum eggs have been stocked into bays along the coast of Texas, and threadfin shad have been successfully stocked using eggs (Maxwell and Easbach 1971). Some agencies are recommending that angler groups plant trout eggs. Harshbarger and Porter (1982) found that direct intergravel plants produced twice the hatch and 3.5 times more swim-up fry than eggs planted in Whitlock-Vibert boxes. Two of five plantings of muskellunge eggs in Wisconsin resulted in measurable year-classes. Lake trout eggs have been hatched (73%) between layers of artificial turf suspended over a reef in Lake Superior (Swanson 1982).

One of the best known put-grow-and-take fisheries is the salmon fishery that has been made possible by smolt stocking in the Great Lakes (Chapter 23). Salmon that are stocked as smolts imprint and return to the streams where they were released. In put-grow-and-take fisheries in smaller inland waters, there are only a few species that are routinely stocked as larvae with any degree of success. These include the striped bass, hybrid striped bass, walleye, and saugeye. The larvae of all of these species are pelagic, and their survival may be due to reduced predation in the open area of a lake as opposed to the vegetated littoral area. Even so, in many lakes larval stocking of walleye and striped bass has not been successful, whereas the stocking of fingerlings has developed and maintained a sport fishery.

Channel catfish are frequently stocked into farm ponds and small lakes that contain largemouth bass (see Chapter 21). In general, there is a positive correlation between size of channel catfish stocked and relative survival (Dudash and Heidinger 1996).

Two-story (that is, a warmwater fishery above a coldwater fishery) rainbow trout populations in reservoirs may have to be maintained by stocking large, less vulnerable rainbow trout. Often 40–50% of the stocked fish are creeled when 20- to 23-cm rainbow trout are stocked, and no fishery may develop if 5- to 10-cm fish are stocked. By its nature a put-and-take fishery requires stocking of what is usually called a catchable-size fish. The operational definition of catchable is a fish large enough to be acceptable to the angling public.

Put-and-take trout fisheries are often based on fish larger than 178 mm in total length (Figure 14.2). In 1983, states stocked 54,000 km of streams and 550,000 ha of lakes with more than 53 million catchable-size trout that cost almost $37 million to produce. Rainbow trout was the most frequently stocked species (78%), followed by brook trout (11%), brown trout (10%), and other species (1%). Some

Figure 14.2 Catch from a put-and-take rainbow trout fishery in the White River, Arkansas.

warmwater urban put-and-take fisheries are based on 0.2-kg bullheads, whereas others use 1.1-kg common carp and 0.5-kg channel catfish. The fisheries manager always has to guard against superimposing his or her personal recreational bias upon the angling public.

14.4.4   What Number?

Obviously, the number of fish stocked should be based on biologically sound principles and the objectives associated with a specific management plan. In reality, few hatchery systems produce enough fish of the needed size to meet the demand. The management decision then becomes, should these fish be distributed at low density into all the waters for which they have been requested or should only a relatively few of the bodies of water be stocked at the desired density? Many state agencies address this problem by prioritizing the bodies of water. The priority order in many states is based on ownership; that is, state over federal over private bodies of water. State biologists then proceed down the priority list, stocking fish until the supply has been exhausted. A priority list can lead to stocking when conditions are less than optimal, such as during periods of low zooplankton or ichthyoplankton abundance.

When maintenance stocking is involved and the broodfish are collected from the wild, the bodies of water from which the broodfish are being taken often are given first priority. Normally, broodfish of only a relatively small number of species are kept in

Table 14.8 Stocking densities of selected taxa in two urban put-and-take fisheries.

Taxa	Number/hectare	Reference
Carps, bullheads, and channel catfish	373 [a]	Haas 1984
Bullheads	415–1,532	Lange 1984

[a] Stocked 10 times during the year at this density.

hatcheries. Typically, bluegill, redear sunfish, channel catfish, crappies, and trout are kept in the hatchery; largemouth bass and smallmouth bass may be; and the rest, such as salmon, walleye, striped bass, shads, northern pike, and muskellunge, are usually collected on their spawning run. Some recommended stocking densities for warmwater and coolwater fishes are listed in Tables 14.1 and 14.8–14.11. Many states stock more fish per hectare in a small body of water than in a large body of water. This sliding scale reduces the absolute number of fish required. The tremendous range of stocking densities for various species reflects the biology of the species in terms of normal standing stock, life stage at which fish are stocked, the management program being followed, and the productivity of the water. Obviously, more fry than fingerlings are stocked per hectare, which reflects the relatively low survival of the fry versus fingerlings. This general pattern of decreasing mortality when larger fish are released has been demonstrated among stream trout populations (Hume and Parkinson 1988).

High trophic level species, such as the muskellunge, may have a low-density standing stock, such as one legal size fish for every 0.4–4.0 ha (Brege 1986). Fishes that produce a trophy fishery are usually stocked at a relatively low density. Is it possible to stock too many fish? Currently, a major concern of biologists working on the Great Lakes is that the forage base may be depleted if additional salmon are stocked annually. Hume and Parkinson (1987) have shown that mortality increases when steelhead are stocked at high densities in British Columbia streams. One concern in trout stocking is that the fry do not move extensively from the stocking site. This fish behavior can lead to overstocking when the number to be stocked has been calculated for the total area of the stream or river. The only solution to this problem is to disperse the fish at the time of stocking. Unfortunately, it is not always economically or physically feasible to do so.

Table 14.9 Part of Embody's revised planting table for trout streams. Table modified from Everhart et al. (1975).

Stream width (m)	Pool grade A Food grade			Pool grade B Food grade			Pool grade C Food grade		
	1	2	3	1	2	3	1	2	3
0.3	89	72	56	73	56	29	56	39	22
1.5	447	364	280	364	280	196	280	196	112
3.0	895	727	559	727	559	391	559	391	224
4.6	1,342	1,091	839	1,091	839	587	839	587	336
6.1 [b]	1,611	1,309	1,087	1,309	1,007	705	1,007	705	403

Number of 7.6-cm fingerlings per kilometer [a]

[a] The number of other size trout to stock can be calculated by multiplying the number of 7.6-cm fish indicated in the table by 12, 1, 0.75, 0.6, and 0.3 for 2.5-, 7.5-, 10.1-, 15.2-, and 25.4-cm fish, respectively.
[b] For streams over 6.1 m in width, the formula is $X = 1.642WN_1 + 8N_1$, where N = number of 7.6-cm fingerlings for 0.3-m wide stream, W = average width of stream in meters, and X = number to be stocked per kilometer.

Table 14.10 Partial summary of policy for stocking trout in streams in Pennsylvania. When the stocking program does not vary with trout biomass or human density, NA is given in column. Table taken from Pennsylvania Fish Commission (1986).

Wild trout biomass (kg/ha)	Stream's mean width (m)	Human population (persons/km²)	Recreational use potential [a]	Annual stocking intensity (trout/ha)		
				Preseason	Inseason	Total
High yield strategy						
Brook trout plus brown trout <20	4–20	NA	High	80	30-30-30	170
Optimum yield strategy						
Brook trout 20–30	4–20	NA	High	50	30–30	110
Brook trout plus brown trout 20–40						
20–40	4–20	≥125	Good	70	50–40	160
20–40	4–20	40–125	Good	40	40–40	120
20–40	4–20	<40	Good	40	40	80
20–40	<4	NA	High	30	30	60
Low yield strategy						
NA	<4	NA	Good to low	30	0	30
Rivers strategy						
NA	>30	>125	High	75		75
NA	>30	40–125	High	65		65
NA	>30	<40	High	60		60
NA	>30	>125	Good	45		45
NA	>30	40–125	Good	35		35
NA	>30	<40	Good	30		30
NA	>30	NA	Low	20		20

[a] Based on percentage of streams in private or public ownership, stream's proximity to road, and available parking.

The biology of the species is not the only criterion that influences the number of fish to be stocked. The actions of the fishing public can be as important. A phenomenon known as "truck following" is often associated with put-and-take trout stocking and put-and-take urban fisheries of catfishes (e.g., channel, blue, and white), common carp, and bullheads. Anglers go to extremes to learn when and where fish will be released, and large numbers of anglers deplete the vulnerable fish almost as soon as they are stocked. This situation is aggravated when large numbers of fish are stocked infrequently. Frequent stockings of small numbers of fish reduce truck following and tend to level out the catch per unit effort.

One of the earliest systems to estimate the number of trout that should be stocked in streams was proposed by Embody in 1927 (Everhart et al. 1975). In Embody's system, the number of trout to be stocked depends upon stream size, suitability of the stream to support trout, food supply, and size of trout (Table 14.9). Streams that have high pool grade are characterized by frequent pools with water depth greater than 0.75 m and abundant shelter (e.g., overhangs, logs, and roots). Streams that have high food grade are characterized by extensive areas of aquatic vegetation, such as watercress, and swift-water substrate suitable for high numbers of macroinvertebrates. The stream margin is also vegetated, which concentrates large numbers of terrestrial insects. Embody's tables have been modified by various managers to reflect their specific goals and environmental conditions (Smith and Moyle 1944; Cooper 1948).

Table 14.11   Partial summary of policy for stocking trout in lakes in Pennsylvania. Table taken from Pennsylvania Fish Commission (1986).

Human population density rank [a]	Annual stocking rate (trout/ha)	
	Optimum	Alternate [b]
Lake ≤8 ha		
1	250	120
2	240	110
3	220	100
4	210	80
Lake >8 but ≤20 ha		
1	200	70
2	190	60
3	170	50
4	160	40
Lake >20 but ≤40 ha		
1	120	60
2	110	50
3	100	40
4	80	30
Lake >40 but ≤81 ha		
1	80	40
2	70	30
3	60	20
4	50	10
Lake >81 hectares		
1	16	
2	12	
3	8	
4	4	

[a] Ranges from urban (1) to rural (4).
[b] Reduced stocking rates when access or some social, biological, or chemical factor limits suitability for intensive management.

A number of states have written detailed trout stocking procedures for the management of stream and lake fisheries. Pennsylvania's plan is an example; it is detailed and includes goals and many testable objectives. In addition, it integrates the available resources with management techniques and user pressure to protect and maintain the resource (Tables 14.10 and 14.11).

14.4.5   When and Where?

Trout, especially catchable-size individuals, must be stocked frequently to avoid truck following, to equalize fishing effort, and to maintain an acceptable harvest rate. In southern states that have two-story fish populations in reservoirs with trout as the coldwater species, usually the trout are not stocked in the summer to avoid the possibility of temperature shock. Attempts have been made to stock trout in the summer through a tube that extends into the metalimnion. One reason for summer stocking is to give a hatchery system more flexibility. In a steelhead stream, Hume and Parkinson (1988) confirmed a pattern of decreasing mortality with later release dates.

One of the most common stocking combinations for small lakes and ponds includes largemouth bass, bluegill, redear sunfish, and channel catfish (see Chapter 21). In many states, the bluegill, redear sunfish, and channel catfish are stocked in the fall, and the largemouth bass are stocked the following spring. To a great extent, this facilitates efficient use of hatchery space and does not necessarily reflect biological restraints. Dillard (1971) concluded that in Missouri there was little difference in the results of stocking farm ponds with largemouth bass in the summer and bluegill and channel catfish in the fall compared with stocking all three species in the fall.

Most warm- and coolwater species spawn in the spring; therefore, the fry must be stocked in the spring. Ideally, fry should be stocked when their natural zooplankton food is abundant. The density of zooplankton can change drastically within 2 weeks. Realistically, because the fry cannot be held for more than a few days without food, they are stocked when they are available.

When the hatchery and the body of water to be stocked are geographically separated the difference between water temperatures may be troublesome. If fry are moved north from a southern hatchery, the water temperature at the stocking site is often below optimum. Similarly, if they are moved south, water temperature at the stocking site may be too warm for the fry to survive. Water temperature can be an extremely acute problem when power plant cooling lakes are stocked. In some cases hatching temperatures can be adjusted to produce fry for stocking at conditions more similar to the receiving body.

It would seem that the hatchery manager would have much more timing flexibility when stocking fingerlings, and this is true especially if the fingerlings are being fed a prepared diet. But most sport species, except channel catfish, some sunfishes, and salmonids, do not accept prepared diets without an intensive training period. Ideally, walleye, striped bass, and hybrid striped bass fingerlings should be stocked into a strong ichthyoplankton pulse. In most cases, these fishes tend to be raised in fertilized rearing ponds. At stocking densities of 125,000 fry/ha and a survival of 50%, walleye deplete the zooplankton by the time the fish reach 5 cm in total length (Nickum 1986). If the fish are not harvested within a few days, cannibalism will rapidly reduce the density. Thus, the fish culturist who is saddled with a numbers quota will want to move the piscivorous fishes out of the hatchery as soon as the zooplankton community is depleted to avoid extensive cannibalism.

The optimum location for stocking and the best time of day or night to stock various species have not been thoroughly investigated. We know zooplankton density is not uniform throughout a reservoir and that they are patchy within any area of the reservoir. Also higher densities of ichthyoplankton and young-of-the-year fish tend to occur in the back end of large coves. Because the larvae of walleye, striped bass, and hybrid striped bass inhabit open water, it may be beneficial to stock them away from shore instead of at convenient boat docks. Alternatively, fingerling northern pike, largemouth bass, and muskellunge are probably better adapted for shoreline stocking than for pelagic stocking.

In Nebraska, Miller (1971) stocked highly vulnerable 5- to 8-cm northern pike at a density of 62–155/km into small streams that emptied into larger bodies of water. Of 15 such stockings, 5 were judged successful, 5 were partially successful, and 5 were considered failures, demonstrating that in at least some streams predator pressure is low.

As indicated earlier, some salmon and trout fry and fingerlings do not disperse well in streams if stocked in one or few spots. Evidently this is not the case for all salmonid fingerlings. Marshall and Menzel (1984) found no difference in survival or growth of spring brown trout fingerlings (9 cm) undergoing concentrated stocking versus dispersed stocking in six northeast Iowa streams (put-grow-and-take fisheries). Summer survival ranged from

3 to 22%, and overwinter survival ranged from 2 to 14%. Similar overwinter survival (7%) of brown trout has been reported in a Wisconsin stream by Brynildson et al. (1966) and in a New York stream (6%) by Schuck (1943).

In some cases fry or fingerlings are not stocked directly into the body of water where the fishery occurs. For example, in the north-central states, northern pike have been reared in managed marshes (Gregory et al. 1984). Production is quite variable, but 1,000–3,000 fingerlings (75–150 mm total length) per hectare is normal. Both fry and adult broodfish have been stocked into the marshes. Royer (1971) found that adult broodfish stockings were much more successful than fry stocking. However, fry stockings have been successful in other marshes (Fago 1977). A variation of the managed spawning marsh is the use of winterkill lakes to rear fish for stocking. In 1967–1968, Minnesota reared about 500,000 northern pike weighing 136,000 kg in shallow winterkill lakes by stocking adult fish in the spring and harvesting them in the late fall or winter (Johnson and Moyle 1969). Walleye can also be raised in winterkill lakes.

Some states construct nursery ponds adjacent to large reservoirs to produce fingerlings of various species for stocking into the reservoir. Arkansas built a number of these nursery ponds during the 1960s (Keith 1970). Production of fishes in such nursery areas can be quite high (Table 14.12).

Certain species are being raised to larger sizes in cages or pens suspended in the body of water in which they are to be released. Channel catfish have been raised to larger sizes in cages to reduce their vulnerability to largemouth bass predation when they are stocked (Collins 1971). Alternate crops of catfishes (blue and channel) and rainbow trout have been raised to large size (0.2–0.4 kg/fish) in net-pens. Arkansas's Bull Shoals net-pen facility produced 44% (by number) of all catchable-size catfishes and 15% of all trout stocked in the state in 1986 (Oliver and Rider 1986). Trout stocking density in the cages ranges from 106 to 176 fish/m^3, and catfishes were usually stocked at approximately 140 fish/m^3. One prerequisite of this technique is that the cultured species will accept a prepared diet.

14.4.6   What Quality?

In addition to factors discussed above, other parameters contribute to the success of a stocking program, including genetic characteristics, parasite infestation, physiological integrity, and handling history (Murphy and Kelso 1986).

14.4.6.1   Genetic Characteristics

Kutkuhn (1981) defined the term stock as a randomly mating group of individuals having temporal, spatial, or behavioral integrity. This is somewhat similar to the ecological definition of a population. Accordingly, the stock, and not the species or subspecies, has been considered the operational unit of interest. More recently there has been an effort to establish the "evolutionary significant unit" as the population unit that should be protected under the Endangered Species Act (Nielsen 1995). All of these concepts may be quite difficult to apply.

Table 14.12   Fish produced in four Arkansas nursery ponds (size of pond given in parentheses after its name). Data are from Keith (1970).

Taxa and year	Number/ha	Length (mm)
Norfork (3.2 ha)		
Walleye		
1964	12,500	63
1965	23,400	63
Channel catfish		
1966	12,500	228
Northern pike		
1967	1,200	253
Striped bass		
1967	1,600	114
Muskellunge		
1969	400	228
Bull Shoals (4.0 ha)		
Walleye		
1966	30,000	63
1967	75,000	63
1968	2,200	177
1969	25,000	101
Channel catfish		
1967	4,500	177
Largemouth bass		
1967	1,200	88
Beaver (10.9 ha)		
Walleye		
1968	45,900	63
Northern pike		
1969	22,900	114
Greeson (8.1 ha)		
Channel catfish		
1968	2,800	[a]
Striped bass		
1968	2,100	126
Walleye		
1969	37,000	63

[a] Mean weight 0.23 kg.

Many subspecies are recognized, and related subspecies can exhibit differences in behavior and physiology. For example, the Florida subspecies of the largemouth bass is not as tolerant of cold water as is the northern subspecies of largemouth bass. Undoubtedly, such differences can exist even among populations of a subspecies, but such differences may or may not have biological, social, economic, or political significance.

The existence of genetic stocks of salmon and trout has been recognized for many years. Initially, biologists attempted to identify stocks by means of classical meristic techniques, but they achieved only limited success. Stocks can be identified with morphological characters, cytogenetic analysis, electrophoretic characteristics of specific protein molecules, immunological analyses of specific protein molecules, mitochondrial DNA analysis, and nuclear DNA analysis (Philipp et al. 1986). Even though the genetic stock concept is important, care must be taken that it is not overemphasized. With sufficient analytical sensitivity, every population could probably be shown to be distinctive. In fact, every individual is distinctive. From a practical point of view, that is, relocating genetic stocks through stocking, the unit that makes up a given stock must have measurable, meaningful differences in biological parameters such as growth rate, reproductive rate, tolerance of environmental conditions, and behavior. The argument over what is a meaningful difference will continue for a long time because in many cases the point is philosophical rather than biological.

Aquaculturists have developed strains (group having distinctive genetic characteristics) of fishes that are often held in hatcheries. They may also use so-called wild strains that have not been altered by controlled breeding. Kincaid and Berry (1986) identified 143 genetic strains of rainbow trout, 43 of brook trout, 42 of brown trout, 39 of cutthroat trout, and 26 of lake trout. These numbers include both natural and domestic strains. Strains can have considerable differences in biological parameters, such as growth rate, fecundity, disease resistance, stress resistance, age to maturity, time of spawning, and spawning location.

The degree of domestication (number of generations in a hatchery) of a species can change the characteristics of a species. For example, in 1959 Boles and Borgeson (1966) stocked equal numbers of brown trout from the Convict Lake strain (wild), the Mount Whitney strain (several generations in a hatchery), and the Massachusetts strain (many generations in a hatchery). After 2 years, the return to creel of the Massachusetts, Mount Whitney, and Convict Lake strains were 63%, 37%, and 25%, respectively. Other studies have shown that hatchery-reared salmonids, including brook, brown, and rainbow trout, may be more vulnerable to angling than are wild fish (Boles 1960; Hunt 1979). Other differences have also been noted between wild and domestic strains of salmonids. Symons (1969) determined that wild strains of Atlantic salmon smelts dispersed more widely than did domestic strains, and Flick and Webster (1976) found that hatchery-reared wild brook trout lived longer than did domestic strains of brook trout when released into natural ponds. Reisenbichler and McIntyre (1977) found that a hatchery strain of steelhead had a higher growth rate than did wild steelhead in hatchery ponds but not in streams; however, the wild trout × hatchery trout hybrid had the highest growth rate in streams.

14.4.6.2   Parasite Infestation

Goede (1986) discussed the potential effects of stocking fish infected with parasites. He pointed out that stocking diseased or carrier fish can have serious consequences, including imminent mortality or reduced performance and increased sensitivity to stressors. An equally serious problem is the possible establishment of a reservoir of infection. Most fisheries agencies have attempted to deal with whether or not to stock infected fish through a set of internally developed policy guidelines. For example, salmon infected with hematopoietic necrosis virus or the protozoan *Myxosoma cerebralis,* which induces whirling disease, are not usually stocked, especially in waters containing salmonid populations with no known history of these diseases. On the other hand, fish that have been infected with bacteria such as *Aeromonas* or with protozoans such as *Ichthyopthirius multifilia* are frequently stocked. The latter category of parasites is found in almost all bodies of water. In fact, *Aeromonas* is a nonobligatory fish parasite.

14.4.6.3   Physiological Integrity

The physiological integrity of hatchery-reared fishes can be compromised during the rearing process. For example, it is possible to produce large numbers of fish that do not have inflated swim bladders. These fish do not survive when released into the wild. This is known to occur in a number of both saltwater and freshwater species (e.g., snook, striped bass, and walleye). Under certain conditions, cultured fishes have been shown to have overinflated swim bladders.

A phenomenon even less understood than noninflated swim bladders is whether or not piscivorous sport fishes that have been raised on a prepared diet have a survival rate as high after they are released as the same taxa reared on natural forage. Biologists have mixed opinions on this subject. If fish do not survive as well if raised on a prepared diet, the problem may relate to the quality of the diet or to a change in the behavior of the fish.

14.4.6.4   Handling History

Even if all of the previously discussed criteria, such as species, genetic stock, size, health, and inflated swim bladder, have been met, fish must be handled correctly during harvest and hauling to the stocking site, or high mortality will result. Some of the fish may be dead when they are stocked (initial mortality), or they may die within several days or weeks after stocking (delayed mortality). Many management failures have probably resulted from fish that were predestined to die from handling, hauling, and stocking stress. Often the people who stock fish are not trained biologists. If the fish are moved too fast from one temperature to another, thermal shock could occur. For information on this subject see Piper et al. (1982).

14.5   ECONOMIC EVALUATION OF STOCKINGS

In a review of the economic benefits and costs associated with stocking fishes, Weithman (1986) indicated that factors usually considered to determine success or failure of stocking programs include (1) survival and reproduction of fish, (2) angler acceptance, (3) days and quality of fishing provided, (4) percentage of stocked fish that are harvested, and (5) economic benefits to anglers and state and local governments. An objective evaluation of stocking is not always easy. It can be measured in terms of survival rate, year-class strength, and contribution to the harvest; in actuality, public perception plays a major part in determining the success of a program.

Post-stocking fish surveys are usually conducted to evaluate survival and reproduction. Days of fishing provided and percentage of stocked fish harvested are obtained from creel surveys. The fish community is sampled to document the presence of stocked fish, obtain fish for age and growth analyses, and determine population structure.

Economic evaluation of the success of a stocking is usually based on a benefit-cost ratio. Monetary cost of stocking is theoretically easy to calculate, but biological cost is often hard to identify and more difficult to quantify. Likewise, the biological, social, and economic benefits of stocking may be difficult to quantify. Various methods are used to calculate economic cost and benefits, including (1) agency cost per

Table 14.13   Selected examples of four types of economic cost analyses of stocking programs. Table based on Weithman (1986).

Taxa and state	Cost/angler-day ($)	Cost/fish creeled ($)	Benefit-cost ratio		Reference
			Dollar value of fishing day	Yearly value to anglers	
Channel catfish					
Kansas			39:1		Stevens (1982)
Missouri	0.24				Haas (1984)
Virginia		18.00			Bryson et al. (1975)
Coho and chinook salmon					
Oregon			8:1		Brown and Hussen (1976)
Largemouth bass, bluegill, and channel catfish					
Missouri			90:1		Novinger (1977)
Muskellunge					
Missouri (Pomme de Terre Lake)				9:1	Belusz and Witter (1986)
Rainbow trout					
Virginia	0.69				Applegate (1963)
New Mexico	0.54				Ferkovich (1969)
Salmon and trout					
Great Lakes				18:1	Talheim and Ellefson (1973)
Walleye					
Wisconsin		1.46			Hauber (1983)
Michigan		4.00			Laarman (1981)

angler-day, (2) agency cost to provide fish in the creel, (3) agency cost for stocking versus days of fishing multiplied by the dollar value of a day's fishing (expressed as benefit-cost ratio), and (4) agency cost versus benefit to anglers and state/province and local economies (expressed as benefit-cost ratio) (Weithman 1986). Examples of cost analyses are compared in Table 14.13.

It is questionable whether benefit-cost ratios have included all of the costs. Attempts have been made to compare the cost of fish produced by government agencies to the cost of those produced by the private sector. A comparative cost analysis (USFWS 1982) showed that fingerling rainbow trout, largemouth bass, and bluegill were probably more expensive to buy from federal hatcheries than from private hatcheries. Salmon were not available from the private sector, and channel catfish were probably cheaper to buy from private producers. Catchable-size rainbow trout were cheaper to buy from commercial producers in some regions of the United States but more expensive in others.

The daily cost of fishing, if it is not known for a specific fishery, is often based on national or state averages. Some economists feel that the true value of a fishing day is how much a person is willing to spend to go fishing; others argue that the dollars spent on recreational fishing are "displaced." This latter philosophy concludes that if the money were not spent on fishing, it would be spent for other forms of recreation and, therefore, it would really not be lost to the total economy.

14.6   THE FUTURE

Radonski and Martin (1986) considered the history of stocking and concluded that:

> The arc of the fish culture pendulum has come full swing: from early consideration as a universal fisheries management panacea through a transitional period of questioning and disrepute, to final recognition as an indispensable tool when appropriately integrated with other equally essential fisheries management protocol. (Radonski and Martin 1986:12–13)

Due to the continued increase in the human population and loss of habitat, stocking will continue to be used as a tool in sport fisheries management. In particular, it will be used more frequently in large rivers and coastal areas. Such examples as transplanting the snail darter (Hickman 1981) and the progress made in recovery of native fishes in the southwestern United States (Rinne et al. 1986) will lead to stocking as an acceptable tool in the management of both endangered species and nonsport fishes.

Through no fault of their own, managers of hatchery systems often become locked into production based on numbers, which may have little to do with the success of stocking programs. An alternative to the "head count" is needed. Quotas (hatchery benefit units) can be set in more biologically significant terms once the relative survival and cost of different size individuals of various species is known and a weighting system is set up among species (Heidinger et al. 1987).

The economics of rearing various taxa to larger sizes will change as more biological information is accumulated and as more fish are reared for stocking by private fish culturists. The considerable resistance by state/province and federal agencies to buying fish from private fish culturists for stocking will eventually be reduced due to legal and political pressure and the maturing of the private sector.

No matter how it is done, the actual cost of stocking makes it an expensive management tool. In some cases, part of the cost is recovered by requiring a special license or stamp to fish for the stocked fish. This is most frequently done for salmon and trout put-and-take and put-grow-and-take fisheries. The requirement for a special fishing license or stamp will likely be extended to other sport fisheries.

One of the most important problems associated with evaluating the effectiveness of stocking programs in artificial lakes and reservoirs is the uncontrolled loss of stocked species over the spillway. We need to start thinking of total mortality as equaling fishing mortality plus natural mortality and spillway emigration.

Fish genetics, in terms of identifying stocks, hybridization, sterilization, and strain improvement, will increasingly become more important. With the advent of gene transfer technology, there is tremendous potential for engineering fish communities well suited for perturbed or artificial bodies of water. In the future, genetic stocks will be preserved in gene banks. The philosophical difference between those who advocate no stocking and those who advocate complete laissez faire will initially widen, but as more information is obtained, these views will be reconciled to a much greater degree than they are today.

14.7   REFERENCES

Applegate, R. L. 1963. Evaluation of trout stocking practices. Virginia Commission of Game and Inland Fisheries, Federal Aid in Fish Restoration, Project F-13-R-1, Completion Report, Richmond.

Belusz, L. C., and D. J. Witter. 1986. Why are they here and how much do they spend? A survey of muskellunge angler characteristics, expenditures, and benefits. Pages 39–45 in G. E. Hall and M. J. Van Den Avyle, editors. Reservoir fisheries management: strategies for the 80's. American Fisheries Society, Southern Division, Reservoir Committee, Bethesda, Maryland.

Blankenship, H. L., and K. M. Leber. 1995. A responsible approach to marine stock enhancement. Pages 167–175 in Schramm and Piper (1995).

Boles, H. D. 1960. Experimental stocking of brown trout in a California Lake. Proceedings of the Annual Conference Western Association of State Game and Fish Commissioners 40:334–349.

Boles, H. D., and D. P. Borgeson. 1966. Experimental brown trout management of Lower Sardine Lake. California Fish and Game 52:166–172.

Brege, D. A. 1986. A comparison of muskellunge and hybrid muskellunge in a southern Wisconsin lake. Pages 203–207 in G. E. Hall, editor. Managing muskies. American Fisheries Society, Special Publication 15, Bethesda, Maryland.

Brooks, R. C., R. C. Heidinger, and C. C. Kohler. 1994. Mass-marking otoliths of larval and juvenile walleyes by immersion in oxytetracycline, calcein, or calcein blue. North American Journal of Fisheries Management 14:143–150.

Brown, W. G., and A. Hussen. 1976. A production economic analysis of the Little White Salmon and Willard national fish hatcheries. Oregon State University, Agricultural Experiment Station, Special Report 428, Corvallis.

Brynildson, O. M., P. E. Degurse, and J. W. Mason. 1966. Survival, growth, and yield of stocked domesticated brown and rainbow trout fingerlings in Black Earth Creek. Wisconsin Conservation Department Research Report (Fish) 18, Madison.

Bryson, W. T., R. T. Lackey, J. Cairns, Jr., and K. L. Dickson. 1975. Restocking after fish kills as a fisheries management strategy. Transactions of the American Fisheries Society 104:256–263.

Chandler, W. J. 1985. Inland fisheries management. Pages 93–129 in R. L. Di Silvestro, editor. Audubon Wildlife Report 1985.

Clepper, H. 1970. A century of fish conservation. American Forests 76(11):16–19, 54–56.

Collins, R. A. 1971. Cage culture of catfish in reservoirs and lakes. Proceedings of the Annual Conference Southeast Association of Fish and Wildlife Agencies 24(1970):489–496.

Conover, M. C. 1986. Stocking cool-water species to meet management needs. Pages 31–39 in Stroud (1986).

Cooper, G. P. 1948. Fish stocking policies in Michigan. Transactions of the North American Wildlife Conference 13:187–198.

Davis, J. T. 1986. Role of private industry in increasing production capabilities. Pages 243–248 in Stroud (1986).

Department of Commerce and Labor. 1906. The propagation and distribution of food fishes in 1905. Bureau of Fisheries Document 602, Washington, D.C.

Dillard, J. 1971. Evaluation of two stocking methods for Missouri farm ponds. Pages 203–204 in R. J. Muncy and R. V. Bulkley, editors. Proceedings of the north central warmwater fish culture-management workshop. Iowa State University, Ames.

Dudash, M., and R. C. Heidinger. 1996. Comparative survival and growth of various size channel catfish stocked into a lake containing largemouth bass. Transactions of the Illinois State Academy of Science 89(1 and 2):105–111.

Embody, G. C. 1927. An outline of stream study and the development of a stocking policy. Cornell University Aquaculture Laboratory, Ithaca, New York.

Everhart, W. H., A. W. Eipper, and W. D. Youngs. 1975. Principles of fishery science. Cornell University Press, Ithaca, New York.

Fago, D. M. 1977. Northern pike production in managed spawning and rearing marshes. Wisconsin Department of Natural Resources Technical Bulletin 96.

Ferkovich, P. 1969. Cost:harvest ratios from southeastern area and northeastern area trout waters. New Mexico Department of Game and Fish, Federal Aid in Fish Restoration, Project F-22-R-10, Completion Report, Santa Fe.

Flick, W. A., and D. A. Webster. 1976. Production of wild, domestic and interstrain hybrids of brook trout *(Salvelinus fontinalis)* in natural ponds. Journal of the Fisheries Research Board of Canada 33:1525–1539.

Goede, R. W. 1986. Management considerations in stocking of diseased or carrier fish. Pages 349–355 *in* Stroud (1986).

Gregory, R. W., A. A. Elser, and T. Lenhart. 1984. Utilization of surface coal mine waste water for construction of a northern pike spawning/rearing marsh. U.S. Fish and Wildlife Service FSW/OBS-84103.

Haas, M. A. 1984. The Missouri urban fishing program. Pages 275–279 *in* L. J. Allen, editor. Urban fishing symposium, proceedings. American Fisheries Society, Fisheries Management Section and Fisheries Administrators Section, Bethesda, Maryland.

Harshbarger, T. J., and P. E. Porter. 1982. Embryo survival and fry emergence from two methods of planting brown trout eggs. North American Journal of Fisheries Management 2:84–89.

Hauber, A. B. 1983. Two methods for evaluating fingerling walleye stocking success and natural year class densities in Severn Island Lake, Wisconsin, 1977–1981. North American Journal of Fisheries Management 3:152–155.

Heidinger, R. C., J. H. Waddell, and B. L. Tetzlaff. 1987. Relative survival of walleye fry versus fingerlings in two Illinois reservoirs. Proceedings of the Annual Conference Southeastern Association of Fish and Wildlife Agencies 39(1985):306–311.

Hendricks, M. L. 1995. The contribution of hatchery fish to the restoration of American shad in the Susquehanna River. Pages 329–336 *in* Schramm and Piper (1995).

Hickman, G. D. 1981. Is the snail darter transplant a success? Pages 338–344 *in* L. A. Krumholz, editor. The warmwater streams symposium. American Fisheries Society, Southern Division, Bethesda, Maryland.

Hume, J. M. B., and E. A. Parkinson. 1987. Effect of stocking density on the survival, growth and dispersal of steelhead trout fry *(Salmo gairdneri)*. Canadian Journal of Fisheries and Aquatic Sciences 44:271–281.

Hume, J. M. B., and E. A. Parkinson. 1988. Effects of size at and time of release on the survival and growth of steelhead fry stocked in streams. North American Journal of Fisheries Management 8:50–57.

Hunt, R. L. 1979. Exploitation, growth, and survival of three strains of domestic brook trout. Wisconsin Department of Natural Resources, Report 99, Madison.

Johnson, F. H., and J. B. Moyle. 1969. Management of a large shallow winter-kill lake in Minnesota for the production of pike *(Esox lucius)*. Transactions of the American Fisheries Society 98:691–697.

Keith, W. 1970. Preliminary results in the use of a nursery pond as a tool in fishery management. Proceedings of the Annual Conference Southeastern Association of Game and Fish Commissioners 23(1969):501–511.

Keith, W. E. 1986. A review of introduction and maintenance stocking in reservoir fisheries management. Pages 144–155 *in* G. E. Hall and M. J. Van Den Avyle, editors. Reservoir fisheries management: strategies for the 80's. American Fisheries Society, Southern Division, Reservoir Committee, Bethesda, Maryland.

Kent, D. B., M. A. Drawbridge, and R. F. Ford. 1995. Accomplishments and roadblocks of a marine stock enhancement program for white seabass in California. Pages 492–498 *in* Schramm and Piper (1995).

Kincaid, H. L., and C. R. Berry, Jr. 1986. Trout broodstocks in management of national fisheries. Pages 211–222 *in* Stroud (1986).

Kutkuhn, J. H. 1981. Stock definition as a necessary basis for cooperative management of Great Lakes fish resources. Canadian Journal of Fisheries and Aquatic Sciences 38:1476–1478.

Laarman, P. W. 1978. Case histories of stocking walleyes in inland lakes, impounds, and the Great Lakes—100 years with walleyes. Pages 254–260 *in* R. L. Kendall, editor. Selected coolwater fishes of North America. American Fisheries Society, Special Publication 11, Bethesda, Maryland.

Laarman, P. W. 1981. Vital statistics of a Michigan fish population with special emphasis on the effectiveness of stocking 15-cm walleye fingerlings. North American Journal of Fisheries Management 1:177–185.

Lange, R. E. 1984. Fishing in the Big Apple: a demonstration program for New York City. Pages 263–274 *in* L. J. Allen, editor. Urban fishing symposium proceedings. American Fisheries Society, Fisheries Management Section and Fisheries Administrators Section, Bethesda, Maryland.

Leber, K. M., N. P. Brennan, and S. M. Arce. 1995. Marine enhancement with striped mullet: are hatchery releases replenishing or displacing wild stocks? Pages 376–387 *in* Schramm and Piper (1995).

Marshall, S. A., and B. A. Menzel. 1984. Recovery, growth and habitual utilization of spring-stocked fingerling brown trout *(Salmo trutta)* in six northeast Iowa streams. Proceedings of the Annual Conference Western Association of Fish and Wildlife Agencies 64:411–421.

Maxwell, R., and A. R. Easbach. 1971. Eggs of the threadfin shad successfully transplanted and hatched after spawning on excelsior mats. Progressive Fish-Culturist 33:140.

McEachron, L. W., C. E. McCarty, and R. R. Vega. 1995. Beneficial uses of marine fish hatcheries: enhancement of red drum in Texas coastal waters. Pages 161–166 *in* Schramm and Piper (1995).

McHugh, J. L. 1970. Trends in fishery research. Pages 25–56 *in* N. G. Benson, editor. A century of fisheries in North America. American Fisheries Society, Special Publication 7, Bethesda, Maryland.

Miller, E. 1971. Some success and failures using very small streams as northern pike stocking areas. Pages 84–88 *in* R. J. Muncy and R. V. Bulkley, editors. Proceedings of the north central warmwater fish culture-management workshop. Iowa State University, Ames.

Moring, J. R. 1986. Stocking anadromous species to restore or enhance fisheries. Pages 59–74 *in* Stroud (1986).

Murphy, B. R., and W. E. Kelso. 1986. Strategies for evaluating fresh-water stocking programs: past practices and future needs. Pages 306–313 *in* Stroud (1986).

Nickum, J. G. 1986. Walleye. Pages 115–126 *in* R. R. Stickney, editor. Culture of nonsalmonid freshwater fishes. CRC Press, Boca Raton, Florida.

Nielsen, J. L., editor. 1995. Evolution and the aquatic ecosystem: defining unique units in population conservation. American Fisheries Society, Symposium 17, Bethesda, Maryland.

Noble, R. L. 1986. Stocking criteria and goals for restoration and enhancement of warmwater and coolwater fisheries. Pages 139–159 *in* Stroud (1986).

Novinger, G. D. 1977. An assessment of Missouri's pond program. Missouri Department of Conservation, Federal Aid in Fish Restoration, Project F-I-R-25, Study 1-19, Job 1, Completion Report, Columbia.

Oliver, M. L., and L. L. Rider. 1986. Net-pen aquaculture in Bull Shoals Reservoir. Pages 287–300 *in* Stroud (1986).

Pennsylvania Fish Commission. 1986. Management of trout fisheries in Pennsylvania. Pennsylvania Fish Commission, Division of Fisheries, Harrisburg.

Philipp, D. P., J. B. Koppelman, and J. L. Van Orman. 1986. Techniques for identification and conservation of fish stocks. Pages 323–338 *in* Stroud (1986).

Piper, R. G., and five coauthors. 1982. Fish hatchery management. U.S. Fish and Wildlife Service, Washington, D.C.

Potter, B. A., and B. A. Barton. 1986. Stocking goals and criteria for restoration and enhancement of coldwater fisheries. Pages 147–159 *in* Stroud (1986).

Radonski, G. C., and R. G. Martin. 1986. Fish culture is a tool, not a panacea. Pages 7–13 *in* Stroud (1986).

Reisenbichler, R. R., and J. D. McIntyre. 1977. Genetic differences in growth and survival of juvenile hatchery and wild steelhead trout, *Salmo gairdneri*. Journal of the Fisheries Research Board of Canada 34:123–128.

Richards, W. J., and R. D. Edwards. 1986. Stocking to restore or enhance marine fisheries. Pages 75–80 *in* Stroud (1986).

Rinne, J. N., J. E. Johnson, B. L. Jensen, A. W. Ruger, and R. Sorenson. 1986. The role of hatcheries in the management and recovery of threatened and endangered fishes. Pages 271–285 *in* Stroud (1986).

Royer, L. M. 1971. Comparative production of pike fingerlings from adult spawners and from fry planted in a controlled spawning marsh. Progressive Fish-Culturist 33:153–155.

Schramm, H. L., Jr., and R. G. Piper, editors. 1995. Uses and effects of cultured fishes in aquatic ecosystems. American Fisheries Society, Symposium 15, Bethesda, Maryland.

Schuck, H. A. 1943. Survival, population density, growth, and movement of wild brown trout in Crystal Creek. Transactions of the American Fisheries Society 73:209–230.

Smith, B. W., and W. C. Reeves. 1986. Stocking warmwater species to restore or enhance fisheries. Pages 17–29 *in* Stroud (1986).

Smith, L. L., Jr., and J. B. Moyle. 1944. A biological survey and fishery management plan for the streams of the Lake Superior north shore watershed. Minnesota Department of Conservation Technical Bulletin 1.

Stevens, V. 1982. Channel catfish fishery maintenance for public fishing waters. Kansas Fish and Game Commission, Federal Aid in Fish Restoration, Project F-25-D, Completion Report, Pratt.

Stroud, R. H., editor. 1986. Fish culture in fisheries management. American Fisheries Society, Fish Culture Section and Fisheries Management Section, Bethesda, Maryland.

Swanson, B. L. 1982. Artificial turf as a substrate for incubating lake trout eggs on reefs in Lake Superior. Progressive Fish-Culturist 44:109–111.

Swingle, H. S. 1970. History of warmwater pond culture in the United States. Pages 95–106 in N. G. Benson, editor. A century of fisheries in North America. American Fisheries Society, Special Publication 7, Bethesda, Maryland.

Symons, P. F. K. 1969. Greater dispersal of wild compared with hatchery-reared juvenile Atlantic salmon released in streams. Journal of the Fisheries Research Board of Canada 26:1867–1876.

Talheim, D. R., and P. V. Ellefson. 1973. Michigan's Great Lakes trout and salmon fishery (1969–1972). Bureau of Sport Fisheries and Wildlife, and the National Marine Fisheries Service, Fisheries Management Report 5, Great Lakes Fish Resource Development Study AFSC-8, Ann Arbor, Michigan.

Thompson, P. E. 1970. The first fifty years-the exciting ones. Pages 1–11 in N. G. Benson, editor. A century of fisheries in North America. American Fisheries Society, Special Publication 7, Bethesda, Maryland.

USFWS (U.S. Fish and Wildlife Service). 1975. Propagation and distribution of fishes from national fish hatcheries for fiscal year 1975. USFWS, Fish Distribution Report 10, Washington, D.C.

USFWS (U.S. Fish and Wildlife Service). 1982. Comparative costs of alternative sources of fish for federal management needs. USFWS, Division of Hatcheries and Fishery Resource Management, Washington, D.C.

USFWS (U.S. Fish and Wildlife Service). 1986. Fish and fish egg distribution report 21. USFWS, Washington, D.C.

USFWS (U.S. Fish and Wildlife Service). 1995. Fish and fish egg distribution report 30. USFWS, Washington D.C.

Weithman, A. S. 1986. Economic benefits and costs associated with stocking fish. Pages 357–364 in Stroud (1986).

Wydoski, R. D. 1986. Informational needs to improve stocking as a coldwater fisheries management tool. Pages 41–57 in Stroud (1986).

Chapter 15

Management of Undesirable Fish Species

RICHARD S. WYDOSKI AND ROBERT W. WILEY

15.1 INTRODUCTION

Beginning in the nineteenth century, many species of fish were intentionally stocked throughout the United States to establish food fish populations, create new fisheries, or restore populations that had declined from overharvest or loss of habitat (Bowen 1970; Ross 1997). The success of stocking (i.e., good growth, survival, and recruitment) depended on suitable habitat and adequate food resources, and federal and state conservation agencies began to control fish species that competed with or preyed upon fishes that were considered desirable by society.

Occasionally, fishes that were introduced became more deleterious than beneficial (Magnuson 1976; Moyle 1986; Courtenay and Robins 1989) and required control. Early fisheries administrators in the United States introduced the common carp into North America because it was considered a good food fish; it was introduced into the Hudson River, New York, in 1831–1832 (Cole 1905). However, common carp was not used for food and failed to provide satisfactory sport for anglers (Cumming 1975). By 1891, the first measures to control common carp were taken in Lake Merced, California (Cole 1904). Chemicals were first used to control fish in the United States in the early 1900s, when copper sulfate was applied in a Vermont lake (Titcomb 1914). From the 1930s (Hubbs and Eschmeyer 1938) through the 1950s, conservation agencies increased efforts to control undesirable fishes by means of chemical toxicants, biomanipulation, electricity, and mechanical removal. Clearly, as the demand for desirable sport and commercial fishes increased so did the perceived need to control undesirable fishes. Presently, management of undesirable fish species in North America appears to be evolving from control of a single species toward an approach that considers fish assemblages or communities.

15.1.1 Definition and Examples of Undesirable Fish Species

Based on questionnaires mailed to them in 1988 and 1996, state and provincial conservation agencies defined an undesirable fish as virtually any species not considered beneficial to humans, analogous to weeds as defined by gardeners. Because of the implicit subjectivity of this term, it is imperative to establish explicit, biologically sound criteria for determining whether a species is undesirable and warrants control in a given situation. A given species (or population) of fish is considered undesirable be-

---

Box 15.1   Criteria upon Which Fish Are Determined Undesirable

1. The species is detrimental to the biological balance of an aquatic system (e.g., large gizzard shad sometimes constitute most of the biomass in reservoirs).

2. The species adversely affects native fishes that are listed as threatened or endangered.

3. The species may serve as a potential reservoir for pathogenic organisms in natural waters or a hatchery water supply.

4. The species does not contribute to a commercial or sport fishery (i.e., may not be available or acceptable to anglers) or provide a forage base.

5. The species inhibits development or maintenance of desirable sport or commercial fishes through predation or direct competition.

6. The species may interfere with other wildlife management practices (e.g., common carp cause turbidity, thus inhibiting the growth of aquatic plants that are used as food by waterfowl).

---

cause it may (1) be detrimental to the biological balance of an aquatic ecosystem, (2) interfere with other wildlife management practices, (3) adversely affect threatened or endangered species, or (4) not contribute to a commercial or sport fishery, as well as for other reasons (Box 15.1).

Often nongame fishes are targeted for control because they compete with or prey upon sport fishes. In addition, stunted stocks of game fishes are sometimes controlled because the fish are too small to be acceptable to anglers. For example, largemouth bass stocked in ponds sometimes become stunted because they become overabundant and deplete their food resources. They then must be reduced in numbers before they can be effectively managed. At other times, largemouth bass do not have to be reduced in number because they have become stunted due to a lack of forage fishes (i.e., quantity) or from food preference for soft-rayed fishes (i.e., quality) over spiny-rayed fishes (Holton 1977). Other examples of undesirable populations of game fishes include transplanted trout species such as brook trout and rainbow trout. Transplanted eastern brook trout compete with native western cutthroat trout (Gresswell 1991), and transplanted western rainbow trout compete with native eastern brook trout (Larson et al. 1986).

In the past, native fishes that did not provide either commercial or sport fisheries and were not used as forage by species with commercial or sport value were considered undesirable. Today, most state wildlife and fisheries conservation agencies have responsibility for all aquatic species and consider the entire fish community in natural waters. Introductions of new fish species and their effects on native fishes are carefully considered in preparing a management plan for a particular water.

In the past the introduction of fishes was believed to be generally benign. It was believed that introduced fishes were compatible with other vertebrates using aquatic habitats (e.g., birds), however, various studies have demonstrated that this assumption is often incorrect (Bouffard and Hanson 1997). Predation by introduced fishes can reduce the abundance and mean sizes of pelagic, benthic, and littoral invertebrates, changes that alter food webs (Murdoch and Bence 1987). Changes in aquatic invertebrate abundance and diversity may affect various vertebrate species that use aquatic habitats and depend on invertebrates for food. Introduced fishes can degrade aquatic systems and reduce the abundance of endemic fishes or other aquatic species. Wilcove et al. (1992) reported that 34% of native North American fishes, 65% of crayfishes, and 73% of unionid mussels may be at risk due to introductions of nonnative fish species (see also Miller et al. 1989). Because fish introductions are often irreversible (Magnuson 1976; Courtenay and Robins 1989), strict protocol must be followed in intentional introductions of new fish species to ensure that such fishes do not become undesirable (Kohler and Courtenay 1986). Some native North American fish species that were once considered to be undesirable are being reevaluated as to their roles in aquatic ecosystem function and stability (Holey et al. 1979; Scarnecchia 1992).

Accidentally introduced fish species (see Chapter 13) can become undesirable and require control. Prevention of such introductions is generally the best safeguard to avoid environmental catastrophes. Accidental fish introductions, such as the sea lamprey in the Great Lakes, have required expensive control programs (Lennon et al. 1970; Cumming 1975). In the 1980s, the Eurasian ruffe was accidentally introduced into the harbor at Duluth, Minnesota, by means of ballast water discharge from seagoing freighters. Although small, the adaptable ruffe is now the most abundant species in the fish community in Duluth Harbor, and most endemic minnows have declined sharply (Boogaard et al. 1996).

15.1.2 Families of Fish Targeted for Control

Various control programs have targeted virtually all species of freshwater fishes (Lennon et al. 1970; Dunst et al. 1974). Over half (54%) of current control efforts are directed at species in three families of fishes: herrings (primarily gizzard shad), minnows (almost exclusively common carp), and stunted sunfishes (Centrarchidae). An additional 22% of the control efforts involve suckers and bullhead catfishes (*Ameiurus* spp.).

15.2 THE FISH CONTROL PROJECT

The assumptions upon which a fish species is deemed undesirable and the problems it is suspected of causing should be thoroughly evaluated before deciding if control is warranted. A fish control project can be ineffective or short lived if it treats the symptom rather than the cause of the problem (Meronek et al. 1996). If control measures are warranted, the best alternative to accomplish management goals must be determined. State and provincial conservation agencies use a variety of criteria to evaluate and select a control method, including the cost, size of water body, characteristics of the water body such as temperature and water quality, the target species, public opinion, water body ownership, environmental concerns, and location of the water body.

15.2.1 Preparation of an Environmental Assessment

Federal agencies or state agencies that propose actions which will be paid by federal funds (e.g., through the Federal Aid in Sport Fish Restoration Act) are required to prepare an environmental assessment to determine if an action may result in adverse effects. Proper planning and assessment of a proposed action will aid in preparation of an environmental assessment (Box 15.2) If a potential for adverse environmental effects exist, then the agency must describe how impacts will be minimized or the kind of mitigation that will be conducted to offset those impacts. Environmental assessments provide a mechanism by which to plan a project and select the best alternative to accomplish agency goals and objectives (Box 15.2). In the United States, environmental assessments are required by the National Environmental Policy Act (42 U.S.C. Sections 4321–4361; see Chapter 4).

Environmental assessments should include the purpose and need for the proposed treatment, a description of the environment, alternatives that were considered to accomplish the proposed work, environmental impacts of the proposed treatment, mitigating measures to offset adverse effects of the proposed treatment, discussion of unavoidable adverse effects, discussion of irreversible and irretrievable commitments of resources, and documentation of public and agency interest or concern.

Environmental assessments are reviewed by the U.S. Fish and Wildlife Service, other federal and state agencies, and the public. If no adverse comments are received or if benefits clearly exceed adverse effects, the control project can proceed. If there are significant adverse environmental effects or potential public health hazards, an environmental impact statement must be prepared. Pretreatment studies may be necessary to obtain information requisite for an environmental assessment. A comparison of alternative actions and potential outcomes and impacts is vital to the success of any project. The advantages and disadvantages of all alternative actions for controlling

---

Box 15.2   Planning a Fish Control Project

1. Establish that a fish control project is warranted by clearly and succinctly describing the purpose and need for the action.

2. Review advantages and disadvantages of various alternative control methods.

3. Determine the potential adverse environmental impacts for each fish control method.

4. Carefully review the potential for any irreversible or irretrievable adverse environmental impacts from the various alternatives.

5. Obtain public input on the proposed action to determine if such action is sensitive or controversial.

6. Prepare a draft environmental assessment that explains the purpose and need, alternative fish control methods considered, and rationale for the proposed action selected following guidelines for the National Environmental Policy Act.

7. Distribute to appropriate agencies, organizations, and individuals for review and comment; place a public notice in newspapers published near to the location of the proposed project; conduct one or more public meetings if warranted because of sensitive or controversial issues.

8. Prepare a final environmental assessment making appropriate revisions from public review and comment or determine if an environmental impact statement is required under the National Environmental Policy Act.

undesirable fishes should be carefully considered before deciding upon the most effective method that will result in the least adverse environmental impact(s). Such planning should be axiomatic in any fish control program.

15.2.2 Basic Methods of Fish Population Control

Three methods are used to control undesirable stocks of fishes: chemical, biological, and mechanical. Chemical and biological means are favored over mechanical controls for most families of fishes. As of the late 1990s, North Carolina, Rhode Island, and the Canadian Province of Prince Edward Island do not control any undesirable fish species, and Texas relies solely on biological control.

Chemical control (Lennon et al. 1970; Eschmeyer 1975; Schnick et al. 1986) is the favored control method of 60% of state and provincial conservation agencies. Biological control of fish stocks are used by 35% of the agencies, and, perhaps as a result of increased consideration of fish communities, use of this method appears to be increasing. North American fisheries managers find mechanical control methods labor intensive and generally ineffective (Dunst et al. 1974). Sometimes integrated pest management principles (using a combination of methods) may be warranted (Marsden 1993). All control programs require an awareness of the entire fish community because targeting a single species without regard to its relationship to others in the community may trade one problem for another.

15.3 CHEMICAL METHODS FOR MANAGING UNDESIRABLE FISH SPECIES

Chemical control is the most popular method to control undesirable fishes (Box 15.3) because of the ease of application, the short time required to achieve lasting results, and the lower cost when compared with other control methods (Lennon et al. 1970; Eschmeyer 1975). Complete removal of all undesirable fish species was the general goal of chemical treatment projects by 86% of the state and provincial conservation agencies. Partial treatments with chemicals include (1) treatment of fish spawning sites, (2) treatment of particular sections of lakes where undesirable fish are aggregated for reasons other than spawning, or (3) thinning of stunted stocks of game fishes (Lennon et al. 1970; Bradbury 1986). Partial chemical treatments have met with varying success depending on size of the water body treated, water quality, water temperature, and target species (Bradbury 1986). Partial treatments are often followed by introductions of predatory game fishes to feed on forage species or to control less desirable fishes. Available food sources should be considered prior to any restocking efforts. For example, stocked fish may exhibit poor growth and low survival if the abundance of zooplankton or other food organisms is low (see Section 15.3.4.4).

After chemical treatment fishing pressure on formerly marginal trout waters was observed to equal the pressure on the best trout waters in Michigan (Trimberger 1975). Chemical treatments of warmwater lakes consistently result in more angler days per acre than are attributable to other control strategies (Lennon et al. 1970). Periods of effectiveness of chemical treatments range from less than 1 to up to 30 years (Figure 15.1). When benefits are substantial, treatment programs may be scheduled at regular intervals.

Box 15.3   Assessment of a Chemical Fish Control Project

1. Based on pretreatment surveys of the fish population, determine the efficacy of chemical treatment for restoring the sport fishery. Review all potential control methods to ensure selecting the most effective and practical method.

2. Obtain and evaluate complete water quality and fishery statistics.

3. Determine the volume (lake or pond) or length and volume stream) of water to be treated.

4. Determine the amount of toxicant required for the desired treatment (amounts of toxicant may be decreased if lake levels can be lowered or the flow of regulated streams reduced).

5. Determine if the chemical must be detoxified (some break down to nontoxic components quickly depending on factors such as water temperature and sunlight); accurately determine the amount of material required to detoxify the specific concentration of the toxicant.

6. Provide an opportunity for the public to become informed and allow them an opportunity to comment on the treatment.

7. Ensure that the treatment will not contaminate potential sources of drinking water.

8. Evaluate the potential adverse effects on environmentally sensitive species, including threatened and endangered species.

9. Develop detailed standard operating procedures that completely cover all aspects of the fish control project.

10. Carefully consider fish species to be used in restocking waters to ensure that suitable environmental conditions are present.

In addition to controlling undesirable fish species, chemicals can also be used to determine the structure and health of fish communities and to determine if other management actions are required (Bettoli and Maceina 1996). Lake cove sampling began in the 1950s and was used widely by the mid-1960s as a reservoir management tool. Rotenone has also been used to evaluate the efficiency of gears for sampling estuarine fishes (Allen et al. 1992). Generally, the objectives of inventories or surveys conducted using chemicals are to determine the species composition, total number and weight by species, and size structure of selected fishes in particular waters. However, results can be biased because animals such as birds (Matlock et al. 1982) or crayfishes (Boccardy and Cooper 1963), which are not affected by toxicants at the dosages used (see Section 15.3.4.4), may feed on the dead and dying fishes and cause underestimates of certain size- or age-classes.

The surfacing of fishes killed with chemical toxicants may be important to estimate or document numbers or growth rates of fishes or species composition (Bettoli and Maceina 1996) or to remove fishes that may add obnoxious odor or taste to municipal water supplies (Cohen et al. 1960, 1961). Surfacing may be related to the resistance of fish species to rotenone, water temperature, fish size, water depth, and density

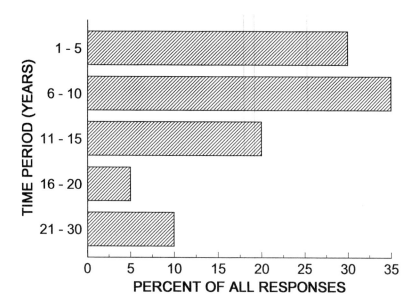

Figure 15.1 Length of time that a chemical treatment has been found to be effective by state and provincial conservation agencies (based on 1988 and 1996 surveys).

of rooted aquatic vegetation. Parker (1970) reported that following application of rotenone, minnows surfaced first, followed by centrarchids, and finally bullhead catfishes. A higher percentage of fish surfaced at warmer water temperatures, and warmwater fish species surfaced slower from deep water and with increased density of rooted aquatic vegetation (Parker 1970; Bradbury 1986).

Prior to chemical treatment of waters, conservation agencies may liberalize regulations to allow harvest of desirable sport fishes because U.S. regulations prohibit human use of fish killed by toxicants. For example, the Utah Division of Wildlife Resources liberalized regulations for trout prior to a 1990 large-scale chemical treatment to control Utah chub in Strawberry Reservoir (4,900 ha) and 259 km of tributary streams in north-central Utah.

Bettoli and Maceina (1996) surveyed 55 fishery agencies in North America and found that 64% have decreased or ceased using chemical fish toxicants in the past 10 years. Fewer streams are treated with chemicals today because several case histories indicate that control of toxicants in running waters is often impossible. If the chemical is not controlled, an agency may be liable to compensate for environmental damages, such as the loss of sport fisheries, under the Comprehensive Environmental Response, Compensation, and Liability Act (42 U.S.C. 9601 et seq.).

15.3.1   Properties of an Ideal Fish Toxicant

Chemical fish toxicants must have properties that are consistent with the needs of fisheries managers while meeting the requirements of government regulations. An ideal toxicant should be specific to the species of fish targeted, easy and safe to apply, harmless to nontarget organisms (plant and animal), effective over a broad range of water quality conditions, and registered for use in the aquatic environment; it also should degrade to harmless constituents in a limited time without the aid of a detoxicant (Lennon et al. 1970). No currently registered fish toxicant meets all of these criteria. Therefore, fisheries managers must carefully evaluate the benefits of using toxicants with potential adverse environmental effects.

15.3.2 Chemicals Registered for Use as Piscicides

Since 1980, the Fisheries Act of Canada has not permitted the application of any deleterious substance (such as toxicants) in fish-bearing waters. The Pesticide Chemicals Act of Canada allows use of a registered powdered form of rotenone but in only small. landlocked lakes. The Canadian Province of Quebec placed a complete ban on the use of chemicals for fish control from 1994 through 1997.

Four toxicants are registered for use as piscicides in the United States (Schnick et al. 1986): rotenone, antimycin, TFM (3-triflouromethyl-4-nitrophenol), and Bayluscide. Registration of new chemical fish toxicants in the near future is unlikely because of the high cost of the registration process. Registration will have to be done by private industry, but the low market potential for a minor use piscicide may not be profitable (Marking 1992).

In the United States, regulations governing piscicide use are administered by the federal government (Federal Insecticide, Fungicide, and Rodenticide Act of 1972, as amended, 7 U.S.C. Sections 136–136y) and by the respective states. Most states require that pesticides (including piscicides) be used by only certified applicators.

Two of the fish toxicants, rotenone and antimycin, are registered for general use and two, Bayluscide and TFM, are registered as lampricides. Rotenone is the common choice and is used by 97% of state conservation agencies. Antimycin is used by 25% of those state agencies. Bayluscide and TFM are used for selective control of sea lampreys by 11% and 8% of the agencies, respectively, in northeastern and central North America. Canada does not consider TFM and Bayluscide deleterious under the Fisheries Act of Canada.

15.3.3 Safe Use of Chemical Toxicants in Fisheries Management

Safety precautions must always be carefully followed when using chemicals (Sowards 1961; Binning et al. 1985). Following proper protocol and using common sense when planning and executing chemical control projects will ensure the safety of workers, the public, and the environment. Labels contain safety precautions to be taken with specific toxicants (Box 15.4), and agencies are liable for damages if full compliance is not made with the label instructions.

15.3.4 Rotenone

Rotenone was first used in North America in 1934 (Lopinot 1975) and is the most commonly used fish toxicant. It is a natural chemical compound found in the roots of several tropical plants of the genera *Derris*, *Lonchocarpus*, and *Tephrosia*, which occur in Asia, Central, and South America (Bradbury 1986). Rotenone is presently registered for use as a general fish toxicant for nonfood fishes. It is readily absorbed through fish gills and inhibits the oxygen transfer system in cell mitochondria (Lindahl and Oberg 1961), which results in physiological suffocation. However, rotenone has been found to be safe to humans when it is applied by certified applicators who follow necessary precautions (Sousa et al. 1987).

Box 15.4   Handling Toxicants in Fisheries Management

1. Always read and follow directions on the container label (information on the label includes product name, type of chemical, formulation, statement of ingredients, net contents, name and address of manufacturer, Environmental Protection Agency [EPA] registration number, number identifying where the product was produced, reference to toxicity to humans, cautionary statements, whether product is for general or restricted use, directions for use, statement on restricted use, statement on misuse, instructions for storage and disposal, and limitations or restrictions on use).

2. Store chemical toxicants in only the original labeled containers.

3. Avoid smoking in areas where toxicants are handled or used.

4. Avoid inhaling fumes or dusts from toxicants and wear protective clothing if so instructed.

5. Avoid toxicant spills; wash immediately if skin contact occurs.

6. Dispose of empty toxicant containers following EPA guidelines.

7. Always wash thoroughly, including clothing, following use of toxicants.

8. Prevent access to any toxicants by children, pets, or irresponsible adults.

15.3.4.1   Rotenone Application

Rotenone is available in both powder or liquid forms. Powder is less expensive than is the liquid, and when powder is mixed into a slurry before application it is very effective. Powdered rotenone can be mixed with sand (1 part to 3 parts), a small amount of gelatin, and water to form a thick paste for use in spring seeps, heavily vegetated shorelines, and deep waters (R. L. Spateholts, Utah Division of Wildlife Resources, personal communication).

Both powdered and emulsified forms of rotenone cause avoidance reactions in fishes (Bradbury 1986). Rainbow trout exhibited an avoidance response to powdered rotenone in a laboratory study (Courtney 1991); however, the response by this species was much stronger to the liquid formulation.

The shelf life of the active ingredient in liquid rotenone formulations is guaranteed for 1 year when stored in sealed containers at 70°F. However, liquid formulations have remained stable for 5 years (R. L. Fisher, AgrEvo, personal communication). Several biologists have estimated that powdered rotenone degrades about 1–2% per year during storage. Rotenone powder, intended for agricultural use, was found to be about 4.5% active ingredient after being stored in a warehouse for about 20 years (Wyoming Game and Fish Department, unpublished data). Over that time, powdered rotenone generally contained about 5% active ingredient, indicating that powdered rotenone may remain stable for a long time. Storing powdered rotenone out of direct sunlight and in a cool location helps preserve its efficacy when stored for a long time.

Rotenone is commonly dispersed by means of pump sprayers, mixed with water in the propwash of boats, pumped into deep waters in lakes, and applied from aircraft. Constant-flow drip stations are used in treating streams.

Biologists should always consult the container label when calculating the amount of toxicant to be used. Rotenone is applied at various rates, depending on water chemistry, that may affect the success of fish removal. In formulations with 5% active ingredient, application rates of rotenone to kill fish range from 1 to 5 mg/L. For freshwater treatments, the Utah Division of Wildlife Resources uses between 1.5 and 2.0 mg/L for treatments of most species, 2.5–3.0 mg/L to eradicate bullheads and carps, and 5.0 mg/L for preimpoundment treatments. Hegen (1985) reported that a rotenone concentration of 2 mg/L was effective on fishes in estuaries. However, he emphasized that it was essential to consider water movement via wind, tide, and currents for proper distribution and confinement of the toxicant.

Rotenone applications at water temperatures of 20°C and above result in optimum fish kills and facilitate detoxification because rotenone kills fish quickly and breaks down rapidly in warm water. Biologists in Wisconsin (Gilderhus et al. 1986; Roth and Hacker 1988) reported application of rotenone shortly before ice formation (water temperature 4.4°C or less) to obtain complete kills of common carp and bullhead catfishes. At low water temperatures (especially if ice cover forms shortly after application), rotenone can remain toxic for up to 3 months. The advantages of applying rotenone in cold water (particularly just before ice formation) are (1) the elimination of odor from decomposing fish, (2) reduction of labor costs associated with disposal of dead fish, and (3) natural detoxification is usually complete by ice breakup so that restocking can proceed at that time. These advantages may be important or mandatory for health reasons (e.g., odor and taste problems in domestic water supplies; Cohen et al. 1960, 1961) or aesthetic reasons (e.g., residential and recreational areas with heavy human use).

Although the toxicity of rotenone-treated bait was similar to the toxicity via water applications, various studies have reported limited success in controlling undesirable fishes such as common carp because they do not readily eat the treated bait (Buck et al. 1960; Fait and Grizzle 1993). However, baiting with corn concentrated common carp, which were then killed with rotenone. This strategy was effective in small- to medium-size lakes, but several treatments may be required to achieve partial control among widely scattered areas of large lakes because common carp readily re-entered chemically treated areas within 36–48 (Buck et al. 1960). Baiting with nontoxic food dispensed from automatic fish feeders was used to concentrate common carp in tributaries of a reservoir by Bonneau et al. (1995).

15.3.4.2   Efficacy of Rotenone

Toxicity of rotenone is affected by the dosage used and its dilution and characteristics of the body of water, such as water temperature, light, dissolved oxygen, turbidity, organic materials, aquatic plants, bottom sediments, and alkalinity. The efficacy of rotenone, and also of antimycin, is probably not significantly affected by aquatic vegetation but is greatly reduced by turbidity (Gilderhus 1982). Some species of fish, such as goldfish and bullhead catfishes, which can tolerate low oxygen levels, are relatively resistant to rotenone whereas salmonids are least resistant (Marking and Bills 1976). In addition, some fish species may develop a tolerance to rotenone in waters that are treated repeatedly with rotenone or other chemicals (Fabacher and Chambers 1972; Orciari 1979).

The dose and exposure time of fishes to toxicants (rotenone and antimycin) are especially critical for effective control of undesirable fishes (Gilderhus 1982). Fish exposed to rotenone will recover if the dose, exposure time, or both are too short (Tate et al. 1965).

There are differences in resistance to rotenone among different life stages and species of fishes. Salmonid eggs are 10–100 times more resistant to rotenone than are fry (Garrison 1968; Olson and Marking 1975; Marking and Bills 1976). Rotenone toxicity to eggs of nonsalmonid fishes is variable. For example, Hester (1959) reported that rotenone toxicity to common carp and fathead minnow eggs was similar in toxicity to fingerlings. However, Marking et al. (1983) reported that eyed common carp eggs were about 50 times as resistant to rotenone as were larvae.

15.3.4.3 Detoxification of Rotenone

Rotenone is environmentally nonpersistent and will detoxify naturally within 2 d to 2 weeks in late summer or early fall (California Department of Fish and Game 1985; Bradbury 1986). Natural detoxification is accelerated by warm water temperatures, high alkalinity, and sunlight in clear waters. Detoxification is inhibited by turbidity and deep water because of decreased light penetration.

Generally, rotenone detoxifies rapidly in lakes, ponds, and reservoirs. Where chemically induced detoxification is necessary, such as in potable water supplies or in delimited instream treatments, potassium permanganate is usually added in an amount equal to the rotenone used plus the permanganate demand of the water (Figure 15.2; Pfeifer

Figure 15.2 A potassium permanganate drip station used to detoxify rotenone in a stream at the downstream end of a reach treated to control undesirable fishes.

1985). Chemical detoxification of rotenone with potassium permanganate in potable water supplies or streams is described in detail by Loeb and Engstrom-Heg (1971), Engstrom-Heg (1972), and Pfeifer (1985). It is important to consider the toxicity of potassium permanganate to fishes when chemical detoxification is warranted. Marking and Bills (1975) reported that 3 mg/L is toxic to trout in soft water (44–170 mg/L total hardness) and 1.5 mg/L is toxic to trout in hard water (300 mg/L total hardness). In an estuary, potassium permanganate successfully neutralized 2 mg/L rotenone at an application rate of 2–4 mg/L (Hegen 1985).

15.3.4.4   Effects of Rotenone Treatment on Other Fauna and Flora

Rotenone is extremely toxic to many species of invertebrates in addition to many fishes (Bradbury 1986). Zooplankton communities are drastically reduced following rotenone treatments; cladocerans and copepods are very sensitive whereas rotifers are more tolerant (Kiser et al. 1963). Most open-water zooplankton die within 1 h after treatment, and 95–100% are dead within a day (Kiser et al. 1963; Schnick 1974; Bradbury 1986). Zooplankton can rebound from cove sampling with rotenone within 1 week by means of recolonization from adjacent untreated water (Neves 1975). Most zooplankton communities exhibit resiliency and are completely recovered within 2–12 months following rotenone treatments (Bradbury 1986). However, some zooplankton communities, such as those in cold, oligotrophic, alpine waters, may not recover for 2–3 years (Wrenn 1965; Anderson 1970; Walters and Vincent 1973). Although temporary shifts occur in zooplankton communities following rotenone treatment, all species recover following treatment, and new zooplankton species may sometimes appear but never become dominant (Bradbury 1986).

Benthic macroinvertebrates vary in their response to rotenone treatments. Generally, immediate reductions in numbers of benthic organisms range from 0 to 70% (mean, 25%) in freshwater with rotenone treatments of 1–2 mg/L (Bradbury 1986). The diversity of benthic organisms is not affected by rotenone treatment, and most benthic communities recover within 2 months (Bradbury 1986). Of the benthic fauna, crustaceans, such as crayfishes, are most tolerant to rotenone treatments, whereas midges and mayflies are least tolerant (Bradbury 1986). Marine crustaceans such as blue crabs survived rotenone concentrations of 4 mg/L (Hegen 1985).

Phytoplankton and rooted aquatic plants are unaffected by rotenone treatments at concentrations used to kill fishes. Phytoplankton blooms occur in some lakes following rotenone treatment in response to nutrients released from decomposing fishes (Bonn and Holbert 1961).

Bradbury (1986) stated that some amphibians and reptiles are affected by rotenone and may be killed. He reported that amphibians with gills were most affected, some alligators were made visibly ill, and some turtles were killed during rotenone treatments. Rotenone is relatively nontoxic to birds and mammals (Bradbury 1986; Marking 1988) due primarily to the entry sites and oxidizing enzyme systems (Fukami et al. 1969, 1970).

15.3.4.5   Effects of Rotenone on Humans

Humans can be exposed to rotenone through dermal exposure, inhalation, and ingestion. No human fatalities have been reported from exposure to rotenone. Symptoms of short-term exposure (a few days) to rotenone include numbing sensation in

mouth and lips, mild sore throat, eye irritation, runny nose, mild headache, and derma-
titis of sensitive tissues in armpits and genital region. Long-term exposure (up to 3
weeks) to rotenone can result in skin rashes, severe eye and nasal inflammation, severe
irritation of tongue and lips, and loss of taste and appetite (California Department of
Fish and Game 1985; Bradbury 1986).

15.3.5 Antimycin

Antimycin (Fintrol is the registered name) is an antibiotic and the only other compound registered as a general fish toxicant. Antimycin was first used as a fish toxicant in
1963 (Lopinot 1975), and it was registered in the United States and Canada in 1966 (Lennon
et al. 1970). Antimycin is less commonly used than rotenone because of its limited availability from a single supplier (Aquabiotics Corporation, Bainbridge Island, Washington).

Fish absorb antimycin through the gills and are killed by interruption of cellular
respiration. Antimycin is an effective fish toxicant in softwater streams, lakes, and
ponds and will kill fertilized fish eggs. Antimycin does not elicit an avoidance response in fishes, and it is highly toxic to some rotenone-resistant fish species. Antimy-
cin-impregnated fish meal bait produced only limited success, 21–51% mortality in 96
h (Rach et al. 1994), when used on common carp that were congregated and actively
feeding. Because antimycin is now available in only a liquid formulation, it is not as
effective in deep ponds and lakes. A formerly available sand-based compound was
effective in deeper waters, and some biologists have applied the liquid formulation to
sand for treating ponds and lakes.

Antimycin is readily detoxified by sunlight, warm water temperatures ($>10°C$),
and alkalinity ($pH> 8.0$) (Tiffan and Bergersen 1996). It is also naturally degraded in
stream habitats that have an elevation drop of 80 m or more, apparently through aeration. In stream habitats, degradation of antimycin associated with elevation drop can
be compensated for by adding more drip stations. Chemical detoxification of antimycin with potassium permanganate was described by Marking and Bills (1975).

15.3.6 Lampricides

Two lampricides, TFM and Bayluscide (Bayer 73 is the registered name), are used to
control sea lamprey, largely by the Great Lakes Fishery Commission (see Chapter 23).

15.3.6.1 TFM

This selective toxicant, first tested as a lampricide in Michigan in 1958 (Cumming
1975), is applied as a liquid or soluble-bar formulation to control larval sea lampreys
(ammocetes) in spawning and nursery areas. The mode of action is not well understood, but TFM appears to cause general collapse of the circulatory system. The toxicant TFM is restricted for nonfood use as a lampricide and can be used by only applicators certified by the Fish and Wildlife Service (Schnick et al. 1986). The effectiveness
of TFM is depressed at water temperatures near freezing and at pH values above 8.0.
The chemical is environmentally nonpersistent, does not affect birds, mammals, or aquatic
plants (though photosynthesis may be reduced), has a varied effect on invertebrates depending upon treatment location and species, and may depress associated fish populations.

15.3.6.2 Bayluscide

Bayluscide, a granular or powdered toxicant, was developed for the control of mollusks (snails) in tropical areas and was first used as a lampricide in 1963 (Cumming 1975). Although the exact mode of action of Bayluscide is unknown, it is believed to be similar to that of TFM (National Research Council of Canada 1985). Bayluscide is registered as a tool to survey sea lamprey populations, and the synergistic combination of Bayluscide and TFM (not more than 2% TFM by weight) is registered as a lampricide. A formulation applied to sand has been found effective for control of larval sea lamprey, especially in alluvial fans of silt and sand at the mouths of rivers entering lakes. Bayluscide is highly toxic to associated fishes and must be used with care. The compound is environmentally nonpersistent, sensitive to pH, moderately toxic to mammals, and very toxic to mollusks and aquatic annelid worms.

15.3.7 Costs Associated with Chemical Control of Undesirable Fishes

Cost-benefit data on the chemical control of fishes is not readily available. Numerous states report spending large sums annually to control undesirable fish stocks, but few have summarized actual cost-benefit data. In the late 1980s, the cost to treat a hectare of water in the United States with rotenone ranged from US$2 to $240 (mean, $40). Bettoli and Maceina (1996) reported that 3 mg/L of powdered rotenone (5% active ingredient) cost one-third that of a liquid formulation with 5% active ingredient. Rotenone is such an important fish toxicant that the expenditure of nearly $3 million and investment of 6 years of testing were justified to reregister the toxicant for fisheries use in the United States (Sousa et al. 1987).

Information on costs associated with other registered fish toxicants is not adequate to permit comparisons, but much time and money have been invested in the control of sea lamprey in the Great Lakes. The increase in the number of lake trout is at least partly attributable to chemical control of the sea lamprey. Chemical control methods are effective, and cost should not serve as the sole measure in making decisions on control methods.

15.3.8 Role of the Public in Use of Chemicals

Throughout the United States public opinion has become important in agency decisions about fisheries management. Many states have jurisdiction for management of *all* aquatic resources and must consider the potential adverse effects of fish control on nontarget aquatic species, such as rare or imperiled native fishes, and terrestrial species that depend upon the aquatic organisms for food. For example, the California Fish and Game Code states that it is the "policy of the state to encourage the conservation, maintenance, and utilization of the living resources of the ocean and other waters under jurisdiction and influence of the state for the benefit of all citizens of the state" (California Department of Fish and Game 1985:9). A programmatic environmental impact report on the use of rotenone in fisheries management was prepared by California to ensure that irreversible, adverse environmental impacts would not occur in aquatic habitats (California Department of Fish and Game 1985). California biologists also prepare site-specific environmental assessments for each chemical treatment project;

the assessments describe potential effects and provide for mitigation, if necessary. In addition, biologists in California completed studies of the residual breakdown products of rotenone because of public concern for the use of chemicals and pesticides in water (B. J. Finlayson and J. M. Harrington, California Department of Fish and Game, data presented at the American Fisheries Society, Western Division meeting, 1991). In Texas, public opposition to the use of chemical toxicants (rotenone) in sampling estuarine habitats has restricted chemical application to areas without high public use (Matlock et al. 1982).

Public concern over the use of chemicals that may affect human health and adversely affect the environment has increased in recent years. The increased concern is due to a higher degree of public awareness about health and the environment through various media. However, Fagin and Lavelle (1996) stated that the regulatory process benefits chemical manufacturers and does not adequately inform the public about potential hazards. They emphasized that very few Americans realize that most chemicals are not screened for human safety before being placed on the market.

15.4 BIOLOGICAL METHODS FOR MANAGING UNDESIRABLE FISH SPECIES

Biological control methods can be grouped into three categories: (1) predation by piscivorous animals including fishes, turtles, birds, and mammals; (2) use of pathogens (viruses, bacteria, and fungi); and (3) biomanipulation, in which interrelationships among plants, animals, and their environment are adjusted to achieve the desired control and ecological balance (Schuytema 1977). Use of predator species is the most promising biological control technique for managing abundant, nongame, forage fish populations because it can establish balanced predator–prey populations. The use of pathogens for fish control is risky; therefore, pathogens are not generally used in aquatic environments.

Biomanipulation is defined as the deliberate adjustment of the biota and habitat (e.g., reservoir water level and streamflow rate) and is based on ecological theory of food web structure and dynamics (Reynolds 1994). It is especially applicable to shallow lakes where the biomass of benthic invertebrates is great enough that benthic-feeding planktivorous fishes are less dependent on zooplankton prey in contrast to fishes occupying deepwater lakes. The key to stability in the fish community is the ratio of piscivorous to planktivorous fish species. Biological control of undesirable benthic-feeding planktivorous fishes is achieved by carefully tailoring the species of piscivorous fish stocked, its size at stocking, and the stocking rate (Reynolds 1994).

15.4.1 Introduction of Predatory Fishes

Generally, predatory fishes used for biological control of nongame fish species in inland waters are selected to provide recreational fishing as well as to control populations of abundant forage fishes or stunted sport fishes (Box 15.5). To be effective as a management tool, predator stocking should (1) effectively utilize abundant forage fish populations or control stunted sport fish populations; (2) be cost-effective with regard

---

**Box 15.5   Factors to Consider in Choosing a Suitable Predator
or Prey Species**

1. Numbers of undesirable or unexploited fishes.

2. Size of the prey fish species.

3. Potential for use of forage organisms by sport fish predators.

4. Size of the water body.

5. Species and strains of predators available.

6. Size of predators at stocking and estimated survival based on size.

7. Stocking rates as well as timing and frequency of stocking.

8. Habitat suitability for the predator.

9. The desirability of the predator as a sport species.

10. Angler harvest of stocked sport fish.

11. Potential control of the predator in the stocked water, if necessary.

12. Native fish species (especially threatened and endangered fishes) that may be af-
    fected by the predator through hybridization, competition, and predation.

13. Potential for transmission of disease or parasites.

14. Fisheries management plan for the basin in which introduction is to take place.

15. Public preference of the predatory fish as a sport fish.

---

to culturing costs, stocking rates, and survival of predators after stocking; (3) produce
consistent fisheries with a balanced predator–prey composition in the fish community;
and (4) result in sufficient growth and survival of stocked predators to provide an
acceptable recreational fishery. Careful consideration of the ecological requirements
of target species for control or utilization and the predatory fish species is a prerequisite for success. For example, the number of a predator such as largemouth bass and
number and size of a forage fish such as bluegill that will produce a balanced predator-
prey fish community are much different in northern waters than in southern waters
(Chapter 21).

Other criteria that are considered important in evaluation of a potential introduction include the forage potential of prey species, sport value of the predatory species,
the target fish species to be controlled, habitat suitability for the fish predator, and
angler preference for the selected sport fish, in declining order of importance. Although several other criteria are not rated highly, they are also very important and can
be critical in influencing the final decision to use biological control. For example, the
introduction of parasites or diseases into a water could result in irreversible consequences; potential introduction of parasites or disease would become the most important criterion in making decisions related to biological control.

Although simultaneous consideration of multiple factors is made by agencies in making decisions related to the use of predatory fish species for biological control programs, high importance is given to the existing fish community by 75% of state and provincial agencies. Nearly half of the agencies using predators for biological control are concerned about the effects of introduction on native fish species (Courtenay 1993). Approximately 66% of the extinctions of 40 North American fishes (27 species and 13 subspecies) during the past century were at least partially related to introduced fishes, both exotic and North American transplants (Miller et al. 1989).

Almost all states and Canadian provinces use predators for biological control of abundant forage fishes, and Texas uses this control method exclusively. Some problems encountered include (1) the predator is not stocked at an adequate density; (2) the predator reduces the desirable forage fish populations, which result in reduced growth of the predator; (3) the predator becomes overpopulated and stunted; (4) the predator causes declines of other sport species; (5) the predator species is so exploited as a sport fish that it never becomes abundant or large enough to control the targeted forage fishes; and (6) the predator is ineffective for control of the target forage fishes.

Quantitative results of predator stocking for biological control are scarce (Hall 1985). Most studies of predator–prey interactions in large reservoirs have been confined to single species or a few species (Kendall 1978). Complex interactions of total fish communities have not been adequately studied. Recent studies strongly suggest that introduced fishes can eliminate native species, reduce survival and growth rates of established species, or change the structure of the fish community (see Chapter 13). Introductions of nonnative species may adversely affect native fishes that are threatened, endangered, or of special concern (see Chapter 16).

Gizzard shad, yellow perch, sunfishes, common carp, and various minnows are the primary fishes controlled through stocking predatory fishes. The primary species used as predators are largemouth bass, walleye, northern pike, striped bass, hybrid striped bass, and muskellunge.

15.4.2   Introduction of Forage Fishes

Prey introductions are sometimes made in efforts to increase the growth of stunted predators that are too small to be acceptable by anglers. However, problems have been caused by well-intentioned decisions to improve the forage base for game fishes. An example is the competitive interaction that resulted from the introduction of redside shiner as possible forage for rainbow trout in British Columbia lakes (Johannes and Larkin 1961). The introduction failed because the behavior of the redside shiner resulted in spatial separation from large rainbow trout while the redside shiner competed directly for food with small rainbow trout. Similarly, cisco stocked as prey for lake trout in Lake Opeongo, Ontario, competed with young smallmouth bass (Emery 1975). Rainbow smelt is considered to be better forage for landlocked Atlantic salmon in New England than is alewife because rainbow smelt is more available spatially as prey.

Fishes considered undesirable in one location may be acceptable in another. For example, Tui chub provides the primary forage for a rainbow trout subspecies in Eagle Lake, California (Burns 1966). In Pyramid Lake, Nevada, Lahontan cutthroat trout prey upon Tui chub, and the Tui chub is not a problem there. However, in other waters of the intermountain and western states, minnows such as the Tui chub and Utah chub compete for food with stocked fingerling rainbow trout, resulting in poor growth of the rainbow trout.

Therefore, the biology of forage species to be introduced should be carefully reviewed. An ideal forage organism should be prolific, stable in abundance, trophically efficient, vulnerable to predators, nonemigrating, and innocuous to other species (Ney 1981).

15.4.3   Control of Reproduction of Introduced Fishes

Natural reproduction of introduced fish species is sometimes considered undesirable. In such cases, controlling reproduction by genetic manipulation is sometimes warranted. One method is to stock hybrids that have little or no potential to reproduce. In farm ponds, hybrid sunfishes (limited reproductive capability) can be established by two methods: (1) stocking F_1 hybrid sunfishes directly into the ponds or (2) stocking parental fishes of the correct sex to produce the desired hybrids (Kurzawski and Heidinger 1982). In such situations, largemouth bass can control F_2 hybrid sunfishes because recruitment of F_2 hybrids is minimal (Brunson and Robinette 1986).

Another method to control reproduction is to alter the sex of fish by means of chemicals or heat to produce fish that will never develop sexually or that will produce monosex populations (Shelton 1986). Some biologists have concern about using this method because some fish may be reproductively viable and establish undesired self-sustaining populations.

15.4.4   Use of Underexploited Fishes

Some conservation agencies have liberalized regulations or initiated programs to encourage the use of less desirable fishes for food and sport. The common carp is a problem fish species in North America because it is very prolific and sought by few anglers. Miller (1972) offered suggestions on how to fish for common carp, when to fish for them, and how to dress and cook them as well as baits which are effective. The American Fisheries Society published a book to stimulate greater interest in angling for common carp (Cooper 1987). Similar programs may be developed for other underexploited species. For example, the Utah Division of Wildlife Resources published brochures on fishing techniques, fishing locations, and cooking tips for mountain whitefish and bullhead catfishes. Angling for mountain whitefish provides stream fishing opportunities in winter when the trout fishery is closed. Angler preference becomes important in the effectiveness of liberalized regulations to harvest underexploited fish species because little or low harvest may result if the size or species is not desired by anglers (Chapter 17). For example, liberalized bag limits did not result in the harvest of numerous small (<60 cm total length) northern pike from small (<200 ha) Minnesota lakes (Goeman et al. 1993), so bag limits held little promise as a management tool for altering the size structure of northern pike. In contrast, angling was considered a cost-effective method for removing introduced rainbow trout from streams in the Great Smoky Mountains National Park where park management philosophy is to manage for native brook trout. Larson et al. (1986) reported removal of 66% of a rainbow trout population over a 9-week angling season; half were removed during the first 2 weeks.

15.5   MECHANICAL METHODS FOR MANAGING UNDESIRABLE FISH SPECIES

Mechanical control methods are used on a variety of species in North America; common carp, other cyprinids, sunfishes, and suckers are the primary target species. The most frequently used mechanical control methods include installation of barriers, commercial fishing with various gears, water level manipulation, and use of traps to remove undesirable fishes (Figure 15.3). Other less used methods include use of electricity, weirs, gill nets, seines, and even dynamite.

All mechanical control methods are selective because of differences in fish behavior or susceptability to the gear or structure used. For example, Van Den Avyle et al. (1995) found that trawling, gillnetting, shoreline seining, and electrofishing differed in the size distribution of fishes captured, relative proportions of fish species present, and labor required to use the gear. They stated that none of the gears was superior to capture gizzard and threadfin shads in reservoirs.

Mechanical control methods are generally effective for 1–5 years. The effectiveness of mechanical controls depends upon the behavior of the target species, the location of undesirable fish concentrations, the bottom of the aquatic habitat to be sampled with active gears, and labor requirements to implement the control. Mechanical control methods are generally temporary because the remaining fishes exhibit compensatory survival, increased growth, and increased fecundity, all of which result in rapid resiliency of populations.

Combinations of control methods may be more effective in capturing or eliminating target species than application of a single method. For example, copper sulfate was used to increase fish movement in ponds and thereby increase fyke-net catches (Brown 1964). Similarly, electrofishing can be used in streams to force fishes into gill, trammel, or trap nets.

Figure 15.3   Trap net set to capture undesirable fish species.

15.5.1   Installation of Barriers

Barriers are the most commonly used mechanical control method because a one-time expense to construct barriers may provide effective control for years, possibly indefinitely. Barriers are installed on streams to prevent colonization by undesirable fishes from downstream reaches. If colonization occurs upstream of the barrier, it often occurs through human transplant of fishes or structural failure of the barrier. The plunge pool below some barriers such as gabions may concentrate some species of undesirable fishes (e.g., suckers) on spawning runs; these fishes then can be readily captured by other mechanical methods (e.g., electrofishing).

15.5.2   Commercial Fisheries

Game fish populations may benefit from commercial fishing of undesirable species (including competitors or predators) if the numbers are reduced drastically. For example, when 80 million kilograms of commercial fishes (mostly freshwater drum) were removed by trap nets and trawls from Lake Winnebago, Wisconsin, between 1955 and 1966 (Priegel 1971), an increase in populations of walleye, sauger, white bass, yellow perch, and black crappie (all desirable species to anglers) was observed. Commercial fishing is an acceptable way of removing less-desirable fish species at no cost to the managing agency; however, most inland commercial fisheries do not remove enough of the undesirable species to benefit game fisheries (Grinstead 1975). Intensive harvest of preferred commercial species usually results in reduction of average size and catch rate of those species, which makes commercial fishing unprofitable. Other undesirable species may remain unharvested because of low market value. The main limitation of commercial fishing is that nets are seldom designed to catch specific species or sizes of fish.

15.5.3   Manipulation of Water Level

Water level manipulations in reservoirs can be achieved by drawdowns, by water level elevation, or a combination of the two (see Chapter 11). Planned drawdowns can be used (1) to increase predator use of forage fishes, which results in faster growth of the predators, (2) to release nutrients from bottom sediments, and (3) to stimulate increased reproduction and survival of young fishes following reflooding (Groen and Schroeder 1978). For spring spawners, significant increases in water levels timed to occur just before, during, or for a brief period after the spawning season can be used to provide more high-productivity littoral habitat. However, water level manipulations may conflict with other reservoir uses such as domestic or irrigation water supply, power generation, flood control, and recreation. Drawdowns can reduce the invertebrate biomass in the littoral zone and may result in winterkill if done in late summer or fall.

15.5.4   Netting and Trapping

Netting or trapping of fishes is used effectively to control fishes in small ponds, lakes, and streams or for partial control in larger waters such as reservoirs and rivers. Seines are active gears (netting) that must be moved to capture fish (Figure 15.4) whereas trap nets are passive gears (trapping) that depend on fish movement to be effective (Figure 15.3). Netting must occur regularly and continually to be effective. Vulnerability of the target species is an important consideration in netting and trapping efforts. Spawning aggregations of common carp, buffalos, freshwater drum, and bullhead catfishes can be effectively harvested with nets and traps. Certain species of fish (such as common carp) assemble as huge schools just before ice formation and can be effectively removed with nets at that time. Nevertheless, netting and trapping programs have produced varying results. Rose and Moen (1953) reported a sixfold increase in game fish stocks when competing species were reduced to about 35% of former abundance in East Lake Okoboji, Iowa. However, removal of longnose suckers over a 7-year period did not improve the rainbow trout fishery in Pyramid Lake, Alberta (Rawson and Elsey 1950). Removal of small northern pike (<60 cm total length) by trap-netting

Figure 15.4   Seining operation to remove undesirable fish species from a lake.

did not result in increased growth of remaining fish in small (<200 ha) Minnesota lakes (Goeman et al. 1993). Successful use of nets and traps has been limited because typically an inadequate portion of the offending population is removed. Mechanical controls are effective only if a drastic decrease in the target species is achieved. The main drawbacks of trapping and netting are that they are labor intensive, they may cause mortality of desirable game fishes, and the equipment is expensive to repair or replace.

15.5.5   Electricity

Electricity is a method sometimes considered to control or guide (Section 15.5.7) fishes. Albertson et al. (1965) and McLain (1956) reported limitations in controlling sea lampreys with electricity. Electrical fields have been used to repel fishes such as gizzard shad and alewife from water intakes of hydroelectric plants. Depletion electrofishing has been used to reduce abundance of introduced brook trout in streams that contained populations of native Colorado River cutthroat trout (Thompson and Rahel 1996) and to reduce rainbow trout in a stream where native brook trout occurred (Larson et al. 1986). There are several limitations to the use of electricity. Portable and boat electroshockers can be used to remove fishes from confined areas but are ineffective in open, large, or deep waters. A voltage gradient sufficient to control small fishes would narcotize, injure, or even kill larger fishes in the electrical field. Water resistance, concentrations of dissolved substances, and water temperature influence the effectiveness of electricity.

15.5.6   Weirs, Fishways, and Screens

Weirs can be used effectively to guide undesirable fish species into traps when those fish are moving (e.g., during spawning runs). However, weirs require continuous maintenance to clean waterborne debris that reduce their effectiveness. Devices such as fishways and screens have potential for controlling fishes. Fishways are usually used to pass desirable fish species over obstacles such as dams, falls, or rapids (Bates 1993). Careful design of fishways may prevent passage of undesirable species (Broach 1968). Fish screens are used to keep fishes out of particular reaches of streams or to keep game fishes or endangered species from entering irrigation diversion canals (Huber 1974). Designs for screens, bypasses, and fishways must take into account the amount of water diverted, water velocity required to attract fish, the swimming ability of the fish of interest, the behavior and sizes of that fish, and the quantity and size of debris that may affect the performance of the device (OTA 1995).

15.5.7   Artificial Guidance

Artificial guidance may be used to direct undesirable fishes from entering certain waters. Bell (1986) identified natural factors that influence the presence or movement of fishes including light (or its absence), velocity, channel shape, depth, sound, odor, and temperature. These factors can be altered to guide fishes to or from aquatic habitats. In addition, artificial mechanical stimuli such as bubbles, electric fields, high water velocities, chemicals, and light (Bell 1986), can be used to guide and thereby control some undesirable fishes. Various techniques have been applied with varying success (Bell 1990).

15.6 CONCLUSION

Successes and failures have been reported for various biological, chemical, and mechanical fish control projects (Burns 1966; Lennon et al. 1970; Schuytema 1977; Meronek et al. 1996). The lesson is that numerous factors must be considered in making decisions about fish control methods. Because there are uncertainties in making natural resource management decisions (Ludwig et al. 1993), it is extremely important to use a systems approach for the design and implementation of a control program for undesirable fishes. The problem that resulted in a fish species being considered undesirable—perhaps habitat alteration, exploitation rate, or biological imbalance of the aquatic system—should be identified so that the corrective action can be directed at the problem rather than at treating the symptom. Long-term ecosystem stability may be required before sound techniques can be implemented to control undesirable fishes (Marsden 1993). Adaptive management allows decisions to be made based on the best available information and actions to be modified or refined as new information becomes available through evaluation of the results from any action (Walters and Hilborn 1978; Walters 1986).

If a fish species is considered undesirable, a variety of management methods (e.g., chemical, biological, or mechanical) should be considered to achieve carefully planned and predetermined objectives. The goals and objectives (Chapter 2) for control of undesirable species must be realistic in terms of the success that may be achieved and the costs related to the benefits. Based on a comprehensive literature review, Meronek et al. (1996) considered only 43% of 250 fish control projects successful. They stressed that comprehensive planning should include (1) critical evaluation of assumptions, (2) suspected causes of problems, (3) explicit rationale for goals and objectives, and (4) pretreatment study followed by long-term posttreatment study.

Fisheries managers must be flexible to increase or decrease management efforts depending on the response of the target fish population. Control, using the most effective method(s), should be applied at a time and location when undesirable fish are most vulnerable. Integrated management that uses a combination of methods may be most effective in controlling undesirable fish species. For example, single control measures of water level manipulation, aquatic vegetation control, stocking of predatory fish, and mechanical fish removal were implemented in a prairie lake over a 50-year period. Little success in producing a stable sport fishery led Scarnecchia and Wahl (1992) to recommend implementation of an integrated fishery management plan.

Partial control that is temporary may be justified even though the target undesirable fish species may have the potential for rapid resiliency through movement from other areas or compensatory survival, increased growth, and increased fecundity. For example, even temporary removal or suppression of nonnative predators or competitors may enhance native fish populations that are threatened or endangered. Although biologists are often intimidated by the inefficient and labor-intensive methods of fish control and possible negative public reaction, there is a sometimes a necessity to reduce (or eradicate if possible) nonnative fish species (Temple 1990) because of their adverse effects on fish communities (Courtenay and Robins 1989).

Control actions should be implemented only when there is the least risk of adverse effects on desirable game fishes or native, threatened, and endangered species (Berryman 1972). Finally, all fish control programs should be based on the biology and habitat of the species, a consideration of all effective methods of control, and an understanding of the level of control that is needed and possible (Binning et al. 1985).

15.7 REFERENCES

Albertson, L. M., B. R. Smith, and H. H. Moore. 1965. Experimental control of sea lampreys with electricity on the south shore of Lake Superior, 1953-60. Great Lakes Fishery Commission Technical Report 10.

Allen, D. M., S. K. Service, and M. V. Ogburn-Matthews. 1992. Factors influencing the collection efficiency of estuarine fishes. Transactions of the American Fisheries Society 121:234–244.

Anderson, R. S. 1970. Effect of rotenone on zooplankton communities and a study of their recovery patterns in two mountain lakes in Alberta. Journal of the Fisheries Board of Canada 27:1335–1356.

Bates, K., compiler. 1993. Fish passage policy and technology. American Fisheries Society, Bioengineering Section, Bethesda, Maryland. (Available from G. Kindschi, 4050 Bridger Canyon Road, Bozeman, Montana 59715, USA.)

Bell, M. C. 1986. Artificial guidance of fish. Pages 185–199 in M. C. Bell, editor. Fisheries handbook of engineering requirements of biological criteria. U.S. Army Corps of Engineers, Fish Passage Development and Evaluation Program, North Pacific Division, Portland, Oregon.

Bell, M. C. 1990. Fisheries handbook of engineering requirements and biological criteria. U.S. Army Corps of Engineers, North Pacific Division, Fish Passage Development and Evaluation Program, Portland, Oregon.

Berryman, J. H. 1972. The principles of predator control. Journal of Wildlife Management 36:395–400.

Bettoli, P. W., and M. J. Maceina. 1996. Sampling with toxicants. Pages 303–333 in B. R. Murphy and D. W. Willis, editors. Fisheries techniques. 2nd edition. American Fisheries Society, Bethesda, Maryland.

Binning, L., and six coauthors. 1985. Pest management principles for the commercial applicator, aquatic pest control. University of Wisconsin, Cooperative Extension Service, Madison.

Boccardy, J. A., and E. L. Cooper. 1963. The use of rotenone and electrofishing in surveying small streams. Progressive Fish-Culturist 23:26–29.

Bonn, E. W., and L. R. Holbert. 1961. Some effects of rotenone products on municipal water supplies. Transactions of the American Fisheries Society 90:287–297.

Bonneau, J., D. Scarnecchia, and E. Berard. 1995. Better fishing means less carping at Bowman-Haley Reservoir. North Dakota Outdoors 10:18–20.

Boogaard, M. A., T. D. Bills, J. H. Selgeby, and D. A. Johnson. 1996. Evaluation of piscicides for control of ruffe. North American Journal of Fisheries Management 16:600–607.

Bouffard, S. H., and M. A. Hanson. 1997. Fish in waterfowl marshes: waterfowl managers' perspective. Wildlife Society Bulletin 25:146–157.

Bowen, J. T. 1970. A history of fish culture as related to the development of fishery programs. Pages 71–93 in N. G. Benson, editor. A century of fisheries in North America. American Fisheries Society, Special Publication 7, Bethesda, Maryland.

Bradbury, A. 1986. Rotenone and trout stocking: a literature review with special reference to Washington Department of Game's lake rehabilitation program. Washington Department of Game, Fisheries Management Report 86–2, Olympia.

Broach, D. 1968. A small-capacity spillway modified to prevent re-entry of undesirable fishes. Progressive Fish-Culturist 30:38.

Brown, E. H., Jr. 1964. Fish activation with copper sulfate in relation to fyke-netting and angling. Ohio Department of Natural Resources, Division of Wildlife, Publication W-71, Columbus.

Brunson, M. W., and H. R. Robinette. 1986. Evaluation of male bluegill x female green sunfish hybrids for stocking Mississippi farm ponds. North American Journal of Fisheries Management 64:156–167.

Buck, D. H., M. A. Whitacre, and C. F. Thoits, III. 1960. Some experiments in the baiting of carp. Journal of Wildlife Management 24:357–364.

Burns, J. W. 1966. Rough fish management. Pages 492–498 in A. Calhoun, editor. Inland fishery management. California Department of Fish and Game, Sacramento.

California Department of Fish and Game. 1985. Rotenone use for fisheries management. California Department of Fish and Game, Final Programmatic Environmental Impact Report, Sacramento.

Cohen, J. M., L. J. Kamphake, A. E. Lemke, C. Henderson, and R. L. Woodward. 1960. Effect of fish poisons on water supplies. Part 1 - removal of toxic materials. Journal of the American Water Works Association 52:1551–1566.

Cohen, J. M., G. A. Rourke, and R. L. Woodward. 1961. Effects of fish poisons on water supplies. Part 2 - odor problems. Journal of the American Water Works Association 53:49–62.

Cole, L. J. 1904. The German carp in the United States. Pages 592–594 in Report to the Commissioner of Fisheries, U.S. Bureau of Fisheries, Washington, D.C.

Cole, L. J. 1905. The status of carp in America. Transactions of the American Fisheries Society 34:131–138.

Cooper, E. L., editor. 1987. Carp in North America. American Fisheries Society, Bethesda, Maryland.

Courtenay, W. R., Jr. 1993. Biological pollution through fish introductions. Pages 35–61 in B. N. Knight, editor. Biological pollution: the control and impact of invasive exotic species. Indiana Academy of Sciences, Indianapolis.

Courtenay, W. R., Jr., and C. R. Robins. 1989. Fish introductions: good management, mismanagement, or no management? Chemical Rubber Corporation Critical Reviews in Aquatic Sciences 1:159–170.

Courtney, C. C. 1991. Behavioral responses of Utah chub and rainbow trout to rotenone. Utah Division of Wildlife Resources, Publication 91-5.

Cumming, K. B. 1975. History of fish toxicants in the United States. Pages 5–21 in Eschmeyer (1975).

Dunst, R. C., and nine coauthors. 1974. Survey of lake rehabilitation techniques and experiences. Wisconsin Department of Natural Resources Technical Bulletin 75.

Emery, A. R. 1975. Stunted bass: a result of competing cisco and limited crayfish stocks. Pages 154–164 in H. Clepper, editor. Black bass biology and management. Sport Fishing Institute, Washington, D.C.

Engstrom-Heg, R. 1972. Kinetics of rotenone-potassium permanganate reactions as applied to the protection of trout streams. New York Fish and Game Journal 19:47–58.

Eschmeyer, P. H., editor. 1975. Rehabilitation of fish populations with toxicants: a symposium. American Fisheries Society, North Central Division, Special Publication 4, Bethesda, Maryland.

Fabacher, D. L., and H. Chambers. 1972. Rotenone tolerance in mosquitofish. Environmental Pollution 3:139–141.

Fagin, D., and M. Lavelle. 1996. Toxic deception: how the chemical industry manipulates science, bends the laws, and endangers your health. Carol Publishing Group, Secaucus, New Jersey.

Fait, J. R., and J. M. Grizzle. 1993. Oral toxicity of rotenone to common carp. Transactions of the American Fisheries Society 122:302–304.

Fukami, J.-I., T. Mitsui, K. Fukunaga, and T. Shishido. 1970. The selective toxicity of rotenone between mammal, fish, and insect. Pages 159–178 in R. D. O'Brien and I. Yamamoto, editors. Biochemical toxicity of insecticides. Academic Press, New York.

Fukami, J.-I., T. Shishido, K. Fukunaga, and J. E. Casida. 1969. Oxidative metabolism of rotenone in mammals, fish, and insects, and its relation to selective toxicity. Journal of Agricultural and Food Chemistry 17:1217-1226.

Garrison, R. L. 1968. The toxicity of Pro-Noxfish to salmonid eggs and fry. Progressive Fish-Culturist 30:35–38.

Gilderhus, P. A. 1982. Effects of aquatic plant and suspended clay on the activity of fish toxicants. North American Journal of Fisheries Management 2:301–306.

Gilderhus, P. A., J. L. Allen, and V. K. Dawson. 1986. Persistence of rotenone in ponds at different temperatures. North American Journal of Fisheries Management 6:129–130.

Goeman, T. J., P. D. Spencer, and R. B. Pierce. 1993. Effectiveness of liberalized bag limits as management tools for altering northern pike population size structure. North American Journal of Fisheries Management 13:621–624.

Gresswell, R. E. 1991. Use of antimycin for removal of brook trout from a tributary of Yellowstone Lake. North American Journal of Fisheries Management 11:83–90.

Grinstead, B. G. 1975. Response of bass to removal of competing species by commercial fishing. Pages 475–479 in H. Clepper, editor. Black bass biology and management. Sport Fishing Institute, Washington, D.C.

Groen, C. L., and T. A. Schroeder. 1978. Effects of water level management on walleye and other coolwater fishes in Kansas reservoirs. Pages 278–283 in R. L. Kendall, Editor. Selected coolwater fishes of North America. American Fisheries Society, Special Publication 11, Bethesda, Maryland.

Hall, G. E. 1985. Reservoir fishery research needs and priorities. Fisheries 10(2):3–5.

Hegen, H. E. 1985. Use of rotenone and potassium permanganate in estuarine sampling. North American Journal of Fisheries Management 5:500–502.

Hester, F. E. 1959. The toxicity of Noxfish and Pro-Noxfish to eggs of common carp and fathead minnows. Proceedings of the Annual Conference Southeastern Association of Game and Fish Commissioners 13(1959):325–331.

Holey, M., and six coauthors. 1979. Never give a sucker an even break. Fisheries 4(1): 2–6.

Holton, D. C. 1977. Predatory behavior of largemouth bass on soft and spiny- rayed forage species. Doctoral dissertation. University of Florida, Gainesville.

Hubbs, C. L., and R. W. Eschmeyer. 1938. The improvement of lakes for fishing. University of Michigan, Institute for Fishery Research, Bulletin 2, Ann Arbor.

Huber, E. E. 1974. Fish protection at intake structures and dams: guidance, screens, and collection devices: a selected bibliography with abstracts. Oak Ridge National Laboratory, ORNL-EIS-74–67, Oak Ridge, Tennessee.

Johannes, R. E., and P. A. Larkin. 1961. Competition for food between redside shiners (Richardsonius balteatus) and rainbow trout (Salmo gairdneri) in two British Columbia lakes. Journal of the Fisheries Research Board of Canada 18:203–221.

Kendall, R. L., editor. 1978. Selected coolwater fishes of North America. American Fisheries Society, Special Publication 11, Bethesda, Maryland.

Kiser, R. W., J. R. Donaldson, and P. R. Olson. 1963. The effect of rotenone on zooplankton populations in freshwater lakes. Transactions of the American Fisheries Society 92:17–24.

Kohler, C. C., and W. R. Courtenay, Jr. 1986. American Fisheries Society position on introduction of aquatic species. Fisheries 11(2):39–42.

Kurzawski, K. F., and R. C. Heidinger. 1982. The cyclic stocking of parentals in a farm pond to produce a population of male bluegill X female green sunfish F_1 hybrids and male redear sunfish x female green sunfish F_1 hybrids. North American Journal of Fisheries Management 2:188–192.

Larson, G. L., S. E. Moore, and D. C. Lee. 1986. Angling and electrofishing for removing nonnative rainbow trout from a stream in a national park. North American Journal of Fisheries Management 6:580–585.

Lennon, R. E., J. B. Hunn, R. A. Schnick, and R. M. Burress. 1970. Reclamation of ponds, lakes, and streams with fish toxicants: a review. FAO (Food and Agriculture Organization of the United Nations) Fisheries Technical Paper 100.

Lindahl, P. E., and K. E. Oberg. 1961. The effect of rotenone on respiration and its point of attack. Experimental Cell Research 23:228–237.

Loeb, H. A., and R. Engstrom-Heg. 1971. Estimation of rotenone concentration by bio-assay. New York Fish and Game Journal 18:129–134.

Lopinot, A. C. 1975. Summary of the use of toxicants to rehabilitate fish populations in the midwest. Pages 1–4 in Eschmeyer (1975).

Ludwig, D., R. Hilborn, and C. Walters. 1993. Uncertainty: resource exploitation, and conservation: lessons from history. Science 260:17, 36.

Magnuson, J. J. 1976. Managing with exotics – a game of chance. Transactions of the American Fisheries Society 105:1–9.

Marking, L. L. 1988. Oral toxicity of rotenone to mammals. U.S. Fish and Wildlife Service, Investigations in Fish Control, 94.

Marking, L. L. 1992. Evaluation of toxicants for the control of carp and other nuisance fishes. Fisheries 17(6): 6–12.

Marking, L. L., and T. D. Bills. 1975. Toxicity of potassium permanganate to fish and its effectiveness for detoxifying antimycin. Transactions of the American Fisheries Society 104:579–583.

Marking, L. L., and T. D. Bills. 1976. Toxicity of rotenone to fish in standardized laboratory tests. U.S. Fish and Wildlife Service, Investigations in Fish Control 72.

Marking, L. L., T. D. Bills, and J. J. Rach. 1983. Chemical control of fish and fish eggs in the Garrison Diversion Unit, North Dakota. North American Journal of Fisheries Management 3:410–418.

Marsden, J. E. 1993. Responding to aquatic pest species: control or management? Fisheries 18(1): 4–5.

Matlock, G. C., J. E. Weaver, and A. W. Green. 1982. Sampling nearshore estuarine fishes with rotenone. Transactions of the American Fisheries Society 111:326–331.

McLain, A. L. 1956. The control of the upstream movement of fish with pulsated fish current. Transactions of the American Fisheries Society 86:269–284.

Meronek, T. G., and eight coauthors. 1996. A review of fish control projects. North American Journal of Fisheries Management 16:63–74.

Miller, G. 1972. Time out for carp. Nebraska Game and Parks Commission, Lincoln.

Miller, R. R., J. D. Williams, and J. E. Williams. 1989. Extinctions in North American fishes during the past century. Fisheries 14(6): 22–38.

Moyle, P. B. 1986. Fish introductions in North America: patterns and ecological impact. Pages 22–43 in H. A. Mooney and J. A. Drake, editors. Ecology of biological invasions of North America and Hawaii. Springer-Verlag, New York.

Murdoch, W. W., and J. Bence. 1987. General predators and unstable populations. Pages 17–30 in W. C. Kerfoot and A. Sih, editors. Predation: direct and indirect impacts on aquatic communities. University of New England Press, Hanover, New Hampshire.

National Research Council of Canada. 1985. TFM and Bayer 73: lampricides in the aquatic environment. Environmental Secretariat, Publication NRCC 22433, Ottawa.

Neves, R. J. 1975. Zooplankton recolonization of a lake cove treated with rotenone. Transactions of the American Fisheries Society 104:390–393.

Ney, J. J. 1981. Evolution of forage-fish management in lakes and streams. Transactions of the American Fisheries Society 110:725–728.

OTA (Office of Technology Assessment). 1995. Fish passage technologies: protection at hydropower facilities. OTA-ENV-641, U.S. Government Printing Office, Washington, D.C.

Olson, L. E., and L. L. Marking. 1975. Toxicity of four toxicants to green eggs of salmonids. Progressive Fish-Culturist 37:143–147.

Orciari, R. D. 1979. Rotenone resistance of golden shiners from a periodically reclaimed pond. Transactions of the American Fisheries Society 108:641–645.

Parker, J. O., Jr. 1970. Surfacing of dead fish following application of rotenone. Transactions of the American Fisheries Society 99:805–807.

Pfeifer, B. 1985. Potassium permanganate detoxification following lake rehabilitation: procedures, costs, and two case histories. Washington Department of Game, Fishery Management Report 85–5.

Priegel, G. R. 1971. Evaluation of intensive freshwater drum removal in Lake Winnebago, Wisconsin, 1955-1966. Wisconsin Department of Natural Resources Technical Bulletin 47.

Rach, J. J., J. A. Luoma, and L. L. Marking. 1994. Development of an antimycin- impregnated bait for controlling common carp. North American Journal of Fisheries Management 14:442–446.

Rawson, D. S., and C. A. Elsey. 1950. Reduction in the longnose sucker population of Pyramid Lake, Alberta, in an attempt to improve angling. Transactions of the American Fisheries Society 78:13–31.

Reynolds, C. S. 1994. The ecological basis for successful biomanipulation of aquatic communities. Archiv fuer Hydrobiologie 130:1–33.

Rose, E. T., and T. Moen. 1953. The increase in game fish populations in East Okoboji Lake, Iowa, following intensive removal of rough fish. Transactions of the American Fisheries Society 82:104–114.

Ross, M. R. 1997. Fisheries conservation and management. Prentice Hall, Upper Saddle River, New Jersey.

Roth, J., and V. Hacker. 1988. Fall use of rotenone at low concentrations to eradicate fish populations. Wisconsin Department of Natural Resources, Research and Management Findings 9, Madison.

Scarnecchia, D. L. 1992. A reappraisal of gars and bowfins in fishery management. Fisheries 17(5): 6–12.

Scarnecchia, D. L., and J. R. Wahl. 1992. Fifty years of fisheries management in an obstinate prairie lake. Journal of the Iowa Academy of Science 99:7–14.

Schnick, R. A. 1974. A review of the literature on the use of rotenone in fisheries. U.S. Fish and Wildlife Service, Report LR-74-15.

Schnick, R. A., F. P. Meyer, and D. L. Gray. 1986. A guide to approved chemicals in fish production and fishery resource management. University of Arkansas, Cooperative Extension Service Bulletin MP-241, Little Rock.

Schuytema, G. S. 1977. Biological control of aquatic nuisances-a review. U.S. Environmental Protection Agency, Corvallis Environmental Research Laboratory, Corvallis, Oregon.

Shelton, W. L. 1986. Reproductive control of exotic fishes – a primary requisite for utilization in management. Pages 427–434 in R. H. Stroud, editor. Fish culture in fisheries management. American Fisheries Society, Fish Culture Section and Fisheries Management Section, Bethesda, Maryland.

Sousa, R. J., F. P. Meyer, and R. A. Schnick. 1987. Re-registration of rotenone: a state/federal cooperative effort. Fisheries 12(4): 9–13.

Sowards, C. L. 1961. Safety as related to the use of chemicals and electricity in fishery management. U.S. Fish and Wildlife Service, Washington, D.C.

Tate, B., T. Moen, and B. I. Severson. 1965. The use of rotenone for recovery of live fish. Progressive Fish-Culturist 27:158–160.

Temple, S. A. 1990. The nasty necessity: eradicating exotics. Conservation Biology 4:113–115.

Thompson, P. D., and F. J. Rahel. 1996. Evaluation of depletion-removal electrofishing of brook trout in small Rocky Mountain streams. North American Journal of Fisheries Management 16:332–339.

Tiffan, K. F., and E. P. Bergersen. 1996. Performance of antimycin in high-gradient streams. North American Journal of Fisheries Management 16:465–468.

Titcomb, J. W. 1914. The use of copper sulfate for the destruction of obnoxious fishes in ponds and lakes. Transactions of the American Fisheries Society 44:20–26.

Trimberger, E. J. 1975. Evaluation of angler-use benefits from chemical reclamation of lakes and streams in Michigan. Pages 60–65 in Eschmeyer (1975).

Van Den Avyle, M. J., J. Boxrucker, P. Michaletz, B. Vondracek, and G. R. Ploaskey. 1995. Comparison of catch rate, length distribution, and precision of six gears used to sample reservoir shad populations. North American Journal of Fisheries Management 15:940–955.

Walters, C. J. 1986. Adaptive management of renewable resources. MacMillan, New York.

Walters, C. J., and R. Hilborn. 1978. Ecological optimization and adaptive management. Annual Review of Ecological Systems 9:157–188.

Walters, C. J., and R. E. Vincent. 1973. Potential productivity of an alpine lake as indicated by removal and reintroduction of fish. Transactions of the American Fisheries Society 102:675–697.

Wilcove, D., M. Bean, and P. C. Lee. 1992. Fisheries management and biological diversity. Transactions of the North American Natural Resource Conference 57: 373–383.

Wrenn, W. B. 1965. Effect of removal of the fish population on the invertebrate fauna and phytoplankton of Emmaline Lake, Colorado. Master's thesis. Colorado State University, Fort Collins.

Chapter 16

Endangered Species Management

FRANK J. RAHEL, ROBERT T. MUTH, AND
CLARENCE A. CARLSON

16.1 INTRODUCTION

Native fish faunas have declined throughout North America as a result of habitat degradation, overfishing, and introduction of nonnative species (Williams et al. 1989; Nehlsen et al. 1991; Campbell 1998; Lyons et al. 1998). As human alterations of aquatic systems continue, more fish species are likely to face population declines and range reductions that may lead to extinction. The decline in native faunas has resulted in increased interest in the value of biotic diversity (biodiversity) and the ethics of species extinctions (Winter and Hughes 1997). As a result of this increased interest, fisheries managers increasingly must consider the welfare of nongame species, a goal not traditionally pursued when development of sport or commercial fisheries was the major focus of fisheries management (Rahel 1997). Because many state and federal management agencies have legal mandates to prevent the loss of native species, including nongame species, recovery of endangered fishes has become an important activity of fisheries biologists.

The terminology used in discussing declining species can sometimes be confusing. Based on Section 16.2.1 of the U.S. Endangered Species Act (ESA) of 1973 (16 U.S.C. Section 1531 et seq.), a taxon in danger of extinction throughout all or a significant portion of its geographic range is designated as *endangered*, and one that is likely to become endangered in the foreseeable future is designated as *threatened*. Some management agencies have developed additional categories such as *species of special concern* or *sensitive species* to designate taxa that could become threatened or endangered by relatively minor disturbances to their habitat or for which additional information is required to determine their status. To avoid any legal connotations of these terms, conservation biologists often use the more general term *imperiled* to refer to declining taxa regardless of their official designation by state, provincial, or federal agencies.

Despite its name, the ESA affords protection to taxonomic categories below the species level (e.g., subspecies). For legal purposes, the ESA defines a species to include any distinct population segment of any species that interbreeds when mature. Thus, several subspecies of cutthroat trout have been proposed for listing as threatened or endangered (see Section 16.3.1.4 of this chapter).

16.1.1 Accelerated Extinction

Extinction usually occurs when populations are unable to persist as their environments change. Local extinction (often referred to as extirpation) occurs when a given population disappears. Global extinction occurs when a species is eliminated from its entire range. Extinction is a normal process, and global extinction has been the fate of most species during the Earth's history (Ehrlich and Ehrlich 1981). Estimates of the rate of species loss have been made in spite of difficulties inherent in such estimation and lack of data on current and past biotic diversity. Myers (1985) estimated that at least 90% of all species that have ever existed have become extinct, mostly as a result of natural processes. By 1600, humans became capable of driving animals to extinction. From 1600 to 1900, the human-induced extinction rate increased to about one species per year and held relatively constant. By 1979, as humans exploited ecologically diverse tropical forests, it increased again to perhaps one species per day. Continued destruction of tropical forests has pushed the estimated extinction rate to as many as 75 species per day (Myers 1997). Myers (1985) compared these rates with the maximum rate of extinction of the dinosaurs (one species every 10,000 years) to emphasize that there has never been such a period of massive and compressed extinction as that which we now face. It has been predicted we could lose 20% of all species within 30 years and up to half of all species before the close of the twenty-first century (Myers 1997).

Habitat disruption by humans is the foremost cause of plant and animal extinctions. Habitat destruction in tropical forests, where up to half the species on our planet reside, is widely recognized as the greatest threat to the Earth's biodiversity (Myers 1997). Fragmentation of once-continuous habitat into island-like refuges increases the risk of extinction. Humans have also contributed to extinction through commercial harvest, predator and pest control, and collection of organisms for medical research, zoos, and ornamental sales. Extinction also can be the result of predation or competition from introduced species. The effects of alien organisms introduced by humans may combine with effects of human exploitation to devastate native species. Ehrlich and Ehrlich (1981) reviewed how species endangerment by humans occurs directly (e.g., by overexploitation and predator control) or indirectly (e.g., through paving over, plowing under, cutting down habitat, discharging pollutants, transporting organisms, and degrading habitat through recreational uses).

16.1.2 The Importance of Preserving Biodiversity

Arguments for conserving biodiversity fall into two general categories: utilitarian and ethical considerations (Callicott 1997). Utilitarian considerations focus on the value of biodiversity to humans. These values include goods such as food, fuel, and medicines; ecosystem services such as pollination, water purification, and regulation of atmospheric gases; and the psycho-spiritual value of enjoying nature. According to a utilitarian perspective, protection of biodiversity is important because human health and well-being are tied to environmental health. All species are considered worthy of saving because we may someday find a use for them and because they contribute to the infrastructure of healthy ecosystems that are the life-support systems of our planet (Ehrlich and Mooney 1983; Naeem 1998). Utilitarian arguments are anthropocentric (human centered) because nature is valued in terms of its benefit to humans.

A second category of arguments for preserving biodiversity is termed biocentric (nature centered) because species are considered to have intrinsic value apart from their usefulness to humans. In a biocentric viewpoint, humans are not the masters of nature but merely another species inhabiting the Earth. Because all species are considered to have an inherent right to exist, extinction of any species is unethical. Ehrenfeld (1976) coined the term "Noah principle" for the notion that communities and species should be conserved simply because they exist and have existed for a long time.

16.2   LAWS AND TREATIES

16.2.1   The U.S. Endangered Species Act

As the environmental movement of the 1960s progressed, the North American public became increasingly concerned about accelerated extinction. The U.S. Congress responded by passing the Endangered Species Preservation Act of 1966 and the Endangered Species Conservation Act of 1969. These laws did not provide comprehensive authority for protecting endangered wildlife because they failed to prohibit the taking of endangered species or to require federal agencies to comply with the laws' intent, so in 1973 these acts were replaced by the more comprehensive and prohibitive ESA. The ESA was an attempt to slow the rate of extinction by singling out animals and plants thought to be near extinction and giving them and their habitats special protection. Its stated purposes were to provide (1) conservation of the ecosystems upon which endangered and threatened species depend and (2) a program for the conservation of such species. The ESA declared that it was the policy of Congress that all federal departments and agencies seek to conserve endangered and threatened species and use their authorities in furtherance of the ESA.

In response to complaints about the ESA, a set of amendments was added in 1978 that weakened the ESA by making it more difficult to list species as endangered. Before a species could be listed, it was necessary to describe economic impacts and boundaries of habitat and to hold public hearings within a 2-year period. Also, disputes between conservationists and developers were to be settled by an Endangered Species Committee, whose power to decide the fate of species earned it the nickname of "the God Squad." Many conservationists considered creation of the Endangered Species Committee an invitation to agencies to avoid seeking acceptable alternatives to agency actions that might jeopardize endangered species. However, the 1978 amendments may fairly be said to have made the ESA more flexible and the consultation and public participation processes stronger. The ESA amendments of 1982 streamlined the listing and exemption processes, and the Endangered Species Committee was replaced by the Secretary of the Interior as the overseer of disputes. The 1982 amendments also facilitated more active management of listed species by allowing experimental populations to be established, especially by introduction to apparently suitable or formerly occupied habitats.

As of this writing, reauthorization of the ESA was still being debated in Congress. Perhaps the biggest challenges facing reauthorization are deciding how to handle private property concerns and how to deal with declining but not yet listed species (e.g., species of special concern). Much of the land containing the nation's fish and wildlife habitat is owned by nonfederal entities, and the future of many declining species is at least partially dependent on conservation efforts on such lands. Further, conservation efforts for declining species are most effective and efficient when initiated early, and, in some cases, early conservation can remove the need to list federally a species as threatened or endangered. Preventing a species from being listed through early conservation efforts can maintain land use and developmental flexibility for property owners.

Interested individuals can keep abreast of changes in the ESA through several sources. The *U.S. Code* contains general and permanent laws of the United States as amended, including the ESA. Changes in the ESA and in lists and regulations are reported in the *Federal Register* and the *Endangered Species Bulletin*. The *Endangered Species Update* published by the University of Michigan and the *Environmental Reporter* of the Bureau of National Affairs are other sources of up-to-date information on the ESA.

The ESA is administered primarily by the U.S. Fish and Wildlife Service (USFWS), although the National Marine Fisheries Service has responsibility for some marine species. Recovery plans are the primary tools for restoring listed species to self-sustaining components of their ecosystems. In recovery planning, priority is given to species most likely to benefit from such plans. Priorities are based on the degree of threat facing a species, its taxonomic uniqueness and recovery potential, and its likelihood to encounter conflicts with development. Plans are intended to guide various conservation programs of federal, state, and local agencies and other organizations. Regional directors of the USFWS are responsible for preparing plans for species in their region, but they may assign the preparation to USFWS personnel, a volunteer recovery team, a state or federal agency, a conservation organization, or knowledgeable individuals. Approved plans are reviewed periodically and revised as new information is compiled or the status of listed species changes. One of the most important things to remember about recovery plans is that they are only guides to conservation of listed species. They do not mandate action, and no one is legally obligated to carry out the conservation measures they contain. Therefore, citizen action, in the form of a lawsuit, is often required to accomplish recovery of a species. In that sense, the ESA affords imperiled species far less protection than most people assume. Nongovernmental environmental organizations with interests in conserving endangered or threatened species can and do play important roles in enforcing the ESA.

Kohm (1991) reviewed what has been learned since the ESA was passed in 1973 and discussed how future efforts should be directed. Ehrlich and Ehrlich (1981) described the ESA, as originally written, as a powerful weapon on the environment's behalf and considered the modified version (prior to 1982) a potentially strong weapon in defense of species and environmental integrity. Despite its shortcomings, such as an initial tendency to favor listing of vertebrates over invertebrates or plants (Yaffee 1982), the ESA has prevented the extinction of some species and slowed the downward trends of others (Bender et al. 1998). At least half of all species listed for a decade or more are now either stable or improving in status. Bender et al. (1998) summarized actions taken in recent years to make the ESA more flexible and to facilitate participation by private landowners in the recovery process. These actions include such things as a Safe Harbor Program, which protects landowners from additional ESA restrictions when they implement conservation practices that benefit listed species on their property.

16.2.2   State Programs

States began enacting laws and enforcing protection of endangered species and natural habitats in the late 1960s. Some states protect only species on federal lists, whereas other states prepare their own lists. California was first to develop a state list of endangered and threatened species. Many states have potentially excellent programs to aid listed species, but the success of such programs is strongly tied to funding levels. The ESA encourages state participation in conservation of endangered species through cooperative agreements that allow federal matching funds for state projects involving listed species. However, in most states the funding for endangered species management has been much less than that for game species management. The basis of this inequity is that management agencies derive the bulk of their funding through hunting and fishing license sales. Those buying licenses expect their fees to be spent on species that can be hunted or fished. Although agencies have tried innovative ways to generate funds for nongame management, such as sales of special automobile license plates or income tax checkoffs, most agencies lack sufficient funding for nongame preservation efforts. A potential source of funding for management of nongame species in the United States is the Teaming with Wildlife initiative (Brouha 1998). As originally conceived, the initiative would create a tax on various outdoor recreation items and the revenue would be used to enhance habitat, create recreational viewing and interpretive facilities, and educate the public about nongame species and their habitats. In a later version, funding for nongame conservation efforts would come from federal royalties from offshore oil and gas leases through legislation named the "Conservation and Reinvestment Act" (Anonymous 1999). As of this writing, this legislation was being debated in the U.S. Congress. Many wildlife management agencies and conservation organizations support the act as a way to provide sustained funding for the conservation of nongame species.

16.2.3   Canadian and Mexican Programs

Canada has no legislative equivalent of the ESA but has a national committee that evaluates and assigns status to species at risk. Established in 1977, the Committee on the Status of Endangered Wildlife in Canada includes representatives of federal, provincial, and territorial governments and national conservation organizations such as the Canadian Nature Federation, Canadian Wildlife Federation, and World Wildlife Fund Canada (Cook and Muir 1984). The committee has no legislative or management role but has close connections to the Convention on International Trade in Endangered Species of Wild Fauna and Flora (CITES). Scientific subcommittees that correspond to major taxonomic categories arrange for and approve status reports on species that are candidates for a national list of species regarded as extinct, extirpated, endangered, threatened, rare, or not in any category (Cook and Muir 1984). Status reports for Canadian fishes are updated periodically and reported in the *Canadian Field-Naturalist* (Campbell 1998).

The conservation status of fishes in Mexico is summarized in the *Norma Oficial Mexicana* (1994). Additional information on threatened and endangered fishes in Mexico can be found in Contreras Balderas (1987) and Lyons et al. (1998). Contreras Balderas (1991) described legislation to protect the environment and ecological equilibrium, which indirectly affords protection to freshwater habitats and fishes.

16.2.4   International Treaties

Because exploitation for consumptive, medicinal, or ornamental uses is a cause of imperilment for many species, laws and treaties that restrict shipments of wildlife between countries are an important component of conservation efforts. The Lacey Act of 1900 was the first U.S. law applied to control wildlife imports, and the 1940 Convention on Nature Protection and Wildlife Preservation in the Western Hemisphere was the first international treaty that protected all animals and plants. As international trade in wildlife and wildlife products increased in the 1960s and early 1970s, broader and more effective controls were needed (Ehrlich and Ehrlich 1981). In 1975, CITES was implemented to regulate international trade in endangered, rare, and protected species among signatory nations. Appendix I of CITES lists species or other taxa that are in danger of extinction. Before a listed species can be traded, an export permit from the country of origin and an import permit from the destination country must be approved by management and scientific authorities in both countries.

Trade in endangered wildlife has flourished in spite of CITES. Countries vary significantly in their ability to enforce the convention, and only signatory nations are bound by it. However, the United States forbids imports of endangered species without regard to their country of origin.

16.3   ENDANGERED FISHES OF NORTH AMERICA

The imperiled status of many North American fishes has been recognized since the early 1960s (Miller 1972). In 1979, Deacon et al. listed 251 taxa as being endangered, threatened, or of special concern in North America. The percentages of listed taxa affected by particular threats were given by Deacon et al. as habitat modification (98%), other natural or human-induced factors (37%), restricted range (16%), overexploitation (3%), and disease (2%). The authors considered this to be evidence that well over 90% of the endangered and threatened fishes of North America could be restored by nationwide programs of habitat restoration and protection. Wilcove et al. (1998) also considered habitat modification to be the primary cause of species declines. They reviewed causes of imperilment for 213 fish species in the United States and reported that 87% were negatively affected by habitat loss and degradation, 53% by introduction of nonnative fishes, 13% by overexploitation, and 1% by disease.

Williams et al. (1989) listed 364 fishes endemic to North America that warrant protection because of their rarity. Twenty-two of the listed fishes occur in Canada, 254 in the United States, and 123 in Mexico, 35 of these fishes occur in more than one country. Fifty-six of the taxa added since Deacon et al.'s (1979) listing were Mexican fishes. These were added because of new information on their status and the degradation of aquatic habitats in that country. No fish was removed from the 1979 list because of successful recovery efforts. Recovery plans have been approved for 79 of the 108 species of U.S. fishes listed as threatened or endangered by USFWS (1998); however, the American Fisheries Society lists far more fishes than does the USFWS. Williams et al. (1989) concluded that recovery efforts had been locally effective for some species, but that North America's fish fauna has generally deteriorated.

Warren and Burr (1994) reported on the taxonomic and geographic distribution of imperiled fishes in the United States. About 51% of nonanadromous salmonids were considered imperiled, along with 37% of ictalurids, 35% of catostomids, 31% of cyprinids, 27% of percids, and 7% of centrarchids. Imperilment was most severe in areas of high diversity or endemicity, that is, the southern and western states.

Compilers of data on endangered and threatened species generally do not attempt to chronicle fish extinctions, probably because it is difficult to demonstrate that a fish species is extinct. Miller et al. (1989) reported extinction of 3 genera, 27 species, and 13 subspecies of fishes from North America during the past 100 years. Physical habitat alteration and detrimental effects of introduced species were the most commonly cited factors contributing to these extinctions. McAllister et al. (1985) discussed 37 Canadian fishes classified provisionally as rare, endangered, extirpated in Canada, or extinct.

16.3.1 Case Histories

16.3.1.1 The Snail Darter

No discourse on endangered fishes would be complete without considering the celebrated case of the snail darter, an early test of the ESA that resulted in significant changes in the act in 1978. In 1973 the snail darter (Figure 16.1) was discovered in a part of the lower Little Tennessee River known as Coytee Springs (Ono et al. 1983). Shortly thereafter, the ESA was signed into law by President Nixon. Convinced that the snail darter was new to science and very rare, Dr. David Etnier, one of the discoverers, submitted a status report on the fish to the USFWS. The report suggested that the 7- to 8-cm-long fish, which lives in fast waters and feeds on snails, was endangered and its existence was jeopardized by a Tennessee Valley Authority (TVA) dam. The snail darter may have lived throughout much of the Little Tennessee River, but dams had apparently restricted its habitat to the last 24 km of the river before it joined the Tennessee River. By the time the snail darter was described, construction of the TVA's controversial Tellico Project had begun (in 1967) (Ono et al. 1983). Controversy stemmed from the TVA's assertion that the benefits of the project would outweigh any disadvantages. Environmentalists countered that the dam would flood productive farmland, destroy an excellent trout stream, and inundate sacred places of the Cherokee Indians. The snail darter's habitat, above the site of the proposed Tellico Dam, was certain to be destroyed by closure of the dam.

The USFWS was petitioned to list the snail darter as endangered in January 1975, and the fish was listed in October 1975. That same year, the TVA began transplanting snail darters to the Hiwassee River, another tributary of the Tennessee River, without informing state authorities or the USFWS (Ono et al. 1983). Clearly the TVA was in violation of the ESA as it rushed to complete construction of the dam. In 1976, the Environmental Defense Fund sued the TVA for violating the ESA, and the USFWS declared the lower portion of the Little Tennessee River critical habitat for the snail darter. The district court in which the trial was held found the snail darter would be eradicated by the dam but denied a request for a permanent injunction on completion

Figure 16.1   Snail darter. Photograph shown by courtesy of R. Behnke.

of construction. In early 1977, a circuit court of appeals reversed the district court's decision and terminated work on the dam. News media began to criticize the decision to stop a multimillion-dollar project because of a small, inconsequential fish; the story was often presented as a case of environmentalism carried to extremes (Ehrlich and Ehrlich 1981).

After apparent defeat, the TVA asked the USFWS to delist the snail darter and requested permission to move all snail darters from the Little Tennessee River. Both requests were denied. The TVA also appealed the circuit court of appeal's decision to the U.S. Supreme Court, which in June 1978 upheld the injunction against the dam but virtually invited Congress to amend the law to allow exceptions (Ehrlich and Ehrlich 1981). The Congress did just that, passing the 1978 amendments to the ESA (section 7) which allowed the Endangered Species Committee to resolve conflicts between endangered species and projects that threatened them. The committee established to resolve the snail darter–Tellico Project conflict considered evidence of the dam's benefits, alternatives to completing the dam, values that would be lost if it were finished, and new economic analyses that showed that electricity would be produced at a deficit (Ehrlich and Ehrlich 1981). In early 1979 the committee voted unanimously not to exempt the Tellico Project because it was ill conceived and uneconomical (Yaffee 1982; Ono et al. 1983). Later that year, the committee's decision was reversed, and the conflict finally ended when an amendment exempting the Tellico Project from provisions of all federal laws was attached to the water projects appropriation bill in the U.S. House of Representatives and passed without opposition. The U.S. Senate tried without success to kill the amendment. President Carter cited political problems and difficulties in vetoing a multipurpose bill, and he signed the bill (with regret, he said) in September 1979 (Ehrlich and Ehrlich 1981; Yaffee 1982). The Tellico Dam was completed in 1980. The snail darter was later reported to be doing well in the Hiwassee River and several other places where new populations had been found (Etnier and Starnes 1993). Its status was changed from endangered to threatened in 1984.

The snail darter case has been called a classic boondoggle and a prime example of how pork barrel politics can threaten the environment and efforts to preserve species. However, Yaffee (1982) cited the case as an example of how the judiciary, Congress, media, and administrative agencies interact to determine how the ESA is implemented.

16.3.1.2   Death Valley Fishes

Another case of vanishing fishes that resulted in intervention by the Supreme Court involved the fishes of Death Valley. During the Pleistocene epoch, much of the Death Valley region of southwestern Nevada and southeastern California was covered by large lakes interconnected by rivers. Today, Death Valley is one of the hottest and driest places in the United States, and its waters have largely shrunk to a few small spring pools and intermittent streams. The Owens and Mohave rivers receive runoff from within the basin, but much of the basin's water enters through two groundwater systems, the Pahute Mesa and Ash Meadows groundwater basins. Only a few native fishes existed in Death Valley at the turn of the century. Among them were killifishes (several species of *Cyprinodon* and two of *Empetrichthys),* minnows (speckled dace and tui chub), and the Owens sucker (Pister 1981, 1985). Many lived in very restricted habitats subject to destruction by a single catastrophic event (Ono et al. 1983).

Human activities have harmed fishes in Death Valley. Early in the 1900s, construction of the Los Angeles aqueduct greatly modified habitats in the Owens River drainage. New dams reduced flooding that had created killifish habitat, marshes were drained, and game fishes (e.g., brown trout and largemouth bass) and mosquitofish were introduced. The game fishes used small native fishes as food, and the mosquitofish competed with some of them. Use of groundwater for irrigation disrupted fish habitat and lowered water tables on which fishes depended in the Amargosa Valley and Ash Meadows. Introduction of nonnative fishes to waters of Ash Meadows also presented problems for its native fishes, particularly killifishes. The Devils Hole pupfish, which exists only in a small spring pool called Devil's Hole in Ash Meadows, was particularly affected by groundwater withdrawals (Figure 16.2). Lowering water levels gradually exposed a shelf on which the species depended for spawning substrate and its algal food supply.

In 1969 wildlife agency personnel met and considered several strategies to preserve Death Valley fishes. Several fishes were transferred to natural or artificial refugia. The Devils Hole pupfish, for example, was first transferred to five refugia in California and Nevada and later to an artificial refugium built below Hoover Dam by the U.S. Bureau of Reclamation. The latter transfer resulted in phenotypic changes in the fish (Williams et al. 1988). Subsequently, a fiberglass shelf was suspended below the surface of Devil's Hole, and it was artificially lighted to enhance algal production. Efforts to rear the fish in aquaria were unsuccessful. Because it was apparent that only maintenance of groundwater levels would protect the Devils Hole pupfish and other Ash Meadows fishes, legal action to prohibit removal of groundwater in the area was initiated. In 1972, before the ESA was passed, the people of the United States, through the U.S. Department of Justice, sued the land developer and the state of Nevada as codefendants on a point of water law (Pister 1985). A district court issued a temporary injunction to prohibit pumping, and the injunction was made permanent by a unanimous ruling of the Supreme Court in 1976. Because land ownership and land use

Figure 16.2   Biologists touring the Devil's Hole pool, home of the Devils Hole pupfish. Photograph shown by courtesy of J. Hawkins.

control best assure habitat integrity, attempts were made to buy Ash Meadows. In 1984, Ash Meadows National Wildlife Refuge was established (Deacon and Williams 1991). In 1982, Fish Slough in California, the primary refuge of the endangered Owens pupfish and Owens tui chub, was saved from development by an act of Congress that allowed a land exchange between the U.S. Bureau of Land Management and a developer (Pister 1985).

What remains of the native fishes of Death Valley? Of the endemic Owens River fishes, the Owens pupfish was nearly wiped out, speckled dace populations were greatly diminished, and Owens tui chub hybridized with bait minnows. Only the Owens sucker was largely unaffected. In Ash Meadows, interactions with introduced fishes led to extinction of the Ash Meadows killifish and the Raycraft Ranch killifish. The Pahrump killifish survives in artificially maintained refuges, where it is threatened by bullfrogs *Rana catesbeiana* and introduced fishes. The Ash Meadows speckled dace and Ash Meadows Amargosa pupfish have been seriously threatened by groundwater pumping

Figure 16.3 Razorback sucker. Photograph shown by courtesy of E. Wick.

(Ono et al. 1983). Of the surviving native fishes of Death Valley, Mohave tui chub, Owens tui chub, Ash Meadows speckled dace, Devils Hole pupfish, Owens pupfish, Ash Meadows Amargosa pupfish, and Pahrump killifish are endangered. Recent habitat preservation measures have led to limited optimism about their future, but most native fishes of Death Valley remain imperiled.

16.3.1.3 Colorado River Basin Fishes

The indigenous fish fauna of the Colorado River basin includes some of the most unique freshwater fishes in North America. Because the basin has been isolated for millions of years, its native fishes are morphologically different and evolutionarily distant from their nearest relatives. Cyprinids, catostomids, and cyprinodontids are the dominant native fishes. Among the better-publicized natives are the Colorado pikeminnow, a large predatory minnow, and humpback chub and razorback sucker (Figure 16.3), both distinguished by a prominent dorsal hump at the nape. In all, about 54 fishes (presently recognized species and subspecies) are native to the basin, and most (83% of the total number) are endemic. Stanford and Ward (1986a) and Carlson and Muth (1989) summarized information on within-basin distributions of native fishes.

Closure of Hoover Dam in 1935 marked the end of the free-flowing Colorado River. Since then, the basin has become one of the most altered and controlled river systems in the United States. Dam construction drastically altered physical and biological features of the river system (Stanford and Ward 1986b) and severely affected native fishes inhabiting the larger river channels. Two native species are extinct (Pahranagat spinedace and Las Vegas dace), 16 are federally listed as endangered, 6 are federally listed as threatened, 1 is a candidate for federal listing, and several of the remaining native fishes are protected by one or more basin states. Researchers have attributed the decline of native fishes to the cumulative and synergistic effects of habitat modification caused by water development and predation or competition by introduced nonnative fishes (Stanford and Ward 1986a).

Since the late 1800s, approximately 67 nonnative fish species have been introduced into the basin, raising the total number of species to over 100 (Carlson and Muth 1989). Most introductions resulted from efforts to establish sport fisheries in and downstream of newly constructed reservoirs. Invoking the island biogeography theory, Molles

(1980) proposed that, because the Colorado River basin is an insular system, the native fish fauna is especially vulnerable to invasion by nonnative fish species. Using local extirpations of native fishes in the lower basin as examples, he suggested that successful invasions can cause loss of native fishes through competitive replacement or predation. Many researchers now consider nonnative fishes to pose the most serious threat to the continued existence of native fishes in the basin.

Efforts to recover the federally protected Colorado River fishes by direct management of populations or habitats have been implemented. A program was initiated in 1987 to recover, delist, and manage humpback chub, bonytail, Colorado pikeminnow, and razorback sucker in the upper basin while allowing for additional water development (Wydoski and Hamill 1991). Objectives of the recovery program are to (1) provide instream flows to protect and recover the species; (2) increase the amount of favorable habitat for reproduction and overwintering; (3) supplement wild populations by artificial propagation and stocking; (4) regulate stocking of nonnative fishes, selectively remove nonnative species, and regulate sportfishing in areas where the endangered fishes occur; and (5) educate the fishing public about these rare and endangered fishes. Management of flow releases from major impoundments of the Colorado River Storage Project to help restore natural conditions for protection and recovery of the endangered fishes and other native species is a high priority. In 1996 an experimental flood was released from Glen Canyon Dam on the Colorado River above Grand Canyon solely for environmental benefit (Collier et al. 1997). This precedent-setting event has potential implications for future management of rivers controlled by dams across North America.

Even with the above conservation strategies in place, the future of the Colorado River's imperiled native fishes is uncertain. Carlson and Muth (1989) concluded that conflicts between development and natural ecosystems in the basin will continue and probably worsen as demand for water increases. Stanford and Ward (1986a) stated that the river's future depends on whether (1) there will be enough water to maintain desirable ecosystem values and (2) native and nonnative fishes can coexist. They concluded that survival of the endangered fishes is incompatible with stream regulation and the persistence of nonnative species and that future water shortages will preclude allocations for native species.

16.3.1.4   Western Salmonids in the Genus *Oncorhynchus*

One of the most publicized crises in fisheries management involves Pacific salmon in western North America (see Chapter 24). Seven species of anadromous salmonids have experienced drastic population declines: chinook, chum, coho, pink salmon, sockeye, and steelhead, and sea-run cutthroat trout. Nehlsen et al. (1991) reported that of 214 naturally spawning stocks of these species, 101 were at high risk of extinction and 58 at moderate risk of extinction. By 1998, many populations of chinook, coho, and sockeye salmon, steelhead, and sea-run cutthroat trout in the Pacific Northwest were officially listed as threatened under the ESA. The status of Pacific salmon populations in terms of the ESA is likely to be dynamic, and current information about particular stocks is available through the Northwest Fisheries Science Center of the National Marine Fisheries Service.

Major factors responsible for the decline of Pacific salmon populations include overfishing, habitat degradation due to agriculture, logging, and urbanization, and migration barriers related to hydropower and flood control dams (Ross 1997). Overfishing has proven especially difficult to control because much of the harvest has occurred on the high seas outside the jurisdiction of any one country. Controlling the nearshore harvest has involved the challenge of getting state, provincial, and tribal governments to cooperate in reducing catch limits.

Habitat degradation associated with agriculture and logging has involved loss of woody debris and increased sedimentation in spawning and nursery streams (see Chapters 9, 10, and 18). A commonly proposed remedy is the establishment of natural buffer zones along streams where land use activities such as farming or logging would be restricted.

Blockage of migration paths by dams is one of the most controversial aspects of Pacific salmon management (Ross 1997). Many western rivers have been dammed to produce electricity, control flooding, and store water for agricultural and urban uses. Unfortunately, dams prevent upstream migration of adult spawners and downstream migration of smolts. Furthermore, the reservoirs created by dams often are unfavorable habitat for young salmonids because of unsuitable thermal conditions and the host of nonnative predator fishes that such reservoirs contain. To mitigate these effects, fish ladders have been installed to allow upstream migration of spawning adults, and various bypass systems have been developed to prevent smolts from entering turbine intakes as they attempt to migrate downstream to the ocean. In some cases, smolts are captured in reservoirs above dams and transported downstream in trucks or barges to continue their seaward migration. Although these mitigation efforts have helped, removal of dams may be the only way to recover some salmonid populations completely. As Nehlsen et al. (1991) noted, we need management approaches that emphasize habitat restoration and ecosystem function rather than hatchery production if we are to achieve recovery of many imperiled stocks.

In addition to the anadromous species of salmonids, many inland trout species have declined in western North America, and some species have been pushed to near extinction (Ono et al. 1983). Behnke (1992) classified native western trout into four species and 22 subspecies. Six species or subspecies are federally listed as threatened (Apache trout, bull trout, greenback cutthroat trout, Lahontan cutthroat trout, Little Kern golden trout, and Paiute cutthroat trout), and one is federally listed as endangered (Gila trout). Other native western trout are protected by several western states and often appear on unofficial lists of threatened and endangered fishes. Some of these, such as Bonneville cutthroat trout, Colorado River cutthroat trout, and westslope cutthroat trout, are under consideration for federal listing.

Several factors have contributed to the decline of inland western trout. One major threat has been extensive introduction of nonnative salmonids that has resulted in hybridization, increased competition for food and space, and increased predation on young trout. Allendorf and Leary (1988) considered introgression, that is, the dilution of native alleles, to be the most important effect of introducing nonnative trout. Another factor blamed for the decline of native western trout is habitat destruction caused by water diversion and removal, channelization, water pollution, and overgrazing of rangeland and damage to streambanks by livestock. Federal recovery plans have been approved for several native western trout, and management programs have been successful in rehabilitating certain populations (Gresswell 1988). Recovery efforts have involved elimination of nonnative trout, reintroduction of genetically pure native stocks, habitat protection and improvement, and restricted angler harvest.

16.4   MANAGEMENT OF ENDANGERED FISHES

When conservation of fishes is needed, Moyle and Cech (1988) recommended the following steps: (1) inventory fishes; (2) monitor habitats, communities, and species; (3) conduct research on the best management of resource and nonresource fishes; (4) plan regional management following natural, rather than political, boundaries; and (5) manage fish communities and habitats as units. We see a special need for biological surveys to determine the status of rare fishes and gain legal protection for those near extinction, particularly in Mexico (Contreras Balderas 1991; Lyons et al. 1998). Suggestions of Soule and Kohm (1989) regarding research directions for conservation biology should be applied to rare fishes; emphasis on ecosystem fragmentation, the biology of small systems, reproduction requirements of given species, and effects of stress and disease is justified. Population dynamics and habitat needs, with particular emphasis on early life stages, must be understood for effective management. Single species approaches to recovery efforts must give way to considering endangered fishes as components of complex ecosystems.

Conservation of native fishes also will benefit other aquatic organisms. Numerous species of amphibians, mussels, and crayfishes are imperiled, usually by the same anthropogenic alterations that affect fishes (Wilcove et al. 1998). For example, 65% of crayfishes and 73% of unionid mussels in North America are either extinct or at risk (Winter and Hughes 1997).

Approaches used in management of endangered fishes include (1) maintaining and enhancing historic populations; (2) protecting, expanding, or restoring habitat; (3) moving specimens to refuges; (4) rearing and stocking in new or formerly occupied areas; (5) minimizing introductions and undesirable effects of nonnative organisms; and (6) controlling exploitation. We have listed maintaining and enhancing historic native populations first as a reminder of their importance.

16.4.1   Management of Fishes and Habitat

Habitat should be a primary consideration in recovery of rare fishes. Land acquisition as accomplished in the Death Valley drainage is an effective means of protecting habitat for endangered species. Other habitat protection approaches involve participation in pollution control efforts, land use planning, and legal actions to prevent habitat destruction. Johnson and Rinne (1982) noted that protection and enhancement of habitats is the almost universal goal of fish recovery plans, but recovery teams have generally left implementation to the discretion of land managers. Johnson and Rinne suggested that existing habitats be protected through increased federal management, prioritization of habitats, and use of ecosystem-based recovery plans.

Restoration of damaged or depleted habitat is also possible. Earlier, we discussed attempts to restore the habitat of the Devils Hole pupfish by shelf construction (Section 16.3.1.2). Re-creating backwaters, wetlands, embayments, and other habitat types has been suggested to aid recovery of the protected native fishes of the Upper Colorado River system. Use of fishways and other remediation measures has also been suggested to preserve those species. Rinne (1982) described effects of instream habitat improvement structures on Gila trout growth and dispersal. Restoration of aquatic habitat might also include removal of dams, return of channelized streams to their original courses, silt removal, and restoration of riparian vegetation (see Chapter 9). In some cases, artificial barriers have been used to separate endemic rare fishes from potential predators, competitors, or sources of hybridization (Ono et al. 1983; Thompson and Rahel 1998).

Rearing and reintroducing endangered fishes have generally met with limited success. Ono et al. (1983) discussed several failed attempts to propagate rare fishes. But, for some species such as the endangered cui-ui of Pyramid Lake, Nevada, artificial propagation is the key to survival. State and federal agencies at first seemed reluctant to embrace reintroduction because of its potential to limit other water uses. Nonetheless, reintroductions of rare fishes began in the southwestern United States in the early 1980s. The 1982 amendments to the ESA initially stimulated increased interest in reintroduction by eliminating the requirement that critical habitat be designated when species were listed and by providing for experimental populations. However, regulations finalized in 1984 were so complicated that many reintroduction plans were disrupted again (Rinne et al. 1986). Williams et al. (1988) examined recovery plans for 39 endangered and threatened U.S. fishes and found that 82% of the plans called for reintroductions to establish new populations, begin artificial propagation, or create educational exhibits.

Our case histories included examples of transferring threatened fishes to refuges. One of the earliest transfers was that of the Owens pupfish. Thought to be extinct when it was described in 1948, it was later rediscovered in Fish Slough in Mono County, California (Pister 1981, 1985). To protect the remnants of the species from introduced predatory fishes and competition from mosquitofish, the Owens Valley Native Fish Sanctuary was built by the California Department of Fish and Game in 1969. Two other refuges were later constructed to receive the Owens pupfish and other native Owens Valley fishes, and all have contributed to the fishes' continued existence. Although selection pressures in artificial refuges may alter a species' gene pool, Turner (1984) found no apparent loss of average genetic heterozygosity in refugium populations of desert pupfish. He considered use of refuges to be a valid practice for endangered fish management and conservation. Nonetheless, there is general agreement that fishes should be transplanted to refugia only when their extinction is imminent, and they should be reintroduced to native habitats as soon as possible.

Genetic considerations should be an important part of any recovery efforts. Maintaining genetic variability and evolutionary flexibility for a species depends on maximizing within-population and among-population variance. To accomplish this for endangered fishes, Meffe (1986) recommended the following: (1) genetic monitoring should be done to determine how variation is distributed within a species and how to preserve it; (2) the largest-feasible genetically effective population size should be maintained in captive breeding programs; (3) if a large population cannot be maintained, inbreeding may be avoided by selective mating; (4) stocks should be kept in hatchery environments for as short a time as possible; and (5) separate stocks of isolated populations should be maintained to preserve interpopulation genetic variability. Maintenance of interpopulation variation and evolutionary flexibility is also the basis behind adapting the evolutionary significant unit as the basic unit of species conservation (Waples 1995). An evolutionarily significant unit is a population (or group of populations) that is substantially reproductively isolated from other conspecific populations and represents an important component in the evolutionary legacy of the species.

The concept of the evolutionarily significant unit has been used to identify populations of anadromous Pacific salmon in need of conservation (Waples 1995) and is evident in several studies of genetic diversity in fishes. Echelle et al. (1989) observed that the greatest genetic diversity in Pecos gambusia was due to differences among populations in the four primary areas where the fish occurred. Maintaining the species in all four areas would ensure maximum allelic diversity for the species. Allendorf and Leary (1988) found that conservation of genetic diversity in some subspecies of cutthroat trout requires maintenance of many populations, but genetic diversity in other subspecies can be conserved by protecting only a few populations. Their data suggested that some subspecies of cutthroat trout have long evolved separately from others and that taxonomic revision of the species is needed to allow development of good conservation plans for the diverse groups now considered a single species.

Many fishes have been introduced to North America to provide sport and food, and others have been moved across North America to areas outside their native ranges (see Chapter 13). Effects of such introductions may include habitat alteration, introduction of parasites or diseases, reduction in growth or survival (or even elimination) of native fishes, and changes in community structure (Moyle et al. 1986). Adverse effects of introduced species have contributed to 66% of fish extinctions in North America in the past century (Miller et al. 1989). When an introduction seems the only solution to a problem, guidelines to facilitate sound decision making should be followed (see Chapter 13).

16.4.2 Multispecies Management and Conservation of Biodiversity

Management of declining species is generally done under the auspices of the ESA. But working within the framework of the ESA poses several problems. First, recovery efforts typically begin only after a species is extremely imperiled. By waiting until extinction is imminent, costly and heroic efforts often are required to save a species. Second, a growing backlog of endangered species threatens to overwhelm fiscal resources and society's will to deal with all the species separately. Third, recovery often focuses on quick fixes such as transplants and captive rearing rather than more difficult solutions, such as ecosystem rehabilitation and protection. Fourth, the focus on saving species ignores protection of the landscapes in which species occur and interact. Abiotic disturbance (e.g., floods) and interactions with other species (e.g., predator–prey relations) can be important for the long-term persistence of a species. For example, native species are more likely to resist replacement by introduced species in streams with natural flood regimes. Because the native species evolved in flood-prone systems, they possess behavioral and life history adaptations that allow them to survive floods better than nonnative species (Meffe 1984; Baltz and Moyle 1993).

To avoid the drawbacks of managing endangered species one at a time, conservation biologists advocate a community or ecosystem approach to preserving the Earth's biotic diversity (Moyle and Yoshiyama 1994; Angermeier and Schlosser 1995). A focus on saving species is considered inadequate because biodiversity encompasses genetic diversity, population and species diversity, and community and ecosystem diversity. The types of biodiversity form a nested hierarchy and, therefore, conserving higher levels of the hierarchy such as ecosystems or communities also will conserve diversity at lower levels, such as species and genes (Figure 16.4). All levels of biodiversity within the hierarchy need to be conserved. It does little good to save a species if its genetic heritage and the environment in which it evolved have been lost. Also, the community–ecosystem approach should be proactive rather than reactive. This means that steps should be taken to rehabilitate communities or ecosystems during early phases of decline instead of waiting until systems are severely degraded.

How does one manage for biodiversity? The first step is to inventory aquatic habitats and map species distributions (Moyle and Yoshiyama 1994; Angermeier and Schlosser 1995). Surveys should collect all fish species, especially nongame species often overlooked when monitoring sport fisheries. In the absence of historical data, areas with high species richness, several rare species, or a unique assemblage would be likely candidates for conservation efforts. When historical data are available, current species distributions can be compared with earlier distributions to identify declining species (Patton et al. 1998; Shaffer et al. 1998).

Armed with knowledge about which species are declining, the next step is to identify groups of species or ecosystem types in danger of being lost or drastically altered in the absence of conservation efforts. Biologists familiar with the natural history and regional distribution of the fish fauna can identify clusters of declining species in key watersheds. These areas would then become the focus of conservation efforts. Such areas have been termed preserves or refuges, but Moyle and Yoshiyama (1994) recommended calling them "aquatic diversity management areas" to emphasize the protection of biotic communities rather than individual species.

A variety of methods have been proposed for ranking areas on the basis of their conservation value (Winston and Angermeier 1995). One method that has received considerable attention is termed gap analysis because it attempts to identify geographic areas that contain the most species not covered in existing protected areas. Incorporating these areas into the preserve network could close gaps in the protection of a regional fauna (Flather et al. 1997). The procedure involves overlaying maps of individual species' distributions to find hotspots where several declining species co-occur. These hotspots are then overlain with a map of nature preserves to identify areas with clusters of species not contained within existing protected areas. Given several potential sites as preserves, the one that contains the greatest number of currently unprotected species would be a logical choice. For aquatic systems, watersheds appear to be the most practical unit for conserving biodiversity because habitat conditions within stream reaches or lakes are the integration of factors operating throughout the watershed (Hynes 1975; also see Chapter 9). An application of a gap analysis at the watershed level is shown in Figure 16.5.

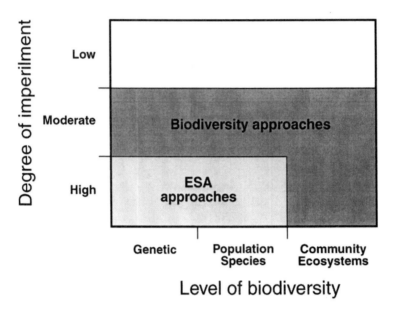

Figure 16.4 Comparison of conservation policy as represented by the Endangered Species Act of 1973 (ESA) and a more holistic approach to protecting biodiversity at the community or ecosystem level prior to extreme imperilment. Biodiversity exists at several levels in a nested hierarchy such that protection of higher levels in the hierarchy (e.g., communities and ecosystems) also will protect diversity at lower levels (e.g., populations and species). The ESA has tended to focus on population or species showing a high degree of imperilment. Most conservation biologists favor protecting entire communities or ecosystems and implementing conservation measures prior to extreme imperilment. Figure modified from Angermeier and Schlosser (1995).

In large water bodies, biodiversity preserves could take the form of no-fishing zones such as those designed to protect commercially important fish. Because so many commercial species are being overfished and because enforcing harvest regulations and eliminating mortality of nontarget species is so difficult, fisheries managers have promoted the establishment of protected areas where no fishing would be allowed (Allison et al. 1998). Excess fish produced in these areas would migrate to nearby harvest areas, but the reserves would ensure that healthy populations remain in at least some locations throughout a species range. In addition to protecting commercially valuable species, reserves would protect a wide range of other species negatively affected either as bycatch or through destruction of benthic habitats by trawling. Reserves have been promoted primarily in marine environments, but the concept has also been suggested to help recovery of Pacific salmon in freshwater habitats of North America (Rahr et al. 1998).

Conserving biodiversity at the watershed scale can be an elusive goal because it requires cooperation among many stakeholders with conflicting interests (Williams et al. 1997). Examples of a watershed approach to protecting native species include reestablishment of natural flow regimes in streams (Poff et al. 1997), controlling sediment inputs and grazing impacts (Binns and Remmick 1994), and maintaining groundwater levels for desert springs (Moyle and Yoshiyama 1994). An interesting example of protecting biodiversity by taking a watershed perspective involves Putah Creek in California (Moyle et al. 1998). A citizen's group, working in conjunction with fisher-

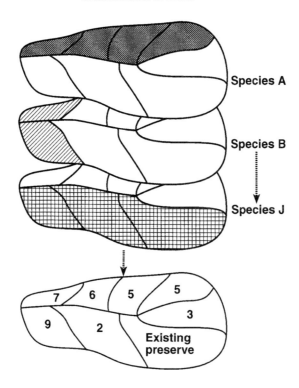

Figure 16.5 An example of gap analysis for conservation planning. Maps for 10 declining species not protected in an existing preserve are superimposed to identify watersheds that contain the largest number of these species. The resultant map (bottom) shows how many of the 10 declining species occur in each watershed and also indicates the location of the existing protected watershed. The two watersheds at the far left contain the largest number of declining species, and including them in a preserve network would help close gaps in the protection of biodiversity across the drainage.

ies biologists, won a court case to improve the flow regime in Putah Creek. Previously, most of the creek's water was diverted for agricultural and municipal use, resulting in low flows and occasional dewatering of the stream channel (Figure 16.6). State law required that sufficient water must pass over a dam to keep any fish that exist downstream in "good condition." The citizen's group argued that good condition included maintenance of an assemblage of native, nongame fishes. The judge recognized that Putah Creek had high value for the preservation of biodiversity and ordered increased flows to keep water in the creek at all times. Also, flows in February and March were to be enhanced to favor reproduction of native fishes. What is distinctive about this case is that it involved preservation of biodiversity in the form of an entire assemblage of native fishes, even though none of the component species was endangered. This represents the goal of many conservation biologists who want to protect species assemblages and ecosystems before individual species become endangered.

16.4.3 The Social Context of Endangered Species Management

Economic, political, and social constraints often have strong bearing on decisions related to managing endangered fishes. The snail darter controversy and other case histories illustrate this point. Conflicts between the protection of endangered fishes

Figure 16.6 Putah Creek, California, August 1990, when water diversion coupled with prolonged drought left only a few remnant pools in the stream channel. Preservation of a native fish assemblage was the main argument for managing water to return year-round flows to the Putah Creek channel. Photograph shown by courtesy of P. Moyle.

and construction projects or other human activities with significant economic, political, and social support are certain to continue. Gaining funding for research on and management of endangered fishes requires consideration of these factors. Fish and game agencies continue to struggle to support nongame programs with funding gained primarily from license fees. However, the people have mandated management and preservation of all fishes and wildlife (Pister 1976). Fisheries management programs need to recognize the inherent values of native fishes, whether the fishes are harvestable or not. Programs will evolve toward that recognition only if practicing fisheries professionals are willing to broaden their perspectives and fisheries students are exposed to new ideas. Meffe (1987) stated that education of the public and professionals in values of biotic diversity and conservation of biological resources would help to launch educational programs in fish conservation genetics. O'Connell (1992) stressed the need for scientists to increase their involvement in legislating sound science and to develop a bit of political sophistication. Biologists should collaborate more with social scientists, be prepared to present social justification of their work to the public, and quibble less with one another.

16.5   CONCLUSION

As the major causal agents of the current extinction spasm, humans have a responsibility to reduce the rate of extinction. Fostering and applying a conservation or evolutionary ethic may be a good starting point, but more immediate practical approaches should also be used. The extinction rate may be slowed by developments such as work of The Nature Conservancy's preservation of selected land areas; the United Nations Educational, Scientific, and Cultural Organization's system of biosphere reserves; and the emergence of the Society for Conservation Biology (Ehrlich 1987). Movement toward an ecosystem paradigm for management of public lands is also encouraging, although there is debate about whether ecosystem management reflects a new, biocentric emphasis on preserving biodiversity or merely an extension of the anthropocentric, multiple-use philosophy to larger spatial scales and longer time frames (Grumbine 1997; Lackey 1998).

Zoos, aquariums, and botanical gardens are useful for preserving organisms that cannot survive in the wild. The major disadvantages of depending on zoos or aquariums as reservoirs of organic diversity are their limited capacity and the difficulties of maintaining large populations under natural selection regimes so as to avoid loss of genetic diversity. Possible reduction of genetic diversity leads to questions about the value of such innovations as gene banks, frozen sperm banks, seed banks, and embryo banks.

Protection of biotic diversity is obviously dependent on future research and political action. National and international conservation programs must share knowledge about biological diversity, extinction rates, and the location and size of reserves and use this information as a basis for legislation. Extinction rates also can be slowed by treaties and laws that protect rare organisms. Biologists must recognize their obligations as informed citizens to see that laws which protect species or their habitats are passed, modified as necessary, and enforced. Finally, nearly all suggestions for preserving the Earth's biodiversity recognize the importance of controlling human population growth and resource consumption if the extinction crisis is to be averted or lessened.

16.6   REFERENCES

Allendorf, F. W., and R. F. Leary. 1988. Conservation and distribution of genetic variation in a polytypic species, the cutthroat trout. Conservation Biology 2:170–184.

Allison, G. W., J. Lubchenco, and M. H. Carr. 1998. Marine reserves are necessary but not sufficient for marine conservation. Ecological Applications 8(Supplement):S79–S92.

Angermeier, P. L., and I. J. Schlosser. 1995. Conserving aquatic biodiversity: beyond species and populations. Pages 402–414 in J. L. Nielsen, editor. Evolution and the aquatic ecosystem: defining unique units in population conservation. American Fisheries Society, Symposium 17, Bethesda, Maryland.

Anonymous. 1999. Teaming with wildlife legislation has been reintroduced. Fisheries 24(3):46.

Baltz, D. M., and P. B. Moyle. 1993. Invasion resistance to introduced species by a native assemblage of California stream fishes. Ecological Applications 3:246–255.

Behnke, R. J. 1992. Native trout of western North America. American Fisheries Society, Monograph 6, Bethesda, Maryland.

Bender, M., K. D. Boylan, and E. L. Smith. 1998. Turning the corner towards recovery. Endangered Species Bulletin 23(2–3):4–9, U.S. Fish and Wildlife Service, Washington, D.C.

Binns, N. A., and R. Remmick. 1994. Response of Bonneville cutthroat trout and their habitat to drainage-wide habitat management at Huff Creek, Wyoming. North American Journal of Fisheries Management 14:669–680.

Brouha, P. 1998. Teaming with wildlife: let's support a great idea. Fisheries 23(1):4.

Callicott, J. B. 1997. Conservation values and ethics. Pages 29–56 in G. K. Meffe and C. R. Carroll, editors. Principles of conservation biology. Sinauer, Sunderland, Massachusetts.

Campbell, R. R. 1998. Rare and endangered fish and marine mammals of Canada: COSEWIC Fish and Marine Mammal Subcommittee status reports XII. Canadian Field-Naturalist 112:94–97.

Carlson, C. A., and R. T. Muth. 1989. The Colorado River: lifeline of the American southwest. Canadian Special Publication of Fisheries and Aquatic Sciences 106:220–239.

Collier, M .P., R. H. Webb, and E. D. Andrews. 1997. Experimental flooding in Grand Canyon. Scientific American 276(1):82–89.

Contreras Balderas, S. 1987. Threatened and endangered fishes of Mexico. Proceedings of the Desert Fishes Council 16–18:58–65.

Contreras Balderas, S. 1991. Conservation of Mexican freshwater fishes: some protected sites and species, and recent federal legislation. Pages 191–197 in Minckley and Deacon (1991).

Cook, F. R., and D. Muir. 1984. The Committee on the Status of Endangered Wildlife in Canada (COSEWIC): history and progress. Canadian Field-Naturalist 98:63–70.

Deacon, J. E., G. Kobetich, J. D. Williams, and S. Contreras. 1979. Fishes of North America endangered, threatened, or of special concern: 1979. Fisheries 4(2):29–44.

Deacon, J. E., and C. D. Williams. 1991. Ash Meadows and the legacy of the Devils Hole pupfish. Pages 69–87 in Minckley and Deacon (1991).

Echelle, A. F., A. A. Echelle, and D. R. Edds. 1989. Conservation genetics of a spring-dwelling desert fish, the Pecos gambusia (Gambusia nobilis, Poeciliidae). Conservation Biology 3:159–169.

Ehrenfeld, D. W. 1976. The conservation of non-resources. American Scientist 64:648–656.

Ehrlich, P. R. 1987. Population biology, conservation biology, and the future of humanity. BioScience 37:757–763.

Ehrlich, P. R., and A. Ehrlich. 1981. Extinction: the causes and consequences of the disappearance of species. Random House, New York.

Ehrlich, P. R., and H. A. Mooney. 1983. Extinction, substitution, and ecosystem services. BioScience 33:248–254.

Etnier, D. A., and W. C. Starnes. 1993. The Fishes of Tennessee. The University of Tennessee Press, Knoxville.

Flather, C. H., K. R. Wilson, D. J. Dean, and W. C. McComb. 1997. Identifying gaps in conservation networks: of indicators and uncertainty in geographic-based analyses. Ecological Applications 7:531–542.

Gresswell, R. E., editor. 1988. Status and management of interior stocks of cutthroat trout. American Fisheries Society, Symposium 4, Bethesda, Maryland.

Grumbine, R. E. 1997. Reflections on "what is ecosystem management?" Conservation Biology 11:41–47.

Hynes, H. B. N. 1975. The stream and its valley. Internationale Vereinigung fur theoretische und angewandte Limnologie Verhandlungen 19:1–15.

Johnson, J. E., and J. N. Rinne. 1982. The Endangered Species Act and southwest fishes. Fisheries 7(4):2–8.

Kohm, K. A., editor. 1991. Balancing on the brink of extinction—the Endangered Species Act and lessons for the future. Island Press, Washington, D.C.

Lackey, R. T. 1998. Ecosystem management: desperately seeking a paradigm. Journal of Soil and Water Conservation 53:92–94.

Lyons, J., G. Gonzalez-Hernandez, E. Soto-Galera, and M. Guzman-Arroyo. 1998. Decline of freshwater fishes and fisheries in selected drainages of west-central Mexico. Fisheries 23(4):10–18.

McAllister, D. E., B. J. Parker, and P. M. McKee. 1985. Rare, endangered and extinct fishes in Canada. National Museums of Canada, National Museum of Natural Sciences Syllogeus 54, Ottawa.

Meffe, G. K. 1984. Effects of abiotic disturbance on coexistence of predator–prey fish species. Ecology 65:1525–1534.

Meffe, G. K. 1986. Conservation genetics and the management of endangered fishes. Fisheries 11(1):14–23.

Meffe, G. K. 1987. Conserving fish genomes: philosophies and practices. Environmental Biology of Fishes 18:3–9.

Miller, R. R. 1972. Threatened freshwater fishes of the United States. Transactions of the American Fisheries Society 101:239–252.

Miller, R. R., J. D. Williams, and J. E. Williams. 1989. Extinctions of North American fishes during the past century. Fisheries 14(6):22–38.

Minckley, W. L., and J. E. Deacon, editors. 1991. Battle against extinction—native fish management in the American West. University of Arizona Press, Tucson.

Molles, M. 1980. The impacts of habitat alterations and introduced species on the native fishes of the upper Colorado River basin. Pages 163–181 *in* W. O. Spofford, A. L. Parker, and A. V. Kneese, editors. Energy development in the southwest: problems of water, fish and wildlife in the upper Colorado River basin, volume 2. Resources for the Future, Washington, D.C.

Moyle, P. B., and J. J. Cech. 1988. Fishes: an introduction to ichthyology, 2nd edition. Prentice-Hall, Englewood Cliffs, New Jersey.

Moyle, P. B., H. W. Li, and B. A. Barton. 1986. The Frankenstein effect: impact of introduced fishes on native fishes in North America. Pages 415–426 *in* Stroud (1986).

Moyle, P. B., M. P. Marchetti, J. Baldrige, and T. L. Taylor. 1998. Fish health and diversity: justifying flows for a California stream. Fisheries 23(7):6–15.

Moyle, P. B., and R. M. Yoshiyama. 1994. Protection of aquatic biodiversity in California: a five-tiered approach. Fisheries 19(2):6–18.

Myers, N. 1985. A look at the present extinction spasm and what it means for the future evolution of species. Pages 47–57 *in* R. J. Hoage, editor. Animal extinctions: what everyone should know. Smithsonian Institution Press, Washington, D.C.

Myers, N. 1997. Global diversity II: losses and threats. Pages 123–158 *in* G. K. Meffe and C. R. Carroll, editors. Principles of conservation biology. Sinauer, Sunderland, Massachusetts.

Naeem, S. 1998. Species redundancy and ecosystem reliability. Conservation Biology 12:39–45.

Nehlsen, W., J. E. Williams, and J. A. Lichatowich. 1991. Pacific salmon at the crossroads: stocks at risk from California, Oregon, Idaho, and Washington. Fisheries 16(2):4–21.

Norma Oficial Mexicana. 1994. Peces. Mexican Federal Government, Mexico, D.F., Rule NOM-ECOL-059-94.

O'Connell, M. 1992. Response to: "six biological reasons why the Endangered Species Act doesn't work and what to do about it. Conservation Biology 6:140–143.

Ono, R. D., J. D. Williams, and A. Wagner. 1983. Vanishing fishes of North America. Stone Wall Press, Washington, D.C.

Patton, T. M., F. J. Rahel, and W. A. Hubert. 1998. Using historical data to assess changes in Wyoming's fish fauna. Conservation Biology 12(5):1120–1128.

Pister, E. P. 1976. A rationale for the management of nongame fish and wildlife. Fisheries 1(1):11–14.

Pister, E. P. 1981. The conservation of desert fishes. Pages 411–445 *in* R. J. Naiman and D. L. Soltz, editors. Fishes in North American deserts. Wiley, New York.

Pister, E. P. 1985. Desert pupfishes: reflections on reality, desirability, and conscience. Environmental Biology of Fishes 12:3–11.

Poff, N. L., and seven coauthors. 1997. The natural flow regime. BioScience 47:769–784.

Rahel, F. J. 1997. From Johnny Appleseed to Dr. Frankenstein: changing values and the legacy of fisheries management. Fisheries 22(8):8–9.

Rahr, G. R., III., J. A. Lichatowich, R. Hubley, and S. M. Whidden. 1998. Sanctuaries for native salmon: a conservation strategy for the 21st century. Fisheries 23(4):6–7, 36.

Rinne, J. N. 1982. Movement, home range, and growth of a rare southwestern trout in improved and unimproved habitats. North American Journal of Fisheries Management 2:150–157.

Rinne, J. N., J. E. Johnson, B. L. Jensen, A. W. Ruger, and R. Sorenson. 1986. The role of hatcheries in the management and recovery of threatened and endangered fishes. Pages 271–285 *in* Stroud (1986).

Ross, M. R. 1997. Fisheries conservation and management. Prentice-Hall, Englewood Cliffs, New Jersey.

Shaffer, H. B., R. N. Fisher, and C. Davidson. 1998. The role of natural history collections in documenting species declines. Trends in Ecology and Evolution 13:27–30.

Soule, M. E., and K. A. Kohm, editors. 1989. Research priorities for conservation biology. Island Press, Washington, D.C.

Stanford, J. A., and J. V. Ward. 1986a. Fishes of the Colorado system. Pages 385–402 *in* B. R. Davies and K. F. Walker, editors. The ecology of river systems. Dr. W. Junk, Dordrecht, The Netherlands.

Stanford, J. A., and J. V. Ward. 1986b. The Colorado River system. Pages 353–374 *in* B. R. Davies and K. F. Walker, editors. The ecology of river systems. Dr. W. Junk, Dordrecht, The Netherlands.

Stroud, R. H., editor. 1986. Fish culture in fisheries management. American Fisheries Society, Fish Culture Section and Fisheries Management Section, Bethesda, Maryland.

Thompson, P. D., and F. J. Rahel. 1998. Evaluation of artificial barriers in small Rocky Mountain streams for preventing the upstream movement of brook trout. North American Journal of Fisheries Management 18:206–210.

Turner, B. J. 1984. Evolutionary genetics of artificial refugium populations of an endangered species, the desert pupfish. Copeia 1984:364–369.

USFWS (U.S. Fish and Wildlife Service). 1998. Box score-listings and recovery plans. Endangered Species Bulletin 23(2–3):44.

Waples, R. S. 1995. Evolutionarily significant units and the conservation of biological diversity under the Endangered Species Act. Pages 8–27 in J. L. Nielsen, editor. Evolution and the aquatic ecosystem: defining unique units in population conservation. American Fisheries Society, Symposium 17, Bethesda, Maryland.

Warren, M. L., Jr., and B. M. Burr. 1994. Status of freshwater fishes of the United States: overview of an imperiled fauna. Fisheries 19(1):6–18.

Wilcove, D. S., D. Rothstein, J. Dubow, A. Phillips, and E. Losos. 1998. Quantifying threats to imperiled species in the United States. BioScience 48:607–615.

Williams, J. E., D. W. Sada, and C. D. Williams. 1988. American Fisheries Society guidelines for introductions of threatened and endangered fishes. Fisheries 13(5):5–11.

Williams, J. E., and seven coauthors. 1989. Fishes of North America endangered, threatened, or of special concern: 1989. Fisheries 14(6):2–20.

Williams, J. E., C. A. Wood, and M. P. Dombeck. 1997. Watershed restoration: principles and practices. American Fisheries Society, Bethesda, Maryland.

Winston, M. R., and P. L. Angermeier. 1995. Assessing conservation value using centers of population density. Conservation Biology 9:1518–1527.

Winter, B. D., and R. M. Hughes. 1997. Biodiversity. Fisheries 22(1):22–29.

Wydoski, R. S., and J. Hamill. 1991. Evolution of a cooperative recovery program for endangered fishes in the upper Colorado River basin. Pages 123–139 in Minckley and Deacon (1991).

Yaffee, S. L. 1982. Prohibitive policy: implementing the federal Endangered Species Act. MIT Press, Cambridge, Massachusetts.

Chapter 17

Managing Fisheries with Regulations

RICHARD L. NOBLE AND T. WAYNE JONES

17.1 INTRODUCTION

To meet defined objectives, successful fisheries management must incorporate the use of appropriate and effective regulations along with other management tools. Although we generally think of regulations as being imposed upon the angler, they are broadly applicable to anyone who might influence fisheries, whether directly or indirectly.

Today's managers must be able to identify fisheries that will respond to either more restrictive or more liberal regulations to meet public expectations. In identifying fisheries that could benefit from regulations, managers must consider ecological data, angler opinions, enforcement ability, judicial systems, and economic impacts of regulations.

Regulations are tools to be used in conjunction with other management practices, such as population, community, and habitat manipulation. Regulations may appear to be an appealing management tool because, in comparison with other management approaches, they seem relatively inexpensive to implement. Unlike other management approaches, however, publicity and enforcement can become major costs. Recent studies have shown that lack of knowledge and understanding of regulations contributes to noncompliance and that noncompliance can have significant effects on achievement of management objectives (Gigliotti and Taylor 1990; Schill and Kline 1995).

17.1.1 Regulatory Authority

Responsibility for regulating fisheries in public waters rests primarily with state or provincial fisheries agencies. However, other agencies may have authority over habitat protection or alterations that are critical to fish communities. Consequently, a major responsibility of fisheries agencies is to review proposed regulations of other agencies and comment on applications for permits by potential resource users other than anglers. Depending on locale, fisheries on private property—particularly small impoundments—may be exempt from agency regulations. If so, owners and managers are free to establish their own regulations to achieve specific objectives.

The role of the federal government is typically limited to managing endangered species, anadromous fishes, federal waters, and international fisheries. Much of the management of the Great Lakes is undertaken through an international commission that coordinates regulations among the bordering states and provinces (see Chapter 23).

17.2 OBJECTIVES ATTAINED THROUGH REGULATIONS

Generally the goal of regulations is to protect or enhance a fishery for the benefit of the users. Consequently, regulations may be implemented for biological, ecological, sociological, and economic reasons. Sound management should be based upon a strong scientific base, and regulations may be used to obtain information that strengthens that base. In any case, regulations should be enacted to meet established management goals through specific measurable objectives (Figure 17.1; see Chapter 2).

17.2.1 Biological and Ecological Objectives

Since the late nineteenth century, harvest regulations in the United States have been used to protect fish populations from overexploitation and to distribute the catch among anglers. As fishing pressure increased, stocks typically declined and fishing success decreased. Therefore, regulations became increasingly stringent on the assumption that reduced harvest would prevent overfishing. Concern that fishing was removing too many adult fish, thereby reducing the spawning population to inadequate levels, frequently was the justification for protection of spawners, particularly during the spawning period when they might be more vulnerable to capture.

As the dynamics of fish populations and communities have become better understood, regulations have been seen as a means of enhancing, as well as protecting, stocks. Fishing can be regulated to adjust the size composition of stocks so that more fish are in the desirable size range. Our understanding of predator–prey systems has improved to the point that we can use regulations to manipulate predator size and abundance, thereby precipitating a top-down or trophic cascade effect (Carpenter et al. 1985) on prey populations and sometimes upon populations of undesirable species (Chapter 15).

The productivity of fisheries depends on quality of habitat. Although most regulations directed at the user have little effect upon the habitat, a series of regulations, based primarily upon fishery requirements, have been promulgated for protection of aquatic habitat. These regulations primarily affect agricultural, industrial, and municipal users of surface waters and watersheds. Habitat protection or enhancement may be the single most important regulatory process with which the professional fisheries manager will deal.

17.2.2 Sociological Objectives

Sociological objectives are primarily aimed at providing the resource user—typically the angler—with a valid expectation for good fishing success. In most inland waters, the fishery resource is preferentially reserved for recreational fishing; consequently, sociological fishery objectives tend to be qualitative. Experiences are measured by satisfaction rather than catch rates or value of catch; therefore, regulations may be aimed at achieving optimal sustainable yield of inland recreational fisheries by incorporating biological, sociological, and economic considerations (Anderson 1975). There is generally an excess of demands upon the fishery resource and a variety of types of anglers to satisfy. Regulations can be used to divide the resource among users; without such regulations a few anglers harvest most of the allowable catch. Regula-

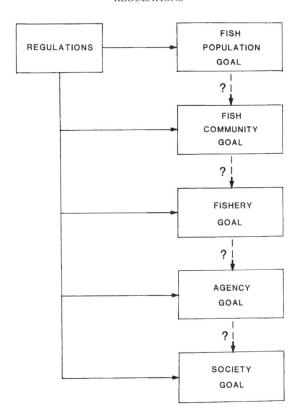

Figure 17.1 Fishery regulations should be aimed at achieving established management goals. Figure is modified from Noble (1986a).

tions may also be used to provide options for anglers according to their desired experiences. Such regulations reduce conflicts among user groups while simultaneously protecting or enhancing the aesthetic qualities of the fishing experience. Finally, it may be necessary to protect the anglers from themselves. In the fervor of the quest for fishing success, anglers may continue their pursuit even though the circumstances may be precarious.

The professional biologist has traditionally been trained to deal with population dynamics and community ecology and is readily prepared to address ecological objectives. Today's fisheries manager must also develop an appreciation for sociological objectives. In instances in which anglers are not the sole users of a resource, sociological objectives are likely to become complex and broadly inclusive.

17.2.3 Informational Objectives

A rather ubiquitous regulation requires that most anglers have a license to fish, at least in public waters. People involved with the sale and transport of fishes are also typically licensed. Although sale of such licenses has economic justification, the angler statistics generated by such sales are equally important. In many states, recreational fisheries management programs are supported primarily with federal aid funds.

Statistics are used to calculate federal aid allotments, which are allocated partially on the basis of number of licensed anglers. Lists of licensed anglers can define human populations to be surveyed for fishing effort and harvest. To replace costly creel census and fish population sampling in some closely monitored or experimental fisheries, a special regulation may require that each angler report all fish caught or harvested as well as information on the size of each fish.

The exceptions to licensing create severe limitations on the validity of statistics generated. Exemptions for youth, senior citizens, Native American anglers, and local residents differ among localities, making statistical comparisons among jurisdictions and over time difficult. Nevertheless, statistics based on licensed anglers can provide important trends that can be useful in evaluating changes, such as effects of license fee increases.

17.3   TYPES OF REGULATIONS

Many types of regulations exist, many of which are unfamiliar to biologists entering the profession. Regulations that affect harvest are usually emphasized. Ability to use regulations effectively depends upon understanding fish population dynamics and how populations can be expected to respond to particular regulations under existing habitat conditions to meet fishery objectives.

17.3.1   Licenses and Permits

Licensing varies with the issuing agency but generally includes basic fishing rights for a period of time. The sale of annual licenses provides revenue for the agency as well as a database for angler surveys. Additional privilege licenses or stamps may be issued for certain bodies of water or species. The sale of special licenses pays for specific programs that are too expensive for an agency to provide from the sale of basic licenses. In most instances the cost of special licenses is considered a user fee, and proceeds are applied to the specific resource whose use generates the fees. For example, trout stamp revenues may be dedicated to supporting trout hatchery operations. Some special fees are short term, such as a one-time surcharge to pay for hatchery construction or to acquire angler easements.

Most agencies issue short-term licenses to both residents and nonresidents. Short-term licenses provide angling opportunities for users who wish to participate only occasionally. Short-term licenses add revenue to an agency and provide it information about visitation by nonresidents, particularly tourists. In some cases, daily permits are available. Daily stocking rates for intensive put-and-take fisheries, such as in Virginia's Big Tumbling Creek trout program, may be determined on the basis of daily permits sold.

In most situations, the privilege of fishing is available to all that are willing to obtain proper licenses and abide by established regulations. However, limited entry— the limitation of the total number of participants in a fishery—has occasionally been applied in inland fisheries. Where commercial fishing is allowed in inland fisheries, the number of commercial participants may be limited to encourage profitability. Recently, the opportunities for application of limited entry to inland recreational fisheries, for purposes of improving the quality of recreational experiences and increasing catch rates, have received attention (e.g., Griffith 1989; Luebke and Betsill 1999).

Fisheries biologists are subject to regulations that govern their work. Scientific collecting permits are usually required of teachers and researchers who use special equipment to collect fishes or who collect fishes protected by law. Detailed reporting is required so that the numbers of each species that are removed for scientific purposes are documented.

17.3.2 Size Limits

Historically, minimum size limits were implemented to prevent overharvest and depletion of fish stocks and were frequently set to protect juvenile fish until maturity. Today a variety of size limits, including both minimum and maximum size limits as well as protected size ranges, are employed to maintain favorable fish populations, community structure, and quality of fishing.

Minimum size limits, which prohibit harvest of fish below some specified length, are generally imposed to lower both angling and total mortality in highly vulnerable populations and to reduce exploitation of fish before they reach sexual maturity. For example, Hunt (1970) stated that a minimum size limit for brook trout, if wisely applied, is the best single regulation for preventing excessive harvest. Clark et al. (1981) demonstrated that as size limits increased, the following general relationships occurred for quality trout fisheries: (1) harvest of trout in terms of numbers and weight decreased, (2) catch and release of trout in numbers and weight increased, (3) total catch in numbers and total yield in weight increased, and (4) numbers of large trout harvested increased. In contrast, Austen and Orth (1988) found that a 305-mm minimum size limit on smallmouth bass in Virginia provided no benefits, apparently due to high exploitation of new recruits, slow growth, and illegal harvest of sublegal-size fish. Minimum size limits have been effective in many fisheries in which harvest is high or recruitment is low. Wilde (1997), who intensively reviewed studies of largemouth bass minimum size limits, concluded that minimum size limits increased abundance and catch rates but did not adjust size distribution and harvest. In some heavily fished, productive waters with minimum size limits, however, the size composition may shift to a modal size just below the minimum limit, a situation termed stockpiling (e.g., Johnson and Anderson 1974). Intraspecific competition then can cause stunting and reduced recruitment to legal size. In waters where harvest has little effect on total mortality, minimum size limits may be of little value.

Maximum size limits are rare, but in situations with relatively few sexually mature adults or where large numbers of smaller fish exist and the manager intends to increase growth rates, maximum size limits can be beneficial. The Yellowstone National Park's cutthroat trout and Idaho's Snake River sturgeon are two examples for which maximum size limits have been imposed. Care must be taken to clarify implementation of maximum size in areas where minimum size limits have been promoted previously. A proposal for a maximum size limit may be misunderstood by constituents, law enforcement officials, and the judicial system. Consider the problem of an agency that for years has promoted minimum size limits for a species and then imposes a maximum size limit to protect a dwindling stock of spawning fish. Although this change may have biological validity, judicial systems may be reluctant to convict violators, who simply may have misunderstood the regulation change.

Slot limits are regulations that prohibit harvest from a designated intermediate size range (Anderson 1980). Harvest of small fish below the protected slot may take advantage of a surplus of recruits and allow additional energy to be channeled into midsize fish, which then survive and grow out of the protected range and into the harvestable size range. In addition to the increased harvest of large fish, the fishery provides harvest of some relatively young fish and substantial catch and release of protected fish. Wilde (1997), in his review of the responses of largemouth bass fisheries to size limits, found that protected slots increased population size and restructured size distributions but did not consistently increase catch rates or harvest. For slot limits to be successful, harvest of the surplus of young must be high enough to sustain the desired size structure of the population (Figure 17.2). The amount of harvest needed depends upon the body of water and the species' vulnerability to angling, longevity, and rates of reproduction and recruitment. Satisfactory growth rates must be sustained so that fish will not accumulate in the slot range.

Inverse slot limits, that is, establishment of a harvest slot rather than a protected slot, combine a minimum and a maximum size limit, thereby protecting small fish to recruitment and large fish for reproduction and attainment of trophy size. An example of the successful use of this approach is the white sturgeon fishery in the lower Columbia River. Recreational and commercial fisheries for this long-lived species are subject to differing minimum size limits, 91 and 122 cm, respectively, and both are restricted from harvesting fish over 183 cm. Combined with other regulations on the commercial fishery, these harvest regulations are credited with protecting, restoring, and stabilizing the white sturgeon fishery (Galbreath 1985).

Minimum size limits may be combined with protected slot regulations and may be associated additionally with limits on the numbers of individuals allowed to be harvested. For example, North Carolina established a 178-mm minimum size limit for salmonids on a stream and imposed a protected slot of 254–406 mm, allowing only one fish of the four-fish creel (catch) to be harvested above the slot. Although designed to achieve a specific objective, this regulation was so complex and compliance so difficult that it was subsequently dropped. This experience illustrates that managers must be certain that there is backing from constituents and law enforcement officials. The general fishing public is often unwilling to remove small fish and is inclined not to return larger fish within the slot size. A complex regulation has a reduced chance of accomplishing its objectives without an effective education program and high angler compliance.

17.3.3   Creel Limits

Creel limits are usually defined as the allowable fish, either in number or weight, that can be harvested during a particular time period. Generally this limit is in the form of number of fish per day. Creel limits may be established for individual species or for species in aggregate, such as salmonids, black bass, crappies, and sunfishes. Creel limits are instituted so that harvest may be more equitably divided; they have an additional sociological value in that they give the angler a specific target and the satisfaction of catching a limit. Although creel limits sometimes may be justified as a means of preventing overexploitation, they are usually not effective in reducing total exploita-

Black Bass Slot Limit

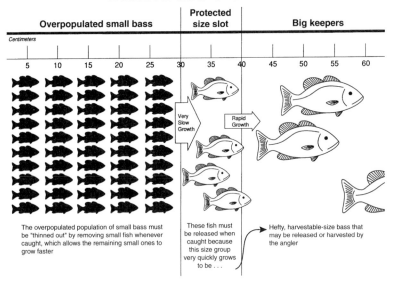

The overpopulated population of small bass must be "thinned out" by removing small fish whenever caught, which allows the remaining small ones to grow faster

These fish must be released when caught because this size group very quickly grows to be . . .

Hefty, harvestable-size bass that may be released or harvested by the angler

Figure 17.2   Managers should attempt to educate anglers about slot limits, which are often difficult to understand.

tion because only a small proportion of anglers actually harvest their limit regularly. In fisheries characterized by limits being regularly attained, managers are increasingly concerned about "high grading," that is, the release of small fish when larger fish are captured after the creel limit has been reached. Prohibition of high grading in paddlefish snag fisheries is an example of steps taken by many states to regulate this practice.

Creel limits are readily understood by constituents, law enforcement officers, and judicial systems. Daily creel limits have a high degree of voluntary compliance and are easily enforced. Agencies that have personnel and monies available to monitor creel limits on specific bodies of water have a distinct advantage in managing fish populations.

Possession limits, usually one or two times the daily creel limit, are established at the request of law enforcement officials. Possession limits aid enforcement of daily creel limits, for example, when anglers fish all night. Creel limits usually require fish to be kept in a manner so that number, species, and length (when applicable) can be verified.

Seasonal or annual creel limits may be developed to limit total harvest by individual anglers, but such regulations are ineffective unless checking stations are available to record catches. Due to lack of enforcement, voluntary compliance is usually nonexistent, and such regulations only serve as additional red tape.

Catch-and-release regulations have widespread applicability and popularity (Barnhart and Roelofs 1977). The principal objective of these regulations is to reduce fishing mortality, thereby maintaining high catch rates and increasing the catch of larger fish. Achieving this objective depends upon minimization of hooking mortality, so some gear restrictions may be incorporated Muoneke and Childress (1994), in a review of hooking mortality studies in the United States and Canada, described the great variation in posthooking mortality among taxa. This variation indicates the variable applicability of catch and release to the spectrum of fisheries being managed with such regulations. The greatest increase in adoption of catch-and-release management generally has been for those species with lowest hooking mortality (Quinn 1996). Catch and release may sometimes be prohibited because of high hooking mortality. Because anglers must fish only for fun, catch-and-release regulations have a sociological component. Participants in catch-and-release fisheries must be satisfied with fish caught rather than fish creeled. Catch-and-release regulations may be enacted on waters where contaminant concentrations make the fishes inedible. Rather than completely closing such waters to fishing, agencies post health advisories and keep the water open to catch-and-release fishing, thereby providing angling opportunities on what otherwise might be a lost resource.

The ideal catch regulation is one that incorporates knowledge of the population dynamics of the target species and controls total harvest by fish size-group. Such a regulation is termed a quota, which is typically applied on an annual basis. After the established annual quota has been harvested, the fishery for that species is closed to further harvest. Unfortunately, most agencies do not possess the necessary personnel to collect up-to-date harvest information from which to determine when a quota has been reached. It is also difficult to inform the public of closure. In North Carolina's Roanoke River striped bass fishery, a combination of quota and protected slot is used effectively to limit total harvest. Because the quota is reached during the early part of the spawning run, closure of harvest also differentially protects females which tend to migrate later than do males. On private waters with controlled access, an annual quota is readily applicable and can be effective in preventing overfishing. For vulnerable species, an annual quota can be filled quickly. For example, Ming (1974) reported that annual quotas of largemouth bass in a 3.8-ha public fishing lake were taken in an average of 4.2 d.

17.3.4   Seasons

Although underexploited fisheries are often managed with continuous open seasons, many agencies close seasons during peak spawning activities to protect spawning fishes. Closure during this period can have the additional effect of conserving the limited stock for later harvest by a larger clientele. If timing of spawning runs is predictable, it may be possible to open the season after the spawning is well under way. This timing allows sufficient reproduction to sustain the population while still giving anglers access to a readily available resource. Such an approach has been used successfully for rainbow trout runs from the New York Finger Lakes, where closure through March protects the early spawners. Their reproduction is adequate to produce enough fingerlings for the limited nursery habitat.

Seasonal partitioning of regulations has also been used, particularly on trout streams. Restrictive early-season regulations, such as catch and release, are eventually liberalized to allow harvest. This delayed harvest approach to management allows some protection of the population during a period of high growth and mortality and allows high exploitation later in the season when conditions may be more marginal for production.

Seasons are sometimes established for the safety of users. In high latitudes, fishing seasons are frequently closed on lakes and reservoirs during springtime ice breakup. This regulation helps keep people off the ice during this dangerous period. These closures may be species specific or pertain to only certain areas.

Diel seasons, that is, closed hours, may also have some justification on the basis of protecting the resource and the user. More likely, however, the justification is enforceability. If fishing is closed during certain hours, such as at night, any suspicious activity during that period can more easily draw the attention of enforcement officers.

Although the purpose is not generally recognized, season closures for fisheries are sometimes established to divert enforcement personnel to monitor traditionally high periods of hunting activity. On the surface this may appear unreasonable, but if adequate protection cannot be guaranteed for the fishery resource, closure may be the only logical solution to the problem.

17.3.5   Closed Areas

Areas are sometimes closed to fishing, generally to protect the user. Fishing is often prohibited within a certain distance below dams to protect anglers from extreme turbulence and sudden releases of water. In some instances, waters have been closed to fishing to reduce human disturbance in areas where threatened or endangered plants and animals exist. Other areas are closed when contaminant accumulations make the fishes inedible. As previously discussed, a prohibition on harvest may make more sense than complete closure for the latter situation.

To a limited extent, areas may be closed to protect or enhance the resource (Hill and Shell 1975). During periods of fish concentration on spawning or nursery grounds, fish may be very vulnerable to harvest. Ontario has used seasonal sanctuaries to restrict anglers from black bass spawning areas for over 60 years (Cholmondeley 1994). In addition, some sanctuaries have been established in Ontario with year-round prohibition of any fishing to enhance largemouth bass populations. However, consequences other than the reduction of nesting-bass disturbance have not been evaluated.

Management of put-and-take fisheries may include closed areas where stocking is done. Exclusion of anglers from the immediate area of stocking prevents harvest of the concentrated fish and provides opportunity for a steady dispersal of fish into adjacent waters. Pennsylvania has used this method in some of its trout waters. This approach compromises maximization of catch but extends benefits to a great number of users.

It is questionable whether closure of an area is justified to protect a few species. If enforcement is adequate, a regulation prohibiting harvest of species of concern would appear to be more viable than closure of the area to all anglers.

Closure of areas may be the best tool to allow initial establishment of fish populations or reestablishment following reclamation. Closure will allow a fish community to develop and subsequently to be managed effectively. Fishing during the initial establishment and growth period, when fish are particularly vulnerable to angling, could upset the community balance and preclude sustained management success.

When managing waters that are open to both recreational and commercial fishing, segregation of the two types of users can be attained by restricting one or both to certain areas. Such zoning can be an effective approach to minimizing conflicts between users, although segregation itself is unlikely to have any direct effects on the resource.

17.3.6   Gear Restrictions

Restrictions on gear are imposed to promote a diversity of fishing experiences, to decrease angling efficiency, and to reduce hooking mortality. Gear restrictions, unless used to decrease mortality, should only be considered to meet social, as opposed to biological, objectives in sportfishing. If proper size limits, creel limits, and seasons are established, most restrictions on conventional recreational fishing gears are not usually necessary for biological purposes. However, great controversy exists among users and managers alike as to whether use of certain gears should be considered to be sporting.

Although gear restrictions are used to decrease user conflicts and promote a diversity of angling opportunities, this approach carries some liability. Implementation of new gear restrictions usually creates a great furor. On the surface it would appear that gear types that result in increased mortality or overexploitation should be regulated. Consider, however, individuals who have been using these gear types for a number of years and suddenly discover that their favorite fishing spots are off limits unless they change gear. Anticipated public resistance does not mean that managers should not consider gear restrictions. A manager has the responsibility to meet social as well as biological objectives. The important point is to realize the difference between the two.

Fly-fishing is a form of angling that has gained widespread popularity in recent years. Because of some special requirements of fly-fishing, especially adequate spatial segregation from other anglers, some areas are designed for fly-fishing only (Figure 17.3). This type of regulation must be considered social in nature and provides no biological gain (Latta 1973). Public demand may be sufficient to warrant gear restriction; however, such a regulation should be weighed against diversification of fishing to determine whether angling opportunity would increase or decrease (e.g., see Wright 1992).

Gear restrictions are also applicable when a majority of fish are to be released after capture, such as when slot limits, minimum size limits, or catch-and-release regulations apply. Single, barbless hooks are frequently required on the premise that minimal physical damage is done to the fish, thereby increasing their chances of survival. Multiple points on hooks (e.g., treble hooks) may also be prohibited in fisheries where high catch rates of protected fish are anticipated. The value of such regulation should not be overestimated. Studies (see Muoneke and Childress 1994) have shown that hooking mortality is minimal unless fish are deeply hooked, such as results from use of single, baited hooks.

Number of hooks per line and number of lines tended by each angler are commonly limited. Likewise, regulations may require continuous tending of lines. These regulations reduce losses of hooked fish and are consistent with an ethic of not wasting or maiming our fishery resources.

Restriction on types of bait can also be considered a form of gear regulation. Bait restrictions are implemented to reduce hooking mortality, to decrease angling efficiency, and to prevent contamination of the habitat by unwanted species. A widespread bait restriction to reduce vulnerability is prohibition of the use of salmon eggs. During spawning runs this bait is so effective for salmonids that its use is considered to give

Figure 17.3 Designated areas for specialized fishing can add to the quality of the fishing experience.

anglers an unfair advantage. Use of exotic and otherwise undesirable fishes is prohibited to prevent their spread into new areas. Despite the effectiveness of species such as common carp, goldfish, blue tilapia, and sea lamprey as live bait, their use can only contribute to further expansion of the ranges of these potentially deleterious fishes. For different reasons, use of game fishes or other recreational species as bait is normally prohibited. Although it is conceivable that collection of sport fishes for bait could have detrimental effects on stocks, their prohibition as bait makes enforcement of creel and size limits much more explicit.

Liberalization of gear restrictions may be used to adjust community composition. Although gear restrictions are frequently applied to recreational fisheries to ensure inefficiency, efficient gear may be used to harvest undesirable species selectively. In North Carolina, where the introduced flathead catfish has reduced other riverine sport fish stocks through predation, local laws allow electrofishing for flathead catfish. Catfishes are extremely susceptible to certain forms of electricity, and their populations can be electrofished recreationally without significant bycatch.

Inland fisheries are primarily reserved for recreational rather than commercial use. When commercial fishing is allowed, regulations carefully limit the gears and harvest (Fritz and Wight 1986). If nets are used, mesh sizes are regulated to minimize catches of sport fishes. Agencies sometimes contract with commercial fishers to harvest underexploited species on a limited-entry basis. Some agencies have gone exclusively to such programs for commercial fishing on inland waters. Although studies have shown that the incidental catch of sport fishes in commercial catches is negligible, public reaction is typically negative if commercial fishers catch game fishes.

17.3.7   Reporting

Total catch and effort data are valuable to fisheries managers and can be provided by the angler. Nevertheless, regulations that mandate anglers to keep catch and effort records are tenuous at best. No matter what the objective, the effectiveness of such a regulation depends upon voluntary compliance unless the agency has the personnel and funds for enforcement. For example, all Atlantic salmon anglers in the Canadian provinces are required to report catch and effort information. To obtain complete information, the Canadian authorities send follow-up letters to those anglers who did not report their catch as required. Even if total response is obtained, the accuracy of the information relies ultimately upon the integrity of the angler.

Record-keeping regulations appear to have their greatest effectiveness for waters with a single point of access. At that location, the value of total catch records can be publicized to encourage accurate reporting. Reporting forms should be readily available, and a secure, well-marked depository for reports should be established. Single-access points that have limited access hours also can be staffed, either by a creel clerk or, as is frequently the case with single-access sites, by a concessionaire.

In some fisheries, record keeping in the form of diaries kept by selected anglers may provide valuable information on annual trends in size composition and catch rates. Such "avid angler" records have the benefit of measuring changes in the fish population based on data from a fixed subset of anglers who use similar fishing techniques each year. Integrity of these anglers can be anticipated to be such that more reliable data are acquired from them than can be expected from the general public. Individual record keeping has been used frequently by private pond owners and fishing clubs and can be a cost-effective method of assessment. It is important to communicate regularly with the record keepers to show them the results of their efforts and to reinforce the value of their data.

Throughout most of the United States, procedures for voluntary registration of state record fishes have been established. Additional citation programs for trophy fish are also common (Quinn 1987). Regulations governing the procedures to follow when submitting an application for a state record fish or a citation fish are usually simple. Most agencies require identification of the fish by a professional and verification of weight on certified scales. This type of regulation is a service to the angler; standardization is designed to control cheating. In addition, such programs may provide valuable information to the fisheries agency.

17.3.8   Regulation of Commercial Ventures

Because inland fisheries are primarily recreational, commercial fisheries are highly regulated, if not totally prohibited, by fisheries agencies. Gear restrictions are set to minimize catch of sport fishes; size limits, season closures, and designated areas are also likely to be stringently regulated. A regulation that prevents sale of sport fish species, whether caught by recreational or commercial fishing, is frequently implemented to keep recreational fishing from becoming commercial and to limit commercial fishing to nonsport fishes.

Federal treaty rights for Native Americans frequently exempt tribal fisheries from agency regulations. Commercial harvest from fisheries that are otherwise designated by agencies as recreational has been a source of conflict between recreational anglers and Native Americans. This is a good illustration of the conflict that can arise when sociological, economic, and ecological objectives are incompatible.

Commercialization of fish production (aquaculture) and recreational fishing has occurred because of the economic values of fishes beyond those associated with traditional commercial fishing. To avoid conflicts with conventional fisheries and to facilitate better enforcement, agencies may develop specific regulations on commercial activities such as fishing tournaments and aquaculture.

One aspect of the commercialization of sportfishing has resulted in what can be described as tournament mania. Competitive fishing has spurred the growth of organizations that conduct regular series of tournaments with high stakes, which have ranged from large cash prizes to expensive fishing equipment, accessories, and lucrative product endorsements. Private businesses and civic organizations have used fishing tournaments as promotional events. The debate over the effects of fishing tournaments on fish communities is ongoing. Some agencies have enacted regulations that require permits for tournaments, but most agencies do not regulate tournaments in any fashion (Duttweiler 1985). However, size and creel limits imposed on certain waters have been so restrictive that tournament organizers have sought exemption for tournaments on the basis of tournaments' stringent, self-imposed policies on release of all fish. Both California and Florida now allow such exemptions for bass tournaments. Many well-organized tournament sponsors develop and impose their own regulations governing handling of fish and release of fish after check-in. The intent of such regulations is good; however, only the highly organized tournaments have such regulations. Tournaments between local clubs and among groups of anglers typically have few self-imposed restrictions. Because these tournaments are so numerous, spontaneously organized, and rarely publicized, their regulation by agencies is impractical. Until a feasible method is found to regulate this type of tournament, the effectiveness of tournament regulations is in doubt. Without regulation, however, this commercial use of public fisheries resources will continue to be a source of conflict with the general sportfishing public.

Most agencies require commercial production and distribution operations to be licensed. Licensing provides information regarding the number of hatcheries and the species raised and allows monitoring to prevent diseases and undesirable species from entering local fisheries. Generally, regulations prohibit the sale, transport, and release of any undesirable species, nonindigenous strains, or any fish not certified to be free of pathogens. Further regulations govern quantity and quality of water intake and release by production facilities. As demand for stocking private waters for recreation has grown, private production and distribution of sport fishes has increased. Licensing of commercial operations that deal with sport fishes minimizes enforcement problems related to discrepancies in possession and size limits otherwise applicable to anglers. Some complications have arisen as a result of bans on the sale of sport fishes where the primary intent was to protect wild stocks. When aquaculture enterprises have gotten into production and sale of sport fish species, either for stocking private waters or as food fish, special provisions have become necessary.

17.3.9   Regulation Evaluation

The effects of changing regulations should always be evaluated. The extent of evaluation may vary from a complex study to simply polling law enforcement officers or constituents on the effectiveness of the regulation. The results of evaluations, whether positive or negative, should be made available to the public. Communication of regulatory successes can increase angler appreciation of the utility of such regulations, and reinforce confidence in the procedures followed in planning and implementing the changes. The manager should have valid data for periods before and after implementation to evaluate the effectiveness of the change. Because results may not be immediate, postimplementation data should be collected over an ample time period for effects to be detected.

Regulations should always be evaluated relative to the specific biological, sociological, informational, or economic objectives that were established (see Chapter 2). Evaluation methodologies should be developed and incorporated into the initial action plan when the regulatory action is designed. Too often an agency's decision to evaluate is an afterthought; consequently there is likely to be little comparative data collected before implementation of the new regulation.

17.4   REGULATIONS FOR SPECIFIC FISHERIES

Early in the twentieth century as more governmental authorities became established, the number and complexity of regulations increased. With this expansion came a shift from specific regulations to more general regulations enacted to cover entire systems. This generic approach had two advantages: it was easily administered by agencies and easily understood by constituents. However, failure to tailor regulations to specific situations and objectives often led to resources being managed for mediocrity. For some bodies of water, generalized regulations would be too restrictive and negative results would be obtained. Consider a statewide minimum size limit imposed on salmonids in infertile waters where density is high and growth rates are slow—such a regulation would essentially prevent harvest.

Eschmeyer's (1945) study of Norris Reservoir provided the impetus toward more liberal regulations. Agencies began removing closed seasons on reservoirs and dropping restrictions on other types of water (Fox 1975; Redmond 1986). This trend continued into the early 1960s, when acquisition of data on fish populations and harvest substantially improved due to widespread use of assessment techniques such as electrofishing, tagging, and creel census. Subsequent experimentation with slot limits, higher minimum size limits, and more restrictive creel limits led to the imposition of water-specific and species-specific regulations.

17.4.1   Coldwater Fisheries

Regulations are particularly applicable to oligotrophic, cold waters, where even moderate fishing effort can lead to exploitation that exceeds annual surplus production. Consequently, we see the most widespread use of regulations in such habitats. Due to low fertility and cold temperatures, growth rates are slow, thereby extending the time to maturation and recruitment to the fishery. Furthermore, low productivity and standing crops of fishes make their populations vulnerable to the effects of fishing. Therefore, strict regulations are frequently promulgated to protect highly accessible coldwater fisheries from overexploitation. However, recognizing that many coldwater streams are in remote mountainous areas that are not readily accessible to anglers, some agencies have liberalized regulations for such areas.

Management of coldwater streams that are marginal for trout or are subject to excessive fishing pressure frequently includes put-and-take or put-grow-and-take stocking, which may require management through regulations. In these cases of more intensive management, regulations are usually set with the objectives of maximizing catch and dividing the resource among users.

It is also in cold waters that much emphasis has been placed upon the diversification of fishing experiences. Therefore, regulations are frequently established for sociological needs. Catch and release and fly-fishing only are commonly used regulations in these waters.

Coldwater habitats are so sensitive to disturbance that emphasis is placed on strict regulations governing habitat degradation and modification. In lakes, nutrient discharge regulations are imposed to minimize eutrophication, which could easily upset the fragile food chains. In streams, regulations emphasize minimal instream disturbances of flow and substrate, as well as limited riparian disturbances by forestry practices, road building, and livestock grazing (see Chapter 9).

17.4.2 Coolwater and Warmwater Streams and Rivers

With some exceptions, coolwater and warmwater streams are seldom overexploited as recreational fishing resources due to limitations in access. Consequently, regulations to protect or divide the resources are unnecessary. Large waters are often opened to commercial fishing, even for commercial harvest of sport fishes, and regulations are apt to be imposed to minimize conflicts.

One exception to liberalized regulations is the smallmouth bass fishery in streams. Much like trout, smallmouth bass have fairly stringent habitat requirements. As a result, a variety of size and catch limits are imposed for this species. Minimum size limits have successfully increased catch rates and harvests in some situations (Fajen 1981). However, because of slow growth rates in some areas, regulations have often been counterproductive except for providing increased numbers of strikes due to release rather than harvest of fish.

Another exception to liberal regulations in stream fisheries applies to anadromous stocks such as striped bass. Until recently, liberalized regulations applied to anadromous fishes in inland waters; the spawning runs allowed an opportunity for inland exploitation of a perceived inexhaustible, ocean-produced resource. However, most fisheries have declined due to degradation of water quality, spawning grounds, and migration routes and overharvest of stocks before becoming riverine. Stringent harvest regulations are now in place for most anadromous fisheries; these apply to both recreational and commercial users, although allocation of the limited resource usually is controversial. Included in this controversy are the treaty rights of Native Americans to exploit certain fisheries. Although these treaties are not subject to limitation by state and provincial agencies, local efforts typically emphasize making tribal harvest compatible with state agency regulations.

Decisions on habitat regulations commonly confront the stream fisheries manager. Instream industrial activities, such as gravel dredging or diversion of irrigation waters, are subject to regulation on the basis of fisheries and water quality. With the widespread construction of dams on rivers and streams, instream flow requirements have increasingly become recognized as a basis for regulation of consumptive users of flowing water resources. Although agencies other than the state or provincial fisheries management agency will likely be responsible for promulgation of habitat regulations, the biologist must be prepared to contribute objectively and quantitatively to development of such regulations on the basis of impacts upon fisheries.

17.4.3   Coolwater and Warmwater Lakes

Natural lakes with moderate productivity, which characterizes most coolwater and warmwater lakes, have highly evolved and complex communities. Consequently, the principal approach to management of these fisheries has been through regulations. A common objective is to prevent overharvest of game fishes, particularly top carnivores. However, because eutrophication, introduced or invading species, and overharvest have altered community dynamics, wide annual fluctuations in harvest have become more common. These fluctuations are usually associated with erratic recruitment due to variations in year-class strength. Effective use of regulations under such conditions is a challenge to the fisheries manager; although restrictive regulations may be desirable to rebuild depleted stocks, flexibility to change regulations is needed to allow harvest of surpluses when they occur.

Although most management and regulatory attention in lake fisheries has been directed at game fishes, effort is also directed at panfish—yellow perch, crappies, and sunfishes (*Lepomis* spp.)—which frequently serve as prey for game fishes. Until recently, liberal harvest regulations were applied to these fishes. As fishing pressure has increased and data on exploitation rates have been obtained, the need for more restrictive harvest regulations has been recognized.

Because there may be more than one piscivore of interest to sport fisheries, it may be desirable to direct management at one species to optimize its fishery. In such cases, stringent regulations are likely to be imposed on the primary fishery while liberal regulations govern the secondary fishery. An example is New York's liberalized regulations on walleye in Chautauqua Lake, where primary interest is in a trophy muskellunge fishery (Mooradian et al. 1986).

17.4.4   Reservoirs

Reservoir management has been characterized by liberal harvest regulations (Redmond 1986). In contrast to natural lakes, reservoirs have poorly evolved predator–prey systems and frequently have large populations of potential prey such as gizzard shad (Noble 1986b). In addition, some reservoirs are subject to extreme habitat variations, such as water level fluctuations, that limit the effectiveness or applicability of harvest regulations as management tools. As fishing effort has increased, exploitation rates of some reservoir sport fishes have been excessive, especially during the early postimpoundment years and on smaller reservoirs. Recent years have brought more restrictive harvest regulations with the objective of reducing fishing mortality and improving size structure. In particular, size and catch limits have been implemented in reservoirs with limited recruitment in attempts to improve size structure and thereby provide higher quality fish to the creel. Missouri has implemented such regulations on crappie to accomplish this objective.

Because of their size and dendritic shape, reservoirs may offer an opportunity for achieving varied fisheries objectives through spatial within-reservoir zoning. Closed areas have been shown to serve as nurseries that can supply catchable-size fish to open areas while preventing lakewide overfishing. Likewise, it should be possible to apply different regulations to different areas, thereby creating differing size structures and varied fishing experiences. Occasionally lakes have been zoned to achieve specific management objectives. An example is Missouri's Table Rock Lake, where one arm of the reservoir is managed with a more restrictive regulation for crappies. Such an approach is applicable only to relatively sedentary species (Noble et al. 1994).

Although many reservoirs have been built for hydropower generation, fish habitat regulations are applicable to such reservoirs under the U.S. Federal Power Act. Consequently, shoreline management plans are developed to protect natural habitat, provide public access, and minimize erosion. The plan may encompass regulations on dredging for access, placement of piers, docks, and boathouses, and specifications for bulkheads. Unfortunately, enforcement of these regulations usually does not rest with the fishery management agencies but rather with the hydropower company, thereby requiring that agencies petition the Federal Energy Regulatory Commission if they believe more stringent enforcement is needed.

17.4.5 Private Ponds

Privately owned ponds are conducive to management with harvest regulations. Fish populations of small impoundments are particularly susceptible to harvest (e.g., Funk 1974) but small pond size, coupled with control over users, allows implementation of a variety and complexity of regulations. These can be closely integrated with other forms of management to meet well-defined objectives. Although enforcement through conventional agency personnel is limited, if at all applicable, the limited fishing constituency of almost any pond can be readily reached with educational programs and materials. Coupled with education, control over access ensures a high degree of compliance to self-imposed regulations (Ashley and Buff 1988). If regulations are developed with the objective of adjusting size structure of fish populations through fishing, adequate fishing pressure must be exerted. In large ponds with limited access, it may be necessary to promote a directed fishery to attain adequate fishing pressure. Likewise, total control of access and compliance with special regulations is necessary. A few uninvited guests or renegade members of a fishing club can quickly offset the accomplishments achieved through the dedicated efforts of the remaining anglers.

Increased demands for urban fishing opportunities have been addressed through a variety of programs (Allen 1984). Among these is intensive put-and-take fishing in small, closely controlled ponds. Although there is no need for regulations to protect the stocks under these highly artificial conditions, special regulations can serve important sociological purposes. Limits on gear can improve human safety and minimize injury of unharvested fish. Regulations may include sociological zoning, providing independent fishing opportunities for youth, senior citizens, physically impaired individuals, and the general public. Such highly structured regulations are unlikely to be possible in any other setting. Educational programs associated with urban fisheries can also communicate information about natural resource management to nontraditional anglers, who have a low awareness of regulations and their purposes (Schramm and Dennis 1993).

17.5   REGULATORY PROCESS

The regulatory process varies with each agency yet follows a generalized procedure. In response to an identified objective, a proposal is formulated. A subsequent period of review by the biological staff will include evaluation of effects of the regulation on the fish stock and the fishery and feasibility of implementation and enforcement (Figure 17.4). This review may involve a formal presentation or preparation of a written document. The next step varies by agency, but some period of public comment, either before or following evaluation by a board of commissioners or directors, is usually established. Depending upon the nature of the regulation, ratification by the state legislature may be necessary.

An important component of the regulatory process is the public comment period (Figure 17.5). Regulations, especially new or radically different ones, often meet with opposition regardless of their merit. The manager must be prepared to justify an idea to peers, the agency governing board, and constituents. In many instances, justifying a controversial regulation requires a carefully planned educational process in which the manager must play the key role. The manager must anticipate questions and concerns that will be expressed by the public and others so that sound, logical answers can be rapidly provided. Depending upon the degree to which the audience is informed about fisheries management, a number of different versions of the same presentation may be required. Passage of regulations in many instances depends more on the "selling" done by the biologist than on the validity or merit of the regulation (see Chapter 3).

The responsibility of today's professional does not end with passage of the regulation. Only by working with the public for acceptance and compliance with the regulation can the objective be met. Professionals cannot rely solely on law enforcement officers to ensure compliance with the enacted regulation. Acceptance and compliance evolve after passage, so the manager may need to continue to work for years toward that goal.

This brings us to one of the most complicated aspects of regulations—enacting unpopular regulations. Unpopular regulations have been proposed and enacted many times. In many instances, these regulations were necessary to protect the resource; however, cases exist where the regulations were of dubious merit. The following scenario will serve as an example.

In a statewide angler opinion survey, a large percentage of largemouth bass anglers indicate they would prefer to catch larger bass, even if it means a decrease in creel limits and an increase in size limits. A management agency having jurisdiction

Figure 17.4   Effective enforcement is critical to the success of fishing regulations. Photograph is courtesy of North Carolina Wildlife Resources Commission.

over a reservoir that meets all the biological requirements for a trophy bass fishery proposes a regulation to increase the minimum size limit and reduce the creel limit. This proposal meets with opposition from the anglers who regularly fish the reservoir and are satisfied with the status quo. Should that regulation be enacted? The professional will face issues like this continually. The answer is found only by evaluating each case on its own merits and with the knowledge that mistakes can and will be made (see Chapter 2).

17.6   THE FUTURE ROLE OF REGULATIONS

There can be little doubt that the role of regulations in fisheries management will increase in the future. As leisure time and human population size increase, we are faced with intensified fishing effort on a fixed inland aquatic resource. In addition, the efficiency of angling will continue to increase as anglers become more highly informed and more sophisticated methods are used. Large, powerful boats now propel many anglers to the most remote coves of our biggest lakes and reservoirs. Fish finders not only chart the bottom topography but also the distribution of fishes. To increase their probability of success, some anglers are monitoring physical conditions such as temperature, oxygen, pH, and color. Most recently, the use of underwater video equipment by anglers to locate trophy fishes has stimulated strong public sentiment for regulation of such highly effective technology. Without appropriate regulations, overharvest will

Figure 17.5 Public hearings offer managers the opportunity to learn the concerns of anglers and other resource constituents to proposed regulation changes and simultaneously to inform the public of anticipated benefits. Photograph is courtesy of North Carolina Wildlife Resources Commission.

occur on most of our major fisheries. Likewise, without appropriate regulations, the resource will be allocated inadvertently to the best-equipped individuals rather than divided among the various users.

Pressure on habitat will also continue to increase. The public sector demands water for many uses, not just recreation. Although legislation protects our water from pollution and our waterways from alteration more than ever before, water withdrawals will continue to become demanding influences on our fisheries. The fisheries manager must be able to provide objective inputs to decision making on the basis of impacts on fisheries. In particular, effects of water flow and water level fluctuations will need to be considered. With better knowledge and some imagination, fisheries managers may develop regulations that allow net benefits to fisheries from industrial and municipal use of water.

To meet the increased demand for fishing, in terms of quantity, quality, and diversity, it will be necessary to tailor regulations more closely to specific situations. That will mean abandonment of many regional regulations in favor of resource-specific regulations that are based upon factors such as fish population dynamics, characteristics of the habitat, proximity to population centers, alternative fishing opportunities in the area, angler preferences, and enforcement limitations (e.g., Brousseau and Armstrong 1987). The benefit of this resource-specific approach will be establishment of more precise management objectives and higher chances of attaining those objectives through integration of all available management tools. There is general resistance to site-specific regulations from the public and those responsible for enforcement, who understandably prefer the simplicity of standardized regulations. The major obstacle, however, is likely to be sufficient personnel who can adequately assess fisheries to determine management needs. Site-specific regulations already are common in highly populated states with popular trout fisheries. With the recent recognition that overexploitation of warmwater fisheries can commonly occur, and the increasing interest in providing a greater diversity and better quality of fishing, greater emphasis on site-specific regulations is justified. In those areas where regional standard regulations have been employed, public education will be an integral part of the development of site-specific regulations. Jakus et al. (1996) found that educational level, club membership, and income affected angler receptivity to new regulations, including those specific to individual fisheries. The uncertainties associated with moving toward site-specific management have recently prompted a position statement by the American Fisheries Society (AFS 1997).

Regardless of whether new regulations are being proposed or old ones defended, the fisheries manager will be more accountable. Justification for regulations will be measured increasingly by their anticipated effectiveness in the achievement of defined objectives in management plans, where regulations are incorporated with other management approaches. Although biologists must start from the standpoint of what is good for the resource, they must be cognizant of the fact that fisheries are often a part of a multiple-use resource. Therefore, public decisions are made according to the political process. Consequently, the manager will need to be able to justify and evaluate sociologically based as well as biologically based objectives. Furthermore, attention will need to be given to public education, thereby developing a better appreciation by the public of objectives, an appreciation that will, in turn, raise the level of manager accountability.

The process by which regulatory change is developed is through proper assessment of need and opportunity by the fisheries manager. The process by which implementation is effective is through understanding and compliance by the user and the ability of the law enforcement officers to enforce the change.

17.7 REFERENCES

AFS (American Fisheries Society, Resource Policy Committee). 1997. Special fishing regulations for managing freshwater sport fisheries. Pages 63–65 in Resource policy handbook, 1st edition. AFS, Bethesda, Maryland.

Allen, L. J., editor. 1984. Urban fishing symposium, proceedings. American Fisheries Society, Fisheries Management Section, and Fisheries Administrators Section, Bethesda, Maryland.

Anderson, R. O. 1975. Optimum sustainable yield in inland recreational fisheries management. Pages 29–38 in P. M. Roedel, editor. Optimum sustainable yield as a concept in fisheries management. American Fisheries Society, Special Publication 9, Bethesda, Maryland.

Anderson, R. O. 1980. The role of length limits in ecological management. Pages 41–45 in S. Gloss and B. Shupp, editors. Practical fisheries management: more with less in the 1980s. American Fisheries Society, New York Chapter, Ithaca.

Ashley, K. W., and B. Buff. 1988. Pond owner utilization of management recommendations made by state fishery personnel. Fisheries 13(6):12–14.

Austen, D. J., and D. J. Orth. 1988. Evaluation of a 305-mm minimum length limit for smallmouth bass in the New River, Virginia and West Virginia. North American Journal of Fisheries Management 8:231–239.

Barnhart, R. A., and T. D. Roelofs, editors. 1977. A national symposium on catch-and-release fishing. Humboldt State University, Arcata, California.

Brousseau, C. S., and E. R. Armstrong. 1987. The role of size limits in walleye management. Fisheries 12(1):2–5.

Carpenter, S. R., J. F. Kitchell, and J. R. Hodgson. 1985. Cascading trophic interactions and lake productivity. BioScience 35:634–639.

Cholmondeley, R. 1994. An historical review of minimum size limits and sanctuaries in Division 10. Pages 81–95 *in* S. J. Kerr and R. Cholmondeley, editors. Bass management in Ontario workshop proceedings, WP-004. Queen's University Biological Station, Ontario.

Clark, R. D., Jr., G. R. Alexander, and H. Gowing. 1981. A history and evaluation of regulations for brook trout and brown trout in Michigan streams. North American Journal of Fisheries Management 1:1–14.

Duttweiler, M. W. 1985. Status of competitive fishing in the United States: trends and state fisheries policies. Fisheries 10(5):5–7.

Eschmeyer, R. W. 1945. The Norris Lake fishing experiment. Tennessee Department of Conservation, Division of Fisheries, Nashville.

Fajen, O. F. 1981. Warmwater stream management with emphasis on bass streams in Missouri. Pages 252–265 *in* L. A. Krumholz, editor. The warmwater streams symposium. American Fisheries Society, Southern Division, Bethesda, Maryland.

Fox, A. C. 1975. Effects of traditional harvest regulations in bass populations and fishing. Pages 392–398 *in* H. Clepper, editor. Black bass biology and management. Sport Fishing Institute, Washington, D.C.

Fritz, A. W., and H. L. Wight. 1986. Commercial fishing as a reservoir management tool. Pages 196–202 *in* Hall and Van Den Avyle (1986).

Funk, J. L., editor. 1974. Symposium on overharvest and management of largemouth bass in small impoundments. American Fisheries Society, North Central Division, Special Publication 3, Bethesda, Maryland.

Galbreath, J. L. 1985. Status, life history, and management of Columbia River white sturgeon, *Acipenser transmontanus*. Environmental Biology of Fishes 14:119–125.

Gigliotti, L. M., and W. W. Taylor. 1990. The effect of illegal harvest on recreational fisheries. North American Journal of Fisheries Management 10:106–110.

Griffith, J. 1989. Is limited entry needed in catch-and-release trout fisheries? Pages 112–125 *in* R. A. Barnhart and T. D. Roelofs, editors. Catch-and-release fishing—a decade of experience. California Cooperative Fisheries Research Unit, Humboldt State University, Arcata.

Hall, G. E., and M. J. Van Den Avyle, editors. 1986. Reservoir fisheries management: strategies for the 80's. American Fisheries Society, Southern Division, Reservoir Committee, Bethesda, Maryland.

Hill, T. K., and E. W. Shell. 1975. Some effects of a sanctuary on an exploited fish population. Transactions of the American Fisheries Society 104:441–445.

Hunt, R. L. 1970. A compendium of research on angling regulations for brook trout conducted at Lawrence Creek, Wisconsin. Wisconsin Department of Natural Resources, Research Report 54, Madison.

Jakus, P. M., J. M. Fly, and J. L. Wilson. 1996. Explaining public support for fisheries management alternatives. North American Journal of Fisheries Management 16:41–48.

Johnson, D. L., and R. O. Anderson. 1974. Evaluation of a 12-inch length limit on largemouth bass in Philips Lake, 1966–1973. Pages 106–113 *in* Funk (1974).

Latta, W. C. 1973. The effects of a flies-only fishing regulation upon trout in the Pigeon River, Otsego County, Michigan. Michigan Department of Natural Resources, Research Report 1807, Ann Arbor.

Luebke, R. W., and R. K. Betsill. 1999. Limited entry in recreational fisheries—has its time come? Proceedings of the Annual Conference Southeastern Association of Fish and Wildlife Agencies 51(1997):184–191.

Ming, A. 1974. Regulation of largemouth bass harvest with a quota. Pages 39–53 *in* Funk (1974).

Mooradian, S. R., J. L. Forney, and M. D. Staggs. 1986. Response of muskellunge to establishment of walleye in Chautauqua Lake, New York. Pages 168–175 *in* G. E. Hall, editor. Managing muskies, a treatise on the biology and propagation of muskellunge in North American. American Fisheries Society, Special Publication 15, Bethesda, Maryland.

Muoneke, M. I., and W. M. Childress. 1994. Hooking mortality: a review for recreational fisheries. Reviews in Fisheries Science 2:123–156.

Noble, R. L. 1986a. Stocking criteria and goals for restoration and enhancement of warm-water and cool-water fisheries. Pages 139–159 *in* R. H. Stroud, editor. Fish culture in fisheries management. American Fisheries Society, Fish Culture Section and Fisheries Management Section, Bethesda, Maryland.

Noble, R. L. 1986b. Predator-prey interactions in reservoir communities. Pages 137–143 *in* Hall and Van Den Avyle (1986).

Noble, R. L., J. R. Jackson, E. R. Irwin, J. M. Phillips, and T. N. Churchill. 1994. Reservoirs as land-scapes: implications for stocking programs. Transactions of the North American Wildlife and Natural Resources Conference 59:281–288.

Quinn, S. P. 1987. The status and usefulness of angler recognition programs in the United States. Fisheries 12(2):10–16.

Quinn, S. P. 1996. Trends in regulatory and voluntary catch-and-release fishing. Pages 152–162 *in* L. E. Miranda and D. R. DeVries, editors. Multidimensional approaches to reservoir fisheries management. American Fisheries Society, Symposium 16, Bethesda, Maryland.

Redmond, L. C. 1986. The history and development of warmwater fish harvest regulations. Pages 186–195 *in* Hall and Van Den Avyle (1986).

Schill, D. J., and P. A. Kline. 1995. Use of random response to estimate angler noncompliance with fishing regulations. North American Journal of Fisheries Management 15:721–731.

Schramm, H. L., Jr., and J. A. Dennis. 1993. Characteristics and perceptions of users and nonusers of an urban fishery program in Lubbock, Texas. North American Journal of Fisheries Management 13:210–216.

Wilde, G. R. 1997. Largemouth bass fishery responses to length limits. Fisheries 22(6): 14–23.

Wright, S. 1992. Guidelines for selecting regulations to manage open-access fisheries for natural populations of anadromous and resident trout in stream habitats. North American Journal of Fisheries Management 12:517–527.

COMMON MANAGEMENT
PRACTICES

Chapter 18

Coldwater Streams

J. S. GRIFFITH

18.1 INTRODUCTION

Although coldwater streams come in a variety of sizes and shapes and hold numerous fish species, for the purposes of this chapter they will be defined as streams where game fish populations are predominantly salmonids. Furthermore, they are defined as streams that maintain salmonid populations by sustaining spawning and rearing rather than just serving as migratory pathways or receptacles for hatchery-produced fish.

Fluvial salmonids in North America may be divided into three categories.

1. Anadromous species that move out of streams and into the sea (chum salmon and pink salmon) or into lakes (sockeye salmon) immediately after emergence (see Chapter 24).

2. Anadromous species that spend 1–4 years in streams before moving to the sea. These include chinook salmon, coho salmon, and Atlantic salmon, as well as anadromous races of rainbow trout (steelhead), cutthroat trout, brown trout, brook trout, and Dolly Varden.

3. Resident species that spend their lives in streams, such as some populations of bull trout and resident races of rainbow trout, cutthroat trout, brown trout, and brook trout. Residents may either rear and spawn in a small stream segment or move periodically throughout a large river system.

All of these fishes, except brown trout, are native to North America. In addition to those mentioned above, other species, such as Arctic char and Arctic grayling, may be locally important. The mountain whitefish is abundant in western rivers but differs from other salmonids in life history characteristics and is not considered in detail in this chapter.

18.1.1 Characteristics of Coldwater Streams

Fish communities in coldwater streams are the product of a highly dynamic environment. From their primary headwaters through increasingly higher orders downstream, stream systems provide a continuous gradient of physical variables such as width, depth, velocity, and temperature. The river continuum concept (Vannote et al. 1980) has been proposed as a framework for describing the structure and function of communities along a river system and for describing community change in response to fluvial geomorphic processes (see Chapter 20).

Salmonids are present at predictable positions along a stream continuum. Their environmental requirements vary considerably among species and life stages, but a defined range of tolerable physical, chemical, and biological characteristics exists. Water temperature may be the primary factor dictating the lower limit of salmonid distribution in terms of elevation and latitude. Although some populations—like those in the Firehole River in Yellowstone National Park—may spend periods at 26°C, summer stream temperatures for most coldwater fishes do not exceed 22°C. The scope for activity (the difference between active metabolism and standard metabolism that indicates the amount of oxygen physiologically available for activity) is greatest for wild rainbow trout at 20°C and for hatchery cutthroat trout at 15°C. Growth for most salmonids rapidly declines above 20°C. Dissolved oxygen concentration, typically near saturation in unpolluted coldwater streams, should remain at a minimum of 8 mg/L for rearing and 10 mg/L for egg and larval development (Davis 1975).

Flow in coldwater streams may range from a few liters per second to hundreds of cubic meters per second. Although salmonids may be found in streams that range from first-order to sixth-order or larger, salmonid abundance and diversity in the Salmon River drainage, Idaho, was greatest in fourth- and fifth-order streams (Platts 1979). A similar relationship holds for stream gradient. Despite the fact that some salmonid populations may exist in streams with gradients exceeding 20%, a range of 0.5–6.0% is typical.

As stream size increases, the influence of riparian vegetation declines. Shading of the stream lessens, and the relative contribution of allochthanous input in the form of litter fall (and terrestrial insects) is reduced. Water temperatures usually increase as stream size increases, providing more thermal units for fish production within temperature limits specified above. The maximum diel temperature range increases from a few degrees in second-order headwaters to a maximum of about 10°C in some fifth-order streams; the range then decreases with further increases in stream order.

With these concepts of physical and biological changes along the length of a stream in mind, it is important to consider some basic differences between one stream system and another. Fluvial environments vary greatly in their basic nature, and the contrast, for example, between a glacial stream and one that is spring fed is striking. Conditions at high latitudes are marked by both extreme photoperiod and temperature, and as such these streams (reviewed by Reynolds 1997) represent special cases.

One key characteristic of a coldwater stream is its flow pattern (Poff and Ward 1989). An annual hydrograph depicts average daily flows as a time series and provides information regarding the extent and timing of minimum, maximum, and average stream flows. Three sample hydrographs (Figure 18.1) show a range of typical patterns. For a high-elevation, unregulated stream, as much as 75% of the precipitation in the drainage basin may come from snow. Its hydrograph has a peak flow in spring that may be from ten to several hundred times that of the average summer low flow. A spring-fed stream may show a much flatter hydrograph. Coastal systems, such as those on the west side of the Cascade Mountains in Oregon and Washington, may be characterized by a hydrograph that reflects 250 cm of rainfall, most of which falls between October and March. Another type of hydrograph may be typical of a regulated stream below an impoundment, where flows are drastically modified by reservoir storage.

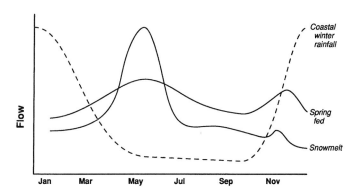

Figure 18.1 Typical seasonal flow patterns of coldwater streams. The coastal winter rainfall hydrograph represents the Alsea River, Oregon; the spring-fed hydrograph, Silver Creek, Idaho; and the snowmelt hydrograph, Falls River, Idaho, where it exits Yellowstone National Park.

The geology of a drainage basin determines water chemistry parameters such as dissolved solids and conductivity. In a similar manner, the geomorphic characteristics of the watershed dictate critical features of a stream, such as its ability to transport sediment. Eight major variables interact in a process such as the transport of sediment: channel width, depth, and slope, roughness of bank and bed, discharge, shape of the bed, sediment concentration, and sediment size. A change in any of these factors may bring a response from the others as the system moves toward a new equilibrium. For example, channel substrate size systematically changes with channel gradient (Figure 18.2).

The transport and deposition of fine sediment have particularly important consequences for fluvial salmonids. Excessive turbidity may impair the health of fish, and levels of sediment that heavily embed the substrate reduce the survival of embryos (Chapman 1988) and juvenile salmonids that overwinter in the substrate. Also, sediment may reduce critical pool volume.

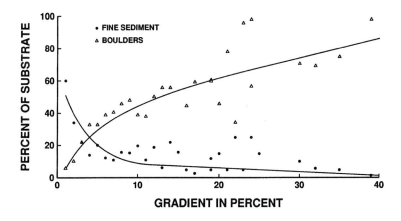

Figure 18.2 Change in stream channel substrate with gradient in tributaries of the Salmon River, Idaho. Data are from Platts 1979.

Interactions with other fish species have shaped the niches of coldwater salmonids. Biological diversity of native species in coldwater streams ranges from simple systems with limited diversity (such as in some western streams where only six to eight species of *Cottus, Catostomus,* and *Rhinichthys* are found with one or two salmonids) to more elaborate biological communities typical of streams in eastern North America (with perhaps 20 species of cyprinids, cottids, catostomids, and percids [darters] interacting with salmonids).

18.1.2   Economic Value of Coldwater Stream Fisheries

Coldwater streams are valuable resources that are in demand by current and potential users, some of whose interests conflict. Streams are highly valued by anglers, a majority of whom prefer streams over lakes. In Idaho, for example, a 1986 survey (V. Moore, Idaho Fish and Game Department, unpublished data) indicated that twice as many trout anglers preferred stream fishing over lake and reservoir fishing even though streams constitute only about 20% of the state's surface water.

The economic value of coldwater stream fisheries is significant. In 1985, the additional amount anglers were willing to pay over and above actual expenditures (e.g., travel and licenses) for a day of stream fishing in Montana was US$102. Angling effort is predicted to reach an average of 300 h/ha for public lakes, streams, and reservoirs in the contiguous 48 states by the year 2000 (McFadden 1969). If so, each hectare of coldwater habitat (a 400-m length of a 25-m-wide stream) will provide $7,500 of net economic value per year. Some popular trout fisheries provide more than 2,500 h of angling per hectare each year; effort in some doubled in the 1980s. Many land use practices directly compete with salmonid fishes and anglers for coldwater stream habitat. Examples are described in Chapters 9 and 10.

18.2   LIFE HISTORY AND BEHAVIORAL PATTERNS OF SALMONIDS

18.2.1   Population Dynamics

Fish production is defined as the total elaboration of fish tissue during any time interval, including what is formed by individuals that do not survive to the end of the interval. The components of production are growth and survival.

Growth of salmonids in streams is typically less than the maximum possible because of periods of suboptimal temperature, limited food and space, and other factors. Growth is most rapid during the first summer of life, when fish may reach lengths of 100–150 mm (Figure 18.3). In subsequent years, growth rate declines but may continue at 20–40 mm per year during the last few years of life. When carrying out investigations of population dynamics, care must be taken in assigning ages (Beamish and McFarlane 1983). Studies using otoliths and daily growth rings have shown that age determination based on traditional scale reading may underestimate true age (Casselman 1987). Furthermore, in severe environments, salmonids that grow little during their first year may not develop first-year annuli (Lentsch and Griffith 1987). Biologists conducting age and growth studies should be aware of these sources of bias.

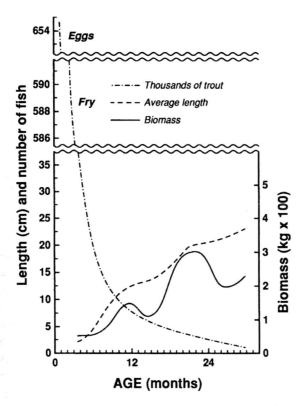

Figure 18.3 Summary of some vital statistics of a typical year-class of brook trout for the first 31 months of life in spring-fed Lawrence Creek, Wisconsin. Data are from McFadden (1961).

Natural mortality of stream-dwelling salmonids varies considerably with age. Mortality during the first few months of life may exceed 90%, although part of this loss may actually reflect emigration. Factors such as intraspecific competition and starvation, which cause mortality of age-0 fish, are density dependent. Other factors, such as spring floods and high summer temperatures, operate regardless of fish density. Predation may be substantial but is difficult to assess accurately. An extensive study on the Au Sable River in Michigan could not identify the causes of mortality during the first year of life for brook and brown trout (Alexander 1979). Predation on yearling and older trout was significant by great blue herons, common mergansers, otters, mink, and brown trout and negligible by belted kingfishers (Table 18.1). These predators killed 79% and 45% of the age-1 brook trout and brown trout present, respectively, and 25% and 44% of the older brook and brown trout, respectively. In a Vancouver Island stream in British Columbia, common merganser broods consumed sufficient wild coho salmon fry during summer months to account for 24–65% of the potential smolt production (Wood 1987).

Overwinter mortality of salmonids in streams appears to approximate 50% of the age-0 fish present in the fall (Smith and Griffith 1994). Death may result from starvation, predation, or physical damage. Because of the decrease in metabolic activity at cold temperatures, starvation of wild salmonids is unlikely. Healthy trout survive several months of fasting in cold water. Predation on juvenile salmonids in winter needs careful evaluation.

Table 18.1   Percentage of total mortality of brook and brown trout due to predation in the Au Sable River, Michigan. Mortality estimates were made from the fall of one year to the fall of the next for the age-groups indicated. Data are from Alexander 1979.

Predator	Brown trout		Brook trout	
	Age 0–1	Age 1–8	Age 0–1	Age 1–4
Anglers	6.8	46.0	6.6	43.7
Belted kingfisher	3.2	0	4.0	0.1
Brown trout	16.1	0.2	58.0	5.1
Common merganser	10.7	12.8	3.6	10.5
Great blue heron	6.8	14.1	7.7	3.3
Mink	7.7	11.4	5.0	4.8
Otter	0.2	5.4	0.6	1.2
Unknown	48.5	10.1	14.5	31.3

Physical injury appears to be the major cause of winter mortality in stream reaches where snow bridging does not occur. During nights when clear skies maximize radiant heat loss from streambed and water and when air temperatures are low enough to supercool the stream, crystals of ice, called frazil ice, form in the water column. The frazil ice drifts downstream and attaches to the first object it contacts, where it may accumulate as a mat of anchor ice. Anchor ice tends to build up where velocity is greatest and flow is most turbulent and where there is no surface ice. As the mats build, dams form, and the stream becomes elevated and is forced out of its bed. Fish may lose their orientation and swim aimlessly into the impounded areas. During the day, heat from the sun melts the ice dams, and the stream quickly returns to its former bed, often stranding fish and other aquatic organisms in the overflow areas. This sequence may be repeated frequently throughout the winter. In coastal systems, winter floods often are major causes of mortality.

Salmonids display one or more of a range of responses to the onset of winter, depending upon fish species and life stage, minimum temperatures experienced, and habitat quality. Two basic options are to move elsewhere or to remain in the same areas that are used in summer. Fish that move may enter thermal refuges such as groundwater seeps or move to off-channel habitat such as ponds and side channels. Movements generally begin when water temperatures drop below 7–10°C. Chapman and Bjornn (1969) suggested that in many Idaho streams 50% or more of the juvenile chinook salmon and steelhead that left the stream did so in fall and winter months. Adults of resident species characteristically move to the deepest pools available, leaving small tributaries to enter main river areas.

Once in the winter-use area, juvenile salmonids typically either aggregate in the water column in deep, slow water beneath cover or conceal themselves in the substrate. Chapman and Bjornn (1969) described movement by juvenile chinook salmon and steelhead into or beneath cobble, small boulders, or large woody debris to depths of 30 cm. Similar behavior has been described for most other fluvial salmonids, as summarized by Cunjak (1996). Where cobble was not available in heavily sedimented portions of an Idaho stream, juvenile chinook salmon overwintered in low-velocity areas among dense growths of sedges and grasses that draped over the bank (Hillman et al. 1987). Fish that are in the substrate during the day may emerge at night. Campbell and Neuner (1985) found that in small Washington streams juvenile and adult resident rainbow trout were concealed during the day in winter, but these fish moved inshore into shallow water at night.

Overwintering within the substrate appears to be adaptive for high elevations and latitudes, where winters are characteristically severe (Cunjak and Power 1986). Here streams freeze over early and remain frozen throughout the winter, and fish survival might be maximized by remaining in the substrate without expending energy for swimming. At lower latitudes and elevations, streams display alternating freezing and thawing, variable discharge, and anchor ice formation. Under such conditions, an active existence might be most adaptive for survival. Life history strategy is an additional determinant of winter behavior for (anadromous) age-0 Atlantic salmon. Some members of a cohort maintain active feeding levels and continue growing throughout the winter; these fish then immediately undergo the physiological, behavioral, and morphological changes of smolting. Other individuals reduce winter feeding and arrest winter growth, delaying smolting for a later year (Metcalfe et al. 1988).

After trout have reached sexual maturity, natural mortality may reduce fluvial populations of trout by 50% or more annually; males typically show higher mortality than do females. Therefore, odds are against extended longevity, especially if the population is heavily fished. "Old" fish might be 4–5 years, depending on the species and the environment. For brook trout in Lawrence Creek, Wisconsin, some year-classes never had members surviving past age 3 (McFadden 1961).

18.2.2 Spawning

The attainment of sexual maturity appears to be more related to fish size than to age. In some populations of brook trout, males become sexually mature at the end of their first growth season, when less than 100 mm in length, and females become mature as yearlings. Other species, and other populations of brook trout that grow more slowly, may not mature until age 4 or older. Salmonids may reproduce over several years, or they may reproduce only once. Egg production varies with size of female and ranges from a few hundred to several thousand eggs. Spawning occurs as water temperature approaches 10°C, either in the fall for species such as brown trout and brook trout or in the spring for rainbow trout and cutthroat trout. Spawning habitat is predictable in terms of water velocity, water depth, and substrate size. Eggs develop in the gravel until they accumulate approximately 300 degree-days over 0°C for resident trout species. During the winter the eggs of fall-spawning species remain in the gravel, where they are prone to damage from intergravel ice formation under severe conditions. However, fall-spawned fry emerge from the gravel before spring-spawned fry.

18.2.3 Territoriality

As juvenile salmonids absorb their yolk sacs and emerge from the gravel in which they developed, three changes in behavior occur. They become positively phototactic, they start to nip at and chase neighboring fish, and they begin to take particles of food from the drift. A territory of adequate size increases the probability that sufficient food (aquatic insects and other macroinvertebrates) will be carried past them by the current.

Agonistic (threat and defense) interactions between fish become more subtle with age; nipping and chasing are replaced with ritualized threat postures and changes in body shading that serve the same purposes. These behavioral signals appear to be effectively communicated among species of fluvial salmonids. The ability to establish and hold a territory is largely dependent on body size, and a fish that is even slightly larger than another will be dominant in most encounters. There is growing evidence (i.e., Brown and Brown 1993) that salmonids recognize their kin and behave more passively toward them.

In a conceptual sense, individual fish remain at a point in the territory referred to as the focal point. Here they may spend the majority of their time during daylight hours. The focal point may be near the center of a teardrop-shaped territory, and the fish darts out from the focal point to feed or to threaten intruders. The distance moved to feed is usually equivalent to about twice the fish's body length. Because the amount of drift that passes a given point in the water column is approximately linear with water velocity (Figure 18.4), the fish scans a high-velocity area for drift. However, because the energy it expends increases exponentially as its swimming speed increases, the fish selects a focal point with reduced water velocity, often 10–20 cm/s, to maximize net energy gain (Fausch 1984; Hughes 1992). Territory size increases with fish size: a trout 10 cm long may defend a 0.1 m² area and a 30-cm-long fish may defend a territory of 1.0 m².

In reality, there are numerous variations on the basic theme described above. Juvenile coho salmon, for example, were territorial in riffles but not territorial in pools, although aggression was displayed toward other pool inhabitants (Puckett and Dill 1985). Other

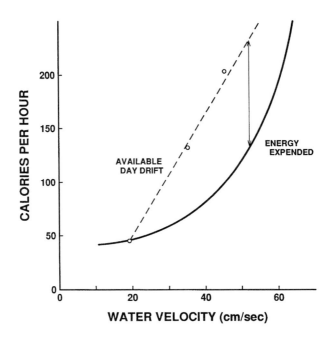

Figure 18.4   Energy available from drifting invertebrates and energy expended by a 100-g trout at 15°C by swimming at a range of water velocities. Net energy gain is maximized at the velocity shown by the arrow. Data are from Feldmeth and Jenkins (1973) and J. S. Griffith (unpublished).

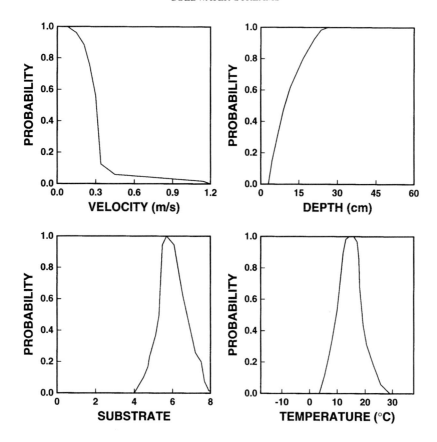

Figure 18.5 Probability-of-use curves for adult brook trout. Typical substrate categories are mud (2), sand (4), cobble (6), and bedrock (8). Data are from Bovee (1978).

individual coho salmon, called floaters, never defended an area and remained in spaces between the territories of other coho salmon. Territorial coho salmon had a net energy intake advantage because of reduced costs of searching for and pursuing food and reduced agonistic activity costs (Puckett and Dill 1985). In a study of wild adult brown trout, Bachman (1984) developed a size-dependent linear dominance hierarchy of individuals with overlapping home ranges. No fish had exclusive use of any home range and no clearly defined territories were observed. The brown trout spent 86% of daylight hours in a sit-and-wait state, searching the water column for drifting food.

Practical implications of these patterns of salmonid behavior are that only a portion of a stream may actually be used by a given life stage of a particular species and an ontogenetic shift of the habitat used occurs with fish growth. The microhabitat of age-0 rainbow trout, for example, is defined by a combination of habitat characteristics such as water depth and velocity, substrate, and cover. Sets of probability-of-use curves (e.g., Bovee 1978; Figure 18.5) developed for many fluvial salmonids illustrate differences in habitat used by different species and different life stages.

Recent studies using radio telemetry indicate that individual salmonids often move long distances (tens or hundreds of kilometers) during their lives (Gowan et al. 1994) and that some may repeatedly return to specific homesites within a stream system (Clapp et al. 1990).

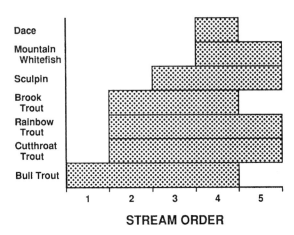

Figure 18.6 Relationship between stream order and fish species composition for tributaries of the Salmon River, Idaho. Dace refers to longnose dace and sculpins to likely more than one species. Data are from Platts (1979).

18.3   FACTORS INFLUENCING SALMONID ABUNDANCE

18.3.1   Species Composition

The present distribution of salmonid species in North America bears little resemblance to original ranges. Efforts began over 100 years ago to introduce virtually every species into every drainage across the continent. Now few coldwater stream systems exist that do not hold brook trout, brown trout, or rainbow trout. The ranges of coho salmon, chinook salmon, and steelhead have been increased dramatically by introductions into the Great Lakes and their tributaries beginning in 1966 (Chapter 23). Only Atlantic salmon, cutthroat trout, bull trout, and Dolly Varden have not been spread widely out of their original ranges. Cutthroat trout have been transplanted around the world, but nearly all of these efforts have been unsuccessful.

Within the continuum of physical variables in a coldwater stream it is possible that salmonid species might occupy separate stream portions that do not overlap. Some segregation has been documented, such as in Idaho's Salmon River drainage, where first-order streams were used by only bull trout and fifth-order streams were not used by brook trout and bull trout (Platts 1979; Figure 18.6). Rainbow trout were the only salmonids found in stream reaches with gradients over 16%. Bull trout were most abundant in reaches with gradients of 6–10%, cutthroat trout between 8 and 14%, and brook trout between 2 and 5%. Stream gradient has been shown to have a strong negative influence on brook trout abundance in Wyoming streams (Chisholm and Hubert 1986); however, stream-dwelling salmonids are generalists, both in terms of habitat use and diet. Their coexistence in multispecies situations depends upon their ability to specialize when influenced by interspecific interactions such as predation, hybridization, and competition.

Competition occurs when a number of individuals (of the same or different species) use common resources that are in short supply or, if the resources are not in short supply, individuals harm one another in the process of using shared resources. Competition among fluvial salmonids usually translates into attempts by individuals of the

Table 18.2 Frequency of annual production values for resident salmonids in streams in North America, Europe, and New Zealand. Data are modified from Waters 1977.

Species	Production (kg/ha)				
	<50	51–100	101–150	151–200	>200
Brook trout	3	3	2	1	3
Brown trout	5	4	6	0	2
Rainbow trout	4	0	2	0	0
Multiple species	0	2	2	2	1

same or different species to secure adequate space and, therefore, food and cover. The outcome of such competition may be temperature specific (DeStaso and Rahel 1994). Over time, continued interactions between two species should either produce a niche shift in one or both species, the local extinction of one species, or fluctuating coexistence as the environment alternately favors one species over the other. Coexistence of sympatric species may result from interspecific differences in habitat use that reflect species differences in agonistic behavior, innate habitat preference, timing of emergence, morphology, or a combination of these factors. There is evidence, though, that the presence of introduced salmonids has detrimentally affected native species (Chapter 24). Examples include the decline of inland subspecies of cutthroat trout following range expansion of brook trout (Griffith 1988), decline of brook trout in the southeastern portion of its range following invasion by rainbow trout (Kelly et al. 1980), and replacement of brook trout by brown trout in Minnesota (Waters 1983).

18.3.2 Density, Biomass, and Production

Annual production of salmonids in coldwater streams, either in mixed- or single-species situations, usually falls within the range of 15–50 kg/ha for softwater streams or for those at higher latitudes and 100–150 kg/ha for more productive streams (Waters 1977; Chapman 1978; Kwak and Waters 1997; Table 18.2). Often more than half of the production is contributed by the youngest age-group. The turnover ratio, or ratio of production to mean standing stock over a time interval, is commonly 1.0–1.5 for fluvial salmonids (Chapman 1978) but ranges from about 0.5–1.0 for small, high-elevation Colorado streams (Scarnecchia and Bergersen 1987).

Platts and McHenry (1988) summarized midsummer estimates of salmonid biomass and density for 313 interior streams in the western United States with pristine or lightly altered habitat. Biomass exhibited great variability, ranging from 0 to 819 kg/ha; the average was 54 kg/ha. Salmonid density was less variable, ranging from 0 to 420 fish per 100 m^2; the average was 25 fish per 100 m^2. Brown trout density and biomass were significantly greater than those of other species, and multispecies communities had densities and biomasses that were not significantly different from those of streams occupied by single species.

Densities of juvenile salmon and steelhead in fully seeded stream-rearing habitat may exceed 700 fish per 100 m^2 for brief periods. By late summer, however, densities of 50–370 fish per 100 m^2 are typical for age-0 Atlantic salmon, and such a range is also realistic for Pacific salmon and steelhead. Average annual smolt production for Atlantic salmon in North American streams may reach 6–10 fish per 100 m^2 (Bley 1987).

Table 18.3   Models similar to Binns and Eiserman's (1979) habitat quality index that predict trout biomass in streams based on measurement of habitat. See Fausch et al. (1988) for a thorough review.

Study	Location	Trout species	Biomass range (kg/ha)	Significant variables	r^2
Wesche et al. (1987)	Southeast Wyoming	Brown	2–211	Overhead bank cover	0.31
Scarnecchia and Bergersen (1987)	Northern Colorado	Brook, cutthroat, rainbow, and brown	39–282	Width and depth, width, alkalinity	0.82
Bowlby and Roff (1986)	Southern Colorado	Brook, brown, and rainbow	0.5–150	Suspended microcommunity, biomass, mean maximum temperature, benthic biomass, pool area, piscivore presence	0.62

18.3.3   Predictors of Abundance

A number of models have been developed to predict the biomass of resident trout in streams based on measurements of habitat. Binns and Eiserman's (1979) habitat quality index predicted trout biomass in Wyoming streams based on nine attributes: late-summer streamflow, annual streamflow variation, maximum summer stream temperature, nitrate-nitrogen, cover, eroding streambanks, submerged aquatic vegetation, water velocity, and stream width. The model explained 96% of the variation in trout biomass for the 36 streams from which it was developed and 87% for 16 Wyoming streams in a follow-up study by Conder and Annear (1987). Attempts to apply the habitat quality index to populations of salmonids in streams outside Wyoming have generally not been successful, indicating that trout populations in different areas respond to different sets of factors. Correlative models based on Binns and Eiserman's model have been assembled for areas ranging from Colorado to Ontario (Table 18.3).

More recently, geomorphic variables such as basin relief and drainage density have been shown to correlate with trout biomass in Wyoming streams (Lanka et al. 1987; Kruse et al. 1997). This linkage may enable the use of simple measures of drainage basin geomorphology, perhaps in combination with stream habitat variables, to predict potential habitat quality for salmonids.

18.3.4   Limiting Factors

An examination of the stability of fluvial salmonid populations over time provides insight into their limiting factors. Populations in spring-fed streams, where flows are relatively uniform throughout the year and floods and droughts are uncommon, are most stable. An example is Lawrence Creek, Wisconsin, where brook trout populations in September varied from 55 to 111 kg/ha over a 5-year period in the mid-1950s (McFadden 1961), and annual production varied only 20% over 11 consecutive years (Hunt 1974).

At the opposite extreme are Great Basin (Nevada) and Rocky Mountain streams between 1975 and 1985, when both near-record low flows and high flows and the most severe winter on record occurred (Platts and Nelson 1988). Biomass of bull trout in the

South Fork Salmon River, Idaho, fluctuated fourfold (range 11–39 kg/ha) in those 11 years, and brook trout in three small Idaho streams also fluctuated fourfold in an 8-year period. Over the same period in the Great Basin, biomass of cutthroat, brown, and rainbow trout also showed a three- to fourfold fluctuation (cutthroat trout range 33–98 kg/ha, brown trout range 20–80 kg/ha, rainbow trout range 4–14 kg/ha). Trout number fluctuated more than did biomass, often eightfold. Platts and Nelson suggested that for these and many other western streams, abiotic factors may be the causal factors limiting salmonid abundance.

Most salmonid streams are intermediate between the two extremes described above. Abiotic events such as floods and severe winters may commonly occur but not frequently enough to affect every generation of trout. When such events do occur, populations would be depressed by the climatic factors so that resources such as food should be abundant and competition between individuals minimal. However, interspersed between these periods of lessened competition would be "ecological crunches," as described by Wiens and Rotenberry (1981), during which populations would increase and resources would become limited. Interspecific interactions then become important, and the niches of one or more species may be modified as a result. For fluvial trout, interspecific competition should be most significant in late summer of those years when population levels are at or above normal. At these times, space and food are minimal, and niche shifts have been observed as summarized by Hearn (1987). Also, when population levels are at or above normal levels, density-dependent mortality of salmonids during the first few months of life is an important population-regulating process.

In a practical evaluation of limiting factor theory, Mason (1976) provided juvenile coho salmon in a small Vancouver Island stream with supplemental food in the form of marine invertebrates. As a result, the density effect on survival was eliminated and growth increased, producing a coho salmon biomass that was six- to sevenfold above natural levels at the end of the summer. However, coho salmon remain in the rearing stream through their first winter of life. By February, Mason's expanded populations had declined to densities typical of most natural coho salmon populations, and the net result was output of about the same number of smolts that the stream would have produced without summer feeding. To accomplish a successful "lifting of the lid," Mason would have had to increase appropriate winter habitat in addition to providing summer food.

18.4 COLDWATER STREAM MANAGEMENT

18.4.1 History

A movie reel depicting the highlights of coldwater stream fisheries management during the past 100 years would show the numerous shifts in emphasis discussed in Chapter 1. From 1870 through the early 1900s was the golden era of introductions. During this time exotics such as brown trout (MacCrimmon and Marshall 1968) and common carp were widely distributed. At the same time, most salmonids native to North America were transplanted throughout the continent. Brook trout, for example, were brought into 12 central and western states before 1900 and into 5 other states and three provinces 30–50 years later (MacCrimmon and Campbell 1969).

In an attempt to maintain high catch rates while stream habitat was noticeably shrinking in the mid-twentieth century, hatchery programs proliferated (Chapters 1, 14). Not until the early 1970s were questions being raised about the effects of stocked trout on wild populations. A gradual change in attitude, placing increased emphasis on wild fish in a quality setting, began and continues to grow. Currently, efforts are being made to maintain or restore species or subspecies of special concern (Chapters 13, 16), adopt special regulations (Chapters 14, 17), and protect and rehabilitate instream and riparian habitat through partnerships with private landowners and watershed advisory groups (Williams et al. 1997).

18.4.2 Management Goals

For anadromous salmonids rearing in freshwater the management goal is simple: produce as many smolts as possible. The issue becomes the quality of the smolts; in particular, whether those of hatchery origin are adequate.

For resident salmonids that support sport fisheries, the traditional goal has been maximization of harvest. However, this goal is shifting toward the more challenging one of optimum sustainable yield. Yield includes fish harvested as well as fish caught and released because the latter also contribute to the quality of the angling experience. It is important that managers understand what constitutes a successful angling experience for their constituents. To establish what made a trip satisfying for anglers throughout Yellowstone National Park, the U.S. Fish and Wildlife Service interviewed over 20,000 anglers (Varley 1984). The percentage of satisfied individuals increased—but at a decreasing rate—as the number and size of trout landed increased (Figure 18.7). About one person in five was satisfied even though no fish were landed. Satisfaction increased greatly when one fish was caught, and about 80% of the anglers were satisfied after catching three fish. In Wisconsin, trout anglers indicated that their satisfaction was based more on aesthetics and solitude than on fish yield (Jackson 1989).

A standardized length categorization system (Gabelhouse 1984) enables a manager to compare the length frequency distribution of a population with the sizes of fish desired by anglers. Minimum lengths for preferred, memorable, and trophy categories correspond to certain percentages of the world record length for that species. For example, preferred rainbow trout must exceed 49 cm, and memorable and trophy individuals must exceed 65 and 81 cm, respectively, whereas trophy brook trout would exceed 59 cm.

It has been facetiously said that management of coldwater streams boils down to three basic components: protect habitat, preserve habitat, and restore habitat. Actually, there are other components—specifically, manipulating the biotic community and regulating the catch. Nonetheless, habitat is the key. The need to maintain quality habitat for coldwater streams is equivalent to the overriding need to maintain a predator–prey balance in ponds and reservoirs. Predator–prey balance in salmonid streams is of lesser concern because invertebrate food sources are generally abundant (but see Cada et al. 1987), and the overall quantity of invertebrates is typically not affected by trout to a significant extent.

The limiting factor concept may help coldwater stream managers in allocating their limited resources. It is important to distinguish between deleterious signs and actual problems. Deleterious signs can be seen (perhaps too easily), whereas the underlying problems may be difficult to detect. By examining the situation from a basinwide perspective and

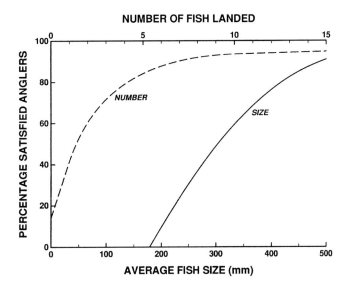

Figure 18.7 Satisfaction of anglers in Yellowstone National Park as determined by number of fish landed and average size of fish landed. Data are from Varley (1984).

critically reviewing the requirements of each life stage of the species concerned, it may be possible to identify the factors that most urgently require management action. For example, an evaluation of steelhead and coho salmon populations and their habitat in Fish Creek, Oregon, provided information on factors limiting their production in the basin (Everest et al. 1985). For steelhead, spawning habitat was more than adequate, but rearing habitat was inadequate. For coho salmon, smolt production was limited only by the numbers of adults returning to the basin. Following this analysis, specific management plans could be implemented to improve steelhead rearing habitat and to increase escapement of adult coho salmon (Reeves et al. 1997).

Another management tool (for situations where requisite data are available) is provided by individual-based modeling analysis. Clark and Rose (1997) illustrated its use in evaluating options to enhance native brook trout negatively affected by rainbow trout in southern Appalachian streams. The model simulated results for four management options (electrofishing removal of rainbow trout, brook trout stocking, habitat improvement, and change in angler harvest) and provided stimulus for future research and management efforts. Another management tool is the estimation of a stream's potential maximum biomass of trout (Hubert et al. 1996). This potential can be compared with current biomass levels to help managers evaluate a range of future management activities.

18.4.3 Habitat Management

When considering management actions, the general strategy is to document existing habitat conditions and then to assess habitat improvement potential (or document habitat degradation) based on a model such as the habitat quality index (Binns and Eiserman 1979). Chapter 10 describes techniques for the restoration of coldwater stream habitat. White (1996) and Cunjak (1996) provide excellent reviews of summer and winter habitat-related issues.

The placement of instream and streambank structures, so in vogue in the 1980s, requires careful consideration. More than $100 million was spent on enhancement of salmonid habitat in the western United States during the 1980s. The ability of this type of enhancement to provide cost-effective results must be carefully examined; in many cases, placing structures to increase fish numbers may not be justifiable, but implementing measures to improve watershed conditions or to deal with specific habitat problems may have large payoffs. Follow-up evaluation is critical, and it is important to realize that the benefits from instream structures may not be detectable for 5 years or more.

If streams have been damaged by livestock, they are usually rehabilitated more easily and cheaply by relieving the grazing pressure (by fencing or other compatible strategies). Streambanks then restore themselves, a process preferable to building artificial stream structures. Also, it is unlikely that habitat structures would survive in the presence of continued heavy grazing.

Habitat management should be based on a need for habitat that is as complex as possible. The need for large woody debris in salmonid nursery streams is clearly recognized. In Oregon, House and Boehne (1986) examined stream segments in old-growth coniferous forests where large woody debris was abundant and compared these segments with previously logged segments that held little large debris. The presence of large woody debris had profound effects on channel morphology. Although channel gradient was the same in test and control segments, large woody debris "stair stepped" the channel, caused the formation of secondary channels, meanders, and undercut banks, and caused trees to tip over and expose rootwads. Spawning gravel was trapped in the stair stepped sections, and pool quality and quantity were increased. Significantly greater salmonid biomass was found in the old-growth segments, and the number of coho salmon was positively correlated with the number of large woody debris intrusions into the stream channel.

As Gowan et al. (1994) pointed out, good management depends upon correct assessment of the extent to which fish move. Different stream reaches should not automatically be considered as independent units of habitat.

In the absence of complete ecological understanding, management for habitat diversity will usually keep the stream in sound condition for biological resources. The manager hopes to approach an optimal relationship between this diversity and year-long trophic and habitat requirements. The best channel has plunges, backwaters, large woody debris and streamside vegetation, cobble, gravel, even sand and silt, shear lines, depth, pockets, undercuts. Beaver often play a critical role in salmonid habitat by providing deep pools for salmonids in summer and winter; however, it is often required to control beaver numbers.

18.4.4 Management of Community Structure

A commonly used management approach is to modify the community structure of a stream by introducing nonnative salmonids, stocking hatchery trout (see also Chapter 14), or removing nonsalmonid competitors. Hatchery-reared fishes have been used either to supplement wild trout populations or to provide angling where no wild trout existed. In 1983, 43 states stocked over 50 million catchable-size trout at a production cost of about $37 million (Hartzler 1988). The fraction of these fishes that were stocked into streams is unknown, but at least 30 states planted catchable-size fishes into more than 54,000 km of streams in the United States. Rainbow trout accounted for 77% of all catchable-size salmonids stocked, and brook and brown trout made up 11% and 10% of the total, respectively.

The cost of producing a catchable-size hatchery trout is typically about $0.60 if only feed, supplies, and salaries are considered. However, if fixed costs such as capital replacement and administrative overhead are included, the total cost is approximately $1.50 per catchable-size trout (Johnson et al. 1995). This might be a good investment of funds if return of these fish to anglers' creels is high. If, however, predators such as cormorants and pelicans (Derby and Lovvorn 1997) or striped bass in reservoir tailwaters (Walters et al. 1997) kill more hatchery trout than do anglers, one hatchery fish in an angler's creel may cost more than an annual fishing license.

In addition, recent studies (such as DeWald and Wilzbach 1992) show potential detrimental effects of hatchery-reared trout on wild salmonids. Populations of 2-year-old and older wild brown and rainbow trout in two Montana streams declined significantly in numbers and total weight following introduction of catchable-size hatchery rainbow trout (Vincent 1987). As Vincent points out, such stocking can reduce the number of wild trout available to anglers and may cause some genetic alteration of the wild stocks. For native cutthroat and bull trout, which have hybridized with rainbow and brook trout, respectively, reproduction by hybrids has been a major factor in the decline of the native species (Behnke 1992; Rieman and McIntyre 1993). In these situations, management of self-sustaining wild trout streams would be better directed at maintaining or enhancing riparian habitat, maintaining adequate water flows, and applying appropriate catch regulations. Another approach would be to stock only fish raised from sterile triploid eggs. Purchase of triploid eggs from commercial suppliers adds only a few cents to the costs of production. In streams in western North America and the southern Appalachian Mountains, exact mechanisms of competition between native and naturalized trout are still not completely understood. However, proactive solutions such as removal of nonnative trout are a feasible management tool (Thompson and Rahel 1996; Clark and Rose 1997).

Experience with an extensive system of hatcheries for anadromous salmonids in the western United States and Canada has shown that hatchery production may reduce natural production through competition, predation, interbreeding, and transmission of disease. Procedures have been developed to mitigate the reduction of natural production by hatchery production, but most efforts are directed toward hatchery production of smolts, and involve some separation of hatchery and wild fish. Recently, interest has expanded to an alternative approach called outplanting—releasing hatchery fish to rear or spawn in streams, usually remote from the hatchery—which may have merit in some situations (Reisenbichler and McIntyre 1986).

Control of nonsalmonid fishes that appear to have a negative effect on trout and salmon in streams has been a focus of attention for several decades (see Chapter 15). In streams, sculpins have been thought to affect salmonid populations through predation on their eggs and fry and through competition for food in the form of benthic invertebrates. However, Moyle (1977) suggested that sculpin effects would be expected only under conditions in which salmonid populations had been badly damaged by overharvest or environmental disturbance. After trapping and removing nontrout species from a small New York stream for 13 years, Flick and Webster (1975) could detect no change in growth of brook trout. They suggested that removal of nontrout species did not seem to be a viable management technique.

From 1964 through 1966, a program to improve the quality of angling by removing predators from a portion of the Au Sable River, Michigan, included reducing numbers of large brown trout by 40–66% and harassing common mergansers (Shetter and Alexander 1970). No significant differences in size or catch of smaller brook and brown trout were subsequently observed, and it was concluded that predator reduction would have to be conducted at a much more intensive level to produce the desired results.

18.5   REGULATION OF THE FISHERY

The traditional objective of regulating catch in freshwater sport fisheries has been to maximize harvest by shifting as much of total mortality as possible into angling-induced mortality (see Chapter 17). Typical annual harvest for some stream fisheries may approach 50% of the biomass present. Regulations are used to protect the fish population from depletion and to maintain an equitable distribution of the resource among the fishing public. Fairly simple regulations, usually creel limits, are set to provide the greatest allowable harvest without overexploitation. Mathematical models have been developed to evaluate the response of trout populations to these and other regulations.

As Anderson and Nehring (1984) have pointed out, however, above some level of fishing pressure, the maximum yield approach sacrifices the quality aspects of the fishery because larger and older fish exist in greatly reduced numbers. This threshold of fishing pressure has been surpassed on the more popular trout streams in Colorado and perhaps across the continent.

More specialized regulations (see Chapter 17) include slot limits: only fish smaller than and larger than a length interval can be killed. Other regulations include restricting harvest to a few fish per day and limiting terminal gear to flies or artificial lures only. The expected mortality upon release of bait-caught resident salmonids is 30%; therefore, the use of bait is generally not compatible with regulations that involve release. Tight-line techniques and new hook designs intended to reduce hooking mortality may change the situation. Another approach is delayed harvest, in which harvest is prohibited for part of the year when environmental conditions are ideal for trout growth and survival. When conditions deteriorate, such as during low flow and high water temperatures, some harvest is allowed.

Catch and release has become an accepted management technique after emerging from a period of controversy in the early 1980s. The strategy here is to recycle individual medium-size and large trout as many times as possible, such as in the Yellowstone River in Yellowstone National Park where cutthroat trout were each caught an average of nearly 10 times in a season (Schill et al. 1986). Catch and release would be expected to be successful for populations having good growth potential, low natural mortality rates, and substantial longevity. Although often called no-kill regulations, catch and release typically results in 4–6% mortality from each capture for salmonids caught with artificial lures or flies (Wydoski 1980). This per-capture mortality could cause significant cumulative mortality for populations whose members are repeatedly caught and released, especially when water temperatures exceed the preferred range for the species. For the Yellowstone Park cutthroat trout fishery, single capture mortality at cool temperatures was estimated to be only 0.3% (Schill et al. 1986); however, Atlantic salmon caught at 22°C showed 40% delayed mortality (Wilkie et al. 1996).

Where they are biologically effective in "stockpiling" larger trout, catch-and-release regulations provide opportunities for high-quality angling. For example, the catch rate of trophy-size trout in the South Platte River in Colorado was 28 times greater in the catch-and-release area than in the harvest area. Rainbow trout dominated the population in the catch-and-release area; over 500 kg/ha rainbow trout were present and 50% were longer than 30 cm. In the harvest area, brown trout were most abundant, biomass was about one-third of that in the catch-and-release area, and only 17% of the population exceeded 30 cm (Anderson and Nehring 1984). This example also illustrates the differential vulnerability of salmonid species to angling, a phenomenon that must be considered in any regulation of harvest.

An index of the susceptibility of salmonids to angling is the amount of angling effort that results in removal of a certain fraction of the catchable-size population. Varley (1984) developed a preliminary model for Yellowstone National Park trout (Figure 18.8). Approximately half of the catchable-size cutthroat trout in a population would be removed annually by about 35 h of fishing per hectare. Corresponding values for rainbow and brown trout were approximately 285 h and over 400 h, respectively.

18.6   THE FUTURE OF COLDWATER STREAM FISHERIES

The future of fisheries management for coldwater streams lies with managing habitat, as discussed previously, managing the angler, and learning to deal with whirling disease (see below). As Larkin (1988) points out, the manager of the future may well conduct a market analysis of latent demand to help develop a management strategy. The value of this technique is shown by Sorg et al. (1985), who determined that the value of a coldwater fishing trip would increase by $8 if the number of fish caught doubled and would increase by $13 if the size of fish caught increased by 50%.

The manager of the future will have to attract new anglers willing to do more (actively support their streams and fishes) but who will be satisfied with less (in terms of fish harvest). The manager will also have to call for limiting entry to the more popular publicly owned streams. Most of the current interest in limited entry comes from anglers who are concerned that the quality of their experience is reduced by high angler density. To date, there is only minimal biological evidence that catch-and-release anglers are negatively affecting stream resources, but this may change as pressure increases.

In 1994, whirling disease, caused by the myxosporean *Myxosoma cerebralis*, was detected in wild trout in several western states. Adult rainbow trout populations in some infected river reaches have declined by more than 90% but have declined little, if at all, in others. Research on this disease and its primary host, the worm *Tubifex tubifex*, is expanding so that appropriate management can be implemented.

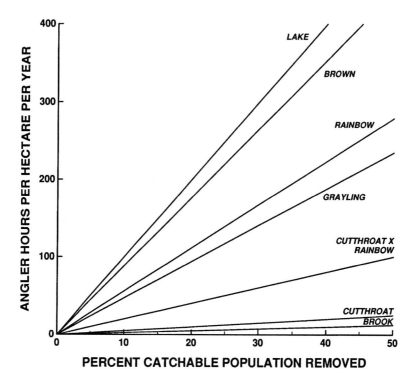

Figure 18.8 Differential vulnerability of resident trout species to angling, as indicated by the number of angling hours required to catch certain fractions of each population. Grayling refers to Arctic grayling. Data are from Varley (1984).

18.7 REFERENCES

Alexander, G. R. 1979. Predators of fish in coldwater streams. Pages 153–170 *in* H. Clepper, editor. Predator-prey systems in fisheries management. Sport Fishing Institute, Washington, D.C.

Anderson, R. M., and R. B. Nehring. 1984. Effects of a catch-and-release regulation on a wild trout population in Colorado and its acceptance by anglers. North American Journal of Fisheries Management 4:257–265.

Bachman, R. A. 1984. Foraging behavior of free-ranging wild and hatchery brown trout in a stream. Transactions of the American Fisheries Society 113:1–32.

Beamish, R. J., and G. A. McFarlane. 1983. The forgotten requirement for age validation in fisheries biology. Transactions of the American Fisheries Society 112:735–743.

Behnke, R. J. 1992. Native trout of western North America. American Fisheries Society, Monograph 6, Bethesda, Maryland.

Binns, N. A., and F. M. Eiserman. 1979. Quantification of fluvial trout habitat in Wyoming. Transactions of the American Fisheries Society 108:215–228.

Bley, P. W. 1987. Age, growth, and mortality of juvenile Atlantic salmon in streams: a review. U.S. Fish and Wildlife Service Biological Report 87(4).

Bovee, K. D. 1978. Probability-of-use criteria for the family Salmonidae. U.S. Fish and Wildlife Service, Cooperative Instream Flow Service Group, Information Paper 4, Fort Collins, Colorado.

Bowlby, J. N., and J. C. Roff. 1986. Trout biomass and habitat relationships in southern Ontario streams. Transactions of the American Fisheries Society 115:503–514.

Brown, G. E., and J. A. Brown. 1993. Social dynamics in salmonid fishes; do kin make better neighbors? Animal Behaviour 45:863–871.

Cada, G. F., J. M. Loar, and M. J. Sale. 1987. Evidence of food limitation of rainbow and brown trout in southern Appalachian soft-water streams. Transactions of the American Fisheries Society 116:692–702.

Campbell, R. F., and J. H. Neuner. 1985. Seasonal and diurnal shifts in habitat utilized by resident rainbow trout in western Washington Cascade Mountain streams. Pages 39–48 *in* F. W. Olson, R. G. White, and R. H. Hamre, editors. Symposium on small hydropower and fisheries. American Fisheries Society, Western Division and Bioengineering Section, Bethesda, Maryland.

Casselman, J. M. 1987. Determination of age and growth. Pages 209–242 *in* A. H. Weatherly and H. S. Gill, editors. The biology of fish growth. Academic Press, London.

Chapman, D. W. 1978. Production in fish populations. Pages 5–25 *in* S. D. Gerking, editor. Ecology of freshwater fish production. Wiley, New York.

Chapman, D. W. 1988. Critical review of variables used to define effects of fines in redds of large salmonids. Transactions of the American Fisheries Society 117:1–21.

Chapman, D. W., and T. C. Bjornn. 1969. Distribution of salmonids in streams, with special reference to food and feeding. Pages 153–176 *in* T. G. Northcote, editor. Salmon and trout in streams. H. R. MacMillan Lectures in Fisheries, University of British Columbia, Vancouver.

Chisholm, I. M., and W. A. Hubert. 1986. Influence of stream gradient on standing stock of brook trout in the Snowy Range, Wyoming. Northwest Science 60:137–139.

Clapp, D. F., R. D. Clark, and J. S. Diana. 1990. Range, activity, and habitat of large, free-ranging brown trout in a Michigan stream. Transactions of the American Fisheries Society 119:1022–1034.

Clark, M. E., and K. A. Rose. 1997. An individual-based modeling analysis of management strategies for enhancing brook trout populations in southern Appalachian streams. North American Journal of Fisheries Management 17:54–76.

Conder, A. L., and T. C. Annear. 1987. Test of weighted usable area estimates derived from a PHABSIM model for instream flow studies on trout streams. North American Journal of Fisheries Management 7:340–350.

Cunjak, R. A. 1996. Winter habitat of selected stream fishes and potential impacts from land-use activity. Canadian Journal of Fisheries and Aquatic Sciences 53(Supplement 1):267–282.

Davis, J. C. 1975. Minimal dissolved oxygen requirements of aquatic life with emphasis on Canadian species: a review. Journal of the Fisheries Research Board of Canada 32:2295–2332.

DeStaso, J., and F. J. Rahel. 1994. Influence of water temperature on interaction between juvenile Colorado River cutthroat trout and brook trout in a laboratory stream. Transactions of the American Fisheries Society 123:289–297.

Derby, C. E., and J. R. Lovvorn. 1997. Predation on fish by cormorants and pelicans in a cold-water river: a field and modeling study. Canadian Journal of Fisheries and Aquatic Sciences 54:1480–1493.

DeWald, L., and M. A. Wilzbach. 1992. Interactions between native brook trout and hatchery brown trout: effects on habitat use, feeding, and growth. Transactions of the American Fisheries Society 121:287–296.

Everest, F. H., and five coauthors. 1985. Abundance, behavior, and habitat utilization by coho salmon and steelhead trout in Fish Creek, Oregon, as influenced by habitat enhancement. Annual Report to Bonneville Power Administration, Project 84-11, Portland, Oregon.

Fausch, K. D. 1984. Profitable stream positions for salmonids: relating specific growth rate to net energy gain. Canadian Journal of Zoology 62:441–451.

Feldmeth, C. R., and T. M. Jenkins. 1973. An estimate of energy expenditure by rainbow trout in a small mountain stream. Journal of the Fisheries Research Board of Canada 30:1755–1759.

Flick, W. A., and D. A. Webster. 1975. Movement, growth, and survival in a stream population of wild brook trout *Salvelinus fontinalis* during a period of removal of non-trout species. Journal of the Fisheries Research Board of Canada 32:1359–1367.

Gabelhouse, D. W. 1984. A length-categorization system to assess fish stocks. North American Journal of Fisheries Management 4:273–285.

Gowan, C., M. K. Young, K. D. Fausch, and S. C. Riley. 1994. Restricted movement in resident stream salmonids: a paradigm lost? Canadian Journal of Fisheries and Aquatic Sciences 51:2626–2637.

Griffith, J. S. 1988. Review of competition between cutthroat trout and other salmonids. Pages 134–140 *in* R. E. Gresswell, editor. Status and management of interior stocks of cutthroat trout. American Fisheries Society, Symposium 4, Bethesda, Maryland.

Hartzler, J. R. 1988. Catchable trout fisheries: the need for assessment. Fisheries 13(2):2–8.

Hearn, W. E. 1987. Interspecific competition and habitat segregation among stream-dwelling trout and salmon: a review. Fisheries 12(5):24–31.

Hillman, T. W., J. S. Griffith, and W. S. Platts. 1987. The effects of sediment on summer and winter habitat selection by juvenile chinook salmon in an Idaho stream. Transactions of the American Fisheries Society 116:185–195.

House, R. A., and P. L. Boehne. 1986. Effects of instream structures on salmonid habitat and populations in Tobe Creek, Oregon. North American Journal of Fisheries Management 6:38–46.

Hubert, W. A., T. D. Marwitz, K. G. Gerow, N. A. Binns, and R. W. Wiley. 1996. Estimation of potential maximum biomass of trout in Wyoming streams to assist management decisions. North American Journal of Fisheries Management 16:821–829.

Hughes, N. F. 1992. Selection of positions by drift-feeding salmonids in dominance hierarchies: model and test for Arctic grayling (*Thymallus arcticus*) in subarctic mountain streams, interior Alaska. Canadian Journal of Fisheries and Aquatic Sciences 49:1999–2008.

Hunt, R. L. 1974. Annual production by brook trout in Lawrence Creek during eleven successive years. Wisconsin Department of Natural Resources Technical Bulletin 82.

Jackson, R. M. 1989. Why, why, why?: the human dimensions of trout angling motivations and satisfactions. Pages 186–192 *in* F. Richardson and R. H. Hamre, editors. Wild trout IV: proceedings of the symposium. Trout Unlimited, Arlington, Virginia.

Johnson, D. M., R. J. Behnke, D. A. Harpman, and R. G. Walsh. 1995. Economic benefits and costs of stocking catchable rainbow trout: a synthesis of economic analysis in Colorado. North American Journal of Fisheries Management 15:26–32.

Kelly, G. A., J. S. Griffith, and R. D. Jones. 1980. Changes in distribution of trout in Great Smoky Mountains National Park, 1900-1977. U.S. Fish and Wildlife Service Technical Paper 102.

Kruse, C. G., W. A. Hubert, and F. J. Rahel. 1997. Geomorphic influences on the distribution of Yellowstone cutthroat trout in the Absaroka Mountains, Wyoming. Transactions of the American Fisheries Society 126:418–427.

Kwak, T. J., and T. F. Waters. 1997. Trout production dynamics and water quality in Minnesota streams. Transactions of the American Fisheries Society 126:35–48.

Lanka, R. P., W. A. Hubert, and T. A. Wesche. 1987. Relations of geomorphology to stream habitat and trout standing stock in small Rocky Mountain streams. Transactions of the American Fisheries Society 116:21–28.

Lentsch, L. D., and J. S. Griffith. 1987. Lack of first-year annuli on scales: frequency of occurrence and predictability in trout of the western United States. Pages 177–188 *in* R. C. Summerfelt and G. E. Hall, editors. Age and growth of fish. Iowa State University Press, Ames.

MacCrimmon, H. R., and J. S. Campbell. 1969. World distribution of brook trout, *Salvelinus fontinalis*. Journal of the Fisheries Research Board of Canada 26:1699–1725.

MacCrimmon, H. R., and T. L. Marshall. 1968. World distribution of brown trout, *Salmo trutta*. Journal of the Fisheries Research Board of Canada 25:2527–2548.

Marcus, M. D., M. K. Young, L. E. Noel, and B. A. Mullan. 1990. Salmonid-habitat relationships in the western United States: a review and indexed bibliography. U.S. Forest Service General Technical Report RM-188.

Mason, J. C. 1976. Response of underyearling coho salmon to supplemental feeding in a natural stream. Journal of Wildlife Management 40:775–788.

McFadden, J. T. 1961. A population study of the brook trout *Salvelinus fontinalis*. Wildlife Monographs 7.

McFadden, J. T. 1969. Trends in freshwater sport fisheries of North America. Transactions of the American Fisheries Society 98:136–150.

McHenry, M. L., P. R. Turner, and W. S. Platts. In press. An evaluation of the Habitat Quality Index in southwestern New Mexico. North American Journal of Fisheries Management.

Metcalfe, N. B., F. A. Huntingford, and J. E. Thorpe. 1988. Feeding intensity, growth rates, and the establishment of life-history patterns in juvenile Atlantic salmon *Salmo salar*. Journal of Animal Ecology 57:463–474.

Moyle, P. B. 1977. In defense of sculpins. Fisheries 2(1):20–23.

Platts, W. S. 1979. Relationships among stream order, fish populations, and aquatic geomorphology in an Idaho river drainage. Fisheries 4(2):5–9.

Platts, W. S., and M. L. McHenry. 1988. Density and biomass of trout and char in western streams. U.S. Forest Service General Technical Report INT-241.

Platts, W. S., and R. L. Nelson. 1988. Fluctuations in trout populations and their implications for land-use evaluation. North American Journal of Fisheries Management 8:333–345.

Poff, N. L., and J. V. Ward. 1989. Implications of streamflow variability and predictability for lotic community structure: a regional analysis of streamflow patterns. Canadian Journal of Fisheries and Aquatic Sciences 46:1805–1818.

Puckett, K. J., and L. M. Dill. 1985. The energetics of feeding territoriality in juvenile coho salmon (*Oncorhynchus kisutch*). Behaviour 92:97–111.

Reeves, G. H., and six coauthors. 1997. Fish habitat restoration in the Pacific Northwest: Fish Creek of Oregon. Pages 335–359 in J. E. Williams, C. A. Wood, and M. P. Dombeck, editors. Watershed restoration: principles and practices. American Fisheries Society, Bethesda, Maryland.

Reisenbichler, R. R., and J. D. McIntyre. 1986. Requirements for integrating natural and artificial production of anadromous salmonids in the Pacific Northwest. Pages 365–374 in R. H. Stroud, editor. Fish culture in fisheries management. American Fisheries Society, Fish Culture Section and Fisheries Management Section, Bethesda, Maryland.

Reynolds, J. B. 1997. Ecology of overwintering fishes in Alaska freshwaters. Pages 281–302 in A. M. Miller and M. W. Oswood, editors. Freshwaters of Alaska. Springer-Verlag, New York.

Rieman, B. E., and J. D. McIntyre. 1993. Demographic and habitat requirements for conservation of bull trout. U.S. Forest Service, Intermountain Research Station, General Technical Report INT-302.

Scarnecchia, D. L., and E. P. Bergersen. 1987. Trout production and standing crop in Colorado's small streams, as related to environmental features. North American Journal of Fisheries Management 7:315–330.

Schill, D. J., J. S. Griffith, and R. E. Gresswell. 1986. Hooking mortality of cutthroat trout in a catch-and-release segment of the Yellowstone River, Yellowstone National Park. North American Journal of Fisheries Management 6:226–232.

Shetter, D. S., and G. R. Alexander. 1970. Results of predator reduction on brook trout and brown trout in 4.2 miles of the North Branch of the Au Sable River. Transactions of the American Fisheries Society 99:312–319.

Smith, R. W., and J. S. Griffith. 1994. Survival of rainbow trout during their first winter in the Henrys Fork of the Snake River, Idaho. Transactions of the American Fisheries Society 123:747–756.

Sorg, C., J. Loomis, D. M. Donnelly, G. Peterson, and L. J. Nelson. 1985. Net economic value of cold and warm water fishing in Idaho. U.S. Forest Service Resource Bulletin RM-11.

Thompson, P. D., and F. J. Rahel. 1996. Evaluation of depletion-removal electrofishing of brook trout in small Rocky Mountain streams. North American Journal of Fisheries Management 16:332–339.

Vannote, R. L., G. W. Minshall, K. W. Cummins, J. R. Sedell, and C. E. Cushing. 1980. The river continuum concept. Canadian Journal of Fisheries and Aquatic Sciences 37:130–137.

Varley, J. D. 1984. The use of restrictive regulations in managing wild salmonids in Yellowstone National Park, with particular reference to cutthroat trout, *Salmo clarki*. Pages 145–156 in J. M. Walton and D. B. Houston, editors. Proceedings of the Olympic wild fish conference. Peninsula College, Fisheries Technology Program, Port Angeles, Washington.

Vincent, E. R. 1987. Effects of stocking catchable-size hatchery rainbow trout in two wild trout species in the Madison River and O'Dell Creek, Montana. North American Journal of Fisheries Management 7:91–105.

Walters, J. P., T. D. Fresques, and S. D. Bryan. 1997. Comparison of creel returns from rainbow trout stocked at two sizes. North American Journal of Fisheries Management 17:474–476.

Waters, T. F. 1977. Secondary production in inland waters. Advances in Ecological Research 10:91–164.

Waters, T. F. 1983. Replacement of brook trout by brown trout over 15 years in a Minnesota stream: production and abundance. Transactions of the American Fisheries Society 112:137–147.

Wesche, T. A., C. M. Goertler, and W. A. Hubert. 1987. Modified habitat suitability index model for brown trout in southeastern Wyoming. North American Journal of Fisheries Management 7:232–237.

White, R. J. 1996. Growth and development of North American stream habitat management for fish. Canadian Journal of Fisheries and Aquatic Sciences 53(Supplement 1):342–363.

Wiens, J. A., and J. T. Rotenberry. 1981. Habitat associations and community structure of birds in shrub steppe environments. Ecological Monographs 51:21–42.

Wilkie, M. P., and six coauthors. 1996. Physiology and survival of wild Atlantic salmon following an-
    gling in warm summer waters. Transactions of the American Fisheries Society 125:572–580.

Williams, J. E., C. A. Wood, and M. P. Dombeck, editors. 1997. Watershed restoration: principles and
    practices. American Fisheries Society, Bethesda, Maryland.

Wood, C. C. 1987. Predation of juvenile Pacific salmon by the common merganser (*Mergus merganser*)
    on eastern Vancouver Island. II: predation of stream-resident juvenile salmon by merganser broods.
    Canadian Journal of Fisheries and Aquatic Sciences 44:950–959.

Wydoski, R. S. 1980. Relation of hooking mortality and sublethal hooking stress to quality fishery
    management. Pages 43–87 *in* R. A. Barnhart and T. D. Roelofs, editors. Catch-and-release fishing
    as a management tool. California Cooperative Fishery Research Unit, Humboldt State University,
    Arcata.

Chapter 19

Warmwater Streams

CHARLES F. RABENI AND ROBERT B. JACOBSON

19.1 INTRODUCTION

Warmwater streams are those streams in which temperature becomes too warm to support a self–sustaining trout population. This designation was developed because management needs differ for different species of fish (see Krumholz 1981). A large number of North American streams meet this warmwater criterion and account for almost one-half million kilometers of fishable waters. Fishing for warmwater species predominates some portion of stream fisheries in 49 states and provinces of the United States and Canada and is the only type of stream fishing in over half of them (Funk 1970).

Numerous species are sought by anglers in warmwater streams. The most popular are largemouth bass, smallmouth bass, spotted bass, striped bass, muskellunge, northern pike, walleye, channel catfish as well as other ictalurids, and common carp. Important species regionally are rock bass, pumpkinseed, bluegill, white and black crappies, other sunfishes (*Lepomis* spp.), various bullheads (*Ameiurus* spp.), white perch, yellow perch, chain pickerel, buffalos, various suckers, and freshwater drum. Overall, 32 species of game fishes, as defined by state and provincial fisheries programs, occur in warmwater streams (Table 19.1).

19.1.1 Characteristics of Warmwater Streams

Warmwater streams occur in parts of North America that have very different climatic and physiographic characteristics. All warmwater streams share some common characteristics (Winger 1981), however, such as occurring in mild climates and having low gradients, higher turbidity, and a wide range of transported substrate particle sizes.

Native fish assemblages of warmwater streams are directly related to the geographic setting of the stream. Geologic and climatic events have influenced speciation and distribution, resulting in species richness that ranges from less than half a dozen species in some western and southwestern drainages to over 70 species in some streams of the Mississippi River drainage. Species introductions, both intentional and unintentional, have expanded the ranges of many warmwater fishes. Geology and climate also influence local stream factors, such as temperature, gradient, bottom type, flow characteristics, dissolved oxygen, depth, cover, food abundance, and the presence of other fish, all of which interact to influence the distribution and abundance of individual fish species. This is why we have "catfish streams" in Kansas and "smallmouth bass streams" in Virginia (Table 19.2).

Table 19.1   Fishes occurring in warmwater streams that U.S. state or Canadian provincial agencies (61 total) list as game fishes.

Fish species	Number of states and provinces in which fish is listed as a game fish
Largemouth bass	41
Smallmouth bass	38
Walleye	36
Crappies	30
Northern pike	30
Striped bass	24
Yellow perch	24
White bass	23
Bluegill	22
Sauger	20
Muskellunge	19
Channel catfish	18
Bullheads	17
Rock bass	17
Tiger muskellunge	16
Spotted bass	14
Blue catfish	12
Pickerel	12
Black and white bass (8 variations)	10
Common carp	10
Flathead catfish	10
Suckers	8
Green sunfish	7
Paddlefish	7
Warmouth	7
Redear sunfish	6
White perch	6
Pumpkinseed	5
Bowfin	3
Redbreast sunfish	3
Gar	2
Longear sunfish	2

Although warmwater streams may differ from each other in physical appearance and resident fish faunas, they have common patterns of fish abundance and distribution (Figure 19.1) that are related to longitudinal zonation (Pflieger 1988).

Headwater streams tend to have higher gradients and a succession of short pools and well-defined riffles. Substrates are usually of a larger mean particle size than in downstream areas. Headwaters of warmwater streams in forested biomes tend to be well shaded by overhanging tree canopies. Because trees that fall into headwaters are infrequently transported, large woody debris contributes substantial scour-pool habitat and fish cover. During dry periods, headwaters may become a series of isolated pools or be entirely dry for long stretches. Fishes inhabiting headwaters are small and possess the ability to survive extremes of temperature, dissolved oxygen, and current velocities. The fish assemblage tends to be controlled primarily by the inherent variability of prevailing physical factors (Matthews 1998). Headwater species rapidly recolonize reaches that periodically become dry. Because the biota of wadable headwaters are more efficiently sampled, much more has been learned about this community compared with downstream areas.

Table 19.2 An idealized classification of warmwater stream types based upon physical characteristics and some representative game fishes.

Physical characteristics	Type I	Type II	Type III
Predominant substrate type	Cobble	Fine	Fine sand and silt
Gradient	Medium	Medium–low	Low
Mean velocity	Medium	Medium–low	Low
Channel type (sinuosity)	Low	Intermediate	High
Riffle–pool development	High	Medium	Low
Turbidity	Low	Low–medium	Medium–high
Edaphic factors	Coarse	Intermediate	Fine
Stream size	Small–large	Small–large	Medium–large
Representative game fishes	Smallmouth bass, rock bass, walleye, northern pike	Largemouth bass, warmouth, buffalo, bullhead catfishes	Channel catfish, common carp,

Midreach areas are more diverse in their habitat conditions. Pools tend to be longer and deeper, and riffles are short and well defined. Fluctuations in flow regularly occur, but cessation of flow usually is not a problem. Substrate particle size tends to be relatively large, and total fish cover in the form of undercut banks, rootwads, and downed trees is more prevalent than in headwater areas. Because stream channels are wider, tree canopies tend to be more open in midreach sections than farther upstream, and midreach sections are more susceptible to heating from sunlight. Fish assemblages are more diverse than in headwater areas. Biological factors, such as spawning habitat requirements and trophic interactions, become more important as regulators of fish assemblage structure.

Main-stem reaches are characterized by long pools separated by deep runs; riffles are usually absent. Backwaters and meander cutoffs are more common than they are upstream. Substrate particle size tends to be small, and turbidity is usually at its highest. The trend of increasing species richness, species diversity, and body size continues in this region.

Predictable changes in fish assemblages occur on a longitudinal basis regardless of the type of warmwater stream. Species richness and species diversity both increase in a downstream direction. Headwater species are found in other areas, but their relative abundance is drastically decreased downstream. Figure 19.1 compares data from two stream types found in two different ecoregions of Missouri, the Ozark and Prairie (comparable to types I and II in Table 19.2, respectively). Although absolute values are greater for the Ozark stream, the longitudinal patterns are similar.

The among-stream-type comparison (Table 19.2) and the within-stream continuum (Figure 19.1) indicate the potential of each stream type and stream segment for productivity, a specific assemblage structure, and the ability to support a particular fishery. Managers of warmwater streams strive to understand these potentials in order to formulate management objectives.

19.1.2 A History of Disturbance

Warmwater streams have historically been foci for human habitation and as a result are susceptible to disturbances such as sedimentation, erosion, water pollution, and altered hydrologic conditions. A major role of management is controlling distur-

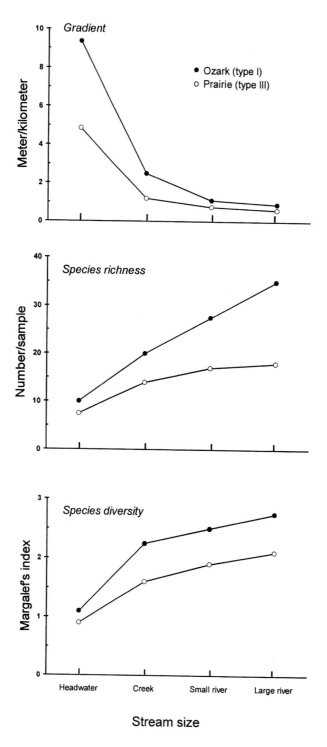

Figure 19.1 Longitudinal patterns of two warmwater stream types found in Missouri. Data are from Pflieger (1988).

bance and perhaps restoring streams from the consequences of past disturbance. To accomplish this role, managers must understand the processes that link physical factors (hydrologic characteristics, sediment supply, and geologic constraints), land use, and fish habitat at a variety of spatial and temporal scales. Many natural sources of disturbance—storms, floods, and wildfires—influence streams and their watersheds. Hence it can be difficult to document how land use changes are linked to a stream's biotic community. An historical perspective of a stream and its drainage basin can provide warmwater stream managers with an understanding of historical baseline conditions, factors that are outside historical norms, and factors that are amenable to management.

The history of fishes in warmwater streams in Ohio has been well documented (Trautman 1981) and probably typifies the magnitude of changes that have occurred throughout North America since the first Europeans arrived. Major damages to warmwater streams occurred rapidly following European settlement, and the situation generally has been exacerbated since then. Around 1800 there were reports of "huge populations of the larger and better food fishes and gamefishes; pikes, walleye, catfishes, suckers, drum, sturgeon and black basses" (Trautman 1981). Within 50 years, however, extensive draining and filling of marshes had irrevocably altered headwater conditions such that many permanent springs and brooks became intermittent, and the shallow, ephemeral wetlands necessary for reproduction of many species were lost. By 1900 increasing agricultural activity and channel and riparian modifications created numerous clay bottom streams, and siltation and sedimentation became the most important pollutants affecting the fish fauna.

In Ohio's Maumee River system, 17 species of fish have completely disappeared in the Twentieth Century and 26 species have become less abundant (Karr et al. 1985). About half of all species in headwaters and midreaches have disappeared. The species that were eliminated tended to be specialists with a narrow range of acceptable conditions for spawning or water quality. Especially affected were invertivores and herbivores because their foods require silt-free surfaces. Top carnivores, including northern pike, walleye, and smallmouth bass, have also become less abundant in the Twentieth Century, and less-specialized species have increased in abundance (e.g., gizzard shad, quillback, and bigmouth buffalo). The river system has shifted from being dominated by predatory species to one of predominantly omnivores. This change in the relative abundance of feeding guilds is important to a warmwater stream manager whose objective may be to increase the quality of the fishery for top predators.

Since the earliest European settlement, warmwater streams were seen as opportunities to produce power with dams, transport wastes, or create lakes and were considered to be obstacles to agricultural production. From the fishery perspective, they were hard to sample and evaluate, and there was a prevailing view that they were "self managing"; that is, there were fish for those few anglers who cared to take advantage of them. Most ecological degradation of warmwater streams is undocumented. We now have the problem of determining the potential of these systems after they have been degraded and the important factors limiting their potential.

19.1.3   Legislation and Policy

Attitudes have been changing gradually, and several events since the late 1960s have encouraged wiser use and greater appreciation of warmwater fisheries resources. First was a realization by state and provincial agencies that fishing in warmwater streams was popular. For example, in Iowa warmwater streams account for about one-quarter of all fishing trips. In Missouri, a state with extensive farm pond and reservoir resources, about 20% of all fishing occurs in warmwater streams. Oregon and Kentucky both initiated warmwater stream management programs after discovering that 25% and 35%, respectively, of all their fishing was on warmwater streams.

Federal legislation in the United States has affected warmwater stream management through mandates for improved water quality, increased funding for management, and increased protection for streams and their watersheds. The Federal Water Pollution Control Act Amendments of 1972 (Clean Water Act) served as a major stimulus (US$30 billion) for abatement of point-source pollution. There are numerous cases in which, once point source pollutants were removed, fishes rapidly returned to the affected waterway, the stream or river became aesthetically acceptable, and fishers were once again attracted there.

The Wallop–Breaux Aquatic Resources Trust Act of 1984, which imposed a federal excise tax on sportfishing equipment, boats, motors, trailers, and motor boat fuels, supplemented the established Dingell–Johnson Federal Aid in Sport Fish Restoration Act so that almost $200 million are available annually for state fish and game agencies. This funding has allowed several states to initiate programs on warmwater streams. The Farm Bill (Food Security Act of 1985), with its key conservation easement and swampbuster sections, gives farmers financial incentives to stop cultivating highly erodible fields and put them into cover crops that may ultimately slow watershed level impacts and protect millions of acres of stream riparian habitat.

Although there has been progress in the abatement of point source pollution, much less has been done to address the major problems associated with flow modification, physical habitat degradation (e.g., alteration of channel shape, substrate size, and amount of woody structure), and other nonpoint source problems. Additionally, an understanding of population dynamics of individual species and interactions among species in warmwater streams lags behind comparable knowledge in other aquatic systems.

19.2   WARMWATER STREAMS MANAGEMENT

The role of fisheries biologists responsible for warmwater streams is different from that of their coldwater counterparts in several respects: (1) warmwater streams tend to be more subject to human modification and often have severe habitat and water quality problems, (2) multispecies recreational fisheries often exist, and (3) much basic ecological information on fishes and the associated biota is not available.

19.2.1 Management Goals

Fishery management goals in warmwater streams fall under three broad categories.

Restoration and rehabilitation—Bringing a degraded fish population or assemblage back toward its former potential. Examples of rehabilitation activities are removing dams, unchannelizing streams, eliminating severely eroding streambanks, and improving water quality. Restoration involves reestablishing the structure and function of the ecosystem in a more holistic process (Williams et al. 1997).

Maintenance—Conserving a fishery's status quo. Common maintenance activities include assessing effects of actual or planned water diversions (instream flow) and defining the consequences of planned development in the watershed or along the stream. Environmental impact analysis, evaluating endangered or threatened species habitats, defining reference reaches, monitoring fishes and habitat, assessing long-term cumulative habitat changes, and facilitating general basinwide planning are all potential components of maintenance management.

Enhancement—Increasing the abundance or changing the size structure of a population or altering existing species assemblages.

Each goal requires particular management activities, but all must occur in an ecological context that takes into account complex interrelationships of the biota with its environment. A holistic view that can serve as the framework for all management goals is the concept of biological integrity. For streams, biological integrity depends upon five primary components: water quality, flow regime, habitat structure, energy source for biota, and biotic interactions (Karr and Dudley 1981; Figure 19.2). Altering the processes associated with any of these components can have a major effect on stream fishes. Additionally, efforts to rehabilitate or maintain a stream fishery by attention to only one component will be disappointing if more than one factor has been affected. The manager must be able to identify processes influencing each primary component.

Management of warmwater stream fisheries is currently divided between efforts to manage a particular species and efforts to manage the entire fish assemblage. The species approach holds that sport fishes, being mostly top carnivores, reflect an integration of conditions in the stream environment, and maintaining their well-being will benefit the entire stream biota. Often the species approach involves an endangered or threatened species, and the primary goal is to save the species from extinction; a secondary goal is to improve the environment for other species. The assemblage management approach is predicated on concepts of biological integrity or biodiversity, and it stresses that an appropriate balance of either taxonomic or ecological groups reflects the quality of the stream environment. The two approaches are not necessarily contradictory. It is unlikely that managing for a native sport fish or an endangered species will be detrimental to other species of fish in that stream, and it is likely that improving conditions for the fish assemblage will help species of particular interest. Managers should view their goal, be it the integrity of the fish assemblage or the well-being of a particular species, in light of the five primary components of stream integrity (Figure 19.2).

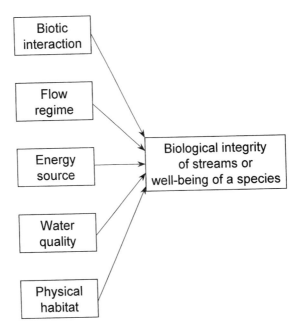

Figure 19.2   Classes of factors that influence aquatic biota and determine either the biological integrity of the stream, as evidenced by the fish assemblage, or the well-being of a particular species. Figure adapted from Karr and Dudley (1981).

19.2.2   The Spatial Scale of Management

Understanding the connection between habitat and the well-being of fish populations is the first step in an effective management strategy. Habitat for a stream-dwelling fish may be defined as the local physicochemical and biological features of a site that constitute the daily environment of fish (Milner et al. 1985). Of importance to a manager is that whereas fish respond to local conditions, some of these conditions originate far from the stream. Some locally expressed elements of the habitat—such as water temperature, nutrients, dissolved oxygen, and flow variation and regime—are reflections of broader-scale watershed influences (Table 19.3). These watershed influences may determine the overall livability of a stream and its potential capacity to support fish. Intense local efforts to remedy particular fisheries problems may be ineffective if watershed influences are the overriding control.

Managers of warmwater stream fisheries must understand that the potential and limitations for fishes in a particular stream depend upon conditions within the stream channel, in the riparian zone, and in the watershed. Whereas watershed-scale management activities expressly directed at fisheries are rare, an understanding of land use and its probable influence on stream biota is crucial to an understanding of a stream's biotic condition.

19.3   WATERSHED-SCALE CONSIDERATIONS

At the scale of the watershed, characteristics of fish habitat are linked to the quantity, timing, and quality of rainfall runoff and sediment supply (see Chapter 9). Responses of fish habitat to land use changes are not instantaneous and are often difficult

Table 19.3 Factors that influence fish distribution and abundance delineated into those influenced primarily at the local site level, requiring local management, and those influenced at the watershed level, requiring watershed-scale management.

	Appropriate scale of management	
Variable	Watershed	Local site
Water quantity		
Mean flow	x	
Extremes of flow	x	
Variation in flow	x	
Water quality		
Nutrients	x	
Dissolved solids	x	
Temperature	x	x
Turbidity	x	x
Dissolved oxygen	x	x
Organic matter	x	x
Channel quality		
Gradient	x	x
Substrate size	x	x
Depth		x
Riffle–pool sequence		x
Width		x
Bank erosion		x
Overhanging vegetation		x
Woody debris		x

to predict. In a study of the relationship between land use and fish assemblages in Wisconsin streams (Wang et al. 1997), the health of fish assemblages, as measured by a fish index of biotic integrity, was negatively correlated with the amount of upstream urban development and agricultural land. Fish assemblage health was positively related to the amount of upstream forest in the watershed. There appeared to be threshold levels of the extent of particular land uses before the fish assemblage was influenced. A decline in the condition of the fish assemblage was noted when 20% or more of the watershed was urbanized. Declines in the index of biotic integrity attributed to agriculture were noted when more than 50% of the watershed was used for this purpose. Similar results were obtained in a Michigan study (Roth et al. 1996) in which regional land use was the prime determinant of stream conditions, even more important than local site vegetation, in supporting high-quality habitat and fish assemblages. Studies that describe correlations among stream biota and land use activities are an important first step in watershed management for fisheries purposes. However, to make sound management decisions the mechanisms behind the associations need to be elucidated. There are many possible reasons for any of the relationships found, and blaming agriculture or urbanization for fisheries problems does not help to solve the problems.

In many areas, a significant consequence of land use change has been alteration of the water table and runoff patterns, a problem that is especially widespread in agricultural areas. Many landscapes in formerly glaciated regions or on floodplains of large rivers were originally wetter, and headwater and bottomland marshes and wetlands were common. Considerable draining and dredging was carried out to prepare the land for agriculture. Encouragement came from legislation such as the Illinois Farm Drainage Act of 1879, which formed drainage districts and enabled widespread installation

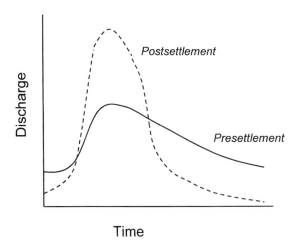

Figure 19.3 Alteration of discharge pattern for streams in watersheds affected by urbanization or row crop agriculture.

of drainage systems. For example, 74% of Champaign County was in drainage districts by 1928 and 82% by 1959. These systems eliminated standing water, created new channels, delivered large quantities of fine sediment to streams, and created a flashy hydrograph (Larimore and Bayley 1996).

The consequences of an altered hydrologic regime to the fish fauna can be substantial. Historically, fish assemblages were adapted to the regular, slow springtime rises in discharge during spawning and other life history activities and were able to complete the year with moderate but adequate flows (Figure 19.3). Many intensive land use activities have reduced the water storage capacity of the earth and have drastically lowered water levels. Species must now cope with much greater flows during limited periods of the year and lesser flows the rest of the time. It is encouraging that in some instances where land use activities have become more compatible with natural processes, there has been a concomitant improvement in flow patterns. Beginning in about 1940, southwestern Wisconsin has undergone substantial changes in basinwide agricultural practices. Steep-sloping pastures were abandoned, and they reverted to forest. Contour plowing and deep plowing were initiated on a wide scale. These changes increased rainfall infiltration and decreased overland flow. Long-term flow records on streams show a corresponding decrease in peak flows and increased and more sustained base flows (Gebert and Krug 1966). Gebert and Krug's study is an indication of the long-range benefits to fisheries from best management practices on agricultural watersheds.

A given land use change can result in unpredicted alterations in runoff and sediment yield; the effect on fish habitat depends upon the balance between the processes. For example, a study (Jacobson and Coleman 1986) of stream responses to agricultural activity in Maryland piedmont watersheds identified a sequence in which streams aggraded (the streambed elevated) due to increased sediment from poor agricultural practices in the eighteenth and nineteenth centuries. With the retirement of marginal land from row crop agriculture and the institution of soil conservation practices, sediment yield decreased. Yet runoff continued to be higher than natural, and the streams became incised (cut down because of bed erosion) and their bed materials coarsened (Figure 19.4).

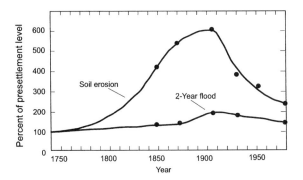

Figure 19.4 Estimated magnitude of soil erosion and the 2-year flood for the Maryland piedmont preceding and following widespread agriculture. Figure adapted from Jacobson and Coleman (1986).

Thresholds of flow conditions that initiate sediment movement, lagged transport of sediment through drainage basins, and downstream movement of sediment to and from floodplain storage can result in disturbances that propagate slowly through drainage networks in unpredictable ways, or what geomorphologists call a complex response (Schumm 1977). Moreover, in larger drainage basins many different land uses and natural climatic variations may take place simultaneously. Thus, it becomes very difficult to link accurately the effect of land use in the watershed to fish habitat in a stream. The complexity of land use effects on the stream channel presents a challenge to management at the watershed scale.

19.4 RIPARIAN ZONE MANAGEMENT

19.4.1 Natural Processes

Fish habitats are also affected by factors that operate in the riparian zone. Examples of riparian zone factors are channel interactions with valley walls, riparian vegetation, riparian land use, and gully inflows. These factors can be described generally as those that affect hydraulic resistance (ability of bank materials or vegetation to slow velocities and dissipate energy), erosional resistance (ability of bank materials and vegetation to resist erosion), and channel morphology. These factors vary longitudinally along most streams and control the spatial variability of habitats useful to fishes.

Some riparian zone factors, such as valley geometry and tributary junctions, influence fish habitat independent of land use. In many warmwater streams, places where the channel impinges on a valley wall are characterized by the input of cobbles and boulders from erosion on adjacent slopes, increased scour, and consequent pool formation. Interactions of the channel and the valley wall also can cause nonuniform distributions of lateral channel erosion along a valley (Figure 19.5). Unlike the standard textbook example of a meandering stream in which lateral erosion is essentially uniform along the stream valley, streams in confining valleys tend to have alternating disturbed and stable reaches. Although unstable, disturbed reaches in Ozark streams tend to have some of the highest hydraulic diversity (variation of depth and velocity at a given discharge) (McKenney 1997). Stable reaches, on the other hand, have low hydraulic diversity but greater canopy coverage. Recognition of these distinct, natural sources of variation among reaches helps a manager target riparian management to

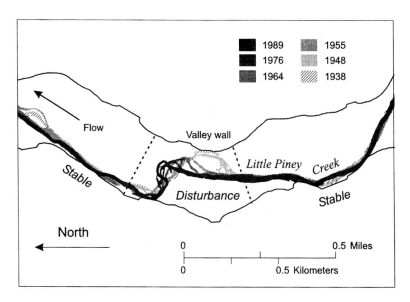

Figure 19.5 Record of channel changes in a dynamic stream system over 50 years, Little Piney Creek, Missouri. Distinct longitudinal differences in inherent channel stability within the river segment distinguish stable and disturbance reaches.

reaches where maximum benefit and success can be achieved. For example, in the case of the river shown in Figure 19.5, riparian woody vegetation was ineffective in slowing lateral channel erosion because much erosional energy was concentrated in disturbed reaches. Thus, some sections of stream, because of their physical characteristics, are much more prone to natural migration. To view these areas as in need of management activity to stabilize them would be a mistake.

Tributary junctions often create stable riffles because large bed materials are deposited there. Furthermore, tributary junctions are sites of greater water and sediment discharges. Management of fisheries resources within a drainage basin can be optimized if conditions at tributary junctions are recognized and management actions are tailored to them.

19.4.2 Buffer Strips

Because riparian corridors have been extensively altered due to channelization, cultivation, livestock grazing, and timber harvest, reestablishment or enhancement of riparian vegetation is a common management action in warmwater streams. Although few professionals dispute the value of riparian management, there is much discussion of appropriate dimensions, both length and width, of buffer strips. Determining the appropriate size of riparian buffer strip requires taking into account (1) the intended function of the buffer strip, (2) the natural situation (especially topography and soil characteristics), and (3) human activities.

Riparian vegetation serves a number of functions depending upon its position relative to the channel. If the banks are not too steep or if the bank height is not greater than the rooting depth (Thorne 1990), vegetation adjacent to and overhanging the channel can contribute to bank stabilization. Vegetation is most effective at stabilizing banks of smaller streams. Riparian vegetation moderates water temperature by shading the channel during the summer, and it increases allochthonous energy in the form of leaf fall and terrestrial insects. Trees that fall into the stream act as retention devices for organic matter as well as increasing the diversity of habitat for fishes and invertebrates. Riparian vegetation that is not adjacent to or overhanging the stream channel also functions to reduce the amount of sediments, pesticides, herbicides, and excessive nutrients reaching the stream. It also increases infiltration and stabilizes the water table, both of which reduce the magnitude of hydrograph fluctuations.

The few studies that have addressed appropriate riparian zone widths on warmwater streams have concentrated on reducing sediment. Biologists of the U.S. Forest Service have quantified soil erodibility, soil drainage, and stream gradient in a model that determines the appropriate distance from the channel that particular activities should be allowed in southeastern forests. Other published guidelines or regulations are often estimates by professionals who are familiar with local conditions. For example, planning documents developed by various states recommend vegetation strip widths ranging from 6 m in the lower coastal plain of Georgia to 152 m in some areas of Alabama.

As important as the area of riparian vegetation is its quality. The ideal riparian buffer zone would have mature stands of trees spaced sufficiently to allow an understory of grasses and shrubs. Trees would be close enough to the stream so that exposed roots would supply cover and allow stable undercut banks to develop. Grasses and shrubs would have deep root systems, dense well-ramified top growth, a resistance to flooding and drought, and an ability to recover and grow subsequent to inundation by sediment.

19.4.3 Sedimentation

Sediment produced by human activities is considered the major pollutant in waters of the United States (Waters 1995). Sedimentation has caused changes in fish faunas as agricultural and other human activities have increased (Smith 1971; Muncy et al. 1979). Excessive siltation from erosion has occurred in 46% of all U.S. streams and is considered the most important factor limiting fish habitat (Judy et al. 1984).

Fine sediments impair the stream biota in a number of ways. While in suspension, fine sediments limit light penetration and reduce primary production by algae and macrophytes. Turbidity affects fish behaviors such as feeding and social interactions. When sediment comes out of suspension and settles on and within the substrate, pools fill in and the channel environment tends toward homogeneity and away from discrete riffle–pool divisions. As sedimentation increases in riffles, fish assemblage structure changes from riffle-specific species to more run- and pool-specific species, thus decreasing overall fish diversity (Berkman and Rabeni 1987).

Sedimentation often degrades spawning areas and may cause behavioral changes in spawning fish, an increase in egg mortality, and a decrease in larval growth and development. The species most affected are those requiring gravel or vegetation for spawning substrate, those with complex spawning behaviors and courtship rituals, and those that construct nests from bottom materials (Bulkley et al. 1976). Sedimentation disrupts normal trophic relations because primary production and invertebrate secondary production can be drastically reduced by siltation. Consequently, piscivores and invertivores, which constitute most sport fishes, are often replaced by omnivores and detritivores.

Managers must determine where sediment originates, but there has been no easy or acceptable way to measure the origin and movement of fine sediments that enter streams. A major sediment contributor in physically degraded streams is the channel itself. Several studies in Illinois (summarized by Roseboom and Russell 1985) show that about 50% of the annual sediment load of streams is from the eroding streambed and banks. This indicates that stream channel stabilization should be a management priority.

Documenting movement of sediment from the upper watershed into floodplains and streams is difficult. One innovative study (Wilkin and Hebel 1982) tracked erosion within a 526-ha watershed in Illinois by examining concentrations of radioactive cesium 137, an isotope evenly deposited onto the landscape as a consequence of the 1950s atmospheric testing of atomic weapons. Concentrations of the isotope in different parts of the watershed 30 years after deposition were due to erosion and redeposition. Wilkin and Hebel's study concluded that sediment delivery to a stream can be controlled by management. A forested buffer zone between upland row crops and the floodplain prevented almost all delivery of eroded soil onto the active floodplain. Floodplains planted in row crops eroded at high rates and delivered eroded soils to downstream flow. Forested floodplains on these small streams, even those grazed, served as depositional areas during overbank flows and removed significant amounts of sediments from upstream. Thus, we can conclude that floodplain management should be a priority in attempts to control excess sediment.

19.5   CHANNEL MANAGEMENT

The way in which a reach of river will respond to disturbance depends upon many factors, including the history of disturbance and adjustment, position of the reach within the channel network, riparian vegetation dynamics, and physical factors that determine watershed hydrology and sediment load. How a river will respond to disturbance also depends upon whether the stream reach is in a state of equilibrium, that is, a conceptual state of balance between water discharge and sediment supply such that channel morphology and physical habitat vary about a mean condition (Figure 19.6). If not, the stream may be in the midst of a disturbance or recovery. Over time, some systems work their way back to an original equilibrium condition (Schumm 1977). Alternatively, some stream systems will react to disturbance by attaining and persisting in an alternative equilibrium state. Once a stream is disturbed, three conceptual pathways exist (Bull 1991; Figure 19.6). (1) The channel returns after some lag to its original form. (2) The channel remains in a new form. This is common when a threshold is surpassed, for example, when streambanks are oversteepened. (3) In a disturbed state, it is possible that streams are more sensitive to disturbances than they were previously; in other words, once disturbed, streams may have a lower threshold to geomorphic change.

The potential that stream channels will exhibit complex responses requires that managers develop an historical understanding of where the channel has been and where it is likely to go. In many cases, historic biological and physical conditions are difficult to establish with certainty but may be estimated from maps, oral accounts, climatic records, alluvial stratigraphy, highway construction records, and aerial photography (Kondolf and Larson 1995; Jacobson and Pugh 1997). Even with a lack of this historical context, the manager may apply management activities in areas that will continue to be overwhelmed by ongoing physical changes.

Only after recognizing the limitations imposed on the biota by activities occurring in the watershed—be these limitations alteration of flows, increases in nutrients or sedimentation, or a change in the temperature regime—are management activities at smaller spatial scales warranted. If watershed limitations are not severe, more localized activities addressing problems affecting fish habitat within a reach of stream may be considered. These activities are interrelated and focus on (1) reducing streambank erosion and instream sedimentation, (2) modifying channel morphology and alignment, and (3) increasing instream cover (Lyons and Courtney 1990) .

19.5.1   Fish Habitat Improvements

Activities carried out to improve the local habitat of fishes are popular because they are usually limited in scope and effort and can be readily evaluated, although evaluation is often lacking. Given that the immediate physical structure of the channel can be manipulated, what is the best strategy? What are the changes that can be made locally that will accomplish a given objective for a species or an entire fish assemblage? Much is now known about how warmwater fishes relate to their immediate environment (see Matthews 1987 for a review). Warmwater stream fishes, both single populations and the entire assemblage, are known to segregate by size within species along gradients of depth (Mendelson 1975), current velocity (Mendelson 1975; Probst et al. 1984), and bottom type (Paine et al. 1982). They are influenced by debris and instream cover, light levels, and aquatic vegetation. The greater the breadth of gradients of these factors, the more opportunity there is for a fish to find its optimal surroundings. Although we have a good idea about how many species relate to each other and to their physical habitat, however, we do not understand how important most habi-

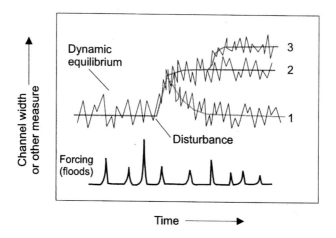

Figure 19.6 Response patterns for streams following a disturbance. See text for explanation of three dynamic equilibria modeled. Figure adapted from Bull (1991).

tat elements are to the well-being of fishes. Knowing a species associates with a habitat element, such as submerged logs, does not mean that altering the amount of downed logs will in any way change the growth, reproduction, or survivorship of that species.

The importance of particular habitat elements to fishes is difficult to evaluate. For example, a particular depth may be used by a species but may not be essential to its well-being. Another habitat element, say a riffle that provides food as drift, may be essential, but the fish may not be found there. Some habitat conditions may have short-term detriment but long-term benefit, such as a spate that washes away young fishes and scours substrate but also removes interstitial silt and creates or maintains undercut banks. Some habitat conditions may be important for only short periods of the day, vary seasonally, or change as fishes grow.

Determining preferences for important habitat variables for multiple life stages of every species in a warmwater stream is not feasible. A more appropriate strategy may be to concentrate on a single assemblage-level objective. Although relationships between fish assemblages and their environment are complex, these relationships can be reduced to fairly manageable categories of fish types (e.g., guilds) and sizes and the relationship of these categories to broad, similar habitat conditions (Bain and Boltz 1989). Habitat conditions most commonly measured are current velocity, depth, substrate type, and structure (fish cover). A promising management strategy aims at increasing the diversity of hydraulic conditions, depth, structure, and substrate.

An initial emphasis of fish habitat improvement is to reestablish natural sinuosity to the stream whenever possible. Such efforts decrease gradient and increase secondary currents, thereby decreasing the tendency of the stream to incise and increasing the diversity of many habitat elements. Next, attention should be turned to achieving appropriate depths and adequate structure.

Depth is an important factor for many sport fishes because it (1) serves as cover, (2) provides refuge during low water or the winter, and (3) increases the optimal habitat for many forage fishes. Greater depth can be obtained in several ways (see Chapters 9 and 10). If natural meandering of the channel and bank stabilization can be facilitated, increased depth will be a natural by-product. Depth also increases as bank strength and scour opportunities increase with establishment of some types of riparian vegetation. Central Indiana streams that were flanked by trees had pools 50% deeper than did streams adjacent to residential areas, grazed meadows, or cultivated fields (Gammon et al. 1983). Additionally, wherever cultivated fields were separated from the stream by strips of grass or trees, the streams were considerably deeper. The installation of instream structures such as gabions (rock-filled baskets) or rock structures may result in scouring and pool deepening; they also act to retain water, which will also serve to increase depth.

Structure can be defined as any physical obstruction in the stream that provides fishes with camouflage, shade, or relief from the current. Structure in warmwater streams can be achieved by many of the methods detailed in Chapter 10. Here, we will concentrate on wood. Wood has many advantages as a structure type: it is inexpensive and natural looking, and it is continually produced on the streambanks. The importance of woody structure (logs, branches, rootwads, snags, and log jams) to coldwater stream habitat is well documented (see Chapters 10 and 18). It may be that wood is even more important in warmwater streams because there is typically less of other classes of objects (e.g., boulders, cobble substrate, and bedrock ledges) that serve similar functions. Wood provides three important functions in low-gradient, small-substrate streams: (1) it interacts with the hydraulics to increase diversity of velocity and depth; (2) it provides fish sanctuary; and (3) it supplies habitat for invertebrates.

Smallmouth bass and rock bass in Ozark streams associate more closely with woody structure—individual logs, log jams (consisting of trunks and branches), and rootwads—than with other habitat types (Probst et al. 1984). The highest densities of muskellunge in 14 streams in Kentucky were found in reaches with the greatest number of fallen trees. This factor was determined to be such an important ingredient in muskellunge management that a program to preserve riparian vegetation is now in place (Axon and Kornman 1986). In a medium-size stream in an agricultural area of northern Missouri, stretches without appreciable woody structure had densities of game fishes (common carp, catfishes, green sunfish, bluegill, smallmouth bass, and crappies) that were 25% lower than were densities in areas with woody structure (Hickman 1975). In areas without wood, there were 51% fewer catchable-size fish than in areas where wood was present.

The adaptive significance of the associations between fishes and woody debris has been attributed to wood's function as camouflage rather than to its functions of increasing food availability or protecting fishes from current velocity. A small stream in Illinois was longitudinally divided for a distance of 30.5 m and woody debris was removed from one side (Angermeier and Karr 1984). On the cleared side, bottom substrate became more homogeneous and of a finer particle size. Depth of the altered channel was reduced, as was organic litter on the bottom. These changes were due to the way in which wood functions to alter the flow regime, consequently assisting in the scouring and removal of fine particles. Sixty percent of the fish species of the experimental reach were more abundant on the side of the stream with woody debris, and larger individuals were found there.

In streams with low gradients and fine-particle substrates, woody structure provides an important habitat for invertebrates used as food by fishes. The Satilla River in southeast Georgia has three major benthic habitats: (1) shifting sand in the main channel, (2) muddy depositional backwaters, and (3) submerged wood (Benke et al. 1985). Woody structure made up 4% of total habitat surfaces yet supported 60% of invertebrate biomass and 16% of invertebrate production. All fish species used woody structure to some extent, and four of eight major species obtained at least 60% of their prey biomass from woody structure.

19.5.2   Rehabilitation of Channelized Streams

Channelization is highly detrimental to stream fishes. The act of straightening a stream is, by itself, harmful to aquatic life, but it is often accompanied by removal of instream cover and replacement of streamside vegetation with grass berms or row crops. The devastating effect of channelization on resident fish fauna is well documented (e.g., Schneberger and Funk 1973; Stern and Stern 1980). It is not unusual for the numbers and biomass of sport fishes to be reduced more than 90% in a channelized area. Channelization is widespread in North America, especially in agricultural areas. For example, Iowa has lost over 2500 km of streams capable of supporting permanent fish populations due to channelization (Bulkley et al. 1976).

The few evaluations of mitigation or restoration efforts in channelized streams indicate that a variety of approaches can improve the habitat and the fishery. One approach is to leave the stream alone and let the stream rehabilitate itself; however, this process requires substantial time. For example, responses to channelization in some western Tennessee streams involved a consistent sequence toward recovery of some of the channel sinuosity and took up to 50 years to complete (Simon 1994). Minimal efforts can accelerate the recovery process. For example, short reaches of channelized streams (<0.5 km) in Iowa could support the same biomass of ictalurid catfishes as unchannelized reaches as long as brush piles and trees accumulated in the channelized stream (Bulkley et al. 1976).

More extensive efforts, such as installing permanent instream structures, can be successful in improving a fishery, but they can decrease the intended drainage performance of channelized reaches. The Styx and Chippewa rivers in central Ohio have been extensively altered. The stream channels of these rivers are bounded by high banks. Instream structures (mainly wing deflectors) that were installed occupied a comparatively small area. Reaches with structures held significantly more species and higher numbers and biomasses than did reaches without structures (Carline and Klosiewski 1985). Nevertheless, important game species did not respond as intended. Although the deflectors appeared to create favorable conditions for centrarchids, low densities and small sizes of these fishes provided minimal fishery benefits. Even though some habitat conditions were improved, erratic flows caused reproductive failures: because of the high-banked channel morphology, rising waters during rains were confined to the channel, and high current velocities caused destruction of nests and eggs.

The Olentangy River in Ohio is an example of a more successful rehabilitation effort (Edwards et al. 1984). It had only a small proportion of its length channelized, and an unaltered section separated the mitigated area from a channelized section. The floodplain adjacent to the channelized and unaltered reaches was forested. When high flows occurred, water spread out into the floodplain, and in-channel velocities were moderated. Artificial riffles and pools were created by constructing from bank to bank five equally spaced riffles with boulders over earthen fill. Game fishes in the mitigated area responded, and relative abundance and body condition increased. Diversity of macroinvertebrates and fishes in the artificial riffles and pools was intermediate when compared with natural and channelized sites. Not only was adequate physical cover provided, but flow conditions were such that high flows were moderated, and structures retarded the rate of dewatering during dry periods.

Warmwater stream habitat can be managed successfully if emphasis is placed on stabilizing hydraulic conditions that improve pool–riffle development. The ideal fishery management response to channelization is to rehabilitate the effected stretch to near its original physical condition. In 1948, Big Buffalo Creek in west-central Missouri was extensively channelized, causing the predictable loss of stream habitat diversity and channel stability. In 1966, the stream was reclaimed by replacing 1.3 km of channelized stream with about 2.9 km of restored channel. About half of the rehabilitated length was a new channel, and the rest consisted of abandoned channels formerly used by the creek. High water rearranged some of the channel morphology during the first 5 years, but soon stability was established, and the stream came into dynamic equilibrium. Little erosion or deposition has since occurred. The standing crop of fishes in the new channel increased from zero in 1966 to 67.5 kg/ha in 1972 (Fajen 1975), and fish populations have remained stable.

19.6   MANAGEMENT OF FISH POPULATIONS

Effective fisheries management must address the triad of fisheries users, aquatic organisms, and habitat. Warmwater stream fisheries managers have focused on the problem of habitat degradation. However, there are thousands of kilometers of fishable warmwater streams where habitat restoration or rehabilitation as a management objective has been achieved or is unnecessary. It is in these waters where benefits may result from manipulation of fish populations and fisheries users.

19.6.1   Stocking

Stocking of common carp by the U.S. Fish Commission in the early 1800s was one of the first management activities to affect the fisheries of warmwater streams. Today, stocking of warmwater streams is not a general policy of most states or provinces because either the need has not been perceived or has been shown to be unnecessary.

Stocking of reclaimed or restored streams in which a given species previously existed may have merit. Stocking of native species into existing populations is usually not necessary or warranted except in special circumstances, for example, in situations in which inconsistent reproductive success is a problem. In Iowa, careful examination of low production of walleye in streams has shown a possible problem with young-of-year survival, and a program of stocking fingerling walleye is underway with encouraging results (Paragamian and Kingery 1992). Walleye were stocked in Wisconsin's lower Fox River to compensate for the lack of suitable spawning substrate (Auer and Auer 1990). A cooperative program in New York State has angler associations culturing walleye in earthen ponds for stocking into tributaries of Lake Ontario (Buttner et al. 1991), providing a fishery that had been absent for over 2 decades. Similar cooperative efforts between the state conservation agency and angler groups are underway in Michigan. Kentucky manages several native muskellunge streams and recommends an annual supplemental stocking of 17–22-cm fish to maintain more constant recruitment (Kornman 1983). Stocking is expensive, and any effort should be based on biological need, should be well planned, and should be closely evaluated. For example, in Missouri, smallmouth bass introductions into established populations did not significantly alter densities or biomass.

Special environmental situations may require an unusual approach. For example, Kentucky has many streams degraded by acid mine drainage, which has devastated the local fish fauna. Introduction of redbreasted sunfish into eight affected drainage basins has proven highly successful because these fish are more tolerant of acid conditions and, perhaps because of reduced competition, have shown excellent growth and size characteristics.

19.6.2   Introductions and Endangered Species

Whereas stocking a species in its historical range is usually considered appropriate, introducing a species outside its historical geographic area is usually problematical. Many western states have dramatically increased fishing opportunities by introducing sport fishes from farther east. Oregon, for example, plans to expand stocking of warmwater stream fishes into waters not suited for salmonids, to introduce new warmwater species into areas where present warmwater species are not meeting objectives, and to develop a mixed trout–warmwater species fishery. However, introductions of warmwater sport fishes have been implicated as a significant factor in the reduction or extirpation of many threatened or endangered native species (Moyle et al. 1986). Extinctions are not commonly recorded for warmwater stream species, probably because of their high tolerance for many environmental conditions. Still, introductions for fishery management purposes were considered an important factor for 7 out of 13 warmwater stream species that have become extinct in North America in the Twentieth Century (Miller et al. 1989). Many species considered to be endangered or threatened are so classified because of adverse interactions with introduced species (Lassuy 1995). Introduced species were cited in 70% of cases as a factor in the decline or as a continuing threat to federally endangered or threatened species in warmwater streams. The majority of these introductions involve management activities to improve sportfishing. The most often introduced species was the largemouth bass, but green sunfish, bluegill, crappies, smallmouth bass, channel catfish, and bullheads also have been commonly introduced. Introduced baitfishes include the red shiner and the fathead minnow. Often these introductions were into lentic environments, especially reservoirs, but individuals invariably escaped into connecting streams and became widespread. Clearly, there is a need to evaluate fully the consequences of introducing a species where it has never occurred.

Of the approximately 110 federally listed endangered and threatened fish species in the United States, the majority are found in warmwater streams. Most states have endangered species programs and list many more species. For example, a survey of prairie streams in all or portions of 13 states in the central United States found 109 warmwater species in need of some protection: 27 species were considered endangered, 29 threatened, and the others in need of conservation or monitoring (Rabeni 1996). These 109 species represent over one-third of all fish species in the prairie region of the United States.

19.6.3   Regulation of the Fishery

States and provinces generally regulate the fisheries of coldwater streams on a stream-by-stream or at least a region-by-region basis, but fishes in warmwater streams are generally either ignored or given treatment identical to that given populations in lentic waters. No fisheries biologist doubts that fish population dynamics are affected by the basic differences between lentic and lotic systems, yet only four states and provinces have extensive statewide or provincewide warmwater stream regulations. About one-quarter of the states or provinces have warmwater stream regulations applying to a very restricted set of streams, usually less than three, and these regulations are often associated with a tailwater. The most common fishes regulated are the black bass, walleye, and esocids. Regulations specific to warmwater stream fishes have the potential to improve fishing in situations in which habitat is adequate. For example, Missouri experimented with a catch–release regulation for smallmouth bass on an Ozark stream and saw the biomass of this species double. Missouri then instituted a 305-mm minimum length limit. The numbers of legal-size smallmouth bass increased substantially within 3 years of implementation. Angler effort remained equal to preregulation times. The same number of fish was harvested, while twice that number was caught and released. A 356-mm minimum length limit on smallmouth bass in streams located in the southern two-thirds of Wisconsin has resulted in greater numbers of sublegal-length smallmouth bass, increased total lengths of fish, and improvement in overall length structure (Lyons et al. 1996).

Proposed regulations for smallmouth bass in Arkansas are an example of biologically based management on a large spatial scale (S. Filipek, Arkansas Game and Fish Commission, unpublished data, 1995). Harvest regulations were proposed based on the known interests of anglers and differences in stream productivity in different ecoregions. These ecoregions were known to support populations with differing density, growth, mortality, and recruitment. The Ozark ecoregion comprises roughly the northern one-fourth of the state and is considered highly productive due to a combination of soil type, hydrology, and topography. This region was selected for a minimum length limit of 305 mm and a creel limit of four fish per day. Three other ecoregions in the state were similar enough in productivity and smallmouth bass characteristics to be considered as a single management zone; a 254-mm minimum length limit and creel limit of four fish per day was proposed. A few streams, known for exceptional populations of smallmouth bass, were designated as exceptional waters. Those in the northern zone will have a 356-mm minimum length limit and a two fish per day creel limit, whereas those in the southern zone will have a 305-mm minimum length limit and a two fish per day creel limit. One stream will be dedicated to catch and release only. Arkansas' approach is unusual; the majority of states with warmwater stream fisheries resources have no specific regulations for their resident fishes, even though many are listed as game fishes (Table 19.1).

19.7 CONCLUSIONS

Most state and provincial agencies have been slow to acknowledge that warmwater stream fisheries deserve or require the type of management accorded other aquatic habitats. Recent efforts show that warmwater stream managers have great opportunities to make significant contributions to society in the protection of endangered species, the maintenance of biological integrity and biodiversity, and the promotion of sport fisheries. Management must be based upon a sound informational base, and appropriate management strategies will follow advances in our understanding of the population dynamics, environmental requirements, trophic relationships, and assemblage level dynamics of warmwater fishes.

19.8 REFERENCES

Angermeier, P. L., and J. R. Karr. 1984. Relationships between woody debris and fish habitat in a small warmwater stream. Transactions of the American Fisheries Society 113:716–726.

Auer, M. T., and N. A. Auer. 1990. Chemical suitability of substrates for walleye egg development in the lower Fox River, Wisconsin. Transactions of the American Fisheries Society 119:871–876.

Axon, J. R., and L. E. Kornman. 1986. Characteristics of native muskellunge streams in eastern Kentucky. Pages 263–272 in G. E. Hall, editor. Managing muskies. American Fisheries Society, Special Publication 15, Bethesda, Maryland.

Bain, M. B., and J. M. Boltz. 1989. Regulated streamflow and warmwater stream fish: a general hypothesis and research agenda. U.S. Fish and Wildlife Service Biological Report 89(18).

Benke, A. L., R. L. Henry, III, D. M. Gillespie, and R. J. Hunter. 1985. Importance of snag habitat for animal production in southeastern streams. Fisheries 10(5):8–13.

Berkman, H. E., and C. F. Rabeni. 1987. Effect of siltation on stream fish communities. Environmental Biology of Fishes 18:285–294.

Bulkley, R. V., and seven coauthors. 1976. Warmwater stream alteration in Iowa: extent, effects on habitat, fish and fish food, and evaluation of stream improvement structures. Summary report for Office of Biological Services, U.S. Fish and Wildlife Service by Iowa Cooperative Fishery Research Unit, Ames.

Bull, W. B. 1991. Geomorphic responses to climatic change. Oxford University Press, New York.

Buttner, J. K., D. B. MacNeill, D. M. Green, and R. T. Colesante. 1991. Angler associations as partners in walleye management. Fisheries 16(4):12–17.

Carline, R. F., and S. P. Klosiewski. 1985. Responses of fish populations to mitigation structures in two small channelized streams in Ohio. North American Journal of Fisheries Management 5:1–11.

Edwards, C. J., B. L. Griswold, R. A. Tubb, E. C. Weber, and L. C. Woods. 1984. Mitigating effects of artificial riffles and pools on the fauna of a channelized warmwater stream. North American Journal of Fisheries Management 4:194–203.

Fajen, O. F. 1975. The fish population of Big Buffalo Creek as affected by stream improvements. Federal Aid in Fish Restoration Project F-1-R-24, Study S-13, Job 2, Progress Report, Missouri Department of Conservation, Jefferson City.

Funk, J. L. 1970. Warmwater streams. Pages 141–152 in N. G. Benson, editor. A century of fisheries in North America. American Fisheries Society, Special Publication 7, Bethesda, Maryland.

Gammon, J. R., and five coauthors. 1983. Effects of agriculture on stream fauna in central Indiana (project summary). Environmental Research Laboratory, U. S. Environmental Protection Agency, Corvallis, Oregon.

Gebert, W. A., and W. R. Krug. 1966. Streamflow trends in Wisconsin's driftless area. Water Resources Bulletin 32:733–744.

Hickman, G. D. 1975. Value of instream cover to the fish populations of Middle Fabius River, Missouri. Missouri Department of Conservation, Aquatic Series 14, Jefferson City.

Jacobson, R. B., and D. J. Coleman. 1986. Stratigraphy and recent evolution of Maryland Piedmont flood plains. American Journal of Science 286:617–637.

Jacobson, R. B., and A. L. Pugh. 1997. Riparian vegetation controls on the spatial pattern of stream–channel instability, Little Piney Creek, Missouri. U.S. Geological Survey Water-Supply Paper 2494.

Judy, R. D. Jr., and five coauthors. 1984. 1982 national fisheries survey, volume 1. Technical report: initial findings. U. S. Fish and Wildlife Service, FWS/OBS-84/06.

Karr, J. R., and D. R. Dudley. 1981. Ecological perspective on water quality goals. Environmental Management 5:55–68.

Karr, J. R., L. A. Toth, and D. R. Dudley. 1985. Fish communities of midwestern rivers: a history of degradation. BioScience 35:90–95.

Kondolf, G. M., and M. Larson. 1995. Historical channel analysis and its application to riparian and aquatic habitat restoration. Aquatic Conservation: Marine and Freshwater Ecosystems 5:109–126.

Kornman, L. E. 1983. Muskellunge stream investigation at Kinnicanick and Tygarts creeks. Kentucky Department of Fish and Wildlife Research Bulletin 68, Frankfort.

Krumholz, L. A., editor. 1981. The warmwater streams symposium. American Fisheries Society, Southern Division, Bethesda, Maryland.

Larimore, R. W., and P. B. Bayley. 1996. The fishes of Champaign County, Illinois, during a century of alterations of a prairie ecosystem. Illinois Natural History Survey Bulletin 35.

Lassuy, D. 1995. Introduced species as a factor in extinction and endangerment of native fish species. Page 391–396 in H. L. Schramm, Jr. and R. G. Piper, editors. Uses and effects of cultured fishes in aquatic ecosystems. American Fisheries Society Symposium 15, Bethesda, Maryland.

Lyons, J. P., and C. C. Courtney. 1990. A review of fisheries habitat improvement projects in warmwater streams, with recommendations for Wisconsin. Wisconsin Department of Natural Resources, Technical Bulletin 169, Madison.

Lyons, J. P., D. Kanehl, and D. M. Day. 1996. Evaluation of a 356-mm minimum-length limit for smallmouth bass in Wisconsin streams. North American Journal of Fisheries Management 16:952–957.

Matthews, W. H. 1987. Physiochemical tolerance and selectivity of stream fishes as related to their geographic ranges and local distribution. Pages 111–120 *in* W. H. Matthews and D. C. Heins, editors, Community and evolutionary ecology of North American stream fishes. University of Oklahoma Press, Norman.

Matthews, W. H. 1998. Patterns in freshwater fish ecology. Chapman and Hall, New York.

McKenney, R. 1997. Formation and maintenance of hydraulic habitat units in the streams of the Ozark plateaus. Doctoral dissertation. Pennsylvania State University, State College.

Mendelson, J. 1975. Feeding relations among species of Notropis (Pisces: Cyprinidae) in a Wisconsin stream. Zoological Monographs 45:199–230.

Miller, R. R., J. D. Williams, and J. E. Williams. 1989. Extinctions of North American fishes during the last century. Fisheries 14(6):27–32.

Milner, N. J., R. J. Hemsworth, and B. E. Jones. 1985. Habitat evaluation as a fisheries management tool. Journal of Fish Biology 27(Supplement A):85–108.

Moyle, P. B., H. W. Li, and B. A. Barton. 1986. The Frankenstein effect: impact of introduced species on native fishes of North America. Pages 415–426 *in* R. H. Stroud, editor. Fish culture in fisheries management. American Fisheries Society, Fish Culture Section and Fish Management Section, Bethesda, Maryland.

Muncy, R. J., and five coauthors. 1979. Effects of suspended solids, and sediment on reproduction, and early life of warmwater fishes: a review. U.S. Environmental Protection Agency, Environmental Research Laboratory, EPA-600/3–79–042, Corvallis, Oregon.

Paine, M. D., J. J. Dodson, and G. Power. 1982. Habitat and food partitioning among four species of darters (Percidae: *Etheostoma*) in a southern Ontario stream. Canadian Journal of Zoology 60:1635–1641.

Paragamian, V. L., and R. Kingery. 1992. A comparison of walleye fry and fingerling stockings in three rivers in Iowa. North American Journal of Fisheries Management 12:313–320.

Pflieger, W. L. 1988. Aquatic community classification system for Missouri. Missouri Department of Conservation, Columbia.

Probst, W. E., C. F. Rabeni, W. G. Covington, and R. E. Marteney. 1984. Resource use by stream-dwelling rock bass and smallmouth bass. Transactions of the American Fisheries Society 113:283–294.

Rabeni, C. F. 1996. Prairie legacies—fish and aquatic resources. Pages 111–124 *in* F. Samson and F. Knopf, editors. Prairie conservation. Island Press, Washington, D.C.

Roseboom, D., and K. Russell. 1985. Riparian vegetation reduces stream bank and row crop flood damages. Proceedings of the North American riparian conference. U.S. Forest Service, Rocky Mountain Forest and Range Experiment Station, Fort Collins, Colorado.

Roth, N. E., J. D. Allan, and D. L. Erickson. 1996. Landscape influences on stream biotic integrity assessed at multiple spatial scales. Landscape Ecology 11:141–156.

Schneberger, E., and J. L. Funk. 1973. Stream channelization: a symposium. American Fisheries Society, North Central Division, Special Publication 2, Bethesda, Maryland.

Schumm, S. A. 1977. The fluvial system. Wiley-Interscience, New York.

Simon, A. 1994. Gradation processes and channel evolution in modified west Tennessee streams: process, response, and form. U.S. Geological Survey Professional Paper 1470.

Smith, P. W. 1971. Illinois streams: a classification based on their fishes and an analysis of factors responsible for disappearance of native species. Illinois Natural History Survey, Biological Notes 76.

Stern, D. H., and M. S. Stern. 1980. Effects of bank stabilization on the physical and chemical characteristics of streams and small rivers: an annotated bibliography. U.S. Fish and Wildlife Service, FWS/OBS-80/12.

Thorne, C. R. 1990. Effects of vegetation on riverbank erosion and stability. Pages 125–144 *in* J. B. Thornes, editor. Vegetation, and erosion. Wiley, New York.

Trautman, M. B. 1981. The fishes of Ohio, 2nd edition. Ohio State University Press, Columbus.

Wang, L., J. Lyons, P. Kanehl, and R. Gatti. 1997. Influences of watershed land use on habitat quality and biotic integrity in Wisconsin streams. Fisheries 22(6):6–12.

Waters, T. F. 1995. Sediment in streams: sources, biological effects, and control. American Fisheries Society, Monograph 7, Bethesda, Maryland.

Wilkin, D. C., and S. J. Hebel. 1982. Erosion, redeposition and delivery of sediment to midwestern streams. Water Resources Research 18:1278–1282.

Williams, J. E., C. A. Wood, and M. P. Dombeck, editors. 1997. Watershed restoration: principles and practices. American Fisheries Society, Bethesda, Maryland.

Winger, P. V. 1981. Physical and chemical characteristics of warmwater streams: a review. Pages 32–44 in L. A. Krumholz, editor. The warmwater streams symposium. American Fisheries Society, Southern Division, Bethesda, Maryland.

Chapter 20

Large Rivers

ROBERT J. SHEEHAN AND JERRY L. RASMUSSEN

20.1 INTRODUCTION

Most large rivers in North America have been modified to accommodate navigation, flood control, hydroelectric power generation, and other human uses. Large Arctic rivers, such as the Mackenzie and Churchill, still exhibit long reaches where constructed barriers do not limit fish movement, but these are exceptions. Dams have transformed long river reaches into impoundments, and many of these, such as the six large Missouri River main-stem reservoirs, exhibit few riverine characteristics. Due to dams, many large river reaches in the United States, as well as some in Canada (e.g., the Peace River), can no longer be classified as riverine based on structure or function.

It is difficult to define a large river. Characteristics used to describe the relative size of rivers, such as length, width, average annual discharge at mouth, and size of drainage basin, are highly correlated for pristine rivers, but anomalies appear among rivers modified by humans. One of the world's largest rivers based on length and drainage area—the Colorado River—has been subjected to numerous flow diversions and thereby reduced to a mere trickle in its lower reaches. It is no longer permanently connected to the sea. Even the Mississippi River, North America's largest river, is dwarfed in size by the Amazon River: its discharge is an order of magnitude less (Table 20.1).

By the world's standards, the rivers of Mexico are not very large, with the exception of the Rio Grande. Although other of the larger Mexican rivers can be relatively long (e.g., Lerma River at 857 km, Balsas at 676 km, and Yaqui at 619 km), their discharges increase little along their lengths due to evaporation, a paucity of tributaries, and water diversions for irrigation.

Lotic systems are arranged into a hierarchy by a concept known as stream order. Stream order positions a stream in relation to its tributaries, the land mass it drains, and its total length. First-order streams are the terminal branches of a stream system—they have no tributaries. Second-order streams are formed when two first order streams join. At least two streams of any given order must join to form a stream of the next order (Cole 1979). Based on stream order, we place lotic systems into three groups: small or headwater streams (orders 1–3), medium rivers (orders 4–6), and large rivers (orders > 6). The Mississippi River is the only river in North America that ranks as high as a 10th-order stream (Cole 1979).

Table 20.1 Drainage area, length, and discharge for large North American rivers. Data are from Czaya (1981) and Patrick (1995). The Amazon River is provided for comparison.

River	Drainage area (1,000 km²)	Length (km)	Discharge at mouth (m³/s)
Amazon	7,180	6,516	180,000
Mississippi	3,221	6,019	17,545
Mackenzie	1,805	4,250	7,500
Missouri	1,370	4,076	2,152
Nelson	1,072	2,575	2,300
St. Lawrence	1,030	3,100	10,400
Yukon	885	3,185	7,000
Columbia	669	2,229	6,650
Colorado	629	3,200	168
Rio Grande	570	2,870	82
Ohio	528	2,102	7,307
Arkansas	417	2,334	1,274
Snake	282	1,671	1,416
Red	241	2,044	1,756
Fraser	225	1,360	3,750

Large rivers are the least understood of the three groupings above and for good reasons. The size, raw power, and inherent complexity of large rivers make them difficult to study. Also, most North American large rivers have been and continue to be modified extensively by humans. Thus, river scientists are burdened with needing to understand large rivers as they once were, as they are now, and as they may someday be. Human manipulations of large rivers and the resulting ecological effects frame the overriding issues in the management of fisheries in most large U.S. rivers. Thus, we open the chapter by describing the nature of large rivers and their ecology. We subsequently attempt to use these concepts to illustrate how human uses and modifications of large rivers have affected their ecology. We then introduce the reader to the fisheries resources of large rivers, contemporary issues in river management, and potential management strategies. We give little coverage to anadromous fishes (see Chapter 18 and 24) or the estuarine environment (see Chapter 12).

20.2   NATURE AND ECOLOGY OF LARGE RIVERS

20.2.1 Diverse, Dynamic Systems

Large rivers are diverse systems that feature a variety of lentic and lotic habitats and biotic community assemblages. Large rivers are also dynamic systems. Natural rivers are not fixed in space because hydraulic processes continuously create new habitats and eliminate old ones. Rivers can be highly interactive with other aquatic systems, such as tributaries and wetlands, as well as with riparian and floodplain habitats (horizontal interactions; Figure 20.1A). Important physical, chemical, and biological interactions occur between river waters, riverbeds, and groundwaters (vertical interactions), as well as between upstream and downstream river reaches, and between rivers and main-stem reservoirs, estuaries, and marine systems (longitudinal interactions; Figure 20.1B) (Ward and Stanford 1989). Important ecological interactions occur between

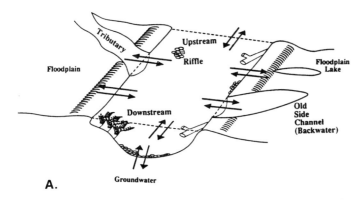

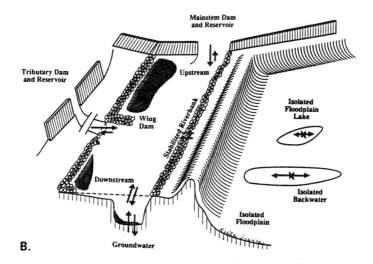

Figure 20.1 Horizontal (horizontal arrows), vertical (vertical arrows), and longitudinal (oblique arrows) physical, chemical, and biological interactions within the river complex and its drainage basin for (**A**) pristine and (**B**) modified large rivers. Smaller arrows indicate diminished interactions in modified rivers. Channel stabilization also fixes the river in its bed, decreasing spatiotemporal interactions. Crosshatched areas indicate deposition. Panel (**A**) taken from Ward and Stanford (1989).

the river and its floodplain where the river is not tightly confined by canyon walls (e.g., portions of the Colorado River) or flood control levees. Many human activities interfere with one or more of these interactive pathways. The interactive nature of large rivers suggests a holistic approach to understanding and managing resources.

20.2.2   Ecology

Conceptual frameworks of how large rivers function remain debatable, but current ecological concepts have been shaped by work in lotic systems dating to the turn of the twentieth century. In the early 1900s, studies of flowing-water systems focused on the biology and distribution of stream organisms. This approach developed into studies of trophic relationships and ecosystem energy flows in streams in the mid-1900s (Johnson et al. 1995). At about the same time, the seminal works of L. B. Leopold and colleagues on fluvial processes in streams were published (e.g., Leopold et al. 1964). Their work suggested that the hydrodynamic and geomorphic features of a stream were predictable along its length. This concept of predictable features and findings on physical and biological characteristics within streams evolved and were merged to form a more holistic model of stream organization and function, the river continuum concept (Vannote et al. 1980).

The river continuum concept proposes that ecosystem structure and function varies in a predictable manner from headwaters to estuary. It is based on the idea that natural watersheds are in a state of dynamic equilibrium and that stream channel characteristics (e.g., width, depth, velocity, and sediment load) tend toward an equilibrium determined by climate, geology, and watershed. Downstream processes are linked to upstream processes, creating resource gradients and biotic communities that are organized and function predictably along these gradients. The river continuum concept stresses upstream–downstream, or longitudinal, relationships in lotic systems, and it implies that the position of aquatic communities along these resource gradients can be predicted along the longitudinal axis of a stream as measured by stream order (Vannote et al. 1980).

The river continuum concept holds that small streams of lower stream orders provide a relatively constant environment due to groundwater inputs. Primary production is limited due to shading by riparian vegetation, so the primary energy input is from allochthonous organic matter washed or blown into the stream from the watershed. The invertebrate community consists chiefly of shredders and collectors of fine particulate matter that are well suited to the available energy source. Medium-size streams are thought to be the most variable in environmental conditions such as temperature and hydrology, thereby creating habitat diversity. Therefore, environmental variability in medium-size streams is thought to provide for niche diversity and, thus, the greatest biological diversity along the continuum. The increased width of the river opens the forest canopy and allows sunlight penetration, and primary production becomes the most important energy source. The primary invertebrate groups collect fine particles or graze on organisms, such as algae, attached to surfaces.

Biological production in large rivers, according to the river continuum concept, is rooted in the fine particulate organic matter delivered from upstream reaches and from periodic inundations of the floodplain. The greater depths and higher turbidity typical in large rivers diminish primary productivity (Sedell et al. 1989), so collectors predominate in the invertebrate community.

The river continuum concept predicts ecological function well in some large rivers (see Sedell et al. 1989). However, energy flows from upstream sources appear to be relatively small in large rivers such as the Amazon that periodically and extensively inundate their floodplain. Bayley (1989) estimated that 93% of the energy in a partially deforested area of the Amazon came from riparian vegetation and tree litter captured during flooding. These kinds of observations led to the development of the flood pulse concept for large river–floodplain ecosystems (Junk et al. 1989; Bayley 1995).

Floodplain organic matter becomes available to aquatic herbivores and detritivores when the river spills into the floodplain during floods. Many assemblages of South American large-river fishes, such as those found in the Amazon River, include numerous species that directly feed on seeds, fruits, and other vegetative matter that become available to herbivorous fishes during floods. North American large-river fish communities have few herbivorous species. Other than this difference, large-river fish populations on both continents derive similar benefits during floodplain inundation. The floodplain is used for spawning and nursery areas by many species. Organic matter in the floodplain stimulates plankton blooms and the production of invertebrates, resulting in more food for fishes.

The river continuum concept emphasizes a longitudinal structuring for rivers, so its predictiveness relates to river channels and not backwaters, marshes, and floodplain lakes (Johnson et al. 1995). The flood pulse concept emphasizes lateral interactions between the river and floodplain. It predicts that organic matter produced and consumed in the floodplain is far more important to higher trophic levels in the river than are organic inputs from upstream sources. It further differs from the river continuum concept in that it predicts that biotic diversity may be higher in large rivers than in medium-size rivers when the entire large river ecosystem, which includes the floodplain, is taken into consideration (Statzner 1987).

Which is the better theory for developing a conceptual framework for addressing large river fisheries? The river continuum concept appears to apply especially well to large rivers that are largely confined to their channels (Johnson et al. 1995), so perhaps this model is more applicable to many large U.S. rivers as they function today. For example, the Mississippi River floodplain is estimated to have been reduced by 90% by levees (Gore and Shields 1995). On the other hand, the longitudinally structured river continuum concept is ill suited for integrating or portraying lateral interactions between the river and its floodplain. The flood pulse concept may be more applicable in more pristine, large river–floodplain systems. It is much more useful for depicting how periodic flooding affects, creates, destroys, and maintains wetlands and associated land forms in large rivers.

There is general agreement that vertical, horizontal, longitudinal, and spatiotemporal interactions are disrupted by human modification of rivers, thereby disturbing and shifting the dynamic equilibria in and between the river and its watershed (Figure 20.1B). The establishment of a main-stem reservoir provides a good illustration of how dams disrupt longitudinal interactions and affect energy flow pathways. For example, 30,000 metric tons of planktonic crustaceans are swept into the Missouri River annually via releases from one of its main-stem reservoirs (Ward and Stanford 1989). Such inputs can lead to increased fish production but also can lead to shifts in fish species composition and relative abundance. The serial discontinuity concept (Ward and Stanford 1989) has been proposed for impounded rivers in an attempt to account for the interruptions in longitudinal interactions and gradients imposed by dams. However, dams create pooled water upstream of them and gradients within these pools; lotic conditions below dams grade into lentic conditions above the next dam. Thus, dams convert downstream reaches to conditions found in more upstream reaches (increased velocities, narrower channels, and streambed degradation). It may be appropriate to consider dams as "retrogressive discontinuities" in rivers.

20.3   LARGE-RIVER HABITATS

Fish habitat use and requirements are important concerns to fisheries managers because human activities tend to diminish habitat diversity in rivers. Factors typically used in describing fish habitats are velocity, depth, substrate (composition, particle size, and stability), structure (woody debris or snags and rock outcroppings), bank coefficient (the ratio of riverbank length to river length), dissolved oxygen, salinity, and the presence of other biotic communities (e.g., macrophytes and aquatic invertebrates). Habitat use by fishes varies by life stage, over a 24-h period, with the seasons, and with changes in discharge, making descriptions of fish habitat requirements difficult.

At a larger scale, river habitats can be divided into either channel or extrachannel habitats (e.g., side channels and backwaters). Large channels demonstrate repeating segments of river bend–crossover–river bend (Leopold et al. 1964), much as small streams show repeating segments of riffle–pool–riffle. Unlike smaller lotic systems, in large rivers islands typically predominate. Islands divide large rivers into multiple channels. The frequency and amount of longitudinal overlap among islands determines the number of channels. For example, a single island results in a split channel. In large river reaches where islands are more frequent and overlap each other, the channel becomes more finely divided into multiple channels. Collectively, the multiple channels can take on a braided or anastomosing appearance (Church 1992).

No single habitat classification system is suitable for all large rivers. Extensively modified rivers show habitats not found in pristine ones; for example, Rasmussen (1979) proposed a classification scheme for the upper Mississippi River that includes habitats such as tailwater, which is found in only impounded rivers. Floodplain rivers demonstrate habitats that are not found in rivers whose channels are confined in rocky canyons, and some habitats, such as salt marches, are near only estuaries. Cowardin et al. (1979) classified wetlands and deepwater habitats of the United States. Their riverine system classification scheme (Figure 20.2) is applicable to large floodplain rivers in forested watersheds; because floodplain lacustrine (lake) and palustrine (marsh) systems also can be inundated during flooding, habitats found in those systems (see Cowardin et al. 1979) are also included.

Channel meandering creates and eliminates riverine and floodplain lacustrine and palustrine systems. Common to all of these systems are the influences that water level fluctuation exerts on floral communities. Each plant community comprises species with certain soil–moisture tolerances and requirements. Plant communities occupy elevations and locations amenable to where the component species fall along the soil–moisture tolerance and requirement spectrum. This spectrum ranges from species that require total immersion to species that tolerate only periodic, short-duration flooding (see Figure 20.2).

20.4   MULTIPLE USES OF LARGE RIVERS

The greatest challenges facing large-river fisheries managers are related to habitat losses and alterations due to nonfishery human uses of river resources. For the most part, large-river fisheries managers in the United States came too late to the table to

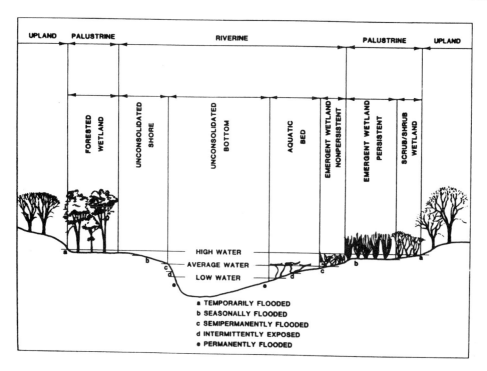

Figure 20.2 Examples of habitats and plant communities found in riverine systems. Figure taken from Cowardin et al. (1979).

make a serious claim to their piece of the river pie. Agricultural, navigational, and hydropower interests were, by and large, there first. What are the consequences? Vast expanses of large-river floodplain habitat no longer exist in the United States. Flood-plains have been leveed and drained, primarily for agriculture. River waters are also diverted to agricultural lands for irrigation or to municipalities for drinking water, further reducing habitat for fishes in large rivers. Navigation has been designated the primary use in what waters remain, principally the main channel, in many large U.S. rivers.

Below we describe how multiple use of and competition for river resources have transformed many large rivers in the United States. But first, history provides a worthy lesson. Compared with fisheries managers, wildlife managers have much more effec-tively provided for their waterfowl interests in large rivers. Using fees paid by water-fowl hunters (1934 Migratory Bird Hunting Stamp Act; 1937 Federal Aid in Wildlife Restoration Act) wildlife agencies have purchased vast tracts of land, much of it large-river extrachannel habitat, in waterfowl flyways in the Mississippi and Missouri rivers. These lands, isolated from large rivers to a great extent by levees, form portions of the National Wildlife Refuge System and are managed by the U.S. Fish and Wildlife Ser-vice. Management of National Wildlife Refuge System lands has been a source of contention between fisheries and waterfowl biologists primarily because ducks can fly over levees whereas fishes cannot. There is little doubt that the acquisition of large-river floodplain lands through outright purchases or through conservation easements, so these lands can be reconnected to their rivers, is a pressing need in many large U.S. river systems.

20.4.1   Navigation Impoundments

The U.S. inland navigation system can be roughly divided into four interconnecting systems (Figure 20.3): (1) Great Lakes and connecting rivers; (2) Atlantic seaboard; (3) Gulf states; and (4) at the center, the Mississippi River and its tributaries. The Mississippi River system comprises more than one-third of the 40,000 km of navigable inland waters (Nielsen et al. 1986). The agency charged with the maintenance of these waters is the U.S. Army Corps of Engineers.

Establishment of the Mississippi River navigation system illustrates the historical use pattern in large rivers. The Ohio River became a traffic route during the earliest days of settlement in the Mississippi River valley. The first steamboat appeared in 1811. Rapidly swelling river traffic provided impetus for the Rivers and Harbors Act of 1824, which authorized snag removal from the Ohio River and marked the beginning of sanctioned habitat modifications.

Shoals and riffles of the natural river impeded boats except during times of high waters, so river-training structures were designed to provide deeper navigation channels in the Mississippi River system. By 1828 a series of wing dams were constructed on the Kanawha River, an Ohio River tributary, to establish a 1-m deep navigation channel. These dams were followed in 1898 by the completion of a system of 10 low-lift locks and dams that created a 2-m deep channel for 142 km upstream from the Kanawha's confluence with the Ohio. Between 1931 and 1937, the lock system was replaced by four high-lift structures, which established a 2.75-m deep navigable depth.

On the Ohio River, 47 back-channel dikes and 111 training dikes were constructed between 1837 and 1866. Then between 1875 and 1900, five low-lift locks and dams were constructed to provide a 2-m deep navigation channel. The completion between

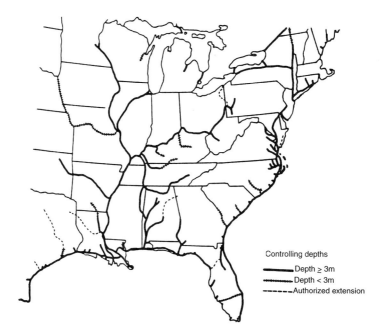

Figure 20.3   Rivers modified for navigation in the eastern United States.

Figure 20.4   Navigation lock (foreground) and dam. Dams such as this one are operated to maintain navigable depths in upstream pools and to control floods. Locks permit vessel passage. Upstream fish movements are restricted to "locking through." Photograph provided by S. Scherck.

1900 to 1930 of 51 low-lift locks and dams increased the Ohio River navigation channel to 2.75 m deep. The dams permitted fish and vessel passage during periods of high water but segmented the river into a series of stairstepped pools. Since 1930, modernization of the Ohio system has involved replacing the 51 low-lift structures with 13 high-lift locks and dams. These high-lift dams elevated navigation pools as much as 10.5 m above downstream channels, thereby restricting fish passage.

Modification of the Mississippi River followed later but paralleled that on the Ohio River. Beginning about 1930, a 2.75-m deep channel was established with the construction of 28 high-lift locks and dams (Figure 20.4).

Lock-and-dam systems convert rivers from a free-flowing condition to a series of impoundments. Impoundments widen some reaches and greatly increase habitat complexity due to inundation of floodplain habitat. The immediate effect of impoundment on some fishes, such as centrarchids, is beneficial by providing more habitat. Because impoundment makes conditions more favorable for lentic species, it can be assumed that conditions become less favorable for lotic species.

Channel deepening occurs downstream of dams because water velocities can be greater due to upstream impoundment and constriction of the channel. Streambed degradation can occur for hundreds of kilometers downstream of dams, as on the Missouri River. Deepening of the channel causes adjacent extrachannel habitats (side channels and backwaters) to become dewatered.

Impoundment enhances sediment deposition, especially in extrachannel habitats, by reducing hydraulic gradient and velocities. Consequently, the boon to river fish populations from inundation of floodplain habitats can be relatively short lived. Most backwaters of the Mississippi River will be lost to sedimentation in the next 50–100 years, a very short time on the geological timescale.

Roughly one-third of the Missouri River has been transformed to reservoirs and one-third has been channelized or stabilized. Impoundments and river-training structures have reduced most of the Missouri River to a single channel. Islands, side channels, backwaters, and sloughs have been eliminated. Hesse et al. (1989) estimated that 41,000 ha of Missouri River habitats have been lost. Impoundment has also led to a substantial loss of submergent vegetation in Pool 8 of the Mississippi River as well as a 79% loss of islands in the downstream reaches of the pool due to erosion (Fischer and Claflin 1995).

Immediately after impoundment, the Illinois River—near its confluence with the Mississippi River—consisted of a main channel and a network of side channels and backwaters. River fishes no longer have access to much of that habitat due to sedimentation and levee construction. Soon all that will be left in many reaches will be the main channel. Aquatic habitat loss, together with the effects of pollution, has led to a decline in the Illinois River commercial fishery harvest from 10% of the entire U.S. freshwater catch in 1908 to virtually nothing in the 1980s (Karr et al. 1985).

The immediate effects of navigation dams and their associated floodplain levees on fishes can be highly variable, but some generalizations are possible: (1) habitats upstream of dams become favorable to lentic species, (2) riffles become inundated so fishes dependent on this habitat are negatively affected or eliminated, (3) species dependent on seasonal flooding of terrestrial habitats for reproduction are selected against, (4) the abundance of fishes increases above dams because surface area is increased, and (5) longitudinal fish migrations are blocked by navigation dams except for limited movement through navigational locks. Many of the positive effects are short lived, and fish species diversity and abundance can be expected to decline eventually below preimpoundment levels.

Fishes do not have the opportunity to lock through dams if those dams are not designed for navigation. Dams have severely reduced survival of Atlantic salmon in eastern rivers and Pacific salmon in western rivers, both during the upstream spawning migration and during the downstream migration of smolts. Although fish passage systems have been somewhat successful in allowing fishes to pass dams, actively transporting migrating fishes past dams by means of trucks and barges has often been most successful (see Chapter 24).

Dams also affect fish populations in ways that would be difficult to anticipate. For example, regulated flow from the Glen Canyon Dam has reduced ponding in Colorado River tributary mouths, such as in the Little Colorado River. Thus, native larval fishes drift from tributaries into the colder waters of the Colorado River more quickly and at a less mature state than before, making it doubtful that they survive well (Robinson et al. 1998).

20.4.2 Navigation Aids, Flood Control Systems, and Wetland Drainage

The lower Mississippi River has been modified for navigation by the construction of deflection dikes to accommodate navigation during low flows. Wing dikes constructed perpendicular to the river's flow slow water down, which promotes sedimentation in nearshore areas instead of the navigated main channel. Wing dikes also promote scouring in the navigation channel (Figure 20.5). Shorelines opposite wing dams are often armored to prevent erosion.

Dikes and channel stabilization structures protect the streambank from erosion, maintain channel alignment, and fix the river in its bed. The lower Mississippi River has been channelized and stabilized by more than 1,300 km of revetments. The effect has been to halt the creation of new river habitat and to promote the loss of extrachannel habitat through reductions in flow and accompanying sediment deposition.

Most rivers modified for navigation are extensively leveed as a flood control measure (Welcomme 1979). Levees block horizontal interactions, such as transfer of organic matter from the floodplain to the river. Levees can directly limit reproduction in fishes that depend on inundated floodplain habitats for spawning. They can also limit recruitment by reducing floodplain habitat, thereby diminishing the diversity of current velocities, depths, and temperatures available to fishes. Levees also can cut off many of the river's natural meanders and isolate floodplain lakes that provide spawning and nursery areas.

Ecological succession is a natural process that eventually leads to the conversion of extrachannel aquatic habitats to dry land. However, river training and bank stabilization have accelerated the process. Also, agricultural practices and construction in large-river watersheds, along with dredge disposal programs, have accelerated aggradation in large rivers. Unfortunately, many large rivers have been purposefully fixed in their beds, and natural fluvial processes that create new aquatic habitats have been all but eliminated.

Much of the floodplain behind levees has been converted to agricultural land, and wetland plant communities have been lost. This loss has several ramifications. The annual use of nitrogen fertilizer is about 9 million metric tons per year in the United States (much of it in the Corn Belt), and 50% of the fertilizer is thought to be lost to agricultural drainage water as nitrate. Nitrogen is of particular concern in the pollution

Figure 20.5 An L-head wing dam. Wing dams divert water to the channel to facilitate navigation. Wing dams increase channel border habitat and create slackwater habitat on their downstream sides but often diminish flows through side channels. Most wing dams jut out perpendicularly from the bank in a straight line, creating a scour hole at their tips. With the L-head configuration the scour hole is lost, but thermal stratification becomes more stable on the downstream side. Photograph provided by S. Scherck.

of estuaries and diminished water quality in major offshore fisheries zones, such as in the Gulf of Mexico. Wetland plant communities, on the other hand, are considered sinks for nitrogen because they create the environments for microbial communities that denitrify nitrates and facilitate nitrogen loss to the atmosphere (Sheehan and Konikoff 1998). Wetlands trap or break down other pollutants as well. The drainage of river wetlands for agriculture has thus had the duel deleterious effects on river and estuarine water quality of greatly increasing the unnatural addition of pollutants as well as eliminating the biotic communities promoting water purification.

In the United States, the 1990 Farm Bill's Agricultural Wetland Reserve Program provides funding to support the restoration of as much as 405,000 ha of former wetlands through the purchase of conservation easements from landowners. Much of the draining of wetlands for agriculture has occurred in the Mississippi River basin (Figure 20.6).

20.4.3 Navigation Traffic

Many river biologists believe that virtually every biological component of the river ecosystem is affected by barge traffic, yet it is held by others that the effects of navigation are imperceptible (Wright 1982). Towboats (which actually do more pushing than pulling) and barges (Figure 20.7) displace water during passage. One effect of boat induced wave action and turbulence on lotic ecosystems is to increase suspended solids. A study of the Illinois River during and after a 2-month period when boat traffic was greatly reduced indicated that the return of boat traffic increased suspended solids by 30–40%. The suspension of sediments during vessel passage may affect fishes.

Figure 20.6 Wetlands of the United States artificially drained for agriculture. Each dot represents approximately 8,100 ha. Map taken from Dahl (1990).

Alabaster and Lloyd (1980) concluded that any increase in suspended solids above a low level may cause a decline in the value and status of a freshwater fishery and that the risk of damage increases with concentration.

Most of the sediment suspended by boat passage originates in the watershed, but some may be generated by the erosional forces exerted by boats. Wave heights created by both recreational and commercial boats are often sufficient to cause erosion. Towboats produce three successive hydraulic effects during passage: (1) a slight rise in water as the bow passes, (2) a dewatering of the shoreline and small tributaries, and (3) the return of the water in a series of waves as the stern of the towboat passes. The vertical drop in water level can be substantial (on the order of 0.5 m), and if shore or extrachannel areas have a shallow slope, a considerable portion of aquatic habitat can be dewatered. Nearby backwaters can also be dewatered if the channel is small. Dewatering of shoreline and extrachannel habitats due to boat passage can kill larval fishes. The proportion of damaged fish eggs has been shown to increase subsequent to barge passage in the upper Mississippi River. Barges traveling upstream in the river can double or triple water velocities, whereas downstream barges can reverse the river's flow.

20.4.4   Electric Power Generation

The electric power industry also affects fish populations and river ecology. Use of rivers by this industry can be divided into two broad categories, direct use of a river's energy for generating hydroelectric power and withdrawal of river water for the dissipation of waste heat from fossil and nuclear fuel power plants. Water is either returned to the river at an elevated temperature or passed to cooling towers and lost as steam.

In the 1960s and 1970s, environmental concerns over power plants focused on effects of heated effluents. These effects are less important in large rivers compared with other surface waters because of the large volume of water available for heat dissi-

Figure 20.7   Towboat pushing a barge string. The specific effects of navigation on river biota remain controversial. Photograph provided by S. Scherck.

pation. However, power plant shutdowns can lead to thermal shock of fishes wintering in heated effluents. Increases in river temperatures can also lead to supersaturation of dissolved gases and increase the toxicity of pollutants such as ammonia.

A long-term debate over the construction of six power stations on the Hudson River culminated in the conclusion that the major detriments to fishes would be the impingement and entrainment of fishes (primarily eggs and larvae) in the cooling water intake system (Limburg et al. 1989). Impingement refers to fish being forced up against screens used to keep debris out of the intake system. Fish that are small enough to pass through the screens and the cooling system are considered entrained; they are subjected to temperature and pressure changes and shear forces that may kill them. Mitigation of these impacts has taken the form of hatchery production and stocking of striped bass fingerlings in the Hudson River.

Water releases from hydroelectric dams to downstream reaches often coincide with peak electrical demands, which have little to do with the needs of the river biota. Peak demands occur daily (morning and evening) or seasonally (midwinter for heating and midsummer for air conditioning). Hydrographs in rivers such as the Columbia, which in pristine condition were primarily influenced by spring runoff, are now predominantly driven by electric power consumption, often with disastrous effects on early life stages of fishes (Figure 20.8). Young salmonids can become stranded on land due to dewatering when releases from dams are curtailed. Flows are completely halted during the night in some reaches of the Columbia River system, having effects on migratory fishes that orient to currents. These types of situations have stimulated interest in determining and establishing minimum instream flows to protect fish populations and channel morphology.

Water quality in hydropower dam releases is also a concern. Releases from 10 Tennessee Valley Authority dams failed to meet the Tennessee water quality standard of 5 mg/L dissolved oxygen, and 5 were at less than 1 mg/L (Voigtlander and Poppe 1989). Toxic metals, such as iron and manganese, which precipitate as oxides when dissolved oxygen is high, reenter solution when dissolved oxygen is low, also harming fishes. When the falling water entrains air, high-head hydropower dams can also cause gas embolism disease in fishes. Gases from entrained air bubbles enter solution due to the increased hydrostatic pressure, and these gases become supersaturated when the water leaves the plunge basin. Other effects of dams on large-river fishes and habitats are discussed in Section 20.4.1.

20.4.5 Pollution

Water pollution can be divided into five general classes: (1) suspended solids (2) nutrient enrichment, (3) toxic chemicals (ammonia, metals, acids and bases, biocides, detergents, and petroleum hydrocarbons), (4) putrecible organic matter (e.g., sewage wastes) that causes dissolved oxygen depletion, and (5) thermal pollution. The first four categories have profound influences on large rivers, although water quality related to putrecible organic matter and toxic chemicals has generally improved since the 1960s.

The most significant pollutant in surface waters in North America is suspended solids. Suspended solids have had greater influence on fish diversity and abundance in the midwestern United States than any other factor. Decades of short-sighted land use

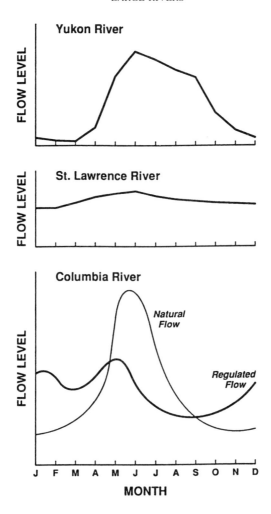

Figure 20.8 Hydrographs (average annual discharge) of three large river systems. The Yukon River demonstrates the effect of season on discharge; this river is an extreme case because its drainage basin thaws over much of the year. The Saint Lawrence River is another extreme case; fluctuations in discharge are reduced by the massive storage capacity of the Great Lakes, the river's source (figure taken from Czaya 1981). For the Columbia River generalized discharge patterns are shown for natural and regulated flow regimes (figure taken from Ebel et al. 1989). Regulated releases from main-stem and tributary storage reservoirs affect discharge in the Columbia River much as the Great Lakes do for the Saint Lawrence River.

practices resulting in erosion and sedimentation culminate today in the near total loss of extrachannel fish habitats in many rivers. The source of most suspended solids is soil erosion, and soils become highly erodible when terrestrial vegetation is denuded due to agriculture, intensive forestry, or construction.

Sewage treatment plants contribute suspended solids to streams, but primary environmental concerns related to sewage center around nutrient enrichment, elevated ammonia, high biological oxygen demand, and disinfectants such as chlorine, which combines with ammonia to form toxic chloramines. Biologists have noted the near elimination of mollusks, such as mussels, for long distances downstream of sewage treatment plants (Goudreau et al. 1993). Sewage effluents can account for a significant portion of total discharge and allochthonous organic inputs (e.g., 24% of the total organic carbon inputs to the Hudson River; Limburg et al. 1989).

North American industry produces more than 55,000 different chemicals (Owen 1985), many of which eventually find their way into rivers. Heavy metals (e.g., mercury and selenium) and halogenated hydrocarbon pesticides have reduced exploitation of several large-river fisheries. The discovery of PCBs in the Hudson River ecosystem below Hudson Falls in 1976 forced closure of the freshwater fishery indefinitely (Limburg et al. 1989). The Food and Drug Administration detected PCB concentrations exceeding acceptable levels in common carp from the Mississippi River, necessitating the destruction of 27,000 kg of presumably contaminated fillets. Despite the ban on sales and manufacture of PCBs since 1977, 60% of upper Mississippi River shovelnose sturgeon examined in the mid-1990s exceeded the acceptable level of 2 mg PCB/kg fish. The long-term trend of PCB contamination appears to be on the decline, but just as the DDT contamination era was supplanted by that of PCBs, other toxic substances are on the increase.

The threat of catastrophic pollution events (spills) is greater in large rivers than other freshwater aquatic systems. Large rivers attract industry to their shores because of the availability of water, inexpensive shipping, and hydroelectric power and because these same factors promote urbanization, providing a workforce and support facilities. Consequently, there is a continuous threat of spills from industrial storage facilities and barges that transport the raw materials, products, and by-products of industry.

There is, however, cause to be encouraged by trends in water quality in some large rivers in the 1980s and 1990s. Many river reaches have followed a scenario similar to the Kanawha River, West Virginia, once considered one of the most polluted streams in the United States. Little information is available regarding the historical effects of pollution, but Addair (1944) noted that fishes were absent in the pool of the Kanawha most severely affected by pollution. As late as 1970, the biological oxygen demand was extremely high, dissolved oxygen ranged as low as 0 mg/L in some reaches, ammonia concentrations and turbidity were elevated, lead and zinc were at toxic levels, and pH fluctuated from 4.2 to 9.5. A trend toward improved water quality in the Kanawha began in the 1960s due to the enforcement of wastewater treatment requirements set by the Federal Water Pollution Control Administration, founded in 1963. Recent data indicate that only fecal coliform and iron exceed West Virginia standards. Average mainstem dissolved oxygen concentrations exceed 8 mg/L, and during a recent 3-year study by the West Virginia Environmental Protection Agency dissolved oxygen minima did not fall below 4 mg/L. The fish community now appears typical of large rivers in terms of standing crop and diversity.

Historically, pollution abatement in streams emphasized laboratory toxicity tests and chemical monitoring in surface waters. Today, biotic assessment methods have been developed to identify waters exhibiting severe anthropogenic impacts. These methods examine the severity of anthropogenic impacts where they matter most—in nature (referred to as instream assessment) rather than in laboratory animals held in aquaria.

The underlying principle of biotic assessment is that environmental disturbances will be reflected in the structure and function of biotic communities. Karr (1981) developed the index of biotic integrity (IBI) for this purpose. For biotic assessment tools to be usefully applied to a stream, standardized sampling methods must be used and the fish community norm must be established for streams in the region that are of similar stream order and relatively free of human impacts. The structural and functional attributes of the fish community norm, referred to as metrics, include measures such as the abundance of tolerant and intolerant species, predator abundance, and disease incidence.

Biotic assessment methods are now routinely used in small streams by state and federal agencies to identify reaches affected by anthropogenic forces, such as pollution. Only recently have IBI methods for large rivers been successfully developed and tested. Simon and Emery (1995), using boat electrofishing to collect fish samples in the Ohio River, found their IBI method was sensitive to differences in land use and municipal and industrial pollutant loadings.

20.4.6 Flow Diversions

River waters are diverted to supply irrigation for croplands and municipalities and to augment the needs of the electrical power industry. Flow diversions have also been used to scour and deepen harbors. The Colorado River provides an example of the effects of flow diversions. Entire species can become eliminated when discharge is reduced below critical levels, and spawning migrations of anadromous fishes can also become jeopardized.

In the western United States arid lands have been greatly developed for agricultural use by means of irrigation. Much of this water is lost to the atmosphere via transporation and evaporation and is not returned to its origin. This loss, in essence, constitutes a loss of fish habitat. The story, however, does not end there. In many cases it would be better if no irrigated water returned. Selenium is an element that is an essential animal nutrient, but it becomes toxic when available to organisms in more than minute quantities. Selenium leaches into irrigation runoff and then becomes more concentrated due to evaporation. The extent of this problem became apparent to all in the 1980s when the media graphically reported selenium-caused deformities and die-offs in ducks and other waterfowl in, ironically, National Wildlife Refuge lands.

Leaching and evaporation also lead to salinization of irrigation waters, a problem more common than selenium poisoning. After all of the flow diversions, the discharge remaining in the Colorado River at the United States–Mexican border is so saline that the U.S. government was required by treaty to build a desalinization plant (costing more than US$300 million) to deliver some semblance of usable water to Mexico. The effects of irrigation and salinization on the Colorado Rivers' freshwater fish community are obvious.

20.5 LARGE-RIVER FISHERIES

Large-river fisheries managers deal with multiple-use conflicts because fisheries resources have been considered secondary to other uses. We know relatively little about large-river fisheries in North America because anthropogenic alterations in hydrogeomorphology and ecology have come at a faster rate than has scientific knowledge. Large Canadian rivers, though relatively free of human modifications, have not received a great deal of attention from fisheries researchers.

20.5.1   Sport Fisheries

Sport fisheries are often underexploited in large rivers due to several factors. Boat ramps are often under water during high river stages, unusable at low river stages, or nonexistent in many large river reaches, all of which diminish angler access. In many large river reaches, high concentrations of contaminants in fish flesh reduce the appeal for consumptive fisheries resource use (see Section 20.4.5). Also, directed angling effort has, in general, been slow to shift with changes in fish communities that have been brought about by the invasion of exotic species (such as the Asian carps—common, grass, bighead, and silver carp) or by changes in habitat. Even so, fishing is a leading river recreational activity. Angling is the foremost recreational use of the Missouri River in Missouri (Weithman and Fleener 1988) and accounts for more than 35% of the total recreational activity in Pools 11–22 of the upper Mississippi River.

Major sport fishes of large rivers vary predictably across the climates of North America. The annual thermal cycle dictates what groups will be abundant, so the temperature-altering effects of impoundments have led to deviations from the natural pattern. Coldwater species dominate the large Arctic rivers of Canada, but the commercial fishery reaps most of the harvest. Impoundment has led to an increase in the abundance of predators of young salmonids (e.g., northern pikeminnow, largemouth and smallmouth bass, and walleye) in the Columbia River. Introduced species, such as the walleye, are gaining importance in the Columbia River sport fishery. Other introduced fishes now enter the Columbia River creel also, such as largemouth bass, smallmouth bass, yellow perch, crappies, and catfishes (*Ictalurus* spp.) (Ebel et al. 1989).

Warmwater species are important in rivers at lower latitudes and elevations. The Missouri River supports 13 sport species: channel catfish, freshwater drum, and common carp are frequently harvested. The sport fishery on the upper Mississippi River consists of about 30 species. Bluegill and crappies are often the most sought-after fishes in warmwater rivers, but a number of other fishes are important, such as channel catfish, bullhead catfishes, largemouth bass, freshwater drum, northern pike, and common carp. Walleye and sauger are important in the northern reaches of the river.

Season, river stage, flow, and habitat affect harvest rates and species harvested. The sport fisheries in unchannelized reaches of the Missouri River tend to be about 2–2.5 times more productive than they are in channelized reaches (Groen and Schmulbach 1978). Tailwaters below main-stem dams and spillways, wing dikes, and backwaters generally produce the greatest yields in impounded rivers. Tailwater fisheries are generally good year-round, especially for walleye and sauger, even when much of the river is covered with ice. Backwaters with adequate depth and habitat diversity can provide year-round fishing.

20.5.2   Commercial Fisheries

The designation commercial fishery is more of a management concept than a biological one. Commercial fisher preferences and effort, market demand, and management agency philosophy are key factors determining whether commercial harvest is permitted. There are about 20 commercially harvested fishes in the Mississippi River and its tributaries (Table 20.2). Some of these species support commercial fisheries in other large warmwater rivers of North America. Although data are not available in many cases, the untapped commercial fisheries resources appear to be considerable in many large warmwater rivers.

Approximately 95% of the catch and 99% of the value of the upper Mississippi River commercial fishery consists of common carp, buffalos, catfishes, and freshwater drum. Total commercial landings in 1975 from the Mississippi River (including tributaries) totaled 4,651 metric tons (an average of 22 kg/ha of river at low flow) and had an exvessel value of over $1.9 million and a processed value of $5 million (Fremling et al. 1989). In 1983, more than 84,000 kg of channel catfish, 180,000 kg of buffalos, and 180,000 kg of goldeye were taken commercially from the Missouri River (Hesse et al. 1989).

There is little doubt that anthropogenic factors greatly influence exploitation of commercial fisheries. The majority of the commercial catch consists of fishes that are largely benthophagic. They consume prey that are in close contact with sediments, and they themselves are often in proximity to sediments that they disturb and mix. Many commercial species are at high risk of uptaking contaminants that are associated with sediments. Thus, a major responsibility of management agencies is to monitor contaminant concentrations in commercial species. In the 1990s, chlordane and PCB contamination forced management agencies to close the commercial harvest of certain species in rivers of the midwestern United States.

Human activities dramatically affect the abundance of commercial species, as reflected in commercial catch. Commercial catch in the upper Mississippi River appears to be related to total water surface area in navigation pools, and commercial catch per unit area appears to be positively correlated with the proportion of shallow marshy habitat and negatively correlated with barge traffic (Fremling et al. 1989). The commercial catch from the Illinois River was estimated at 77–200 kg/ha in 1908, and catch per unit area was highest in reaches with the highest proportions of backwater lakes connected to the river (Richardson 1921). Catch per unit area declined to 45.6 kg/ha by the 1950s and to 8.4 kg/ha in the 1970s (Fremling et al. 1989). The decline in the commercial catch in the Illinois River during this century can be directly attributed to habitat loss.

Highly sought game fishes are routinely harvested commercially in large rivers. It seems inevitable that the number of species on commercial lists will decline. If the trend toward increased sportfishing effort continues, recreational fishing demand for many river fishes will increase to the point that it will become uneconomical to permit

Table 20.2 Major commercially harvested fishes of the upper Mississippi River and its tributaries.

Bighead carp	Freshwater drum	Quillback
Bowfin	Gar	Shads (*Dorosoma* spp.)
Buffalos	Grass carp	Shovelnose sturgeon
Bullheads	Lake whitefish	Suckers (*Moxostoma* spp.)
Burbot	Mooneye	Walleye
Catfishes (*Ictalurus* spp.)	Northern pike	White bass
Common carp	Paddlefish	Yellow perch
Crappies	Pickerels	

their commercial harvest. The relative value of sport fishes to local economies greatly outweighs their value as commercial fishes, especially when direct angler expenditures and the resulting economic ripple effects are considered.

Two groups of invertebrates, freshwater mussels and crayfishes, are harvested in significant numbers from large rivers. In the early part of the twentieth century, mussels were taken in great numbers from several eastern and midwestern U.S. rivers for the pearl button industry. This industry declined with the invention of plastic buttons, and interest in the resource faded. However, demand for mussels increased in the 1970s when the Japanese cultured pearl industry began using mussel shells as seeds for pearl formation. Commercially important mussels are limited to species that have heavy shells and reach a fairly large size. Suitable pearl "blanks" can be cut only from the thick shells of relatively large species. The commercial mussel fishery is currently highly profitable, with high-quality shells selling for more than $1 per pound—which puts considerable pressure on the resource. The majority of the harvest comes from the Mississippi and Tennessee river systems.

With the rekindling of the commercial mussel industry came renewed concern for mussels and mussel communities, also sparking interest in protection of endangered mussels.

Loss of habitat, unexplained die-offs in the 1980s, and overharvest threaten the continued existence of commercially harvested mussels. Mussel sanctuaries have now been established below all Cumberland River and Tennessee River dams in Tennessee, Kentucky, and Alabama. Parallel efforts are underway for some historically important mussel beds in the upper Mississippi River.

Crayfishes are harvested for use as bait and food. The two primary species taken as food are the red swamp crawfish and the white river crawfish. They are harvested primarily from the Atchafalaya River basin to satisfy the growing demand for Cajun food delicacies. Harvests are greatest in high-water years when the floodplain is maximally inundated and provides ample decaying vegetation, which is fundamental to the food web supporting crayfish production. Most of the harvest is in Louisiana (more than 20 million kilograms annually), where the unique topography results in huge overflow swamps.

20.6   LARGE-RIVER FISHERIES MANAGEMENT

Fisheries management can be accomplished through manipulating (1) fish stocks and other biotic communities, (2) anglers, and (3) the physical and chemical environment. All three of these options are greatly affected by anthropogenic factors associated with other primary uses in most large North American rivers.

Large rivers remain among the few inland waters where sportfish harvest can be significantly increased. Harvest manipulation and habitat management can be effective and economical management approaches. Also, a number of important river fisheries already depend primarily on fish culture and stocking, and this trend may increase in the future unless natural habitats are restored.

20.6.1   Stock Assessment

The first step in managing a fishery, before any consideration should be given to manipulation, is to gather information about the fish community. Harvest will affect the size structure and abundance of the exploited population. This, in turn, will affect the entire fish community due to changes in trophic relationships among species. Specific information about the exploited population, such as growth, abundance, recruitment, mortality, yield, age and size structure, and habitat requirements is needed. In practice, however, large-river fisheries managers rarely can acquire all of the above information due to factors such as ingress and egress of fishes in the managed river reach, expense, equipment and personnel requirements, and the challenging sampling conditions found in large rivers.

A multiple-gear approach is usually warranted in sampling fish communities in large rivers because of biases associated with various gear types and because of strong interactions between the environment and sampling efficiency (e.g., sampling gears that work well in flowing water may not be effective where velocities are low). Commonly used river sampling tools include gill nets, trammel nets, seines, trawls (bottom and midwater), hoop nets, fyke nets, frame nets, larval fish nets, and electrofishing gear.

Shoreline seining is often effective for collecting smaller specimens. For collection of juveniles and adults of many large river species, we and others (e.g., Simon and Emery 1995; Pugh and Schramm 1998) have found boat electrofishing to be an effective sampling method, if only one method is to be used. Electrofishing works well along channel borders and in off-channel habitats but is not very effective in the channel proper.

Biologists increasingly are using trawls to sample within large channels. Great precaution must be taken when trawling (or conducting any activity, for that matter) in large rivers because it can be dangerous! The entire boat and all its occupants can be forced under water by the river's current if the trawl becomes entangled in debris and cannot be freed or cut loose in time. A colleague of ours sunk a boat in the Amazon River when an ichtyoplankton net became entangled. We recommend primary and backup systems to detach the trawl from the boat in an emergency and floats to facilitate trawl recovery. Trawling in an upstream direction allows more time to react to an entanglement situation. The boat can be drifted back over the trawl if it becomes caught, which often results in the trawl coming free. Trawling requires a boat with a powerful enough motor to overcome the drag from the trawl and the river's current and to attain an effective trawling speed.

20.6.2 Fish Population Manipulations

Population reductions in large rivers are generally considered infeasible due to the magnitude of the resource and the expense involved. When needed, such reductions are usually handled through the establishment of a commercial fishery. Managers can also enhance populations of large piscivorous fishes through stocking or catch restrictions. To our knowledge, reductions of overly abundant species, such as shads and alewife, by manipulation of predatory fish populations has never been effectively demonstrated in large rivers, but this approach still merits further attempt.

Just as they receive all the point and nonpoint source pollution in their watersheds, large rivers are also the receptacle of downstream drift from every upstream minnow bucket, aquarium release, hatchery, pond, and reservoir. Thus, the potential for species introductions is constantly present. Introduced warmwater and coolwater fishes have become serious threats to young salmonids in the Columbia River. There is also concern that the history of the common carp in North America may be repeated by other carp species such as the grass carp (stocked for vegetation control in small impoundments). The grass carp, introduced originally in Arkansas, has found its way as far upstream in the Mississippi River as Pool 5a on the Minnesota–Wisconsin border. Natural reproduction of grass carp has not been reported in the Upper Mississippi River, but there is evidence of reproduction in the lower Mississippi, Arkansas, Red, Black, Missouri, and Oachita rivers. Bighead carp have become very abundant in the Mississippi River, and silver and black carp are now found there as well.

Reservoir stocking programs and species introductions can have dramatic effects on river fisheries due to spillway escapement that can sometimes create quality tailwater fisheries and sometimes severely compromise endemic river fishes. The early assumption was that reservoirs would not support self-reproducing fish populations, so initial stocking focused on indigenous species. Today, more nonnative species, stocks, and hybrids are being introduced, such as striped bass and its hybrids, rainbow trout and brown trout, kokanee, cisco, rainbow smelt, alewife, and saugeye. Brown trout and rainbow trout introductions have been highly successful in Tennessee River main-stem reservoirs (Voigtlander and Poppe 1989). Striped bass and hybrid striped bass introductions have resulted in high-profile trophy fisheries. The striped bass will not successfully reproduce in most reservoirs, but inability to reproduce is often desirable for an introduced species because it gives the manager more control over the predator–prey balance.

Stocking may be the only remedy for native species if spawning has been disrupted, spawning grounds have been destroyed, or exploitation is extensive. Stocking is now the mainstay in many salmonid fisheries because of dams that block spawning migrations, the degradation of spawning grounds, and other factors. In 1977, about 85% of the coho salmon harvested off the Columbia River coast came from hatchery production (Ebel et al. 1989). Stocking can also be used to reintroduce and enhance threatened and endangered species (Chapter 16). Several U.S. states along the Missouri and Mississippi rivers are participating in a program for stocking artificially propagated pallid sturgeon, an endangered species. Missouri's paddlefish stocking program was stimulated by inundation of this species' spawning grounds by the construction of the Harry S. Truman Dam on the Osage River. The effectiveness of stocking to mitigate habitat loss in large rivers is uncertain at best; habitat loss should be avoided whenever possible. Also, stocking is being discouraged in many rivers due to the threat of disrupting the genetic integrity of wild populations (see Chapter 24).

20.6.3 Habitat Manipulations

Habitat manipulation can be cost effective for enhancing river fish communities. However, most habitat manipulation projects on large rivers are complex and expensive. Input from biologists and engineers with a variety of specialties is required to ensure that a project has the desired effects and endures beyond the next period of high water.

Habitat restoration and enhancement for the purpose of benefiting native fish communities is a relatively new science to large warmwater rivers, but a handbook of potential mitigation and enhancement techniques for large rivers has been compiled

(Schnick et al. 1982). Most of the earlier work on habitat manipulation was directed at large coldwater rivers in the western United States to ensure passage and continued propagation of anadromous species (Chapter 24). The Upper Mississippi River System Master Plan (USACE 1985) initiated a major environmental management program that began in 1987; the program includes habitat rehabilitation and enhancement projects. Projects include backwater dredging, dike and levee construction, island creation, bank stabilization, side channel openings and closures, wing and closing dike modification, aeration and water control systems, waterfowl nesting cover, acquisition of wildlife lands, and forest management. This environmental management program should generate useful information on habitat enhancement and restoration in large coolwater and warmwater rivers.

Modification of wing and closing dikes to reduce the negative effects of navigation channel maintenance and improve fish habitat has been used extensively on the upper Mississippi River and Missouri River. Many dike designs have been tried, including notched wing dikes. The usual objective is to reduce riverbed aggradation and to encourage the river itself to develop aquatic habitats usable at various river stages (Hesse et al. 1989).

Advances have been made in developing techniques to predict the physical effects of towboat traffic (Bhowmik et al. 1982; Chen et al. 1984). Reduction of navigation effects may require improved regulations (including possible speed limits and establishment of protected areas), enforcement of regulations, and perhaps redesign of boat propellers.

The armoring of bends in riverbanks causes erosional forces to be directed downward, deepening the channel. Deposition occurs on the inside of the bend, causing it to become narrower. This results in narrow, high-velocity river bends that pose hazards to shipping. A new river-training structure, the bendway weir, is being used in the Mississippi River to correct this problem. These weirs are submerged throughout deep river bends and are oriented obliquely to the current. They are designed to armor and prevent further scouring and deepening of the channel as well as to promote widening of the channel by deflecting erosional forces toward the inside of the bend. The Army Corps of Engineers is currently collecting data to determine whether more fish use river bends outfitted with bendway weirs.

Blockage of fish movements in warmwater rivers has not been considered a major problem because spawning has not generally been prevented. However, ongoing degradation of the Missouri River caused by upstream reservoirs and maintenance of flood control and navigation projects may be threatening catfish spawning in major tributaries. Grade stabilization structures, installed to prevent streambed degradation in Missouri River tributaries, have blocked catfishes from reaching spawning areas. A project implemented on the Little Sioux River in Iowa to provide for catfish passage is currently being evaluated.

During the winter, young salmon congregate at high densities in lentic habitats of river complexes, such as channel and extrachannel ponds, gravel pits, and cut-off channel meanders. These habitats, as compared with flowing-channel habitats where temperatures fall to 0°C for extended periods, provide at least two conditions that probably promote overwinter survival: (1) warmer temperature due to stratification, and (2) lack of flow. The importance of this habitat has led to calls for creation and enhancement of lentic habitats in salmon rivers. Lentic habitats are aggrading and disappearing in large, aging, modified rivers, also leading to the need to create and enhance these habitats.

Water level management for flood control and navigation projects often affects river fisheries. Public Law 697, passed in 1948 and known as the anti-drawdown law, prevents winter drawdown of slackwater navigation pools for flood control in the upper Mississippi River in the interest of preserving wintering habitat for fishes. The law ordered the Army Corps of Engineers to maintain pools "as though navigation was carried on throughout the year."

Increases in water level and flow rarely meet with favor from riparian property owners. Consequently, flow augmentation and water level manipulation on a large scale is rarely, if ever, practiced for large-river fisheries. However, control of water levels and flows on a smaller scale in isolated backwaters could be a useful tool for large-river fisheries and wildlife managers. Several of these types of flow manipulation projects have been completed with varying degrees of success on the upper Mississippi River.

On a larger scale, water level manipulation is being tried in Pools 24–26 of the Mississippi River to promote the growth of nonpersistent wetland vegetation. Pool levels are drawn down 0.3–0.6 m for about 1.5 months in summer, a period long enough to allow lush vegetative production (mostly smartweed and wild millet), and then are returned to normal levels in the fall (K. L. Dalrymple, Missouri Department of Conservations, E. Atwood, Illinois Department of Natural Resources, and D. Busse and C. N. Strauser, U.S. Army Corps of Engineers, data presented during a meeting of the Upper Mississippi River Basin Committee, 1996).

Sparks (1995) pointed out that dams greatly affect hydrological regimes. The spring flood pulse immediately below some Mississippi River dams is similar to predam conditions, but it is progressively attenuated downstream, and it can even become inverted just above, or upstream of, the next downstream dam (Figure 20.9). Sparks and most other river scientists (e.g., Bayley 1991) recognize that restoration of a more natural hydrological regime in regulated rivers would greatly benefit river fisheries. The productivity of the floodplain and its importance to fish production became clear in the Mississippi River during the flood of 1993, which produced exceptionally large year-classes in many species, such as sauger and blue sucker (Fisheries News 1993).

Water releases from dams can be planned to promote beneficial effects in downstream reaches. Guidelines for planning releases can be developed from the study of minimum instream flow requirements of key river fishes. Water releases could also be used to simulate seasonal inundation of the floodplain (Ward and Stanford 1989), stimulating fish production and reproduction. Water and sediment releases could possibly be managed to maintain natural morphology below dams (Ligon et al. 1995). Sedimentation in the upper Mississippi River is a major problem. Interestingly, it is the lack of sediment in the lower Mississippi River floodplain that is thought to cause the loss of wetland plant communities there. The levee system causes most of the sediments to be transported to the river mouth and lost over the continental shelf. The sediment deposition rate in the floodplain currently appears to be less than sediment submergence rate (due to sea level rise and subsidence). The excess water levels are thought to stress and kill plants by exposing them to hypoxic conditions, high salinities, and high sulfide (Sheehan and Konikoff 1998). Clearly, better management of sediments is a significant need in many large rivers.

Main-stem reservoirs can improve water quality in downstream reaches by trapping sediments. However, reservoir releases of hypolimnetic waters are often devoid of oxygen, and toxic metals will enter hypolimnetic waters under anoxic conditions. A program to alleviate these and other problems in reservoirs on the Tennessee River includes techniques such as reservoir destratification and hypolimnetic aeration, hydroelectric turbine modification to increase dissolved oxygen, aeration of release waters, and weirs that automatically provide minimum instream flows (Voigtlander and Poppe 1989).

Although most large river reaches in the United States may not be restored to near-pristine conditions in the foreseeable future, plans for partial restoration of more natural river processes are being developed nationwide. There is growing sentiment, as well as funding from public interest groups and governments, to move people and activities such as agriculture from floodplains. This relocation would afford opportunities to move flood control levees further from the river. There is probably no better way at the present time to improve river ecosystems and fish populations in many U.S. rivers. Stanford et al. (1996) suggested a general protocol for the restoration of regulated rivers that includes: (1) restoring peak flows needed to reconnect and reconfigure channel and floodplain habitats, (2) stabilizing base flows to revitalize food webs in shallow-water habitats, (3) reconstituting seasonal temperature patterns, (4) maximizing dam passage to allow recovery of fish metapopulation structure, and (5) instituting a management belief system that relies upon natural habitat restoration and maintenance.

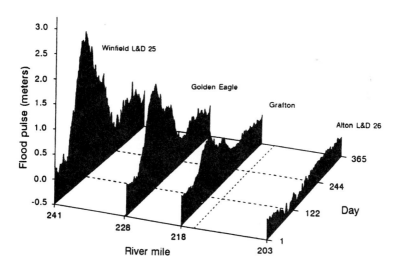

Figure 20.9 The spring flood pulse is inverted at river mile 203, upstream of the lock and dam (L&D) at river mile 201; the spring flood increases in amplitude upstream of the dam in Pool 26 of the Mississippi River. Alton L&D 26 could be operated in a manner that promotes a more natural flood pulse throughout the pool. Inundation of the floodplain near the dam would increase the river's productivity and allow fishes access to the floodplain. Figure taken from Sparks (1995).

20.6.4   Management of Fisheries Users

Commercial fishing for salmonids is often highly regulated, and extensive information is kept on catch. Catch can be allocated and partitioned into harvest by fishing gear types and even into various user groups (e.g., Native Americans). Although there are large standing stocks of large salmonids in some large rivers, such as the Churchill and Mackenzie (Bodaly et al. 1989), the stocks cannot tolerate much fishing pressure. Salmon in Arctic rivers are slow growing but long lived (due to low exploitation), permitting them to attain large size.

On the other hand, most state agencies have regarded fisheries in large warmwater rivers as limitless resources requiring little regulation. Consequently, less restrictive regulations have been imposed. However, advances in angling technology, organized fishing contests, and increased competition among anglers for use of inland waters have led to a rediscovery of large-river fishing.

There is a growing desire on the part of river managers to protect populations from overexploitation, which has resulted in more restrictive harvest regulations. Consequently, many inland fisheries regulations (including some creel limits and closed seasons) presently also apply to certain large-river reaches. For example, a 38-cm minimum length limit for largemouth bass was recently imposed on Pools 16–19 of the upper Mississippi River.

Mandated harvest restrictions combined with voluntary changes in harvest practices are especially effective in improving the quality of a fishery. Smallmouth bass catch rates have doubled in the Tennessee River since the establishment of a 36-cm minimum length limit for smallmouth bass. The improvement probably did not come about solely as a direct result of the length limit. A concomitantly developed greater respect for the resource on the part of anglers (anglers release 98% of their catch) has led to an order of magnitude decrease in harvest since the length limit was instituted (Slipke et al. 1998).

Commercial fishing is a traditional practice on almost every large river, constituting a way of life for some individuals. The commercial fishery on the upper Mississippi River can be described as a limited-entry, territorial fishery (fishers lay claim to areas they have historically fished). The commercial fisheries of the basin as a whole are being exploited at nearly optimal levels; this conclusion is based on the observation that catch per fisher declined and yields remained stable during the 1950s to 1970s when effort increased substantially. Unfortunately, good catch records are not often kept for most large rivers, and the reporting procedures are usually voluntary, which causes most harvest estimates to be low. However, even if underestimated, valuable trend information can be obtained from these data if one is willing to assume that underestimates have been similarly made each year. For example, declines in commercial fishing catch from 1955 to 1984 in the upper Mississippi River suggested that channel catfish above the legal length limit of 33 cm were being overharvested. An increase to a 38 cm minimum length limit resulted in an upward trend in both commercial and recreational fishing catch (Pitlo 1997).

20.7   CONCLUSION

It has been estimated that 60% of the world's total streamflow will be regulated by dams by the year 2000 (Petts 1980), and many (e.g., Welcomme 1989) believe that restoration of large rivers to near-pristine conditions is incompatible at present human population levels. There have been attempts to return modified rivers to more natural conditions; the

Kissimmee River in Florida (Toth et al. 1997) and the Colorado River are examples. However, from a pragmatic standpoint, impoundment and many other modifications for navigation and flood control will probably persist long into the future in many large river systems. It is unlikely that there will be large-scale dismantling of the inland waterway and flood control system. Therefore, biologists should consider using innovative approaches to managing large rivers under existing conditions. Management efforts should be directed at maintaining a diverse ecosystem and optimizing exploitation of renewable resources, such as sport and commercial fishes. Some management goals may necessitate artificial manipulations, such as environmental engineering of habitats or manipulations of species composition. These measures are appropriate because most large rivers no longer behave entirely as natural systems. Emphasis on conservation, rather than preservation, thus appears to be the more realistic management approach in rivers modified by humans. However, protection of unique natural habitats (such as the wild and scenic reach of the Missouri River in Montana) and threatened and endangered species must be considered viable management practices for these resources.

20.7.1   Directions for Fisheries Management

River managers should consider the maintenance of a healthy ecosystem as their primary goal. Supplemental fish stocking, reintroductions, and habitat restoration and enhancement projects may be required to meet this goal. In general, fisheries management objectives in large rivers should be set according to the following criteria.

1. Threatened and endangered species and remaining unique habitats should be preserved, and reintroductions of extirpated species should be contemplated.

2. Fish communities should be managed to maximize species diversity.

3. Resources, such as commercial and sport fisheries, should be managed for optimum sustained yield.

4. At a minimum, habitat diversity should be maintained at its current level through sound conservation practices. In altered systems, habitat diversity should be increased through management projects directed at restoration, enhancement, or creation of fish habitat.

Management of river fisheries requires an improved understanding of large-river fisheries biology and ecology. This, in turn, requires collection of long-term data sets on population dynamics, movements, and habitat requirements. Because of the scale of these efforts, and the fact that rivers cross political boundaries, fisheries agencies should develop long-term management plans in cooperation with fisheries agencies in other states and other agencies involved with rivers. Several states acted together to implement new minimum length limits to enhance commercial and recreational fishing harvest of channel catfish in the upper Mississippi River (Pitlo 1997).

20.7.2   Basinwide Management

As the appellation implies, large rivers require large efforts to manage them effectively. Only through ecosystem management can all large-river resources, not just fisheries, be optimally used and conserved. All large-river users (Figure 20.10) need to be aware of the importance of ecosystem level management. Broad-scope and long-term management plans need to be developed at the multijurisdictional and multiple-stakeholder levels. This kind of effort is underway in the Atchafalaya River basin. The Army Corps of Engineers is planning to spend more than $250 million (about $90 million from the state of Louisiana and the remainder allocated by the U.S. Congress) to purchase or secure conservation easements on approximately 160,000 ha of land for the purpose of management for wetlands.

Such holistic management starts with a vision, developed and shared by all user groups, of what we want our large rivers to be in the next century. In our experience, the consensus-building process itself reaps palpable benefits (see Williams et al. 1997 for examples). As stakeholders begin to become more aware and, thereby, more sensitive to each other's needs and desires, they begin to show more flexibility in their own needs and desires. A common purpose develops, as well as a sense that big things can get done.

Figure 20.10 Mississippi River paper plant and hydroelectrical station at Sartell Dam in Saint Cloud, Minnesota. Large rivers attract industries and municipalities to their banks because they provide water, inexpensive transportation, recreational opportunities, and other resources, such as hydroelectrical power. Anthropogenic impacts and periodically harsh natural conditions, such as ice-covered channels, low temperatures, and high discharge, combine to provide a challenging environment for river fishes. Photograph provided by D. Logsdon, Minnesota Department of Natural Resources.

20.8 REFERENCES

Addair, J. 1944. Fishes of the Kanawha River system and some factors which influence their distribution. Doctoral dissertation. Ohio State University, Columbus.

Alabaster, J. S., and R. Lloyd. 1980. Water quality criteria for freshwater fish. Butterworth, Stoneham, Massachusetts.

Bayley, P. B. 1989. Aquatic environments in the Amazon Basin, with an analysis of carbon sources, fish production, and yield. Special Publication of the Canadian Journal of Fisheries and Aquatic Sciences 106:399–408.

Bayley, P. B. 1991. The flood pulse advantage and the restoration of river-floodplain ecosystems. Regulated Rivers Research & Management 6:75–86.

Bayley, P. B. 1995. Understanding large river-floodplain ecosystems. BioScience 45:153–158.

Bhowmik, N. G., M. Demissie, and C. Y. Guo. 1982. Waves generated by river traffic and wind on the Illinois and Mississippi rivers. University of Illinois, Water Resources Center, Illinois State Water Survey, UILU-WRC-82-0167 Research Report 167, Champaign.

Bodaly, R. A., J. D. Reist, D. M. Rosenberg, P. J. McCart, and R. E. Hecky. 1989. Fish and fisheries of the Mackenzie and Churchill river basins, northern Canada. Canadian Special Publication of Fisheries and Aquatic Sciences 106:128–144.

Chen, Y. H., D. B. Simons, R. Li, and S. S. Ellis. 1984. Investigation of effects of navigation traffic activities on hydrologic, hydraulic, and geomorphic characteristics in the upper Mississippi River system. Pages 229–324 in J. G. Wiener, R. V. Anderson, and D. R. McConville, editors. Contaminants in the upper Mississippi River. Butterworth, Stoneham, Massachusetts.

Church, M. 1992. Channel morphology and typology. Pages 126–143 in P. Calow and G. E. Petts, editors. The rivers handbook. Blackwell Scientific Publications, Cambridge, Massachusetts.

Cole, G. A. 1979. Textbook of limnology. C. V. Mosby, St. Louis, Missouri.

Cowardin, L. M., V. Carter, F .C. Golet, and E. T. LaRoe. 1979. Classification of wetlands and deepwater habitats of the United States. U.S. Fish and Wildlife Service, Office of Biological Services, FWS/OBS-79/31, Washington, D.C.

Czaya, E. 1981. Rivers of the world. Cambridge University Press, New York.

Dahl, T. E. 1990. Wetland losses in the United States: 1780s to 1980s. U.S. Fish and Wildlife Service, Washington, D.C.

Ebel, W. J., C. D. Becker, J. W. Mullan, and H. L. Raymond. 1989. The Columbia River-toward a holistic understanding. Canadian Special Publication of Fisheries and Aquatic Sciences 106:205–219.

Fischer, J. R., and T. O. Claflin. 1995. Declines in aquatic vegetation in navigation pool no. 8, upper Mississippi River between 1975 and 1991. Regulated Rivers Research & Management 11:157–165.

Fisheries News. 1993. Mississippi flooding could benefit fisheries. Fisheries 18(10):48–49.

Fremling, C. R., and five coauthors. 1989. Mississippi River fisheries: a case history. Canadian Special Publication of Fisheries and Aquatic Sciences 106:309–351.

Gore, J. A., and F. D. Shields. 1995. Can large rivers be restored? BioScience 45:142–152.

Goudreau, S. E., R. J. Neves, and R. J. Sheehan. 1993. Effects of sewage treatment plant effluents on mollusks in the upper Clinch River, Virginia. Hydrobiologia 252:211–230.

Groen, C. L., and J. C. Schmulbach. 1978. The sport fishery of the unchannelized and channelized middle Missouri River. Transactions of the American Fisheries Society 107:412–418.

Hesse, L. W. 1982. The Missouri River channel catfish. Nebraska Game and Parks Commission, Nebraska Technical Series 11, Lincoln.

Hesse, L. W., and six coauthors. 1989. Missouri River fishery resources in relation to past, present, and future stresses. Canadian Special Publication of Fisheries and Aquatic Sciences 106:352–371.

Johnson, B. L., W. B. Richardson, and T. J. Naimo. 1995. Past, present, and future concepts in large river ecology. BioScience 45:134–141.

Junk, W. J., P. B. Bayley, and R. E. Sparks. 1989. The flood-pulse concept in river floodplain ecosystems. Canadian Special Publication of Fisheries and Aquatic Sciences 106:110–127.

Karr, J. R. 1981. Assessment of biotic integrity using fish communities. Fisheries 6(6):21–27.

Karr, J. K., L. A. Toth, and D. R. Dudley. 1985. Fish communities of midwestern rivers: a history of degradation. BioScience 35:90–95.

Leopold, L. B., M. G. Wolman, and J. P. Miller. 1964. Fluvial processes in geomorphology. Freeman, San Francisco.

Ligon, F. K., W. E. Dietrich, and W. J. Trush. 1995. Downstream ecological effects of dams. BioScience 45:183–192.

Limburg, K. E., S. A. Levin, and R. E. Brandt. 1989. Perspectives on management of the Hudson River ecosystem. Canadian Special Publication of Fisheries and Aquatic Sciences 106:265–291.

Nielsen, L. A., R. J. Sheehan, and D. J. Orth. 1986. Impacts of navigation on riverine fish production in the United States. Polish Archives of Hydrobiology 33:377–394.

Owen, O. S. 1985. Natural resource conservation. Macmillan, New York.

Petts, G. E. 1980. Long-term consequences of upstream impoundment. Environmental Conservation 7:325–332.

Pitlo, J., Jr. 1997. Response of upper Mississippi River channel catfish populations to changes in commercial harvest regulations. North American Journal of Fisheries Management 17:848–859.

Pugh, L. L., and H. L. Schramm, Jr. 1998. Comparison of electrofishing and hoopnetting in lotic habitats of the lower Mississippi River. North American Journal of Fisheries Management 18:649–656.

Rasmussen, J. L. 1979. A compendium of fishery information on the upper Mississippi River. Upper Mississippi River Conservation Committee, Rock Island, Illinois.

Richardson, R. E. 1921. The small bottom and shore fauna of the middle and lower Illinois River and its connecting lakes, Chillicothe to Grafton: its valuation; its sources of food supply; and its relation to the fishery. Illinois Natural History Survey Bulletin 13:363–522.

Robinson, A. T., R. W. Clarkson, and R. E. Forrest. 1998. Dispersal of larval fishes in a regulated river tributary. Transactions of the American Fisheries Society 127:772–786.

Toth, L. A., D. A. Arrington, and G. Begue. 1997. Headwater restoration and reestablishment of natural flow regimes: Kissimmee River of Florida. Pages 425–442 in J. E. Williams, C. A. Wood, and M. P. Dombeck, editors. Watershed restoration: principles and practices. American Fisheries Society, Bethesda, Maryland.

USACE (U.S. Army Corps of Engineers). 1985. Upper Mississippi River system environmental management program—general plan. USACE, North Central Division, Chicago.

Schnick, R. A., J. M. Morton, J. C. Mochalski, and J. T. Bailly. 1982. Mitigation and enhancement techniques for the upper Mississippi River system and other large river systems. U.S. Fish and Wildlife Service Resource Publication 149.

Sedell, J. R., J. E. Richey, and F. J. Swanson. 1989. The river continuum concept: a basis for the expected ecosystem behavior of very large rivers? Canadian Special Publication of Fisheries and Aquatic Sciences 106:49–55.

Sheehan, R. J., and M. Konikoff. 1998. Wetlands and fisheries resources of the Mississippi River. Pages 628–647 in S. K. Mujumdar, E. W. Miller, and F. J. Brenner, editors. Wetlands and associated systems. Pennsylvania Academy of Science, Harrisburg.

Simon, T. P., and E. B. Emery. 1995. Modification and assessment of an index of biotic integrity to quantify water resource quality in great rivers. Regulated Rivers Research & Management 11:283–298.

Slipke, J. W., M. J. Maceina, V. H. Travnichek, and K. C. Weathers. 1998. Effects of a 356-mm minimum length limit on the population characteristics and sport fishery of smallmouth bass in the shoals reach of the Tennessee River, Alabama. North American Journal of Fisheries Management 18:76–84.

Sparks, R. E. 1995. Need for ecosystem management of large rivers and their flood-plains. BioScience 45:168–182.

Stanford, J. A., and six others. 1996. A general protocol for restoration of regulated rivers. Regulated Rivers Research & Management 12:391–413.

Statzner, B. 1987. Characteristics of lotic ecosystems and consequences for future research directions. Pages 365–390 in E. D. Schulze and H. Zwolfer, editors. Potentials and limitations of ecosystem analysis. Springer-Verlag, New York.

Vannote, R. L., G. W. Winshall, K. W. Cummins, J. R. Sedell, and C. E. Cushing. 1980. The river continuum concept. Canadian Journal of Fisheries and Aquatic Sciences 37:130–137.

Voigtlander, C. W., and W. L. Poppe. 1989. The Tennessee River. Canadian Special Publication of Fisheries and Aquatic Sciences 106:372–384.

Ward, J. V., and J. A. Stanford. 1989. Riverine ecosystems: the influence of man on catchment dynamics and fish ecology. Canadian Special Publication of Fisheries and Aquatic Sciences 106:56–64.

Weithman, A. S., and G. G. Fleener. 1988. Recreational use along the Missouri River in Missouri. Pages 67–78 *in* N. G. Benson, editor. The Missouri River, the resources, and their uses and value. American Fisheries Society, North Central Division, Special Publication 8, Bethesda, Maryland.

Welcomme, R. L. 1979. Fisheries ecology of floodplain rivers. Longman Group, London.

Welcomme, R. L. 1989. Floodplain fisheries management. Pages 209–233 *in* J. A. Gore and G. E. Petts, editors. Alternatives in regulated river management. CRC Press, Boca Raton, Florida.

Williams, J. E., C. A. Wood, and M. P. Dombeck, editors. 1997. Watershed restoration: principles and practices. American Fisheries Society, Bethesda, Maryland.

Wright, T. D. 1982. Potential biological impacts of navigation traffic. U.S. Army Corps of Engineers, South Atlantic Division, Environmental and Water Quality Operation Studies Miscellaneous Paper E-82, Atlanta.

Chapter 21

Small Impoundments

STEPHEN A. FLICKINGER, FRANK J. BULOW, AND DAVID W. WILLIS

21.1 INTRODUCTION

Small impoundments are ponds made by constructing dams to impound water from springs, streams, wells, direct precipitation, and surface runoff or by digging depressions or pits to hold water (Dendy 1963). What constitutes a small impoundment varies regionally, but the designation generally includes impoundments ranging from less than 0.4 ha to more than 40 ha. The American Fisheries Society's Central States Pond Management Work Group defined a pond as an impoundment with a surface area of 0.2–2.4 ha (Anderson 1978). Manageability is a key feature of most ponds that are constructed for fish production, and such ponds are typically furnished with a drain. Another common feature of ponds is that they are usually privately owned. Farm or ranch ponds are often built for livestock water and irrigation (Figure 21.1).

Ponds are most numerous in the central and southeastern United States (Figure 21.2). The Canadian provinces and some northern states have a large number and variety of natural ponds and potholes that are primarily the result of relatively recent glaciation. Thousands of small, often temporary sloughs in the southern prairie provinces of Canada are important to waterfowl and agriculture of the semiarid prairie. In addition, many small reservoirs have been constructed in Canada. Severe winter stagnation is a common limnological characteristic that impedes sustained fish production in smaller, northern-latitude impoundments. Alaskan pond management is very limited, but an active program does exist for stocking borrow pits along the highway system. There is a paucity of small impoundments in Mexico, and those that exist are used primarily to store water for livestock.

In many areas, farm ponds make substantial contributions to sportfishing. Surveys indicate that farm ponds support a large portion of the recreational fishing days in states with good pond resources. Arid western states and northern states with potential for winterkill do not have many ponds.

Development of farm pond management in North America has a long history (see overviews in Regier 1962; Dendy 1963; Swingle 1970; Bennett 1971). The present management philosophy for recreational pond fisheries is sustained or improved fishing quality and favorable benefit-cost ratios; the ultimate goals are high catch rates and above-average sizes. Many of the principles and techniques developed for manage-

Figure 21.1   Farm ponds and other small impoundments are constructed for livestock water, irrigation, fish production, field and orchard spraying, domestic water supply, fire protection, energy conservation, wildlife habitat, recreation, and landscape improvement. Small impoundments are typically more easily managed for increased fish production than are large, multipurpose reservoirs.

ment of farm ponds are being applied to small natural lakes, flood control impoundments, and impoundments created as a by-product of surface material or mineral mining (Bulow 1967; Noble et al. 1979; Noble 1988).

Although management techniques have been developed by fisheries biologists, it is usually the pond owner who is responsible for implementation. Literature and personal assistance are often available to the pond owner, and specific topics not covered in this chapter, such as fish parasites, are typically addressed in booklets and extension literature. Guidance can often be obtained from Cooperative Extension Service, the Natural Resources Conservation Service, or various fisheries agencies. In addition, pond management services are available from private consulting firms.

21.2   ECOLOGICAL PRINCIPLES RELATED TO PONDS

Current recommendations for pond construction, species and numbers of fish to stock, and subsequent management techniques are based on an understanding of basic ecological principles relating to trophic relationships and population dynamics. Carrying capacity when applied to pond fish populations may be defined as the maximum weight of a given species of fish that a pond will support during a stated interval of time; standing stock is the actual weight of a given species or complex of species present in a pond at a specific moment (Bennett 1971). The growing season for warmwater species may be 2 months in Manitoba, 3 months in Minnesota, 6–7 months in Illinois, and 11–12 months in Louisiana or Mexico. Consequently, new or reclaimed ponds initially stocked with largemouth bass and bluegills may reach carrying capacity within 1 year at southern latitudes but may take two or more years at northern latitudes.

Fishes are consumers, being dependent upon the production of phytoplankton and a series of lower-level consumers such as zooplankton, benthos, and other smaller fishes. Because the amount of usable energy is reduced with each transfer in the food chain, planktivorous fishes are expected to yield more biomass per unit area for a given

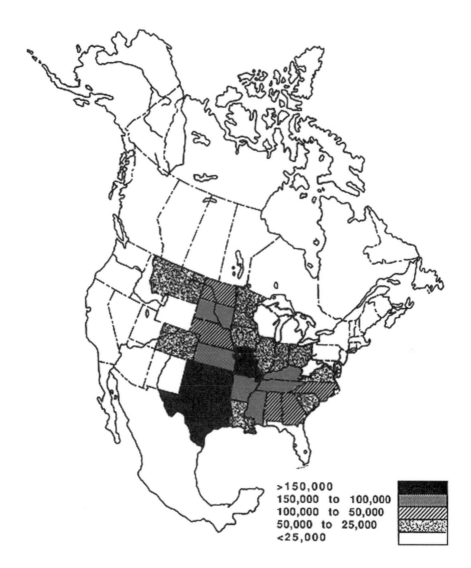

>150,000
150,000 to 100,000
100,000 to 50,000
50,000 to 25,000
<25,000

Figure 21.2 Distribution by state or province of the total number of ponds constructed in North America. Figure adapted from Modde (1980).

level of primary production than are carnivores at the end of a long food chain. Standing crop tends to increase with number of species because of greater use of available resources; however, simple species combinations are more manageable in producing fishes considered useful in a sport fishery.

Relationships between carrying capacity, population density, and individual sizes have long been recognized. For example, at a given level of productivity, the average size of bluegills will be inversely related to population density (Bennett 1971). The carrying capacity of a pond may be increased by fertilization, supplemental feeding, or stocking different species.

21.3 HISTORICAL OVERVIEW OF POND MANAGEMENT

The drought of the 1930s led to the Soil Conservation Service (presently Natural Resources Conservation Service) farm pond building program in the United States and the Prairie Farm Rehabilitation Act Pond Construction Program in Canada. The demands for protein and inexpensive recreation during World War II intensified the interest in pond construction and management. Questions of what species to stock initiated a search for a stocking combination that would establish balanced predator–prey populations and sustain quality fishing. Two schools of thought developed concerning farm pond stocking strategies, one centered in Alabama and one in the Midwest (Regier 1962). The former considered the largemouth bass-bluegill combination as optimal, and the latter sought other prey species to stock with largemouth bass.

21.3.1 Warmwater Ponds

Research by Swingle and Smith (1938) at Auburn University established the largemouth bass and bluegill as a suitable stocking combination. The basic premise of this combination was that largemouth bass and bluegill provide an effective predator–prey system and both are desirable sport fishes. The bluegill would convert invertebrate production into bluegill flesh, small bluegills would serve as food for the largemouth bass, largemouth bass predation would control excessive numbers of bluegills, and the few bluegills surviving would grow to a large average size because of low population density. Although Swingle experimented with alternative forage species, he found the bluegill to be most desirable because it met all of his criteria for a good prey fish for largemouth bass (Dillard and Novinger 1975): (1) in adult form it must not be too large for largemouth bass to eat; (2) in adult form it must not be too small or largemouth bass will eliminate its population; and (3) to be of continuing value as largemouth bass forage, the bluegill population must be reduced annually by angling or predation to stimulate spawning. After years of experimentation with fish size and stocking density, Swingle (1951) concluded that stocking fingerlings of each species was the most desirable technique. He obtained the highest average standing crop of harvestable bass by stocking 250 bass and 2,500 bluegill fingerlings per hectare in fertilized ponds (Dillard and Novinger 1975).

A symposium on the management of farm ponds included evidence that stocking and management procedures applicable to Alabama did not necessarily apply elsewhere (Meehean 1952). The largemouth bass–bluegill combination frequently leads to an overpopulation of bluegills that interferes with successful reproduction of largemouth bass. This situation is more common in northern latitudes where interest in bluegill fishing is not as great as in the South. The shorter growing season and slower growth rate in northern regions mean that largemouth bass and bluegill may not reach harvestable size until 3 years after stocking. The faster-growing, southern-latitude largemouth bass populations also exert greater predatory control of bluegills. Northern-latitude largemouth bass populations may require an additional year to reach sexual maturity in contrast to southern populations (Ball 1952; Stone and Modde 1982), and northern-latitude bluegills often spawn 1 year before largemouth bass spawn (James 1946; Saila 1952). These differences in growth and spawning, along with a high mortality rate within the population before the attainment of harvestable-size fish, result in a lack of predictability for a given stocking. The larger, deeper (and less manageable) ponds necessary to escape winterkill in northern climates also contribute to lower success rate with the largemouth bass–bluegill combination. An additional contributory factor is the protection afforded to bluegills by excessive aquatic macrophytes; in southern but not in northern latitudes, fertilization of ponds promotes phytoplankton growth, which, in turn, shades out macrophytes. Despite these limitations, Guy and Willis (1990) found population structures of largemouth bass and bluegills in South Dakota ponds that were similar to those of midwestern and southeastern ponds. Additionally, minimum (Lindgren and Willis 1990a) and slot length limits (Neumann et al. 1994) were as effective for largemouth bass in South Dakota ponds as they were in small impoundments farther south and east.

21.3.2 Coldwater Ponds

There are considerably fewer coldwater ponds than warmwater ponds; however, lower elevation warmwater ponds in the central and northern Rocky Mountain states often can also support trout. Fall stocking of catchable-size trout for winter and spring fishing is an option for southern ponds. There are no coldwater pond research counterparts to the works of Swingle, Bennett, and Regier. Rainbow, brook, cutthroat, and brown trout have been stocked in ponds, rainbow trout being the most common choice. Trout are usually not stocked in combination with other species; in fact, if other fish species are present in the pond, they are removed before trout stocking. Ponds of average fertility will produce enough natural food to support an average of 250 kg of trout per hectare, but 10- to 20-fold increases can be produced with supplemental feeding of pelleted foods (Marriage et al. 1971) as long as dissolved oxygen depletion from excessive feeding does not become a problem. Repeated stockings are necessary because trout reproduction seldom occurs in ponds.

21.3.3 Coolwater Ponds

The term coolwater fishes has been introduced to denote an intermediate category of water temperatures and species suited to those temperatures. Despite the validity of a coolwater concept, the few papers published on stocking these species in ponds have reported poor to limited success. More research is needed in this area. Coolwater fishes include walleye, yellow perch, northern pike, striped bass, and striped bass hybrids. These fishes have usually been introduced to increase predation on prey populations and provide an additional sport fish.

21.4   STOCKING AND HARVEST STRATEGIES

Stocking recommendations vary according to purpose and region. Experimentation with alternatives to the standard largemouth bass–bluegill stocking combination (Dillard and Novinger 1975; Modde 1980) is ongoing. Dillard and Novinger (1975) stressed the need to develop a stocking strategy that will provide a desirable pond fishery in the shortest time possible within physical and fiscal limits. This philosophy dictates a diversity of stocking strategies rather than one standard.

Another approach to stocking is the replacement of the traditional standardized recommendations for each region with selected stocking combinations that most appropriately meet specific environmental situations and the angling preferences or other goals of the pond owner. Regier (1962) suggested that the kind of fish and fishing that the pond owner desires should be given greater consideration. Dillard and Novinger (1975) explored this concept in detail and suggested that pond owners could be presented with several stocking and harvest options and, if necessary, could buy fish from private suppliers. Along these lines, Lewis and Heidinger (1978a) presented several stocking combinations to satisfy specific conditions and objectives, and Gabelhouse et al. (1982) presented five optional stocking and management strategies to meet specific objectives. Several stocking and management options are described below.

21.4.1 All-Purpose Option

This option allows the harvest of largemouth bass, bluegill, and channel catfish of a variety of sizes. It employs a protected slot limit of 30–38 cm for largemouth bass 4 years after stocking. To reduce competition and allow largemouth bass to grow consistently over 38 cm, about 75 largemouth bass per hectare (length range 20–30 cm) should be harvested annually after the fourth year following largemouth bass fingerling stocking (Gabelhouse et al. 1982). These guidelines are based on data from Kansas; initiation of harvest can be sooner or later depending on growth rates at a particular latitude. Release of all 30–38 cm largemouth bass will ensure that at least 10% of the catchable-size largemouth bass survive to lengths of 38 cm and longer. A larger population of 30–38 cm largemouth bass will help reduce densities of intermediate-size bluegills and allow some bluegills to reach 20 cm. Bluegills and channel catfish can be harvested as desired, but channel catfish must be replaced with 20 cm or longer channel catfish. With this option, if anglers do not obey the slot limit, overharvest of largemouth bass can result in overpopulation of bluegills. On the other hand, if anglers release largemouth bass in the slot limit but do not harvest 20–30 cm largemouth bass under the slot, overpopulation and stunting of largemouth bass can occur.

21.4.2 Harvest Quota Option

Because of problems associated with length limits, quotas have been proposed as an alternative way to regulate harvest of largemouth bass. This option involves the harvest of a given number or weight of largemouth bass annually, regardless of size. Catch-and-release fishing is possible after the quota has been reached. Few or no largemouth bass should be harvested for the first 4 years, followed by an annual harvest of about 125 individuals or 22 kg of largemouth bass per hectare without regard for length. This is a general guideline that should be adjusted for the fertility of a particular pond. Also, the harvest needs to be spread out over the entire fishing season. With this option, there is a tendency to overharvest large largemouth bass and underharvest small ones. Channel catfish harvest is unrestricted, but those harvested should be replaced with 20 cm or larger individuals. There is no limit for bluegill harvest, but with this option various state recommendations call for the harvest of 3–10 kg of bluegill for each kilogram of largemouth bass harvested. Although this option is generally expected to keep the pond in balance, Gabelhouse et al. (1982) questioned its effectiveness. Obviously, very accurate harvest records would have to be kept. Application of harvest quotas in management of small impoundments has greatly restricted harvest opportunities in some cases (Ming 1974) but has met with success in others (Powell 1975).

21.4.3 Panfish Option

This option emphasizes the harvest of big panfishes instead of largemouth bass by imposing a 38-cm minimum length limit for largemouth bass. Few largemouth bass will grow beyond 38 cm because of overpopulation of the largemouth bass. The high density of 20–38 cm largemouth bass will reduce bluegill densities and allow growth of bluegills to 20 cm and longer. Other panfish species include channel catfish, black and white crappies, black bullhead, and yellow perch (Gabelhouse 1984b; Boxrucker 1987; Saffel et al. 1990; Guy and Willis 1991). Because crappies, black bullhead, and yellow perch tend to overpopulate, addition of these species requires release of nearly all largemouth bass and maintenance of water clarity to depths greater than 46 cm to encourage predation by the sight-feeding largemouth bass. With adherence to the length restriction on harvest of largemouth bass, the panfish option is unlikely to fail. However, pond owners may be disappointed in the small size of largemouth bass caught. It is also possible to have too many small largemouth bass, which will then compete with bluegills (Gabelhouse 1987). In such a case it would be desirable to remove some small largemouth bass periodically or temporarily impose a slot limit of 30–38 cm (Gabelhouse 1987; Neumann et al. 1994).

An alternative approach for panfish fishing is to stock hybrid sunfish and largemouth bass. Several sunfish crosses (bluegill × green sunfish and redear sunfish × green sunfish) produce offspring that are mostly male (Childers 1967; Lewis and Heidinger 1978b). With limited reproductive potential, hybrid sunfishes are less likely to overpopulate, and thereby they achieve large sizes. Hybrid sunfishes also exhibit hybrid vigor and higher vulnerability to angling. Ellison and Heidinger (1978) reported that most pond owners who had tried some type of hybrid sunfish and largemouth bass would do it again, but a majority of pond owners who had tried hybrid sunfish without largemouth bass would not stock them again. Hybrid sunfishes are of little value in ponds containing other sunfishes. Hybrids, or the correct parental sexes to produce hybrids, must be restocked every few years.

21.4.4 Big Bass Option

The objective of this option is the consistent production of trophy largemouth bass without regard to bluegill size. Catch rate for largemouth bass will be low with this option, but the largemouth bass caught should be large. After 4 years, densities of 20- to 38-cm largemouth bass should be greatly reduced to allow for rapid growth of those remaining. In a pond of average fertility, 75 largemouth bass 20–30 cm long and about 12 largemouth bass 30–38 cm long should be harvested per hectare each year. Unless a trophy largemouth bass is caught, all bass over 38 cm should continue to be released.

This option includes the stocking of 50 adult gizzard shad per hectare 2 years after stocking of fingerling largemouth bass. The gizzard shad increase the chances of producing trophy largemouth bass because they serve as food for large largemouth bass whereas bluegills serve as stable food for small largemouth bass. Gabelhouse et al. (1982) cautioned that this option is largely unevaluated and may be practical in only larger ponds because maintenance of largemouth bass over 1.5 kg may be limited to 25 per hectare. In addition to trophy largemouth bass, this option is expected to produce a high catch rate of small bluegills. With this option, overpopulation of bluegills may reduce recruitment of largemouth bass through predation on largemouth bass eggs and fry. The catchability of large largemouth bass may also be low because of their low density in the presence of abundant prey. The use of gizzard shad is unique to this option, and its use should be avoided in ponds with other management strategies. Bluegill and crappie populations can be adversely affected by gizzard shad introduction in small impoundments.

21.4.5 Catfish Only Option

The catfish only option is especially desirable for channel catfish in muddy ponds where sight-feeding largemouth bass and bluegill would not do well or in ponds less than 0.2 ha where overharvest of largemouth bass would be likely. The pond should be free of any structures, such as hollow logs and rock ledges, that would provide the seclusion required for channel catfish spawning (section 21.6.3). Fathead minnows can be stocked as prey, and harvest of channel catfish is unrestricted. Replacement stocking of channel catfish over 20 cm should take place during the cool of spring or fall; the number harvested plus an additional 10% to account for natural mortality should be stocked. A density of 250–500 channel catfish per hectare should be maintained in ponds receiving no supplemental feeding.

21.4.6 Black Bass Only Option

For those ponds whose owners are not interested in panfishes or in very shallow ponds where excessive aquatic vegetation provides too much cover for bluegill, black bass alone is a viable option. Bennett (1952) found that spotted, smallmouth, or largemouth bass, when stocked alone in Illinois ponds, were able to do well by feeding on crayfish, large aquatic insects, and their own young. Bennett (1971) reported that such populations produced more kilograms of black bass per hectare than did populations in which black bass were stocked with other species. Bennett cautioned that it was necessary to stock several year-classes to prevent the development of a dominant year-class, which might become stunted. Buck and Thoits (1970) also found that standing crops of either largemouth bass or smallmouth bass stocked by themselves were comparable to those with prey species; however, monocultured bass grew more slowly. Swingle (1952) experimented with stocking largemouth bass alone but considered this to be an inefficient use of available food resources of the pond. Also, stocking largemouth bass alone resulted in about one-third of the angling pleasure as did the largemouth bass–bluegill combination. Overpopulation can occur with the black bass only option, but this can be corrected with a slot length limit (Gabelhouse 1987; Neumann et al. 1994). The largemouth bass–golden shiner combination is a slight variation to the black bass only option that can be used in ponds that contain sufficient macrophytes to protect some adult golden shiners. Regier (1963) suggested this combination for the pond owner who is not interested in bluegill and whose pond is small and shallow and has surface water temperatures that rise above 23°C in summer.

21.4.7 Trout Options

Rainbow trout will achieve catchable size in coldwater ponds. Because rainbow trout will not spawn in standing water, growth rates and ultimate sizes are easily controlled by stocking rates. In addition, rainbow trout readily accept formulated feeds if supplemental feeding is desired. However, many strains of hatchery-reared rainbow trout have become so domesticated that stocked rainbow trout do not live much beyond 2–3 years.

Interviews with pond owners have revealed that knowing which species and approximately what size fish are going to be caught reduces the anticipation of going fishing. Thought should be given to adding a few large fish or to stocking more than one species or a color variation. The occasional catch of something different will provide excitement. Hatchery broodstock become available for stocking from time to time. Several state and private trout hatcheries have golden-colored strains of rainbow trout that would provide a surprise catch when stocked in low numbers. Brown trout generally live longer and grow larger than do rainbow trout, but they can be cannibalistic and hard to catch.

21.5 STOCKING RATES

Biologists sometimes become too concerned with adjustments to stocking rates to achieve desired results. In most cases, harvest has more to do with a pond's success than does stocking.

Ponds should not be stocked with fish caught elsewhere by anglers. In some states, it is illegal for anglers to transport live fishes. In addition, stocking a few individuals of sometimes erroneously identified species is not likely to produce a balanced community. Also, shipments of fish should be checked for unwanted species before stocking. Most warmwater fish-rearing facilities raise many species, and it is not uncommon for a few stray fish to find their way into a shipment of another species. Sometimes a shipment contains desirable species like fathead minnows mixed with largemouth bass, but other times the mixture may contain undesirable species such as goldfish with channel catfish.

21.5.1 Warmwater Ponds

Stocking rates vary geographically, but the traditional 250 largemouth bass and 2,500 sunfish (either all bluegill or a combination of bluegill and redear sunfish at southern latitudes) per hectare remains the most common. Successive, or split, stocking involves stocking fingerling sunfishes in the fall and fingerling largemouth bass the following spring. In some northern regions this sequence is reversed or both are stocked at the same time. Due to the slow growth of largemouth bass, bluegills stocked before largemouth bass may become too big to be prey for the stocked largemouth bass. However, by stocking bluegills before largemouth bass, the bluegills have reached sexual maturity by the time the largemouth bass are stocked. Bluegill spawning produces appropriate size prey for stocked fingerling largemouth bass.

When channel catfish are desired, they should be stocked at a rate of 250 fish per hectare when stocked with other species and up to 500 fish per hectare when stocked alone. Periodic restocking is necessary for perpetuation of this species in most ponds. Alternate-year stocking ensures a steady supply of channel catfish for harvest. If adult largemouth bass are present, fingerling channel catfish longer than 20 cm should be stocked to avoid predation.

Because little research has been conducted on coolwater ponds, no stocking rates are suggested here for coolwater pond species.

21.5.2 Coldwater Ponds

Marriage et al. (1971) recommended fall stocking with 5–10 cm trout at rates of 600 per hectare in western states and 1,500 per hectare in eastern states. If larger trout are used, smaller numbers should be stocked. Fingerlings less than 10 cm are adequate for initial stocking, but subsequent stockings should be done with fish greater than 10 cm. Record keeping will aid in determining the need for more or fewer trout in subsequent stockings. Some owners of coldwater ponds prefer to stock in alternate years to reduce cost of transportation. However, such practice results in uneven fishing success.

21.6   MANAGEMENT OF STOCKED PONDS

21.6.1 Influence of Predation

In a predator–prey fish community, not only do the predators gain from eating prey, but also the prey population benefits by having fewer individuals competing for a limited food supply. However, there are many problems associated with maintaining a predator population in a pond. For some species, such as northern pike and muskellunge, the carrying capacity is often one adult or less per hectare. Most predators are highly prized game fishes in North America, and anglers actively seek these species, which are also the least abundant. Northern pike and largemouth bass are notorious for their emigration over spillways. Although there are many desirable attributes of cover, too much cover in ponds, especially aquatic vegetation, makes it difficult for predators to capture prey. High turbidity has the same detrimental effect. Finally, some predators have rigid spawning requirements or do not recruit well in the forced interaction that occurs in the small area typical of ponds.

21.6.2 Regulation of Harvest

As mentioned in many of the stocking options, regulation of largemouth bass harvest is essential in the success of those options (Lindgren and Willis 1990b). It is common for 70% of adult largemouth bass to be caught in the first few days of fishing in a new pond (Redmond 1974). If that many largemouth bass are harvested, the predator–prey balance will be ruined. Minimum length limits, if obeyed, will protect against overharvest. If largemouth bass recruitment is high, a minimum length limit will result in overpopulation and reduced growth, a condition that is desirable in some management options but not in others. A protected range, or slot limit, would be a better choice when larger largemouth bass are desired (Novinger 1990). However, anglers must harvest smaller largemouth bass below the slot; otherwise the regulation would function like a minimum length limit. Harvest recommendations for other species are given throughout section 21.4.

21.6.3 Lack of Natural Spawning

For most stocking options, reproduction is desired and necessary. However, the use of hybrid sunfishes (section 21.4.3) is an example of reducing spawning potential. In the catfish only option (section 21.4.5), no reproduction is desired because there would be no predation to control population size. Measures should be taken to eliminate spawning sites for channel catfish. Brook trout are the only trout species likely to spawn in a pond. Therefore, most trout ponds can be stocked with appropriate numbers of fish to achieve desired growth.

21.6.4 Supplemental Feeding

Use of formulated feeds is usually not recommended for fishing ponds because of the potential problem of dissolved oxygen depletion and because natural food production will support normal harvest rates. Nevertheless, supplemental feeding can be useful where harvest is high or where large fish are desired. Channel catfish and trout are most commonly fed formulated feeds, but there is also interest in feeding bluegill, hybrid sunfishes, and other species. Floating feed is advisable because the fish can be observed feeding and feeding rates can be appropriately adjusted. Schmittou (1969) found that supplemental feeding increased fishing success and total production in largemouth bass–bluegill ponds.

The seasonal timing, quantity, and frequency of feeding vary with fish species, fish sizes, water temperature, water quality, weather conditions, and quality of feed. For example, seasonal feeding of channel catfish typically commences when water temperatures are over 15°C. They may be fed at 3% of their body weight per day with a maximum daily food input of 22 kg/ha. The addition of large amounts of feed increases the danger of oxygen depletion, and feeding is discontinued when surface dissolved oxygen levels drop below 5.0 mg/L.

21.6.5 Fertilization and Liming

In many areas where soil fertility is low, fertilizers are added to ponds to increase fish production by stimulating phytoplankton growth. Ponds in fertile watersheds, well-managed pastures with high densities of cattle, northern latitudes where winterkill is a problem, and areas where clay turbidity is a persistent problem, as well as ponds that receive very little fishing pressure generally should not be fertilized. Fertilization programs are used in Alabama, Arkansas, some parts of northern Florida, Georgia, Kentucky, Mississippi, North Carolina, South Carolina, Tennessee, Texas, and Virginia. With a few exceptions, fertilization and liming are generally not recommended elsewhere.

Box 21.1   Standard Pond Fertilization Program for Southeastern United States

This procedure is given by Boyd (1979b). Once a pond fertilization program is begun, it should be continued each year or stunted fish and the emergence of nuisance aquatic macrophytes may result. The standard fertilization practice will require 8–12 applications per year.

1. In mid-February or early March (when water temperatures rise to 16–18°C) apply 45 kg/ha of 20–20–5 fertilizer (equivalent amounts of other formulations are also used). Follow with two additional applications at 2-week intervals.

2. Make three more applications of 45 kg/ha of 20–20–5 fertilizer at 3-week intervals.

3. Continue applications of 45 kg/ha of 20–20–5 fertilizer at monthly intervals or whenever the water clears so that a Secchi disk or equivalent is visible to a depth of 45 cm.

4. Discontinue applications for the year by the last week in October.

Swingle and Smith (1938) found that fertilized ponds in Alabama supported four to five times the weight of largemouth bass and bluegill as did unfertilized ones. The shading effect of a phytoplankton bloom also aids in control of submerged aquatic macrophytes, and this beneficial turbidity increases fishing success. However, most fertilizers are expensive, and the pond owner should determine whether the increased production justifies the expense. Bennett (1971) cautioned about the dangers of summerkill at any latitude and winterkill in northern latitudes when abundant nutrients result in excessive plant growth.

Inorganic fertilizers are most commonly used as pond fertilizers in the United States. A common fertilizer is granular 20–20–5: 20% nitrogen (as N), 20% phosphorus (as P_2O_5), and 5% potassium (as K_2O; also called potash). Secondary nutrients in fertilizers include calcium, magnesium, and sulfur. Minor or trace nutrients such as copper and zinc may also be present. In recent years, powdered fertilizers and liquid formulations have become popular with pond owners because these forms are easier to apply.

Boyd (1979b) described the chemistry of fertilizers, relationships between chemical fertilization and fish production, and the early experiments of Auburn University professors Swingle and Smith; Boyd modified Swingle and Smith's work into a standard fertilizer application procedure for the southeastern United States (Box 21.1). Additional accounts of early research on pond fertilization were presented by Dendy (1963), Swingle (1970), and Bennett (1971).

Swingle et al. (1965) found that old ponds which had been fertilized for many years had adequate supplies of nitrogen and potassium through nitrogen fixation by bacteria or algae and through decomposition of bottom organic materials. For such ponds, they recommended a standard fertilization rate with phosphorus only fertilizer: 45 kg/ha of superphosphate (16–20%) or 20 kg/ha of triple (or treble) superphosphate (44–54%). If a satisfactory phytoplankton bloom cannot be maintained, use of a complete fertilizer should be resumed.

Granular fertilizers applied to waters over 1 m deep may sink into the bottom mud, be locked into the bottom water and sediment, and be unavailable to phytoplankton. Therefore, fertilizers should be broadcast over shallow water or be placed on underwater plat-

Figure 21.3 With a pontoon barge and pump sprayers, 2 metric tons of agricultural limestone can be applied in several minutes. Lime increases production in impoundments with acidic bottom muds and soft water (less than 20 mg/L total alkalinity). Lime is especially important in making more phosphorus available for phytoplankton production.

forms in the euphotic zone (zone with sufficient light penetration to support plant growth) (Lawrence 1954). A platform at a depth of about 0.3 m and with an area of 4 m² is adequate for 2–4 ha of pond area (Boyd 1979b). The mixing of liquid fertilizers, after first diluting, in the prop wash of a boat motor is another method of distributing fertilizer.

In certain regions, pond waters have low alkalinity and may be very acidic. Such waters may not respond to fertilization unless some form of liming is used. Lime has been routinely used to increase fish production in ponds in many other nations and in areas of the southeastern United States where natural shortages of calcium result in low pH and low buffering capacity. Liming has several effects on water quality and productivity including the following: (1) increases the pH of bottom mud and thereby increases the availability of phosphorus added via fertilizer; (2) increases benthic invertebrate production by increasing nutrient availability; (3) increases alkalinity and thereby the availability of carbon dioxide for photosynthesis; (4) increases microbial activity by means of increasing pH; (5) buffers against pH shifts common in soft-water ponds; and (6) may clear water of humic stains of vegetative origin, thereby increasing sunlight penetration. The net effect is an increase in phytoplankton productivity that leads to increased food production.

Liming needs are determined by total alkalinity and hardness measurements of water. Boyd (1979b) described methods used in analysis of agricultural soils and pond muds to determine the lime requirement. Agricultural limestones, $CaCO_3$ or $CaMg(CO_3)_2$, are the materials most commonly used in fish ponds. New ponds can be limed by spreading limestone over the pond bottom and disking it with farm equipment. Old ponds can be limed by spreading limestone over the entire pond surface (Figure 21.3). Addition of limestone or other liming materials is best accomplished during the winter before the fertilization season. The length of time a treatment is effective will depend upon the rate of water loss through overflow and seepage.

21.6.6 Control of Turbidity

Clay turbidity can be a problem when it limits light penetration and reduces primary production (see Chapter 11). Clay particles can also cover fish eggs and benthic macroinvertebrates. The best management practice is limiting soil erosion in the watershed (see Chapter 9), but it may be necessary to clear the water of colloidal soil particles. The introduction of positively charged electrolytes, such as aluminum sulfate, has been effective in clearing pond water of clay turbidity (Boyd 1979a).

21.6.7 Importance and Management of Aquatic Vegetation

Primary production by aquatic plants, especially phytoplankton, is the means by which a portion of the sun's energy is fixed into forms available to other aquatic organisms. A variety of invertebrates and a few vertebrates feed on living and dead aquatic plants. Many aquatic organisms use plants for shelter and attachment. Isolated weed beds become fish attractors. However, abundant vegetation can create unpleasant conditions for anglers, swimmers, and boaters. Aquatic plant respiration and decomposition depletes oxygen and can result in dead fish. Certain plant species can contribute to taste and odor problems in fish.

Vegetation may need to be controlled through reduction, but control is not synonymous with elimination. Durocher et al. (1984) found a positive relationship between submerged vegetation (up to 20% of total lake coverage) in Texas reservoirs and both standing stock of largemouth bass and numbers being recruited to harvestable size. Any reduction in submerged vegetation below 20% coverage resulted in a decrease in both recruitment and standing stock of largemouth bass. The optimal vegetation coverage for largemouth bass production in Illinois ponds was 36% (Wiley et al. 1984).

There are times when growth of aquatic vegetation should be encouraged. Fertilizing to promote phytoplankton production increases fish production and discourages more noxious plants. If a pond is to be used for waterfowl, at least a portion of the pond should have shallow water to encourage growth of vascular plants; fertilization may or may not be needed. An interesting plan for interspersing plants was outlined by Belusz (1986): high and low spots in shallow areas of a pond will encourage and discourage, respectively, growth of aquatic plants.

Publications on control of aquatic vegetation list three general methods: mechanical, chemical, and biological (see Flickinger and Satterfield 1995 for an overview). Mechanical methods such as raking and cutting are labor intensive and produce short-term results. Properly timed drawdowns can temporarily reduce aquatic macrophytes, but there is also a risk of creating conditions that allow macrophytes to colonize areas that were heretofore too deep. When herbicides work, they give results quickly, but sometimes, such as with fall applications, they show no detectable effect. Hard waters also reduce or negate the effectiveness of herbicides, and some plant species are more difficult to kill than are others. There are three major disadvantages to the use of herbicides: (1) only a few compounds have been cleared, at least in the United States, for use with food fishes (Schnick et al. 1989); (2) all approved herbicides except crystalline copper sulfate are expensive; and (3) decomposition of vast amounts of dead plants depletes dissolved oxygen. Fish pond publications, extension agents, and fisheries agencies should be consulted for herbicides recommended for a particular area.

Biological control of aquatic vegetation is receiving considerable attention largely because of environmental concerns over chemicals. However, biological control, specifically the use of grass carp, has generated other environmental concerns, such as nontarget effects, downstream migration of the grass carp, and change from a macrophyte-dominated plant community to a phytoplankton community that is often dominated by blue-green algae. Use of grass carp is illegal in some states, and other states require the use of triploid (sterile) grass carp or have other restrictions. Swanson and Bergersen (1988) have created a model for grass carp stocking that incorporates the major variables important to the success or failure of grass carp. The variables of the model are daily temperature units, vegetation density and distribution, feeding preference, disturbance, management objectives, fish size, and genetic makeup. Because grass carp eat proportionately less as they get larger and will not spawn in standing water, it is necessary to stock additional fish periodically. However, grass carp are long lived (Hill 1986). Grass carp may be too effective by completely eliminating vegetation. Blackwell and Murphy (1996) reported success in partial vegetation control in Texas.

21.7   BALANCE AND POPULATION ANALYSIS

The concept of balanced fish populations was proposed by Swingle (1950) and further developed by Anderson (1973). Ponds are usually stocked with species combinations that are likely to attain balance. A balanced system is a dynamic one characterized by continual reproduction of predator and prey species, diverse size composition of prey species so that food is available for all sizes of predators, high growth rates of predators and prey, and an annual yield of harvestable-size fish in proportion to basic fertility.

Bennett (1971) refuted the concept of balance largely on the basis that ponds represent artificial ecosystems which cannot be expected to show great stability. More recently, Gabelhouse et al. (1982) and Gabelhouse (1984a) proposed panfish (section 21.4.3) and big bass (section 21.4.4) management options that are out of balance but are useful. The all-purpose option (section 21.4.1) does fit the concept of balance, and that is the management strategy that Swingle and Anderson had in mind.

No one has applied the concept of balance to coldwater ponds. To be sure, lack of reproduction and no prey fish are major deviations from Swingle's (1950) thinking. However, the objective of sustained crops of harvestable-size fish in proportion to pond fertility is equally valid for coldwater systems. Perhaps the relationship between zooplankton size and fishing quality for rainbow trout will emerge as an assessment of balance in coldwater ponds (Galbraith 1975).

21.7.1 Biomass Indices

In an attempt to describe balanced and unbalanced populations, Swingle (1950) analyzed biomass data from 55 balanced and 34 unbalanced ponds and computed certain biomass ratios or indices. The most commonly used biomass indices are the $F{:}C$ ratio, the $Y{:}C$ ratio, and the A_T value.

The $F{:}C$ ratio is the total weight of all forage fishes (F) divided by the total weight of all carnivorous fishes (C). Ratios from 3.0 to 6.0 are considered the most desirable in the balanced range (Figure 21.4).

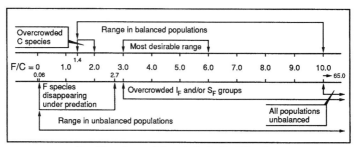

Conditions in populations indicated by F/C ratios

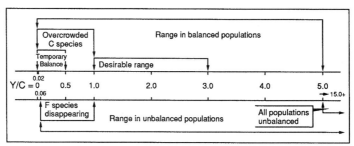

Conditions in populations indicated by Y/C ratios

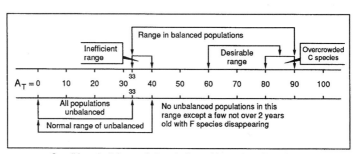

Conditions in fish populations indicated by various A_T values

Figure 21.4 Conditions of fish populations as indicated by $F{:}C$ ratios, $Y{:}C$ ratios, and A_T values (see text). Additional abbreviations are I_F (intermediate forage) and S_F (small forage) groups. Figure taken from Swingle (1950).

The $Y{:}C$ ratio is the total weight of all forage fishes that are small enough to be eaten by the average-size individual in the C group divided by the C value. The most desirable range for $Y{:}C$ ratios in balanced populations is between 1.0 and 3.0.

The A_T value (total availability value) is the percentage of the total weight of a fish population composed of fishes of harvestable size. In calculating the A_T value, it is necessary to define minimum weights suitable for harvest, and Swingle (1950) did this for common pond species (e.g., 45 g for sunfishes, 180 g for largemouth bass, and 230 g for channel catfish). Swingle suggested a range of 60–85% as most desirable for balanced populations.

Swingle (1956) presented a method of analysis that was based on use of a minnow seine to sample for evidence of reproduction and a larger seine to sample intermediate and harvestable-size fishes (Figure 21.5). His description of possible catches and the

Figure 21.5 The minnow seine method of pond analysis is a simple method for determining the state of balance of a population within a short time and without destruction of the population. The state of balance is deduced from the degree of success of reproduction of largemouth bass and bluegills and the abundance of intermediate size (8–13 cm) bluegills (see Box 21.2).

interpretations that should be drawn are summarized in Box 21.2. Because so much of the analysis is based on presence or absence of young fishes, sampling must be done after spawning has occurred.

21.7.2 Length-Frequency Indices

It is difficult to sample with equal effort the different species and sizes of fishes needed to obtain the biomass figures required for Swingle's (1950) pond analysis system. Consequently, Anderson (1976) introduced a different approach called proportional stock density (PSD). The PSD and related length-frequency indices can be calculated from subsamples of pond populations (Anderson and Neumann 1996). These indices reflect an interaction of rates of reproduction, growth, and mortality of the age-groups present.

The PSD index is calculated by dividing the number of fish of a given species greater than or equal to quality length by the number of fish greater than or equal to stock length and then multiplying by 100 (Chapter 7). Quality length is the size most anglers like to catch, and stock length has been defined as the size at or near which fish often reach sexual maturity, the minimum length typically sampled by traditional gears, or the minimum length of fish that provides recreational value (Gabelhouse 1984a). Stock and quality categories are based on a percentage of world record length (Table 21.1), and the lengths are standard for anyone using them (Anderson and Neumann 1996). In contrast, Swingle's (1950) notations of small (S), intermediate (I), and large (A) had enough latitude that different biologists could put different lengths of a given species into the categories. Such a practice hampers discussion of results from different geographical regions.

Box 21.2   Swingle's (1956) Method of Pond Analysis Based on Seining

1. No young largemouth bass present:

   A. Many recently hatched bluegills; no or very few intermediate bluegills. (Temporary balance with bass overcrowded.)

   B. No recent hatch of bluegills; many intermediate bluegills. (Unbalanced population with overcrowded bluegills and insufficient bass.)

   C. No recent hatch of bluegills; many intermediate bluegills; many tadpoles or minnows or crayfish. (Unbalanced population with overcrowded bluegills and very few bass.)

   D. No recent hatch of bluegills; few intermediate bluegills. (Unbalanced population, crowding due to species competitive with bluegills.)

   E. No recent hatch of bluegills; few intermediate bluegills; many intermediate fish of a species competitive with bluegills. (Unbalanced population due to crowding by competitive species.)

   F. No recent hatch of bluegills; no intermediate bluegills. (Unbalanced population; possible no fish present or water unsuitable for bass–bluegill reproduction.)

2. Young largemouth bass present:

   A. Many recently hatched bluegills; few intermediate bluegills. (Balanced population.)

   B. Many recently hatched bluegills; very few or no intermediate bluegills. (Balanced population with slightly crowded bass.)

   C. No recent hatch of bluegills; no intermediate bluegills. (Unbalanced population; bluegills prevented from spawning by low water temperature or salinity, etc.)

   D. No recent hatch of bluegills; few intermediate bluegills. (Temporary balance with possibility of imbalance developing due to a reduction of the food available to the bluegill or overcrowding by a species growing to a competitive size.)

   E. No recent hatch of bluegill; many intermediate bluegills. (Unbalanced population similar to 1.B., but less severely overcrowded.)

The PSD is similar to A_T (Swingle 1950) in that both represent a percentage of fish that are an attractive size to anglers. However, PSD is based on length frequency instead of total weight. In addition, fish below stock length are not included in PSD; therefore, no special effort is needed to sample small fish. For balanced ponds, largemouth bass PSD should be between 40 and 70 (Gabelhouse 1984a) and bluegill PSD should be between 20 and 60 (Anderson 1985). Sequential sampling for PSD may shorten the amount of time spent sampling (Weithman et al. 1980). Calculation of confidence limits for PSD estimates was developed by Gustafson (1988).

The relative stock density (RSD) is the proportion of fish of any designated length-group in the stock length and longer portion of a population (Chapter 7). Anderson and Neumann (1996) considered RSD to be a more sensitive indicator of potential fishing quality than is PSD alone because RSD provides opportunity to divide a population into more than just stock and quality lengths. Gabelhouse (1984a) described the basis of a five-cell analysis of fish lengths. Gabelhouse did not use RSD to assess balance, but he proposed desirable ranges in the five cells for different management options.

Table 21.1 Proposed maximum total length (cm) of minimum stock and quality, preferred, memorable, and trophy lengths for selected species based on percentages of world record lengths. Data are from Anderson and Neumann (1996).

| Species | Size designation | | | | |
	Stock	Quality	Preferred	Memorable	Trophy
Largemouth bass	20	30	38	51	63
Smallmouth bass	18	28	35	43	51
Spotted bass	18	28	35	43	51
Bluegill	8	15	20	25	30
Green sunfish	8	15	20	25	30
Redear sunfish	10	18	23	28	33
Black crappie	13	20	25	30	38
White crappie	13	20	25	30	38
Yellow perch	13	20	25	30	38
Black bullhead	15	23	30	38	46
Channel catfish	28	41	61	71	91

21.7.3 Abundance and Weight Indices

All of the indices presented so far do not provide any indication of fish abundance. Desirable ratios might exist with very low abundance, giving the impression that good fishing is available when catch rates probably would be poor. A measure of fish abundance that is sometimes used is the number of stock length and longer fish captured per hour of electrofishing. General consensus of biologists using this expression is that under good sampling conditions 100 stock length and longer largemouth bass captured per hour of electrofishing is a dense largemouth bass population. Other catch-per-unit-effort methods, such as number of target species per angler-hour, can be used as indices of population density given that gear, methods, and sampling design are standardized (Hubert 1996; Malvestuto 1996).

Relative weight (W_r) is the actual weight of a fish divided by a standard weight for the same length for that species multiplied by 100 (Wege and Anderson 1978; Anderson and Neumann 1996; Chapter 7). Equations for calculating standard weights have been developed for selected species (Table 21.2). A W_r of 100 is not average, but typically represents the 75th percentile of weights attained by that species across its range. Fish with a W_r close to 100 are in balance with their food supply, whereas fish with values below 85 are underweight and may be too abundant for their food supply. On the other end, fish with W_r above 105 are more plump than necessary, reflecting an overabundant food supply. In such a situation, the pond could support more fish without detrimental effects.

Average W_r should not be calculated for an entire population. Different size individuals of the same species often have different food habits; consequently, fish of different lengths could have considerably different W_r. A plot of W_r against length will likely show where food is abundant or limiting for certain length ranges. It would be acceptable to calculate mean W_r values for fish grouped by the five-cell length categories proposed by Gabelhouse (1984a).

The progression of material presented here may lead readers to the conclusion that Swingle's concepts of balance have been outdated. This is not at all true, and there is no better discussion of the structure and dynamics of warmwater pond fish populations than Swingle's 1950 publication. Furthermore, Swingle provides an understanding from a biomass perspective rather than one of relative numbers.

Table 21.2   Equations for calculating standard weight (W_s) for a given length of given species. Standard weights are then used in relative weight (W_r) computations. Table is from Anderson and Neumann (1996).

Species	Standard weight equation[a]	Source
Largemouth bass	$\log_{10}W_s = -5.528 + 3.273 \log_{10}L$	Henson (1991)
Smallmouth bass	$\log_{10}W_s = -5.329 + 3.200 \log_{10}L$	Kolander et al. (1993)
Bluegill	$\log_{10}W_s = -5.374 + 3.316 \log_{10}L$	Hillman (1982)
Black crappie	$\log_{10}W_s = -5.618 + 3.345 \log_{10}L$	Neumann and Murphy (1991)
White crappie	$\log_{10}W_s = -5.642 + 3.332 \log_{10}L$	Neumann and Murphy (1991)
Channel catfish	$\log_{10}W_s = -5.800 + 3.294 \log_{10}L$	Brown et al. (1995)
Rainbow trout[b]	$\log_{10}W_s = -4.898 + 2.990 \log_{10}L$	Simpkins and Hubert (1996)

[a] Total length (L) is in millimeters and weight (W) is in grams.
[b] Equation based on lentic populations.

21.7.4 Angler-Collected Data

Swingle (1956) also developed a method of distinguishing balanced and unbalanced ponds from angler catches. Assuming that anglers have been fishing for both largemouth bass and bluegill, the following interpretations can be made: (1) in balanced ponds most bluegills caught will be larger than 15 cm, and the average largemouth bass will be 500–1,000 g, although smaller and larger ones also will be caught; (2) in unbalanced ponds the catch will be principally bluegills ranging from 7.5 to 12.5 cm, and the few largemouth bass caught will be larger than 1,000 g (note the similarity to the big bass option, section 21.4.4); and (3) in ponds crowded with largemouth bass (unbalanced condition), bluegills will average more than 150 g and largemouth bass will be less than 500 g and in poor condition (note similarity to the panfish option, section 21.4.3). New York biologists used angler-collected data to provide recommendations to private pond owners (Green et al. 1993). In fact, the pond management bulletin for New York (Eipper et al. 1997) provides information that pond owners can use to assess their pond fish community based on angling catch rates and size structure of fishes that are caught.

However, we caution that biologists consider size selectivity when angler-collected data are used to assess populations. The type of lure or bait used (e.g., Payer et al. 1989) and the size of lure used (e.g., Gabelhouse and Willis 1986) can influence the resulting size structure of the sample and could lead to improper recommendations. For example, the use of large lures to capture largemouth bass could result in a sample dominated by larger (e.g., >30 cm) individuals, even if a population was dominated by smaller individuals. Biologists can alternatively use angling gear themselves to obtain useful samples from ponds. Isaak et al. (1992) purposely varied the lure sizes used to collect largemouth bass in small impoundments and found that catch per unit effort and size structure were highly correlated for angling and electrofishing samples.

21.8   MANAGEMENT OF PROBLEM PONDS

Fisheries biologists working on ponds today will be heavily involved with assessing old ponds and devising management plans that will make the best out of what presently exists. Management strategies for old ponds will fall into the categories of do nothing, use a corrective stocking, or reclaim the pond.

21.8.1 The Do Nothing Option

The first question that a fisheries biologist should ask after assessing a pond community is, what will happen if nothing is done? Both short- and long-term answers should be assessed. Sometimes nothing needs to be done, especially in the short term. Sometimes nothing can be done to improve a pond's condition. With limits on time and money, it is better to focus on ponds that can be managed relatively easily.

21.8.2 Corrective Stocking

Often when pond fish communities are out of balance, there are too few predators. Through the use of formulated feeds, advanced sizes of largemouth bass and tiger muskellunge are available from hatcheries. In theory, stocking predators should help reduce abundant prey, but results have been variable (Bennett 1971; Boxrucker 1987). Success may be improved if standing stock of prey is reduced below potential carrying capacity—based on available nutrients—and enough predator fish are stocked to approach the carrying capacity for that species (Bennett 1971).

21.8.3 Reclamation

For older ponds that are far out of predator–prey balance, often the best remedy is to eliminate the fish community and start over. Sometimes unwanted species of fish have detrimentally altered the pond's dynamics. Elimination is a possible solution to that problem, but first consider the sources of species contamination and decide if it is practical to avoid reoccurrence. If the pond has a bottom drain, it is a simple matter to drain the pond. Sometimes a siphon can be used to remove most of the water in a pond. If puddles remain, the fish can be poisoned (Chapter 15). The entire pond can also be poisoned but at a greater cost. Although elimination destroys fishing, good fishing can occur in as little as 2 years after reclamation. Without reclamation, some older ponds will never produce good fishing.

21.9 POND CONSTRUCTION

Although construction of ponds has slowed in recent years, it is likely that fisheries biologists will be called upon to advise on construction of ponds. Biologists should focus their advice on aspects such as installing a drain, constructing a proper spillway, and avoiding shallow water, which will adversely affect future management. With a drain, fall drawdowns, reclamation of the fish community (section 21.8.3), and renovation of the pond itself are simple matters. A considerable number of fish can be lost over emergency spillways (Louder 1958; Lewis et al. 1968). The spillway should be wide so that flows will be shallow, or a parallel bar barrier (Figure 21.6) should be installed (Powell and Spencer 1979).

Water depth of less than 1 m is undesirable from an aquatic plant management perspective (section 21.6.7) because sunlight penetration to the bottom encourages rooted macrophytes. For safety and stability, a gradual slope of 3 to 1 is recommended, which means 1 m of depth will be reached about 3 m from shore. Plants can be prevented from growing in an area such as a swimming beach by a process called blanketing (Belusz 1986). Black plastic sheeting is laid over the area during construction, and then 15–20 cm of sand or fine gravel is placed over the plastic.

Figure 21.6   Properly constructed spillway barriers prevent harvestable-size fish from escaping and require little maintenance. Certain species such as grass carp and juvenile largemouth bass are especially vulnerable to losses over emergency spillways.

Ponds can be too small as well as too large. Hooper (1970) found that 80% of warmwater ponds less than 0.1 ha failed to achieve a balance among multiple species. An option to consider for small ponds is single-species stocking and supplemental feeding, such as is done with rainbow trout and channel catfish. Ponds are too large when owners cannot afford to buy the quantity of fish that should be stocked or the amount of chemicals, such as herbicides and fertilizers, that may be necessary to maintain the pond. Pond owners should be aware of such potential expenses before a large pond is designed. They should also be advised on proper watershed to pond surface ratios and also the minimum depths necessary to avoid problems such as winterkill at northern latitudes and higher elevations (SCS 1982).

Older extension bulletins on pond management stated that all brush and trees should be removed from a pond site because they provided too much cover for bluegills. Fish attractors are popular in fish management now (Johnson and Stein 1979; Wege and Anderson 1979), and newer bulletins suggest leaving or placing trees and brush in ponds (Figure 21.7). Also, leaving the pond bottom irregular provides fish habitat. Humps are better than dips so that no puddles remain if the pond has to be drained.

The concept of multiple use, which is more commonly applied to larger impoundments, should also be considered for ponds. A pond that is intended for fishing, swimming, and duck hunting will need a few special provisions during construction, such as

Figure 21.7 Fishing piers can be used in conjunction with various fish attractors. In addition, piers improve fishing opportunities for anglers lacking boats. Piers are especially helpful to handicapped and elderly anglers. Such structures can be provided with open and shaded areas.

fish attractors, a sand beach, and a point or perhaps an island for a duck blind to best accommodate the three recreational activities. Proximity to residences may also vary for different uses. One commonly desired use of ponds is livestock watering. Without carefully planned restricted access or a water tank below the dam, livestock will ruin pond banks. In arid parts of North America, pond owners may want to use pond water for irrigation, but anything more than watering the lawn should be discouraged for impoundments of the size considered in this chapter.

Before any pond is built, it may be necessary to obtain permits from several agencies. Employees of agencies such as the Natural Resources Conservation Service have the technical knowledge and experience to advise on actual construction and on application for federal cost-sharing funds. If a biologist wants to become more conversant about pond construction, a useful reference is *Ponds: Planning, Design, Construction* (SCS 1982).

21.10   CONCLUSION

Although small impoundments are often constructed for purposes other than sport fish production, their great numbers and small sizes do provide convenient angling opportunities for millions of people of all ages. Unbalanced fish populations and poor fishing generally result when the pond owner lacks an understanding of basic fish management concepts or is unwilling to alter practices that degrade fish habitat, such as livestock watering.

Fisheries biologists have too often tried to make an impoundment be all things to all people. Pond owner or angler desires and the production capabilities of the impoundment should be used in establishing management objectives. Balance is best illustrated in ponds with relatively simple communities, but it is a valid concept regardless of impoundment size and community complexity. There are several useful options that differ from the traditional concept of a balanced largemouth bass-bluegill community.

The goal of fisheries management biologists should be to provide annual crops of harvestable-size fish commensurate with fertility of the impoundment. It is the responsibility of the pond owner, however, to implement management procedures designed to achieve fish management objectives. Successful angling opportunities can be provided in a multipurpose pond with minimal impact on other uses, provided that the pond is properly constructed, stocked, and managed.

21.11 REFERENCES

Anderson, R. O. 1973. Application of theory and research to management of warmwater fish populations. Transactions of the American Fisheries Society 102:164–171.

Anderson, R. O. 1976. Management of small warm water impoundments. Fisheries 1(6):5–7, 26–28.

Anderson, R. O. 1978. New approaches to recreational fishery management. Pages 73–78 in G. D. Novinger and J. G. Dillard, editors. New approaches to the management of small impoundments. American Fisheries Society, North Central Division, Special Publication 5, Bethesda, Maryland.

Anderson, R. O. 1985. Managing ponds for good fishing. University of Missouri Extension Division, Agricultural Guide 9410, Columbia.

Anderson, R. O., and R. M. Neumann. 1996. Length, weight, and associated structural indices. Pages 447–482 in B. R. Murphy and D. W. Willis, editors. Fisheries techniques, 2nd edition. American Fisheries Society, Bethesda, Maryland.

Ball, R. C. 1952. Problems in farm fish pond management. Journal of Wildlife Management 16:266–269.

Belusz, L. C. 1986. Aquatic plant management in Missouri. Missouri Conservation Commission, Jefferson City.

Bennett, G. W. 1952. Pond management in Illinois. Journal of Wildlife Management 16:249–253.

Bennett, G. W. 1971. Management of lakes and ponds. Van Nostrand Reinhold, New York.

Blackwell, B. G., and B. R. Murphy. 1996. Low-density triploid grass carp stockings for submersed vegetation control in small impoundments. Journal of Freshwater Ecology 11:475–484.

Boxrucker, J. 1987. Largemouth bass influence on size structure of crappie populations in small Oklahoma impoundments. North American Journal of Fisheries Management 7:273–278.

Boyd, C. E. 1979a. Aluminum sulfate (alum) for precipitating clay turbidity from fish ponds. Transactions of the American Fisheries Society 108:307–313.

Boyd, C. E. 1979b. Water quality in warmwater fish ponds. Alabama Agricultural Experiment Station, Auburn University, Auburn.

Brown, M. L., F. Jaramillo, Jr., D. M. Gatlin, III, and B. R. Murphy. 1995. A revised standard weight (W_s) equation for channel catfish. Journal of Freshwater Ecology 10:295–302.

Buck, D. H., and C. F. Thoits III. 1970. Dynamics of one-species populations of fishes in ponds subjected to cropping and additional stocking. Illinois Natural History Survey Bulletin 30:69–165.

Bulow, F. J. 1967. The suitability of strip-mine ponds for producing marketable channel catfish. The Progressive Fish-Culturist 29:222–228.

Childers, W. F. 1967. Hybridization of four species of sunfishes (Centrarchidae). Illinois Natural History Survey Bulletin 29:159–214.

Dendy, J. S. 1963. Farm ponds. Pages 595–620 in D. G. Frey, editor. Limnology in North America. University of Wisconsin Press, Madison.

Dillard, J. G., and G. D. Novinger. 1975. Stocking largemouth bass in small impoundments. Pages 459–474 in H. Clepper, editor. Black bass biology and management. Sport Fishing Institute, Washington, D.C.

Durocher, P. P., W. C. Provine, and J. E. Kraai. 1984. Relationship between abundance of largemouth bass and submerged vegetation in Texas reservoirs. North American Journal of Fisheries Management 4:84–88.

Eipper, A. W., H. A. Reiger, and D. M. Green. 1997. Fish management in New York ponds. Cornell Cooperative Extension, Information Bulletin 116, Ithaca, New York.

Ellison, D. G., and R. C. Heidinger. 1978. Dynamics of hybrid sunfish in southern Illinois farm ponds. Proceedings of the Annual Conference Southeastern Association of Fish and Wildlife Agencies 30(1976):82–87.

Flickinger, S. A., and J. R. Satterfield, Jr. 1995. Aquatic plant management for fish and wildlife. Habitat Extension Bulletin 21, Wyoming Game and Fish Department, Cheyenne.

Gabelhouse, D. W., Jr. 1984a. A length-categorization system to assess fish stocks. North American Journal of Fisheries Management 4:273–285.

Gabelhouse, D. W., Jr. 1984b. An assessment of crappie stocks in small midwestern private impoundments. North American Journal of Fisheries Management 4:371–384.

Gabelhouse, D. W., Jr. 1987. Responses of largemouth bass and bluegills to removal of surplus largemouth bass from a Kansas pond. North American Journal of Fisheries Management 7:81–90.

Gabelhouse, D. W., Jr., R. L. Hager, and H. E. Klaassen. 1982. Producing fish and wildlife from Kansas ponds. Kansas Fish and Game Commission, Pratt.

Gabelhouse, D. W., Jr., and D. W. Willis. 1986. Biases and utility of angler catch data for assessing size structure and density of largemouth bass. North American Journal of Fisheries Management 6:481–489.

Galbraith, M. G. 1975. The use of large *Daphnia* as indices of fishing quality for rainbow trout in small lakes. Internationale Vereinigung für theoretische und angewandte Limnologie Verhandlungen 19:2485–2492.

Green, D. M., E. L. Mills, and D. J. Decker. 1993. Participatory learning in natural resource education: a pilot study in private fishery management. Journal of Extension 31:13–15, Madison, Wisconsin.

Gustafson, K. A. 1988. Approximating confidence intervals for indices of fish population size structure. North American Journal of Fisheries Management 8:139–141.

Guy, C. S., and D. W. Willis. 1990. Structural relationships of largemouth bass and bluegills in South Dakota ponds. North American Journal of Fisheries Management 10:338–343.

Guy, C. S., and D. W. Willis. 1991. Evaluation of largemouth bass-yellow perch communities in small South Dakota impoundments. North American Journal of Fisheries Management 11:43–49.

Henson, J. C. 1991. Quantitative description and development of a species-specific growth form for largemouth bass, with application to the relative weight index. Master's thesis. Texas A&M University, College Station.

Hill, K. R. 1986. Mortality and standing stocks of grass carp planted in two Iowa lakes. North American Journal of Fisheries Management 6:449–451.

Hillman, W. P. 1982. Structure and dynamics of unique bluegill populations. Master's thesis. University of Missouri, Columbia.

Hooper, G. R. 1970. Results of stocking largemouth bass, bluegill, and redear sunfish in ponds less than 0.25 acre. Proceedings of the Annual Conference Southeastern Association of Game and Fish Commissioners 23(1969):474–479.

Hubert, W. A. 1996. Passive capture techniques. Pages 157–181 *in* B. R. Murphy and D. W. Willis, editors. Fisheries techniques, 2nd edition. American Fisheries Society, Bethesda, Maryland.

Isaak, D. J., T. D. Hill, and D. W. Willis. 1992. Comparison of size structure and catch rate for largemouth bass samples collected by electrofishing and angling. Prairie Naturalist 24:89–96.

James, M. F. 1946. Histology of gonadal changes in bluegill, *Lepomis macrochirus* Raf., and the largemouth bass, *Huro salmoides* (Lac.). Journal of Morphology 79:63–92.

Johnson, D. L., and R. A. Stein, editors. 1979. Response of fish to habitat structure in standing water. American Fisheries Society, North Central Division, Special Publication 6, Bethesda, Maryland.

Kolander, T. D., D. W. Willis, and B. R. Murphy. 1993. Proposed revisions of the standard weight (W_s) equation for smallmouth bass. North American Journal of Fisheries Management 13:398–400.

Lawrence, J. M. 1954. A new method of applying inorganic fertilizer to farm fish ponds. Progressive Fish-Culturist 16:176–178.

Lewis, W. M., and R. C. Heidinger. 1978a. Tailor-made fish populations for small lakes. Southern Illinois University, Fisheries Bulletin 5, Carbondale.

Lewis, W. M., and R. C. Heidinger. 1978b. Use of hybrid sunfishes in the management of small impoundments. Pages 104–108 *in* G. D. Novinger and J. G. Dillard, editors. New approaches to the management of small impoundments. American Fisheries Society, North Central Division, Special Publication 5, Bethesda, Maryland.

Lewis, W. M., R. Heidinger, and M. Konikoff. 1968. Loss of fishes over the drop box spillway of a lake. Transactions of the American Fisheries Society 97:492–494.

Lindgren, J. P., and D. W. Willis. 1990a. Evaluation of a 380-mm minimum length limit for largemouth bass in Lake Alvin, South Dakota. Proceedings of the South Dakota Academy of Science 69:121–127.

Lindgren, J. P., and D. W. Willis. 1990b. Vulnerability of largemouth bass to angling in two small South Dakota impoundments. Prairie Naturalist 22:107–112.

Louder, D. 1958. Escape of fish over spillways. Progressive Fish-Culturist 20:38–41.

Malvestuto, S. P. 1996. Sampling the recreational creel. Pages 591–623 *in* B. R. Murphy and D. W. Willis, editors. Fisheries techniques, 2nd edition. American Fisheries Society, Bethesda, Maryland.

Marriage, L. D., A. E. Borell, and P. M. Scheffer. 1971. Trout ponds for recreation. U.S. Department of Agriculture Farmers' Bulletin 2249.

Meehean, O. L. 1952. Problems of farm fish pond management. Journal of Wildlife Management 16:233-238.

Ming, A. 1974. Regulation of largemouth bass harvest with a quota. Pages 39–53 *in* J. L. Funk, editor. Symposium on overharvest and management of largemouth bass in small impoundments. American Fisheries Society, North Central Division, Special Publication 3, Bethesda, Maryland.

Modde, T. 1980. State stocking policies for small warmwater impoundments. Fisheries 5(5):13–17.

Neumann, R. M., and B. R. Murphy. 1991. Evaluation of the relative weight (W_r) index for the assessment of white crappie and black crappie populations. North American Journal of Fisheries Management 11:543–555.

Neumann, R. M., D. W. Willis, and D. D. Mann. 1994. Evaluation of largemouth bass slot length limits in two small South Dakota impoundments. Prairie Naturalist 26:15–32.

Noble, R. L. 1988. Fish communities of small impoundments of semi-arid areas. Pages 63–78 *in* J. L. Thames and C. D. Ziebell, editors. Small impoundments of semi-arid regions. University of New Mexico Press, Albuquerque.

Noble, R. L., F. H. Sprague, W. C. Hobaugh, and D. W. Steinback. 1979. Wildlife benefits through construction and management of floodwater retarding structures. Rocky Mountain Forest and Range Experiment Station General Technical Report 65:181–185.

Novinger, G. D. 1990. Slot length limits for largemouth bass in small private impoundments. North American Journal of Fisheries Management 10:330–337.

Payer, R. D., R. B. Pierce, and D. L. Pereira. 1989. Hooking mortality of walleyes caught on live and artificial baits. North American Journal of Fisheries Management 9:188–192.

Powell, D. H. 1975. Management of largemouth bass in Alabama's state-owned public fishing lakes. Pages 386–390 *in* H. Clepper, editor. Black bass biology and management. Sport Fishing Institute, Washington, D.C.

Powell, D. H., and S. L. Spencer. 1979. Parallel-bar barrier prevents fish loss over spillways. Progressive Fish-Culturist 41:174–175.

Redmond, L. C. 1974. Prevention of overharvest of largemouth bass in Missouri impoundments. Pages 54–68 *in* J. L. Funk, editor. Symposium on overharvest and management of largemouth bass in small impoundments. American Fisheries Society, North Central Division, Special Publication 3, Bethesda, Maryland.

Regier, H. A. 1962. On the evolution of bass–bluegill stocking policies and management recommendations. Progressive Fish-Culturist 24:99–111.

Regier, H. A. 1963. Ecology and management of largemouth bass and golden shiners in farm ponds in New York. New York Fish and Game Journal 10(2):139–169.

Saffel, P. D., C. S. Guy, and D. W. Willis. 1990. Population structure of largemouth bass and black bullheads in South Dakota ponds. Prairie Naturalist 22:113–118.

Saila, S. B. 1952. Some results of farm pond management studies in New York. Journal of Wildlife Management 16:279-282.

Schmittou, H. R. 1969. Some effects of supplemental feeding and controlled fishing in largemouth bass-bluegill populations. Proceedings of the Annual Conference Southeastern Association of Game and Fish Commissioners 22(1968):311–320.

Schnick, R. A., F. P. Meyer, and D. L. Gray. 1989. A guide to approved chemicals in fish production and fishery resource management. University of Arkansas, Cooperative Extension Service, Little Rock.

SCS (Soil Conservation Service). 1982. Ponds: planning, design, construction. U.S. Department of Agriculture, Agriculture Handbook 590, Washington, D.C.

Simpkins, D. G., and W. A. Hubert. 1996. Proposed revision of the standard weight equation for rainbow trout. Journal of Freshwater Ecology 11:319–326.

Stone, C. C., and T. Modde. 1982. Growth and survival of largemouth bass in newly stocked South Dakota ponds. North American Journal of Fisheries Management 2:326–333.

Swanson, E. D., and E. P. Bergersen. 1988. Grass carp stocking model for coldwater lakes. North American Journal of Fisheries Management 8:284–291.

Swingle, H. S. 1950. Relationships and dynamics of balanced and unbalanced fish populations. Alabama Polytechnic Institute, Agricultural Experiment Station Bulletin 274, Auburn.

Swingle, H. S. 1951. Experiments with various rates of stocking bluegills, *Lepomis macrochirus* Rafinesque, and largemouth bass, *Micropterus salmoides* (Lacepede) in ponds. Transactions of the American Fisheries Society 80:218–230.

Swingle, H. S. 1952. Farm pond investigations in Alabama. Journal of Wildlife Management 16:243–249.

Swingle, H. S. 1956. Appraisal of methods of fish population study, part 4: determination of balance in farm fish ponds. Transactions of the North American Wildlife Conference 21:298–322.

Swingle, H. S. 1970. History of warmwater pond culture in the United States. Pages 95–105 *in* N. G. Benson, editor. A century of fisheries in North America. American Fisheries Society, Special Publication 7, Bethesda, Maryland.

Swingle, H. S., B. C. Gooch, and H. R. Rabanal. 1965. Phosphate fertilization of ponds. Proceedings of the Annual Conference Southeastern Association of Game and Fish Commissioners 17(1963):213–218.

Swingle, H. S., and E. V. Smith. 1938. Fertilizers for increasing the natural food for fish in ponds. Transactions of the American Fisheries Society 68:126–135.

Wege, G. J., and R. O. Anderson. 1978. Relative weight (W_r): a new index of condition for largemouth bass. Pages 79–91 *in* G. D. Novinger and J. G. Dillard, editors. New approaches to the management of small impoundments. American Fisheries Society, North Central Division. Special Publication 5, Bethesda, Maryland.

Wege, G. J., and R. O. Anderson. 1979. Influence of artificial structures on largemouth bass and bluegills in small ponds. Pages 59–69 *in* D. L. Johnson and R. A. Stein, editors. Response of fish to habitat structure in standing water. American Fisheries Society, North Central Division, Special Publication 6, Bethesda, Maryland.

Weithman, S. A., J. B. Reynolds, and D. E. Simpson. 1980. Assessment of structure of largemouth bass stocks by sequential sampling. Proceedings of the Annual Conference Southeastern Association of Fish and Wildlife Agencies 33(1979):415–424.

Wiley, M. J., R. W. Gordon, S. W. Waite, and T. Powless. 1984. The relationship between aquatic macrophytes and sport fish production in Illinois ponds: a simple model. North American Journal of Fisheries Management 4:111–119.

Chapter 22

Natural Lakes and Large Impoundments

DANIEL B. HAYES, WILLIAM W. TAYLOR,
AND PATRICIA A. SORANNO

22.1 INTRODUCTION

Natural and artificial lakes are a valuable resource to our society. One has only to look at the property value of lakeside lots to see the tangible value that we place on the aesthetic and recreational opportunities lakes provide. Although sportfishing contributes to the value we place on lakes, it is important to realize that sportfishing is rarely the only or even the most important use we make of lakes. Because of lakes' multiple values, lake fisheries management needs to take place in the broader context of lake management, which includes such issues as water quality, flood control, suitability for other aquatic recreation, and the use of water for power generation, irrigation, or other human needs.

In this chapter we (1) provide an understanding of the relationship between how lakes function and fishery and water management issues, (2) outline typical fishery management problems in lakes and some of their solutions, and (3) suggest future trends in lake fisheries management. For simplicity, we will classify all bodies of standing water greater than 4 ha as lakes (Noble 1980). Reservoirs are artificial lakes created by damming rivers. In this chapter, we will focus on the management of natural lakes (excluding the Great Lakes) and reservoirs greater than 40 ha. Because reservoirs less than 40 ha encompass their own suite of characteristics (e.g., small reservoirs are often privately owned), their management is covered in Chapter 21. Likewise, the Great Lakes are covered separately because of the unique characteristics of these water bodies (Chapter 23).

22.1.1 Lakes and Their Environment

Lakes are a product of their geologic history, the landscape in which they occur, and the climatic regime of their watershed. On a continental scale, the occurrence of natural lakes is most strongly influenced by landscape morphology and the timing and amount of precipitation. In North America, the effect of glaciers on geomorphology was the most important force creating landscapes rich in natural lakes, including the Great Lakes. Consequently, there are far fewer natural lakes south of the extent of glaciation. In the southern part of the continent, lakes are formed by other processes, such as earthquakes, dissolution of limestone bedrock, and river meanders forming oxbows (Wetzel 1975). The paucity of lakes in the western United States and parts of

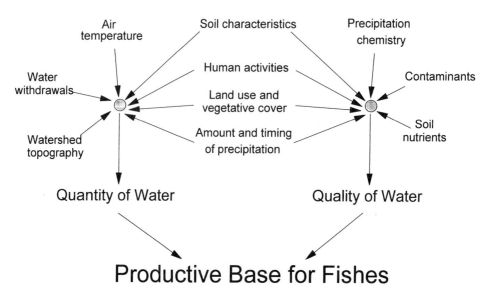

Air temperature

Soil characteristics

Precipitation chemistry

Water withdrawals

Human activities

Contaminants

Land use and vegetative cover

Watershed topography

Amount and timing of precipitation

Soil nutrients

Quantity of Water   Quality of Water

Productive Base for Fishes

Figure 22.1 Interrelationships among watershed features, climate, and human activities on the quantity and quality of water supplied to lakes and reservoirs.

Mexico is a reflection of the relatively dry climate. Although the landscape may be conducive for lake formation, there may not be sufficient precipitation to offset evapotranspiration losses. Hence, some "lakes" in these arid regions are ephemeral and contain water only after heavy rains or snowmelt and are dry otherwise. In the southern and western United States, many reservoirs have been created for power generation and water control, thus offsetting the paucity of natural lakes.

In addition to determining where lakes occur and their size and shape, landscape morphology and climate interact to determine many of the physical traits of lakes. One of the key physical characteristics that determines the fish community supported by lakes and reservoirs is their temperature regime. Although lakes in warmer climates tend to be warmer, whether or not they stratify will determine if a cool hypolimnion (deep layer that does not mix) is present. Lake Casitas in southern California is an example of a lake in a warm climate that supports warmwater fishes such as largemouth bass in its upper waters and also supports coldwater fishes such as rainbow trout in its hypolimnion (Fast 1993).

The natural chemical properties of lakes are largely determined by external factors such as the geology of the watershed, climate, and the chemistry of precipitation falling on the watershed. All of these factors interact to determine the quality as well as quantity of water delivered to lakes and reservoirs (Figure 22.1). The rate of internal cycling also affects the chemical composition of lake water, but internal cycling is often overshadowed by human and natural inputs, especially of nutrients. Humans have a tremendous effect on water chemistry through activities such as the discharge of pollutants into lake water, changes in watershed characteristics (e.g., through agricultural practices), and the dispersal of toxic substances (e.g., pesticides) and acidic compounds (e.g., SO_x and NO_x) into the atmosphere, where they affect large parts of the continent (Chapter 11).

22.1.2   Comparison of Natural Lakes and Large Impoundments

Reservoirs are generally built to generate hydroelectric power, provide flood control and water storage, or control water level in natural lakes. Because of construction costs and the need to provide adequate head (or water height) for power generation, dams are often built where river channels are narrowest and have the steepest gradient. Because of such site selection, reservoirs tend to be longer and narrower than are nearby natural lakes and also tend to have more narrow bays than do nearby natural lakes. Reservoirs also tend to have larger watersheds for their size (i.e., they have a greater watershed to lake surface area ratio) than do natural lakes, resulting in a relatively greater influence of the watershed on their water chemistry and physical properties. Although the dendritic shape of reservoirs tends to create longer shorelines than occur for natural lakes, the littoral zone in reservoirs is often limited by their relatively steep depth contours. The productivity of the littoral zone in reservoirs may be further limited by fluctuations in water level associated with power generation, flood control, or irrigation. When shoreline areas are dewatered for extended periods, the fish, invertebrate, and macrophyte communities that would normally be there can be disrupted.

Because of their dependence on watershed and climatic factors, lakes and reservoirs share many common traits; therefore, fisheries management practices are often similar in natural and artificial lakes. However, reservoirs often differ from natural lakes in some of their physical, chemical, and biological properties because of the strong influence of the inflowing river or stream and the effects of human control of water flows. Inflowing water generally creates a strong longitudinal gradient of conditions in reservoirs from a stream (lotic) environment to a lakelike (lentic) environment (Hayward and Van den Avyle 1986). For example, the upper reaches of large reservoirs often have high sediment inputs from the source river. Excessive sedimentation can reduce the abundance of benthic invertebrates and directly cause mortality of fishes and their eggs. Fine sediment input can also increase turbidity in reservoirs. High turbidity affects light penetration and can reduce algal and macrophyte growth. Additionally, high turbidity decreases the ability of visually orienting predators to detect prey, potentially leading to an unbalanced predator–prey system. Because of high water input from their source river, reservoirs tend to have high nutrient loading rates, which often results in higher productivity provided that light penetration is sufficient.

Another difference between natural lakes and reservoirs is their age. Most natural lakes were formed by the glaciers 10,000–12,000 years ago. Reservoirs are much younger, however; most are less than 100 years old. Because of the young age of new impoundments, fish communities generally have not had time to adapt to the conditions peculiar to these habitats. When a river is impounded, the majority of fish species present before impoundment are often displaced, leaving an impoverished fish community. Therefore, carefully thought out stocking plans may be needed to create an appropriate fish community in new reservoirs (Fry and Hanson 1968). In addition to being relatively young, the expected life span of reservoirs is often shorter than that of most natural lakes because of the greater sediment input reservoirs receive. As an example, Stronach Dam, constructed in 1912 on the Pine River, Michigan, created a 27-ha reservoir. Because of the high sand load carried by the river, the impoundment disappeared within 40 years. Without dredging or some other active sediment removal program, even the largest of our reservoirs face a similar fate, albeit over a longer time horizon. Although the size of most large reservoirs has not been greatly reduced by sedimentation, the process of reservoir sedimentation does pose a limit to how long we can expect reservoirs to function as lakes and support lake fisheries.

Variability in the recruitment of fishes occurs in both natural lakes and reservoirs and is problematic because this variability reduces the ability of these systems to produce sustained high-quality fisheries. This variability also reduces our ability to predict fish abundance over time and the response of these fisheries to management manipulations. The problem of variable recruitment is often intensified in reservoirs through the manipulation of water levels. The reproductive success of many species in reservoirs, including northern pike, largemouth bass, and common carp, depends on the occurrence of high water levels and the occurrence of appropriate, flooded substrate (Aggus 1979). For example, largemouth bass reproduction is enhanced if shoreline vegetation is inundated (Von Geldern 1971). Therefore, dewatering of the littoral zone can significantly affect reproductive success and juvenile survival of largemouth bass. The timing and magnitude of water level fluctuations will affect species differentially, giving managers a tool for controlling undesirable species (e.g., carps) while promoting growth and survival of desired species (Chapter 15).

The ecological immaturity of impoundments is also a source of variability in lake productivity. Kimmel and Groeger (1986) coined the terms trophic upsurge and trophic depression to describe the changes that occur in reservoir productivity with age (Figure 22.2). The magnitude and variability of biological production is greatest in the early phase of an impoundment's existence because of the surplus of available nutrients and large quantities of organic material present in the area that is flooded. Often this initial period of high productivity has produced false hopes for long-term fish production. As reservoirs mature, fish populations usually stabilize (5–10 years) at a lower level of abundance, and productivity tends to decline as competition reduces food availability (Kimmel and Groeger 1986). As mentioned earlier, the relative youth of reservoirs and the dramatic habitat changes that occur when a river is impounded result in a relatively impoverished fish community. Because of the relatively simple trophic structure of most reservoir systems, they are particularly vulnerable to overfishing and environmental variation.

22.2   IN-LAKE HABITATS

A fundamental concept in fisheries management is that fish habitat determines the composition of the fish community and the potential productivity of a system. Fisheries management is an attempt to manipulate people, aquatic populations, and the habitat itself to direct that potential productivity. In the context of lakes and reservoirs, what is fish habitat and what are some of its critical features?

Biotic and abiotic factors both play an important role in determining fish growth and survival. As such, we use a broad definition of habitat that includes both biotic and abiotic components of the environment. Our preferred definition of habitat is:

> Habitat is simply the place where an organism lives. Physical, chemical and biological variables (the environment) define the place where an organism lives. (Hudson et al. 1992:75).

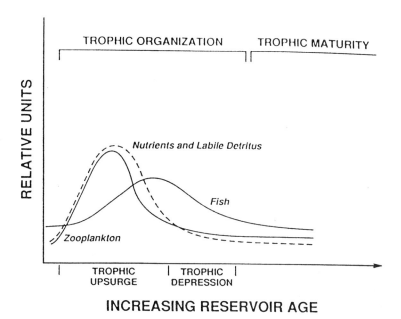

Figure 22.2 A generalized model of trophic dynamics as a reservoir matures and ages. Soon after reservoir construction, nutrient levels increase, leading to increased zooplankton production. This is the trophic upsurge phase. As interactions among organisms develop, trophic maturity is reached. Production often declines (trophic depression), however, as nutrients decline. Figure adapted from Kimmel and Groeger (1986).

In this definition, space is the primary component of fish habitat, and other resources and environmental conditions modify the utility of that space. A fundamental ecological concept that is complementary to the idea of fish habitat is niche. Niche is an expression of a fish's needs as determined by its biology. The boundary of a fish's niche is the combination of environmental conditions it will tolerate. Within these bounds, fish can survive, but their growth, survival, and reproductive success will vary, depending on how well environmental conditions match the fish's needs. Fish abundance and production will be greatest when the environmental conditions within a habitat best fulfill a fish's needs (i.e., the habitat matches the fish's niche requirements).

Lakes and reservoirs provide habitats with a variety of environmental conditions, and every water body has a unique combination of habitat characteristics. Here we will briefly review some of the habitat features that most often limit fish populations in lakes, and we refer the reader to Chapter 11 for a more detailed discussion of lake and reservoir habitat management.

An important set of limnological features that affect fish communities is variation of temperature, oxygen, and light with depth (Figure 22.3). The waters of shallow lakes tend to mix from top to bottom, resulting in constant temperature and oxygen concentrations with depth. Across most of the continent, deeper lakes, or lakes that are protected from the wind, generally stratify (i.e., have layers of different temperature) during the summer months. Typically, the epilimnion (surface layer) occupies the upper several meters of the lake and has a similar temperature throughout (Figure 22.3)

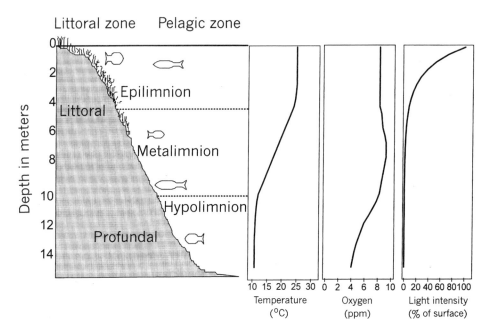

Figure 22.3 Habitat features in a stratified lake and temperature, dissolved oxygen, and light intensity with water depth.

and relatively high concentrations of dissolved oxygen. Warmwater littoral and pelagic fishes are generally found in epilimnetic waters. The metalimnion is the next layer and is characterized by having temperatures that decline by more than 0.5°C for every meter of depth. Thus, the top of the metalimnion is the same temperature as the epilimnion and holds similar species of fishes, but as one progresses deeper the temperature declines and cool- or coldwater fishes tend to predominate. The extent of the metalimnion is highly variable and depends on lake size, wind exposure, and the time of year. The deepest layer in stratified lakes is the hypolimnion. The hypolimnion is the coolest part of the lake and may support coldwater fishes if sufficient oxygen is present. In lakes with low productivity, dissolved oxygen concentrations are often high in the metalimnion and hypolimnion; in more productive lakes, decomposing organic material depletes oxygen, sometimes to the point that there is insufficient oxygen to support any fish species.

The intensity of light declines with water depth, and the rate of decline depends on the amount of suspended and dissolved material present in the water column. Fishes vary tremendously in their ability to see and feed at low light levels. Some fishes such as walleye have a special structure in their eye to enhance their low-light vision, and others rely on special chemosensory organs (e.g., the barbels of catfishes) that allow them to function in low light levels. Many species of fishes, however, lack these specialized structures and require more light to see well. The rate of decline in light intensity with depth is also an important determinant of where plants occur in lakes. In general, most algae and aquatic plants require a light intensity greater than 1% of the light intensity at the surface. Along the shoreline above this depth, rooted aquatic plants (macrophytes) can maintain themselves if the substrate is suitable. The portion of lake bottom and water column containing macrophytes is termed the littoral zone (Figure 22.3). The area where the water is too deep for sufficient light to reach the bottom for macrophyte growth is termed the pelagic zone. The upper part of the pelagic zone that has more than 1% of the light intensity at the surface can support algal growth; it is called the photic zone. Below the photic zone, and along the lake bottom, is the profundal zone (Figure 22.3). The extent of these different zones directly affects the fish community as well as the amount of primary production of macrophytes and algae and the secondary production of zooplankton and benthic invertebrates. Thus, for example, biological production in shallow lakes may be dominated by macrophytes and their associated invertebrate community (e.g., odonates), resulting in a fish community with a predominance of littoral fishes adapted for feeding on these invertebrates.

22.3   FISH SPECIES ASSEMBLAGES IN LAKES AND RESERVOIRS

Across the continent, there is a tremendous variety in the organisms found in lakes. Focusing on fishes, several key factors determine the species present in a given lake. In this section, we will discuss some of these factors and their relationship to fisheries management.

As described earlier, one of the critical factors determining species composition is the habitat conditions that a lake provides. Simply put, if the habitat conditions lie outside a fish's niche (over its entire life history), the species will not be able to maintain a viable population. Although all fish species (and potentially individual stocks within a species) have their own particular niche, there are distinguishable groups of fish species that share common habitat needs. For example, trout, salmon, and graylings are examples of coldwater fishes. Technically, these fishes are all coldwater stenotherms, requiring cold water temperatures and high oxygen levels. In contrast, peacock bass and mosquitofishes are examples of warmwater stenotherms. Because of these commonalities in habitat needs, we often see groups of fish co-occurring on a regular basis. These groups may be indicative of different lake types (e.g., Lake Okeechobee, Florida, is a largemouth bass–bluegill lake, whereas Oneida Lake, New York, is a yellow perch–walleye lake), or these groups may be seen to segregate within a lake (e.g., where they co-occur, adult largemouth bass and bluegill occur primarily in warm littoral areas, whereas rainbow trout are found in the hypolimnion [Hodgson et al. 1991]).

Even where habitat conditions are suitable to support a given species, that species may not occur for other reasons. In particular, the limited dispersal mechanisms fishes have generally restricts their natural range. Because barriers to migration reduce the dispersal of all fish species (although not necessarily equally), the distribution of many fishes fall into separate zoogeographical zones (Figure 22.4). Examples of such zones include the Mississippi province, the Hudson Bay province, and the Mexican transition subregion (Figure 22.4). Before European settlement of North America, movement of fishes among these zones was limited, often taking thousands of years to occur (e.g., the current zoogeographical distribution of fishes is in part the result of the most recent glaciation). Following European settlement, many of the barriers to dispersal

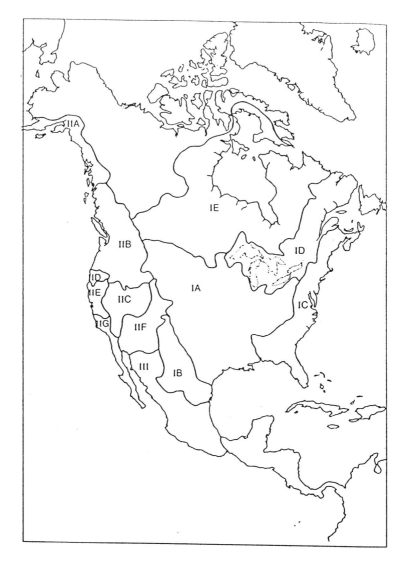

Figure 22.4 Zoogeographic zones of freshwater fishes for North America: (IA) Mississippi province; (IB) Rio Grande province; (IC) Atlantic Coast province; (ID) Great Lakes–St. Lawrence province; (IE) Hudson Bay province; (IF) Arctic province; (IIA) Alaska Coastal province; (IIB) Columbia province; (IIC) Great Basin province: (IID) Klamath province; (IIE) Sacramento province; (IIF) Colorado province; (IIG) South Coastal province; and (III) Mexican transition subregion. Figure taken from Moyle and Cech (1988).

have been broken down, both through the intentional stocking of fishes outside their native range as well as the unintentional introduction of aquatic organisms by human activities such as the building of canals and the release of ballast water from ships.

Although the merits of a particular intentional introduction may be debatable, the unintentional introduction of aquatic organisms outside their natural range is generally more problematic because of the accidental and unplanned nature of such an event.

Aquatic organisms from all over the world crossing virtually all taxonomic groups have been spread across much of the continent (Mills et al. 1993). Some examples include

- the spread of various cyprinid species that are used as bait in recreational fisheries;
- the wide dispersal via boats of aquatic macrophytes such as Eurasian water milfoil and the encroachment of wetland plant species such as purple loose-strife due to their superior dispersal abilities after initial introduction;
- the appearance of the European zebra and quagga mussels in lakes across the eastern seaboard and the Great Lakes; and
- the release of the European predaceous macrozooplankton, the spiny water flea *Bythotrephes cederstroemi*, in the Great Lakes via ballast water.

In addition to the effects of habitat and zoogeography on fish community composition, interactions among species may also limit where a fish species can survive. We will expand on this topic in the next section but here briefly illustrate this concept with an example. In a survey of 195 randomly selected lakes in the northeast United States, Whittier et al. (1997) found that the presence of littoral predators (particularly large-mouth and smallmouth bass) reduced the median number of minnow species from 2 to 0, implying that these predators effectively eliminated minnows in many lakes.

A useful conceptual framework synthesizing the processes that determine fish community composition in lakes is found in Tonn (1990). His conceptual model is based on the idea that the species found in a lake are "filtered" from a larger pool of species (Figure 22.5). These filters initially act at broad geographic scales (e.g., the species pool is determined in part by zoogeographical provinces) and then work on finer and finer scales until reaching the level of a particular lake where the specific combination of physical habitat characteristics, prey community, and other fish species present are the primary factors limiting the occurrence of a species (Tonn 1990).

22.4 REGULATION OF FISH PRODUCTION IN LAKES AND RESERVOIRS

There are a number of different conceptual models regarding fish production in lakes. The dominant paradigms of fish production include the trophic state model and the food web model. Briefly, the trophic state model focuses on the importance of nutrients as a limiting factor for fish production, whereas the food web model focuses on community structure and species interactions as limiting factors. In this section, we will discuss these models and evidence for them in more detail and present what we feel is the current synthesis and future direction of these models.

22.4.1 Trophic State Model

The trophic state model is a widely recognized conceptual framework for describing the natural changes in productivity that occur as lakes age and the changes in productivity that occur through human impacts (i.e., cultural eutrophication; Wetzel 1975). Within this framework, three major classes or trophic states of lakes and im-

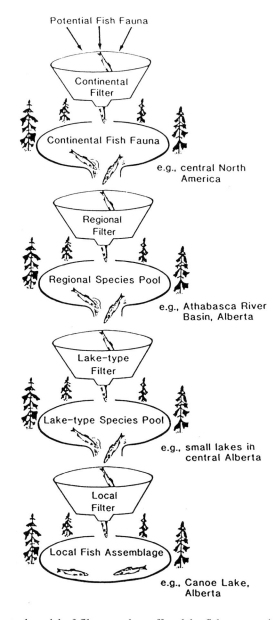

Figure 22.5 Conceptual model of filters as they affect lake fish community composition. Figure taken from Tonn (1990).

poundments are recognized: oligotrophic, mesotrophic, and eutrophic. Trophic state designations are based on nutrient loading and primary production: oligotrophic lakes have the lowest levels of nutrient loading and primary production and eutrophic lakes have the highest. In addition to nutrient inputs, climate plays an important role in determining trophic status of a lake. When lakes are compared globally, primary production is more closely linked to latitude and radiant energy inputs than it is to nutrients (Brylinsky and Mann 1973). For lakes at a similar latitude, or for a single lake over time, the role of nutrient loading becomes paramount.

The foundation for the trophic state model is that productivity at all levels in lakes is constrained by the amount of nutrients present. In this model, nutrient concentrations (particularly phosphorus and nitrogen) are the primary limiting factor for algal production. Limitations on algal production in turn result in limitations on zooplankton production and so on up through the food web to piscivorous fishes.

Many studies have demonstrated the correlation between productivity at various trophic levels (e.g., producers, herbivores, and carnivores) and nutrient concentration. In some cases, the correlational approach has been expanded to include a regression that predicts fish productivity from nutrients or productivity at a lower level. Some examples of such regressions are given here.

$$\text{Yield (kg/ha)} = 0.792 + 0.072 \cdot \text{total P (µg/L)}, \tag{22.1}$$

for which $r^2 = 0.84$ and $N = 21$ (Hanson and Leggett 1982).

$$\text{Yield (g dry weight/m/yr)} = -1.92 + 1.17 \cdot \log_{10} \text{chlorophyll } a \text{ (mg/m}^3\text{)}, \tag{22.2}$$

for which $r^2 = 0.84$ and $N = 19$ (Oglesby 1977).

$$\text{Harvest (kg/ha)} = -1.8 + 2.7 \cdot \text{chlorophyll } a \text{ (mg/m}^3\text{)}, \tag{22.3}$$

for which $r^2 = 0.83$ and $N = 25$ (Jones and Hoyer 1982).

$$\text{Yield (kg/ha)} = 1.293 + 0.012 \cdot \text{macrobenthos (kg/ha)}, \tag{22.4}$$

for which $r^2 = 0.47$ and $N = 26$ (Hanson and Leggett 1982).

$$\log_{10}\text{Biomass (kg/ha)} = 1.55 + 0.321 \cdot \log_{10} \text{total P (µg/L)}, \tag{22.5}$$

for which $r^2 = 0.24$ and $N = 60$ (Bachmann et al. 1996).

$$\log_{10} \text{Production (kg/ha)} = 0.32 + 0.94 \cdot \log_{10} \text{biomass (kg/ha)} - 0.17 \log_{10}\text{maximum weight of individual fish (g)}, \tag{22.6}$$

for which $r^2 = 0.84$ and $N = 100$ fish populations from 38 lakes (Downing and Plante 1993).

In addition to research evaluating the correlation between nutrients and fish production, manipulative experiments have also been conducted to evaluate the response of various trophic levels to nutrient addition. Examples of increased production due to fertilization are available for phytoplankton (e.g., Elser et al. 1990; Gross et al. 1997), zooplankton (e.g., Luecke et al. 1996), benthic invertebrates (e.g., Clarke et al. 1997), and fishes (e.g., Stockner 1987; Luecke et al. 1996). These studies provide direct evidence of the importance of nutrients for lake productivity.

The classic trophic state model focuses on the relationship between nutrients and algae. Recently more attention has been placed on the role nutrient loading plays in supporting macrophyte growth. Dense stands of macrophytes interfere with swimming, boating, and other recreational uses of lakes and can pose particular problems in fisheries management. There is good evidence that macrophytes respond to nutrients as do algae, but the quantitative relationship is less well known. In addition, there appears to be a negative interaction between algal abundance and macrophyte growth through competition for light (Spencer and King 1984).

There are at least two areas for which the evidence for the trophic state model is less conclusive. One area is the relationship between benthic invertebrate productivity and nutrients and the other is the productivity of detrital food webs. There are studies that have examined the correlation between benthic invertebrates and nutrients (e.g., Hanson and Leggett 1982), however detailed information on this relationship is still lacking. Although some studies have detected increases in benthos abundance following nutrient addition, it appears that other factors, including substrate characteristics, dissolved oxygen concentration, fish predation, and benthic invertebrate life cycles, also influence benthic productivity, masking the effect of changes in nutrients (Clarke et al. 1997).

Although the trophic state model is a useful approach for understanding the productivity of lakes and their fishery potential, there are limitations to the model's capabilities. Like most models, the predictions made using the trophic state model represent average responses across a range of observations and may not be accurate for a particular lake. For example, lakes with high phosphorus inputs tend to have high productivity on average, but productivity in a particular lake may be limited by other factors, such as the amount of available nitrogen (Elser et al. 1990). Another important limitation of the trophic state model is that its predictions are confined to total production or yield of fishes and not the species composition of the lake. As such, caution must be used in trying to predict changes in the yield of a particular species as nutrient inputs change.

Beyond issues related to fish production, the trophic state model has played a prominent role in our management of lake water quality. Based on the evidence discussed above, many regulations in the United States and Canada have been passed that limit the discharge of nutrients into lakes and streams. Broadly speaking, these regulations have been successful in reducing point source discharges of nutrients in lakes and have "improved" water quality (i.e., water clarity is greater and blooms of nuisance algae have declined). However, the trophic model suggests that limitations on nutrient input may have the potentially undesirable consequence of reducing fish production. Such tradeoffs between lake productivity and water quality can present problems in managing lakes to satisfy multiple user groups.

22.4.2 Food Web Model

In the food web model, predation is viewed as the dominant force determining the structure and abundance of prey communities (Carpenter et al. 1985). Predation by piscivorous fishes decreases the abundance of planktivorous forage fishes, thereby affecting zooplankton abundance, size structure, and community composition. Alterations in the zooplankton community further lead to changes in algal and nutrient dynamics. Thus, fish predation has direct effects on their prey items, which in turn affects biological, physical, and chemical components of lake ecosystems. A logical extension to this model views anglers as the highest trophic level, controlling the abundance of piscivorous fishes (Johnson and Carpenter 1994). Studies supporting the food web model include the correlation between fish biomass and zooplankton biomass and fish biomass and phytoplankton biomass (McQueen et al. 1986).

Fish introductions provide additional evidence for food web control of lake ecosystems. The classic example is the effect alewives had on the structure of the zooplankton communities in coastal ponds in Connecticut (Brooks and Dodson 1965). The results of their investigations showed that ponds with alewives had zooplankton communities dominated by small species such as *Bosmina longirostris*, *Ceriodaphnia lacustris*, *Cyclops bicuspidatus thomasi*, and *Tropocyclops prasinus.* In contrast, zooplankton communities in ponds where alewives were absent were predominantly large plankters, such as *Daphnia* spp., large *Diaptomus* spp., and *Mesocyclops edax.*

Situations where the evidence for the food web model is less clear is for littoral zone communities and benthic-invertebrate-based and detritus-based food webs. Interestingly, these are also areas for which there is less information supporting the trophic state model. In the littoral zone, interactions between top-level predators (e.g., largemouth bass) and fishes of intermediate trophic level (e.g., bluegill and pumpkinseed) have been extensively investigated; these studies have shown that predation has a strong effect on the abundance, size structure, and distribution of prey fishes (e.g., Werner et al. 1983). Although some studies (Cuker et al. 1992; Diehl 1995; Johnson et al. 1996) have shown that predation by intermediate trophic level fishes can reduce the abundance of their prey, the generality of these findings are not known (Pierce and Hinrichs 1997). Furthermore, there is little evidence that zooplankton and benthic invertebrates in the littoral zone have a substantial effect on littoral algae and macrophytes in natural lakes or large reservoirs. In addition, relatively few studies are available documenting the effect of fish predation on the abundance of benthic invertebrates in natural lakes. In a whole-lake manipulation study, Hayes et al. (1992) demonstrated that removal of 85% of the adult white sucker population resulted in large (5–10-fold) increases in the abundance of most benthic invertebrates. Post and Cucin (1984) likewise demonstrated substantial (30–60%) reductions in the abundance, biomass, and mean weight of benthic invertebrates following the introduction of yellow perch into a small Ontario lake. As in the littoral zone, the effect of changes in the abundance of benthic invertebrates on productivity at lower trophic levels is not well known.

22.4.3 Synthesis of Trophic State and Food Web Models

Although both the trophic state and food web models have been used individually to conceptualize the major controls of lake productivity, it is generally recognized that the processes represented by each model occur simultaneously in aquatic systems. In the alewife example above, it is true that these fish had a dramatic influence on the zooplankton community. However, it can be argued that the biomass of the alewife population was ultimately determined by the rate of nutrient flux into these ponds. A current synthesis of the trophic state and food web models proposes that predation mainly controls community structure and tends to reduce biomass at the next lower trophic level, but competition and availability of nutrients limit the maximum production at each trophic level (Mills and Forney 1988). We would extend this by proposing that although maximal fish production is constrained by nutrient levels, other limiting factors such as the physical habitat conditions (e.g., amount of spawning area available) often reduce fish production below the maximum set by nutrient levels. Another dimension of the synthesis of these models is that fishes act not only as predators but also affect the internal cycling of nutrients within lakes. For example, planktivorous fishes increase algal biomass not only by removing zooplankton (which eat the algae) but also by recycling available nutrients for algal growth (Vanni and Layne 1997).

A critical feature of both models and their synthesis is that productivity of different trophic levels is linked, and management actions imposed on one trophic level will have effects on other trophic levels and on the physical and chemical characteristics of the lake. For example, Spencer and King (1984) showed that manipulation of fish populations affected water clarity, macrophyte growth, and even the species of birds and mammals that made use of eutrophic ponds in southern Michigan. Understanding that all trophic levels in lakes are linked leads to the development of an ecosystem approach to lake management, with consideration given to anglers in addition to other lake users (Chapter 5). Given the diverse demands being placed on our aquatic resources, an ecosystem approach is necessary to evaluate the consequences a given management prescription will have on food web interactions, trophic dynamics, and ultimately fish production.

The specific problem of maximizing fish abundance, fish yield, or the quality of fish produced depends critically on how production is packaged in a food web. If one considers trophic status versus food web effects on lake productivity, there are two alternative approaches to increasing fish yield (Figure 22.6). The first involves enlarging the food resource base without altering energy transfer efficiency among trophic

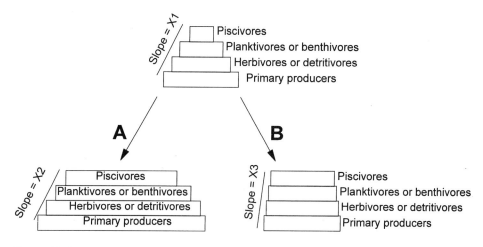

Figure 22.6 Alternate methods of increasing fish yield. Box size represents production at each trophic level. Slopes of lines represent energy transfer efficiency. Method A involves enlarging the food resource base (e.g., via fertilization) without changing the energy transfer efficiency among trophic levels (slope X1 = slope X2). Method B involves a food web manipulation (e.g., introduction of a predator) designed to increase energy transfer efficiency (slope X3 > slopes of X1 or X2) without an increase in food resource base. Diagram adapted from Wagner and Oglesby (1984).

levels. This approach amounts to increasing lake fertility to achieve peak biomass possible at all trophic levels. The second approach increases transfer efficiency without changing the size of the food resource base. This is most commonly achieved by maintaining an "efficient" and "balanced" suite of predators and prey species. For example, gizzard shad are often stocked to provide forage for large predators such as largemouth bass. Because gizzard shad can use detritus as a food source directly (Stein et al. 1995), energy transfer from lower tropic levels to higher trophic levels may be increased. Ideally, these approaches should be viewed as complementary to one another, with the possibility that both need to be considered simultaneously.

22.4.4 Further Refinements and Future Directions

Further refinement of these two models will need to incorporate two additional factors. First is the connectivity between the pelagic and littoral zones. These two zones interact with each other in several ways that affect nutrient concentrations, fish population and community dynamics, and other ecosystem processes. Second is the recognition of the important interactions between the lake and its watershed that strongly determine the lake's trophic state.

The littoral zone is used by many pelagic and littoral fishes at some point in their life, depending on the species. Littoral structure (substrate, macrophytes, or debris) serves three important functions for fishes: (1) as habitat for spawning, (2) as a production source of invertebrate forage, and (3) as refuge from predation (Weaver et al. 1996). Each one of these functions is affected by both biotic (predation and competition) and abiotic (e.g., oxygen concentrations, temperature, and nutrients) factors. However, because many fishes depend on both the littoral and pelagic zones at various stages of their life, it is necessary to consider explicitly both the littoral and pelagic zone when thinking of synthesizing the food web and trophic state models.

Nutrient transport between the littoral and pelagic zones can be important to overall food webs and to nutrient cycles in lakes through a variety of mechanisms. Not only are dissolved nutrients in the water easily mixed between littoral and pelagic zones through physical processes (e.g., wind generated currents), but consumers in lake food webs (such as fishes, zooplankton, and benthic invertebrates) can also transport nutrients between littoral and pelagic zones (Vanni and de Ruiter 1996). For example, fishes that feed in the littoral zone can assimilate nutrients as they feed and grow and then later excrete these nutrients to make them available to phytoplankton in open water habitats (Schindler et al. 1996; Vanni and de Ruiter 1996). This process is also evident for detrivorous fishes that feed on benthic detritus and then excrete some of the consumed nutrients into pelagic habitats (Vanni 1996). The transport of nutrients requires further study, however, before it can be effectively applied to lake and fisheries management.

It is becoming clear that focusing on the lake or reservoir basin alone is not adequate to assess fully a lake and, in particular, its fish community. Lakes and reservoirs are strongly connected to their surrounding watershed through the transport of water, nutrients, and sediments from land to water. In addition, many lakes are not isolated units but in fact have one or several stream inflows and outflows. These stream linkages can influence both abiotic and biotic aspects of lakes and reservoirs. For example, streams flowing into lakes are often the main vector through which nutrients from the watershed reach lakes. Most nutrients entering northern temperate lakes come from nonpoint sources within the watershed and are transported to lakes by connecting streams. Streams are also the main transport mechanisms through which sediments enter lakes and are deposited in littoral zones. The actual amount of both sediments and nutrients depends on watershed characteristics, in particular land use practices and topographical features. Therefore, the amount and timing of sediment and nutrient inputs to a lake can be predicted from information only at the watershed scale.

The presence of streams also influences biotic components in lakes. For example, the distribution of fish species in different water bodies across the landscape will be determined to a certain extent by the presence of streams through which species can migrate between different water bodies. This linkage between different water bodies is also important for the spread of plants and other aquatic organisms. Thus, both abiotic and biotic factors are influenced by the presence of streams that connect lakes to other lakes or to other streams.

The important linkages of surrounding wetlands, stream inputs, and groundwater inputs to lakes and reservoirs become apparent at only the watershed scale (NRC 1992). A true watershed perspective is one that recognizes the strong interrelationships between watersheds, nutrients, water quality, and fisheries. For example, lakes with dense populations of bottom-feeding fishes are usually eutrophic and suffer from algal blooms. To manage these lakes effectively for both water quality and fisheries, it will be necessary to consider where the fish feed, what species are there, and the magnitude of nutrient input from the watershed (Vanni 1996).

22.5   LAKE ASSESSMENT

The process of fisheries management is ideally goal oriented, with a concrete set of objectives in place before taking any management action (Barber and Taylor 1990; Chapter 2). The manager's goals and objectives are ultimately set by the public's demands (see Chapter 8) but are limited by the capability of a lake to produce fishes while maintaining water quality suitable for other uses. Determining the fishery potential of a lake is difficult due to the complexity of factors influencing fish production as described earlier. Both direct and indirect approaches to assessing fisheries potential are presented in this section.

The first step in assessing a lake or reservoir is to evaluate the condition of the watershed. Although we will not go into detail on how to accomplish that here, some of the major considerations are the size of the watershed and its land use patterns (Wang et al. 1997). The next step is to sample the key limnological and fishery characteristics of the lake in question to prescribe a management program to achieve the previously set goals. Some of the key limnological information required to make informed fisheries management decisions include nutrient concentration, lake temperature and stratification pattern, dissolved oxygen, and water clarity.

22.5.1   Direct Assessment of the Fishery

There are two general strategies for assessing fisheries. First is to sample and measure directly the fish populations in terms of their abundance, individual growth, recruitment, and population size structure. The key to successful direct assessment is obtaining a representative sample with sufficient sample sizes to provide statistically reliable estimates of these fish population characteristics. Interpretation of the results from this type of assessment program depends on an understanding of the population ecology and biology of the species present in relation to their habitat. Several indices are available for comparing fish population growth and size structure between lakes. Growth is one of the most sensitive parameters to ecosystem change and is perhaps best indexed by comparing lengths at given ages with statewide or regional averages. Difficulties emerge in this approach when growth stanzas in a fish's life history differ in their response to ecosystem stress or management activities. For example, in a population of stunted yellow perch, young-of-the-year fish had above average growth, whereas older individuals grew very slowly (Hayes 1990). Thus, depending on the life stage being examined, different conclusions regarding the general well-being of the population could be drawn. Fish growth can also be evaluated using condition factors or associated measures such as the relative weight (W_r) index (Wege and Anderson 1978). Higher values of W_r indicate that the fish in the sampled population have a higher weight at a given length than a broad regional mean. Interpretation of W_r is not always straightforward because W_r varies seasonally, especially during periods of gonadal development (see Chapter 7).

To index population size structure, the proportional stock density (PSD) index was developed by Anderson (1976) based on research performed by Swingle (1950). Basically, PSD indicates the proportion of the reproductive segment of a fish population that is of quality size (Chapter 7). Higher values of PSD mean that more of the fish population is of quality size to anglers. Target values for PSD have been established for various species and are usually in the range of 40–80%. Values higher than 80% are generally indicative of a senescent population or a population with low reproductive success. The PSD index is most commonly used in small ponds or reservoirs (see Chapter 21) but has also been used in assessing lake fisheries (e.g., Willis 1987).

22.5.2   Indirect Assessment of Fishery Potential

A complementary strategy for fisheries assessment is to examine other elements of the ecosystem that have an effect on fish production. There are a number of such indirect assessment approaches, including the morphoedaphic index and regressions of fish yield on phosphorus, phytoplankton, and benthos.

The morphoedaphic index (MEI) is a widely used index of limnological and fishery productivity. It is the ratio of total dissolved solids to mean lake depth (Ryder 1982). Although the concentration of total dissolved solids is a less accurate indicator of nutrient availability than are direct measurements, the relative ease of measuring total dissolved solids is attractive for use in an index such as the MEI. Mean lake depth provides an indication of lake morphometry. In general, deeper lakes are cooler and tend to have higher energy losses to deepwater sediments. Deep lakes also have more water volume than do shallow lakes of the same area, and thus nutrient inputs are diluted to a greater degree (Ryder 1982). Finally, deep lakes may have a greater proportion of their water volume below the photic zone, thus reducing total primary production. At a regional level, the MEI can be used to provide fisheries managers with a preliminary estimate of the amount of

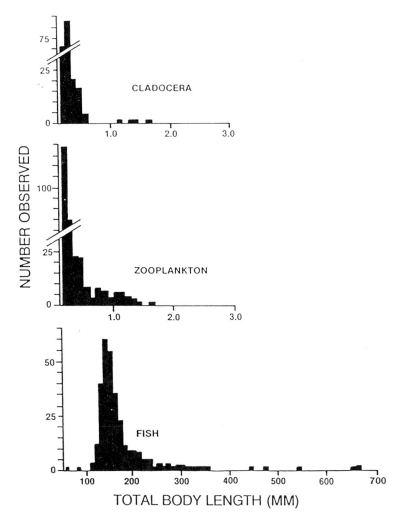

Figure 22.7   Size distribution of Cladocera, zooplankton, and fishes in a warmwater freshwater lake (Black Lake, New York) with a stunted panfish population, few predators, and small-sized zooplankton. Figure adapted from Mills and Schiavone (1982).

fish yield that can be expected from a given lake. When used on a continent-wide or global basis, the MEI should be modified to account for latitudinal differences in climate and growing season.

Although this index provides an indication of the current level of fish production, it cannot predict what level of fish production a lake can achieve through appropriate habitat manipulation (i.e., fertilization or brush piles), nor does it predict the amount of production for particular species. Our abilities to predict the extent and magnitude of the responses to such habitat manipulations are limited due to a lack of mechanistic data. As such, more research is need to develop methods for determining a lake's fishery potential.

Although providing information regarding fish production, none of the indirect assessment approaches require the direct sampling of fish populations and as such are often easier or less expensive to measure. From a fisheries management

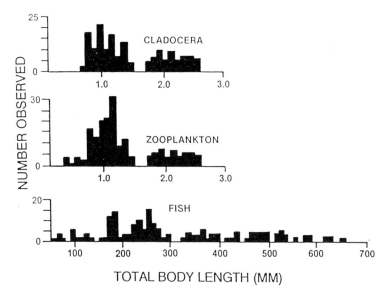

Figure 22.8 Size distribution of Cladocera, zooplankton, and fishes in a warmwater freshwater lake (Clear Lake, New York) with a balanced piscivore–panfish community and numerous large-bodied zooplankton. Figure adapted from Mills and Schiavone (1982).

point of view, these indices may be of limited value because they do not provide information regarding fish species composition, population size structure, growth, or abundance. An indicator that has been successful, however, for indexing the size structure of fish populations is the size structure of the zooplankton community (Mills et al. 1987). This measure follows from work by Galbraith (1967), in which he recognized the importance of large-bodied *Daphnia* to rainbow trout production in lakes. From his research, he suggested that a simple indicator of the probable success of rainbow trout stocking in lakes is the density of large *Daphnia*. Galbraith's work has been further developed by Mills and Schiavone (1982) and Mills et al. (1987), who have shown that the mean body size of the crustacean zooplankton community is correlated with the mean body size of the fish community, particularly in lakes dominated by percids and centrarchids (Figures 22.7 and 22.8). Representative samples of zooplankton can be taken much more easily and quickly than can samples of fishes to provide an indication of the fish community size structure within a lake. Data from Canadarago Lake, an eutrophic lake with a warmwater fishery in New York State, is an example of the use of zooplankton size structure for indirect fishery assessment (Mills et al. 1987). Predators were stocked from 1977 to 1982 to shift a stunted panfish community toward a balanced (see Chapter 21) predator–prey system. Coincident with the establishment of predatory fishes and the suppression of stunted forage species, average zooplankton size increased. The shift toward larger zooplankton coincided with a change toward *Daphnia pulex*. The body size approach to ecosystem analysis has been further generalized to all pelagic trophic levels (Sheldon et al. 1972; 1973; Sprules et al. 1983) and is a step toward a holistic view of lake management.

22.6 LAKE FISHERIES MANAGEMENT

22.6.1 Overview of Management Approaches

Fisheries management prescriptions are often broken into three general classes: people management, aquatic population management, and habitat management. The goal of people management is to provide opportunities for access to aquatic resources for all interested parties and to ensure resource availability and integrity over time. Secondary goals are to provide for specialized opportunities (e.g., quality and catch-and-release fishing) for various user groups when the resource permits. The most common forms of people management are regulations such as length limits, bag limits, closed seasons, gear restrictions, and closed areas (see Chapter 17).

Population management involves the direct manipulation or introduction of aquatic populations. Some examples include stocking desired game species, stocking prey species, removing undesirable fishes, and thinning overcrowded game species. As fisheries management arose from the fish culture discipline, it is not surprising that many of our historical management tools centered on stocking. In fact, a common stereotype of the fisheries manager shows him or her with a bucket of rotenone in one hand and a bucket of fish in the other. Early concepts of fisheries management focused on the idea that the failure of a lake to produce the desired fishery was a function of high exploitation combined with the inability of fish to sustain themselves reproductively (Chapters 13 and 14). Degradation of aquatic habitats early this century also prompted fisheries managers to rely on stocking, largely because the sources of the damage were outside of their control. All these issues continue to be concerns, but successful management today relies more on an ecosystem level approach to management that integrates the people, aquatic population, and habitat components of a lake (Chapter 5).

Habitat management in lakes has matured more slowly than has people or aquatic population management, possibly due to the difficulty in defining and directly observing the critical elements of fish habitat. Basically, there are two distinct categories of habitat manipulations used in lakes. One approach is to alter structure within lakes directly. Examples of this include macrophyte management, formation of spawning areas, and addition of brush as cover (Chapter 11). Perhaps the most extreme example is the building of impoundments by which lotic fish habitats are converted to lentic fish habitats. In effect, the total area of a stream is increased many-fold through impoundment, potentially producing higher total fish production. The desirability of converting a stream to an impoundment depends on the ratio of the cost of impoundment (including loss of recreational opportunities) to the benefits that impoundment would return to society (e.g., flood control).

An alternative habitat management approach is to increase the productivity of a fishery through the direct manipulation of the physical and chemical attributes of a lake or through indirect manipulation by watershed management. Nutrient additions, phosphorus detergents bans, artificial aeration, and destratification are all examples of directly managing limnological characteristics. Regulations on land use or requirements for buffer strips along streams are examples of watershed manipulations that indirectly affect a lake's limnology. Such management actions are common measures for lake management, and as such the effect of such prescriptions on fishery production and the utility of the lake for other uses needs to be carefully evaluated.

Table 22.1 Classes of lake fisheries management problems, their causes, and appropriate management prescriptions.

Fisheries management problems and causes	Management prescriptions
Absence of desired species	
Winterkill or summerkill	Decrease nutrient inputs
	Install artificial aeration
Low pH	Perform mitigative liming
	Stock acid-tolerant strains of fishes
	Reduce emissions of sulphur- and nitrogen-based acids
Current fish species not desirable	Stock desired species
Toxic chemical pollution	Discontinue inputs
Low abundance of desired species	
Inadequate recruitment	Add new spawning sites
	Manipulate water levels
	Close angling season during spawning
	Add cover to increase young-of-the-year survival
	Stock desired species
High adult fishing mortality	Impose bag limits
	Impose length limits
Fishes dispersed over wide areas	Add cover to congregate fishes
Low system productivity	Fertilize
Slow growth	
Intraspecific competition	Thin populations with nets or chemicals
	Add or protect predators for balance
	Reduce weeds to increase predation
	Stock new prey species
	Stock sterile hybrids or triploid fishes
Interspecific competition	Remove competitors with nets or chemicals
	See above management prescriptions for intraspecific competition
Low lake productivity	Fertilize lake
Poor population size structure	
Slow growth	See management prescriptions for improving slow growth
High adult mortality	Impose length limits
	Impose bag limits
	Stock catchable-size fish

22.6.2 Overview of Management Problems

Although the exact nature of the management problem encountered in a given lake will depend to a large degree on its limnology and the fish assemblage present, there are several general classes of management problems that commonly occur. Organizing these problems into a logical scheme is useful in choosing among possible fisheries management prescriptions (Table 22.1). An analogous scheme can be developed for lake management problems (Table 22.2). Many of the fisheries and lake management problems share similar causes and management prescriptions. For example, excessive macrophyte growth can cause stunted populations of fishes as well as create problems for boaters and swimmers. The possible management prescriptions are there-

Table 22.2   Classes of lake management problems, their causes, and appropriate management prescriptions.

Lake management problems and causes	Management prescriptions
Poor water clarity	
Excessive algal biomass	Reduce nutrient inputs to lake
	Manipulate piscivorous fish stocks
Excessive turbidity (suspended sediments)	Reduce sediment inputs by reducing erosion (construction sites and agriculture)
Excessive macrophyte growth	
High nutrient levels	Reduce nutrient inputs to lake
	Reduce water levels to expose submerged macrophytes
	Harvest macrophytes (mechanical and chemical)
Invasion of nonnative macrophyte species	Harvest macrophytes (mechanical and chemical)
	Take preventative measures in lakes where nonnative species absent
Pollutants (in lake and biota, especially fishes)	
Point sources (industrial and municipal)	Reduce point and nonpoint inputs
Nonpoint sources (watershed and atmosphere)	Dredge lake bottom to remove contaminated sediments
	Post fish consumption advisories
Low dissolved oxygen concentrations	
High nutrient levels	Reduce nutrient inputs
	Install artificial aeration
Low water potability	
High nutrient levels	Reduce nutrient inputs
	Treat water

fore similar. The second point we want to emphasize is that sometimes the management prescriptions may be in conflict; the solution for a fishery management problem may exacerbate a lake management problem. For example, in very oligotrophic lakes, fishery productivity may be limited by nutrient inputs. Although the addition of phosphorus and nitrogen may increase fish yields, fertilization will likely result in greater algal abundance and lower lake water clarity. Our final point is that the complexity of lake and reservoir systems, and the specifics of their history, result in a great deal variability among these systems. As such, we need to take into account the individuality of each lake or reservoir.

22.6.3   Classes of Lake Fishery Management Problems

22.6.3.1   Absence of Desired Species

The absence of desired fish species is commonly a result of extremely harsh environmental conditions such as low oxygen concentrations or low lake pH. Winterkill and summerkill occur most frequently in shallow eutrophic lakes in which decomposition and algal respiration decrease oxygen. In stratified reservoirs, hypolimnetic oxygen is often depleted through the decomposition of organic matter associated with newly flooded land and allochthonous material brought in through tributary streams.

Coldwater species such as rainbow trout are particularly sensitive to oxygen deficits: they cannot tolerate low hypolimnetic oxygen concentrations, yet they are unable to use warm epilimnetic waters for prolonged periods of time. The thermal habitat of striped bass in southern reservoirs is similarly limited by low hypolimnetic oxygen. As a consequence, the viability of this species is significantly reduced in these reservoirs (Coutant 1987). Low oxygen levels may be mitigated by imposing restrictions on point and nonpoint sources of nutrients. A side benefit of this management scheme is reduction of algal blooms and possibly macrophyte growth, thus increasing lake desirability for swimming and boating users. In some cases nutrient reduction is not feasible, and more intensive measures will need to be taken. One such action is the installation of artificial aeration or destratification devices. These devices are successful in increasing oxygen levels but do incur substantial installation and operation costs, which may be prohibitive for larger lakes (Cooke et al. 1986). Destratification devices may also have the undesirable effect of reducing the volume of metalimnetic water available for coolwater fishes such as walleye.

Low pH is of international concern because of the wide geographic distribution of lakes that are candidates for acidification. In the long term, the solution to this problem lies in reducing emissions of SO_x and NO_x. Short-term solutions fall primarily into two categories: mitigative liming to increase buffering capacity and pH and development and stocking of acid-tolerant strains of fishes. Both of these measures are effective for individual lakes, but the scope and magnitude of acidification makes treating all affected lakes impractical. Associated with low lake pH are increased heavy metal lability (especially mercury) and decreased productivity of lower trophic levels. These concerns seriously limit the feasibility of using acid-tolerant fish species to mitigate lake acidification.

Inputs of contaminants from the air and watershed can also cause the loss of fisheries. The acute effects of a toxic chemical may be to reduce or eliminate fish populations. Coal mine tailings are one example of a source of toxins that can cause the direct loss of fish species in a lake or river. Contamination of fishes by such chemicals as DDT, PCB, dioxin, mirex, and toxaphene may also cause direct losses of fishes and fish-eating birds and mammals. Additionally, the presence of these contaminants in fish tissue may affect human health, thereby decreasing the value of these fisheries to the public. The effects of toxins may be more subtle than causing direct mortality on a fish species. For example, Bryan et al. (1995) found that juvenile bluegills showed reduced foraging efficiency and lower growth when exposed to cadmium.

Desired species may also be lacking from a lake if public demands are for game species other than the one(s) present. In fact, the definition of desired species and the relative rank of desirability is a regional phenomenon and is dependent on local fishing traditions. An example of a species with variable desirability is the yellow perch. Throughout the upper midwestern states, yellow perch are a desirable panfish and are sought after by many anglers throughout the year. However, in the Owasco Lake Basin (New York), the results of a survey of angler's species preferences rated yellow perch tied for seventh place, with only 6% of anglers preferring that species (Duttweiler 1976). The results of this study highlight the importance of knowing angler preferences when forming management prescriptions. When the current fish community doesn't contain desirable species, a common approach is to stock the desired species.

Spectacular fisheries have been developed using stocking, with the best-known cases probably being the salmonid fishery in the Great Lakes (Chapter 23) and Lake Ohae, South Dakota. Failures with stocking have also occurred. A classic example of the dangers inherent in introducing fish species is the introduction of peacock bass into Lake Gatun, Panama. The introduction of this desired sport fish significantly affected the overall ecosystem structure and function. Within 2 years, predation by this fish reduced the abundance of all other fishes by 99% and eventually eliminated all but one of the species present before introduction (Zaret 1979).

22.6.3.2   Low Abundance of Desired Species

Low fish abundance may occur for a variety of reasons, including inadequate recruitment, overfishing, or low system productivity. In natural lakes, poor recruitment may occur due to limited spawning habitat, extreme or suboptimal environmental conditions, or low parental stock. In reservoirs, recruitment problems are often intensified due to the inherent variability of physical and chemical factors, which often reduces spawning success and survival of young fish. For example, when water levels are maintained above normal summer stage, reproductive success of largemouth bass is enhanced, resulting in a strong year-class (Martin et al. 1981; Rainwater and Houser 1982; Miranda et al. 1984). When water levels are lower than normal, or shoreline vegetation is dewatered quickly, reproductive success of largemouth bass is reduced. Thus, one way to increase largemouth bass abundance is to maintain high summer water levels at 2–3 year intervals. High abundances of undesirable fishes may be managed in a similar fashion by water level manipulation during the appropriate time period (Chapter 15). These management prescriptions are contingent, however, on competing uses for reservoir water during the summer months (e.g., irrigation).

For some species, the lack of suitable reproductive sites has a technological solution. The construction of spawning reefs in Brevort Lake, Michigan, has greatly enhanced walleye reproduction and substantially increased walleye harvest. Another case is Smith Mountain Lake, Virginia, where largemouth bass were observed nesting near tire reefs (Prince et al. 1985). In addition to acting as spawning sites, artificial reefs may serve to concentrate diffuse fish populations, thereby increasing angler success. Production of food items is another potential benefit of artificial structures by providing substrate for periphyton and invertebrates. The success of these habitat devices is contingent upon proper structural design and placement, which necessitates consideration of the target species' ecological requirements.

Low lake productivity has been implicated as being the cause of low fish abundance in certain lakes. We feel that low productivity in itself is not usually the sole reason for low fish abundance; low-productivity lakes in northern Canada support substantial populations of Arctic char. Low abundance more often results from harvest rates that are in excess of system productivity. Where this is a problem, lake fertilization may increase basal fish production and result in greater harvest and higher standing stocks. This has been successfully applied in lakes in British Columbia where salmonid production has been significantly increased, resulting in substantial economic benefits. However, adding excessive nutrients may have negative side effects, including the occurrence of blue-green algal blooms, increased macrophyte growth, decreased water clarity, and decreased hypolimnetic oxygen concentration. Due to the cost and potential for negative side effects of lake fertilization, managers generally regulate fish harvest in relationship to lake productivity rather than increase lake productivity directly.

Recreational angling rarely endangers the persistence of fish species, but decreases in population abundance through overfishing have been observed. Such overexploitation can occur surprisingly quickly. Eagle Lake, a 176-ha lake in Acadia National Park, Maine, was stocked with brook trout to provide an ice fishery. Before the opening of the season, on 1 February 1977, there were an estimated 1,556 brook trout present in the lake. A thorough creel census of anglers indicated that a minimum of 1,344 of these fish were caught in 742 angler trips by 6 February (Havey and Locke 1980). Despite the near elimination of the stocked population, an additional 706 angling trips were taken during the winter season, yielding only 121 brook trout. Clearly these fish were highly vulnerable to anglers and despite drastically reduced catch rates were subjected to continued high fishing pressure. To reduce the impact of fishing on the population, regulations such as length limits or bag limits could have been imposed (Paragamian 1982). However, compliance with these regulations must be high (>80%) to achieve significant population improvement (Gigliotti and Taylor 1988).

22.6.3.3   Slow Growth of Desired Species

Slow fish growth is probably the most common fishery problem experienced and perhaps the most difficult to manage. Growth is highly plastic in fishes. Many factors determine growth rates, but the primary determinants are generally food and temperature. The amount of food available to a fish may depend on intraspecific competition, interspecific competition, lake productivity, and limnological characteristics of the lake such as turbidity. Food availability is further complicated by the presence of predators, which influences habitat choice of vulnerable size-classes of prey fishes. Habitats that provide protection against predators frequently have less available food (often because of food competition); thus, fishes using these habitats experience increased survival but at a cost of substantially reduced growth. For example, Werner et al. (1983) found that in weedy lakes young bluegills lived almost entirely in weed beds. Predation rates on young bluegills were low, but food resources in the weed beds became depleted. Use of weed beds led to relatively high survival of these bluegills, but their growth was poor.

To manage the growth of fish populations effectively, one must be able to identify the key factors limiting growth. Because of the diverse set of circumstances that can lead to poor growth (Table 22.1), a number of management options are available to managers to attack this problem. Where intraspecific competition or low lake productivity has led to slow growth, reductions in fish abundance have led to higher growth rates. The most desirable means of decreasing fish abundance is to increase predation pressure on the species in question. Yellow perch and bluegills are two species that show a tendency to grow very slowly or stunt when large predators are low in abundance. Increasing the appropriate predator populations by stocking can be an effective method of thinning the population and increasing growth. An appropriate predator must be large enough to consume the size prey available and be effective in consuming them in the habitats they use. As these predators are often game fishes, the manager should limit the number and size of predators harvested to allow them to control their prey species effectively.

When a balanced predator–prey system cannot be achieved by stocking and regulations, other methods of population reduction might be needed. These include removal of all or part of the population using chemicals (e.g., rotenone or antimycin) or mechanical (e.g., nets or electrofishing) means (Chapter 15). A successful application of fish removal through mechanical means was reported by Hanson et al. (1983). Removal of black crappie and black bullhead with fyke nets resulted in a higher abundance of large individuals through increased growth rates. The effectiveness of fish removal projects depends on maintaining balanced population density over time. Doing so is difficult because most fishes have a high reproductive potential, and their populations can quickly rebound after artificial thinning (Rutledge and Barron 1972).

Interspecific competition may result in the depletion of food resources and reduced growth of desired species. Reducing the abundance of competing species, especially if they are considered socially undesirable (Chapter 15), has led to enhanced sport fish populations. An example of a fish removal project that has been successful is the removal of white suckers to improve yellow perch fisheries. In Big Bear Lake, Michigan, yellow perch harvest increased from 500 per year to over 12,000 per year after a sustained white sucker removal project (Schneider and Crowe 1980). Similarly, Johnson (1977) observed higher recruitment and growth of yellow perch after reducing white sucker abundance in Wilson Lake, Minnesota. However, not all removals have resulted in improved game fish populations (Holey et al. 1979).

When food resources are limited, an intuitive solution is to provide more food by introducing additional prey species. Theoretically, this adds linkages to the food web present and increases system utilization and retention of primary production. Furthermore, production can be channeled toward desired game fishes. In practice, however, complications often arise from introductions of forage species. Results of stocking gizzard shad and threadfin shad as prey for game species in southern reservoirs provide examples of both successes and failures in forage fish management. The addition of these fishes has benefited largemouth bass, walleye, and striped bass in many of these reservoirs (Noble 1981). However, in systems in which predation pressure is insufficient to control numbers of gizzard shad, this species has a tendency to overpopulate at sizes too large for predators to consume. Where this occurs, condition and reproductive success of largemouth bass and centrarchids may suffer (Noble 1981). Threadfin shad, which do not grow as large as gizzard shad, are often perceived as being a better forage species. Despite this advantage, threadfin shad may compete with young game fishes, causing decreased growth and survival (Noble 1981). Thus, the manager needs to evaluate the overall effect of forage introductions on the fish community to ensure that the expected benefits will outweigh any negative interactions.

Food availability is not always the primary factor limiting growth rates. Temperature significantly affects growth, survival, and habitat choice of fishes. The way that the temperature regime influences growth, however, is a function of the genetic composition of the population. For example, the northern subspecies of largemouth bass grows poorly at high temperatures, whereas the Florida subspecies grows rapidly at these warmer temperatures (Philipp et al. 1981). Consequently, by selecting fish species or strains whose growth optima match the thermal regime in a lake, managers can improve the growth efficiency and productivity of the fishery.

To assess the effects of temperature and food availability on fish consumption and growth, bioenergetics models based on the physiology of different fish species have been developed (Kitchell et al. 1977). These models have been used primarily in two ways: (1) to estimate growth from feeding rate and temperature and (2) to estimate feeding rate from observed growth and temperature. This information can then be used by the manager to identify which prey are limiting fish growth and determine the extent that temperature influences growth. When using bioenergetics models, managers should note that the genetic composition of a fish strain affects the growth–temperature response and thermal tolerance portions of these models. As such, bioenergetics models whose parameters are obtained from one strain of fish may not accurately predict the growth of other strains. Accordingly, managers may need to parameterize the model with data from the strain of fish they are managing.

22.6.3.4 Poor Population Size Structure

Management practices to improve population size structure typically focus on balancing predator–prey relationships (Chapter 21). It should be noted here that the "desirable" population size structure varies by species and by region. For example, a yellow perch population with a majority of fish in the 225–300 mm total length size range would be considered highly desirable, whereas a walleye population with the same size structure would not. One of the reasons for poor population size structure is slow growth, and the management prescriptions discussed earlier would be appropriate for fishery improvement. However not all slow-growing populations exhibit a poor population size structure. Even where growth rates are low, sizeable numbers of large fish can be maintained in a population if mortality rates are low enough. Examples of this are found in unexploited lake whitefish and Arctic char populations in Canada (Johnson 1976). Conversely, good growth rates do not ensure abundant large fish because excessive exploitation can reduce the numbers of individuals above the desired size. Thus, the size structure of a fish population results from the interaction between reproduction, growth, and survival rates. When growth rates are low and mortality on recruited individuals high, poor population size structure is virtually guaranteed.

Where macrophytes are overly abundant, undesirable size structure often results for game fishes. This is especially true for panfishes such as bluegill, which generally live in macrophyte beds, presumably to escape predation. Where fish are crowded in weed beds, intraspecific competition for food is often severe and growth is poor, even though ample food resources for growth are available in the limnetic zone of the lake (Werner et al. 1983). Additionally, growth of predators is depressed because they cannot effectively forage where dense weeds are present (Savino and Stein 1982).

Aquatic weeds are often removed by mechanical or chemical means. However, if the limnological conditions resulting in heavy weed growth (i.e., eutrophication) are not treated, macrophytes may quickly return. Biological control of macrophytes with grass carp has been successfully used in some areas of the United States, but the importation of this species is illegal in many states due to concerns regarding their compatibility with other fishes and nonfish prey species.

A specialized aspect of population size structure management occurs in trophy fisheries. Generally all anglers want to catch big fish, but they may also want to be able to keep some fish. It is very difficult to meet these two objectives in a single lake because of the reduction in the abundance of trophy size fish that occurs where significant harvesting takes place. One way to provide both of these opportunities is to manage some lakes for trophy fisheries and other lakes for fish harvest. Anglers can then choose the type of fishing experience they desire most, and no single fishery is pushed to produce conflicting products. An alternative is to impose a slot length limit (Chapter 17), which allows harvest of some fish while protecting larger size-classes, and thereby increase the abundance of quality size fish. Maintenance of large numbers of trophy size fish usually requires low mortality rates, and consequently bag limits are usually highly restrictive.

22.7   FUTURE DIRECTIONS

Technological advances will continue to improve the way we acquire, analyze, and deliver information. The worldwide web, for example, provides access to data and information more quickly and less expensively than does paper communications, but there can be real concerns over the reliability of information provided. Perhaps more importantly, technological advances shape our thinking, allowing us to ask different questions than we did in the past. Such questions form the basis for new areas of scientific inquiry, that will someday generate new approaches and solutions for fisheries and lake management. An example is the emerging trend of taking a spatial, or landscape, approach to natural resource management. The accessibility of technologies such as geographic information systems and satellite imagery has facilitated and encouraged a landscape approach. With these technologies, fisheries managers are better able to place lake ecosystems in the context of the lake's watershed, emphasizing the linkage between the "aquascape" and the landscape. Another emerging trend is the use of adaptive management concepts as applied to whole-ecosystem management. The premise of adaptive management is to view management actions as experiments by which we learn how a system functions; by wisely choosing the timing and site selection for management, we can improve our management in the future (Walters 1986). Linked with new technology, adaptive management has the potential to improve greatly our knowledge of how lake fisheries respond to management actions and which management strategies work best for a particular situation.

Among the greatest conceptual challenges facing fisheries science is how to deal with fish habitat in lakes. In this chapter, we have espoused a view that habitat encompasses all the physical, chemical, and biological characteristics of the environment. We feel that partitioning abiotic from biotic factors is counterproductive to the goal of understanding how the environment influences fish production. In the future, we need to develop a clearer understanding of how the abiotic environment affects individual species and further how the environment shapes biotic interactions between species. Exploring the relationship between abiotic and biotic factors in essence represents a reemphasis on the fundamental question of ecology, what factors limit the distribution and abundance of organisms?

One of the most daunting problems facing fisheries managers is the reality that lakes are truly a multiple-use resource. As such, a continuing need in the future will be a strong commitment to conflict resolution and consensus building. As part of this process, fisheries managers will play a critical role in helping to manage society's expectations of the resource. Too often the constituents of a multiple-use resource have expectations for their interest area that exceed the capability of the system, especially given other demands. One of the ways that we foresee managing user's expectations is through their direct involvement in all phases of lake management. It is already commonplace for lake associations, township boards, or other similar organizations to take a primary role in data collection and management for limnological problems (e.g., many lake associations routinely monitor water clarity and have active programs to help with nutrient abatement). Lake fisheries management in the future will increasingly take a similar self-help approach, with the public being increasingly involved in fish sampling, habitat monitoring, and management actions. Beyond the value of this involvement in developing realistic expectation, directly engaging the public in all phases of management will foster a greater sense of stewardship. Increasing the level of stewardship among all lake users will be the most beneficial action that can be taken.

22.8 REFERENCES

Aggus, L. R. 1979. Effects of weather on freshwater fish predator-prey dynamics. Pages 47-56 *in* H. Clepper, editor. Predator-prey systems in fisheries management. Sport Fishing Institute, Washington, D.C.

Anderson, R. O. 1976. Management of small warmwater impoundments. Fisheries 1(6):5–7.

Bachmann, R. W., and five coauthors. 1996. Relations between trophic state indicators and fish in Florida (U.S.A.) lakes. Canadian Journal of Fisheries and Aquatic Sciences 53:842–855.

Barber, W. E., and J. N. Taylor. 1990. The importance of goals, objectives, and values in the fisheries management process and organization: a review. North American Journal of Fisheries Management 10:365–373.

Brooks, J. L., and S. I. Dodson. 1965. Predation, body size, and composition of plankton. Science 150:28–35.

Bryan, M. D., G. J. Atchison, and M. B. Sandheinrich. 1995. Effects of cadmium on the foraging behavior and growth of juvenile bluegill, *Lepomis macrochirus*. Canadian Journal of Fisheries and Aquatic Sciences 52:1630–1638.

Brylinsky, M., and K. H. Mann. 1973. An analysis of factors governing productivity in lakes and reservoirs. Limnology and Oceanography 18:1–14.

Carpenter, S. R., J. F. Kitchell, and J. R. Hodgson. 1985. Cascading trophic interactions and lake productivity. BioScience 35:634–639.

Clarke, K. D., R. Knoechel, and P. M. Ryan. 1997. Influence of trophic role and life-cycle duration on timing and magnitude of benthic macroinvertebrate response to whole-lake enrichment. Canadian Journal of Fisheries and Aquatic Sciences 54:89–95.

Cooke, G. D., E. B. Welsh, S. A. Peterson, and P. R. Newroth. 1986. Lake and reservoir restoration. Buttersorst Publishing Co., Boston.

Coutant, C. C. 1987. Poor reproductive success of striped bass from a reservoir with reduced summer habitat. Transactions of the American Fisheries Society 116:154–160.

Cuker, B. E., M. E. McDonald, and S. C. Mozley. 1992. Influences of slimy sculpin (*Cottus cognatus*) predation on the rocky littoral invertebrate community in an arctic lake. Hydrobiologia 240:83–90.

Diehl, S. 1995. Direct and indirect effects of omnivory in a littoral lake community. Ecology 76:1727–1740.

Downing, J. A., and C. Plante. 1993. Production of fish populations in lakes. Canadian Journal of Fisheries and Aquatic Sciences 50:110–120.

Duttweiler, M. W. 1976. Use of questionnaire surveys in forming fishery management policy. Transactions of the American Fisheries Society 105:232–238.

Elser, J. J., E. R. Marzolf, and C. R. Goldman. 1990. Phosphorus and nitrogen limitation of phytoplankton growth in the freshwaters of North America: a review and critique of experimental enrichments. Canadian Journal of Fisheries and Aquatic Sciences 47:1468–1477.

Fast, A. W. 1993. Distributions of rainbow trout, largemouth bass and threadfin shad in Lake Casitas, California, with artificial aeration. California Fish and Game 79:13–27.

Fry, J. P, and W. D. Hanson. 1968. Lake Taneycomo: a cold-water reservoir in Missouri. Transactions of the American Fisheries Society 97:138–145.

Galbraith, M. G., Jr. 1967. Size-selective predation on *Daphnia* by rainbow trout and yellow perch. Transactions of the American Fisheries Society 96:1–10.

Gigliotti, L. M., and W. W. Taylor. 1988. The effect of illegal harvest on recreational fisheries. North American Journal of Fisheries Management 10:106–110.

Gross, H. P., W. A. Wurtsbaugh, C. Luecke, and P. Budy. 1997. Fertilization of an oligotrophic lake with a deep chlorophyll maximum: predicting the effect on primary production. Canadian Journal of Fisheries and Aquatic Sciences 54:1177–1189.

Hanson, D. A., B. J. Belonger, and D. L. Schoenike. 1983. Evaluation of a mechanical population reduction of black crappie and black bullheads in a small Wisconsin Lake. North American Journal of Fisheries Management 3:41–47.

Hanson, J. M., and W. C. Leggett. 1982. Empirical prediction of fish biomass and yield. Canadian Journal of Fisheries and Aquatic Sciences 39:257–263.

Havey, K. A., and D. O. Locke. 1980. Rapid exploitation of hatchery-reared brook trout by ice fishermen in a Maine lake. Transactions of the American Fisheries Society 109:282–286.

Hayes, D. B. 1990. Competition between white sucker (*Catostomus commersoni*) and yellow perch (*Perca flavescens*): results of a whole-lake manipulation. Michigan Department of Natural Resources Fish Division, Fisheries Research Report 1972, Ann Arbor.

Hayes, D. B., W. W. Taylor, and J. C. Schneider. 1992. Response of yellow perch and the benthic invertebrate community to a reduction in the abundance of white suckers. Transactions of the American Fisheries Society 121:36–53.

Hayward, R. S., and M. J. Van den Avyle. 1986. The nature of zooplankton spatial heterogeneity in a nonriverine impoundment. Hydrobiologia 131:261–271.

Hodgson, J. R., C. J. Hodgson, and S. M. Brooks. 1991. Trophic interactions and competition between largemouth bass (*Micropterus salmoides*) and rainbow trout (*Oncorhynchus mykiss*) in a manipulated lake. Canadian Journal of Fisheries and Aquatic Sciences 48:1704–1712.

Holey, M, and six coauthors. 1979. Never give a sucker an even break. Fisheries 4(1):2–6.

Hudson, P. L., R. W. Griffiths, and T. J. Wheaton. 1992. Review of habitat classification schemes appropriate to streams, rivers, and connecting channels in the Great Lakes drainage basin. Pages 73–107 *in* W. D. N. Busch and P. G. Sly, editors. The development of an aquatic habitat classification system for lakes. CRC Press, Ann Arbor, Michigan.

Johnson, B. M., and S. R. Carpenter. 1994. Functional and numerical responses: a framework for fish-angler interactions? Ecological Applications 4:808–821.

Johnson, D. M., T. H. Martin, P. H. Crowley, and L. B. Crowder. 1996. Link strength in lake littoral food webs: net effects of small sunfish and larval dragonflies. Journal of the North American Benthological Society 15:271–288.

Johnson, F. H. 1977. Responses of walleye (*Stizostedion vitreum vitreum*) and yellow perch (*Perca flavescens*) populations to removal of white sucker (*Catostomus commersoni*) from a Minnesota lake, 1966. Journal of the Fisheries Research Board of Canada 34:1633–1642.

Johnson, L. 1976. Ecology of arctic populations of lake trout *Salvelinus namaycush*, lake whitefish, *Coregonus clupeaformis*, arctic char, *S. alpinus*, and associated species in unexploited lakes of the Canadian Northwest Territories. Journal of the Fisheries Research Board of Canada 33:2459–2488.

Jones, J. R., and M. V. Hoyer. 1982. Sportfish harvest predicted by summer chlorophyll *a* concentration in midwestern lakes and reservoirs. Transactions of the American Fisheries Society 111:176–179.

Kimmel, G. L., and A. W. Groeger. 1986. Limnological and ecological changes associated with reservoir aging. Pages 103–109 in G. E. Hall and M. J. Van Den Avyle, editors. Reservoir fisheries management: strategies for the 80's. American Fisheries Society, Southern Division, Reservoir Committee, Bethesda, Maryland.

Kitchell, J. F., D. J. Stewart, and D. Weininger. 1977. Applications of a bioenergetic model to yellow perch (Perca flavescens) and walleye (Stizostedion vitreum vitreum). Journal of the Fisheries Research Board of Canada 34:1922–1935.

Luecke, C., W. A. Wurtsbaugh, P. Budy, H. P. Gross, and G. Steinhart. 1996. Simulated growth and production of endangered Snake River sockeye salmon: assessing management strategies for the nursery lakes. Fisheries 21(6):18–23.

Martin, D. B., L. J. Mengel, J. F. Novotny, and C. H. Walburg. 1981. Spring and summer water levels in a Missouri River reservoir: effects on age-0 fish and zooplankton. Transactions of the American Fisheries Society 110:370–381.

McQueen, D. J., J. R. Post, and E. L. Mills. 1986. Trophic relationships in freshwater pelagic ecosystems. Canadian Journal of Fisheries and Aquatic Sciences 43:1571–1581.

Mills, E. L., and J. L. Forney. 1988. Trophic dynamics and development of freshwater pelagic food webs. Pages 11–30 in S. R. Carpenter, editor. Complex interactions in lake communities. Springer-Verlag, New York.

Mills, E. L., D. M. Green, and A. Schiavone. 1987. Use of zooplankton size to assess the community structure of fish populations in freshwater lakes. North American Journal of Fisheries Management 7:369–378.

Mills, E. L., J. H. Leach, J. T. Carlton, and C. L. Secor. 1993. Exotic species in the Great Lakes: a history of biotic crises and anthropogenic introductions. Journal of Great Lakes Research 19:1–54.

Mills, E. L., and A. Schiavone, Jr. 1982. Evaluation of fish communities through assessment of zooplankton populations and measures of lake productivity. North American Journal of Fisheries Management 2:14–27.

Miranda, L. E., W. L. Shelton, and T. D. Bryce. 1984. Effects of water level manipulation on abundance, mortality, and growth of young-of-the-year largemouth bass in West Point Reservoir, Alabama–Georgia. North American Journal of Fisheries Management 4:314–320.

Moyle, P. B., and J. J. Cech, Jr. 1988. Fishes: an introduction to ichthyology, 2nd edition. Prentice Hall, Englewood Cliffs, New Jersey.

NRC (National Research Council). 1992. Restoration of aquatic ecosystems. National Academy Press, Washington, D.C.

Noble, R. L. 1980. Management of lakes, reservoirs, and ponds. Pages 265–295 in R. T. Lackey and L. A. Nielsen. editors. Fisheries management. Blackwell Scientific Publications, Oxford, UK.

Noble, R. L. 1981. Management of forage fishes in impoundments of the southern United States. Transactions of the American Fisheries Society 110:738–750.

Oglesby, R. T. 1977. Relationships of fish yield to lake phytoplankton standing crop, production and morphoedaphic factors. Journal of the Fisheries Research Board of Canada 34:2271–2279.

Paragamian, V. L. 1982. Catch rates and harvest results under a 14.0-inch minimum length limit for largemouth bass in a new Iowa impoundment. North American Journal of Fisheries Management 2:224–231.

Philipp, D. P., W. F. Childers, and G. S. White. 1981. Management implications for different genetic stocks of largemouth bass (Micropterus salmoides) in the United States. Canadian Journal of Fisheries and Aquatic Sciences 38:1715–1723.

Pierce, C. L., and B. D. Hinrichs. 1997. Response of littoral invertebrates to reduction of fish density: simultaneous experiments in ponds with different fish assemblages. Freshwater Biology 37:397–408.

Post, J. R., and D. Cucin. 1984. Changes in the benthic community of a small precambrian lake following the introduction of yellow perch, Perca flavescens. Canadian Journal of Fisheries and Aquatic Sciences 41:1496–1501.

Prince, E. D., O. E. Maughan, and P. Brouha. 1985. Summary and update of the Smith Mountain Lake artificial reef project. Pages 401-430 in F. M. D'Itri, editor. Artificial reefs—marine and freshwater applications. Lewis Publishers, Chelsea, Michigan.

Rainwater, W. C., and A. Houser. 1982. Species composition and biomass of fish in selected coves in Beaver Lake, Arkansas, during the first 18 years of impoundment (1963–1980). North American Journal of Fisheries Management 4:316–325.

Rutledge, W. P., and J. C. Barron. 1972. The effects of the removal of stunted white crappie on the remaining crappie populations of Meridian State Park Lake. Texas Parks and Wildlife Department, Technical Series 12, Austin.

Ryder, R. A. 1982. The morphoedaphic index-use, abuse, and fundamental concepts. Transactions of the American Fisheries Society 111:154–164.

Savino, J. F., and R. A. Stein. 1982. Predator-prey interactions between largemouth bass and bluegills as influenced by simulated, submerged vegetation. Transactions of the American Fisheries Society 111:255–266.

Schindler, D. E., and six coauthors. 1996. Food web structure and littoral zone coupling to pelagic trophic cascades. Pages 96–105 in G. A. Polis and K. O. Winemiller, editors. Food webs: integration of patterns and dynamics. Chapman and Hall, New York.

Schneider, J. C., and W. R. Crowe. 1980. Effect of sucker removal on fish and fishing at Big Bear Lake. Michigan Department of Natural Resources, Institute for Fisheries Research Report 1887, Ann Arbor.

Sheldon, R. W., A. Prakash, and W. H. Sutcliffe, Jr. 1972. The size distribution of particles in the ocean. Limnology and Oceanography 17:327–340.

Sheldon, R. W., W. H. Sutcliffe, Jr., and A. Prakash. 1973. The production of particles in the surface waters of the ocean with particular reference to the Sargasso Sea. Limnology and Oceanography 18:719–733.

Spencer, C. N., and D. L. King. 1984. Role of fish in regulation of plant and animal communities in eutrophic ponds. Canadian Journal of Fisheries and Aquatic Sciences 41:1851–1855.

Sprules, W. G., J. M. Casselman, and B. J. Shuter. 1983. Size distribution of pelagic particles in lakes. Canadian Journal of Fisheries and Aquatic Sciences 40:1761–1769.

Stein, R. A., D. R. DeVries, and J. M. Dettmers. 1995. Food-web regulation by a planktivore: exploring the generality of the trophic cascade hypothesis. Canadian Journal of Fisheries and Aquatic Sciences 52:2518–2526.

Stockner, J. G. 1987. Lake fertilization: the enrichment cycle and lake sockeye salmon production. Pages 198–215 in H. D. Smith, L. Margolis, and C. C. Woods, editors. Sockeye salmon population biology and future management. Canadian Special Publication of Fisheries and Aquatic Sciences 96.

Swingle, H. S. 1950. Relationships and dynamics of balanced and unbalanced fish populations. Alabama Agricultural Experiment Station, Auburn University, Bulletin 274:1-74.

Tonn, W. M. 1990. Climate change and fish communities: a conceptual framework. Transactions of the American Fisheries Society 119:337–352.

Vanni, M. J. 1996. Nutrient transport and recycling by consumers in lake food webs: implications for algal communities. Pages 81–95 in G. A. Polis and K. O. Winemiller, editors. Food webs: integration of patterns and dynamics. Chapman and Hall, New York.

Vanni, M. J., and P. C. de Ruiter. 1996. Detritus and nutrients in food webs. Pages 25–29 in G. A. Polis and K. O. Winemiller, editors. Food webs: integration of patterns and dynamics. Chapman and Hall, New York.

Vanni, M. J., and C. D. Layne. 1997. Nutrient recycling and herbivory as mechanisms in the "top-down" effect of fish on algae in lakes. Ecology 78:21–40.

Von Geldern, C. E., Jr. 1971. Abundance and distribution of fingerling largemouth bass, Micropterus salmoides, as determined by electro-fishing at Lake Nacimiento, California. California Fish and Game 57:228–245.

Wagner, K. J., and R. T. Oglesby. 1984. Incompatibility of common lake management objectives. Pages 97–100 in Lake and reservoir management. U.S. Environmental Protection Agency, 440/5/83-001, Washington, D.C.

Walters, C. 1986. Adaptive management of renewable resources. Macmillan, New York.

Wang, L., J. Lyons, P. Kanehl, and R. Gatti. 1997. Influences of watershed land use on habitat quality, and biotic integrity in Wisconsin streams. Fisheries 22(6):6–12.

Weaver, M. J., J. J. Magnuson, and M. K. Clayton. 1996. Habitat heterogeneity and fish community structure: inferences from north temperate lakes. Pages 335–346 *in* L. E. Miranda and D. R. DeVries, editors. Multidimensional approaches to reservoir fisheries management. American Fisheries Society, Symposium 16, Bethesda, Maryland.

Wege, G. J., and R. O. Anderson. 1978. Relative weight (W_r): a new index of condition for largemouth bass. Pages 79–91 *in* G. D. Novinger and J. G. Dillard, editors. New approaches to the management of small impoundments. American Fisheries Society, North Central Division, Special Publication 5, Bethesda, Maryland.

Werner, E. E., J. F. Gilliam, D. J. Hall, and G. G. Mittelbach. 1983. An experimental test of the effects of predation risk on habitat use in fish. Ecology 64:1540–1548.

Wetzel, R. G. 1975. Limnology. Saunders, Philadelphia.

Whittier, T. R., D. B. Halliwell, and S. G. Paulsen. 1997. Cyprinid distributions in Northeast U.S.A. lakes: evidence of regional-scale minnow biodiversity losses. Canadian Journal of Fisheries and Aquatic Sciences 54:1593–1607.

Willis, D. W. 1987. Reproduction and recruitment of gizzard shad in Kansas reservoirs. North American Journal of Fisheries Management 7:71–80.

Zaret, T. M. 1979. Predation in freshwater fish communities. Pages 135–143 *in* H. Clepper, editor. Predator–prey systems in fisheries management. Sport Fishing Institute, Washington, D.C.

Chapter 23

Great Lakes Fisheries

DAVID J. JUDE AND JOSEPH LEACH

23.1 INTRODUCTION

The Laurentian Great Lakes are the world's largest freshwater system and contain the world's largest and most valuable freshwater fisheries resources. There are 38 million people in the Great Lakes basin and 25 million people obtain drinking water from the lakes. Because the lakes are distributed over such a large area, individual lakes in many cases differ markedly in climatic, physical, and chemical characteristics and consequently in productivity and fish species diversity (Figure 23.1, Table 23.1).

Much has been written about changes in water quality and biota, particularly fish stocks, in the Great Lakes. Early accounts of explorers present a picture of crystal-clear water and abundant fishes. Historical changes in water quality and biota of the Great Lakes are related to cultural development. Increasing environmental degradation paralleled human progression through exploitation of fur, timber, agricultural lands, minerals, and energy resources. The degree of environmental alteration to the lakes from stresses related to cultural development has been and continues to be substantial (Table 23.2).

Smith (1970a) attributed deterioration of fish habitat and stocks partly to piecemeal management resulting from the division of the jurisdiction of the Great Lakes among two nations, eight states, and one province. Because management approaches have not been uniform throughout the lakes, waste disposal, fishing, and nonnative species introductions have been largely uncontrolled.

Recent initiatives in Great Lakes fisheries science have been directed toward rehabilitation. Smith (1970a) considered that the mechanisms for rehabilitation of Great Lakes fisheries had been put in place with ratification of the 1955 Convention of Great Lakes Fisheries and formation of the Great Lakes Fishery Commission (GLFC) (Eshenroder 1987). The GLFC undertook the tasks of sea lamprey control (see section 23.4.5) and coordination of research and regulatory efforts of management agencies. Early problems with water quality led to formation of the International Joint Commission (IJC) under the Boundary Waters Treaty of 1909 (Fetterolf 1988). Water quality degradation in the 1950s and 1960s, particularly in the lower Great Lakes, spurred renewal of the IJC's activities.

Many changes have occurred in the Great Lakes fisheries since 1970. There have been increases in the amount of information accruing from research and assessment efforts, changes in public perceptions and expectations of ecosystem quality in the basin, changes in attitudes of the public toward allocation of fisheries resources, and

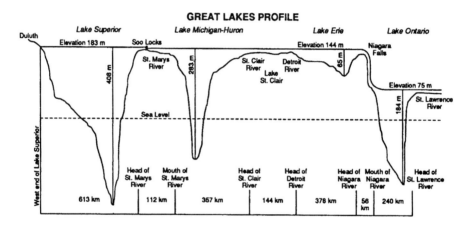

Figure 23.1   Map and profile of the Great Lakes. Figure adapted from *Michigan Department of Natural Resources Magazine*.

reorganization of institutional arrangements. A series of international symposia, supported largely by the GLFC, focused international knowledge on several key areas important to Great Lakes management (Loftus and Regier 1972; Colby 1977; Berst and Simon 1981; Spangler et al. 1987). Reports of the U.S. Federal Water Pollution Control Administration (FWPCA 1968), the IJC, and the Canadian Department of the Environment (Vallentyne 1974) alerted the public and governments to the seriousness of pollution problems in the Great Lakes and the need for remedial efforts. In 1972 and 1978, water quality agreements were signed between the United States and Canada. The 1978 agreement incorporated an ecosystem approach toward Great Lakes management, a concept that was also supported by the GLFC.

Table 23.1 Climatic, morphometric, and limnological characteristics of the Great Lakes. Sources are Ryder (1972); Christie (1974); Dobson et al. (1974); Oglesby (1977); Matuszek (1978); and Herdendorf (1982).

	Lake				
Parameter	Superior	Michigan	Huron	Erie	Ontario
Mean July temperature (°C)[a]	16–18	18–21	18–24	21–24	21–24
Growing season (days)	140–160	160–200	160–240	220–240	200–220
Lake area (km²)	82,100	57,750	59,500	25,657	19,000
Watershed area (km²)	127,700	118,100	133,900	58,800	70,000
Mean depth (m)	149	85	59	18.5	86
Maximum depth (m)	407	282	229	64	245
Volume (km³)	12,230	4,920	3,537	483	1,637
Mean total dissolved solids (mg/L)	60	134	108	146	148
Mean total phosphorus (mg/m)	3.3	6.6	4.4	17.8	19.4
Mean chlorophyll *a* (mg/m)	1.0	2.1	1.0	5.0	3.8
MEI[b]	0.40	1.58	1.83	7.89	1.72
Number of native fish species	67	114	99	114	112

[a] Range represents means for cool to hot summers.
[b] Morphoedaphic index (MEI) = total dissolved solids ÷ mean depth.

In 1980 a major step in cooperation among fisheries agencies occurred with the signing of the Joint Strategic Plan for Management of Great Lakes Fisheries by all parties with jurisdiction for management. Continuing efforts in the 1990s to create lake management plans for each of the Great Lakes (Governments of the United States and Canada 1988a; Smith and Coape-Arnold 1995) and to illuminate the problems and impediments to rehabilitation in the nearshore zone (Governments of the United States and Canada 1997) signal a strong effort to maintain progress in implementing the ecosystem approach in all the Great Lakes. Clearly, the mechanisms for rehabilitation of Great Lakes fisheries have been enhanced and expanded in the past 3 decades.

In this chapter, we outline changes in fish communities of the Great Lakes, particularly since 1970, and provide explanations as to their most likely causes. Rehabilitation efforts and the responses of management agencies to resource crises are consid-

Table 23.2 Degree of environmental alteration of the Great Lakes from cultural development. Table adapted from Ryder (1972).

	Lake				
Perturbation	Superior	Michigan	Huron	Erie	Ontario
Fishing	Intensive	Intensive	Intensive	Intensive	Intensive
Nonnative fish introductions	Many	Many	Many	Many	Many
Pollution	Moderate	Moderate	Extensive	Extensive	Extensive
Other factors[a]	Many	Many	Many	Many	Many

[a] Damming, water diversions, canals, and shoreline restructuring.

ered in detail, and the obvious successes are highlighted. Finally, the major challenges that fisheries managers now face are discussed, and strategies for avoiding future crises are suggested.

23.2   OVERVIEW OF THE GREAT LAKES FISHERIES

23.2.1 Native and Introduced Fish Populations

The original fish species assemblage was similar in Lakes Superior, Michigan, Huron, and Ontario (salmonine complex) but markedly different in Lake Erie (percid complex). The most abundant fishes in Lake Erie were lake whitefish, lake herring, lake sturgeon, blue pike, walleye, sauger, yellow perch, freshwater drum, and channel catfish (Leach and Nepszy 1976). Remaining lakes contained lake trout, lake whitefish, lake sturgeon, lake herring, several species of deepwater ciscoes, burbot, deepwater sculpin, slimy sculpin, emerald shiner, and (in Lake Ontario only) Atlantic salmon. Green Bay of Lake Michigan and Saginaw Bay of Lake Huron also contained large populations of walleye and yellow perch (see review papers in Loftus and Regier 1972). In Lakes Superior, Michigan, and Huron, top predators were lake trout and burbot, of which lake trout were the more abundant; prey were mainly coregonines, consisting of perhaps 11 species in Lake Michigan and somewhat fewer in Lakes Huron and Superior (Koelz 1929), and sculpins. Lake trout became the dominant predator in Lake Ontario after 1850, when Atlantic salmon disappeared from that lake (Christie 1974). Top predators in Lake Erie were walleye and blue pike, except that lake trout may have been the dominant predator of the limnetic zone of eastern Lake Erie, at least until about 1890 (Regier and Hartman 1973). Primary prey were probably yellow perch and freshwater drum. After settlers arrived, nearly all but the smallest native species became heavily fished. Perhaps the most valuable as human food have been lake trout, lake whitefish, walleye, and blue pike.

Most native fish populations in the Great Lakes suffered dramatic declines between the mid-1800s and the mid-1900s (see Loftus and Regier 1972). For example, lake sturgeon had become scarce in all the lakes by 1900, and by the 1950s lake trout were extirpated in Lakes Michigan, Ontario, and Erie, nearly extirpated in Lake Huron, and at low levels of abundance in Lake Superior. Lake herring collapsed in Lake Erie in the 1920s, followed by declines in lake whitefish, blue pike, and walleye. In all the lakes, declines in key predators, such as lake trout, resulted in an explosion of prey species and a constantly changing assemblage of fishes.

One result of the changes outlined here was the emergence of a large biomass of nonnative rainbow smelt and alewife, which provided the opportunity to stock nonnative salmon and trout. Consequently, fish assemblages now largely consist of pelagic species, a condition that probably did not exist 200 years ago when nearshore littoral communities were in dynamic equilibrium with offshore pelagic communities (Regier 1979). Small, fecund pelagic species (alewife and rainbow smelt) now predominate over the larger species that were originally more closely associated with the benthic and coastal habitats of the lakes. Pacific salmon (coho, chinook, pink, and sockeye) have been introduced into all the lakes (Mills et al. 1993a). Self-sustaining populations of pink salmon exist in all the Great Lakes, and remnant stocks of reproducing sockeye salmon persist in Lake Huron. White perch has recently invaded and prospered in Lakes Erie and Huron. In spite of major changes in species composition of the Lake Erie fish assemblages, fish yield (total biomass) has remained approximately the same as it was before the major cultural stresses (Regier and Hartman 1973). The present identity of most of the Great Lakes fish stocks is but faintly recognizable as being derived from the diverse communities of 200 years ago.

23.2.2 Factors Causing Declines in Native Fishes

Changes in native Great Lakes fish stocks have been attributed to overfishing, degradation of habitat, and invasion of nonnative fish species, particularly the sea lamprey, alewife, and rainbow smelt from the Atlantic Ocean (Table 23.3). Lake Ontario was probably the earliest of the lakes to be affected, followed by Lakes Erie, Huron, Michigan, and Superior. Lake Superior has had far less drastic changes than have the other lakes.

Overfishing probably began during the mid-1800s, as European-American settlement of the Great Lakes basin and the consequent demand for fresh fish rapidly increased. Greater exploitation of stocks resulted not only from greater commercial fishing effort but also from more efficient gear. As local stocks were depleted, fishers moved farther away from home and continued the process. A good example of this pattern of overfishing is the history of the lake trout. During early periods of exploitation there existed many ecologically and genetically subspecific populations of lake trout (Brown et al. 1981; Dehring et al. 1981). The most accessible—those of greatest commercial value—were depleted first, often to extinction, before less accessible stocks were affected. Although commercial harvest figures remained fairly constant for a time, overall populations declined as the different subpopulations were successively depleted. Another prime example of fish that were over-exploited were Lake Michigan deepwater ciscoes, which originally included seven species that differed in size at maturity. Use of decreasing mesh sizes as stocks were depleted sequentially decimated the young of the larger species (Smith 1964). This process, coupled with the even more devastating sea lamprey predation (Coble et al. 1990), eventually led to extirpation of all but the smallest species; highly prized species were all overfished in the Great Lakes.

Loss and degradation of habitat in the Great Lakes began soon after settlers arrived and has become widespread. The consequences of habitat alteration, that is, deforestation of watersheds, were demonstrated early in Lake Ontario, where increased water temperature and sedimentation in spawning streams led to extinction of Atlantic salmon by the mid-1800s (Christie 1974). By the late 1800s, many tributary streams and estuaries of Lakes Superior, Michigan, and Huron were useless for spawning due to contamination by sawdust and other sawmill wastes. Habitat degradation has also resulted from agricultural runoff (causing excessive enrichment of waters), damming of spawning streams, pollution from industrial and domestic sources, and filling of marshes that had been used as spawning and nursery grounds. About 60% of the 39 km^2 of marsh originally lining the nearshore areas of Green Bay has been lost to commercial development (Whillans 1982). These watershed changes have set into motion the irreversible changes that precipitated the calamitous disruptions of Great Lakes ecosystems.

Table 23.3 Summary of stresses affecting major fish stocks in the Great Lakes. Given are the approximate dates of population declines: current status of native stocks (extinct [E], population has not recovered [−], and population resurgence [+]); and factors affecting stock's decline (major factor in decline [xx]; minor factor in decline [x]; factor present [P]. Exotic species are sea lamprey (SL); rainbow smelt (RS); alewife (AL); and white perch (WP).

Affected native fish species	Period of major decline	Exploitation	Exotic species				Nutrient loading	Toxic substances	Alteration of spawning habitats	Current status of native stocks
			SL	RS	AL	WP				
Lake Superior										
Lake trout	1930s–1950s	x	xx					P	P	+
Lake whitefish	1890s–1950s	xx	xx					P	P	+
Lake herring	1940s–1970s	xx		x				P		+
Lake Michigan										
Lake trout	1940s–1950s	xx	xx					P		E
Lake whitefish	1890s–1950s	xx	x					P	x	+
Lake herring	1950s	P		x	xx			P	x	−
Ciscoes	1960s	P			xx			P		+
Yellow perch	1960s–1980s	xx			x					+
Lake Huron										
Lake trout	1930s–1950s	xx	xx					P		−
Lake whitefish	1930s, 1950s	x	x					P		+
Lake herring	1940s–1950s	P					x	x	x	−
Ciscoes	1960s	xx	x					P		+
Yellow perch	1970s	x					P	P		+
Walleye	1940s	xx					xx	P		+
Lake Erie										
Lake trout	late 1800s	xx							x	E
Lake whitefish	1950s	x		x			x	P	x	+
Lake herring	1920s	xx					x		x	−
Blue pike	1950s	xx		x			x	P	x	E
Sauger	1920s–1950s	xx					x	P	x	E
Walleye	1950s	xx		x			x	P	x	+
Yellow perch	1970s	xx		P		x	x	P		+
Lake Ontario										
Lake trout	1920s–1950s	xx	x					P	x	E
Lake whitefish	1920s–1960s	xx	x					P		+
Lake herring and ciscoes	1940s–1950s	xx		x	x			P	x	−
Walleye	1960s				x		x	P	x	+

Perhaps the most notable nonnative species to invade the Great Lakes has been the parasitic sea lamprey, which attaches itself to a fish, rasps a hole, and sucks out body juices, often killing the host fish in the process. Although the sea lamprey preys on many species of fishes, its greatest effect has been on lake trout (Lawrie 1970), an effect perhaps exacerbated by commercial fishing. The lake trout fishery tended to remove larger fish, leaving the brunt of sea lamprey attacks to be borne by smaller lake trout, which are less likely to survive such attacks (Christie 1972). Sea lampreys are believed to have entered Lake Ontario via the Hudson River and the Erie Canal (Emery 1985). Their date of arrival in Lake Ontario is not known, but it was probably at least as early as 1819, when the Erie Canal was opened into Lake Ontario. The parasite probably entered Lake Erie via the Welland Canal and was first reported in Lake Erie in 1921 (Van Meter and Trautman 1970); it spread to Lake Superior by 1946 (Figure 23.2).

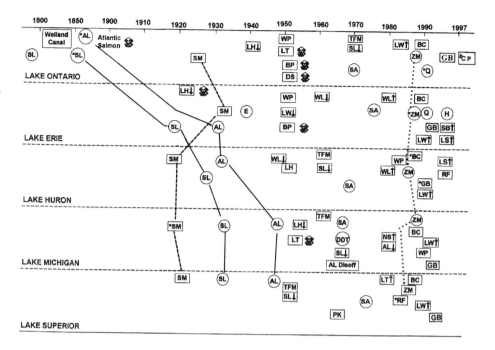

Figure 23.2 Temporal sequence of various critical events affecting fish populations in the Great Lakes. Abbreviations and symbols are as follows: alewife (AL), *Bythotrephes cederstroemi* (BC), blue pike (BP), *Cercopagis pengoi* (CP), pesticide DDT (DDT), deepwater sculpin (DS), eutrophication (E), tubenose (Lake Erie only) and round gobies (GB), *Hexagenia* (H), lake herring (LH), lake sturgeon (LS), lake trout (LT), lake whitefish (LW), native species (NS; yellow perch, bloater, and deepwater sculpin), pink salmon (PK), quagga mussel (Q), ruffe (RF), stocked salmonines (SA), smallmouth bass (SB), sea lamprey (SL), rainbow smelt (SM), lamprey larvicide (TFM), walleye (WL), white perch (WP), zebra mussel (ZM), decline ([downarrow]), resurgence ([up arrow]), and first occurrence in lakes (*). Circled codes indicate major impact, codes in squares designate pertinent information, and skull and crossbones indicate extirpation.

In Lake Ontario the effect of sea lamprey on lake trout was not great until deforestation of watersheds was sufficiently widespread to provide, through siltation of streams, optimum substrate for larval sea lampreys, which reside in streams up to 7 years (Christie 1972). Thus, an environmental stress was first necessary in Lake Ontario before sea lamprey markedly affected the indigenous community. The sea lamprey never did well in Lake Erie until the 1990s because this lake lacks suitable cool, clean spawning tributaries. In Lakes Huron and Michigan, however, the sea lamprey, in concert with overfishing, reduced lake trout to extinction by the late 1940s. In Lake Superior, the sea lamprey was well on its way to exacting the same toll on lake trout as in Lakes Huron and Michigan, but treatment of spawning streams with the selective lampricide TFM (3-trifluoromethyl-4-nitrophenol), under the direction of the GLFC, has reduced sea lamprey numbers and allowed native lake trout to rebound.

A second invader that has had adverse effects on native fish populations is the alewife. Loss of keystone predators allowed proliferation of nonnative species into disrupted fish communities. The alewife was first recorded in Lake Ontario in 1873 (Figure 23.2), then moved into Lake Huron by 1933, Lake Michigan by 1949, and Lake Superior by 1954 (Smith 1970b; Figure 23.2). Prior to the mid-1870s, the alewife was prevented from proliferation in Lake Ontario by cold temperatures (maintained by heavily forested watersheds) that prevented spawning and by predation from the well-established top predators, lake trout and Atlantic salmon. Alewife has never attained great abundance in Lakes Erie and Superior. In Lake Erie, high winter mortality, resulting from low winter water temperatures in that shallow lake, as well as stable predator populations are thought to have prevented the alewife from establishing high populations (Smith 1968; 1970b). In Lake Superior, cold water in spring and summer inhibited reproduction (Smith 1972a; Christie 1974) and growth (Bronte et al. 1991). The story was much different in Lakes Michigan and Huron. During the 1960s, in the absence of terminal predators and with a severely altered coregonine population in these lakes, alewife became the most abundant species, often undergoing massive dieoffs (Brown 1972). With increased stocking of salmonines and successful sea lamprey control, which reduced salmonine mortality, large alewife dieoffs have not been recorded since the 1970s. In fact, alewife has suffered a decline in Lake Michigan (Jude and Tesar 1985; Eck and Wells 1987) and in Lake Ontario during the 1990s (R. O'Gorman, U.S. Geological Survey, personal communication). In Lake Michigan, alewife reached a population low around 1986 and has fluctuated around that mean through the 1990s (D. R. Passino-Reader, R. M. Stedman, and C. P. Madenjian, unpublished data presented during Great Lakes Fishery Commission Lake Michigan Committee Meeting, Duluth, 1996).

Alewife detrimentally affects other species mainly by inhibiting reproduction. In Lake Michigan, alewives, through size-selective predation on zooplankton, reduced the abundance of larger individuals among the plankton (Wells 1970), and created bottlenecks in the food supply to the extent that recruitment of native species was affected (Crowder et al. 1987). Alewives are known to eat pelagic fish larvae, including those of lake trout (Kohler and Ney 1980; Brandt et al. 1987; Kreuger et al. 1995a), and eggs (Wells 1980); larvae and eggs appear to be the most susceptible life stages to alewife predation. The emerald shiner, once extremely abundant in Lake Michigan, has large, 2-mm diameter pelagic eggs (Auer 1982), pelagic larvae, and a short life cycle; it was the first species to decline precipitously as the alewife entered and moved through Lake Michigan (Smith 1970b). Other species, such as the deepwater ciscoes, lake herring, and nonnative rainbow smelt, suffered declines suspected to be related to alewife predation on eggs and larvae (Smith 1970b). Deepwater sculpin and yellow perch, which both have pelagic larvae, were also reduced to low levels in Lake Michigan. As alewife populations declined in the 1980s populations of both these species increased (Jude and Tesar 1985; Eck and Wells 1987). In Lake Ontario, O'Gorman et al. (1987) found that abundant adult alewives suppressed production of age-0 and yearling alewives, thereby depressing growth of young brown trout and coho salmon, which fed on young alewives.

Alewives composed a large part of the diet of all salmonines in Lake Michigan (Jude et al. 1987) and Lake Ontario (Brandt 1986b), probably because alewives are more pelagic, slower swimming, and in larger and denser schools than are native prey species. Heavy predation on alewives has led to an inverse relationship between salmonine and alewife abundance. In addition, Pacific salmon eat mainly alewife because of the propensity of salmonines to feed in midwater (Jude et al. 1987). These trends have continued into the 1990s, despite a shift in dominance of forage species from alewife to bloater.

Rainbow smelt had colonized all five Great Lakes by the 1930s and has been implicated in adversely affecting many native species. At times, rainbow smelt has been abundant in all the lakes; in the 1940s, it suffered catastrophic dieoffs in Lakes Michigan and Huron. Rainbow smelt is believed to suppress recruitment of native coregonines either through food competition (Anderson and Smith 1971) or predation on fry (Selgeby et al. 1978). However, rainbow smelt predation was not believed to be an important factor in the decline of Lake Superior lake herring stocks (Selgeby et al. 1978); predation or some other density-dependent factor may have prevented overexploited stocks from recovering (Selgeby 1982).

23.3 SHIFTS IN GREAT LAKES FISH COMMUNITIES

23.3.1 Theoretical Framework of Fish Population Responses

The importance of natural assemblages of fishes was stressed by Evans et al. (1987) in regards to disruptions of Great Lakes fish communities and their potential for restoration. This view is also part of the current, strongly advocated ecosystem approach to Great Lakes rehabilitation. Fish communities among the oligotrophic lakes of the north temperate zone were similar and have responded similarly to perturbations. They have persisted rather well under moderate exploitation but have changed greatly under extreme disturbances. For example, loss of predators has been followed by major shifts in abundances of prey species and establishment of large stocks of introduced species, often small planktivores. These observations suggest structural conservatism and resistance to change among the original species assemblages and stress the importance of the integrative influence of large piscivores on community structure. Whillans (1979) analyzed the changes in fish communities in three Great Lakes bays and found fishes that survived perturbations were less reliant on specific pathways than were the species they replaced. As a result, stressed communities develop looser structures. Therefore, as Christie et al. (1987a) have propounded, the best insurance against unexpected crises in the distant future is to manage fisheries in the direction of increased system maturity at every opportunity. Reestablishment of a mature system requires movement away from artificial fisheries (stocking and genetic manipulation) toward restoration of native fishes, control over factors that impinge on such restoration (e.g., overfishing, habitat destruction, sea lamprey predation, and pollution), and communication of these goals to user groups, managers, and politicians, who are too often short sighted and driven more by bioeconomics than the long-term goal of ecosystem rehabilitation.

Stresses, particularly cultural stresses, can lead to an array of ecosystem, community, and species responses, many of which may be predictable. Ryder and Edwards (1985) and Francis et al. (1979) discussed stress responses at various hierarchical levels but specifically those occurring within communities in oligotrophic systems (Table 23.4). Stress effects may be perceived first within specific hierarchical levels (e.g., species and populations) due to stress specificity. For example, stress effects at the individual level may be manifested by bioaccumulation of toxic substances or as inordinate sensitivity. Integrated effects can occur among component organisms or between organisms and their abiotic environment (Ryder and Edwards 1985). The hierar-

Table 23.4 Community responses of an oligotrophic system under cultural stress. Adapted from Ryder and Edwards (1985).

Mean size of organisms decreases	Food webs shorten
Life span shortens	Introgression may occur
Pelagic organisms predominate	Reproduction may cease
Benthic organisms decline	Species rations change
Prey species increase	Number of species decreases
Top predators decrease	Mesotrophic forms increase
Eurybionts replace stenobionts	Production declines
Food webs decompose or simplify	

chical level that first demonstrates readily recognizable signs of malaise may not necessarily be the level at which the greatest ultimate stress is experienced. Species with maximum growth or reproductive compensation, such as alewife and rainbow smelt, tend to have survival advantages under stress over those species with minimal density-dependent response flexibility, such as lake trout. Consequently, the species and community compensation characteristics set the stage, at least in part, for the sequential order of species and community demise in a culturally stressed system, which has often led to valuable species being replaced by less valuable ones and the loss of irreplaceable genetic material.

Ecosystem response to stress may also be evaluated by the particle size concept (Sheldon et al. 1972). In marine pelagic communities, organisms, whether they are algae or whales, occur at approximately equal biomass on a $\log_{10}$ size interval basis. Kerr and Dickie (1984) extended this concept to ecosystems, and Ryder et al. (1981) noted similar characteristic community structure in the Great Lakes during early periods of stability. Ryder et al. speculated that particle size changes due to cultural stresses in the Great Lakes should cause the large terminal predators and benthic feeders to disappear first. Ultimate survivors would likely be small opportunistic pelagic species. This theory is in agreement with the observed community structure of the lakes. The lake sturgeon (large benthic feeder) and the lake trout (terminal predator) were highly susceptible to cultural stresses, whereas alewife and rainbow smelt, two small pelagic species, tended to thrive through ecological opportunism, high fecundity, and short turnover times. Intermediate-size species, such as lake herring, lake whitefish, and walleye, were drastically reduced in an orderly sequence that was consistent with the particle size density theory. Even within species, the largest individuals tended to disappear first. Disruption in the particle size relations of organisms, from whatever causal mechanism, may be an indicator of community malaise and is perhaps indicative of a dysfunction in the ecosystem as a whole. This theory was used to predict changes in the shape of the size spectrum of organisms in Lake Ontario after toxic chemical contamination (Borgmann and Whittle 1983) and to predict Lake Ontario production (Borgmann 1982).

Degradation of salmonine communities in response to fishing, cultural eutrophication, and introduced species results in similar changes in fish communities (guild shifts) no matter the perturbation (Christie et al. 1987a). This common stress response suggests that some fundamental principles of ecosystem organization, such as ecological succession, govern community development in response to strong environmental perturbations. Implicit to this perspective is the understanding that degradation of aquatic ecosystems is at least a partially reversible process. A current example of this reversibility is the improvement of Lake Ontario's Bay of Quinte after nutrient loading reductions (Minns et al. 1986). Many important issues facing ongoing management of the Great Lakes fisheries relate to the process of ecological succession in ecosystems and fish communities.

One generalized population response to stress is introgression (the loss of distinct genes through hybridization), which has occurred particularly within the salmonines and coregonines. Before the 1800s, high levels of morphological and functional diversity of phenotypically plastic fish stocks, such as the lake trout and cisco complexes, allowed ready adaptation to the heterogeneous habitats of the lake basins. Losses of particular stocks identified as either genotypically or phenotypically distinct, through extinction processes or introgression, may be a symptom of more insidious environmental stresses. These changes are usually associated with intense harvesting, cultural eutrophication, and physical perturbation or restructuring of the nearshore zone (Regier et al. 1969). Steedman and Regier (1987) made a strong case for the importance of the nearshore zone (nearshore diverse habitats) as "centers of organization," which are often neglected, destroyed, or modified, as in the case of wetlands (Whillans 1982; Jude and Pappas 1992) and harbors. Smith (1964) documented introgression for Lake Michigan coregonines, in which six of seven species were lost as a result of overfishing and sea lamprey predation; only the least valuable and smallest of the species, the bloater, survived. Phillips and Ehlinger (1995) discuss the importance and problems with restoring this historically important component of Great Lakes fish communities. Introgression was also cited as contributing to the demise of blue pike and sauger in Lake Erie (Regier et al. 1969), and in Lake Ontario, blackfin cisco, shortnose cisco, and kiyi were extirpated, leaving only the bloater, which eventually also disappeared (Christie 1972). In Lake Ontario, the deepwater sculpin became extirpated and the slimy sculpin, a species with a more shallow-water distribution, replaced it. In Lake Michigan, the spoonhead sculpin, a sensitive species, was thought to be extinct, but some were found in the late 1980s (Potter and Fleischer 1992). The deepwater sculpin and slimy sculpin are the two more common remaining species. In all of these replacements, premium or highly sensitive species were replaced by a less valuable, more tolerant species. This pattern—a progressive decline in abundance and eventual extirpation of the more sensitive members of a congeneric group—is common.

23.3.2 Lake Superior Fish Community Dynamics

Lake Superior's highly oligotrophic state is reflected by its native fish community, which was once dominated by lake trout, burbot, lake whitefish, lake sturgeon, lake herring, suckers, sculpins, and several species of deepwater ciscoes. The community has been altered since the 1930s by an intense fishery, predation or competition from nonnative species (particularly sea lamprey and rainbow smelt), habitat loss, and water quality deterioration in some nearshore areas and tributaries. Lake Superior is unique in that it has not had the excessive nutrient loadings that the lower lakes have experienced. Lawrie and Rahrer (1972) and Lawrie (1978) summarized changes in native fish populations as follows: nonnative rainbow smelt had become abundant by the 1930s and sea lamprey by 1950; alewife entered the lake in the 1950s but never became abundant. The native lake trout, which was being overfished in the 1930s and 1940s, declined abruptly in abundance in the 1950s as a result of added stress from sea lamprey predation. Several deepwater ciscoe species collapsed and remained at low levels through 1992; predation by the expanding Siscowet lake trout population was cited to explain their low abundance (MacCallum and Selgeby 1987). Lake whitefish populations have fluctuated considerably since the late 1800s; sharp declines have been attributed to overfishing, spawning habitat degradation resulting from lumbering operations, and sea lamprey predation. Lake herring suffered steep declines in abundance between the 1930s and 1960s; these declines have been attributed to overfishing of spawning aggregations (Selgeby 1982) and expansion of rainbow smelt populations (Anderson and Smith 1971). Lake sturgeon and the deepwater ciscoes were also exploited commercially.

Recent trends show a shift in abundance from rainbow smelt, which dominated prey populations since the 1960s, to lake herring, the historical dominant in the prey fish complex (Selgeby 1985; MacCallum and Selgeby 1987). It remains unclear whether historic levels of lake herring abundance have been attained in the 1990s (Hansen 1994). The rainbow smelt decline, more than 90% from 1978 to 1981, was attributed to salmonine predation; lake trout still prefer rainbow smelt over lake herring, even though the lake herring population is increasing. Rainbow smelt stocks have increased since the early 1980s (Selgeby 1985). The lake herring recovery from near extirpation in the 1960s is noteworthy because no other lake herring stocks in the Great Lakes have recovered after reaching such low levels. Reasons for the recovery are not obvious (MacCallum and Selgeby 1987). An immense year-class was produced in 1984 despite the buildup of large populations of predators.

Stocks of lake whitefish have increased since the 1970s, probably due to sea lamprey control measures and favorable abiotic conditions. Harvests in the 1990s have been the largest ever recorded. Lake trout rehabilitation is inhibited due to incidental bycatch in lake whitefish nets.

Since large-scale stocking of lake trout began in 1958, its abundance has increased in most areas, partly as a result of improved natural reproduction believed to be derived mainly from native fish that never became extinct (Krueger et al. 1986). The long-range goal for lake trout rehabilitation calls for self-sustaining populations capable of yielding 1,810 metric tons annually. Since the 1980s, there have been plantings of splake, rainbow trout, brown trout, brook trout (native to Lake Superior), and chinook, sockeye, and coho salmon. Also, pink salmon were inadvertently introduced in 1956, and reproducing populations were established in all the Great Lakes by 1979 (Emery 1985). Natural reproduction by some of these introduced species is believed to be large (MacCallum and Selgeby 1987), although natural reproduction by lake trout continues to be the most substantial. All of these salmonines contribute to sport fisheries in nearshore areas.

Some major controls have been put in place on the Lake Superior commercial fisheries, including quotas, banning of gill nets by Minnesota, sanctuaries, and sportfishing-only zones (MacCallum and Selgeby 1987). Michigan has forced commercial fishers to use impoundment gear. However, tribal groups fish commercially, primarily with gill nets, throughout the lake.

In 1953, strategies for rehabilitation of stocks commenced with attempts to control the sea lamprey, mainly by blocking spawning streams with electrical barriers. In 1958, control measures became much more effective with use of the lampricide TFM (see section 23.4.5). In the 1990s, sea lamprey was at 10% of its precontrol levels, but it still destroys over 30% of the lake trout in U.S. waters (Hansen 1994).

Various management actions in the past 30 years, such as reduction in exploitation, sea lamprey control, and heavy stocking of salmonines, have resulted in conspicuous reversals in downward trends of most native species. However, walleye, lake sturgeon, which remain threatened throughout North America (Williams et al. 1989), and brook trout remain depressed due to overharvest, habitat degradation, and competition with introduced species (Hansen 1994).

23.3.3 Lake Michigan Fish Community Dynamics

In the past 150 years, native fish stocks in Lake Michigan have experienced huge fluctuations (Wells and McLain 1973). Until the 1940s, exploitation was the most important factor affecting the Lake Michigan fish community; the sea lamprey and alewife were most influential in later years (Eck and Wells 1987). Overfishing caused reductions in abundance of lake whitefish, lake trout, lake sturgeon, and the largest species of coregonines. Sea lamprey predation caused the extirpation of lake trout and adversely affected the abundance of lake whitefish. The alewife has been accused of inhibiting reproduction of native species (Eck and Wells 1987; Kreuger et al. 1995b) and, along with rainbow smelt, caused the lake herring collapse in the late 1950s. In the 1960s and 1970s, cultural eutrophication and contaminants damaged the habitat in southern Green Bay and affected populations of walleye and lake herring.

Rainbow smelt became well established in the 1930s, and since then populations have fluctuated. Abundance was low in 1942–1943 due to winter mortality and in the early 1960s, probably due to competition and predation from alewives. Rainbow smelt were abundant in the 1980s and contributed to the diet of salmonines (Jude et al. 1987).

The lake trout collapse in the 1950s, the rise in abundance of alewife in the early 1960s, and implementation of sea lamprey control in the late 1960s created an opportunity for stocking top predators. Lake trout plantings commenced in 1965 and have continued at an annual rate in excess of 2 million yearlings. These fish grew well and contributed to an expanding sport fishery, but there is little evidence of natural reproduction (Jude et al. 1981). Planting of coho and chinook salmon began in 1966 and 1967, respectively, and these species together with lake trout, rainbow trout, and brown trout contribute to a sport fishery evaluated at over US$150 million annually (Rakoczy and Rogers 1987).

Lake whitefish stocks have expanded since the late 1960s, probably due to reduced sea lamprey predation, and in the 1980s harvest reached levels of pre-sea lamprey years (Eck and Wells 1987). On the other hand, lake herring stocks have not recovered and contribute very little to the commercial fishery or the population of prey fish. Of the seven species of deepwater ciscoes originally present, only the bloater remained abundant by the early 1960s (Smith 1964). Bloater stocks declined in the 1960s and early 1970s due to alewife interference with reproduction and an intensive fishery. Harvest restrictions beginning in 1976 and diminished alewife populations have resulted in a bloater recovery. In the late 1990s, bloaters constituted the majority of the biomass of forage fishes, but reproduction has been minimal (Passino-Reader et al., unpublished data).

Changes have occurred in other native stocks as well. Burbot are now common in Lake Michigan; their recovery is attributed to sea lamprey control. Slimy sculpins increased from 1973 to 1976 but, probably due to lake trout predation, declined between 1976 and 1984. Spoonhead sculpins, which were thought to be extinct in Lake Michigan, lake herring, and emerald shiners declined to extremely low levels of abundance before the 1970s and remain scarce (Eck and Wells 1987). Emerald shiners have shown some increases in the 1980s and 1990s.

During the first half of the twentieth century, yellow perch contributed between 450 and 1,350 metric tons annually to the commercial harvest. Production increased briefly in the 1960s (to a peak of 2,700 metric tons) due to increased fishing intensity and then declined in the late 1960s. The primary cause of the decline was poor reproduction probably resulting from competition with, and predation on larval yellow perch by, alewives. Yellow perch populations remained low through the 1970s, but as alewives declined in the early 1980s, yellow perch populations rebounded, particularly in the southeastern (Jude and Tesar 1985) and southern (Shroyer and McComish 1998) part of the lake. Abundance continued to expand through the 1980s, and the yellow perch became the most abundant species in the sport catch in Michigan's waters (Rakoczy and Rogers 1987). However, in 1988, and continuing through 1997, yellow perch suffered recruitment failure because few young of year survived (Francis et al. 1996) and density of spawning adults was low (Shroyer and McComish 1998). Recruitment failure corresponded with the appearance of zebra mussels in Lake Michigan. These mussels compete with filter-feeding zooplankton; hence larval yellow perch, which depend on zooplankton, were probably negatively affected. Further evidence of the impact of zebra mussels on the Lake Michigan ecosystem was the drastic decline in the southern basin of *Diporeia*, an amphipod that feeds on the phytoplankton, which apparently has been considerably diminished by the zebra mussel (Nalepa et al. 1998). The lack of yellow perch recruitment has resulted in banning of commercial harvest of the species and reductions in angler harvest. Although juveniles remained scarce lakewide, in 1997 larval yellow perch densities were comparable to historical densities in eastern Lake Michigan (Perrone et al. 1983; D. J. Jude, unpublished data), where only sportfishing was allowed. Densities of larvae were considerably reduced in southern Lake Michigan (Robillard and Marsden 1997), where a commercial fishery was a major factor in reducing abundance of spawners to extremely low levels (Francis et al. 1996). In 1998, moderate recruitment of juvenile yellow perch was observed lakewide; the only noticeable changes from previous years were related to early and prolonged warming of the water, presumably attributable to El Niño.

23.3.4 Lake Huron Fish Community Dynamics

As victims of overfishing and through predation by sea lamprey, native lake trout are extirpated in Lake Huron except for remnant stocks in Georgian Bay and the North Channel. Populations are maintained through stocking that began in 1973, and some natural reproduction by lake trout has been recorded in Thunder Bay (Eshenroder et al. 1995). Although the first round of stream treatments with lampricides was completed in 1969, control has not been as successful as in the other Great Lakes. There is concern that the St. Marys River, which is difficult to treat effectively with lampricide, is a refugium for ammocetes, larval sea lampreys (Eshenroder et al. 1987, 1995). There are more sea lamprey in the St. Marys River than in all the other Great Lakes combined (Ebener 1995). Lake trout in northern Lake Huron (and Lake Michigan) have experienced 25–75% mortality due to sea lamprey. Backcrosses of splake with lake trout (three-quarters lake trout and one-quarter brook trout), which are much more abundant than are pure lake trout in Canadian waters, are less vulnerable than are lake trout to attacks from sea lampreys because they are smaller and mature earlier. Limited reproduction from splake and backcrosses was first observed in 1985. The Lake Huron Management Plan for Lake Trout Rehabilitation calls for self-sustaining salmonine populations (predominantly lake trout) capable of providing annual harvests of about 900 metric tons by 2010 (Ebener 1995).

Despite sea lamprey predation and an intensive fishery, lake whitefish populations have remained viable. Harvests increased in the 1980s due to increased pressure from commercial fisheries. Lake whitefish year-classes have been strong, and prospects for future harvests are good, with the exception of southern Georgian Bay.

Lake herring, which was an important commercial species in the first 4 decades of the twentieth century, has not recovered from a population collapse that occurred in the 1940s and 1950s. Lake herring is still present in low abundances and could rebound if exotic species (e.g., alewife and rainbow smelt) are reduced (Ebener 1995). The most probable cause of the collapse was deterioration in water quality in Saginaw Bay spawning grounds; however, lake herring were already overfished in the 1920s.

Until about 1940, the deepwater cisco fishery provided relatively stable catches; then it declined to low levels, possibly due to the alewife invasion (Brown et al. 1987) and overfishing. Between 1958 and 1966, harvests increased and populations collapsed. In 1970, the commercial fishery for deepwater ciscoes was closed, and by the late 1970s stocks began to recover in association with predator restorations. In 1985, surveys indicated deepwater ciscoes in Lake Huron were dominated by bloater, which persisted as the five larger species were reduced to low levels (or exterminated) by fishing and sea lamprey predation. Nonnative species (e.g., alewife and rainbow smelt) flourished as a result.

Although the rainbow smelt has been in Lake Huron at least since 1925, it has not been prominent in the commercial fishery. It has been an important prey species and is now the primary prey of salmonines, along with alewife, despite the biomass of bloater eclipsing that of these two exotic species (Ebener 1995). Abundance of rainbow smelt has remained high throughout the 1980s, whereas stocks of alewife have declined. Abundance of alewife has been declining since the early 1970s (Henderson and Brown 1985) and as of the late 1990s is much below pre-1970s levels due to a combination of cold winters and increased salmonine predation.

Yellow perch are harvested mainly from Saginaw Bay and the southern part of Lake Huron. Yields from Ontario waters increased gradually from 1900 to a peak in the 1960s (Spangler et al. 1977). In the 1970s, abundance in Saginaw Bay declined due to overfishing; the increasingly intensive commercial fishery was more prominent in the outer bay (Eshenroder 1977). In the 1980s, production of yellow perch was relatively high in Saginaw Bay and elsewhere in Lake Huron; however, growth in the 1990s has been slow. It has been hypothesized that the extirpated burrowing mayfly *Hexagenia*, which acted as a rich, intermediate-size food source, needs to be restored to promote good growth for this and other fish species (Hayward and Margraf 1987; Ebener 1995).

Historically, the walleye fishery of Saginaw Bay was the second largest in the Great Lakes, occasionally exceeding 700 metric tons annually (Schneider and Leach 1977). In the 1940s, the population collapsed, probably as a result of degradation of spawning habitat and overfishing. Commercial fishing for walleye was banned in Saginaw Bay in 1969; with improved water quality, however, a successful sport fishery, which is supported largely by stocking, has existed since the 1980s. Large numbers of walleye larvae are produced in the Saginaw River, but few survive to contribute to the fishery (Jude 1992). Since the 1980s, the commercial harvest, which is relatively stable, has been mainly from the southern end of the lake.

23.3.5 Lake Erie Fish Community Dynamics

Lake Erie is considerably shallower and warmer than the other Great Lakes, has the highest turnover and sedimentation rate, and has undergone some of the most dramatic changes of any large lake. The lake has three distinct basins (western, central, and eastern) that range from oligotrophic in the east to eutrophic in the west. The mainly mesotrophic lake is best adapted for percid communities (Leach et al. 1977). Most of the deterioration of Lake Erie's environment and fish populations occurred in the twentieth century. In order of their impact, major sources of stress have been exploitation of fishes, cultural eutrophication, introduction of nonnative species, habitat destruction, siltation, and inputs of toxic substances. These changes have resulted in dissolved oxygen depletion, major decline of an important fish food organism, *Hexagenia*, and native fish extinctions.

Lake Erie provides an example of sequential degradation of fish communities due to multiple stresses of cultural origin (Regier and Hartman 1973). Fish stocks in Lake Erie underwent these changes before such changes occurred in the upper Great Lakes. Populations declined starting with lake trout, which were moderately abundant in the eastern basin (Applegate and VanMeter 1970), followed by lake sturgeon, and then by lake whitefish, lake herring, sauger, blue pike, walleye, and yellow perch. By 1970, the exotic and low-valued rainbow smelt, which had been in Lake Erie since 1935, was the predominant species over much of Lake Erie.

Populations of lake trout, lake herring, lake whitefish, and longjaw cisco have all declined to low levels or have been extirpated. The commercial fishery had greatly reduced native lake trout by 1900, and habitat deterioration (mainly eutrophication) virtually eliminated lake trout by the 1930s (Hartman 1972). Stocking efforts in the eastern basin have not established a self-sustaining population. Sea lamprey predation appears to be preventing survival of lake trout over 3 years of age. Rehabilitation of lake trout will require limiting total annual mortality to 40% or less, which will require sustained sea lamprey control measures along with new sea lamprey control initiatives by the GLFC.

Despite dramatic changes within Lake Erie, commercial fish yield from the lake has generally been higher than from all the other Great Lakes combined (Baldwin et al. 1979). Warmwater species such as channel catfish, emerald shiner, white bass, gizzard shad, alewife, spottail shiner, and common carp are maintaining healthy populations. Since the 1950s, rainbow smelt have been fished commercially mainly in the eastern and central basins by an efficient trawl fishery developed in Ontario waters. The yellow perch has been the most valuable commercial species since the 1950s (Hartman 1988). Annual production declined from a peak of more than 15,000 metric tons in 1969 to about 3,000 metric tons in the early 1980s. Since then, production has held steady at around 5,000 metric tons, mostly from Ontario waters. Interestingly, yellow perch harvests in Lake Erie, especially the eastern basin, have tracked the long-term decline in phosphorus loading.

Commercial production of lake whitefish was irregular during the first half of the twentieth century due to occasional strong year-classes, but the stock collapsed in the 1950s due to degradation of summer and spawning habitat and exploitation (Hartman 1988). A remnant population remains in the eastern basin of the lake and spawns regularly, mainly in the western basin. In the 1990s, lake whitefish populations increased dramatically (Ebener 1997).

Lake herring is no longer fished commercially. This species was the most abundant commercial species in the early decades of the twentieth century, but populations collapsed in 1925, probably due to excessive fishing pressure during spawning migrations (Regier et al. 1969).

Alewife entered Lake Erie around 1931 through the Welland Canal from Lake Ontario (Smith 1970a). Predators and low water temperatures in winter have limited its abundance (Hartman 1972).

Harvests of blue pike, another very important commercial species that sustained some of the largest harvests in the Great Lakes, fluctuated widely before the fishery collapsed by 1958. Reasons for the loss of the blue pike, which is now considered extinct, are not clear, but an intensive fishery, loss of summer habitat, and competition and predation from rainbow smelt have been suggested (Regier et al. 1969; Leach and Nepszy 1976). Introgressive hybridization with walleyes may have caused final disappearance of the species. Native sauger in Lake Erie may have experienced the same fate. Because the sauger fishery declined gradually over a long period, environmental deterioration, particularly pollution, siltation, and dam construction, along with increasing fishing pressure, are considered to have caused the demise of native sauger. The state of Ohio began stocking sauger in the western basin in the 1970s, but a self-sustaining population has yet to become established.

Walleye has undergone fluctuations in abundance in western Lake Erie. Declines in the 1950s and 1960s have been attributed to overexploitation, interactions with invading rainbow smelt, and nutrient loading (Schneider and Leach 1977). Nutrient loadings degraded spawning and summer habitats and adversely affected benthic organisms, such as *Hexagenia* (Regier et al. 1969). Since 1970, western basin walleyes have responded favorably to rehabilitation measures (Knight 1997), especially limitations on harvest and reductions in phosphorus loadings, which reached target levels in 1982. In the 1990s walleyes increased to or exceeded former abundance levels.

White perch, an East Coast estuarine species, was first reported in Lake Erie in 1953 but did not become established until the 1970s (Boileau 1985). It has expanded its range to include Saginaw Bay and Green Bay. In Saginaw Bay, this species suffers high winter mortalities. In Lake Erie it is most prevalent in the western basin where it contributes to the commercial fishery.

The Lake Erie ecosystem of the 1990s is in a recovering phase; chlorophyll levels are very low, anoxia has been reduced, and water clarity is at all time high levels (see Makarewicz and Bertram 1993). There has been a resurgence of walleye, smallmouth bass, lake whitefish, burbot, and, to some degree, lake sturgeon; more lake trout are stocked and *Hexagenia* has returned. These signs of change toward more keystone predators, with large and benthic species predominating over a large biomass of non-native pelagic species, portends well for ecosystem integrity of the lake. These recovered species may assist in the process of stabilizing the fish community.

23.3.6 Lake Ontario Fish Community Dynamics

Lake Ontario is unique among the Great Lakes because of its high level of nutrient loading (Beeton 1965), its connection to the Atlantic Ocean, and its native fish assemblage. Its early fish community differed from that of the upper Great Lakes by the presence of Atlantic salmon and American eel (Ryder and Edwards 1985). Major changes to Lake Ontario's stocks due to cultural interventions preceded those of the other Great Lakes by several decades. There have been several fish extirpations in Lake Ontario, including Atlantic salmon around 1900, lake trout, blue pike, and deepwater sculpin (Brandt 1986a). Four coregonines went extinct in the 1940s and 1950s, due to overfishing and possibly rainbow smelt predation. Lake herring and coregonine stocks have not recovered from declines that occurred in the 1940s and 1950s. There have been drastic declines in catches of lake whitefish and lake sturgeon because of overfishing (Christie 1972); however, lake whitefish populations in the east end (Bay of Quinte) appear to be in a state of recovery (Christie et al. 1987b, Casselman et al. 1996). Nonnative species include alewife, white perch, common carp, and rainbow smelt. This fish potpourri is complicated by the stocking of several species of salmon and trout in the 1970s into a lake that was devoid of predators for 20 years (Kerr and LeTendre 1991). Currently, fish stocks are in an unpredictable and continual state of flux (Governments of the United States and Canada 1988b; Jones et al. 1993).

Lake trout rehabilitation in Lake Ontario is being hampered by high mortality rates of adults and lack of natural reproduction. There is evidence that sportfishing mortality may be excessive. From 1985 to 1992, the total number of lake trout harvested by anglers was 2.4 times greater than that killed by sea lampreys (Elrod et al. 1995). Lack of substantial reproductive success (Krueger et al. 1986), despite approximately 10 years of sea lamprey control, indicates that spawning habitat for lake trout may be degraded and alewife predation may be important (Krueger et al. 1995b). Efforts at Atlantic salmon rehabilitation have been hampered by dams and tributary degradation.

Lake Ontario alewife–salmonine interactions appear to be following the same trajectory as those in Lake Michigan, but there are several differences. In Lake Michigan, it is generally believed that salmonine predation (Stewart et al. 1981) and possibly cold winters (Eck and Brown 1985) caused a precipitous decline in alewife populations (Jude and Tesar 1985). Large-bodied *Daphnia* appeared in Lake Michigan after the alewife decline (Evans and Jude 1986), but similar changes in zooplankton have not yet been observed in Lake Ontario (Johannsson 1987). The alewife decline in Lake Michigan was followed by a dramatic increase in yellow perch, bloater, and, to some degree, rainbow smelt populations (Jude and Tesar 1985). These species may now compose a higher proportion of salmonine diets in Lake Michigan. In Lake Ontario, bloater and deepwater sculpin were believed extirpated (some deepwater sculpins were collected in 1996 by J. Casselman, Ontario Ministry of Natural Resources), so it is not clear what species might respond to an alewife decline. O'Gorman et al. (1987) believe that yellow perch, nonnative white perch, and remnant stocks of lake herring may increase if alewife decline further. Lake trout recruitment may also increase if alewife predation on their fry is reduced (Krueger et al. 1995b). Impetus is thus given to reestablishment of the endemic fauna, especially bloater and deepwater sculpin, should the alewife population decline dramatically. The New York Department of Environmental Conservation, however, has a policy of maintaining a reasonable population of alewife to serve as prey for salmonines (Brandt 1986b).

In the 1980s and 1990s, more salmon have been stocked into Lake Ontario on a surface area basis than have gone into Lake Michigan, and anglers are harvesting large numbers of salmonines (O'Gorman et al. 1987). Yet alewife populations in Lake Ontario have been affected more by weather than by predation, although there is some evidence that the first predation-induced alewife declines may have occurred (Jones et al. 1993). Additionally, because of phosphorus controls, the Lake Ontario carrying capacity for alewife will probably be reduced. An alewife population decline will have far-reaching effects throughout the lake and has the potential to affect commercial and sport fisheries. Therefore, Lake Ontario is poised on the threshold of change.

An alewife dieoff in 1976–1977 was followed by increased growth of age-0 alewife and a strong yellow perch year-class. Rainbow smelt populations have increased since then, apparently buffered by predation directed at alewife, whereas slimy sculpin numbers declined due to predation by juvenile lake trout. In 1993, in response to lower phosphorus concentrations, declines in primary productivity and zooplankton densities, a threat of bacterial kidney disease, and a decline in abundance and poor growth of alewife and rainbow smelt, substantial stocking reductions (>50%) were made for chinook salmon and lake trout (Jones et al. 1993).

In the 1990s there were many signs of recovery in Lake Ontario. Contaminants, phosphorus, sea lamprey predation, and angler harvests decreased, water clarity increased, and there was some natural reproduction by lake trout. Alewife populations declined, thereby reducing their predatory effect on lake trout, yellow perch increased (Jones et al. 1993), walleye staged a resurgence (Bowlby et al. 1991), and lake whitefish attained historical high levels (Casselman et al. 1996).

23.4 RESPONSE TO CRISIS: MAJOR FISHERIES MANAGEMENT ACTIONS

23.4.1 Regulation of Exploitation

Overfishing has altered most Great Lakes fish communities. Species resilience, competitive advantage, and resistance to stress can all be affected by the progressive removal of the largest members of a population. Early exploitation by fishers and other perturbations (e.g., pollution, habitat modification, and nutrient and thermal loading) affected the nearshore zone first. Overfishing and the inability of fisheries managers to detect or control it were exemplified by the lake trout fishery (Brown et al. 1981). Lake trout production was a composite of many stocks that represented many ages and sizes of mature and immature fish. This complexity of stock composition gave the population stability and resilience during times of adverse density-independent effects (e.g., bad weather) and overfishing. Stability of the fishery was maintained because commercial fishers were able to move sequentially from depleted to unexploited stocks (Lawrie and Rahrer 1972). As a result, there was little evidence of fluctuations in the total catch data that would indicate marked variations in year-class strength. Although some stocks became greatly depleted or extinct, there was no indication of general depletion (Lawrie 1978). Smith (1968) recognized the problem of sequential stock elimination and provided the basis for future management efforts to curb and control harvest. Efforts to control harvest have resulted in resurgence of bloater in Lakes Michigan and Superior, walleye in Lake Erie, and lake herring in Lake Superior.

23.4.2 Fisheries Resource Allocation

Conflicts have developed between commercial fishing and sportfishing interests (Berkes 1983). Activities of one group can adversely affect those of the other. Each group is opposed to allocation decisions that appear to be unfavorable to its specific interests. Currently, the Great Lakes sport fishery is the largest in North America, and in most of the Great Lakes it possesses economic values that far exceed those of the commercial fisheries. In an economic analysis of the allocation of Lake Michigan alewife to commercial (harvest for pet food) or sport fishery (prey for salmonines) interests, the commercial value was virtually nil, whereas the marginal net social value of the sport fishery was $4.10 per salmonine. This strongly suggests that alewives should not be harvested by commercial fishers (Samples and Bishop 1987). However, there is definitely a place for regulated commercial fisheries in the Great Lakes (Figure 23.3). There are many species, such as the deepwater ciscoes and common carp, which are not exploited by anglers, and others such as rainbow smelt and yellow perch, which are underharvested in some lakes.

Another basic conflict among fisheries researchers and managers is whether ecosystems should be managed toward native species or maintained as a put-and-take fishery with stocked salmonines and the nonnative alewife as a precious forage base. The Lake Michigan fishery management plan recognizes the importance of stocked predators and naturalized species, such as brown trout, and strives to foster a fish community with native species to the degree possible.

23.4.3 Alewife Management

The exploding alewife population in Lakes Ontario, Michigan, and Huron caused damage to native stocks (e.g., yellow perch and emerald shiner; Smith 1970b) and changed zooplankton size and diversity (Wells 1970). Alewives plugged water intakes and littered beaches, causing economic and aesthetic havoc among residents and businesses dependent on tourists. The population explosion was successfully thwarted by restoration of top predators. Tanner (1988) detailed the state of Michigan's plan both to control the alewife by means of large predaceous fishes and to create a sport fishery, thereby relegating commercial fishing to a much lesser role. In 1964–1965, 1 million

Figure 23.3 Commercial fishing has influenced the composition of fish communities in the Great Lakes. Commercial harvests are currently regulated to avoid such problems and to enable sustained harvest. Shown here is a trap net being serviced by a commercial fishing boat in Saginaw Bay, Lake Huron.

coho salmon were obtained from Oregon and raised to smolts; in 1966, 850,000 fish were released into tributaries of Lakes Michigan and Superior. Due to an alewife population explosion in Lake Michigan, a massive alewife dieoff occurred in 1967 (Brown 1972), so additional stocking of chinook salmon and other salmonines was undertaken to control alewife further. As of the 1990s, the number of salmonines stocked into the Great Lakes has been over 30 million annually. In the late 1990s, stocking was restricted in Lakes Michigan and Ontario because of declining alewife stocks. Stocking salmon to control alewives can be viewed as an end in itself that provides the additional economic and recreational benefits of a world-renowned sport fishery. It can also be viewed as a stopgap measure along a road to rehabilitation of the ecosystem. We are already seeing the revival of native species through a substantial decline in alewife populations in Lakes Michigan and Huron.

23.4.4 Maintenance of Trophic Interactions

Heightened predator–prey interactions, whether caused by an increase in the number of predators or by a decrease in the larger members of the predator community (e.g., by overfishing) can cascade through the trophic food web (Carpenter et al. 1985). When the largest terminal predators are harvested, prey fishes such as coregonines may grow beyond optimal size for the unharvested, next largest predators. Species like lake herring, when released from intense predation, may essentially reach a predation refugium (become too large to be prey) and become increasingly abundant. An overly large and abundant prey species may then overgraze the zooplankton food supply, leading to inefficient transfer of energy to the terminal predator.

The reverse of this case occurred in Lake Michigan, where increasing numbers of stocked salmon overgrazed alewife populations (Stewart et al. 1981). This reduction, in concert with a series of cold winters (Eck and Brown 1985), reduced alewife populations in 1983 to their lowest levels since they first entered the lake. Reduced densities of alewife released large zooplankton from predation, and a dramatic increase in large-bodied zooplankton occurred (Evans and Jude 1986). Species that competed with alewife, such as yellow perch, bloater, and rainbow smelt, also increased in abundance (Jude and Tesar 1985). Large-bodied zooplankton cleared the water column more efficiently, and water clarity increased (Scavia et al. 1986). Hence, the maintenance of appropriate relative size relations and numbers between any two trophic levels is important in ensuring a stable fishery and maintaining the community at steady state (Ryder and Edwards 1985).

23.4.5 Sea Lamprey Control

In the late 1940s to 1960s, fisheries management agencies in the Great Lakes were challenged by degraded fish populations and habitats. Populations of lake trout and other large species collapsed sequentially in the lakes due to sea lamprey predation and overfishing. An adult sea lamprey kills many fish (up to 18 kg) during its life (Fetterolf 1988). Following the sea lamprey invasion was that of the alewife, and conditions were further exacerbated by habitat destruction. The GLFC was established in 1955 primarily to develop ways to control sea lamprey (Fetterolf 1980, 1988).

Vernon Applegate was a pivotal figure in applying scientific principles to the study of sea lamprey biology (Applegate 1950) and subsequently testing thousands of chemicals to apply to the ammocete, the early life history stage identified as vulnerable. Thus TFM was discovered (Applegate et al. 1961). Partial sea lamprey control by use of selective toxicants has been accomplished since 1958 on Lake Superior, 1960 on Lake Michigan, 1966 on Lake Huron, and 1971 on Lake Ontario. Control of sea lamprey is expensive; over $7.5 million was spent in 1988 alone. However, the combined worth of the sport and commercial fisheries is $2–4 billion (Fetterolf 1988).

23.4.6 Eutrophication Abeyance

Each of the Great Lakes has experienced cultural eutrophication over the last 150 years; Lake Erie has experienced very high levels whereas Lake Superior has experienced low levels (Loftus and Regier 1972). Eutrophication still poses an environmental problem. Fish community response to nutrient enrichment can take two courses (Evans et al. 1987). If food abundance changes but habitat quality does not, increased growth rates, earlier maturation, and increased production are expected. If nutrient enrichment is severe, habitat is degraded and reduced survival of sensitive species results in a decline in abundance (Leach et al. 1977). If the most sensitive species is also a top piscivore, dramatic secondary shifts in dominance toward increased abundance of small prey species and possibly some benthivores would be expected. Further ramifications would include a decline in large zooplankton and an increase in algae. Overfishing would cause a similar sequence of changes. Minns et al. (1986) concluded that nutrient additions in the Bay of Quinte, Lake Ontario, set the general level of production, but that to a major extent, species interactions explained observed changes in the structure of the fish community.

Responses to cultural enrichment in the Great Lakes have included increased algal populations, hypolimnetic dissolved oxygen depletion, and dramatic changes in fish communities (Ryder et al. 1981). In Lake Erie, nutrient-induced dissolved oxygen depletion in the central basin hypolimnion was cited as contributing to the ultimate extirpation of lake trout (Hartman 1972). Excessive phosphorus inputs resulted in reduction in the abundance of lake trout stocks by limiting the amount of suitable habitat; in some situations, phosphorus loading led to extirpation. In Lake Erie, reduced loadings of phosphorus between 1970 and 1980 have caused ambient concentrations of phosphorus in the west and central basins of the lake to decline 35% and 33%, respectively (Great Lakes Water Quality Board 1981). However, anoxia in the bottom waters of the central basin still occurs in most years. Filtering by zebra mussels, however, has cleared the water column and reduced anoxia even further, to the extent that *Hexagenia* have returned to all basins of the lake. Lake Ontario shows some of the signs of degradation evident in Lake Erie. However, Lake Ontario has an additional problem of hazardous chemicals, particularly those originating from the Niagara River (Allan et al. 1983).

It has been estimated that projected reductions in municipal phosphorus loads will decrease fish yield by 15% or more in lower Green Bay, main Lake Michigan, Saginaw Bay, western and central Lake Erie, and Lake Ontario (Sullivan et al. 1981). Decreased fish yield from reduced phosphorus loads to the lakes is based on a reduction in phytoplankton growth. Such reductions, along with the filtering of zebra mussels, have resulted in increased macrophyte growth because of improved transparency, especially in harbors, embayments such as Saginaw Bay (Skubinna et al. 1995), and nearshore areas of the lower lakes. Fish species associated with these habitats therefore should increase in abundance.

23.5 SUCCESSES IN FISHERIES AND ECOSYSTEM MANAGEMENT

23.5.1 Ecosystem Rebirth

Christie et al. (1987a) are optimistic about the future of the Great Lakes and the opportunities for providing valuable fisheries. As of the late 1980s, the annual regional economic contribution of the Great Lakes fisheries has been estimated at $270 million for commercial fisheries and from $2 to 4 billion for recreational fisheries that provide some 55 million angler-days (Fetterolf 1988). Realization of continued opportunities will depend upon accurate assessment of ecosystem health and development of management measures that will maintain or enhance it.

In Lake Ontario and some of the other Great Lakes, the 1970s saw implementation of sea lamprey control, a quota system for commercial fishers, intensive salmonine stocking programs, and reduced inputs of mercury, DDT, PCBs, and phosphorus (Christie et al. 1987b). The changes wrought by these actions have all been toward earlier system status, favoring a return to the two harmonic communities (percid and salmonine) that characterized the Great Lakes before extensive human influences (Ryder and Edwards 1985). However, considerable effort and money must be committed to maintain current reversals, and an additional commitment to new initiatives to overcome impediments to rehabilitation of these valuable ecosystems must be made.

23.5.2 The Great Lakes Fisheries Management Plan

Since 1955, fisheries managers, with the help of the GLFC and the IJC, have made progress in rehabilitating Great Lakes fish communities and habitat. However, a continuing problem in the Great Lakes has been the fragmented control of and responsibility for management of the fisheries (Francis 1987). In 1978, the IJC and the GLFC nurtured the ecosystem approach to Great Lakes management into general acceptance. This approach recognizes that any perturbation may affect an individual lake, a connecting channel, or even the entire basin. As part of this ecosystem approach, the GLFC promoted a strong, practical Great Lakes fisheries management plan to ensure that the public's fisheries resources receive full recognition and consideration in any plans for development in the Great Lakes basin (GLFC 1980; Fetterolf 1988). The planning process was begun in 1978 through establishment of the Council of Lake Committees (see section 23.6.2). To ensure success, the effort was well funded, and commitments were secured from high-ranking officials of fisheries agencies in Canada and the United States. These officials and their staffs were involved from the beginning of the planning process, and the plan was to be strategic in scope and form a nucleus from which all efforts could be coordinated.

23.5.3 Recovery of Native Stocks

There have been many positive results of the steps taken to curb overfishing, control expanding nonnative fish populations, reverse eutrophication, restore habitat, eliminate toxic substances, and maintain sea lamprey at low levels. As of the 1990s, lake whitefish are thriving in all five Great Lakes. Causes for the dramatic increase include favorable winter conditions, decline of nonnative species due to cold winters and predation by increasing lake trout populations, reduced sea lamprey and angler mortalities, and decreased phosphorus loading (Casselman et al. 1996). Since 1959, commercial harvest of lake whitefish has increased 7% each year, and yield has been consistent (Ebener 1997).

In Lake Superior, native lake trout are reproducing at a near-record rate (90% of the fish are wild in Michigan waters; Schreiner and Schram 1997), lake herring has been resurrected from very low levels, and bloater populations have been restored. In Lake Michigan, alewife populations have declined but still provide considerable prey for salmonines. In response to the alewife decline, yellow perch and bloater populations have rebounded to all-time high levels in the 1980s. The once-depressed deepwater sculpin has recovered, and emerald shiner is showing evidence of a modest return. Lake sturgeon is showing signs of a rebound in the St. Clair–Detroit River corridor evidenced by increased catches of young of year (Ebener 1995). In Lake Huron, bloater has recovered moderately, and as of the late 1990s, walleye provides a renowned sport fishery in Saginaw Bay (through stocking) and Lake Erie (through natural reproduction) (Knight 1997). *Hexagenia*, long extirpated (Britt et al. 1980), is also showing signs of recovery in Lake Erie (Krieger et al. 1996). Unfortunately, the blue pike, now extinct, was best adapted for the unique mesotrophic habitat of the central basin of Lake Erie (Ryder and Edwards 1985); time will tell if the walleye can replace it in this productive zone or if the blue pike will re-emerge.

Declines in major walleye stocks in the Great Lakes due to cultural perturbations have been reviewed by Schneider and Leach (1977). Some stocks have responded favorably to efforts to reduce sources of stress. One of the best examples of success has been the walleye population resurgence in western Lake Erie (Hatch et al. 1987; Knight 1997), which was accomplished mainly because the fishery was closed in 1970 due to mercury contamination (Muth and Wolfert 1986). A charter boat fishery has since been established, and the stock has extended its range into the central basin. Future fisheries management efforts will be directed at increasing walleye harvest (Knight et al. 1984), which should increase walleye growth, decrease age at maturity, and bring prey and predator stocks more nearly into balance, thereby promoting greater yields.

A walleye stock in the St. Louis River, Lake Superior, has been restored because paper mill waste has been diverted (MacCallum and Selgeby 1987). Many fish caught in the 1980s were old individuals that accumulated in the river because their tainted flesh was repugnant to anglers.

Saginaw Bay once supported annual commercial yields of walleye of about 700 metric tons (Schneider and Leach 1977), but by 1950 stocks were reduced due to a deteriorating environment, exploitation, and a moderate amount of sea lamprey predation. Walleyes were reared and stocked around Saginaw Bay by the Michigan Department of Natural Resources and angler groups; these efforts resulted in a spectacular sport fishery in the 1980s (Rakoczy and Rogers 1987) and beyond. A ban on commercial fishing for walleye and limits on commercial harvest of yellow perch are now in effect in the bay. In Lake Ontario, walleye has also rebounded after a long period at low levels (Bowlby et al. 1991). In each of these cases, major impediments to stock revivals were removed through improvements in wastewater treatment and, in the case of Saginaw Bay, stocking to initiate and enhance natural reproduction.

Burbot populations have increased dramatically in some parts of Lakes Michigan, Huron, Ontario, and Erie and may require some control to reduce competition with stocked salmonines. Predation by burbot on deepwater sculpin in Lake Michigan and on lake trout in Lake Huron has also been recorded (Ebener 1995).

23.5.4 Salmonine Stocking Programs

Establishment of salmonine sport fisheries in some of the Great Lakes and rehabilitation of the lake trout fishery, especially in Lake Superior, are notable achievements. The present put-grow-and-take sport fishery gives management agencies flexibility in managing the system (Smith 1972b). This fishery is maintained at a cost though; a self-sustaining natural predator–prey system, would be more consistent with an ecosystem approach to management (Christie et al. 1987a). Ecosystem management issues remain that are intractable with existing fisheries management approaches.

Led by the state of Michigan, management agencies have shifted emphasis and management from commercial to sport fisheries. The first stocking of Pacific salmon occurred in 1966, when over one-half million coho salmon were placed into Lake Michigan. Returns of over 30% in spawning runs and catches were more than had been expected. Salmon have done well in all the Great Lakes, except Lakes Superior and Erie, where alewives are relatively scarce. The dependence of salmon on alewives highlights the importance of this prey fish to the success of the stocking program and also forewarns of problems in Lakes Michigan and Ontario where alewife stocks have declined. Response of the salmonines to changing prey fish populations is still being evaluated in Lake Michigan (Jude et al. 1987) and Lake Ontario (Jones et al. 1993). One of the consequences of stocking so many chinook salmon in Lake Michigan has been the prevalence of bacterial kidney disease, which depleted stocks in the 1990s (Starliper et al. 1997). The bacterial disease had been present for 25 years and had not caused problems until the 1990s; spread of the disease was attributed to too many stocked fish transmitting the disease and poor growth of stocked fish due to reduced alewife densities (Jude and Tesar 1985; Passino-Reader et al., unpublished data).

23.5.5 Control of Nutrient and Toxic Substance Inputs

Nutrient input to the Great Lakes has been reduced, most dramatically in Lake Erie. Improved water quality there has assisted in walleye resurgence (Hatch et al. 1987) and improved beaches. Such improvements have resulted in increased tourist revenues and resource use. In Lake Michigan, total phosphorus declined over the 1970s and early 1980s, especially in 1977 and 1982 (Scavia et al. 1986). Phosphorus reduction in 1977 was attributed to reduced regeneration from the sediments during an unusually cold winter with extensive ice cover; reduction in 1982 was attributed to a salmonine-induced increase in abundance of large-bodied *Daphnia*, which cleared the water column. Similar decreases in phosphorus in Lake Ontario have not shown similar effects, but the alewife and rainbow smelt populations have declined substantially.

Because of the high accumulation rates in bottom-dwelling and top-predator fishes, a major effort has been underway to control the input of toxic substances into the Great Lakes. Many toxic compounds, including DDT and PCBs, have been banned or restricted in use. A decline in toxic contaminants in fishes has been the general response (D'Itri 1988). Studies are now underway to determine bioaccumulation pathways from the sediments through the food chain, sources of these anthropogenic substances, and their rates of entry to and release from the Great Lakes. Large quantities of toxic substances are now coming into the Great Lakes through the air, carried from areas outside the Great Lakes basin (Sweet 1992).

23.6 MANAGEMENT FOR ONGOING FISHERIES PROBLEMS

23.6.1 Public Expectations

Fisheries managers must be cognizant of what Gale (1987) called "resource miracles and rising expectations" of the fishing public. He argued that dismal conditions (e.g., degraded water quality of Lake Erie and dead alewives on Lake Michigan beaches) will not stimulate revolt if those conditions are consistent with public expectation. Instead, the revolution comes when conditions improve and there is a gap between actual improvement and rapidly increasing public expectations. Accelerated public expectations could force managers into an impossible situation, where available resources cannot catch up with escalating expectations. In 1987, which was an unusually warm year, anglers had greater difficulty catching salmonines in Lake Michigan, and there were complaints about the paucity of large fish. There was even some evidence that anglers shifted effort from chinook salmon, which were difficult to catch, to lake trout (Rakoczy and Rogers 1987). Such shifts can potentially inhibit lake trout restoration efforts in Lake Michigan.

Another aspect of public expectations is the perception that through technology fisheries managers can provide artificial fisheries, or "fixes," by stocking more or different strains of fish or by pursuing other massive and costly fishery rehabilitation techniques. Again, if we pursue such strategies we lose sight of the underlying causes of existing problems and the long-term goal of restoring ecosystem integrity, which has many benefits, including self-reproducing populations, clean water, and fish communities that act as a "miner's canary" for the environmental quality of the basin. The challenge for fisheries managers is to keep expectations in line with biological and ecological possibilities for improvement, to try new techniques aggressively, to build new constituencies (involve environmental groups and agencies concerned with water quality and public health), and to begin new research initiatives. Fisheries managers must plan for a future of reduced expectations now.

23.6.2 Institutional Arrangements

There is an exceedingly complex system of institutional arrangements that bears on all aspects of the planning and management of the Great Lakes ecosystem (Francis 1987). These complex institutional interactions have hampered effective management of the Great Lakes (Smith 1970a), contributed to legacies of extinct species, allowed contamination of aquatic organisms, lost wetlands, and irrevocably changed some of the most complex and fragile ecosystems in North America. Francis (1987) stressed that rehabilitation measures must take into account two things for effective design: first, that some areas are local and require local governmental input and responsibility, and second, that basin or lakewide problems (e.g., contamination of fishes) require governmental inputs at a higher level. The IJC and the GLFC are international leaders in these latter efforts. The IJC is currently spearheading efforts to clean contaminated sites throughout the Great Lakes, especially at 42 areas of concern (Hartig and Thomas 1988). The GLFC provides initiatives in sea lamprey control and rehabilitation of native stocks, especially lake trout. Ironically, as water quality of tributary streams has been improved in Lake Erie, more habitat became available for spawning sea lampreys.

The Joint Strategic Plan for Management of the Great Lakes Fisheries (GLFC 1980) provides some mechanisms for overcoming past obstacles. These mechanisms operate within the framework of lake committees, which act as a forum for agency representatives to represent their own interests and negotiate consensus decisions regarding joint concerns. This process has worked very well in the past, especially in planning strategies to control sea lamprey. Much more could be done because the mechanisms for united management are in place. For example, a unified fish contaminants advisory has recently been initiated for Lake Michigan. A similar plan is needed for controlling the number and species of salmonines stocked, based on current forage fish estimates for all the lakes.

23.6.3 Management of Complex Trophic Interactions

In Lake Ontario, the phosphorus management strategy (reducing phosphorus load-ing) and fisheries management strategy (stocking of salmonines) have been successful (Governments of the United States and Canada 1988a). However, these two strategies may promote system instability if they are not managed effectively. Alewife is the main food for salmonines and can also control zooplankton size and, hence, algal popu-lations. Alewife can also depress native fish stocks through predation and competition. By decreasing phosphorus loadings and increasing salmonine predation pressure, we may be strongly depressing the alewife population; controlling the alewife population, in turn, may allow native stocks to increase along with increases in zooplankton sizes. Winter severity may also negatively affect the alewife, further exacerbating these com-plex interactions. Thus Lake Ontario is on the threshold of major changes, which will most likely be keyed on the stochastic influence of weather on alewife mortality. Christie et al. (1987b) maintain that bottom-up forces (nutrient-dominated pathways and inter-actions) prevail in the early stages of rehabilitation and top-down forces (predation effects on lower levels) dominate at later stages of the maturation cycle. Vertical vec-toring of energy through trophic levels may be inhibited in Lake Ontario because of the loss of the deepwater coregonine complex, so consideration should be given to restoration of this component of the fish community to promote increased ecosystem efficiency and energy transfer among trophic levels.

23.6.4 Lake Trout Rehabilitation

Native lake trout are essentially extirpated in all the Great Lakes except Lake Superior. They are victims of many human activities that have affected their stocks and habitat. Populations in the lakes other than Superior are currently maintained through stocking that began in 1958. To date, limited natural reproduction has been noted, and many agencies continue striving to solve the rehabilitation problem (Eshenroder et al. 1984; Journal of Great Lakes Research 21[Supplement 1]).

Ultraoligotrophic conditions optimize lake trout reproduction, whereas early me-sotrophic conditions are best for growth (Ryder and Edwards 1985). In the Great Lakes, Lake Superior may be viewed as the lake with the best conditions for lake trout repro-duction, whereas the central or eastern basins of Lake Erie likely provided maximal growth rates.

Factors that may inhibit lake trout reproduction include contaminants, deterioration of incubation environment, insufficient numbers of spawners due to high fishing mortality (Eshenroder et al. 1984), and predation on fry by alewife (Krueger et al. 1995b) and by other nonnative species. Reproduction in hatchery stocks may be further compromised by the hatchery experience disrupting reproductive biology, inappropriate stocking sites and hence homing to inappropriate spawning areas (Horrall 1981), and poorly adapted genetic strains. Fitzsimons (1995) and Fisher et al. (1995) have documented an early mortality syndrome among piscivores eating alewife and rainbow smelt, both marine species that contain high levels of thiaminase, an enzyme that destroys the B vitamin thiamine. Re-duced consumption of these forage species and administration of thiamine to eggs in hatch-eries have ameliorated this problem in lake trout as well as salmon.

There has been no natural reproduction of lake trout in Lake Erie despite efforts to reestablish populations in the eastern basin; there has been scattered reproduction in Lakes Michigan, Huron, and Ontario. Substantial reproduction occurs in Lake Superior, where native stocks never became extirpated. Krueger et al. (1986) believe that most reproduction in Lake Superior is by wild, not hatchery, fish. Rehabilitation of lake trout invariably involves stocking of hatchery-reared individuals. Selection in hatcheries is usually for maximum production of eggs and survivability, which sometimes results in diminished genetic diversity. Failure to determine causes for poor success with natural reproduction, despite almost 32 years of intensive stocking of yearlings, points to the difficulty in rehabilitating lake trout (Krueger et al. 1995a). Simple restocking without restoration of water quality and co-occurring native species and without reduction in the stresses of overfishing, sea lamprey predation, and interactions with nonnative species has not worked in lakes below Lake Superior. Current attempts to rehabilitate lake trout stocks include establishment of offshore reef sanctuaries in Lakes Michigan and Huron, where some limited reproduction has been documented (Eshenroder et al. 1995).

23.6.5 Introduced Species

Emery (1985) reviewed 34 species of fishes that were introduced (both intentionally and unintentionally) into the Great Lakes basin between 1819 and 1974 and documented their distribution and present status. Only about half of the introduced species became successfully established. Mills et al. (1993a) provided an updated summary. The sea lamprey, alewife, and white perch invaded the Great Lakes through canals built to link the Great Lakes for shipping. Many species of animals and plants have arrived into the Great Lakes in the ballast water of foreign vessels, a recent trend in the introduction of exotic species. In the United States, regulations under the Nonindigenous Aquatic Nuisance Species Prevention and Control Act of 1990 require exchange of original ballast water in the open ocean before arrival in the Great Lakes. Similar regulations are under development in Canada to strengthen existing voluntary guidelines for the exchange of ballast water; these guidelines are administered by the Canadian Coast Guard and achieve over 90% compliance (Mills et al. 1994). Research into alternative strategies (ship design, thermal alteration, microfiltration, ultraviolet treatment, ozonation, or other chemical additives, such as gluteradlehyde) to manage exotic species in ballast water continues (Lubomudrov et al. 1997).

Eurytemora affinis, a copepod introduced by ballast water, and goldfish introduced via hobbyists no longer enamored with their pets occur in all the Great Lakes but have had little impact (Anderson and Clayton 1959). A more recent invader, *Bythotrephes cederstroemi,* is an epilimnetic zooplankton species from Europe, that preys on smaller zooplankton (Evans 1988). It is now found in all the Great Lakes, where it has the potential of disrupting zooplankton communities and adversely affecting bloater recruitment (Warren and Lehman 1988). It has been found in the stomachs of alewife and yellow perch in Lake Michigan and in yellow perch and coho salmon in Lake Erie. Another closely related species, *Cercopagis pengoi,* originating from the Caspian Sea area, has been documented throughout Lake Ontario in high abundances (MacIsaac et al. 1999). *Cercopagis pengoi* has fouled sport fishers' lines and is expected to depress native zooplankton and be eaten by larger fishes, such as alewife and rainbow smelt. *Echinogammarus ischnus*, another exotic amphipod, was discovered in the 1990s in the Detroit River and has spread throughout Lakes Erie and Ontario (Witt et al. 1997). The snail *Potamopyrgus antipodarum* is found in high abundance in selected areas of Lake Ontario (Zaranko et al. 1997).

The zebra mussel, recently found in high densities in Lake St. Clair and Lake Erie, has spread through all the Great Lakes and far outside the basin (New York Sea Grant 1993). It has a planktonic veliger stage that attaches to solid substrates (Figure 23.4). This organism is a nuisance in Europe where it fouls water intakes (Stanczykowska 1978) and has become a nuisance in North America as well. The discovery of another closely related form in 1991, the quagga mussel, which is now found in Lakes Erie and Ontario, portends even more dire consequences because this species can colonize soft substrates, tolerates colder water temperatures, and occurs in deeper waters than does the zebra mussel (Mills et al. 1993b). Although the effects on fisheries have not yet been adequately measured, managers are concerned that these mussel invaders could affect fish production in two ways. First, the fouling of spawning reefs could affect reproduction of shoal spawners, especially lake trout which require the interstices to protect eggs, and the habitat might be altered sufficiently to reduce egg deposition and survival. Results from observations on walleye in the western basin of Lake Erie indicated that reproduction and survival in 1990 were satisfactory (Fitzsimons et al. 1995). Second, shifting of organic matter from pelagic to benthic areas through filtering by zebra mussels could disrupt the food web that links primary production to larval fishes. For example, between 1988 and 1990 zebra mussels and nutrient reductions in western Lake Erie caused chlorophyll *a* to decline to 1–3 μg/L, a 54% reduction and an order of magnitude decline from just 15 years ago (Leach 1993). Extensive studies have also been done in Saginaw Bay, where chlorophyll *a* declined 59%, total phosphorus declined 42%, phytoplankton production declined by 38%, water transparency increased by 60% (Nalepa and Fahnenstiel 1995), and zooplankton declined (Bridgeman et al. 1995). As a result, there was an increased surface area expansion of macrophytes (Skubinna et al. 1995) and increased production of benthic algae (Lowe and Pillsbury 1995). In Lake Erie, walleyes moved farther offshore to avoid light (Ryder 1977), making anglers change traditional fishing techniques.

The threespine stickleback is another nonnative species present in Lake Ontario. In the 1980s it was collected in the St. Marys River, the St. Clair River, and in Lakes Michigan and Huron as well (Stedman and Bowen 1985). The ruffe, a European immigrant, has reached high densities in the St. Louis River, a Lake Superior tributary (Pratt et al. 1992); in the mid-1990s, it spread along the southern shoreline of Lake Superior. In the late 1990s it was transferred via freighters to Thunder Bay, Lake Huron, and it is now poised to enter Saginaw Bay. It should flourish in this eutrophic bay. In Europe, the ruffe feeds on fish eggs and zooplankton. It was suspected of affecting stocks of yellow perch and trout-perch in the St. Louis harbor area probably because of its ability to feed at night (Busiahn and McClain 1995; Gunderson et al. 1998).

The most recent piscine invaders to become established in the Great Lakes are the tubenose and round gobies. They were first discovered in the St. Clair River in 1990 and have increased in abundance and spread throughout the river (Jude et al. 1992). As of 1995, they have been found in all five Great Lakes and are concentrated in the St.

Figure 23.4   Zebra mussels have been introduced to the Great Lakes and are of concern to fisheries managers as well as municipal and industrial users of water.

Clair-Detroit river corridor, Lake Erie, and southern Lake Michigan. In 1996, bait bucket transfers to the Flint and Shiawassee rivers, tributaries of Saginaw Bay, were also documented. The gobies are believed to have entered the Great Lakes basin before 1990, possibly with zebra mussels in the ballast water of foreign tankers from the Black or Caspian seas. The tubenose goby is an endangered species in Russia; the round goby reaches lengths of 400 mm and is harvested commercially in Russia (Jude et al. 1995). Both gobies are benthic, hearty fishes that prefer rocky substrates (Jude and DeBoe 1996); they are prey for walleyes, smallmouth bass, and other predators and have displaced mottled sculpin and possibly small percids, such as logperch and darters, that currently occupy benthic niches (Jude et al. 1995). An electrical barrier in the Chicago Sanitary Canal in southern Lake Michigan is under consideration for thwarting their entry into the Mississippi River.

Nonnative species, especially of marine origin, have at one time or another dominated the fish communities of all the Great Lakes. Since the 1850s, chlorides have increased in the lakes by a factor of three, which probably has enhanced survival of some of these species. The sea lamprey and Pacific salmon are dominant top predators in Lakes Ontario, Michigan, and Huron. White perch, pink salmon, ruffe, threespine stickleback, and gobies are recent invaders. There is ample evidence that we have not established well-integrated fish communities capable of resisting expansion of nonnative species.

23.6.6 Sea Lamprey Conundrums

The sea lamprey replaced the lake trout and burbot as the dominant terminal predator in the Great Lakes. Chemical control of sea lamprey has been successful in allowing re-establishment of top predators, including native lake trout and burbot and introduced salmon and trout (Eshenroder 1987). In Lake Ontario's U.S. waters during

1985–1992, sea lampreys killed 30% as many lake trout as did anglers (Elrod et al. 1995). According to Smith (1973), stocking will likely have to be continued indefinitely to maintain lake trout as long as current levels of sea lamprey predation and overfishing by commercial fishers, anglers, and Native Americans continue.

Average sea lamprey length has doubled over the past 2 decades in direct correspondence with salmonine stocking rates in Lake Michigan and probably the other Great Lakes as well. As a result, potential mortality imposed by a fixed number of sea lamprey may have increased as much as six-fold for small salmonines (Kitchell 1990). In addition, ammocetes may reside in soft sediments off the mouth of rivers, where they are too costly to treat; TFM may lose its potency; and resistant strains of sea lamprey may develop. Accordingly, alternatives to control sea lamprey (e.g., sterilization of males, Hansen and Manion 1980; attraction of adults using ammocetes pheromones, Li et al. 1995) are being implemented.

23.6.7 Toxic Substances

As of the late 1990s, toxic contamination of fishes is a serious problem facing Great Lakes fisheries and environmental agencies, but the problem is lessening. Identification of contaminants in several important fish species has caused disruption of commercial and sport fishing in some lakes. Despite some successes in toxic contaminant reduction, the IJC has identified 42 areas of concern in the basin (Hartig and Thomas 1988). These areas are identified mainly because of serious problems with fish and sediment contamination and nutrient inputs. These areas of concern are in tributaries, connecting channels, embayments, and nearshore areas, some of which are prime nursery, spawning, and fishing grounds.

D'Itri (1988) noted that between 1965 and 1980, PCBs, total DDT, and mercury concentrations in lake trout, bloater, coho salmon, and chinook salmon decreased significantly throughout the Great Lakes. These declines reflect more stringent controls on point discharges of contaminants into the Great Lakes since the 1970s. Since 1980, however, there appears to be a leveling off or even slight increase in contaminant burdens. Declines observed were probably due to easily controlled inputs, and now more intractable sources, such as the atmosphere, spurious industrial and domestic discharges, runoff, landfills, and resuspension of contaminated sediments continue to plague Great Lakes organisms.

Xenobiotic substances never before documented in aquatic systems are now common in some Lake Michigan biota. For example, Willford et al. (1981) demonstrated that water from Lake Michigan was contaminated to such a degree that survival of young lake trout carrying maternally inherited residues was impeded. Hesselberg and Seelye (1982) demonstrated that adult lake trout from Lake Michigan contained some 167 identified organic chemicals as compared with 8 identified compounds in hatchery broodfish reared in well water.

We appear to be at a plateau in the reduction of inputs of toxic substances into the Great Lakes. We now must deal with these anthropogenic substances in the various compartments of the ecosystem, especially sediments and fishes (Camanzo et al. 1987). Contamination of premium sport and commercial fishes persists and confounds fisheries man-

agement efforts to enhance stocks whose consumption may have long-term effects on human health. Tanner (1988) maintains that fisheries managers in the Great Lakes should
seriously consider their goals of restoring the long-lived lake trout and brown trout because
of their known accumulation of toxic substances. However, these fishes are bellwethers of
ecosystem health, and as long as they remain contaminated, we need to redouble our efforts
to control xenobiotic substance entry into and movement within the Great Lakes.
Bioaccumulation of many toxic substances through food webs, including potential endo-
crine disrupters (Colborn et al. 1993), continues to be a source of concern despite successes
in decreasing the input of toxic chemicals to the system.

23.6.8 Global Warming

Potential changes in species and abundances of fishes induced by global warming
may be dramatic (Meisner et al. 1987; Magnuson et al. 1997). Predicted warming of
air temperatures of 3.2–4.8°C could cause a reduction of extant populations of
salmonines and coregonines through reduction in preferred thermal habitat, whereas
range extensions are expected for more warmwater species (Mandrak 1989). Spawning and nursery areas may be reduced along with yields.

23.7   CONCLUSIONS

The Great Lakes exemplify a vast aquatic ecosystem that has been used and abused
by the humans who inhabit the basin. The dark forces of habitat destruction,
overexploitation, and nonnative species have made massive changes to the fishes and
fabric of the ecosystem. Attempts to rehabilitate native fishes have met with mixed
success—hatchery fish lack genetic diversity, incubation habitat may be deteriorated,
nonnative species inhibit native species growth and reproduction, and fishers are unwilling to reduce harvests to allow more spawner escapement.

The Great Lakes ecosystem has been altered for human use, often to the detriment
of the lakes. Water is used for drinking, industrial applications, and dilution of domestic wastes. Commercial and sport fishers ply these waters for premium species. Ships
move large quantities of goods throughout the system, often transporting nonnative
species in ballast water; recreational boats abound in nearshore waters. Demands continue for more fish, more harbors, and more water (within and outside the basin) for
irrigation, industry, and dilution of human wastes. We have not been good stewards.
We have allowed nutrient input because it was expedient and inexpensive; we have
allowed harbor development and shoreline and wetlands destruction for economic
growth; we have allowed toxic contamination to assault the Great Lakes from the land,
tributaries, and the air. We now know that dilution is not the solution, that toxic substances will accumulate somewhere, and that the premium fishes as well as humans are
the terminal receptors of these xenobiotic substances.

The changes in fish habitat and fish populations were incremental, insidious, and
sometimes irrevocable. No other large freshwater resource has undergone the amount
of change in the last 60 years as have the Great Lakes. We have been in a reactive mode
over most of the history of Great Lakes changes, doing postmortems on ecosystem ills,

seldom learning, and less often applying lessons of past failures. However, there is hope. Reconstituting the Great Lakes to their former diversity and elegance is impossible, but some compromise, based on applying the ecosystem approach as a template (Chapter 5) and retaining as many of the native species and as much of their genetic diversity as possible, along with maintaining some suite of introduced predators, may prove to be the best we can do with current levels of political will and technology. We have made improvements; however, we allowed conditions to deteriorate to grossly unacceptable levels before progress in rehabilitation was made. Nutrient enrichment has been lessened and overfishing curbed. We have slowed down toxic substance entry into the Great Lakes and have attained some semblance of control over the pestiferous alewife and the parasitic sea lamprey. There has been a resurgence of several native species, including walleye, lake sturgeon, and lake whitefish.

Anglers now enjoy fishing for a variety of native and introduced predators, from Atlantic salmon, through the various trout species, to walleye. Signs of successful rehabilitation are present throughout the Great Lakes. We are in an era of recovery. We will continue to need the economic and political will of the people and governments in the Great Lakes basin to maintain the Great Lakes and improve them for future residents so they too can catch uncontaminated yellow perch and chinook salmon from piers, swim and enjoy clear waters on beaches, and watch the eternal, ever varying sunrises and sunsets on these jeweled inland seas.

23.8   REFERENCES

Allan, R., A. Murdoch, and A. Sudar. 1983. An introduction to the Niagara River/Lake Ontario pollution problem. Journal of Great Lakes Research 9:249–273.

Anderson, D., and D. Clayton. 1959. Plankton in Lake Ontario. Ontario Department of Lands and Forests, Division of Research, Phytoplankton Section, Phytoplankton Research Note 1, Toronto.

Anderson, E., and L. L. Smith, Jr. 1971. Factors affecting abundance of lake herring (*Coregonus artedii* Lesueur) in western Lake Superior. Transactions of the American Fisheries Society 100:691–707.

Applegate, V. C. 1950. Natural history of the sea lamprey (*Petromyzon marinus*) in Michigan. U.S. Fish and Wildlife Service Special Scientific Report-Fisheries 55:1–237.

Applegate, V., J. Howell, J. Moffett, B. Johnson, and M. Smith. 1961. Use of 3-trifluoromethyl-4-nitrophenol as a selective sea lamprey larvicide. Great Lakes Fishery Commission Technical Report 1.

Applegate, V. C., and H. D. VanMeter. 1970. A brief history of commercial fishing in Lake Erie. U.S. Fish and Wildlife Service, Bureau of Commercial Fisheries Fishery Leaflet 630:1–28.

Auer, N. A. 1982. Identification of larval fishes of the Great Lakes Basin with emphasis on the Lake Michigan drainage. Great Lakes Fishery Commission, Special Publication 82–3, Ann Arbor, Michigan.

Baldwin, N. S., R. Saalfeld, M. Ross, and H. J. Buettner. 1979. Commercial fish production in the Great Lakes 1967–1977. Great Lakes Fishery Commission Technical Report 3.

Beeton, A. M. 1965. Eutrophication of the St. Lawrence Great Lakes. Limnology and Oceanography 10:240–254.

Berkes, F. 1983. Ontario's Great Lakes fisheries: managing the user groups. Institute of Urban and Environmental Studies, Brock University Working Paper 19, St. Catharines, Ontario.

Berst, A. H., and R. C. Simon. 1981. Introduction to the proceedings of the 1980 stock concept international symposium (STOCS). Canadian Journal of Fisheries and Aquatic Sciences 38:1457–1458.

Boileau, M. 1985. The expansion of white perch *Morone americana*, in the lower Great Lakes. Fisheries 10(1):6–10.

Borgmann, U. 1982. Particle-size-conversion efficiency and total animal production in pelagic ecosystems. Canadian Journal of Fisheries and Aquatic Sciences 39:668–674.

Borgmann, U., and D. Whittle. 1983. Particle-size conversion efficiency and contaminant concentrations in Lake Ontario biota. Canadian Journal of Fisheries and Aquatic Sciences 40:328–336.

Bowlby, J. N., A. Mathers, D. A. Hurley, and T. H. Eckert. 1991. The resurgence of walleye in Lake Ontario. Pages 169–205 in P. J. Colby, C. A. Lewis, and R. L. Eshenroder, editors. Status of walleye in the Great Lakes: case studies prepared for the 1989 workshop. Great Lakes Fishery Commission, Special Publication 91–1, Ann Arbor, Michigan.

Brandt, S. B. 1986a. Disappearance of the deepwater sculpin (*Myoxocephalus thompsoni*) from Lake Ontario: the keystone predator hypothesis. Journal of Great Lakes Research 12:18–24.

Brandt, S. B. 1986b. Food of trout and salmon in Lake Ontario. Journal of Great Lakes Research 12:200–205.

Brandt, S. B., D. M. Mason, D. B. MacNeill, T. C. Coates, and J. Gannon. 1987. Predation by alewives on larvae of yellow perch in Lake Ontario. Transactions of the American Fisheries Society 116:641–645.

Bridgeman, T. B., G. L. Fahnenstiel, G. A. Lang, and T. F. Nalepa. 1995. Zooplankton grazing during the zebra mussel (*Dreissena polymorpha*) colonization of Saginaw Bay, Lake Huron. Journal of Great Lakes Research 21:567–573.

Britt, N. W., A. J. Pliodzinskas, and E. M. Hair. 1980. Record low dissolved oxygen in the island area of Lake Erie. Ohio Journal of Science 68:175–179.

Bronte, C. P., J. H. Selgeby, and G. L. Curtis. 1991. Distribution, abundance, and biology of the alewife in U.S. waters of Lake Superior. Journal of Great Lakes Research 17:304–313.

Brown, E. H., Jr. 1972. Population biology of alewives, *Alosa pseudoharengus*, in Lake Michigan, 1949–1970. Journal of the Fisheries Research Board of Canada 29:477–500.

Brown, E. H., Jr., R. L. Argyle, N. R. Payne, and M. E. Holey. 1987. Yield and dynamics of destabilized chub (*Coregonus* spp.) populations in Lakes Michigan and Huron, 1950–84. Canadian Journal of Fisheries and Aquatic Sciences 44(Supplement 2):371–383.

Brown, E. H., G. W. Eck, N. R. Foster, R. M. Horrall, and C. E. Coberly. 1981. Historical evidence for discrete stocks of lake trout (*Salvelinus namaycush*) in Lake Michigan. Canadian Journal of Fisheries and Aquatic Sciences 38:1747–1758.

Busiahn, T. R., and J. R. McClain. 1995. Status and control of ruffe (*Gymnocephalus cernuus*) in Lake Superior and potential for range expansion. Pages 461–470 in M. Munawar, T. Edsall, and J. L. Leach, editors. The Lake Huron ecosystem: ecology, fisheries and management. Ecovision World Monograph Series, SPB Academic Publishing, Amsterdam, The Netherlands.

Camanzo, J., C. Rice, D. Jude, and R. Rossmann. 1987. Organic priority pollutants in nearshore fish from 14 Lake Michigan tributaries and embayments, 1983. Journal of Great Lakes Research 13:296–309.

Carpenter, S., J. F. Kitchell, and J. Hodgson. 1985. Cascading trophic interactions and lake productivity-fish predation and herbivory can regulate lake ecosystems. BioScience 35:634–639.

Casselman, J. M, D. M. Brown, and J. A. Hoyle. 1996. Resurgence of lake whitefish, *Coregonus clupeaformis*, in Lake Ontario in the 1980s. Great Lakes Research Review 2(2):20–28.

Christie, W. J. 1972. Lake Ontario: effects of exploitation, introductions, and eutrophication on the salmonid community. Journal of the Fisheries Research Board of Canada 29:913–929.

Christie, W. J. 1974. Changes in the fish species composition of the Great Lakes. Journal of the Fisheries Research Board of Canada 31:827–854.

Christie, W. J., K. A. Scott, P. G. Sly, and R. H. Strus. 1987b. Recent changes in the aquatic food web of eastern Lake Ontario. Canadian Journal of Fisheries and Aquatic Sciences 44 (Supplement 2):37–52.

Christie, W. J., and six coauthors. 1987a. A perspective on Great Lakes fish community rehabilitation. Canadian Journal of Fisheries and Aquatic Sciences 44 (Supplement 2):486–499.

Coble, D. W., R. E. Bruesewitz, T. W. Fratt, and J. W. Scheirer. 1990. Lake trout, sea lampreys, and overfishing in the upper Great Lakes: a review and reanalysis. Transactions of the American Fisheries Society 119:985–995.

Colborn, T., F. S. vom Saal, and A. M. Soto. 1993. Developmental effects of endocrine-disrupting chemicals in wildlife and humans. Environmental Health Perspectives 101:378–384.

Colby, P. J. 1977. Introduction to the proceedings of the 1976 percid international symposium (PERCIS). Journal of the Fisheries Research Board of Canada 34:1447–1449.

Crowder, L. B., M. E. McDonald, and J. A. Rice. 1987. Understanding recruitment of lake Michigan fishes: the importance of size-based interactions between fish and zooplankton. Canadian Journal of Fisheries and Aquatic Sciences 44 (Supplement 2):141–147.

D'Itri, F. 1988. Contaminants in selected fishes from the upper Great Lakes. Pages 51–84 *in* N. W. Schmidtke, editor. Toxic contamination in large lakes, vol. II, impact of toxic contaminants on fisheries management. Proceedings of a technical session of the world conference on large lakes. Lewis Publishers, Chelsea, Michigan.

Dehring, T., A. Brown, D. Daugherty, and S. R. Phelps. 1981. Survey of the genetic variation among eastern Lake Superior lake trout (*Salvelinus namaycush*). Canadian Journal of Fisheries and Aquatic Sciences 38:1738–1746.

Dobson, H. F. H., M. Gilbertson, and P. G. Sly. 1974. A summary and comparison of nutrients and related water quality in Lakes Erie, Ontario, Huron, and Superior. Canadian Journal of Fisheries and Aquatic Sciences 31:731–738.

Ebener, M. P. 1995. The state of Lake Huron in 1992. Great Lakes Fishery Commission Special Publication 95–2, Ann Arbor, Michigan.

Ebener, M. P. 1997. Recovery of lake whitefish populations in the Great Lakes. Fisheries 22(7):18–20.

Eck, G., and E. Brown, Jr. 1985. Lake Michigan's capacity to support lake trout (*Salvelinus namaycush*) and other salmonines: an estimate based on the status of prey populations in the 1970s. Canadian Journal of Fisheries and Aquatic Sciences 42:449–454.

Eck, G., and L. Wells. 1987. Recent changes in Lake Michigan's fish community and their probable causes, with emphasis on the role of alewife (*Alosa pseudoharengus*). Canadian Journal of Fisheries and Aquatic Sciences 44 (Supplement 2):53–60.

Eisenreich, S., B. Looney, and J. Thornton. 1981. Airborne organic contaminants in the Great Lakes ecosystem. Environmental Science and Technology 15:30–38.

Elrod, J. H., and six coauthors. 1995. Lake trout rehabilitation in Lake Ontario. Journal of Great Lakes Research 21(Supplement 1):83–107.

Emery, L. 1985. Review of fish species introduced into the Great Lakes, 1819–1974. Great Lakes Fishery Commission, Technical Report 45, Ann Arbor, Michigan.

Eshenroder, R. L. 1977. Effects of intensified fishing, species changes, and spring water temperatures on yellow perch, *Perca flavescens*, in Saginaw Bay. Journal of the Fisheries Research Board of Canada 34:1830–1838.

Eshenroder, R. L. 1987. Socioecononomic aspects of lake trout rehabilitation in the Great Lakes. Transactions of the American Fisheries Society 116:309–313.

Eshenroder, R. L., N. R. Payne, J. E. Johnson, C. Bowen II, and M. P. Ebener. 1995. Lake trout rehabilitation in Lake Huron. Journal of Great Lakes Research 21 (Supplement 1):108–127.

Eshenroder, R. L., T. P. Poe, and C. Olver. 1984. Strategies for rehabilitation of lake trout in the Great Lakes: proceedings of a conference on lake trout research. Great Lakes Fishery Commission, Technical Report 40, Ann Arbor, Michigan.

Eshenroder, R. L., and seven coauthors. 1987. Report of the St. Marys River sea lamprey task force to the Great Lakes Fishery Commission. Great Lakes Fishery Commission, Ann Arbor, Michigan.

Evans, D. O., and five coauthors. 1987. Concepts and methods of community ecology applied to freshwater fisheries management. Canadian Journal of Fisheries and Aquatic Sciences 44: (Supplement 2):448–470.

Evans, M. S. 1988. *Bythotrephes cederstroemi*: its new appearance in Lake Michigan. Journal of Great Lakes Research 14:234–240.

Evans, M. S., and D. Jude. 1986. Recent shifts in *Daphnia* community structure in southeastern Lake Michigan: a comparison of the inshore and offshore. Limnology and Oceanography 31:56–67.

Fetterolf, C. M., Jr. 1980. Why a Great Lakes Fishery Commission and why a sea lamprey international symposium. Canadian Journal of Fisheries and Aquatic Sciences 37:1588–1593.

Fetterolf, C. M., Jr. 1988. Uniting habitat quality and fishery programs in the Great Lakes. Pages 85–100 *in* N. W. Schmidtke, editor. Toxic contamination in large lakes, volume 2. Impact of toxic contaminants on fisheries management. Lewis Publishers, Chelsea, Michigan.

Fisher, J. P., and six coauthors. 1995. Reproductive failure of landlocked Atlantic salmon from New York's Finger Lakes: investigation into the etiology and epidemiology of the "Cayuga syndrome." Journal of Aquatic Animal Health 7:81–94.

Fitzsimons, J. D. 1995. The effect of B-vitamins on a swim-up syndrome in Lake Ontario lake trout. Journal of Great Lakes Research 21 (Supplement 1):286–289.

Fitzsimons, J. D., J. H. Leach, S. J. Nepszy, and V. W. Cairns. 1995. Impacts of zebra mussels on walleye (*Stizostedion vitreum*) reproduction in western Lake Erie. Canadian Journal of Fisheries and Aquatic Sciences 52:578–586.

Francis, G. 1987. Toward understanding Great Lakes "organizational ecosystems." Journal of Great Lakes Research 13:233–234.

Francis, G. R., J. J. Magnuson, H. Regier, and D. Talhelm, editors. 1979. Rehabilitating Great Lakes ecosystems. Great Lakes Fishery Commission, Technical Report 37, Ann Arbor, Michigan.

Francis, J., S. Robillard, and J. E. Marsden. 1996. Yellow perch management in Lake Michigan: a multi-jurisdictional challenge. Fisheries 21(2):18–20.

FWPCA (Federal Water Pollution Control Administration). 1968. Pollution of Lake Erie and its tributaries. U.S. Department of the Interior, FWPCA, Great Lakes Region, Cleveland Program Office, Cleveland, Ohio.

Gale, R. P. 1987. Resource miracles and rising expectations: a challenge to fishery managers. Fisheries 12(5):8–13.

GLFC (Great Lakes Fishery Commission). 1980. A joint strategic plan for management of Great Lakes fisheries. Publication of the Great Lakes Fishery Commission, Ann Arbor, Michigan, December, 1980.

Governments of the United States and Canada. 1988a. Rehabilitation of Lake Ontario: the role of nutrient reduction and food web dynamics. Report to the Great Lakes Science Advisory Board. Windsor, Ontario.

Governments of the United States and Canada. 1988b. Revised Great Lakes Water Quality Agreement of 1978 as amended by protocol signed November 18, 1987. International Joint Commission, Washington, D.C., and Ottawa.

Governments of the United States and Canada. 1997. State of the Great Lakes ecosystem conference. Environment Canada, Burlington, Ontario and U.S. Environmental Protection Agency, Chicago.

Great Lakes Water Quality Board. 1981. Report on Great Lakes water quality. Appendix, Great Lakes surveillance. International Joint Commission, Windsor, Ontario.

Gunderson, J. L., M. R. Klepinger, C. R. Bronte, and J. E. Marsden. 1998. Overview of the international symposium on Eurasian ruffe (*Gymnocephalus cernuus*) biology, impacts, and control. Journal of Great Lakes Research 24(2):165–169.

Hansen, L., and P. Manion. 1980. Sterility method of pest control and its potential role in an integrated sea lamprey (*Petromyzon marinus*) control program. Canadian Journal of Fisheries and Aquatic Sciences 37:2108–2117.

Hansen, M. J., editor. 1994. The state of Lake Superior in 1992. Great Lakes Fishery Commission, Special Publication 94–1, Ann Arbor, Michigan.

Hartig, J. H., and R. L. Thomas. 1988. Development of plans to restore degraded areas in the Great Lakes. Environmental Management 132:327–347.

Hartman, W. L. 1972. Lake Erie: effects of exploitation, environmental changes and new species on the fishery resources. Canadian Journal of Fisheries and Aquatic Sciences 29:899–912.

Hartman, W. L. 1988. Historical changes in the major fish resources of the Great Lakes. Pages 103–132 *in* M. E. Evans, editor. Toxic contaminants and ecosystem health: a Great Lakes focus. Wiley, New York.

Hatch, R., S. Nepszy, K. Muth, and C. Baker. 1987. Dynamics of the recovery of the western Lake Erie walleye (*Stizostedion vitreum vitreum*) stock. Canadian Journal of Fisheries and Aquatic Sciences 44(Supplement 2):15–22.

Hayward, R. S., and F. J. Margraf. 1987. Eutrophication effects on prey size and food available to yellow perch in Lake Erie. Transactions of the American Fisheries Society 116:210–223.

Henderson, B. A., and E. H. Brown, Jr. 1985. Effects of abundance and water temperature on recruitment and growth of alewife (*Alosa pseudoharengus*) near South Bay, Lake Huron, 1954–82. Canadian Journal of Fisheries and Aquatic Sciences 42:1608–1613.

Herdendorf, C. E. 1982. Large lakes of the world. Journal of Great Lakes Research 8:379–412.

Hesselberg, R. J., and J. G. Seelye. 1982. Identification of organic compounds in Great Lakes fishes by gas chromatography/mass spectrometry: 1977. Great Lakes Fishery Laboratory, Administration Report 82–1, Ann Arbor, Michigan.

Horrall, R. 1981. Behavioral stock-isolating mechanisms in Great Lakes fishes with special reference to homing and site imprinting. Canadian Journal of Fisheries and Aquatic Sciences 38:1481–1496.

Johannsson, O. E. 1987. Comparison of Lake Ontario zooplankton communities between 1967 and 1985: before and after implementation of salmonid stocking and phosphorus control. Journal of Great Lakes Research 13:328–339.

Jones, M. L., J. F. Koonce, and R. O'Gorman. 1993. Sustainability of hatchery-dependent salmonine fisheries in Lake Ontario: conflicts between predator demand and prey supply. Transactions of the American Fisheries Society 122:1002–1018.

Jude, D. J. 1992. Evidence for natural reproduction by stocked walleyes in the Saginaw River tributary system, Michigan. North American Journal of Fisheries Management 12:386–395.

Jude, D. J., and S. DeBoe. 1996. Possible impact of gobies and other introduced species on habitat restoration efforts. Canadian Journal of Fisheries and Aquatic Sciences 53(Supplement 1):136–141.

Jude, D. J., J. Janssen, and G. Crawford. 1995. Ecology, distribution, and impact of the newly introduced round and tubenose gobies on the biota of the St. Clair and Detroit Rivers. Pages 447–460 in M. Munawar, T. Edsall, and J. L. Leach, editors. The Lake Huron ecosystem: ecology, fisheries and management. Ecovision World Monograph Series, SPB Academic Publishing, Amsterdam, The Netherlands.

Jude, D. J., S. A. Klinger, and M. D. Enk. 1981. Evidence of natural reproduction by planted lake trout in Lake Michigan. Journal of Great Lakes Research 7:57–61.

Jude, D. J., and J. Pappas. 1992. Fish utilization of Great Lakes coastal wetlands. Journal of Great Lakes Research 18:651–672.

Jude, D. J., R. H. Reider, and G. R. Smith. 1992. Establishment of Gobiidae in the Great Lakes basin. Canadian Journal of Fisheries and Aquatic Sciences 49:416–421.

Jude, D., and F. Tesar. 1985. Recent changes in the inshore forage fish of Lake Michigan. Canadian Journal of Fisheries and Aquatic Sciences 42:1154–1157.

Jude, D. J., F. J. Tesar, S. DeBoe, and T. J. Miller. 1987. Diet and selection of major prey species by Lake Michigan salmonines, 1973–1982. Transactions of the American Fisheries Society 116:677–691.

Kerr, S. J., and G. C. LeTendre. 1991. The state of the Lake Ontario fish community in 1989. Great Lakes Fishery Commission Special Publication 91–3, Ann Arbor, Michigan.

Kerr, S. R., and L. M. Dickie. 1984. Measuring the health of aquatic ecosystems. Pages 279–284 in V. W. Cairns, P. V. Hodson, and J. O. Nriagu, editors. Contaminant effects on fisheries. Wiley, New York.

Kitchell, J. F. 1990. The scope for mortality caused by sea lamprey. Transactions of the American Fisheries Society 119:642–648.

Knight, R. L. 1997. Successful interagency rehabilitation of Lake Erie walleye. Fisheries 22(7):16–17.

Knight, R. L., F. J. Margraf, and R. F. Carline. 1984. Piscivory by walleyes and yellow perch in western Lake Erie. Transactions of the American Fisheries Society 113:677–693.

Koelz, W. 1929. Coregonid fishes of the Great Lakes. U.S. Bureau of Fisheries Bulletin 43:1–643.

Kohler, C. C., and J. J. Ney. 1980. Piscivory in a land-locked alewife (Alosa pseudoharengus) population. Canadian Journal of Fisheries and Aquatic Sciences 37:1314–1317.

Krieger, K. M., and six coauthors. 1996. Recovery of burrowing mayflies (Ephemeroptera: Ephemeridae: Hexagenia) in western Lake Erie. Journal of Great Lakes Research 22:254–263.

Krueger, C. C., M. L. Jones, and W. W. Taylor. 1995a. Restoration of lake trout in the Great Lakes: challenges and strategies for future management. Journal of Great Lakes Research 21 (Supplement 1):547–558.

Krueger, C. C., K. L. Perkins, E. L. Mills, and J. E. Marsden. 1995b. Predation by alewives on lake trout fry in Lake Ontario: role of an exotic species in preventing restoration of a native species. Journal of Great Lakes Research 21(Supplement. 1):458–469.

Krueger, C. C., B. L. Swanson, and J. H. Selgeby. 1986. Evaluation of hatchery-reared lake trout for reestablishment of populations in the Apostle Islands region of Lake Superior, 1960–84. Pages 93–107 in R. H. Stroud, editor. Fish culture in fisheries management. American Fisheries Society, Fish Culture Section and Fisheries Management Section, Bethesda, Maryland.

Lawrie, A. H. 1970. The sea lamprey in the Great Lakes. Transactions of the American Fisheries Society 99:766–775.

Lawrie, A. H. 1978. The fish community of Lake Superior. Journal of Great Lakes Research 4:513–549.

Lawrie, A. H., and J. F. Rahrer. 1972. Lake Superior: effects of exploitation and introductions on the salmonid community. Journal of the Fisheries Research Board of Canada 29:765–776.

Leach, J. H. 1993. Impacts of the zebra mussel (*Dreissena polymorpha*) on water quality and fish spawning reefs in western Lake Erie. Pages 381–397 *in* T. F. Nalepa and D. W. Schloesser, editors. Zebra mussels: biology, impacts, and control. Lewis Publishers, Boca Raton, Florida.

Leach, J. H., and five coauthors. 1977. Responses of percid fishes and their habitats to eutrophication. Journal of the Fisheries Research Board of Canada 34:1964–1971.

Leach, J. H., and S. J. Nepszy. 1976. The fish community in Lake Erie. Journal of the Fisheries Research Board of Canada 33:622–638.

Li, W., P. W. Sorensen, and D. S. Gallagher. 1995. The olfactory system of migrating adult sea lamprey (*Petromyzon marinus*) is specifically and acutely sensitive to unique bio acids released by conspecific larvae. Journal of General Physiology 105:569–587.

Loftus, K. H., and H. A. Regier. 1972. Introduction to other proceedings of the 1971 symposium on salmonid communities in oligotrophic lakes. Journal of the Fisheries Research Board of Canada 29:613–616.

Lowe, R. L., and R. W. Pillsbury. 1995. Shifts in benthic algal community structure and function following the appearance of zebra mussels (*Dreissena polymorpha*) in Saginaw Bay, Lake Huron. Journal of Great Lakes Research 21:558–566.

Lubomudrov, L. M., R. A. Moll, and M. G. Parsons. 1997. An evaluation of the feasibility and efficacy of biocide application in controlling the release of nonindigenous aquatic species from ballast water. Final report to Michigan Department of Environmental Quality, Office of the Great Lakes, Lansing.

MacCallum, W. R., and J. H. Selgeby. 1987. Lake Superior revisited 1984. Canadian Journal of Fisheries and Aquatic Sciences 44 (Supplement 2):23–36.

MacIsaac, H. J., I. A. Grigorovich, J. A. Hoyle, N. D. Yan, and V. E. Panov. 1999. Invasion of Lake Ontario by the Ponto-Caspian predatory cladoceran *Cercopagis pengoi*. Canadian Journal of Fisheries and Aquatic Sciences 56:1–5.

Magnuson, J. J., and 11 coauthors. 1997. Potential effects of climate changes on aquatic systems: Laurentian Great Lakes and precambrian shield region. Hydrological Processes 11:825–871.

Makarewicz, J. C., and P. Bertram. 1993. Evidence for the restoration of the Lake Erie ecosystem. Journal of Great Lakes Research 19:197.

Mandrak, N. E. 1989. Potential invasion of the Great Lakes by fish species associated with climatic warming. Journal of Great Lakes Research 15:306–316.

Matuszek, J. E. 1978. Empirical predictions of fish yields of North American lakes. Transactions of the American Fisheries Society 107:385–394.

Meisner, J. D., J. L. Goodier, H. A. Regier, B. J. Shuter, and W. J. Christie. 1987. An assessment of the effects of climate warming on Great Lakes basin fishes. Journal of Great Lakes Research 13:340–352.

Mills, E., J. E. Leach, J. T. Carlton, and C. L. Secor. 1993a. Exotic species in the Great Lakes: a history of biotic crises and anthropogenic introductions. Journal of Great Lakes Research 19:1–54.

Mills, E. L., J. H. Leach, J. T. Carlton, and C. L. Secor. 1994. Exotic species and the integrity of the Great Lakes: lessons from the past. BioScience 44:666–667.

Mills, E. L., and six coauthors. 1993b. Colonization, ecology, and population structure of the "quagga" mussel (Bivalvia: Dreissenidae) in the lower Great Lakes. Canadian Journal of Fisheries and Aquatic Sciences 50:2305–2314.

Minns, E. K., D. A. Hurley, and K. H. Nicholls, editors. 1986. Project Quinte: point source phosphorus control and ecosystem response in the Bay of Quinte, Lake Ontario. Canadian Special Publication of Fisheries and Aquatic Sciences 86.

Muth, K., and D. Wolfert. 1986. Changes in growth and maturity of walleyes associated with stock rehabilitation in western Lake Erie, 1964–1983. North American Journal of Fisheries Management 6:168–175.

Nalepa, T. F., and G. L. Fahnenstiel. 1995. *Dreissena polymorpha* in the Saginaw Bay, Lake Huron ecosystem: overview and perspective. Journal of Great Lakes Research 21:411–416.

Nalepa, T. F., D. J. Hartson, D. L. Fanslow, G. A. Lang, and S. J. Lozano. 1998. Declines in benthic macroinvertebrate populations in southern Lake Michigan, 1980–1993. Canadian Journal of Fisheries and Aquatic Sciences 55:2402–2413.

Oglesby, R. T. 1977. Relationships of fish yield to lake phytoplankton standing crop, production and morphoedaphic factors. Journal of the Fisheries Research Board of Canada 34:2271–2279.

O'Gorman, R., R. Bergstedt, and T. Eckert. 1987. Prey fish dynamics and salmonine predator growth in Lake Ontario, 1978–84. Canadian Journal of Fisheries and Aquatic Sciences 44 (Supplement 2):390–403.

Perrone, M., Jr., P. J. Schneeberger, and D. J. Jude. 1983. Distribution of larval yellow perch (*Perca flavescens*) in nearshore waters of southeastern Lake Michigan. Journal of Great Lakes Research 9:517–522.

Phillips, R. B., and T. J. Ehlinger. 1995. Evolutionary and ecological considerations in the reestablishment of Great Lakes coregonid fishes. Pages 133–144 *in* J. L. Nielsen, editor. Evolution and the aquatic ecosystem: defining unique units in population conservation. American Fisheries Society, Symposium 17, Bethesda, Maryland.

Potter, R. L., and G. W. Fleischer. 1992. Reappearance of spoonhead sculpin (*Cottus ricei*) in Lake Michigan. Journal of Great Lakes Research 18:755–758.

Pratt, D., W. H. Blust, and J. H. Selgeby. 1992. Ruffe, *Gynnocephalus cernuus*: newly introduced in North America. Canadian Journal of Fisheries and Aquatic Sciences 49:1616–1618.

Rakoczy, G. P., and R. D. Rogers. 1987. Sportfishing catch and effort from the Michigan waters of Lakes Michigan, Huron, and Erie, and their important tributary streams. Michigan Department of Natural Resources, Fisheries Division, Fisheries Technical Report 87–6A, Lansing.

Regier, H. A. 1979. Changes in species composition of Great Lakes fish communities caused by man. Proceedings of the 44th North American Wildlife Conference 558–566.

Regier, H. A., V. C. Applegate, and R. A. Ryder. 1969. The ecology and management of the walleye in western Lake Erie. Great Lakes Fishery Commission, Technical Report 15, Ann Arbor, Michigan.

Regier, H. A., and W. L. Hartman. 1973. Lake Erie's fish community: 150 years of cultural stresses. Science 180:1248–1255.

Robillard, S. R., and J. E. Marsden. 1997. Yellow perch population assessment in southwestern Lake Michigan, including evaluation of sampling techniques and identification of the factors that determine yellow perch year-class strength. Annual report to the Illinois Department of Natural Resources, Aquatic Ecology Technical Report 97/9, Springfield.

Ryder, R. A. 1972. The limnology and fishes of oligotrophic glacial lakes in North America (about 1800 AD). Journal of the Fisheries Research Board of Canada 29:617–628.

Ryder, R. A. 1977. Effects of ambient light variations on behavior of yearling, subadult, and adult walleyes (*Stizostedion vitreum vitreum*). Journal of the Fisheries Research Board of Canada 34:1481–1491.

Ryder, R. A., and C. Edwards. 1985. A conceptual approach for the application of biological indicators of ecosystem quality in the Great Lakes Basin. Report to the International Joint Commission's Science Advisory Board, Windsor, Ontario.

Ryder, R. A., S. R. Kerr, W. W. Taylor, and P. A. Larkin. 1981. Community consequences of fish stock diversity. Canadian Journal of Fisheries and Aquatic Sciences 38:1856–1866.

Samples, K., and R. Bishop. 1987. An economic analysis of integrated fisheries management: the case of the Lake Michigan alewife and salmonid fisheries. Journal of Great Lakes Research 8:593–602.

Scavia, D., G. Fahnenstiel, M. Evans, D. Jude, and J. Lehman. 1986. Influence of salmonine predation and weather on long-term water quality trends in Lake Michigan. Canadian Journal of Fisheries and Aquatic Sciences 43:435–443.

Schneider, J., and J. Leach. 1977. Walleye (*Stizostedion vitreum vitreum*) fluctuations in the Great Lakes and possible causes, 1800–1975. Journal of the Fisheries Research Board of Canada 34:1878–1889.

Schreiner, D. R., and S. T. Schram. 1997. Lake trout rehabilitation in Lake Superior. Fisheries 22(7):12–14.

Selgeby, J. H. 1982. Decline of lake herring (*Coregonus artedii*) in Lake Superior: an analysis of the Wisconsin herring fishery, 1936–78. Canadian Journal of Fisheries and Aquatic Sciences 39:554–563.

Selgeby, J. H. 1985. Population trends of lake herring (*Coregonus artedii*) and rainbow smelt (*Osmerus mordax*) in U.S. waters of Lake Superior, 1968–1984. Pages 1–12 *in* R. L. Eshenroder, editor. Great Lakes Fishery Commission, Technical Report 853.

Selgeby, J. H. 1995. Introduction to the proceedings of the 1994 international conference on restoration of lake trout in the Laurentian Great Lakes. Journal of Great Lakes Research 21(Supplement 1):1–2.

Selgeby, J. H., W. R. MacCallum, and D. V. Swedberg. 1978. Predation by rainbow smelt (*Osmerus
mordax*) on lake herring (*Coregonus artedii*) in western Lake Superior. Journal of the Fisheries
Research Board of Canada 35:1457–1463.

Sheldon, R. W., A. Prakash, and W. H. Sutcliffe, Jr. 1972. The size distribution of particles in the ocean.
Limnology and Oceanography 17:327–340.

Shroyer, S. M., and T. S. McComish. 1998. Forecasting abundance of quality-size yellow perch in Indiana waters of Lake Michigan. North American Journal of Fisheries Management 18:19–24.

Skubinna, J. P., T. G. Coon, and T. R. Batterson. 1995. Increased abundance and depth of submersed
macrophytes in response to decreased turbidity in Saginaw Bay, Lake Huron. Journal of Great
Lakes Research 21:476–488.

Smith, I. R., and T. Coape-Arnold. 1995. Lakewide management planning in the Laurentian Great Lakes:
a perspective for Lake Huron. Pages 483–498 *in* M. Munawar, T. Edsall, and J. L. Leach, editors.
The Lake Huron ecosystem: ecology, fisheries and management. Ecovision World Monograph Series, SPB Academic Publishing, Amsterdam, The Netherlands.

Smith, S. H. 1964. Status of the deepwater cisco population of Lake Michigan. Transactions of the
American Fisheries Society 93:155–163.

Smith, S. H. 1968. Species succession and fishery exploitation in the Great Lakes. Journal of the Fisheries Research Board of Canada 25:667–693.

Smith, S. H. 1970a. Trends in fishery management of the Great Lakes. Pages 107–114 *in* N. Benson,
editor. A century of fisheries in North America. American Fisheries Society, Special Publication 7,
Bethesda, Maryland.

Smith, S. H. 1970b. Species interactions of the alewife in the Great Lakes. Transactions of the American
Fisheries Society 99:754–765.

Smith, S. H. 1972a. Factors in ecological succession in oligotrophic fish communities of the Laurentian
Great Lakes. Journal of the Fisheries Research Board of Canada 29:717–730.

Smith, S. H. 1972b. The future of salmonid communities in the Laurentian Great Lakes. Journal of the
Fisheries Research Board of Canada 29:951–957.

Smith, S. H. 1973. Application of theory and research in fishery management of the Laurentian Great
Lakes. Transactions of the American Fisheries Society 101:156–163.

Spangler, G., K. Loftus, and W. J. Christie. 1987. Introduction to the international symposium on stock
assessment and yield prediction (ASPY). Canadian Journal of Fisheries and Aquatic Sciences 44
(Supplement 2):7–9.

Spangler, G. R., N. R. Payne, and G. K. Winterton. 1977. Percids in the Canadian waters of Lake Huron.
Journal of the Fisheries Research Board of Canada 34:1839–1848.

Stanczykowska, A. 1978. Occurrence and dynamics of *Dreissena polymorpha* (Pall.) (Bivalvia).
Verhandlungen Internationale Vereinigung Limnologie 20:2431–2434.

Starliper, C. E., D. R. Smith, and T. Shatzer. 1997. Virulence of *Renibacterium salmoninarum* to salmonids. Journal of Aquatic Animal Health 9:1–7.

Stedman, R. M., and C. A. Bowen, II. 1985. Introduction and spread of the threespine stickleback
(*Gasterosteus aculeatus*) in Lakes Huron and Michigan. Journal of Great Lakes Research 11:508–511.

Steedman, R. J., and H. A. Regier. 1987. Ecosystem science for the Great Lakes: perspectives on
degradative and rehabilitation transformations. Canadian Journal of Fisheries and Aquatic Sciences 44 (Supplement 2):95–103.

Stewart, D. J., J. F. Kitchell, and L. B. Crowder. 1981. Prey fishes and their salmonid predators in Lake
Michigan. Transactions of the American Fisheries Society 110:751–763.

Sullivan, R., T. Heidtke, J. Hall, and W. Sonzogni. 1981. Potential impact of changes in Great Lakes
water quality on fisheries. Great Lakes environmental planning study (GLES). Great Lakes Basin
Commission, Contribution 26, Ann Arbor, Michigan.

Sweet, C. W. 1992. International monitoring of the deposition of airborne toxic substances to the Great
Lakes. Proceedings of the 9th World Clean Air Congress. Air and Waste Management Association,
Pittsburgh, Pennsylvania.

Tanner, H. 1988. Restocking of Great Lakes fishes and reduction of environmental contaminants 1960–1980. Pages 209–228 *in* N. W. Schmidtke, editor. Toxic contamination in large lakes, volume 2. Impact of toxic contaminants on fisheries management. Lewis Publishers, Chelsea, Michigan.

Vallentyne, J. R. 1974. The algal bowl. Lakes and man. Canada Department of the Environment, Miscellaneous Special Publication 22, Ottawa.

Van Meter, H. D., and M. B. Trautman. 1970. An annotated list of the fishes of Lake Erie and its tributary waters exclusive of the Detroit River. Ohio Journal of Science 70:65–78.

Warren, G. J., and J. T. Lehman. 1988. Young-of-the-year *Coregonus hoyi* in Lake Michigan: prey selection and influence on the zooplankton community. Journal of Great Lakes Research 14:420–426.

Wells, L. 1970. Effects of alewife predation on zooplankton populations in Lake Michigan. Limnology and Oceanography 15:556–565.

Wells, L. 1980. Food of alewives, yellow perch, spottail shiners, trout-perch, and slimy and fourhorn sculpins in southeastern Lake Michigan. U.S. Fish and Wildlife Service Technical Papers 98.

Wells, L., and A. L. McLain. 1973. Lake Michigan: man's effects on native fish stocks and other biota. Great Lakes Fishery Commission, Technical Report 20, Ann Arbor, Michigan.

Whillans, T. H. 1979. Historic transformations of fish communities in three Great Lakes bays. Journal of Great Lakes Research 5:195–215.

Whillans, T. H. 1982. Changes in marsh area along the Canadian shore of Lake Ontario. Journal of Great Lakes Research 8:570–577.

Willford, W. A., and eight coauthors. 1981. Chlorinated hydrocarbons as a factor in the reproduction and survival of lake trout (*Salvelinus namaycush*) in Lake Michigan. U.S. Fish and Wildlife Service Technical Paper 105.

Williams, J. E., and seven coauthors. 1989. Fishes of North America, endangered, threatened, or of special concern. Fisheries 14(6):2–20.

Witt, J. D. S., P. D. N. Hebert, and W. B. Morton. 1997. *Echinogammarus ischus*: another crustacean invader in the Laurentian Great Lakes. Canadian Journal of Fisheries and Aquatic Sciences 54:264–268.

Zaranko, D. T., D. G. Farara, and F. G. Thompson. 1997. Another exotic mollusk in the Laurentian Great Lakes: the New Zealand native *Potamopyrgus antipodarum* (Gray 1843) (Gastropoda, Hydrobiidae). Canadian Journal of Fisheries and Aquatic Sciences 54:809–814.

Chapter 24

Anadromous Stocks

JOHN R. MORING

24.1 INTRODUCTION

Effective management of anadromous fishes (fishes that are born in freshwater, travel to the sea to grow, and then return to freshwater to spawn; see Box 24.1) is often more complicated than management of inland species. This is because anadromous stocks typically cross state, provincial, and national boundaries and use international waters. Even if passage by anadromous species between the freshwater and marine domains is limited, the management of such species likely involves more than one principal management agency. Sound management of anadromous sport fisheries is further complicated because portions of the life cycles of such fishes are in oceanic habitats, where managers have few options except control of harvest.

Following a series of court decisions in the 1970s and 1980s, Native American tribes in many areas obtained or reaffirmed legal jurisdiction of fishes, guaranteed portions of the catch, or both. The Boldt decision (*United States v Washington*) of 1974 ordered the catch of steelhead and Pacific salmon in western Washington to be divided evenly between the state and the tribes. As a result, the state of Washington and 21 tribes now determine catch and allocation of these resources; a court-established Fisheries Advisory Board handles disputes (Clark 1985). Similar decisions by Judge Belloni (*United States v Oregon*, 1969 and 1974) affect four tribes and the states of Washington and Oregon in the management of Pacific salmon and steelhead in the Columbia River basin (Marsh and Johnson 1985). Principal fisheries issues for Native American tribes in the United States and Canada have been reviewed in detail by Busiahn (1984).

Fisheries managers responsible for anadromous species must often deal with both sport and commercial interests. They must address the concerns of these two groups as well as ensure that tribal interests are protected and still protect the fishes. Often, management issues are legally as well as biologically complex. For example, federally protected marine mammals have been actively depleting runs of Pacific salmon and steelhead at the Chittenden Locks in Seattle, Washington. Fisheries managers are caught between legal restrictions on the control of protected mammals and the decline of important salmonid fisheries. Therefore, managers must rely on all available tools when making decisions: research, fish culture, regulation, habitat maintenance and restoration, and monitoring.

**Box 24.1   North American Anadromous Fishes Having Commercial
or Sport Fisheries Importance or Endangered Status**

Alewife	Delta smelt	Sea-run cutthroat trout
American shad	Eulachon	Sea-run Dolly Varden
Atlantic salmon	Green sturgeon	Shortnose sturgeon
Atlantic sturgeon	Pink salmon	Sockeye salmon
Blueback herring	Rainbow smelt	Steelhead
Chinook salmon	Sea-run Arctic char	Striped bass
Chum salmon	Sea-run brook trout	White sturgeon
Coho salmon	Sea-run brown trout	

In this chapter, I discuss the roles of fish culture, stock management, and fisheries management, as well as the roles of interagency cooperation and public involvement, in the management of anadromous fishes. I then review the status of and restorative efforts for most North American anadromous species.

24.2   THE ROLE OF FISH CULTURE

24.2.1 Historical Perspective

Fish culture (and subsequent stocking) is a useful management tool, but no amount of stocking can restore or enhance a fishery unless the underlying causes of depleted fish populations are understood and addressed. These causes can include overharvest and habitat degradation, as well as other direct and indirect by-products of human activities. For example, the rapid decline of striped bass populations along the Atlantic coast in the 1980s was a result of several factors, including habitat destruction, pollution, and overfishing. Until each of these factors was addressed, stocking could not have brought about true restoration. In the case of striped bass, a combination of several management choices—including stocking—resulted in increased runs in the mid-1990s.

The culture of anadromous species effectively began in North America in the 1860s after techniques were brought from Europe (Moring 1986). Initial efforts concentrated on Atlantic salmon and American shad, two species that were declining in abundance in New England during that era. Livingstone Stone, a fish culture pioneer, raised Atlantic salmon eggs from the Miramichi River, Canada, and sold the resultant progeny to state fisheries agencies in Massachusetts, Vermont, and New York (Bowen 1970). New Hampshire imported salmon eggs from Ontario as early as 1866, and Stone's partner, J. Goodfellow, sold Atlantic salmon eggs to Vermont. Thus, using fish culture as a tool in anadromous fish enhancement was viewed as a viable management option over 130 years ago.

Efforts to revive declining runs of Atlantic salmon and American shad have continued sporadically since the 1860s. Between 1866 and 1871 almost 200 million American shad eggs were hatched in five northeastern states (Bowen 1970), and the first Atlantic salmon hatchery, at Craig Brook (then known as Craig's Brook), Maine, was constructed in 1871. These Atlantic salmon eggs supplied Maine, Massachusetts, and Connecticut in an ultimately futile attempt to reverse the effects of dams and water pollution on fish populations. In the early 1870s alone, over 6 million Atlantic salmon were released in northeastern streams as far south as New Jersey (Bowen 1970). Other Atlantic coast anadromous species have been cultured, transported, and stocked widely. Rainbow smelt and alewife were distributed to many states during the last 3 decades of the nineteenth century.

The culture of anadromous stocks on the Pacific coast also began in the 1870s. A landmark of that era was the operation of the Crooks Creek egg taking station on the McCloud River, California. During the 1870s, 51 million eggs of chinook salmon were collected and distributed to public and private hatcheries in the West and East. After 2 decades of declining Pacific salmon runs to the Columbia River, the state of Washington constructed its first hatchery in 1894 and had 15 hatcheries by 1902. Alaska's first hatchery was constructed in 1891, and canneries operated numerous private hatcheries that had released 450 million sockeye salmon and coho salmon smolts by 1906 (McNeil 1980). Anadromous fish hatcheries began operating in Oregon in 1877, British Columbia in 1884, and the Maritime Provinces in the 1880s. Over 4.5 million chinook and sockeye salmon fry were stocked in British Columbia by early 1887 (Mowat 1889).

Most early efforts of the U.S. Fish Commission, created in 1871, were directed at stocking Atlantic salmon, American shad, chinook salmon, and a few other species in new waters or waters where fish populations were extirpated. In defining the role of the Fish Commission, the U.S. Congress was strongly influenced by stocking proponents of the American Fish Culturists Association (now the American Fisheries Society). Some of these early introductions of anadromous fishes were successful. Beginning in 1871, American shad were stocked into California waters. By 1880, 654,000 American shad fry had been stocked into the Sacramento River (Bowen 1970), and runs later became established in California, Oregon, and Washington. There were occasional strays into British Columbia waters. Striped bass were initially introduced to San Francisco Bay with shipments of 132 fish in 1879 and 300 fish in 1882; a commercial fishery developed within 10 years. From these beginnings, striped bass were reported from San Diego to Oregon as early as 1887, and fisheries are now established in California and southern Oregon.

24.2.2 Current and Future Directions

Modern culturists can now draw on more than a century of fish culture experience. As a result of this experience, even species with less culture history in North America (e.g., white sturgeon and shortnose sturgeon) can be cultured with some degree of confidence. The current challenge to fish culturists is how to do it better: how to refine these culture techniques so that growth is faster, survival is higher, and fish are healthier.

Perhaps even more important than culturing fishes is to recognize when it is *appropriate* to stock fishes. Today there is more emphasis on reducing widespread stocking and confining such activities to areas with high angler demand and limited natural production as well as concentration on producing higher quality fish that are able to survive or contribute to fisheries. In recent years, fish culture also has become an appropriate tool for rehabilitating populations of threatened or endangered species. Stocking fish produced from wild gametes has helped the recovery of federally listed shortnose sturgeon and several species of threatened trout, among other species.

A continuing concern among fish culturists and managers is the role of fish genetics (see section 24.3). Hatchery-reared fish, particularly of highly inbred stocks, can be radically changed—morphologically, physiologically, and behaviorally—from their wild ancestors. The more that cultured stocks are inbred or are moved geographically, the less adaptable they become (Griffith et al. 1989). Many studies have documented differences between wild and hatchery fish (Schramm and Piper 1995), and the effects of stocking hatchery-reared fish have been modeled (Bryne et al. 1992). Behavioral interactions between wild and stocked fish can have positive as well as negative consequences for resident species (Moring 1993). The role of fish culture is to produce high-quality fish that can compete favorably with wild fish without severely altering the genetic attributes of native stocks. The development of new and innovative techniques will continue to improve culture operations, reduce costs, or expand production capabilities.

24.2.3 Disease Research and Control

Fish diseases affect management in many ways, including reduced production of fish available for stocking, impaired stamina and growth, and altered behavior. Diseases are natural and kill unknown numbers of fishes in the wild, but it is in the hatchery that mortalities are most visible. Diseases can spread quickly through hatchery facilities because fish are held at abnormally high densities. The implications of diseases to management of anadromous fishes include (1) diseased fish may be stressed, may not feed, and may exhibit reduced growth rates, all of which affect release size and rearing costs; (2) diseased fish subjected to therapeutic control may exhibit delayed mortality after release; and (3) management decisions or other consequences of human activities may expose fish to disease zones during their downstream migration.

An example of the effect of management decisions on fish exposure to disease is the release of steelhead smolts into the South Santiam River, Oregon. In the 1970s, biologists documented delays of up to 2 weeks in the downstream migration of smolts caused by Foster Dam. In some years, smolts did not reach the lower Willamette River until water temperatures had increased. This late arrival exposed them to outbreaks of *Ceratomyxa shasta*, a myxosporidian parasite, and biologists believe that some of the variability in adult returns may have been due to the effects of this disease on smolts. Thus, the presence of dams and the timing of hatchery releases ultimately exposed smolts to suboptimal river conditions (Raymond 1979; Bartholomew et al. 1992).

The key aspect of effective disease management, including diseases promoted by nutritional deficiencies (Snieszko 1972), is prevention and early detection. Most states, provinces, and federal agencies have disease policies in effect that regulate importation of eggs and hatched fish; monitoring of hatchery lots; preventative measures; disinfection of hatchery trucks, raceways, nets, and equipment; detection procedures; and control measures. For example, Maine allows the importation of only eggs that have been certified by a reputable pathologist as being free of diseases (a misnomer, because no tests can actually detect the presence of all diseases). The state prohibits the importation of all live fishes.

Despite efforts at prevention and control, disease outbreaks continue to occur. The Atlantic salmon aquaculture industry in New Brunswick recently suffered a Can$10 million loss due to hemorrhagic kidney syndrome. Other recent outbreaks affecting anadromous salmonids have included infectious salmon anemia in Canadian fish farms and hatcheries and sporadic, continuing outbreaks of bacterial kidney disease in several locations in North America. When there is an outbreak of a devastating disease, such as untreatable whirling disease in popular trout waters of the western United States, fisheries managers are placed under considerable public pressure. It is extremely disheartening for anglers as well as fisheries biologists to see a world-renown fishery decline to mediocrity because there are no effective tools to combat a devastating disease.

When treatment for a disease outbreak is necessary, the prognosis is much better when treatment is started early. Unfortunately, there are few legal medicines available to culturists. Over time, various strains of pathogens and bacteria become resistant to tetracycline, sulfa, and other treatment compounds. Only a few such drugs are certified for use under the 1913 Federal Virus-Serum-Toxin Act, and the certification process is long (at least 2 years) and complicated. It is to the advantage of managers and culturists to support continued disease research and drug certification efforts.

24.3 STOCK MANAGEMENT

Ricker's (1972) and Saunders' (1981) definitions of the word stock are still widely used and refer to groups of interbreeding individuals that spawn in a particular stream and do not interbreed with other such groups due to differences in spawning location or timing. Many fish geneticists assert that each stream has its own stock of salmon, steelhead, American shad, striped bass, or other species, and some larger rivers may have more than one stock.

Ricker (1972) concluded that most differences between stocks were due to either environmental or genetic factors, and perhaps as many as 10,000 distinct stocks of Pacific salmon may exist along the north Pacific rim. Saunders and Bailey (1980) conservatively estimated 2,000 stocks of Atlantic salmon along the North Atlantic rim. Each of these stocks has unique characteristics that separate it from other stocks, such as fecundity, growth rate, egg size, survival, age and size at maturity, resistance to disease and low pH, and timing of migrations (Saunders 1981).

24.3.1 Genetic Selection

With the long history of stocking as a management tool, genetic selection—the development of selected traits—is a role taken on by management biologists, fish culturists, and geneticists. Fish managers and culturists may wish to select for growth, food conversion, survival, disease resistance, timing of return, age at maturity, fecundity, timing of spawning, or other characteristics.

In its basic form, genetic selection can be strictly selective breeding—the accentuation of certain traits in fish by selecting for those traits over generations of artificial breeding. The classic example of such selective breeding is the "super trout," or "Donaldson trout," developed over years by Lauren Donaldson at the University of Washington. These fish are actually rainbow trout × steelhead hybrids, for which traits such as weight, fecundity, and age at maturity have been selected. To achieve such results, large rainbow trout were crossed with steelhead to increase growth rates. After 36 generations (years), 2-year-old Donaldson trout averaged 4.5 kg compared with

0.04–0.05 kg for wild rainbow trout. All Donaldson trout females matured in the second year compared with 4–5 years for wild rainbow trout, and fecundity averaged 18,000 eggs in the third year compared with 400–500 eggs for wild 5-year-old wild females (Donaldson 1968; L. Donaldson, personal communication).

Genetic selection for other characteristics, such as resistance to disease in chinook salmon and resistance to low pH levels in Atlantic salmon, also have been useful management approaches. Further, the substantial genetic variability in freshwater growth characteristics of coho salmon has allowed biologists to select fish for appropriate size at release (Iwamoto et al. 1982; Mahnken et al. 1982). Other traits are being explored for their long-term benefits. Scientists are isolating and transferring the antifreeze protein gene in winter flounder to Atlantic salmon in an attempt to produce fish able to withstand extremely low water temperatures for cage culture and other applications.

Studies by the Canadian Department of Fisheries and Oceans Biological Station at St. Andrews, New Brunswick, and the nearby North American Salmon Research Center have shown that genetic selection for certain strains can decrease the proportion of grilse in runs of Atlantic salmon (Saunders et al. 1983; O'Flynn et al. 1990). Grilse are small salmon that return after only one year at sea and form large proportions of many Canadian runs. By decreasing the portion of grilse in the run, the size and potential catch of the run can be increased, and preferred sizes will be available to anglers and commercial fishers.

Despite such culture successes, genetic selection is much more than simply selective breeding. It is a broader concept of planned genetic exploration, and many challenges remain. For example, Hershberger and Iwamoto (1985) have pointed out that some less heritable traits may take considerable time to develop and that inbreeding within a strain can be a problem. Random mating as opposed to assortive (or selected) mating is often required to avoid inbreeding. A selected fish may be large and produce many eggs but it may not be of efficient form to survive in the wild and return. In addition, caution must be exercised when stocking fish from one drainage into another. For example, Reisenbichler et al. (1992) found many genetic differences between steelhead along the Oregon and northern California coasts. Unique site-adapted characteristics could be lost (Thorpe et al. 1981; Nielsen 1995), and some of these characteristics can be helpful to management as well as being essential to local populations.

24.3.2 Homing and Straying of Stocks

Most anadromous fish species exhibit some degree of homing to natal streams or river systems. This trait was observed as least as far back as 1635, when Izaak Walton reported homing of Atlantic salmon in *The Compleat Angler*.

> Much of this has been observed by tying a Ribband or some known tape or thread in the tail of some young Salmons, which have been taken in Wiers as they have swimm'd toward the salt water, and then by taking a part of them again with the known mark at the same place at their return from the Sea,…which has inclined many to think that every Salmon usually returns to the same River in which it was bred.

The degree of such homing behavior depends on the species, but homing can be a mechanism for developing genetic homogeneity, thus leading to the formation of discrete stocks.

A concern among geneticists and fisheries managers is that with increased hatchery output to enhance existing anadromous fish populations there will be greatly increased chances for straying and mixing among different stocks. When two stocks are mixed at spawning, whether by chance or intent, the unique characteristics of a stock can be lost or become diluted. It is common practice to introduce wild fishes to hatchery broodstocks periodically to ensure the continuance of preferred, native traits in the gene pool. Stock diversity and preservation of unique stock characteristics are management goals because such diversity of characteristics provides agencies with the most management options.

How extensive is straying among anadromous fishes? Without adequate surveys or recovery systems in place straying is difficult to assess because it takes the recovery of tags from all potential streams to obtain an accurate estimate. The precise nature of homing has been shown for Atlantic and Pacific salmon. Reviews by Harden-Jones (1968), Hasler and Scholz (1983), and others have concluded that, based on artificially imprinted fish and tag returns for all salmonids, average straying rates are 5% of returning fish. However, there are exceptions, with straying rates of up to 45% reported for coho salmon and of up to 4% for steelhead. Quinn et al. (1991) found wide variation in straying for runs of chinook salmon; some variation apparently was due to the presence of volcanic ash from Mount St. Helens. Quinn (1993) later reviewed the subject of straying in all salmonids and concluded that causes of straying are many and general estimates are misleading.

24.3.3 Effects of Rearing Environment

It is well known among geneticists and fisheries managers that fish survival is normally highest when a stock that is adapted to local conditions is used for stocking rather than a stock from a more distant river system (Reisenbichler and McIntyre 1986). That is why management agencies stocking striped bass into Chesapeake Bay waters try to use local stocks rather than those from the Hudson River. Similarly, agencies stocking Pacific salmon into the Snake River of Idaho will use local, or at least upper Columbia River stocks, rather than stocks from Alaska or the Fraser River. There are, however, other aspects of this in-system versus out-system rearing. Buchanan and Moring (1986) noticed higher recycling (fallback) of adult summer steelhead from groups reared outside the Santiam River system (Oregon) and then released there compared with groups reared and released within the Santiam system. Recycling refers to adult fish returning to a dam, entering a trap, being released into the reservoir forebay, then passing through the turbines in an attempt to move back downstream, confused about their natal origin or release point. This trait is being studied; it may be a byproduct of rearing conditions in the hatchery.

Managers also are concerned with the effect of in-system versus out-system rearing on homing of anadromous fishes. It is a common practice for management agencies to obtain gametes from returning spawners on one river, raise the fish in a hatchery, and then distribute the fry or smolts to several rivers or river systems. However, biologists in Oregon have recently documented significant homing of steelhead stocked in one system back to the river of parental origin (D. V. Buchanan, Oregon Department of Fish and Wildlife, personal communication). Thus, the in-system versus out-system rearing procedures may have several long-term implications for management, such as altering the timing of migrations, achieving habitat saturation, and shifting angler effort and harvest.

24.3.4 Effects of Commercial Aquaculture Operations

Commercial aquaculture operations in North America typically use two rearing procedures that can affect management of natural and hatchery-supplemented stocks. The first procedure—cage culture—is theoretically self-contained, but escapes almost always occur and the proportions are not trivial. Shrinkage, or unknown losses, from 23 cages of chinook salmon in Puget Sound ranged from 2.5–46.5%, without rips or holes in netting (Moring 1989). When rips do occur, much larger numbers of fish can be released into a local area. Eventually, some escapees may return to nearby streams to spawn with native stocks. Worldwide, it is estimated that 10–30% of cage-reared salmonids disappear due to predation or escape (Moring 1989).

The second commercial rearing approach—ocean ranching—imitates the procedures used by federal, state, and provincial hatcheries whereby anadromous salmonids are reared in captivity to the smolt stage, then are imprinted to water at a private collection facility, and then are released. This procedure greatly reduces rearing costs compared with a complete life cycle in captivity. Those fish that return to the commercial facility can be harvested for meat or gametes. With ocean ranching, a portion of returning adults will not enter the collection facility of the commercial company but may stray into other streams along the migration route. If fish are not strongly attracted to the return facility, the portion spawning elsewhere will be higher. To reinforce homing, Hasler and Scholz (1983) artificially imprinted hatchery coho salmon by using the chemical morpholine; 95% of the fish recaptured in the experimental stream contained the chemical attractant. Other such chemicals also have been used.

Fisheries managers are concerned with escapees from commercial facilities because these fish can be of different stocks that originated from other river systems, other states or provinces, or even from other countries. For example, one commercial operation in Oregon imported chum salmon eggs from the former Soviet Union, and another ranching operation in Maine imported Pacific salmon from Alaska and Japan. A commercial cage operation in eastern Maine has received Atlantic salmon from Scotland as well as from North America. Genetic mixing of these stocks with native runs is an ongoing concern (Berg 1981); continuing entry of escapees into coastal rivers has long-term genetic implications. A detailed analysis by Thorpe et al. (1981) indicates these concerns are not without foundation. For example, hundreds of aquaculture-reared Atlantic salmon escapees from salmon farms that have entered rivers of Downeast Maine complicated deliberations to list the species under the Endangered Species Act and the eventual implementation of an inclusive state rehabilitation plan.

24.4 POPULATION MANAGEMENT

Population management includes techniques for enhancing or reducing the size of anadromous fish runs. The application of stock–recruitment and production yield models, is to maintain run sizes necessary for continuation of populations while allocating the surplus of the run. In such models, the term stock, as defined by Gulland (1969), is sometimes different from the genetic definition and may include several populations or a part of a population managed with similar production characteristics. Although maximum sustained yield is a widely known concept—the maximum yield that can be produced year after year—it is also widely criticized in actual application, generally on economic grounds: the economic costs increase rapidly as the maximum sustainable yield is approached. Although other approaches to managing stocks are often promoted, those who study population dynamics differ on a universal concept. The theories and applications of fish population dynamics are complex and are discussed in Chapter 6. Here, management options for maintaining run sizes of anadromous fishes are discussed.

24.4.1 Allocation of Catch

One of the difficult tasks of a fisheries manager is the allocation of catch and escapement of anadromous fishes: how many fish can be harvested from a run while still providing sufficient escapement of potential spawners? To answer this question, managers need reliable information on numbers of adults ascending rivers and loss to sport and commercial catch. They also need predictors of run size based on previous smolt migration numbers and ocean survival. In most cases, all of this information is not available to management agencies. For example, Pacific salmon ascending and descending glacially turbid rivers of Alaska and British Columbia are not visible to human observers. Adult run size, smolt numbers, ocean survival, and other quantitative descriptors of fisheries can be estimated by only indirect means, such as smolt catch indices As a result, the allocation of catch is often an art. If one is too conservative on harvest allotments, commercial fisheries may suffer, particularly if run size is larger than expected. If one is too liberal on harvest allotments, commercial fisheries may boom in one season but be depressed in a subsequent season due to poor escapement and production of fish.

24.4.2 Fish Passage

Impediments to fish passage during migrations can have significant effects on fish mortality and hence the ultimate size of runs.

24.4.2.1 Upstream Fish Passage

Under natural conditions, anadromous fishes have always experienced some mortality during upstream migration to spawning areas, but when there are dams or other artificial obstructions, the mortality rate increases. One of the principal causes of the extirpation of Atlantic salmon from all the major rivers in the eastern United States (e.g., Connecticut, Merrimack, Kennebec, Androscoggin, and Penobscot rivers) was the construction of dams without passage facilities. In the six major rivers undergoing restoration of Atlantic salmon, there are still numerous dams without fish passage facilities. When such passage is available, mortality varies with the size of the dam (particularly height), type of fish ladder or lift, length of fish ladder, water flow, location of dam on the river, and other factors. Passage efficiency can also vary with species; few fishways, such as those on dams of the Columbia River, will pass sturgeon but may still pass Pacific salmon, American shad, and steelhead.

Upstream passage mortalities, even with modern fish ladders, can be significant. Swartz (1985) reported a 4% loss of migrating steelhead at each of eight dams on the main-stem Columbia and Snake rivers. However, when properly designed fish passage facilities are present, losses can be minimal; only 0.1% of adult steelhead died while passing through the short ladders and traps at Foster and Green Peter dams in Oregon (Wagner and Ingram 1973).

Other problems associated with dams include gas supersaturation—a factor for both upstream and downstream migrants—and migration timing. Dams often delay migrations, upstream and downstream, although Bjornn and Peery (1992) found indications that adult Pacific salmon may actually benefit from the presence of Columbia River reservoirs; upstream movement speeds may actually increase in such quieter waters.

24.4.2.2 Downstream Fish Passage

A rule of thumb for many years has been that 10% of salmonid smolts will be killed at each dam during their downstream migration, whether those mortalities are dam related or due to gas supersaturation, increased susceptibility of predators, or other causes. Fish mortalities can be high when fish encounter turbine blades or undergo stress caused by pressure changes. Although some hydropower developers believe mortality can be reduced to 8% by technical improvements, a review of over 70 studies of fish passage indicates that it will be unlikely that most fish passage facilities will reduce the site mortality much below 10%. In some locations, even a 10% value may be conservative (Table 24.1).

In general, the extent of mortality varies with each facility, each type of turbine, and each type of attraction or deflector system used. There are also other considerations. Dams with poorly placed downstream passage facilities can cause delays in migration, and these delays can significantly affect migrating fish. For example, at Foster Reservoir, Oregon, migrating steelhead smolts were delayed by up to 2 weeks, making them susceptible to anglers seeking legal-size rainbow trout. An estimated 19,500–22,800 of these steelhead smolts were taken annually in this relatively small, 494-ha reservoir (Bley and Moring 1988). Further, predators in reservoirs have greater access to migrating prey if the prey are delayed at a dam. Recent studies by Giorgi et al. (1997) have shown that movement rates of downstream migrating juvenile steelhead and Pacific salmon are largely determined by flow volume. Thus, it may be possible to manipulate flows to expedite the migrations of juvenile anadromous salmonids.

Adult fishes migrating downstream (e.g., Atlantic salmon, steelhead, and sea-run cutthroat trout) can be subjected to high mortality at dams because many dams do not have proper downstream adult passage facilities, and survival of adults passing through turbines is low. An estimate made based on mark and recapture of adult American shad in the Susquehanna River suggested that adults migrating downstream suffered 50% mortality (Whitney 1961). Technological advances have improved American shad survival since that estimate was made (Mathur et al. 1994).

Table 24.1 Representative downstream passage mortalities for anadromous fishes.

Location	Species and stage	Mortality (%)	Source
Columbia River, Washington–Oregon	Steelhead smolts	20–25[a]	Swartz (1985)
Connecticut River, Massachusetts–Connecticut	Atlantic salmon smolts	12–14	Stier and Kynard (1986)
Connecticut River, Massachusetts–Connecticut	Juvenile American shad and blueback herring	62–82	Taylor and Kynard (1985)
Five Maine rivers	Atlantic salmon smolts	11–23	Unpublished[b]
Merrimack River, New Hampshire–Massachusetts	Atlantic salmon smolts	17	Unpublished[b]
Penobscot River, Maine	Atlantic salmon smolts	9–11	Unpublished[b]
South Santiam River, Oregon	Steelhead and chinook salmon smolts	7–61	Wagner and Ingram (1973)
St. Croix River, Maine–New Brunswick	Atlantic salmon smolts	52	Watt (1986)

[a] Mortality at each dam.
[b] Unpublished data from the U.S. Fish and Wildlife Service.

There is some innovative research on improving survival of downstream-migrating fishes at dams. Biologists are exploring ways of diverting smolts and adults to passage facilities and away from turbines by means of such features as inclined plane screens, angled drum screens, acoustic transducers, bar racks, and lights. On the Connecticut River, for example, adult American shad are being repelled from undesirable areas by strobe lights. Juvenile American shad can now be passed at 90–100% efficiency with the proper flow velocities (Mecum 1980). A recent development has been the use of electric fields to guide juvenile and spent adult American shad past Holyoke Dam, Massachusetts, and high-frequency sound appears to be effective for redirecting blueback herring (Nestler et al. 1992).

24.4.2.3 Trucking and Barging

Another management technique for reducing mortalities associated with fish passage at dams is to capture fish and transport them by truck or barge them around dams, releasing smolts farther downstream or adults farther upstream. An agreement has been made between hydropower developers and state agencies in Maine to truck American shad and alewife past a series of dams on the Kennebec River. Pacific salmon and steelhead trout are routinely captured at Lower Granite and Little Goose dams on the Snake River, Washington, transported 460 km downstream to below Bonneville Dam on the Columbia River, and released. Trucking and barging circumvent the cumulative mortalities caused by passing multiple dams and minimizes exposure to predators concentrated below dams and in reservoirs.

Some biologists, however, believe that bypassing much of the river can weaken the homing instinct of returning adults. This may be especially so for fish that are trucked; barged fish are still exposed to circulating river water en route. Experiments by Bjornn and Ringe (1984) seem to support this supposition. Although transported fish had higher survival to the mouth of the Columbia River, adult returns to the upper

Columbia River were highest for fish that had migrated the entire river distance as smolts. Additional studies seem to show higher stress levels in transported smolts (Congleton et al. 1984), a factor that can delay migration or induce mortality (Specker and Schreck 1980). Similar results were found with Atlantic salmon in Norway, where behavioral changes and straying in fish without a complete downstream migration experience were observed (Jonsson et al. 1990).

Although trucking and barging operations are much more costly and labor intensive than allowing natural migrations, they can improve the percentage of smolts reaching the lower part of a river or the number of adults reaching spawning grounds. In that respect, they can be effective management tools, and the subject has been reviewed by Ward et al. (1997).

24.4.3 Release Procedures

Since the earliest days of fish culture of anadromous species, biologists have been attempting to refine release procedures to determine the appropriate size of fish at release and the appropriate timing of release. In the nineteenth century, hatchery managers drew subjective conclusions as to the proper time to release smolts, but it was not until fisheries managers began to tag large numbers of fish in the mid- and late-twentieth century and experiment with different release times, fish sizes, and locations that quantitative information became available. This quantification process continues throughout North America today, yet surprisingly little high-quality information has been analyzed and applied (Moring 1986). The problem is the time required to collect the data. A group of marked fish may not return as adults for several years, and replications are needed for each treatment group to provide statistically meaningful data. Thus, because of the life histories of anadromous fishes, decisions regarding optimal size and time of release of a species could easily require 5–10 years of effort.

An example of the evolution of release strategies is provided by the history of steelhead releases. Many steelhead were once released as fry. Later experiments, however, concluded that ocean survival would be increased with increased size at release (Larson and Ward 1955). Studies in the 1950s and 1960s in the Alsea River, Oregon, showed the highest rates of return were for steelhead of the largest release size (83 g); the location of release affected survival as well (Wagner 1969). Sheppard (1972) summarized several time and size-at-release studies and concluded that smolt-to-returning-adult survival is about 2.5% for releases of small, 1-year-old steelhead, 6% for 2-year-old steelhead, and 18% for 3-year-old steelhead. Some more recent experimental releases (D. Buchanan, Oregon Department of Fish and Wildlife, personal communication) in Oregon have indicated that, for the Skamania stock of summer steelhead, a 2-cm increase in length at release can result in up to twice the previous return rate for adults. Most of Oregon's steelhead production today relies on releasing large smolts: 170–180 mm in length and 76–114 g in weight. The success rate drops rapidly when steelhead are stocked below 160–170 mm; scientists are finding that such smaller fish may not migrate at all. The optimal time of stocking for these populations is now considered to be mid-April to mid-June; those fish stocked later often will not migrate.

24.4.3.1 Delayed-Release Programs

Delayed release is a management technique that retains anadromous salmon smolts past their normal migration times before releasing them into saltwater areas. The most successful program to date has been in Puget Sound. The idea was first proposed in the 1950s and 1960s by biologists in Washington and Oregon who had considered the relationship between the distance traveled between hatcheries and saltwater, the distribution of catches of coho salmon, and the timing of releases (Allen 1966). Coho salmon released from hatcheries moved rapidly out of Puget Sound and the Strait of Juan de Fuca, and then northward along the west coast of Vancouver Island. Commercial catches of those fish in British Columbia were 10 times higher than sport catches in Puget Sound (Novotny 1980). It was found that by delaying the release of smolts past their normal migration time—by holding them in saltwater cages—many remained in Puget Sound.

The Washington Department of Fisheries began experimenting with delayed release of coho salmon in 1969 (Haw and Bergman 1972), and the work was expanded to include chinook salmon in the early 1970s (Abbott and Salo 1972; Moring 1976). Similar procedures also have improved local, inshore sport fisheries in Alaska (Novotny 1975).

From the beginning, the delayed-release program in Puget Sound reaped benefits for fish managers (Figure 24.1). Buckley and Haw (1978) found that catches of delayed-release coho salmon were 21 times greater in the Puget Sound sport fishery than were catches of identical lots released at normal times. More than 14% of the delayed-release chinook salmon tagged and released by Abbott and Salo (1973) were ultimately recovered by anglers in or near the sound, and 1.5% of the 24,500 chinook salmon tagged by Moring (1976) were recovered by anglers over a 2-year period. Returns from some lots of delayed-release coho salmon have been as high as 28% (Novotny 1980). These delayed-release salmon, held at various locations, ultimately provide fish for localized sport fisheries. For example, 84% of the tagged fish recovered from a release made in southern Puget Sound were fish from the same portion of the sound (Moring 1976). Delayed release has been shown to be an effective management technique for improving local sport fisheries for Pacific salmon, and there is some evidence that it may also be successful with sea-run cutthroat trout (Johnston and Mercer 1976).

24.4.4 Control of Predators

It seems obvious that the elimination or reduction of potential predators of anadromous fishes would result in higher production, survival, and catches of anadromous species. Harris (1973), for example, reported a five-fold increase in Atlantic salmon fry-to-smolt survival when trout and eel predators were controlled. The effects of predator control were also documented in the classic 1930s study of sockeye salmon in Cultus Lake, British Columbia. Over 20,000 potential predators were removed from the lake during 7 years of gill netting. Survival to migration of sockeye salmon increased from a pretreatment average of 1.8% to a posttreatment level of 7.8%.

Because of the expense of removing predators from a body of water, this procedure should be done as efficiently as possible when it is warranted. Control of predation is possible because, often, predation is significant during only a brief period of the year or in specific areas. For example, Foerster (1968) reported predation on sockeye

Figure 24.1   "Blackmouth," or immature chinook salmon, caught in Puget Sound as part of the delayed-release program. (Photograph provided by Frank Haw, Washington Department of Fisheries.)

salmon by prickly sculpin when newly hatched fry were emerging from the gravel, when fingerlings moved into shallow areas prior to migrating, and following releases from a hatchery. Similarly, Buchanan et al. (1981) suggested that the number of anadromous steelhead smolts consumed is greater in areas directly below dam spillways or near hatcheries. Poe et al. (1991) reached similar conclusions for predation of salmonids in a Columbia River reservoir.

Birds also prey on anadromous fishes, particularly when the fish are confined in hatchery raceways or floating cages (Moring 1982) or when smolts are migrating, which they do in surface waters. Double-crested cormorants have been implicated in consuming 75-80% of migrating Atlantic salmon smolts in some areas in spring (Elson 1962). Such predators place the fisheries manager in a dilemma because this bird is federally protected in the United States. In addition, population numbers of double-crested cormorants in Maine have increased from 0 in 1930 to an estimated 25,000 nesting pairs today, with another 10,000 nesting pairs along the remaining coastline of New England (Krohn et al. 1995).

Fisheries managers have used two techniques to reduce the extent of predation by double-crested cormorants. First, they have attempted to shift, as much as possible, the release times of hatchery Atlantic salmon smolts so that abundant runs of other species, such as rainbow smelt and alewife are available to double-crested cormorants as alternative prey. Second, in the mid-1980s, biologists with the Maine Atlantic Sea Run Salmon Commission (now the Atlantic Salmon Authority), Penobscot Indian Nation, and U.S. Department of Agriculture conducted a double-crested cormorant removal and harassment program along the Penobscot River during the spring smolt migration. Decreased bird population numbers resulted during this critical smolt migration period, at least temporarily (Moring 1987). However, as of the late 1990s, over 7% of hatchery-reared smolts stocked into the Penobscot River each year are consumed by double-crested cormorants, and most of this predation occurs around dams during the smolt migration (Blackwell 1996).

Other techniques used to dissuade predators include overhead bird nets or covered raceways at fish hatcheries, the use of harassing cannons, the use of inflated scarecrows near fish-holding ponds or on aquaculture cages, and electric fences around hatcheries or floating cages to prevent entry by terrestrial predators. Ocean cage-culture operations sometimes use double nets to enclose rearing fish. Growers must check and remove mortalities daily to prevent rips in netting by spiny dogfish, otters, seals, and sea lions attempting to reach dead fish. Ocean ranchers have found that barging salmon smolts offshore can sometimes avoid heavy mortality from fishes and marine mammals in bays and estuaries near salmon release sites.

24.4.5 Habitat Quality

Habitat quantity and quality is probably the most important factor affecting natural production of anadromous fishes. If spawning gravel is limited for salmonids in a stream or lake, egg yield will be limited and smolt and adult numbers will be affected. If nursery habitat (rearing area for juvenile fish) is limited, the carrying capacity of a stream for fry will be reduced. For a population, the quantity and quality of habitat defines the limits within which food and environmental factors are influential. Other chapters of this text discuss habitat issues in great detail (e.g., Chapters 9, 10, 12, and 13), but some aspects are worth noting as they pertain to anadromous fishes.

One of the causes of declines in anadromous fish populations has been habitat destruction. During the 1980s and early 1990s, urban development and pollutants severely reduced the amount of spawning and nursery habitat available for striped bass in the Chesapeake Bay drainage. In part, the decline in many stocks of coho salmon near urbanized areas of Washington has been due to the destruction of small streams, which are used for coho salmon spawning and nursery areas for juveniles. Similar declines in populations of sea-run cutthroat trout have occurred near human population centers of British Columbia. As a result, the province has embarked on a large stocking and habitat improvement program. It is hoped that hatchery fish can absorb demand from anglers while wild runs are allowed to increase through habitat improvement (Moring 1986).

Logging and other land use activities can have a profound effect on habitat of anadromous fishes. While the timber industry is extremely important in several coastal states and provinces, the value of anadromous salmonids can be almost as great. Everest (1977) estimated fisheries as second in value only to timber in national forests of southwestern Oregon and of even higher value in some watersheds. Effects of land use activities potentially include changes in water temperature, suspended sediment, dissolved oxygen, stream flow, and insect and fish populations (Moring et al. 1994). Nevertheless, timber management and anadromous fisheries management can coexist, particularly when riparian (streamside) vegetation is preserved (Moring and Finlayson 1996). Maintenance of riparian vegetation is a key in the protection of instream habitat of anadromous fishes because these buffer strips provide shading and bank stability. The value of riparian zones has been recognized by federal management agencies, and some states, such as Oregon, provide property tax incentives to landowners who preserve and manage riparian zones for fisheries as well as wildlife. In 1999, Washington State and the timber industry there arrived at an agreement to prohibit logging within 46 m of streams used by Pacific salmon.

24.5   INTERAGENCY AND PUBLIC INVOLVEMENT IN MANAGEMENT

Effective management of anadromous fishes often requires interagency as well as intraagency decisions. When different agencies work together in managing stocks, potential conflicts can sometimes be avoided. For example, Native American tribes in the Columbia River basin interact with state agencies in implementing tribal fisheries so that adjustments can be made to management decisions and possible future conflicts can be avoided. Members of the Columbia Basin Fish and Wildlife Authority include all involved state and federal agencies and four lower Columbia River basin tribes. Members confer on major management issues and have completed a Columbia River fish management plan to direct future allocation of catch and set management goals.

In 1976, the Fishery Conservation and Management Act (FCMA) established a series of regional fisheries councils in the United States to manage coastal fisheries beyond the 3-mi limit, that is, out to 200 mi offshore. These regional councils, composed of agency representatives, user group representatives, and private citizens, prepare fisheries management plans and coordinate their actions with those of nearby regional councils. Although foreign vessels can fish for certain species within the zone under jurisdiction of the councils, most anadromous species are taken strictly by home-based fisheries.

Pacific salmon are considered a special case under the FCMA because stocks range widely and have been protected under other treaties, agreements, and agencies covering the high seas, such as the International North Pacific Fisheries Commission (United States, Canada, and Japan). American shad, alewife, and striped bass are generally not covered by the FCMA because they are restricted to coastal waters within the 3-mi limit.

The restoration programs for Atlantic salmon in New England involve numerous agencies, even on a single river system. A technical advisory committee makes recommendations to Maine's Atlantic Salmon Authority on stocking rates and locations, disposition of eggs, fish culture activities, tagging, and other aspects of restoration science for Maine rivers. Voting members of the technical advisory committee include representatives from the U.S. Fish and Wildlife Service, National Marine Fisheries Service, three state fisheries agencies, the University of Maine, and the Penobscot Indian Nation. Regular participants also include the Atlantic Salmon Federation and other environmental groups. On the Connecticut River, a policy and technical committee includes representatives from the states of Connecticut, Massachusetts, New Hampshire, and Vermont, along with the U.S. Fish and Wildlife Service and the National Marine Fisheries Service.

Another example of interagency coordination is the centralized processing of information on Pacific salmon and steelhead from coded wire tags (small pieces of binary-coded wire). Hundreds of thousands of these tags are applied annually in snouts of fishes by agencies along the Pacific coast. The tags are recovered from fishes harvested across their range in foreign and domestic commercial and sport fisheries. The agencies involved in the tag program have formed a central receiving house for tag information to maximize the recovery of information. Similarly, Atlantic salmon tagging information is now coordinated through the National Marine Fisheries Service in Woods Hole, Massachusetts, and there are plans by the North American Salmon Conservation Organization to coordinate worldwide Atlantic salmon tag data at one location.

A growing trend with many state agencies is involving the public in the management of anadromous species (Williams et al. 1997). Generally, this involves the active participation of a sports organization, school, or particular group of individuals with selected, localized projects. For example, in the 1980s California appropriated over US$900,000 annually to the Northcoast Cooperative Salmon and Steelhead Project. These funds were subsequently provided to nonprofit groups to rehabilitate Pacific salmon and steelhead streams on the northern coast of California. A similar program in Oregon, the Salmon and Trout Enhancement Program, employs a full-time biologist to work with nonprofit groups to improve salmonid populations in streams. Projects to date have concentrated on incubating salmonid eggs in instream boxes, rehabilitating streambanks by planting willows, restoring riparian vegetation, creating spawning areas, and stocking steelhead trout fry. Another education program, Salmon Watch, is a cooperative venture between Oregon Trout and the Oregon Chapter of the American Fisheries Society.

Since the mid-1970s, a public program in Washington has involved sports groups, Native American tribes, power companies, and private corporations in rearing Pacific salmon; over 10 million salmon were released by this program in 1983 alone. Financial arrangements vary from donation of eggs by the state and rearing costs covered by private groups to the outright purchase by state agencies from private growers of live fish, which are then reared to smolt size or beyond. This latter arrangement is often made in conjunction with delayed-release programs to improve sportfishing in Puget Sound. A similar, though smaller-scale, cooperative program has also been conducted with Pacific salmon in San Francisco Bay, and the initial success of culture studies with white sturgeon by the University of California-Davis came with the assistance of local anglers in the San Francisco Bay area.

Several salmon clubs in Maine are currently engaged in cooperative projects with state agencies to assist ongoing restoration efforts for Atlantic salmon, and the Atlantic Salmon Federation's Maine Council brings salmon incubation units to over 75 schools in the state so that students can learn about fish biology. Students hatch the eggs and eventually release the fish into local streams. Such private efforts are incubating several hundred thousand eggs and releasing fry into rivers of coastal and interior Maine.

In 1997, public involvement became a major aspect of the management of Maine's seven rivers that never lost their fragile populations of Atlantic salmon. As part of a state management plan accepted by the federal government as an alternative to inclusion of Atlantic salmon on the federal endangered species list, local watershed councils have considerable control over management activities on specific rivers.

Taken individually, these public involvement projects and others may have only minor effects on the overall management of anadromous species, but collectively, and certainly from a public relations standpoint, the effect can be significant. Some management agencies believe that public involvement programs have sufficient merit to warrant full-time biologists who work with private groups and individuals.

24.6   CURRENT STATUS OF FISHERIES AND RESTORATION AND ENHANCEMENT PROGRAMS

24.6.1 Striped Bass

For many years, the enhancement and restoration of striped bass in the Chesapeake Bay region has been a high national priority. Although runs in the Hudson River are increasing, it has been estimated that perhaps 90% of the striped bass along the East Coast originate from tributaries of the Chesapeake Bay. With declining populations in the 1980s and early 1990s, the Chesapeake stock became quite low. The Hudson and Roanoke (North Carolina and Virginia) stocks were about equal in contribution (Deuel 1987).

At one time, striped bass populations were endemic from the Sabine River, Texas, along the Gulf of Mexico coast, and up the Atlantic coast to the St. Lawrence River. Today, however, most of the Gulf stocks have been extirpated, and spawning stocks north of the Hudson River are essentially gone with the exception of (at least) nine rivers in eastern Canada (Rulifson and Dadswell 1995). Spawning populations from Florida to North Carolina are low but apparently stable (a few may even be increasing), whereas the important stocks north of North Carolina have been undergoing dramatic changes. Chesapeake Bay and Hudson River populations are fished from North Carolina to Maine, but environmental factors, particularly water flow and water temperature, pollution, and habitat degradation were apparently responsible for the severe decline of stocks from the late 1970s to the early 1990s (see Box 24.2).

Striped bass, introduced from the East Coast in 1879, are found on the Pacific coast from Baja California to British Columbia. They are concentrated in the Sacramento-San Joaquin river system of California and some smaller coastal runs, most notably the Coos and Umpqua rivers of Oregon. Pacific coast runs have declined to 25% of their former level, and causes of this decline have been somewhat similar to those on the Atlantic coast: low numbers of spawners, pollution, loss of juveniles to water diversions, and reduced plankton densities (Stevens et al. 1985).

The emphasis in restoration and enhancement has been on the Atlantic coastal stocks, and catch restrictions (minimum lengths), sales restrictions, or moratoriums on fishing are now imposed in almost every coastal state. As a consequence of the Atlantic

---

Box 24.2   The Decline of the Striped Bass Harvest in Atlantic Waters
(Data are from Alperin 1987 and Deuel 1987)

1. Average harvest 1958–1976   4.3 million kilograms

   Peak harvest 1973   6.7 million kilograms

   Harvest in 1983   0.8 million kilograms

   Harvest in 1985   0.5 million kilograms

2. The decline in the Maryland harvest alone was 94% between 1973 and 1983.

3. The 1981 Atlantic hatch was the lowest of any season in the 28 years of monitoring.

4. Sport catches also declined from North Carolina to Maine: 2 million fish in 1979; 300,000 fish in 1985.

Striped Bass Conservation Act and amendments to the Anadromous Fish Conservation Act, the Atlantic States Marine Fisheries Commission influenced coastal states (through financial and regulatory pressure) to reduce harvests of striped bass by 55%. Most states today have some type of minimum size limit for commercial and recreational fisheries. In seven states with limited populations, the striped bass is declared to be a sport fish, and commercial fishing is banned.

The management philosophy for improving striped bass populations along the Atlantic coast of the United States involves monitoring, regulation, habitat improvement, research, and stocking. An important tool for managers is an effective monitoring program whereby any successes of other management procedures can be assessed. Since the mid-1950s, Maryland has monitored age-0 striped bass in Chesapeake Bay. Similar sampling programs for juveniles now occur in five other states and the District of Columbia, and adults are being monitored in six states and the District of Columbia.

The imposition of minimum length limits and fishing moratoriums have been designed to reduce fishing mortality while other management measures are taking effect, such as habitat improvement. Six states are conducting research projects on identifying and reducing the effects of pollutants on striped bass. In particular, state and federal researchers have found that striped bass are very susceptible to arsenic, copper, cadmium, aluminum, and malathion (Tatum et al. 1965; O'Rear 1973; Dawson et al. 1977; Mehrle et al. 1982), and there are closures or advisories against catching or eating PCB-laden striped bass in New York, New Jersey, and Rhode Island. Other researchers have shown that chlorine (used to treat sewage effluent, among other things) may kill eggs and larvae of striped bass (Morgan and Prince 1977) and also reduces zooplankton densities and, hence, food for juvenile striped bass. Applied research continues on the role of these potential toxicants and measures to reduce or eliminate their input. Research is also being conducted on feeding and nutritional requirements of striped bass, disease control, and stock identification. In the latter area, considerable research has been conducted on genetic identification of stocks through electrophoretic or mitochondrial DNA techniques (Stellwag et al. 1994; Waldman and Wirgin 1995; Laughlin and Turner 1996), but more recent optimism involves identifying stocks by scale shape (J. Margraf, Maryland Cooperative Fish and Wildlife Research Unit, personal communication).

Management measures have apparently been successful. In the mid-1990s, runs of striped bass rebounded. Chesapeake Bay runs and numbers of migratory fishes traveling along the southern New England coast have all increased substantially. It is hoped that stocking will show increasing success. One success story has occurred in Maine. Maine has had a limited program of stocking striped bass since 1982, with several hundred thousand striped bass stocked in the Kennebec and Androscoggin rivers. Every year since 1987, juveniles have been captured by seine in Merrymeeting Bay, probably the first such natural spawning since populations were destroyed in the nineteenth century. More recently, successful spawning also has been documented in the Sheepscot River (L. Flagg, Maine Department of Marine Resources, personal communication).

Unfortunately, these successes have resulted in some short-term memories by user groups. Some commercial and recreational fishers have petitioned for reduced minimum length limits. Fisheries managers must be careful: casting aside the lessons of recent history may cause the striped bass fishery to repeat its disastrous record of the 1980s.

24.6.2 Atlantic Salmon

The restoration of Atlantic salmon in the United States relies almost entirely on fish stocking. By the 1860s, runs of Atlantic salmon in many industrialized areas of the northeastern United States were depleted or eliminated entirely by a combination of dams and pollution. As stocks declined, commercial fishing offshore and in rivers helped to speed the extirpation. Only runs in eastern Canada and a few small streams in eastern and central Maine maintained wild stocks, even though numerous runs in eastern Canada suffered many of the same consequences of overharvest, pollution (including acid precipitation), and dams.

Restoration was not able to proceed in any area of New England until progress was made on improving water quality in rivers and providing fish passage around and over dams. Current restoration programs are centered in three regions: Maine (primarily the Penobscot River but a dozen other rivers as well), the Merrimack River basin, and the Connecticut River basin. A number of smaller rivers in New England also are undergoing limited restoration efforts. In each case, a multiagency group directs the restoration effort, the federal government is involved in fish culture, and state agencies are involved in monitoring and management.

Although Maine began a restoration program in 1948, it was the model 1965 river program (Moring 1987) for the Penobscot River that was the principal thrust for subsequent programs (Connecticut River in 1967 and Merrimack River in 1969). These programs involved fish passage and stocking. Currently, there are some two dozen state, federal, and private rearing, holding, and kelt (spawned out adult salmon) rejuvenation facilities in New England.

Full restoration of the Penobscot River will require a consistent annual run of at least 6,000 adults, unsupported by stocking (Moring 1986). The Penobscot, the most successful of the Atlantic salmon restoration efforts, receives about one-third of New England's stocked Atlantic salmon smolts, yet it is far from being restored. The run has averaged over 2,500 Atlantic salmon over the past 18 years, including a record return of 4,125 in 1986. However, numbers of adults returning in the 1990s have ranged from

1,000 to 2,500 adults each year (Moring et al. 1995; Maine Atlantic Salmon Authority trap records). Although the proportion of the run originating from naturally constructed redds in the wild is steadily increasing, full restoration will not occur until the twenty-first century and may have to be supplemented by some level of stocking.

The situation on the other two restoration rivers is more dismal. Less than 300 adult Atlantic salmon have returned annually in the mid- to late-1990s to the Connecticut River, and less than 100 fish each year, during the same time frame, to the Merrimack River. Despite the stocking of millions of Atlantic salmon fry each year, less than 0.0007% eventually return as adults to the Merrimack River drainage and 0.003%–0.02% to the Connecticut River drainage (U.S. Fish and Wildlife Service trap and stocking records).

Stocking programs in Canada's Maritime Provinces totaled over 2.4 million in 1993; most of the Atlantic salmon were placed in New Brunswick waters and the remainder in Nova Scotia (Marshall 1994). Wild runs, however, are still the mainstay of much of the fisheries in the Maritime Provinces.

Part of the problem in restoring Atlantic salmon runs to U.S. streams has been the interception of fish on the high seas, primarily in the Davis Strait and as they pass through Canadian fisheries. Some estimates placed this interception figure at half of all U.S. fish. To identify U.S. stocks, there have been some promising studies with scale shape as a distinguishing characteristic and onsite monitoring programs at local Canadian and Greenland canneries. Recent closure agreements have helped reduce commercial harvest, and other agreements are anticipated.

Another constraint to Atlantic salmon restoration is the source of eggs. With the native runs to most New England rivers eliminated, the unique set of adaptive genetic characteristics of those populations have also been lost. The source of eggs for restoration of Maine rivers, typically now called the Penobscot strain, is probably a combination of some eastern Maine stocks that were spared (those populations have now been somewhat diluted by stocking) and some Canadian stocks. As a result, it has taken considerable time to create the new Penobscot strain, which has shown satisfactory, if not spectacular, return rates. The restoration of southern New England rivers will take even longer because those runs have relied on Penobscot River salmon eggs as a source. The genetic and geographic origin of these eggs is even more removed from the extirpated stocks of southern New England, and it will take considerable time to create new Merrimack, Pawcatuck, and Connecticut stocks adapted for those waters. Successes have been minimal, but fisheries managers are optimistic. Over the past several years, Atlantic salmon have returned to the Pawcatuck, Merrimack, and White rivers—in each case for the first time in almost a century.

24.6.3 Pacific Salmon

Pacific salmon are the most valuable of North America's anadromous fishes. Recreational fisheries from northern California to Alaska are multimillion dollar industries that have suffered from declines in the 1990s along with the populations of fishes. Stocking of hatchery-reared fishes is an essential, costly component of anadromous fish management in Pacific coastal states and British Columbia. The numbers stocked along the Pacific coast annually exceed 1 billion salmon.

Despite these large stocking programs, Pacific salmon runs in some areas of the Pacific coast have declined rapidly (Dauble and Watson 1997), particularly in the 1990s. This decline is a difficult problem for fish management. An analysis of population trends by Konkel and McIntyre (1987) concluded that in the mid 1980s, populations of salmon in south-central Alaska increased the most whereas those in the Columbia River basin declined the most. Populations in Alaska that have been generally increasing are sockeye, pink, and chinook salmon; populations in the Pacific Northwest have been either stable (pink salmon) or generally declining (sockeye salmon). Three times more coho and chum salmon populations are declining than are increasing along the Pacific coast. Baker et al. (1996) concluded that the situation is better for stocks in southeastern Alaska: 36% of stocks were increasing and 60% were stable. This is significant because 79% of the North American harvest of Pacific salmon occurs in Alaska, 17% in British Columbia, and the remainder in Washington, Oregon, and California (Baker et al. 1996). The majority of Pacific salmon stocks analyzed in British Columbia are classified as being "at risk" or "at high risk" (Slaney et al. 1996).

Coho salmon numbers in California have declined by 70% since the 1960s; populations in the early 1990s were only 6% of what they were in the 1940s (Brown et al. 1994). Despite these declines, quotas or proposals for outright bans on commercial catches of coho salmon have met with heated opposition. Part of the decline in the 1980s and 1990s has probably been due to an extended El Niño event that may have reduced ocean nutrients and food supplies.

Improvements in water quality in major rivers, such as Oregon's Willamette, have led to increased runs of Pacific salmon and sea-run cutthroat trout in conjunction with stocking. The runs of chinook salmon in Oregon waters broke several records in 1987, including record commercial catches.

There are some management options that can improve the picture for Pacific salmon, including more precise data on the timing of and size at release of hatchery fish, habitat improvement and protection, increased survival and returns of hatchery fish via genetic selection, disease control, and harvest regulations. There is little that management agencies can do about conditions in the open ocean except for cooperative harvest agreements. Hatchery stocking is an important management option, but the effect of large numbers of hatchery fishes on ocean food supplies is still being debated. Other options for managers include implementing delayed release (section 24.4.3.1) and interacting with commercial ocean ranching facilities to expand rearing capabilities. In the latter case, commercial fisheries and sport anglers can benefit from increased catches of privately released salmon, although other potential conflicts can exist regarding the ownership of these fish.

24.6.4 Steelhead

Generally, runs of steelhead have not been totally eliminated except near human population centers (Barnhart 1975). Stocking to supplement runs is a vital component of management programs, even though the commercial harvest has been greatly reduced over the past 70 years. In most states and British Columbia, the steelhead is managed as a sport fish and increasing pressure from anglers has influenced management agencies to increase run sizes.

Steelhead have been transplanted across North America, but this anadromous form of rainbow trout is cultured primarily in northern California, Oregon, Washington, Idaho, and British Columbia and, to a lesser extent, in Alaska. Over 28 million fish are stocked annually. Smolts account for most of the 12.5 million steelhead released in Washington; other state, federal, and provincial agencies are primarily releasing fish larger than fry size.

Stocking is a critical management tool in many waters. At least 80% of the run on the Alsea River, Oregon, is of hatchery origin. Where runs are supplemented by stocking, these hatchery fish are critical: from 75 to 80% of the steelhead in hatchery-supported runs in Oregon are of hatchery origin (only one-third of total steelhead catch consists of wildfish). To improve runs in southeastern and south-central Alaska, steelhead broodstocks have been established at five hatcheries with a combined capacity of over 200,000 fish.

24.6.5 Other Anadromous Salmonids

Sea-run brook trout occur in limited numbers along the northeastern coast of North America and are managed primarily as sport fish. The most abundant populations are in eastern Canada. Some work on sea ranching has been conducted by the Matamek Research Station (Quebec) of Woods Hole Oceanographic Institute, but the principal management effort today is in Massachusetts and the Maritime Provinces. Sea-run populations were almost eliminated in Massachusetts, and by the mid-1980s existed in less than two dozen streams concentrated on Cape Cod (Bergin 1985). To protect these fragile populations, management efforts have been to stock brown trout in nearby waters to absorb angler pressure. Approximately 15,000–25,000 yearling brown trout are stocked annually in coastal streams of Massachusetts. Populations of sea-run brook trout in eastern Canada have been examined in more detail; locations of runs and life history patterns are known, but sea-run populations are localized and may be quite susceptible to human-induced changes. Increasing snippets of information about sea-run brook trout in Maine indicate there may be many small, localized runs of fish along the coast, but these runs are susceptible to overfishing and environmental alteration.

Connecticut has a limited stocking program for sea-run brown trout, derived from eggs obtained from Europe. Brown trout fisheries have been created in the estuaries of the Royal, St. George, Kennebec, and Ogunquit rivers of Maine by stocking nonanadromous brown trout directly in or near estuaries. Although this species is an exotic introduction, estuarine fisheries have become quite popular. Stocked fish do move into coastal waters, but virtually no information is available on this portion of the brown trout life cycle in North American waters.

Sea-run cutthroat trout are found from northern California to Alaska (Figure 24.2). Hatchery production is centered in Oregon, Washington, and British Columbia, and fisheries managers in these areas employ different strategies for stocking this trout. Oregon seeks to protect wild populations of sea-run cutthroat trout and has been stocking fish in major coastal river systems to meet angler demand. At one time, about 900,000 sea-run cutthroat trout were stocked annually to provide spring, summer, and autumn fisheries at four different life stages: a legal-size cutthroat trout fishery prior to fish moving into the estuary in the spring, a fishery for larger cutthroat trout in the

Figure 24.2   Sea-run cutthroat trout from the Willamette River, Oregon. (Photograph provided by Jay Nicholas, Oregon Department of Fish and Wildlife.)

estuary in summer, a fishery on returning spawners in the fall, and a limited fishery on downstream migrating kelts in the following spring. Fisheries managers have now phased out stocking in coastal streams in favor of stocking in standing bodies of water. Beginning in 1997, stream stocking has been discontinued in waters containing wild sea-run cutthroat in an effort to reduce potential genetic mixing (Hooton 1997).

A study in 1972 indicated that only 0.1% of sea-run cutthroat trout smolts stocked in Hood Canal (Puget Sound) were ultimately taken by anglers (Hisata 1973). De-layed-release techniques that have been successful with Pacific salmon (see section 24.4.3.1) have been employed with sea-run cutthroat trout. Juveniles were reared in floating cages in Puget Sound and then released as adults into Hood Canal; the angler harvest rose to 9% (Johnston and Mercer 1976). In 1983–1984, Washington stocked over 329,000 sea-run cutthroat trout.

British Columbia has attempted to reverse a declining trend in sea-run cutthroat trout populations near urban areas. The Salmonid Enhancement Program is intended eventually to stock sufficient sea-run cutthroat trout juveniles to provide annually an additional 19,000 adults in coastal streams (Moring 1986).

Arctic char have both landlocked and anadromous runs in North America, and the species is managed as a sport fish in Alaska and several provinces of Canada. Com-mercial fisheries exist in the Maritime Provinces, where Arctic char is primarily taken by gill net. However, fisheries managers in Labrador have seen a rapid decline in the size of commercially caught Arctic char. Arctic char processed in 1987 averaged 1–2 kg compared with fish of 3.5–4.5 kg in the early 1980s. This indicates an overexploitation problem. Arctic char are also being reared commercially in hatcheries and floating cages along the coast of New Brunswick and may become an important species for aquaculture by the twenty-first century. If population enhancement measures are needed to improve coastal populations, the culture techniques being developed may be effec-tive management tools.

A related char, the Dolly Varden, is distributed from northern California to Alaska. The species is a popular sport fish, particularly in areas without substantial populations of other trout. At one time, there was a $0.025 per fish bounty on Dolly Varden in Alaska, and in the 1930s some $20,000 per year was expended on bounties (McPhail and Lindsey 1970; Scott and Crossman 1973). Like most bounties, this one was ill-conceived, and removing these predators probably had little effect on other salmonid populations. Although there has been some commercial processing of Dolly Varden in the twentieth century, it has been only as a by-product of other fisheries. Today, the species is managed exclusively as a sport fish.

24.6.6 American Shad, Alewife, and Blueback Herring

Despite restoration efforts, some unknown factor—apparently marine-related— has caused clupeid runs to decline in recent years throughout New England. Whatever the cause, it is affecting American shad, alewife, and blueback herring in many rivers. Thus, restoration efforts will be difficult for fisheries managers.

The American shad was probably the first species of anadromous fishes to be aided by North American fish culture programs because it was the first anadromous species to be severely affected by humans. Over 200 million eggs were artificially hatched in northeastern states between 1866 and 1871 (Bowen 1970). Commercial catches in northeastern United States peaked at 3,000 metric tons in 1970, then declined.

American shad are found from the Gulf of Mexico northward to New Brunswick, but it is in the mid- and north Atlantic regions where the populations have declined due to dams, pollution, and overfishing. It is estimated that in Maine only 5% of the habitat formerly accessible to American shad can now be reached. For several years, there have been commercial and sport angling restrictions, but even with these restrictions the annual U.S. harvest declined to only 1,000 metric tons in the early 1990s. As a consequence, enhancement and restoration of runs in the northeastern United States has become a priority with management agencies.

Restoration and enhancement efforts primarily involve transplantation of adults. For example, New Hampshire and Rhode Island have imported mature American shad from the Connecticut River to stock in state waters, and Maine transfers American shad from the Narraguagus River (Maine) and the Merrimack River (New Hampshire) to stock in Maine rivers that have limited or no American shad populations (Moring 1986). Restoration was successful on the Merrimack River until 1993; then, over a 2-year period, American shad runs declined by almost half each year, before increasing in 1995 and 1996. Runs declined by 75% on the Connecticut River during the same time period (B. Kynard, U.S. Geological Survey, personal communication).

Most mid-Atlantic and northeastern states have at least a moderate restoration program, but major efforts are underway in the Susquehanna and Schuylkill rivers of Pennsylvania. In 1984, 7.6 million American shad were released in Pennsylvania. In the Susquehanna, dams have long blocked runs in the upper river. Downstream bay catches in Maryland dropped from 900,000 kg in 1969 to 9,000 kg in 1980 (Moring 1986). Today, states are in the midst of a long-term restoration program involving fish passage, hatchery culture, and research.

Two related anadromous clupeids also are being stocked in limited enhancement programs. Sea-run alewife and blueback herring are commercially important along the Atlantic coast. By the 1980s, over 2,000 metric tons of the two species were harvested annually, a significant decline from the peak catch in 1970 (Mullen et al.1986; National Marine Fisheries Service 1992). While 90% of the catch in Maine is used as lobster bait, the two species are also valued for human consumption, fish meal, and roe. New Hampshire is currently improving fish passage to increase runs of alewife and blueback herring, and Massachusetts, Rhode Island, and Maine have ongoing stocking programs. Following a legislative mandate to restore alewife to its former range, Maine has plans for an ambitious reintroduction program to stock sea-run alewife in numerous lakes. Before implementing the full program, three state agencies and the University of Maine have been conducting a long-term study on the effects of stocking sea-run alewife on freshwater fish populations.

24.6.7 Other Anadromous Species

The endangered shortnose sturgeon has been cultured for over 75 years; the Atlantic sturgeon has a much shorter culture history. In the early 1980s, efforts by the South Carolina Wildlife and Marine Resources Department and the U.S. Fish and Wildlife Service resulted in the culture and later release of juvenile shortnose sturgeon and Atlantic sturgeon. Because the shortnose sturgeon is on the federal list of endangered species, and the Atlantic sturgeon is likely also to receive such legal protection, stocking and harvest restrictions may be the most effective management tools. Fragile populations exist, and more research is needed to address critical constraints (Kieffer and Kynard 1996). For example, shortnose sturgeon of the Merrimack River, Massachusetts–New Hampshire, prefer to spawn in some heavily urbanized sections of the river, such as the concrete-lined sections that flow through downstream Haverhill (Kieffer and Kynard 1996). Thus, management concerns must focus on habitat protection as well as stocking and harvest restrictions. While shortnose and Atlantic sturgeons have been studied in the eastern United States, additional work by researchers at the University of California at Davis have made strides in culturing white sturgeon (Buddington and Doroshov 1984; Doroshov and Lutes 1984).

Researchers on the Pacific and Atlantic coasts have experimented with the culture and release of anadromous smelts. Taken as a group, these fishes are popular with anglers and are particularly valuable as forage species for freshwater, anadromous, and marine game fishes. Biologists in Massachusetts have had difficulty managing the widely fluctuating year-classes of rainbow smelt; this difficulty may be a reflection of biological factors, including increased numbers of predators, low pH in streams, and changes in substrate quality and aquatic vegetation (B. Chase, Massachusetts Division of Marine Fisheries, personal communication). Rainbow smelt populations can be severely affected by overharvest (Murawski and Cole 1978), and biologists in New England states have often transplanted eggs to rehabilitate rainbow smelt populations. Work with the rainbow smelt, an anadromous and freshwater species, has been successful from two standpoints: the culture of fish in tanks to commercial baitfish size and the stocking of eyed-eggs in natural bodies of water (Akielaszek et al. 1985).

On the Pacific coast, progress has been made culturing delta smelt (on the federal endangered species list) and surf smelt, although the popular eulachon may hold the most promise for future management options.

24.7 CONCLUSION

Management of anadromous species can be highly complex, involving the input of many governmental agencies, municipalities, and private groups as well as consideration of biological and political concerns. Fortunately, fisheries managers have numerous tools at their disposal. Fish culture has been an important component of management plans since the mid-1800s, but refinements in diet, disease control, growth, and survival of cultured fishes are needed. A critical concern for anadromous species is fish passage—maximizing the survival for upstream- and downstream-migrating fishes. Modeling simulations suggest that new techniques for diverting downstream fry, smolts, and spent adults may have a significant effect on survival, overall run numbers, and production (J. Marancik, U.S. Fish and Wildlife Service, personal communication).

Genetic selection and improvements in stream habitat are also options available to fisheries managers to improve survival of wild and hatchery-produced fish. Further management options include some more innovative techniques, such as delayed-release programs to promote localized sport fisheries.

Fisheries managers are involved with the complex issue of how resources should be allocated. Along with this responsibility comes questions such as, what is the optimal timing of and size of fish at release? How can the public become involved in habitat improvement and other restoration and enhancement programs? How can fisheries be improved through interagency and other types of cooperative management? These are the present challenges in anadromous fish management.

24.8 REFERENCES

Abbott, R. R., and E. O. Salo. 1973. Contribution to the recreational fishery of pen-reared chinook salmon released as yearlings. Pages 34–35 in D. E. Rogers, editor. 1972 Research in fisheries. University of Washington, College of Fisheries, Seattle.

Akielaszek, J. J., J. R. Moring, S. R. Chapman, and J. H. Dearborn. 1985. Experimental culture of young rainbow smelt Osmerus mordax. Transactions of the American Fisheries Society 114:596–603.

Allen, G. H. 1966. Ocean migration and distribution of fin-marked coho salmon. Journal of the Fisheries Research Board of Canada 23:1043–1061.

Alperin, I. M. 1987. Management of migratory Atlantic coast striped bass: an historical perspective. Fisheries 12(6):2–3.

Baker, T. T., and eight coauthors. 1996. Status of Pacific salmon and steelhead escapements in southeastern Alaska. Fisheries 21(10):6–18.

Barnhart, R. A. 1975. Pacific slope steelhead trout management. Pages 7–11 in W. King, editor. Wild trout management symposium. Trout Unlimited, Vienna, Virginia.

Bartholomew, J. L., J. L. Fryer, and J. S. Rohovec. 1992. Impact of the myxosporean parasite Ceratomyxa shasta on survival of migrating Columbia River salmonids. Pages 33–41 in R. S. Svrjcek, editor. Control of disease in aquaculture. NOAA (National Oceanic and Atmospheric Administration) Technical Report NMFS (National Marine Fisheries Service) 111.

Berg, E. R. 1981. Management of Pacific Ocean salmon ranching: a problem of federalism in the coastal zone. Coastal Zone Management Journal 9:41–76.

Bergin, J. D. 1985. Massachusetts coastal trout management. Pages 137–142 in F. Richardson and R. H. Hamre, editors. Wild trout III. Federation of Fly Fishers and Trout Unlimited, Vienna, Virginia.

Bjornn, T. C., and C. A. Peery. 1992. A review of literature related to movements of adult salmon past dams and through reservoirs in the lower Snake River. U.S. Army Corps of Engineers, Walla Walla District, Technical Report 92-1, Walla Walla, Washington.

Bjornn, T. C., and R. R. Ringe. 1984. Homing of hatchery salmon and steelhead allowed a short-distance voluntary migration before transport to the lower Columbia River. University of Idaho, Idaho Cooperative Fishery Research Unit, Technical Report 84-1, Moscow.

Blackwell, B. F. 1996. Ecology of double-crested cormorants using the Penobscot River and Bay, Maine. Doctoral dissertation. University of Maine, Orono.

Bley, P. W., and J. R. Moring. 1988. Freshwater and ocean survival of Atlantic salmon and steelhead. U.S. Fish and Wildlife Service Biological Report 88(9).

Bowen, J. T. 1970. A history of fish culture as related to the development of fishery programs. Pages 71–93 in N. G. Benson, editor. A century of fisheries in North America. American Fisheries Society, Special Publication 7, Bethesda, Maryland.

Brown, L. R., P. B. Moyle, and R. M. Yoshiyama. 1994. Historical decline and current status of coho salmon in California. North American Journal of Fisheries Management 14:237–261.

Bryne, A., T. C. Bjornn, and J. D. McIntyre. 1992. Modeling the response of native steelhead to hatchery supplementation programs in an Idaho river. North American Journal of Fisheries Management 12:62–78.

Buchanan, D. V., R. M. Hooton, and J. R. Moring. 1981. Northern squawfish (Ptychocheilus oregonensis) predation on juvenile salmonids in sections of the Willamette River basin, Oregon. Canadian Journal of Fisheries and Aquatic Sciences 38:360–364.

Buchanan, D. V., and J. R. Moring. 1986. Management problems with recycling of adult summer steelhead trout at Foster Reservoir, Oregon. Pages 191–200 in Stroud (1986).

Buckley, R., and F. Haw. 1978. Enhancement of Puget Sound populations of resident coho salmon, Oncorhynchus kisutch (Walbaum). Canadian Fisheries and Marine Service Technical Report 759:93–103.

Buddington, R. K., and S. I. Doroshov. 1984. Feeding trials with hatchery produced white sturgeon juveniles (Acipenser transmontanus). Aquaculture 36:237–243.

Busiahn, T. R. 1984. An introduction to native peoples' fisheries issues in North America. Fisheries 9(5):8–11.

Clark, W. G. 1985. Fishing in a sea of court orders: Puget Sound salmon management 10 years after the Boldt decision. North American Journal of Fisheries Management 5:417–434.

Congleton, J. L., T. C. Bjornn, C. A. Robertson, J. L. Irving, and R. R. Ringe. 1984. Evaluating the effects of stress on the viability of chinook salmon smolts transported from the Snake River to the Columbia River estuary. University of Idaho, Idaho Cooperative Fishery Research Unit, Technical Report 84-4, Moscow.

Dauble, D. D., and D. G. Watson. 1997. Status of fall chinook salmon populations in the mid-Columbia River, 1948–1992. North American Journal of Fisheries Management 17:283–300.

Dawson, M. A., E. Gould, F. P. Thurberg, and A. Calabrese. 1977. Physiological response of juvenile striped bass, Morone saxatilis, to low levels of cadmium and mercury. Chesapeake Science 18:353–359.

Deuel, D. 1987. The role of the National Marine Fisheries Service in the Emergency Striped Bass Research Study. Fisheries 12(6):7–8.

Donaldson, L. R. 1968. Selective breeding of salmonid fishes. Pages 65–74 in W. J. McNeil, editor. Marine aquaculture. Oregon State University Press, Corvallis.

Doroshov, S. I., and P. B. Lutes. 1984. Preliminary data on the induction of ovulation in white sturgeon (Acipenser gtransmontanus Richardson). Aquaculture 38:231–227.

Elson, P. F. 1962. Predator–prey relationships between fish-eating birds and Atlantic salmon. Fisheries Research Board of Canada Bulletin 133.

Everest, F. H. 1977. Evaluation of fisheries for anadromous salmonids produced on western national forests. Fisheries 2(2):8–9, 11, 34–36.

Foerster, R. E. 1968. The sockeye salmon, Oncorhynchus nerka. Fisheries Research Board of Canada Bulletin 162.

Giorgi, A. E., T. W. Hillman, J. R. Stevenson, S. G. Hays, and C. M. Peven. 1997. Factors that influence the downstream migration rates of juvenile salmon and steelhead through the hydroelectric system in the Mid-Columbia River Basin. North American Journal of Fisheries Management 17:268–282.

Griffiths, B., J. M. Scott, J. W. Carpenter, and C. Reed. 1989. Translocation as a species conservation tool: status and strategy. Science 245:477–480.

Gulland, J. A. 1969. Manual of methods for fish stock assessment, part 1. FAO (Food and Agriculture Organization of the United Nations) Manuals in Fisheries Science M4.

Harden-Jones, F. R. 1968. Fish migration. Edward Arnold, London.

Harris, G. S. 1973. Rearing smolts in mountain lakes to supplement salmon stocks. International Atlantic Salmon Foundation Special Publication Series 4:237–252.

Hasler, A. D., and A. T. Scholz. 1983. Olfactory imprinting and homing in salmon. Springer-Verlag, Berlin.

Haw, F., and P. K. Bergman. 1972. A salmon angling program for the Puget Sound regions. Washington Department of Fisheries, Information Booklet 2, Olympia.

Hershberger, W. K., and R. N. Iwamoto. 1985. Systematic genetic selection and breeding in salmonid culture and enhancement programs. NOAA (National Oceanic and Atmospheric Administration) Technical Report NMFS (National Marine Fisheries Service) 27.

Hisata, J. S. 1973. Evaluation of stocking hatchery reared sea-run cutthroat trout in streams of Hood Canal. Washington State Department of Game, Job Completion Reports, AFS-44-1 and AFS-44-2, Olympia.

Hooton, B. 1997. Status of coastal cutthroat trout in Oregon. Oregon Department of Fish and Wildlife, Fish Division, Informational Report 97-2.

Iwamoto, R. N., A. M. Saxton, and W. K. Hershberger. 1982. Genetic estimates for length and weight of coho salmon (*Oncorhynchus kisutch*). Journal of Heredity 73:187–191.

Johnston, J. M., and S. P. Mercer. 1976. Sea-run cutthroat in saltwater pens: broodstock development and extended juvenile rearing (with a life history compendium). Washington State Department of Game, Fisheries Research Report, Olympia.

Jonsson, B., N. Jonsson, and L. P. Hansen. 1990. Does juvenile experience affect migration and spawning of adult Atlantic salmon? Behavioral Ecology and Sociobiology 26:225–230.

Kieffer, M. C., and B. Kynard. 1996. Spawning of the shortnose sturgeon in the Merrimack River, Massachusetts. Transactions of the American Fisheries Society 125:179–186.

Konkel, G. W., and J. D. McIntyre. 1987. Trends in spawning populations of Pacific anadromous salmonids. U.S. Fish and Wildlife Service, Fish and Wildlife Technical Report 9.

Larson, R. W., and J. H. Ward. 1955. Management of steelhead trout in the State of Washington. Transactions of the American Fisheries Society 84:261–274.

Laughlin, T. F., and B. J. Turner. 1996. Hypervariable DNA markers reveal high genetic variability within striped bass populations of the lower Chesapeake Bay. Transactions of the American Fisheries Society 125:49–55.

Mahnken, C., and five coauthors. 1982. The application of recent smoltification research to public hatchery releases: an assessment of size/time requirements for Columbia River hatchery coho salmon (*Oncorhynchus kisutch*). Aquaculture 28:251–268.

Marsh, J. H., and J. H. Johnson. 1985. The role of Stevens Treaty tribes in the management of anadromous fish runs in the Columbia River. Fisheries 10(4):2–5.

Marshall, L., C. J. Farmer, and R. E. Cutting. 1994. Atlantic salmon initiatives in Scotia-Fundy Region, Nova Scotia and New Brunswick. Pages 116–133 *in* S. Calabi and A. Strout, editors. A hard look at some tough issues. New England Salmon Association, Newburyport, Massachusetts.

Mathur, D., P. G. Heisey, and D. A. Robinson. 1994. Turbine-passage mortality of juvenile American shad at a low-head hydroelectric dam. Transactions of the American Fisheries Society 123:108–111.

McNeil, W. J. 1980. Salmon ranching in Alaska. Pages 13–27 *in* J. E. Thorpe, editor. Salmon ranching. Academic Press, New York.

McPhail, J. D., and C. C. Lindsey. 1970. Freshwater fishes of northwestern Canada and Alaska. Fisheries Research Board of Canada Bulletin 173.

Mecum, W. L. 1980. The efficiency of various bypass configurations for juvenile striped bass, *Morone saxatilis*, and American shad, *Alosa sapidissima*. California Department of Fish and Game, Anadromous Fisheries Branch, Administrative Report 80-12, Sacramento.

Mehrle, P. M., L. Cleveland, and D. R. Buckler. 1987. Chronic toxicity of an environmental contaminant mixture to young or larval striped bass. Water, Air, and Soil Pollution 35:107–118.

Morgan, R. P., II, and R. D. Prince. 1977. Chlorine toxicity to eggs and larvae of five Chesapeake Bay fishes. Transactions of the American Fisheries Society 106:380–385.

Moring, J. R. 1976. Contributions of delayed-release, pen-cultured chinook salmon to the sport fishery in Puget Sound, Washington. Progressive Fish-Culturist 38:36–39.

Moring, J. R. 1982. Fin erosion and culture-related injuries of chinook salmon raised in floating net pens. Progressive Fish-Culturist 44:189–191.

Moring, J. R. 1986. Stocking anadromous species to restore or enhance fisheries. Pages 59–74 in Stroud (1986).

Moring, J. R. 1987. Restoration of Atlantic salmon to Maine: overcoming physical and biological problems in the estuary. Pages 3129–3140 in O.T Magoon, and five coeditors. Coastal zone '87. American Society of Civil Engineers, New York.

Moring, J. R. 1989. Documentation of unaccounted-for losses of chinook salmon from saltwater cages. Progressive Fish-Culturist 51:173–176.

Moring, J. R. 1993. Effect of angling effort on catch rate of wild salmonids in streams stocked with catchable-size trout. North American Journal of Fisheries Management 13:234–237.

Moring, J. R., and K. Finlayson. 1996. Relationships between land use activities and Atlantic salmon (Salmo salar) habitat: a literature review. Maine Cooperative Fish and Wildlife Research Unit, Final Report to the National Council of the Paper Industry for Air and Stream Improvement.

Moring, J. R., G. C. Garman, and D. M. Mullen. 1994. Effects of logging practices on fishes in streams and techniques for protection: a review of four studies in the United States. Pages 194–207 in I. G. Cowyx, editor. Rehabilitation of freshwater fisheries. Fishing News Books, Oxford, UK.

Moring, J. R., J. Marancik, and F. Griffiths. 1995. Changes in stocking strategies for Atlantic salmon restoration and rehabilitation in Maine, 1871–1993. Pages 38–46 in H. L. Schramm and R. G. Piper, editors. Uses and effects of cultured fishes in aquatic ecosystems. American Fisheries Society, Symposium 15, Bethesda, Maryland.

Mowatt, T. 1889. Statistics of the fisheries of the Province of British Columbia for 1886. U. S. Fish Commission Bulletin 7(1887):24–25.

Mullen, D. M., C. W. Fay, and J. R. Moring. 1986. Species profiles: life histories and environmental requirements of coastal fishes and invertebrates (North Atlantic)-alewife/blueback herring. U.S. Fish and Wildlife Service Biological Report 82(11.56).

Murawski, S. A., and C. F. Cole. 1978. Population dynamics of anadromous rainbow smelt Osmerus mordax in a Massachusetts river system. Transactions of the American Fisheries Society 107:535–542.

National Marine Fisheries Service. 1992. Our living oceans, NOAA (National Oceanic and Atmospheric Administration) Technical Memorandum NMFS (National Marine Fisheries Service)-F/SPD-2.

Nestler, J. M., G. R. Ploskey, J. Pickens, J. Menezes, and C. Schilt. 1992. Responses of blueback herring to high-frequency sound and implications for reducing entrainment at hydropower dams. North American Journal of Fisheries Management 12:667–683.

Nielsen, J. L., editor. 1995. Evolution and the aquatic ecosystem: defining unique units in population conservation. American Fisheries Society, Symposium 17, Bethesda, Maryland.

Novotny, A. J. 1975. Net-pen culture of Pacific salmon in marine waters. Marine Fisheries Review 37(1):36–47.

Novotny, A. J. 1980. Delayed release of salmon. Pages 325–369 in J. E. Thorpe, editor. Salmon ranching. Academic Press, New York.

O'Flynn, F. M., and five coauthors. 1990. Grilsification: genetic and environmental factors. Pages 26–28 in Salmon genetics research program, annual report 1989/90. Atlantic Salmon Federation, St. Andrews, New Brunswick.

O'Rear, C. W., Jr. 1973. The toxicity of zinc and copper to striped bass eggs and fry with methods for providing confidence limits. Proceedings of the Annual Conference Southeastern Association of Game and Fish Commissioners 26(1972):484–489.

Poe, T. P., H. C. Hamsel, S. Vigg, D. E. Palmer, and L. A. Prendergast. 1991. Feeding of predaceous fishes on out-migrating juvenile salmonids in John Day Reservoir, Columbia River. Transactions of the American Fisheries Society 120:405–420

Quinn, T. P. 1993. A review of homing and straying of wild and hatchery-produced salmon. Fisheries Research 18:29–44.

Quinn, T. P., R. S. Nemeth, and D. O. McIsaac. 1991. Homing and straying patterns of fall chinook salmon in the lower Columbia River. Transactions of the American Fisheries Society 120:150–156

Raymond, H. L. 1979. Effects of dams and impoundments on migrations of juvenile chinook salmon and steelhead from the Snake River, 1966 to 1975. Transactions of the American Fisheries Society 108:505–529.

Reisenbichler, R. R., and J. D. McIntyre. 1986. Requirements for integrating natural and artificial production of anadromous salmonids in the Pacific Northwest. Pages 365–374 in Stroud (1986).

Reisenbichler, R. R., J. D. McIntyre, M. F. Solazzi, and S. W. Landino. 1992. Genetic variation in steelhead of Oregon and northern California. Transactions of the American Fisheries Society 121:158–169.

Ricker, W. E. 1972. Heredity and environmental factors affecting certain salmonid populations. Pages 27–160 in R. C. Simon and P. A. Larkin, editors. The stock concept in Pacific salmon. H. R. MacMillan Lectures in Fisheries, University of British Columbia, Vancouver.

Rulifson, R. A., and M. J. Dadswell. 1995. Life history and population characteristics of striped bass in Atlantic Canada. Transactions of the American Fisheries Society 124:466–507.

Saunders, R. L. 1981. Atlantic salmon (*Salmo salar*) stocks and management implications in the Canadian Atlantic provinces and New England, USA. Canadian Journal of Fisheries and Aquatic Sciences 38:1612–1625.

Saunders, R. L., and J. K. Bailey. 1980. The role of genetics in Atlantic salmon management. Pages 182–200 in A. E. J. Went, editor. Atlantic salmon: its future. Fishing News Books, Farnham, UK.

Saunders, R. L., E. B. Henderson, B. D. Glebe, and E. J. Loudenslager. 1983. Evidence of a major environmental component in determination of the grilse:larger salmon ratio in Atlantic salmon (*Salmo salar*). Aquaculture 33:107–118.

Schramm, H. L., Jr., and R. G. Piper, editors. 1995. Uses and effects of cultured fishes in aquatic ecosystems. American Fisheries Society, Symposium 15, Bethesda, Maryland.

Scott, W. B., and E. J. Crossman. 1973. Freshwater fishes of Canada. Fisheries Research Board of Canada Bulletin 184.

Sheppard, D. 1972. The present status of the steelhead trout stocks along the Pacific coast. Pages 519–556 in D. H. Rosenberg, editor. A review of the oceanography and renewable resources of the northern Gulf of Alaska. University of Alaska, Institute of Marine Science Report R72-23 and Sea Grant Report 73-3, Fairbanks.

Slaney, T. L., K. D. Hyatt, T. G. Northcote, and R. J. Fielden. 1996. Status of anadromous salmon and trout in British Columbia and Yukon. Fisheries 21(10):20–35.

Snieszko, S. F. 1972. Nutritional fish diseases. Pages 404–437 in J. E. Halver, editor. Fish nutrition. Academic Press, New York.

Specker, J. L., and C. B. Schreck. 1980. Stress responses to transportation and fitness for marine survival in coho salmon (*Oncorhynchus kisutch*) smolts. Journal of the Fisheries Research Board of Canada 37:765–769.

Stellwag, E. J., E. S. Payne, and R. A. Rulifson. 1994. Mitochondrial DNA diversity of Roanoke River striped bass. Transactions of the American Fisheries Society 123:321–334.

Stevens, D. E., D. W. Kohlhorst, L. W. Miller, and D. W. Kelley. 1985. The decline of striped bass in the Sacramento-San Joaquin estuary, California. Transactions of the American Fisheries Society 114:12–30.

Stier, D. J., and B. Kynard. 1986. Use of radio telemetry to determine the mortality of Atlantic salmon smolts passed through a 17-MW Kaplan turbine at a low-head hydroelectric dam. Transactions of the American Fisheries Society 115:771–775.

Stroud, R. H., editor. 1986. Fish culture in fisheries management. American Fisheries Society, Fish Culture Section and Fisheries Management Section, Bethesda, Maryland.

Swartz, D. 1985. Critical steelhead habitat and habitat problems in Oregon. Pages 105–108 in F. D. Van Hulle, editor. Alaska steelhead workshop, 1985. Alaska Department of Fish and Game, Division of Sport Fish, Juneau.

Tatum, B. L., J. D. Bayless, E. G. McCoy, and W. B. Smith. 1965. Preliminary experiments in the artificial propagation of striped bass, *Roccus saxatilus*. North Carolina Wildlife Resources Commission, Division of Inland Fisheries, Report.

Taylor, R. E., and B. Kynard. 1985. Mortality of juvenile American shad and blueback herring passed through a low-head Kaplan hydroelectric turbine. Transactions of the American Fisheries Society 114:430–435.

Thorpe, J. E., and eight coauthors. 1981. Assessing and managing man's impact on fish genetic resources. Canadian Journal of Fisheries and Aquatic Sciences 38:1899–1907.

Wagner, E., and P. Ingram. 1973. Evaluation of fish facilities and passage at Foster and Green Peter dams on the South Santiam River drainage in Oregon. Fish Commission of Oregon, final report to U.S. Army Corps of Engineers, Contract DACW57-68-C-0013, Portland.

Wagner, H. H. 1969. Effect of stocking location of juvenile steelhead trout, *Salmo gairdnerii* on adult catch. Transactions of the American Fisheries Society 98:27–34.

Waldman, J. R., and I. I. Wirgin. 1995. Comment: mitochondrial DNA stability and striped bass stock identification. Transactions of the American Fisheries Society 124:954–956.

Ward, D. L., R. R. Boyce, F. R. Young, and F. E. Olney. 1997. A review and assessment of transportation studies for juvenile chinook salmon in the Snake River. North American Journal of Fisheries Management 17:652–662.

Watt, W. D. 1986. Report to the St. Croix River Biological Working Party on statistical analysis of St. Croix river tag return data for information on downstream migration mortality of Atlantic salmon smolts. Department of Fisheries and Oceans, Fisheries Research Branch, Halifax, Nova Scotia.

Whitney, R. R., 1961. The Susquehanna fishery study, 1957–1960. A report on a study on the desirability and feasibility of passing fish at Conowingo Dam. Maryland Department of Resource Education, Contribution 169, Solomons.

Williams, J. E., C. A. Wood, and M. P. Dombeck, editors. 1997. Watershed restoration: principles and practices. American Fisheries Society, Bethesda, Maryland.

Symbols and Abbreviations

The following symbols and abbreviations may be found in this book without definition. Also undefined are standard mathematical and statistical symbols given in most dictionaries.

A	ampere	ha	hectare (2.47 acres)
AC	alternating current	hp	horsepower (746 W)
Bq	becquerel	Hz	hertz
C	coulomb	in	inch (2.54 cm)
°C	degrees Celsius	Inc.	Incorporated
cal	calorie	i.e.	(id est) that is
cd	candela	IU	international unit
cm	centimeter	J	joule
Co.	Company	K	Kelvin
Corp.	Corporation		(degrees above absolute zero)
cov	covariance	k	kilo (10^3, as a prefix)
DC	direct current; District of Columbia	kg	kilogram
D	dextro (as a prefix)	km	kilometer
d	day	l	levorotatory
d	dextrorotatory	L	levo (as a prefix)
df	degrees of freedom	L	liter (0.264 gal, 1.06 qt)
dL	deciliter	lb	pound (0.454 kg, 454g)
E	east	lm	lumen
E	expected value	log	logarithm
e	base of natural logarithm	Ltd.	Limited
	(2.71828…)	M	mega (10^6, as a prefix); molar (as
e.g.	(exempli gratia) for example		a suffix or by itself)
eq	equivalent	m	meter (as a suffix or by itself);
et al.	(et alii) and others		milli (10^{-3}, as a prefix)
etc.	et cetera	mi	mile (1.61 km)
eV	electron volt	min	minute
F	filial generation; Farad	mol	mole
°F	degrees Fahrenheit	N	normal (for chemistry); north (for
fc	footcandle (0.0929 lx)		geography); newton
ft	foot (30.5 cm)	N	sample size
ft³/s	cubic feet per second	NS	not significant
	(0.0283 m³/s)	n	ploidy; nanno (10^{-9}, as a prefix)
g	gram	o	ortho (as a chemical prefix)
G	giga (10^9, as a prefix)	oz	ounce (28.4 g)
gal	gallon (3.79 L)	P	probability
Gy	gray	p	para (as a chemical prefix)
h	hour	p	pico (10^{-12}, as a prefix)

Pa	pascal	USA	United States of America (noun)
pH	negative log of hydrogen ion activity	V	volt
		V, Var	variance (population)
ppm	parts per million	var	variance (sample)
qt	quart (0.946 L)	W	watt (for power); west (for geography)
R	multiple correlation or regression coefficient	Wb	weber
r	simple correlation or regression coefficient	yd	yard (0.914 m, 91.4 cm)
		α	probability of type I error (false rejection of null hypothesis)
rad	radian		
S	siemens (for electrical conductance); south (for geography)	β	probability of type II error (false acceptance of null hypothesis)
		Ω	ohm
SD	standard deviation	m	micro (10^{-6}, as a prefix)
SE	standard error	′	minute (angular)
s	second	″	second (angular)
T	tesla	°	degree (temperature as a prefix, angular as a suffix)
tris	tris(hydroxymethyl)-aminomethane (a buffer)	%	per cent (per hundred)
UK	United Kingdom	‰	per mille (per thousand)
U.S.	United States (adjective)		

Index